Lecture Notes in Computer Science 2761
Edited by G. Goos, J. Hartmanis, and J. van Leeuwen

Springer
Berlin
Heidelberg
New York
Hong Kong
London
Milan
Paris
Tokyo

Roberto Amadio Denis Lugiez (Eds.)

CONCUR 2003 - Concurrency Theory

14th International Conference
Marseille, France, September 3-5, 2003
Proceedings

Springer

Volume Editors

Roberto Amadio
Denis Lugiez
Université de Provence
and
Laboratoire d'Informatique Fondamentale de Marseille, UMR 6166
CMI, 39 Avenue Joliot-Curie
13453 Marseille, France
E-mail: amadio,lugiez@cmi.univ-mrs.fr

Cataloging-in-Publication Data applied for

A catalog record for this book is available from the Library of Congress.

Bibliographic information published by Die Deutsche Bibliothek
Die Deutsche Bibliothek lists this publication in the Deutsche Nationalbibliografie; detailed bibliographic data is available in the Internet at <http://dnb.ddb.de>.

CR Subject Classification (1998): F.3, F.1, D.3, D.1, C.2

ISSN 0302-9743
ISBN 3-540-40753-7 Springer-Verlag Berlin Heidelberg New York

Springer-Verlag Berlin Heidelberg New York
a member of BertelsmannSpringer Science+Business Media GmbH

http://www.springer.de

Typesetting: Camera-ready by author, data conversion by PTP Berlin GmbH
Printed on acid-free paper SPIN: 10930809 06/3142 5 4 3 2 1 0

Preface

This volume contains the proceedings of the 14th International Conference on Concurrency Theory (CONCUR 2003) held in Marseille, France, September 3–5, 2003. The conference was hosted by the Université de Provence and the Laboratoire d'Informatique Fondamentale de Marseille (LIF).

The purpose of the CONCUR conferences is to bring together researchers, developers, and students in order to advance the theory of concurrency, and promote its applications. Interest in this topic is continuously growing, as a consequence of the importance and ubiquity of concurrent systems and their applications, and of the scientific relevance of their fundations. The scope of the conference covers all areas of semantics, logics, and verification techniques for concurrent systems. Topics include concurrency-related aspects of: models of computation and semantic domains, process algebras, Petri nets, event structures, real-time systems, hybrid systems, decidability, model-checking, verification and refinement techniques, term and graph rewriting, distributed programming, logic constraint programming, object-oriented programming, types systems and algorithms, case studies, and tools and environments for programming and verification.

Of the 107 papers submitted this year, 29 were accepted for presentation. Four invited talks were given at the conference: on *Distributed Monitoring of Concurrent and Asynchronous Systems* by Albert Beneveniste, on *Quantitative Verification via the MU-Calculus* by Luca De Alfaro, on *Input-Output Automata: Basic, Timed, Hybrid, Probabilistic, Dynamic,...* by Nancy Lynch, and on *Composition of Cryptographic Protocols in a Probabilistic Polynomial-Time Process Calculus* by Andre Scedrov.

Several workshops were organized together with CONCUR:

- EXPRESS, Expressiveness in Concurrency, organized by Flavio Corradini and Uwe Nestmann;
- FOCLASA, Foundations of Coordination Languages and Software Architectures, organized by Antonio Brogi, Jean-Marie Jacquet and Ernesto Pimentel;
- INFINITY, Verification of Infinite State Systems, organized by Philippe Schnoebelen;
- FORMATS, Formal Modelling and Analysis of Timed Systems, organized by Peter Niebert;
- GETCO, Geometric and Topological Methods in Concurrency, organized by Ulrich Fahrenberg;
- CMCIM, Categorical Methods for Concurrency, Interaction, and Mobility, organized by Thomas Hildebrandt and Alexander Kurz;
- BioConcur, Concurrent Models in Molecular Biology, organized by Vincent Danos and Cosimo Laneve;
- SPV, Security Protocols Verification, organized by Michael Rusinowitch.

We would like to thank the Program Committee members and the referees who assisted us in the evaluation of the submitted papers. Also, many thanks to the Local Organization Chair, Peter Niebert, and to the other members of the Local Organization, Rémi Morin, Sarah Zennou and all the members of the MOVE team. Thanks to the Workshop Chair, Silvano DalZilio, and to the workshops' organizers. Thanks to Vincent Vanackère for installing the START conference system. We would also like to thank the invited speakers and the authors of the submitted papers. The program committee was unanimous in considering that the quality of the submissions was unusually high this year.

We gratefully acknowledge support from Conseil Général des Bouches du Rhônes, Région Provence-Alpes-Côte d'Azur, Ville de Marseille, Université de Provence, and Laboratoire d'Informatique Fondamentale de Marseille.

Roberto Amadio, Denis Lugiez

Organization

Steering Committee

Jos Baeten (Technical University of Eindhoven, The Netherlands)
Eike Best (University of Oldenburg, Germany)
Kim G. Larsen (Aalborg University, Denmark)
Ugo Montanari (University of Pisa, Italy)
Scott Smolka (SUNY Stony Brook, USA)
Pierre Wolper (University of Liege, Belgium)

Program Committee

Roberto Amadio, Chair (University of Provence, Marseille, France)
Denis Lugiez, Co-chair (University of Provence, Marseille, France)
David Basin (ETH Zurich, Switzerland)
Julian Bradfield (University of Edinburgh, UK)
Witold Charatonik (University of Wroclaw, Poland)
Alessandro Cimatti (IRST, Trento, Italy)
Philippa Gardner (Imperial College, London, UK)
Patrice Godefroid (Bell Labs, Murray-Hill, USA)
Holger Hermanns (University of Saarland, Germany)
Naoki Kobayashi (Tokyo Institute of Technology, Japan)
Kim Larsen (University of Aalborg, Denmark)
Madhavan Mukund (Chennai Math. Institute, India)
Doron Peled (University of Warwick, UK)
Jean-François Raskin (University of Brussels, Belgium)
Eugene Stark (SUNY, Stony-Brook, USA)
Roberto Segala (University of Verona, Italy)
Peter Van Roy (University of Louvain-la-Neuve, Belgium)
Thomas Wilke (University of Kiel, Germany)
Glynn Winskel (University of Cambridge, UK)

Referees

Erika Abraham-Mumm
Rafael Accorsi
Luca Aceto
Rajeev Alur
Eugene Asarin
Christel Baier
Patrick Baillot
Marek A. Bednarczyk
Massimo Benerecetti
Martin Berger
Lennart Beringer
Marco Bernardo
Piergiorgio Bertoli
Henrik Bohnenkamp
Bernard Boigelot
Alexandre Boisseau
Beate Bollig
Ahmed Bouajjani
Patricia Bouyer
Marco Bozzano
Jeremy Bradley

Linda Brodo
Glenn Bruns
Jean Cardinal
Paul Caspi
Franck Cassez
Ilaria Castellani
Anders B. Christensen
Dave Clarke
Thomas Colcombet
Raphael Collet
Deepak D'Souza
Silvano Dal Zilio
Bram De Wachter
Martin De Wulf
Giorgio Delzanno
Stephane Demri
Josee Desharnais
Raymond Devillers
Sophia Drossopoulou
Kousha Etessami
Harald Fecher
Bernd Finkbeiner
Cormac Flanagan
Emmanuel Fleury
Wan Fokkink
Stefan Friedrich
Sibylle Froschle
Yuxi Fu
Paul Gastin
Gilles Geeraerts
Rosella Gennari
Alain Girault
Andy Gordon
Ole H. Jensen
Tom Henzinger
Jane Hillston
Daniel Hirschkoff
Michael Huth
Radha Jagadeesan
David Janin
Klaus Jansen
Mark Jerrum
Colette Johen
Sven Johr
Jan Jurjens
Joost-Pieter Katoen
Emanuel Kieronski
Felix Klaedtke
Josva Kleist
William Knottenbelt
Gregory Kucherov
Dietrich Kuske
Rom Langerak
Dirk Leinenbach
Stefan Leue
Cédric Lhoussaine
Kamal Lodaya
P. Madhusudan
Sergio Maffeis
Fabio Masacci
Thierry Massart
Paulo Mateus
Richard Mayr
Paul-Andre Mellies
Massimo Merro
Sebastian Moedersheim
Rémi Morin
Almetwally Mostafa
Kedar Namjoshi
K. Narayan Kumar
Peter Niebert
Gethin Norman
David Nowak
Catuscia Palamidessi
Prakash Panangaden
Iain Phillips
Marco Pistore
Pedro R. D'Argenio
R. Ramanujam
Julian Rathke
Arend Rensink
Marco Roveri
Davide Sangiorgi
Luigi Santocanale
Vladimiro Sassone
Alan Schmitt
Philippe Schnoebelen
Roberto Sebastiani
Peter Sewell
Alex Simpson
Arne Skou
Ana Sokolova
Fred Spiessens
Jiri Srba
Ian Stark
Colin Stirling
Marielle Stoelinga
Eijiro Sumii
Jean-Marc Talbot
P.S. Thiagarajan
Simone Tini
Akihiko Tozawa
Mark Tuttle
Yaroslav Usenko
Frits Vaandrager
Laurent Van Begin
Franck van Breugel
Robert van Glabbeek
Björn Victor
Luca Vigano
Adolfo Villafiorita
Mahesh Viswanathan
Walter Vogler
Bogdan Warinschi
Rafael Wisniewski
Pierre Wolper
Nobuko Yoshida
Gianluigi Zavattaro
Sarah Zennou

Table of Contents

Partial Orders and Asynchronous Systems

Process Algebra

Games

Infinite Systems

Probabilistic Automata

Model-Checking

Model-Checking and HMSC

Security

Mobility

Compositional Methods and Real Time

Probabilistic Models

Distributed Monitoring of Concurrent and Asynchronous Systems*

Albert Benveniste[1], Stefan Haar[1], Eric Fabre[1], and Claude Jard[2]

[1] Irisa/INRIA, Campus de Beaulieu, 35042 Rennes cedex, France
Albert.Benveniste@irisa.fr
http://www.irisa.fr/sigma2/benveniste/
[2] Irisa/CNRS, Campus de Beaulieu, 35042 Rennes cedex, France

Abstract. Developing applications over a distributed and asynchronous architecture without the need for synchronization services is going to become a central track for distributed computing. This research track will be central for the domain of autonomic computing and self-management. Distributed constraint solving, distributed observation, and distributed optimization, are instances of such applications. This paper is about distributed observation: we investigate the problem of distributed monitoring of concurrent and asynchronous systems, with application to distributed fault management in telecommunications networks.
Our approach combines two techniques: compositional unfoldings to handle concurrency properly, and a variant of graphical algorithms and belief propagation, originating from statistics and information theory.

Keywords: asynchronous, concurrent, distributed, unfoldings, event structures, belief propagation, fault diagnosis, fault management

1 Introduction

Concurrent and distributed systems have been at the heart of computer science and engineering for decades. Distributed algorithms [18, 25] have provided the sound basis for distributed software infrastructures, providing correct communication and synchronization mechanisms, and fault tolerance for distributed applications. Consensus and group membership have become basic services that a safe distributed architecture should provide. Formal models and mathematical theories of concurrent systems have been essential to the development of languages, formalisms, and validation techniques that are needed for a correct design of large distributed applications.

However, the increasing power of distributed computing allows the development of applications in which the distributed underlying architecture, the function to be performed, and the data involved, are tightly interacting. Distributed optimization is a generic example. In this paper, we consider another instance of

* This work is or has been supported in part by the following projects: MAGDA, MAGDA2, funded by the ministery of research. Other partners of these projects were or are: Alcatel, France Telecom R&D, Ilog, Paris-Nord University.

R. Amadio, D. Lugiez (Eds.): CONCUR 2003, LNCS 2761, pp. 1–26, 2003.

those problems, namely the problem of inferring, from measurements, the hidden internal state of a distributed and asynchronous system. Such an inference has to be performed also in a distributed way. An important application is *distributed alarm correlation and fault diagnosis* in telecomunications networks, which motivated this work.

The problem of recovering state histories from observations is pervasive throughout the general area of information technologies. As said before, it is central in the area of distributed algorithms [18, 25], where it consists in searching for globally coherent sets of available local states, to form the global state. In this case the local states are available. Extending this to the case when the local states themselves must be inferred from observations, has been considered in other areas than computer science. For instance, estimating the state trajectory from noisy measurements is central in control engineering, with the Kalman filter as its most popular instance [16]; the same problem is considered in the area of pattern recognition, for stochastic finite state automata, in the theory of Hidden Markov Models [24]. For both cases, however, no extension exists to handle distributed systems. The theory of Bayesian networks in pattern recognition addresses the problem of distributed estimation, by proposing so-called belief propagation algorithms, which are chaotic and asynchronous iterations to perform state estimation from noisy measurements [19, 20, 23]. On the other hand, systems with dynamics (e.g., automata) are not considered in Bayesian networks. Finally, fault diagnosis in discrete event systems (e.g., automata) has been extensively studied [6, 27], but the problem of distributed fault diagnosis for distributed asynchronous systems has not been addressed.

This paper is organized as follows. Our motivating application, namely distributed alarm correlation and fault diagnosis, is discussed in Sect. 2. Its main features are: the concurrent nature of fault effect propagation, and the need for distributed supervision, where each supervisor knows only the restriction, to its own domain, of the global system model. Our goal is to compute *a coherent set of local views for the global status of the system,* for each supervisor. We follow a true concurrency approach. A natural candidate for this are 1-safe Petri nets with branching processes and unfoldings. Within this framework, we discuss in Sect. 3 a toy example in detail. The mathematical background is recalled in Sect. 4.1. As our toy example shows, two non classical operators are needed: a projection (to formalize what a local view is), and a composition (to formalize the cooperation of supervisors for distributed diagnosis). Occurrence nets, branching processes, and unfoldings will be shown to be not closed under projection, and thus inadequate to support these operations. Therefore, projection and composition are introduced for event structures (in fact, for "condition structures", a variant of event structures in which events are re-interpreted as conditions). This material is developed in Sect. 4.2. It allows us to formally express our problem, which is done in Sect. 5.

With complex interaction and several supervisors, the algorithm quickly becomes intractable, with no easy implementation. Fortunately, it is possible to organize the problem into a high-level *orchestration* of low-level primitives, expressed in terms of projections and compositions of condition structures. This orchestration is obtained by deriving, for our framework, a set of key properties relating condition structures, their projections, and their compositions. Further analysing these key properties shows that they are shared by basic problems such as: distributed combinatorial constraint solving, distributed combinatorial optimization, and Bayesian networks estimation. Fortunately, chaotic distributed and asynchronous iterations to solve these problems have been studied; perhaps the most well-known version of these is that of *belief propagation* in belief networks [19, 20, 23]. Our high-level orchestration is derived from these techniques, it is explained in Sect. 6. In addition, this approach allows us to consider maximum likelihood estimation of fault scenarios, by using a probabilistic model for fault propagation.

Missing details and proofs can be found in the extended version [1].

2 Motivating Application: Fault Management in Telecommunication Networks [3]

Distributed self-management is a key objective in operating large scale infrastructures. Fault management is one of the five classical components of management, and our driving application. Here, we consider a distributed architecture in which each supervisor is in charge of its own domain, and the different supervisors cooperate at constructing a set of *coherent* local views for their repective domains. Of course, the corresponding global view should never be actually computed.

To ensure modularity, network management systems are decomposed into interconnected Network Elements (NE), composed in turn of several Managed Objects (MO). MO's act as peers providing to each other services for overall fault management. Consequently, each MO is equipped with its own fault management function. This involves self-monitoring for possible own internal sources of fault, as well as propagating, to clients of the considered MO, the effect of one of its servers getting disabled.

Because of this modularity, faults propagate throughout the management system: when a primary fault occurs in some MO, that MO emits an alarm to the supervisor, sends a message to its neigbours, and gets disabled. Its neighbouring MOs receive the message, recognize their inability to deliver their service, get disabled, emit alarms, and so on.

Figure 1 shows on the left hand side the SDH/SONET ring in operation in the Paris area (the locations indicated are subsurbs of Paris). A few ports and links are shown. The right diagram is a detailed view of the Montrouge node. The nested light to mid gray rectangles represent the different layers in the SDH hierarchy, with the largest one being the physical layer. The different boxes are the MOs, and the links across the different layers are the paths for upward/downward fault propagation. Each MO can be seen as an automaton

[3] This section has been prepared with the help of with Armen Aghasaryan Alcatel Research & Innovation, Next Generation Network and Service Management Project, Alcatel R&I, Route de Nozay, Marcoussis, 91461 France

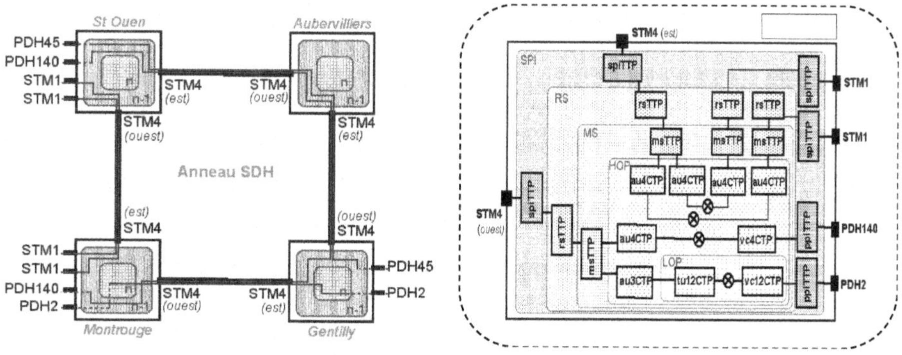

Fig. 1. The Paris area SDH/SONET ring (left), and a detail of the Montrouge node (right). The different levels of the SDH hierarchy are shown: SPI, RS, etc.

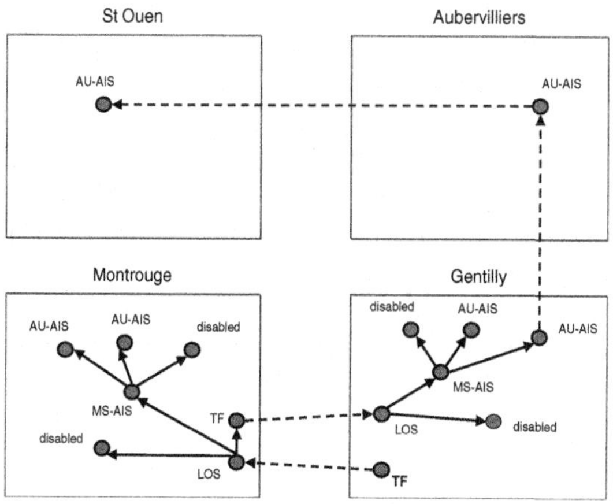

Fig. 2. A fault propagation scenario distributed across the four different sites. The dashed arrows indicate distant propagation. The cryptic names are SDH/SONET fault labels.

reacting to input events/messages, changing its state, and emitting events and alarms to its neighbours, both co-located and distant. Figure 2 shows a realistic example of a fault propagation scenario distributed across the four different sites.

To summarize, the different supervisors are distributed, and different MO's operate concurrently and asynchronously within each supervisor.

3 Informal Discussion of an Example

If all variables involved in the above scenarios possess a finite domain, we can represent these in an enumerative form. This suggests using safe Petri nets with a true concurrency semantics to formalize distributed fault diagnosis.

Presenting the Example, and the Problem. Our example is shown in Fig. 3–1st diagram, in the form of a labeled Petri net, with two interacting components, numbered 1 and 2. Component 2 uses the services of component 1, and therefore may fail to deliver its service when component 1 is faulty. The two components interact via their shared places 3 and 7, represented by the gray zone; note that this Petri net is safe, and that the two places 3 and 7 are complementary.

Component 1 has two private states: safe, represented by place 1, and faulty, represented by place 2. Upon entering its faulty state, component 1 emits an alarm β. The fault of component 1 is temporary, thus self-repair is possible and is represented by the label ρ. Component 2 has three private states, represented by places $4, 5, 6$. State 4 is safe, state 6 indicates that component 2 is faulty, and state 5 indicates that component 2 fails to deliver its service, due to the failure of component 1. Fault 6 is permanent and cannot be repaired.

The failure of component 2 caused by a fault of component 1 is modeled by the shared place 3. The monitoring system of component 2 only detects that component 2 fails to deliver its service, it does not distinguish between the different reasons for this. Hence the same alarm α is attached to the two transitions posterior to 4. Since fault 2 of component 1 is temporary, self-repair can also occur for component 2, when in faulty state 5. This self-repair is not synchronized with that of component 1, but bears the same label ρ. Finally, place 7 guarantees that fault propagation, from component 1 to 2, is possible only when the latter is in safe state.

The initial marking consists of the three states $1, 4, 7$. Labels (alarms α, β or self-repair ρ) attached to the different transitions or events, are generically referred to as *alarms* in the sequel.

Three different setups can be considered for diagnosis, assuming that messages are not lost:

Setup $\mathbf{S}_1$: The successive alarms are recorded in sequence by a single supervisor, in charge of fault monitoring. The sensor and communication infrastructure guarantees that causality is respected: for any two alarms such that α causes α', α is recorded before α'.

Setup $\mathbf{S}_2$: Each sensor records its local alarms in sequence, while respecting causality. The different sensors perform independently and asynchronously, and a single supervisor collects the records from the different sensors. Thus any interleaving of the records from different sensors is possible, and causalities among alarms from different sensors are lost.

Setup $\mathbf{S}_3$: The fault monitoring is distributed, with different supervisors cooperating asynchronously. Each supervisor is attached to a component, records its local alarms in sequence, and can exchange supervision messages with the other supervisors, asynchronously.

A Simple Solution? For setup $\mathbf{S}_1$, there is a simple solution. Call $\mathcal{A}$ the recorded alarm sequence. Try to fire this sequence in the Petri net from the initial marking. Each time an ambiguity occurs (two transitions may be fired explaining the next event in $\mathcal{A}$), a new copy of the trial (a new Petri net) is instanciated to follow the additional firing sequence. Each time no transition can be fired in a trial to explain a new event, the trial is abandoned. Then, at the end of $\mathcal{A}$, all the behaviours explaining $\mathcal{A}$ have been obtained. Setup $\mathbf{S}_2$ can be handled similarly, by exploring all interleavings of the two recorded alarm sequences. However, this direct approach does not represent efficiently the set of all solutions to the diagnosis problem.

In addition, this direct approach does not work for Setup $\mathbf{S}_3$. In this case, no supervisor knows the entire net and no global interleaving of the recorded alarm sequences is available. Maintaining a coherent set of causally related local diagnoses becomes a difficult problem for which no straightforward solution works. The approach we propose below addresses both the Setup $\mathbf{S}_3$ and the efficient representation of all solutions, for all setups.

An Efficient Data Structure to Represent All Runs. Figure 3 shows our running example. The Petri net $\mathcal{P}$ is repeated on the 2nd diagram: the labels α, β, ρ have been discarded, and transitions are i, ii, iii, iv, v, vi. Places constituting the initial marking are indicated by thick circles.

The mechanism of constructing a run of $\mathcal{P}$ in the form of a partial order is illustrated in the 2nd and 3rd diagrams. Initialize any run of $\mathcal{P}$ with the three conditions labeled by the initial marking $(1, 7, 4)$. Append to the pair $(1, 7)$ a copy of the transition $(1, 7) \to i \to (2, 3)$. Append to the new place labeled 2 a copy of the transition $(2) \to iii \to (1)$. Append, to the pair $(3, 4)$, a copy of the transition $(3, 4) \to iv \to (7, 5)$ (this is the step shown). We have constructed (the prefix of) a run of $\mathcal{P}$. Now, all runs can be constructed in this way. Different runs can share some prefix.

In the 4th diagram we show (prefixes of) all runs, by superimposing their shared parts. The gray part of this diagram is a copy of the run shown in the

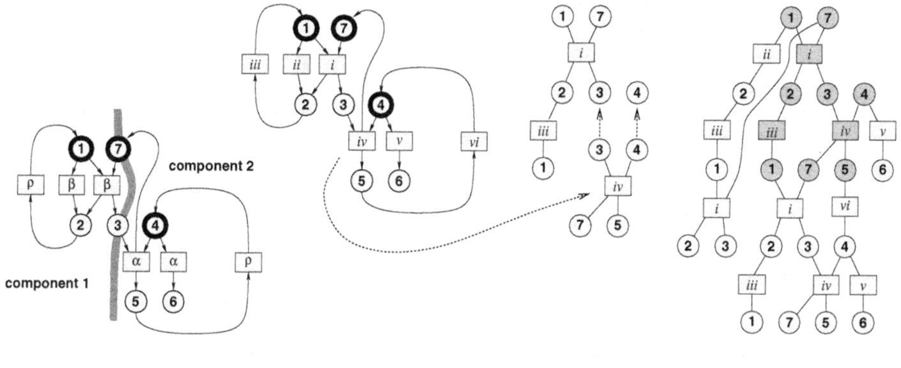

Fig. 3. Running example in the form of a Petri net (left), and representing its runs in a branching process. Petri nets are drawn by using directed arrows; on the other hand, since occurrence nets are acylic, we draw them using nondirected branches which have to be interpreted as implicitly directed toward bottom.

3rd diagram. The alternative run on the extreme left of this diagram (it involves successive transitions labeled *ii*, *iii*, *i*) shares only its initial places with the run in gray. On the other hand, replacing, in the gray run, the transition labeled *iv* by the one labeled *v* yields another run which shares with the gray one its transitions respectively labeled by *i* and by *iii*. This 4th diagram is a *branching process* of $\mathcal{P}$, we denote it by $\mathcal{U}_{\mathcal{P}}$; it is a net without cycle, in which the preset of any condition contains exactly one event. Nodes of $\mathcal{U}_{\mathcal{P}}$ are labeled by places/transitions of $\mathcal{P}$ in such a way that the two replicate each other, locally around transitions.

Terminology. When dealing with branching processes, to distinguish from the corresponding concepts in Petri nets, we shall from now on refer to *conditions/events* instead of places/transitions.

Asynchronous Diagnosis with a Single Sensor and Supervisor. Here we consider setup $\mathbf{S}_1$, and our discussion is supported by Fig. 4. The 1st diagram of this figure is the alarm sequence $\beta, \alpha, \rho, \rho, \beta, \alpha$ recorded at the unique sensor. It is represented by a cycle-free, linear Petri net, whose conditions are not labeled—conditions have no particular meaning, their only purpose is to indicate the ordering of alarms. Denote by $\mathcal{A}' = \beta \to \alpha \to \rho$ the shaded prefix of $\mathcal{A}$.

The 2nd diagram shows the net $\mathcal{U}_{\mathcal{A}'\times\mathcal{P}}$, obtained by unfolding the product $\mathcal{A}' \times \mathcal{P}$ using the procedure explained in the figure 3. The net $\mathcal{U}_{\mathcal{A}'\times\mathcal{P}}$ shows how successive transitions of $\mathcal{P}$ synchronize with transitions of $\mathcal{A}'$ having identical label, and therefore explain them.

Since we are not interested in the conditions originating from $\mathcal{A}'$, we remove them. The result is shown on the 3rd diagram. The dashed line labeled # ori-

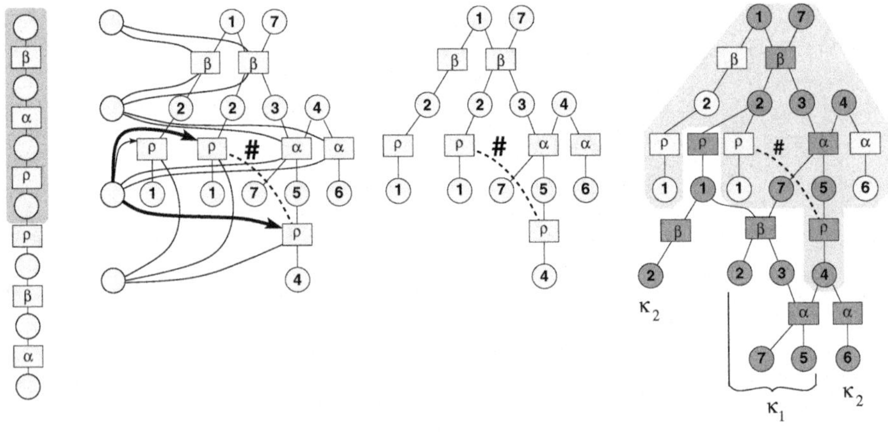

Fig. 4. Asynchronous diagnosis with a single sensor: showing an alarm sequence $\mathcal{A}$ (1st diagram), the explanation of the prefix $\mathcal{A}' = \beta \to \alpha \to \rho$ as the branching process $\mathcal{U}_{\mathcal{A}'\times\mathcal{P}}$ (2nd and 3rd diagrams), and its full explanation in the form of a net $\mathcal{U}_{\mathcal{P},\mathcal{A}}$ (4th diagram). In these diagrams, all branches are directed downwards unless otherwise explicitly stated.

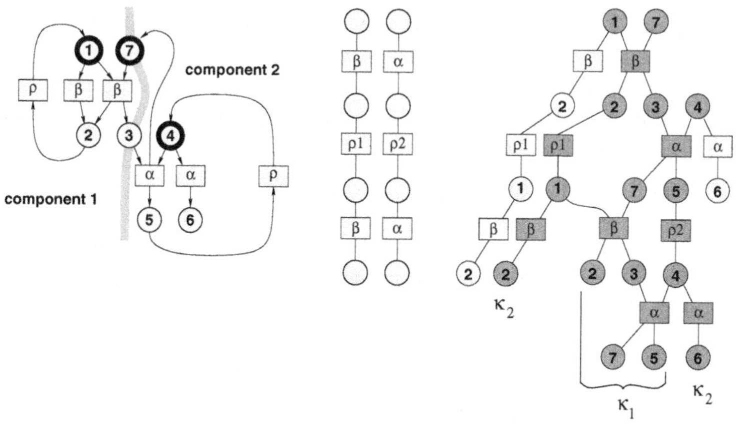

Fig. 5. Asynchronous diagnosis with two independent sensors: showing an alarm pattern $\mathcal{A}$ (middle) consisting of two concurrent alarm sequences, and its explanation in the form of a branching process $\mathcal{U}_{\mathcal{P},\mathcal{A}}$ (right).

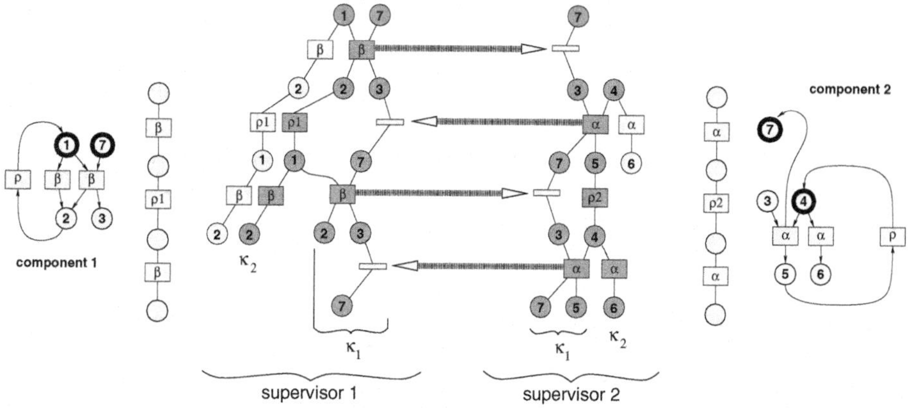

Fig. 6. Distributed diagnosis: constructing two coherent local views of the branching process $\mathcal{U}_{\mathcal{P},\mathcal{A}}$ of Fig. 5 by two supervisors cooperating asynchronously.

ginates from the corresponding conflict in $\mathcal{U}_{\mathcal{A}' \times \mathcal{P}}$ that is due to two different conditions explaining the same alarm ρ. Thus we need to remove, as possible explanations of the prefix, all runs of the 3rd diagram that contain the #-linked pair of events labeled ρ. All remaining runs are valid explanations of the subsequence β, α, ρ.

Finally, the net $\mathcal{U}_{\mathcal{P},\mathcal{A}}$ shown in the 4th diagram contains a prefix consisting of the nodes filled in dark gray. This prefix is the union of the two runs κ_1 and κ_2 of $\mathcal{P}$, that explain $\mathcal{A}$.

Asynchronous Diagnosis with Two Independent Sensors but a Single Supervisor. Focus on setup $\mathbf{S}_2$, in which alarms are recorded by two independent sensors, and then collected at a single supervisor for explanation. Figure 5 shows the same alarm history as in Fig. 4, except that it has been recorded by two independent sensors, respectively attached to each component. The supervisor knows the global model of the system, we recall it in the 1st diagram of Fig. 5.

The two "repair" actions are now distinguished since they are seen by different sensors, this is why we use different labels: ρ_1, ρ_2. This distinction reduces the ambiguity: in Fig. 5 we suppress the white filled path $(2) \to \rho \to (1)$ that occured in Fig. 4. On the other hand, alarms are recorded as two concurrent sequences, one for each sensor, call the whole an *alarm pattern*. Causalities between alarms from different components are lost. This leads to further ambiguity, as shown by the longer branch $(1) \to \beta \to (2) \to \rho_1 \to (1) \to \beta \to (2)$ in Fig. 5, compare with Fig. 4.

The overall result is shown in Fig. 5, and the valid explanations for the entire alarm pattern are the two configurations κ_1 and κ_2 filled in dark gray.

Distributed Diagnosis with Two Concurrent Sensors and Supervisors. Consider setup $\mathbf{S}_3$, in which alarms are recorded by two independent sensors, and processed by two local supervisors which can communicate asynchronously. Figure 6 shows two branching processes, respectively local to each supervisor. For completeness, we have shown the information available to each supervisor. It consists of the local model of the component considered, together with the locally recorded alarm pattern. The process constructed by supervisor 1 involves only events labeled by alarms collected by sensor 1, and places that are either local to component 1 (e.g., $1, 2$) or shared (e.g., $3, 7$); and similarly for the process constructed by supervisor 2.

The 3rd diagram of Fig. 5 can be recovered from Fig. 6 in the following way: glue events sitting at opposite extremities of each thick dashed arrow, identify adjacent conditions, and remove the thick dashed arrows. These dashed arrows indicate a communication between the two supervisors, let us detail the first one. The first event labeled by alarm β belongs to component 1, hence this explanation for β has to be found by supervisor 1. Supervisor 1 sends an abstraction of the path $(1,7) \to \beta \to (2,3)$ by removing the local conditions $1, 2$ and the label β since the latter do not concern supervisor 2. Thus supervisor 2 receives the path $(7) \to [] \to (3)$ to which it can append its local event $(3,4) \to \alpha \to (7,5)$; and so on.

Discussion. The cooperation between the two supervisors needs only asynchronous communication. Each supervisor can simply "emit and forget." Diagnosis can progress concurrently and asynchronously at each supervisor. For example, supervisor 1 can construct the branch $[1 \to \beta \to 2 \to \rho_1 \to 1 \to \beta \to 2]$ as soon as the corresponding local alarms are collected, without ever synchronizing with supervisor 2. This technique extends to distributed diagnosis with several

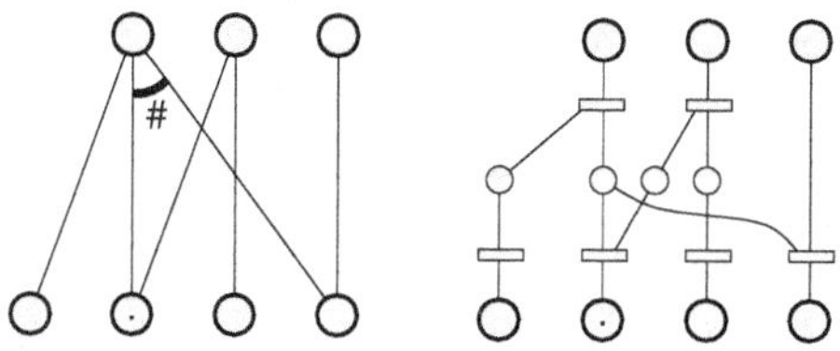

Fig. 7. Illustrating the problem of representating causality, conflict, and concurrency with an occurrence net. The 1st diagram represents a set of conditions together with a causality relation depicted by branches, and a conflict relation whose source is indicated by the symbol #. The 2nd diagram interpolates the 1st one as an occurrence net. (Arcs are assumed directed downwards.)

supervisors. But the algorithm for this general case is indeed very tricky, and its analysis is even more so. Thus, theoretical study is really necessary, and this is the main subject of the remainder of this paper.

As Fig. 4 and Fig. 6 indicate, we need projection and composition operations on branching processes. Focus on projections. Projecting away, from an occurrence net, a subset of conditions and events, can be performed by taking the restriction, to the subset of remaining conditions, of the two causality and conflict relations. An instance of resulting structure is shown in Fig. 7–left. A possible interpolation in the form of an occurrence net is shown in Fig. 7–right. Such an interpolation is not unique. In addition, the introduction of dummy conditions and events becomes problematic when combining projections and compositions.

We need a class of data structures equipped with causality and conflict relations, that is stable under projections. The class of event structures is a candidate. However, since diagnosis must be explained by sets of state histories, we need to handle conditions, not events. For this reason, we rather consider *condition structures,* as introduced in the next section.

4 Mathematical Framework: Nets, Unfoldings, and Condition Structures

4.1 Prerequisites on Petri nets and Their Unfoldings [6, 8, 26]

Petri nets. A *net* is a triple $\mathcal{N} = (P, T, \to)$, where P and T are disjoint sets of *places* and *transitions*, and $\to \subseteq (P \times T) \cup (T \times P)$ is the *flow relation.* Reflexive and irreflexive transitive closures of a relation are denoted by the superscripts $(.)^*$ and $(.)^+$, respectively. Define $\preceq \,=\, \to^*$ and $\prec \,=\, \to^+$. Places and transitions are called *nodes,* generically denoted by x. For $x \in P \cup T$, we denote by ${}^\bullet x = \{y : y \to x\}$ the *preset* of node x, and by $x^\bullet = \{y : x \to y\}$ its *postset.* For $X \subset P \cup T$, we write ${}^\bullet X = \bigcup_{x \in X} {}^\bullet x$ and $X^\bullet = \bigcup_{x \in X} x^\bullet$. A *homomorphism* from a net $\mathcal{N}$ to a net $\mathcal{N}'$ is a map $\varphi : P \cup T \mapsto P' \cup T'$ such that: (i) $\varphi(P) \subseteq P'$, $\varphi(T) \subseteq T'$, and (ii) for every transition t of $\mathcal{N}$, the restriction of φ to ${}^\bullet t$ is a bijection between ${}^\bullet t$ and ${}^\bullet \varphi(t)$, and the restriction of φ to $t^\bullet$ is a bijection between $t^\bullet$ and $\varphi(t)^\bullet$.

For $\mathcal{N}$ a net, a *marking* of $\mathcal{N}$ is a multiset M of places, i.e., a map $M : P \mapsto \{0, 1, 2, \ldots\}$. A *Petri net* is a pair $\mathcal{P} = (\mathcal{N}, M_0)$, where $\mathcal{N}$ is a net having finite sets of places and transitions, and M_0 is an *initial* marking. A transition $t \in T$ is *enabled* at marking M if $M(p) > 0$ for every $p \in {}^\bullet t$. Such a transition can *fire*, leading to a new marking $M' = M - {}^\bullet t + t^\bullet$, denoted by $M[t\rangle M'$. Petri net $\mathcal{P}$ is *safe* if $M(P) \subseteq \{0, 1\}$ for every reachable marking M. Throughout this paper, we consider only safe Petri nets, hence marking M can be regarded as a subset of places.

For $\mathcal{N}_i = \{P_i, T_i, \rightarrow_i\}$, $i = 1, 2$, two nets such that $T_1 \cap T_2 = \emptyset$, their *parallel composition* is the net

$$\mathcal{N}_1 \,\|\, \mathcal{N}_2 =_{\text{def}} (P_1 \cup P_2, T_1 \cup T_2, \rightarrow_1 \cup \rightarrow_2).$$

Petri nets and occurrence nets inherit this notion. For Petri nets, we adopt the convention that the resulting initial marking is equal to $M_{1,0} \cup M_{2,0}$, the union of the two initial markings. We say that $\mathcal{N}_1 \,\|\, \mathcal{N}_2$ *has no distributed conflict* if:

$$\forall p \in P_1 \cap P_2, \exists i \in \{1, 2\} : p^\bullet \subseteq T_i. \tag{1}$$

Note that our example of Fig. 3 satisfies (1). This is a reasonable assumption in our context, since shared places aim at representing the propagation of faults between components. Having distributed conflict would have no meaning in this case. A study of this property in the context of the synthesis of distributed automata via Petri nets is available in [5].

For $\mathcal{N} = (P, T, \rightarrow)$ a net, a *labeling* is a map $\lambda : T \mapsto A$, where A is some finite alphabet. A net $\mathcal{N} = (P, T, \rightarrow, \lambda)$ equipped with a labeling λ is called a *labeled net*. For $\mathcal{N}_i = \{P_i, T_i, \rightarrow_i, \lambda_i\}$, $i = 1, 2$, two labeled nets, their product $\mathcal{N}_1 \times \mathcal{N}_2$ is the labeled net $(P, T, \rightarrow, \lambda)$ defined as follows: $P = P_1 \uplus P_2$, where $\uplus$ denotes the disjoint union, and:

$$T = \begin{cases} \{t =_{\text{def}} t_1 \in T_1 \mid \lambda_1(t_1) \in A_1 \setminus A_2\} & \text{(i)} \\ \cup \{t =_{\text{def}} (t_1, t_2) \in T_1 \times T_2 \mid \lambda_1(t_1) = \lambda_2(t_2)\} & \text{(ii)} \\ \cup \{t =_{\text{def}} t_2 \in T_2 \mid \lambda_2(t_2) \in A_2 \setminus A_1\}, & \text{(iii)} \end{cases}$$

$$p \rightarrow t \text{ iff } \begin{cases} p \in P_1 \text{ and } p \rightarrow_1 t_1 & \text{for case (i)} \\ \exists i = 1, 2 : p \in P_i \text{ and } p \rightarrow_i t_i & \text{for case (ii)} \\ p \in P_2 \text{ and } p \rightarrow_2 t_2 & \text{for case (iii)} \end{cases}$$

and $t \rightarrow p$ is defined symmetrically. In cases (i,iii) only one net fires a transition and this transition has a private label, while the two nets synchronize on transitions with identical labels in case (ii). Petri nets and occurrence nets inherit the above notions of labeling and product.

The *language* $\mathcal{L}_\mathcal{P}$ of labeled Petri net $\mathcal{P}$ is the subset of A^* consisting of the words $\lambda(t_1), \lambda(t_2), \lambda(t_3), \ldots$, where $M_0[t_1\rangle M_1[t_2\rangle M_2[t_3\rangle M_3 \ldots$ ranges over the set of finite firing sequences of $\mathcal{P}$. Note that $\mathcal{L}_\mathcal{P}$ is prefix closed.

Occurrence Nets and Unfoldings. Two nodes x, x' of a net $\mathcal{N}$ are *in conflict*, written $x \# x'$, if there exist distinct transitions $t, t' \in T$, such that ${}^\bullet t \cap {}^\bullet t' \neq \emptyset$

and $t \preceq x$, $t' \preceq x'$. A node x is in *self-conflict* if $x\#x$. An *occurrence net* is a net $\mathcal{O} = (B, E, \rightarrow)$, satisfying the following additional properties:

(i) $\forall x \in B \cup E : \neg[x\#x]$ (no node is in self-conflict);
(ii) $\forall x \in B \cup E : \neg[x \prec x]$ ($\preceq$ is a partial order);
(iii) $\forall x \in B \cup E : |\{y : y \prec x\}| < \infty$ ($\preceq$ is well founded);
(iv) $\forall b \in B : |{}^\bullet b| \leq 1$ (each place has at most one input transition).

We will assume that the set of minimal nodes of $\mathcal{O}$ is contained in B, and we denote by $\min(B)$ or $\min(\mathcal{O})$ this minimal set. Specific terms are used to distinguish occurrence nets from general nets. B is the set of *conditions,* E is the set of *events,* $\prec$ is the *causality* relation. We say that node x is *causally related* to node x' iff $x \prec x'$. Nodes x and x' are *concurrent,* written $x \perp\!\!\!\perp x'$, if neither $x \preceq x'$, nor $x \preceq x'$, nor $x\#x'$ hold. A *conflict set* is a set $\mathcal{A}$ of pairwise conflicting nodes, i.e. a clique of $\#$; a *co-set* is a set X of pairwise concurrent conditions. A maximal (for set inclusion) co-set is called a *cut.* A *configuration* is a sub-net κ of $\mathcal{O}$, which is *conflict-free* (no two nodes are in conflict), *causally closed* (if $x' \preceq x$ and $x \in \kappa$, then $x' \in \kappa$), and contains $\min(\mathcal{O})$. We take the convention that maximal nodes of configurations shall be conditions.

A *branching process* of Petri net $\mathcal{P}$ is a pair $\mathcal{B} = (\mathcal{O}, \varphi)$, where $\mathcal{O}$ is an occurrence net, and φ is a homomorphism from $\mathcal{O}$ to $\mathcal{P}$ regarded as nets, such that: (i) the restriction of φ to $\min(\mathcal{O})$ is a bijection between $\min(\mathcal{O})$ and M_0 (the set of initially marked places), and (ii) for all $e, e' \in E$, ${}^\bullet e = {}^\bullet e'$ and $\varphi(e) = \varphi(e')$ together imply $e = e'$. By abuse of notation, we shall sometimes write $\min(\mathcal{B})$ instead of $\min(\mathcal{O})$. The set of all branching processes of Petri net $\mathcal{P}$ is uniquely defined, up to an isomorphism (i.e., a renaming of the conditions and events), and we shall not distinguish isomorphic branching processes. For $\mathcal{B}, \mathcal{B}'$ two branching processes, $\mathcal{B}'$ is a *prefix* of $\mathcal{B}$, written $\mathcal{B}' \sqsubseteq \mathcal{B}$, if there exists an injective homomorphism ψ from $\mathcal{B}'$ into $\mathcal{B}$, such that $\psi(\min(\mathcal{B}')) = \min(\mathcal{B})$, and the composition $\varphi \circ \psi$ coincides with φ', where $\circ$ denotes the composition of maps. By theorem 23 of [9], there exists (up to an isomorphism) a unique maximum branching process according to $\sqsubseteq$, we call it the *unfolding* of $\mathcal{P}$, and denote it by $\mathcal{U}_\mathcal{P}$. Maximal configurations of $\mathcal{U}_\mathcal{P}$ are called *runs* of $\mathcal{P}$.

4.2 Condition Structures

Occurrence nets give rise in a natural way to *(prime) event structures* [22]: a prime event structure is a triple $\mathcal{E} = (E, \preceq, \#)$, where $\preceq \subseteq E \times E$ is a partial order such that (i) for all $e \in E$, the set $\{e' \in E \mid e' \preceq e\}$ is finite, and (ii) $\# \subseteq E \times E$ is symmetric and irreflexive, and such that for all $e_1, e_2, e_3 \in E$, $e_1\#e_2$ and $e_2 \preceq e_3$ imply that $e_1\#e_3$. Obviously, "forgetting" the net interpretation of an occurrence net yields an event structure, and even restricting to the event set E does. This is the usual way of associating nets and event structures, and explains the name. Below, we will use event structures whose elements are interpreted as *conditions* in the sense of occurrence nets. To avoid confusion, we will speak of *condition structures,* even if the mathematical properties are invariant under this change of interpretation. Restricting an occurrence net to its set of conditions yields a condition structure.

Condition structures are denoted by $\mathcal{C} = (B, \preceq, \#)$. Denote by $\rightarrow$ the successor relation, i.e., the transitive reduction of the relation $\preceq$. For $b \in B$, we denote by ${}^\bullet b$ and $b^\bullet$ the preset and postset of b in $(B, \rightarrow)$, respectively. For $\mathcal{C}' \subseteq \mathcal{C}$ two condition structures, define the preset ${}^\bullet\mathcal{C}' = \bigcup_{b \in \mathcal{C}'} {}^\bullet b$; the postset $\mathcal{C}'^\bullet$ is defined similarly. The prefix relation on condition structures is denoted by $\sqsubseteq$. If $\mathcal{C}' \sqsubseteq \mathcal{C}$ are two condition structures, we write

$$\mathcal{C}' \rightarrow b \text{ iff } b \in \mathcal{C}'^\bullet \text{ but } b \notin \mathcal{C}', \tag{2}$$

and we say that b is an *extension* of $\mathcal{C}'$. Each condition structure $\mathcal{C} = (B, \preceq, \#)$ induces a *concurrency* relation defined by $b \perp\!\!\!\perp b'$ iff neither $b \preceq b'$ nor $b' \preceq b$ nor $b \# b'$ holds. Two conditions b, b' are called in *immediate conflict* iff $b \# b'$ and $\forall b''$ such that $b'' \prec b$, then $\neg[b'' \# b']$, and symmetrically.

For $\mathcal{C} = (B, \preceq, \#)$ a condition structure, a *labeling* is a map $\varphi : B \mapsto P$, where P is some finite alphabet[4]. We shall not distinguish labeled condition structures that are identical up to a bijection that preserves labels, causalities, and conflicts; such condition structures are called

$$\textit{equivalent}, \text{ denoted by the equality symbol } = \tag{3}$$

For $\mathcal{C} = (B, \preceq, \#, \varphi)$ and $\mathcal{C}' = (B', \preceq', \#', \varphi')$ two labeled condition structures,

$$\text{a } \textit{(partial) morphism } \psi : \mathcal{C} \mapsto \mathcal{C}'$$

is a surjective function $B \supseteq dom(\psi) \mapsto B'$ such that $\psi(\preceq) \supseteq \preceq'$ and $\psi(\#) \supseteq \#'$ (causalities and conflicts can be erased but not created), which is in addition label preserving, i.e., $\forall b \in dom(\psi)$, $\varphi'(\psi(b)) = \varphi(b)$. Note that $\psi(\preceq) \supseteq \preceq'$ is equivalent to $\forall b \in B : {}^\bullet\psi(b) \subseteq \psi({}^\bullet b)$. For $X \subset B$, define, for convenience:

$$\psi(X) =_{\mathsf{def}} \psi(X \cap dom(\psi)), \text{ with the convention } \psi(\emptyset) = nil. \tag{4}$$

$\mathcal{C}$ and $\mathcal{C}'$ are *isomorphic*, written $\mathcal{C} \sim \mathcal{C}'$, if there exist two morphisms $\psi' : \mathcal{C} \mapsto \mathcal{C}'$ and $\psi'' : \mathcal{C}' \mapsto \mathcal{C}$. It is not true in general that $\mathcal{C} \sim \mathcal{C}'$ implies $\mathcal{C} = \mathcal{C}'$ in the sense of (3). However:

Lemma 1. *If $\mathcal{C}$ and $\mathcal{C}'$ are finite, then $\mathcal{C} \sim \mathcal{C}'$ implies $\mathcal{C} = \mathcal{C}'$.*

Be careful that $\mathcal{C} = \mathcal{C}'$ means that $\mathcal{C}$ and $\mathcal{C}'$ are *equivalent,* not "equal" in the naive sense—we will not formulate this warning any more. To be able to use lemma 1, we shall henceforth assume the following, where the *height* of a condition structure is the least upper bound of the set of all lengths of finite causal chains $b_0 \rightarrow b_1 \rightarrow \ldots \rightarrow b_k$:

[4] The reader may be surprised that we reuse the symbols P and φ for objects that are different from the set of places P of some Petri net and the homomorphism φ associated with its unfolding. This notational overloading is indeed intentional. We shall mainly use condition structures obtained by erasing events in unfoldings: restricting the homomorphism $\varphi : B \cup E \mapsto P \cup T$ to the set of conditions B yields a labeling $B \mapsto P$ in the above sense.

Assumption 1 *All condition structures we consider are of finite width, meaning that all prefixes of finite height are finite.*

In this paper, we will consider mainly labeled condition structures satisfying the following property, we call them *trimmed* condition structures:

$$\forall b, b' \in B \;:\; \left.\begin{array}{c} {}^\bullet b = {}^\bullet b' \\ \text{and } \varphi(b) = \varphi(b') \end{array}\right\} \Rightarrow b = b'. \tag{5}$$

Condition (5) is similar to the irreducibility hypothesis found in branching processes. It is important when considering the projection operation below, since projections may destroy irreducibility.

A *trimming procedure* can be applied to any labeled condition structure $\mathcal{C} = (B, \preceq, \#, \varphi)$ as follows: inductively by starting from $\min(\mathcal{C})$, identify all pairs (b, b') of conditions such that both ${}^\bullet b = {}^\bullet b'$ and $\varphi(b) = \varphi(b')$ hold in $\mathcal{C}$. This procedure yields a triple $(\bar{B}, \bar{\preceq}, \bar{\varphi})$, and it remains to define the trimmed conflict relation $\bar{\#}$, or, equivalently, the trimmed concurrency relation $\bar{\perp\!\!\!\perp}$. Define $\bar{b} \;\bar{\perp\!\!\!\perp}\; \bar{b}'$ iff $b \perp\!\!\!\perp b'$ holds for some pair (b, b') of conditions mapped to $(\bar{b}, \bar{b}')$ by the trimming procedure. This defines $\bar{\mathcal{C}} = (\bar{B}, \bar{\preceq}, \bar{\#}, \bar{\varphi})$.

It will be convenient to consider the following *canonical form* for a labeled trimmed condition structure. Its conditions have the special inductive form (X, p), where X is a co-set of $\mathcal{C}$ and $p \in P$. The causality relation $\preceq$ is simply encoded by the preset function ${}^\bullet(X, p) = X$, and the labeling map is $\varphi(X, p) = p$. Conditions with empty preset have the form (nil, p), i.e., the distinguished symbol *nil* is used for the minimal conditions of $\mathcal{C}$. The conflict relation is specified separately. Unless otherwise specified, trimmed condition structures will be assumed in canonical form.

For $\mathcal{C} = (B, \preceq, \#, \varphi)$ a trimmed labeled condition structure, and $Q \subset P$ a subset of its labels, the *projection of* $\mathcal{C}$ *on* Q, denoted by $\Pi_Q(\mathcal{C})$, is defined as follows. Take the restriction of $\mathcal{C}$ to $\varphi^{-1}(Q)$, we denote it by $\mathcal{C}_{|\varphi^{-1}(Q)}$. By this we mean that we restrict B as well as the two relations $\preceq$ and $\#$. Note that $\mathcal{C}_{|\varphi^{-1}(Q)}$ is not trimmed any more. Then, applying the trimming procedure to $\mathcal{C}_{|\varphi^{-1}(Q)}$ yields $\Pi_Q(\mathcal{C})$, which is trimmed and in canonical form. By abuse of notation, denote by $\Pi_Q(b)$ the image of $b \in \mathcal{C}_{|\varphi^{-1}(Q)}$ under this operation. The projection possesses the following universal property:

$$\forall \psi : \mathcal{C} \mapsto \mathcal{C}', \text{ and } \mathcal{C}' \text{ has label set } Q \implies \exists \psi' : \quad \begin{array}{ccc} \mathcal{C} & \xrightarrow{\Pi_Q} & \Pi_Q(\mathcal{C}) \\ & {}_{\psi}\searrow & \downarrow \psi' \\ & & \mathcal{C}' \end{array} \tag{6}$$

In (6), symbols ψ, ψ', Π_Q are morphisms, and the diagram commutes.

The *composition* of the two trimmed condition structures $\mathcal{C}_i = (B_i, \preceq_i, \#_i, \varphi_i)$, $i = 1, 2$, where labeling φ_i takes its values in alphabet P_i, is the condition structure

$$\begin{array}{l} \mathcal{C}_1 \wedge \mathcal{C}_2 = (B, \preceq, \#, \varphi), \text{ with two associated morphisms} \\ \psi_i : \mathcal{C}_1 \wedge \mathcal{C}_2 \mapsto \mathcal{C}_i, i = 1, 2, \end{array} \tag{7}$$

satisfying the following universal property:

$$
\forall \psi'_1, \psi'_2 : \quad
\begin{array}{ccc}
 & C' & \\
{}^{\psi'_1}\swarrow & & \searrow^{\psi'_2} \\
C_1 & & C_2
\end{array}
\quad \Longrightarrow \quad \exists \psi' : \quad
\begin{array}{ccccc}
 & & C' & & \\
 & {}^{\psi'_1}\swarrow & \downarrow \psi' & \searrow^{\psi'_2} & \\
C_1 & \xleftarrow[\psi_1]{} & C_1 \wedge C_2 & \xrightarrow[\psi_2]{} & C_2
\end{array}
\qquad (8)
$$

In (8), the ψ's denote morphisms, and the second diagram commutes. The composition is inductively defined as follows (we use the canonical form):

1. $\min(C_1 \wedge C_2) =_{\text{def}} \min(C_1) \cup \min(C_2)$, where we identify $(nil, p) \in C_1$ and $(nil, p) \in C_2$ for $p \in P_1 \cap P_2$. The causality relation and labeling map follow from the canonical form, and $\# =_{\text{def}} \#_1 \cup \#_2$. The canonical surjections $\psi_i : \min(C_1 \wedge C_2) \mapsto \min(C_i), i = 1, 2$ are morphisms.
2. Assume $C' =_{\text{def}} C'_1 \wedge C'_2$ together with the morphisms $\psi_i : C' \mapsto C'_i$ are defined, for C'_i a finite prefix of C_i, and $i = 1, 2$. Then, using (4) we define, for all co-sets X of C':

$$
\left.\begin{array}{r} X \subset dom(\psi_1), p \in P_1 \setminus P_2 \\ C'_1 \rightarrow (\psi_1(X), p) \end{array}\right\} \Rightarrow \quad C' \rightarrow (X, p) \qquad \text{(i)}
$$

$$
\left.\begin{array}{r} X \subset dom(\psi_1), p \in P_1 \cap P_2 \\ C'_1 \rightarrow (\psi_1(X), p) \end{array}\right\} \Rightarrow \quad C' \rightarrow (X, p) \qquad \text{(i')}
$$

$$
\left.\begin{array}{r} X \subset dom(\psi_2), p \in P_2 \setminus P_1 \\ C'_2 \rightarrow (\psi_2(X), p) \end{array}\right\} \Rightarrow \quad C' \rightarrow (X, p) \qquad \text{(ii)}
$$

$$
\left.\begin{array}{r} X \subset dom(\psi_2), p \in P_1 \cap P_2 \\ C'_2 \rightarrow (\psi_2(X), p) \end{array}\right\} \Rightarrow \quad C' \rightarrow (X, p) \qquad \text{(ii')}
$$

$$
\left.\begin{array}{r} p \in P_1 \cap P_2 \\ C'_1 \rightarrow (\psi_1(X), p) \\ C'_2 \rightarrow (\psi_2(X), p) \end{array}\right\} \Rightarrow \quad C' \rightarrow (X, p) \qquad \text{(iii)}
$$

Some comments are in order. The above five rules overlap: if rule (iii) applies, then we could have applied as well rule (i') with $Y = X \cap dom(\psi_1)$ in lieu of X, and the same for (ii'). Thus we equip rules (i–iii) with a set of priorities (a rule with priority 2 applies only if no rule with priority 1 is enabled):

$$
\text{rules (i,ii,iii) have priority 1, rules (i',ii') have priority 2.} \qquad (9)
$$

For the five cases (i,i',ii,ii',iii), extend $\psi, i = 1, 2$ as follows:

$$
\begin{array}{rl}
\psi_1(X, p_1) =_{\text{def}} (\psi_1(X), p_1) & \text{(i, i')} \\
\psi_2(X, p_2) =_{\text{def}} (\psi_2(X), p_2) & \text{(ii, ii')} \\
\text{for } i = 1, 2 : \psi_i(X, p) =_{\text{def}} (\psi_i(X), p) & \text{(iii),}
\end{array}
$$

where convention (4) is used.

Using the above rules, define the triple $(B, \preceq, \varphi)$ as being the smallest[5] triple containing $\min(C_1 \wedge C_2)$, and such that no extension using rules (i,i',ii,ii',iii)

[5] for set inclusion applied to the sets of conditions.

applies. It remains to equip the triple $(B, \preceq, \varphi)$ with the proper conflict relation. The conflict relation # on $\mathcal{C}_1 \wedge \mathcal{C}_2$ is defined as follows:

$$\# \text{ is the smallest conflict relation containing } \psi_1^{-1}(\#_1) \cup \psi_2^{-1}(\#_2). \quad (10)$$

Comments. The composition of labeled event structures combines features from the product of labeled nets (its use of synchronizing labels), and from unfoldings (its inductive construction). Note that this composition is not strictly synchronizing, due to the special rules (i',ii'), which are essential in ensuring (8).

4.3 Condition Structures Obtained from Unfoldings

Consider a Petri net $\mathcal{P} = (P, T, \rightarrow, M_0)$ with its unfolding $\mathcal{U}_\mathcal{P}$. Let $\mathcal{C}_\mathcal{P} = (B, \preceq, \#, \varphi)$ be the condition structure obtained by erasing the events in $\mathcal{U}_\mathcal{P}$. Such condition structures are of finite width.

Lemma 2. *If $\mathcal{P}$ satisfies the following condition:*

$$\forall p \in P \;:\; t, t' \in {}^\bullet p \textit{ and } t \neq t' \;\Rightarrow\; {}^\bullet t \neq {}^\bullet t'. \quad (11)$$

then $\mathcal{C}_\mathcal{P}$ is a trimmed condition structure, i.e., it satisfies (5). Unless otherwise stated, all Petri nets we shall consider satisfy this condition.

Condition (11) can always be enforced by inserting dummy places and transitions, as explained next. Assume that ${}^\bullet t \rightarrow t \rightarrow p$ and ${}^\bullet t' \rightarrow t' \rightarrow p$ with $t' \neq t$ but ${}^\bullet t' = {}^\bullet t$. Then, replace the path ${}^\bullet t \rightarrow t \rightarrow p$ by ${}^\bullet t \rightarrow t \rightarrow q_{t,p} \rightarrow s_{t,p} \rightarrow p$, where $q_{t,p}$ and $s_{t,p}$ are a fresh place and a fresh transition associated to the pair (t, p). Perform the same for the other path ${}^\bullet t' \rightarrow t' \rightarrow p$. This is a mild transformation, of low complexity cost, which does not modify reachability properties.

Important results about condition structures obtained from unfoldings are collected below. In this result, $\mathbf{1} =_{\text{def}} \emptyset$ denotes the empty condition structure, and $\mathcal{C}, \mathcal{C}_1, \mathcal{C}_2$ denote arbitrary condition structures with respective label sets P, P_1, P_2, and label set Q is arbitrary unless otherwise specified.

Theorem 1. *The following properties hold:*

$$\begin{aligned}
\Pi_{P_1}(\mathcal{C}) \wedge \Pi_{P_2}(\mathcal{C}) &= \Pi_{P_1 \cup P_2}(\mathcal{C}) && \text{(a0)}\\
\Pi_{P_1}(\Pi_{P_2}(\mathcal{C})) &= \Pi_{P_1 \cap P_2}(\mathcal{C}) && \text{(a1)}\\
\Pi_{P}(\mathcal{C}) &= \mathcal{C} && \text{(a2)}\\
\forall Q \supseteq P_1 \cap P_2 : \Pi_Q(\mathcal{C}_1 \wedge \mathcal{C}_2) &= \Pi_Q(\mathcal{C}_1) \wedge \Pi_Q(\mathcal{C}_2) && \text{(a3)}\\
\mathcal{C} \wedge \mathbf{1} &= \mathcal{C} && \text{(a4)}\\
\mathcal{C} \wedge \Pi_Q(\mathcal{C}) &= \mathcal{C} && \text{(a5)}
\end{aligned}$$

4.4 Discussion

To our knowledge, compositional theories for unfoldings or event structures have received very little attention so far. The work of Esparza and Römer [10] investigates unfoldings for synchronous products of transition systems. The central issue of this reference is the construction of finite complete prefixes. However, no product is considered, for the unfoldings themselves. To our knowledge, the work closest to ours is that of J-M. Couvreur et al. [7]–Sect. 4. Their motivations are different from ours, since they aim at constructing complete prefixes in a modular way, for Petri nets that have the form $\mathcal{P} = \mathcal{P}_1 \times \ldots \times \mathcal{P}_k$ (synchronous product). It is required that the considered Petri nets are not *reentrant,* a major limitation due to the technique used.

We are now ready to formally state our distributed diagnosis problem.

5 Distributed Diagnosis: Formal Problem Setting [3]

We are given the following labeled Petri nets:

$\mathcal{P} = (P, T, \rightarrow, M_0, \lambda)$ **:** the underlying "true" system. $\mathcal{P}$ is subject to faults, thus places from $\mathcal{P}$ are labelled by *faults,* taken from some finite alphabet (the non-faulty status is just one particular "fault"). The labeling map λ associates, to each transition of $\mathcal{P}$, a label belonging to some finite alphabet A of *alarm labels.* For its supervision, $\mathcal{P}$ produces so-called *alarm patterns,* i.e., sets of causally related alarms.

$\mathcal{Q} = (P^{\mathcal{Q}}, T^{\mathcal{Q}}, \rightarrow, M_0^{\mathcal{Q}}, \lambda^{\mathcal{Q}})$ **:** $\mathcal{Q}$ represents the faulty behaviour of $\mathcal{P}$, as observed via the sensor system. Thus we require that: (i) The labeling maps of $\mathcal{Q}$ and $\mathcal{P}$ take their values in the same alphabet A of alarm labels, and (ii) $\mathcal{L}_{\mathcal{Q}} \supseteq \mathcal{L}_{\mathcal{P}}$, i.e., the language of $\mathcal{Q}$ contains the language of $\mathcal{P}$. In general, however, $\mathcal{Q} \neq \mathcal{P}$. For example, if a single sensor is assumed, which collects alarms in sequence by preserving causalities (as assumed in [3]), then $\mathcal{Q}$ is the net which produces all linear extensions of runs of $\mathcal{P}$. In contrast, if several independent sensors are used, then the causalities between events collected by different sensors are lost. Configurations of $\mathcal{Q}$ are called *alarm patterns.*

Global Diagnosis. Consider the map: $\mathcal{A} \mapsto \mathcal{U}_{\mathcal{A}\times\mathcal{P}}$, where $\mathcal{A}$ ranges over the set of all finite alarm patterns. This map filters out, during the construction of the unfolding $\mathcal{U}_{\mathcal{P}}$, those configurations which are not compatible with the observed alarm pattern $\mathcal{A}$. Thanks to Lemma 2, we can replace the unfolding $\mathcal{U}_{\mathcal{A}\times\mathcal{P}}$ by the corresponding condition structure $\mathcal{C}_{\mathcal{A}\times\mathcal{P}}$. Then, we can project away, from $\mathcal{C}_{\mathcal{A}\times\mathcal{P}}$, the conditions labeled by places from $\mathcal{A}$ (all this is detailed in [3]–Theorem 1). Thus we can state:

Definition 1. *Global diagnosis is represented by the following map:*

$$\mathcal{A} \longmapsto \Pi_P(\mathcal{C}_{\mathcal{A}\times\mathcal{P}}), \tag{12}$$

where $\mathcal{A}$ ranges over the set of all finite configurations of $\mathcal{Q}$.

Modular Diagnosis. Assume that Petri net $\mathcal{P}$ decomposes as $\mathcal{P} = \|_{i\in I}\mathcal{P}_i$. The different subsystems $\mathcal{P}_i$ interact via some shared places, and their sets of transitions are pairwise disjoint. In particular, the alphabet A of alarm labels decomposes as $A = \bigcup_{i \in I} A_i$, where the A_i are pairwise disjoint. Next, we assume that each subsystem $\mathcal{P}_i$ possesses its own local sets of sensors, and the local sensor subsystems are independent, i.e., do not interact. Thus $\mathcal{Q}$ also decomposes as $\mathcal{Q} = \|_{i \in I} \mathcal{Q}_i$, and the $\mathcal{Q}_i$ possess pairwise disjoint sets of places. Consequently, in (12), $\mathcal{A}$ decomposes as $\mathcal{A} = \|_{i \in I} \mathcal{A}_i$, where the $\mathcal{A}_i$, the locally recorded alarm patterns, possess pairwise disjoint sets of places too.

Definition 2. *Modular diagnosis is represented by the following map:*

$$\mathcal{A} \longmapsto [\Pi_{P_i}(\mathcal{C}_{\mathcal{A} \times \mathcal{P}})]_{i \in I}, \tag{13}$$

where $\mathcal{A}$ ranges over the set of all finite prefixes of runs of $\mathcal{Q}$.

Note that, thanks to Theorem 1, we know that $\Pi_P(\mathcal{C}_{\mathcal{A} \times \mathcal{P}}) = \bigwedge_{i \in I} [\Pi_{P_i}(\mathcal{C}_{\mathcal{A} \times \mathcal{P}})]$, i.e., fusing the local diagnoses yields global diagnosis. However, we need to compute modular diagnosis without computing global diagnosis. On the other hand, the reader should notice that, in general, $\Pi_{P_i}(\mathcal{C}_{\mathcal{A} \times \mathcal{P}}) \neq \mathcal{C}_{\mathcal{A}_i \times \mathcal{P}_i}$, expressing the fact that the different supervisors must cooperate at establishing a coherent modular diagnosis.

6 Distributed Orchestration of Modular Diagnosis [12, 13]

This section is essential. It provides a framework for the high-level orchestration of the distributed computation of modular diagnosis. We first link our problem with the seemingly different areas of distributed constraint solving and distributed optimization.

6.1 A Link with Distributed Constraint Solving

Consider Theorem 1, and re-interpret, for a while, the involved objects differently. Suppose that our above generic label set P is a set of variables, thus $p \in P$ denotes a variable. Then, suppose that all considered variables possess a finite domain, and that $\mathcal{C}$ generically denotes a *constraint* on the tuple P of variables, i.e., $\mathcal{C}$ is a subset of all possible values for this tuple. For $Q \subset P$, re-interpret $\Pi_Q(\mathcal{C})$ as the projection of $\mathcal{C}$ onto Q. Then, re-interpret $\wedge$ as the conjunction of constraints. Finally, $\mathbf{1}$ is the trivial constraint, having empty set of associated variables. It is easily seen that, whith this re-interpretation, properties (a0–a5) are satisfied.

Modular constraint solving consists in computing $\Pi_{P_i}(\bigwedge_{j \in I} \mathcal{C}_j)$ without computing the global solution $\bigwedge_{j \in I} \mathcal{C}_j$. Distributed constraint solving consists in computing $\Pi_{P_i}(\bigwedge_{j \in I} \mathcal{C}_j)$ in a distributed way. Thus distributed constraint solving is a simpler problem, which is representative of our distributed diagnosis problem, when seen at the proper abstract level.

A Chaotic Algorithm for Distributed Constraint Solving. We build a graph on the index set I as follows: draw a branch (i,j) iff $P_i \cap P_j \neq \emptyset$, i.e., $\mathcal{C}_i$ and $\mathcal{C}_j$ interact directly. Denote by $\mathcal{G}_I$ the resulting *interaction graph*. For $i \in I$, denote by $\mathrm{N}(i)$ the neighbourhood of i, composed of the set of j's such that $(i,j) \in \mathcal{G}_I$. Note that $\mathrm{N}(i)$ contains i. The algorithm assumes the existence of an "initial guess", i.e., a set of $\mathcal{C}_i, i \in I$ such that

$$\mathcal{C} = \bigwedge_{i \in I} \mathcal{C}_i, \tag{14}$$

and it aims at computing $\Pi_{P_i}(\mathcal{C})$, for $i \in I$, in a chaotic way. In the following algorithm, each site i maintains and updates, for each neighbour j, a message $\mathcal{M}_{i,j}$ toward j. Thus there are two messages per edge (i,j) of $\mathcal{G}_I$, one in each direction:

Algorithm 1

1. *Initialization:* for each edge $(i,j) \in \mathcal{G}_I$:

$$\mathcal{M}_{i,j} = \Pi_{P_i \cap P_j}(\mathbf{1}). \tag{15}$$

2. *Chaotic iteration:* until stabilization, select an edge $(i,j) \in \mathcal{G}_I$, and update:

$$\mathcal{M}_{i,j} := \Pi_{P_i \cap P_j}\left(\mathcal{C}_i \wedge \left[\bigwedge_{k \in \mathrm{N}(i) \setminus j} \mathcal{M}_{k,i}\right]\right). \tag{16}$$

3. *Termination:* for each $i \in I$, set:

$$\mathcal{C}_i^\star = \mathcal{C}_i \wedge \left[\bigwedge_{k \in \mathrm{N}(i)} \mathcal{M}_{k,i}\right]. \tag{17}$$

The following result belongs to the folklore of smoothing theory in statistics and control. It was recently revitalised in the area of so-called "soft" coding theory. In both cases, the result is stated in the context of distributed constrained optimization. In its present form, it has been proved in [13] and uses only the abstract setting of Sect. 6.1 with properties (a0–a5):

Theorem 2 ([13]). *Assume that $\mathcal{G}_I$ is a tree. Then Algorithm 1 converges in finitely many iterations, and $\mathcal{C}_i^\star = \Pi_{P_i}(\mathcal{C})$.*

An informal argument of why the result is true is illustrated in Fig. 8. Message passing, from the leaves to the thick node, is depicted by directed arrows. Thanks to (16), each directed arrow cumulates the effect, on its sink node, of the constraints associated with its set of ancestor nodes, where "ancestor" refers to the ordering defined by the directed arrows. Now, we provide some elementary calculation to illustrate this mechanism. Apply Algorithm 1 with the particular policy shown in Fig. 8, for selecting the successive branches of $\mathcal{G}_I$. Referring to this figure, select concurrently the branches $(9,5)$ and $(8,5)$, and then select the

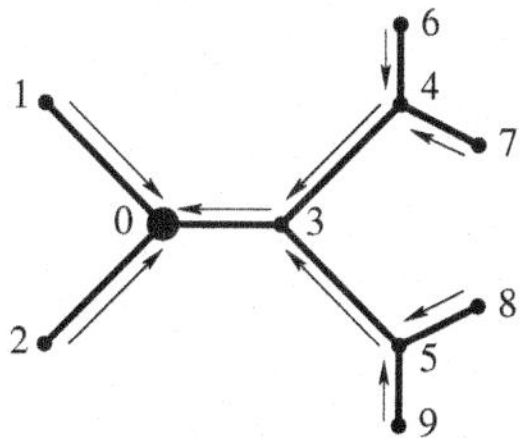

Fig. 8. Illustrating why Algorithm 1 performs well when $\mathcal{G}_I$ is a tree. Messages are generated first at the leaves. They meet at some arbitrarily selected node—the "center", here depicted in thick. Then they travel backward to the leaves.

branch $(5, 3)$. Then, successive applications of formula (16) yield:

$$\begin{aligned} \mathcal{M}_{(9,5)} &= \Pi_{P_9 \cap P_5}\left(\mathcal{C}_9 \wedge \left[\bigwedge_{k \in \mathrm{N}(9) \setminus 5} \mathcal{M}_{(k,i)}\right]\right) \\ &= \Pi_{P_9 \cap P_5}(\mathcal{C}_9) \text{ (since } \mathrm{N}(9) \setminus 5 = \emptyset) \\ \mathcal{M}_{(8,5)} &= \Pi_{P_8 \cap P_5}(\mathcal{C}_8) \\ \mathcal{M}_{(5,3)} &= \Pi_{P_5 \cap P_3}(\mathcal{C}_5 \wedge \mathcal{M}_{(9,5)} \wedge \mathcal{M}_{(8,5)}) \\ &= \Pi_{P_5 \cap P_3}(\mathcal{C}_5 \wedge \mathcal{C}_9 \wedge \mathcal{C}_8) \end{aligned} \tag{18}$$

The calculations can be further continued. They clearly yield the theorem for the center node $i = 0$, for the particular case where the branches are selected according to the policy shown in Fig. 8. A back-propagation to the leaves yields the result for all nodes. In this explanation, the messages were partially ordered, from the leaves to the thick node, and then vice-versa. Due to the monotonic nature of the algorithm, a chaotic and concurrent emission of the messages yields the same result, possibly at the price of exchanging a larger number of messages. Here is a formal proof borrowed from [13], we omit details.

Proof. Write $\mathcal{C} \leq \mathcal{C}'$ iff $\mathcal{C} = \mathcal{C}' \wedge \mathcal{C}''$ for some $\mathcal{C}''$. Using this notation, note that (16) is monotonic in the following sense: if $\forall k : \mathcal{M}_{k,i} \leq \mathcal{M}'_{k,i}$ in the right hand side of (16), then $\mathcal{M}_{i,j} \leq \mathcal{M}'_{i,j}$ in the left hand side of (16). Next, mark $\mathcal{M}_{i,j}$ in formula (16) with a running subset $J_{i,j} \subseteq I$, initialized with $J_{i,j} = \emptyset$:

$$\begin{aligned} \mathcal{M}_{i,j} &:= \Pi_{P_i \cap P_j}\left(\mathcal{C}_i \wedge \left[\bigwedge_{k \in \mathrm{N}(i) \setminus j} \mathcal{M}_{k,i}\right]\right) \\ J_{i,j} &:= J_{i,j} \cup \{i\} \cup \left[\bigcup_{k \in \mathrm{N}(i) \setminus j} J_{k,i}\right] \end{aligned} \tag{19}$$

Then, by using properties (a0–a5) and the monotonicity of (16), we get (left as an exercise to the reader):

$$\mathcal{M}_{i,j} = \Pi_{P_i \cap P_j}\left(\bigwedge_{k \in J_{i,j}} \mathcal{C}_k\right). \tag{20}$$

Hint: compare with (18). Verify that the assumption that $\mathcal{G}_I$ is a tree is used for applying repeatedly axiom (a3): this assumption guarantees that $(\bigwedge_{k \in J_{i,j}} \mathcal{C}_k)$ and $\mathcal{C}_j$ interact only via $P_i \cap P_j$. From (20) the theorem follows easily. $\diamond$

6.2 A Link with Distributed Constrained Optimization

As announced before, Theorem 2 generalizes to distributed constrained optimization. For this case, $\mathcal{C}$ becomes a pair $\mathcal{C}$ =(constraint, cost)$=_{\text{def}}$ $(\mathbf{C}, \mathbf{J})$. Constraints are as before, and costs have the following additive form:

$$\mathbf{J}(x) = \sum_{p \in P} \mathbf{j}_p(x_p), \tag{21}$$

where $x_p \in dom(p)$, the domain of variable p, $x = (x_p)_{p \in P}$, and the local costs $\mathbf{j}_p$ are real-valued, nonnegative cost functions. We require that $\mathbf{J}$ be normalized:

$$\sum_{x \models \mathbf{C}} \exp(\mathbf{J}(x)) = 1, \tag{22}$$

where $x \models \mathbf{C}$ means that x satisfies constraint $\mathbf{C}$, and exp(.) denotes the exponential. In the following, the notation $\mathbf{Cst}$ will denote generically a normalization factor whose role is to ensure condition (22). Then, define:

$$\Pi_Q(\mathbf{J})(x_Q) =_{\text{def}} \frac{1}{\mathbf{Cst}} \max_{x : \Pi_Q(x) = x_Q} \mathbf{J}(x), \tag{23}$$

In (23), $\Pi_Q(x)$ denotes the projection of x on Q. Then, define $\Pi_Q(\mathbf{C})$ to be the projection of constraint $\mathbf{C}$ on Q, i.e., the elimination from $\mathbf{C}$, by existential quantification, of the variables not belonging to Q. Finally, define $\Pi_Q(\mathcal{C}) =_{\text{def}} (\Pi_Q(\mathbf{C}), \Pi_Q(\mathbf{J}))$. Next, we define the composition $\wedge$. To this end, take for $\mathbf{C}_1 \wedge \mathbf{C}_2$ the conjunction of the considered constraints. And define:

$$(\mathbf{J}_1 \wedge \mathbf{J}_2)(x) =_{\text{def}} \frac{1}{\mathbf{Cst}} \left(\mathbf{J}_1(\Pi_{P_1}(x)) + \mathbf{J}_2(\Pi_{P_2}(x))\right). \tag{24}$$

It is easily checked that the properties (a0–a5) are still satisfied. Thus, Algorithm 1 solves, in a distributed way, the following problem:

$$\max_{x \models \mathbf{C}} \mathbf{J}(x), \tag{25}$$

for the case in which $\mathcal{C} =_{\text{def}} (\mathbf{C}, \mathbf{J})$ decomposes as $\mathcal{C} = \bigwedge_{i \in I} \mathcal{C}_i$.

Problem (25) can also be interpreted as maximum likelihood constraint solving, to resolve nondeterminism. In this case, the cost function $\mathbf{J}$ is interpreted as the logarithm of the likelihood (loglikelihood)—whence the normalization constraint (22). Then, the additive decomposition (21) for the loglikelihood means that the different variables are considered independent with respect to their prior distribution (i.e., when ignoring the constraints). In fact, the so defined systems $\mathcal{C}$ are Markov random fields, for which Algorithm 1 provides distributed maximum likelihood estimation. This viewpoint is closely related to belief propagation in belief networks [19, 20, 23].

6.3 Application to Off-Line Distributed Diagnosis

As said above, our framework of Sects. 4 and 5 satisfies properties (a0–a5). We can therefore apply algorithm 1 to perform off-line diagnosis, i.e., compute $\Pi_{P_i}(\mathcal{C}_{\mathcal{A}\times\mathcal{P}})$, cf. (13), for $\mathcal{A}$ being fixed. Now, we need an initial guess $\mathcal{C}_i$ satisfying (14), i.e., in our case: $\mathcal{C}_{\mathcal{A}\times\mathcal{P}} = \bigwedge_{i\in I} \mathcal{C}_i$. As we said, $\mathcal{C}_{\mathcal{A}_i\times\mathcal{P}_i}$ is not a suitable initial guess, since $\mathcal{C}_{\mathcal{A}\times\mathcal{P}} \neq \bigwedge_{i\in I} \mathcal{C}_{\mathcal{A}_i\times\mathcal{P}_i}$. Thus we need to find another one. Of course, we want this initial guess to be "cheap to compute", meaning at least that its computation is purely local and does not involve any cooperation between supervisors.

This is by no means a trivial task in general. However, in our running example, we can simply complement $\mathcal{P}_1$ by a new path $(3) \to [] \to (7)$, and $\mathcal{P}_2$ by a path $(7) \to [] \to (3)$, and the reader can check that (14) is satisfied by taking $\mathcal{C}_i =_{\text{def}} \mathcal{C}_{\mathcal{A}_i\times\bar{\mathcal{P}}_i}$, where $\bar{\mathcal{P}}_i, i = 1, 2$ denote the above introduced completions. In fact, this trick works for pairs of nets having a simple interaction involving only one pair of complementary places. It is not clear how to generalize this to more complex cases—fortunately, this difficulty disappears for the on-line algorithm, which is our very objective.

Anyway, having a correct initial guess at hand, Algorithm 1 applies, and yields the desired high-level orchestration for off-line distributed diagnosis. Each primitive operation of this orchestration is either a projection or a composition. For both, we have given the detailed definition above. All primitives are local to each site, i.e., involve only its private labels.

Again, since only properties (a0–a5) are required by Algorithm 1, we can also address the maximum likelihood extension discussed before (see [4, 17] for issues of randomizing Petri nets with a full concurrency semantics). This is the problem of maximum likelihood diagnosis that our prototype software solved, in the context described in Sect. 2.

6.4 An On-Line Variant of the Abstract Setting

Handling on-line diagnosis amounts to extending the results of Sect. 6.1 to "time-varying" structures in a certain sense [14]. We shall now complement the set of abstract properties (a0–a5) to prepare for the on-line case. Equip the set of condition structures with the following partial order:

$$\mathcal{C}' \sqsubseteq \mathcal{C} \text{ iff } \mathcal{C} \text{ is a prefix of } \mathcal{C}', \tag{26}$$

please note the inversion! To emphasize the analogy with constraint solving, $\mathcal{C}' \sqsubseteq \mathcal{C}$ reads: $\mathcal{C}'$ *refines* $\mathcal{C}$. Note that

$$\mathcal{C}' \sqsubseteq \mathcal{C} \text{ holds in particular if}: \mathcal{C}' = \mathcal{C}_{\mathcal{A}'\times\mathcal{P}}, \mathcal{C} = \mathcal{C}_{\mathcal{A}\times\mathcal{P}}, \mathcal{A} \sqsubseteq \mathcal{A}', \tag{27}$$

this is the situation encountered in incremental diagnosis. The following result complements Theorem 1:

Theorem 3. *The following properties hold:*

$$\begin{array}{rr} \mathcal{C} \sqsubseteq \mathbf{1} & \text{(a6)} \\ \mathcal{C}_1 \sqsubseteq \mathcal{C}_2 \Rightarrow \quad \mathcal{C}_1 \wedge \mathcal{C}_3 \sqsubseteq \mathcal{C}_2 \wedge \mathcal{C}_3 & \text{(a7)} \\ \mathcal{C}' \sqsubseteq \mathcal{C} \Rightarrow \forall Q : \Pi_Q(\mathcal{C}') \sqsubseteq \Pi_Q(\mathcal{C}) & \text{(a8)} \end{array}$$

Distributed Constraint Solving with Monotonically Varying Constraints. Consider again our re-interpretation as distributed constraint solving. Now, instead of constraint $\mathcal{C}_i$ being given once and for all, we are given a *set of constraints* $\mathbf{C}_i$ *ordered by* $\sqsubseteq$. More precisely, for $\mathcal{C}_i, \mathcal{C}'_i \in \mathbf{C}_i$, write $\mathcal{C}_i \sqsubseteq \mathcal{C}'_i$ iff $\mathcal{C}_i \Rightarrow \mathcal{C}'_i$, i.e., the former refines the latter. We assume that $\mathbf{C}_i$ is a lattice, i.e., that the supremum of two constraints exists in $\mathbf{C}_i$. Since all domains are finite, then $\mathcal{C}_i^\infty = \lim_{\sqsubseteq}(\mathbf{C}_i)$ is well defined. Algorithm 1 is then modified as follows, [14]:

Algorithm 2

1. *Initialization:* for each edge $(i,j) \in \mathcal{G}_I$:
$$\mathcal{M}_{i,j} = \Pi_{P_i \cap P_j}(\mathbf{1}). \tag{28}$$
2. *Chaotic nonterminating iteration:* Choose nondeterministically, in the following steps:
CASE 1: select a node $i \in I$ and update:
$$\text{read } \mathcal{C}_i^{\mathsf{new}} \sqsubseteq \mathcal{C}_i^{\mathsf{cur}}, \text{ and update } \mathcal{C}_i^{\mathsf{cur}} := \mathcal{C}_i^{\mathsf{new}}. \tag{29}$$
CASE 2: select an edge $(i,j) \in \mathcal{G}_I$, and update:
$$\mathcal{M}_{i,j} := \mathit{\Pi}_{P_i \cap P_j}(\mathcal{C}_i^{\mathsf{cur}} \wedge [\textstyle\bigwedge_{k \in \mathrm{N}(i) \setminus j} \mathcal{M}_{k,i}]).. \tag{30}$$
CASE 3: Update subsystems: select $i \in I$, and set:
$$\mathcal{C}_i^\star = \mathcal{C}_i^{\mathsf{cur}} \wedge [\textstyle\bigwedge_{k \in \mathrm{N}(i)} \mathcal{M}_{k,i}]. \tag{31}$$

In step 2, $\mathcal{C}_i^{\mathsf{cur}}$ denotes the current estimated value for $\mathcal{C}_i$, whereas $\mathcal{C}_i^{\mathsf{new}}$ denotes the new, refined, version. Algorithm 2 is *fairly* executed if it is applied in such a way that every node i of CASE 1 and CASE 3, and every edge (i,j) of CASE 2 is selected infinitely many times.

Theorem 4 ([14]). *Assume that $\mathcal{G}_I$ is a tree, and Algorithm 2 is fairly executed. Then, for any given $\mathcal{C} = \bigwedge_{i \in I} \mathcal{C}_i$, where $\mathcal{C}_i \in \mathbf{C}_i$, after sufficiently many iterations, one has $\forall i \in I : \mathcal{C}_i^\star \sqsubseteq \Pi_{P_i}(\mathcal{C})$.*

Theorem 4 expresses that, modulo a fairness assumption, Algorithm 2 refines, with some delay, the solution $\Pi_{P_i}(\mathcal{C})$ of any given intermediate problem $\mathcal{C}$.

Proof. It refines the proof of Theorem 2. The monotonicity argument applies here with the special order $\sqsubseteq$. Due to our fairness assumption, after sufficiently many iterations, each node i has updated its $\mathcal{C}_i^{\mathsf{cur}}$ in such a way that $\mathcal{C}_i^{\mathsf{cur}} \sqsubseteq \mathcal{C}_i$. Select such a status of Algorithm 2, and then start marking the recursion (30) as in (19). Applying the same reasoning as for the proof of Theorem 2 yields that $\mathcal{M}_{i,j} \sqsubseteq \Pi_{P_i \cap P_j}(\bigwedge_{k \in J_{i,j}} \mathcal{C}_k)$, from which Theorem 4 follows. ◇

How Algorithms 1 and 2 behave when $\mathcal{G}_I$ possesses cycles is discussed in [1, 15].

Application to On-Line Distributed Diagnosis. Here we consider the case in which a possibly infinite alarm pattern is observed incrementally: $\mathcal{A}^\infty = \|_{i\in I}\, \mathcal{A}_i^\infty$, meaning that sensor i receives only finite prefixes $\mathcal{A}_i \sqsubseteq \mathcal{A}_i^\infty$, partially ordered by inclusion. Now, by (27):

$$\forall i \in I : \mathcal{A}'_i \sqsubseteq \mathcal{A}_i \Rightarrow \mathcal{C}_i = \mathcal{C}_{\mathcal{A}_i \times \mathcal{P}_i} \sqsubseteq \mathcal{C}_{\mathcal{A}'_i \times \mathcal{P}_i} = \mathcal{C}'_i. \quad (32)$$

Thus on-line distributed diagnosis amounts to generalizing Algorithm 1 to the case in which subsystems $\mathcal{C}_i$ are updated on-line while the chaotic algorithm is running. Theorem 4 expresses that, if applied in a fair manner, then Algorithm 2 explains any finite alarm pattern after sufficiently many iterations.

For the off-line case, we mentioned that obtaining an "initial guess" for the $\mathcal{C}_i$ was a difficult issue. Now, since Algorithm 2 progresses incrementally, the issue is to compute the increment from $\mathcal{C}_i^{\mathsf{cur}}$ to $\mathcal{C}_i^{\mathsf{new}}$ in a "cheap" way. As detailed in [1], this can be performed locally, provided that the increment from $\mathcal{A}_i^{\mathsf{cur}}$ to $\mathcal{A}_i^{\mathsf{new}}$ is small enough.

Back to Our Running Example. Here we only relate the different steps of Algorithm 2 to the Fig. 6. Initialization is performed by starting from empty unfoldings on both supervisors. CASE 1 of step (a) consists, e.g., for supervisor 1, in recording the first alarm β ($\mathcal{A}_i^{\mathsf{cur}} = \emptyset$ and $\mathcal{A}_i^{\mathsf{new}} = \{\beta\}$), and then explaining β by the net $(1) \to \beta \to (2) \cup (1,7) \to \beta \to (2,3)$. CASE 2 of step (a) consists, e.g., for supervisor 1, in computing the abstraction of this net, for use by supervisor 2, this is shown by the first thick dashed arrow. Step (b), e.g., consists, for supervisor 2, in receiving the above abstraction and using it to append $(3,4) \to \alpha \to (7,5)$ as an additional explanation for its first alarm α; another explanation is the purely local one $(4) \to \alpha \to (6)$, which does not require the cooperation of supervisor 1.

7 Conclusion

For the context of fault management in SDH/SONET telecommunications networks, a prototype software implementing the method was developed in our laboratory, using Java threads to emulate concurrency. This software was subsequently deployed at Alcatel on a truly distributed experimental management platform. No modification was necessary to perform this deployment.

To ensure that the deployed application be autonomous in terms of synchronization and control, we have relied on techniques from true concurrency. The overall distributed orchestration of the application also required techniques originating from totally different areas related to statistics and information theory, namely belief propagation and distributed algorithms on graphical models. Only by blending those two orthogonal domains was it possible to solve our problem, and our work is a contribution in both domains.

Regarding concurrency theory, we have introduced a new compositional theory of modular event (or condition) structures. These objects form a category equipped with its morphisms, with a projection and two composition operations; it provides the adequate framework to support the distributed construction of unfoldings or event structures. It opens the way to using unfoldings or event structures as core data structures for distributed and asynchronous applications.

Regarding belief propagation, the work reported here presents an axiomatic form not known before. Also, time-varying extensions are proposed. This abstract framework allowed us to address distributed diagnosis, both off-line and on-line, and to derive types of algorithms not envisioned before in the field of graphical algorithms.

The application area which drives our research raises a number of additional issues for further investigation. Getting the model (the net $\mathcal{P}$) is the major one: building the model manually is simply not acceptable. We are developing appropriate software and models for a small number of generic management objects. These have to be instanciated on the fly at network discovery, by the management platform. This is a research topic in itself. From the theoretical point of view, the biggest challenge is to extend our techniques to dynamically changing systems. This is the subject of future research. Various robustness issues need to be considered: messages or alarms can be lost, the model can be approximate, etc. Probabilistic aspects are also of interest, to resolve nondeterminism by performing maximum likelihood diagnosis. The papers [4, 17] propose two possible mathematical frameworks for this, and a third one is in preparation.

References

1. extended version, available as IRISA report No 1540 http://www.irisa.fr/bibli/publi/pi/2003/1540/1540.html
2. A. Aghasaryan, C. Dousson, E. Fabre, Y. Pencolé, A. Osmani. Modeling Fault Propagation in Telecommunications Networks for Diagnosis Purposes. XVIII World Telecommunications Congress 22–27 September 2002 – Paris, France. Available: http://www.irisa.fr/sigma2/benveniste/pub/topic_distribdiag.html
3. A. Benveniste, E. Fabre, C. Jard, and S. Haar. Diagnosis of asynchronous discrete event systems, a net unfolding approach. *IEEE Trans. on Automatic Control,* 48(5), May 2003. Preliminary version available from http://www.irisa.fr/sigma2/benveniste/pub/IEEE_TAC_AsDiag_2003.html
4. A. Benveniste, S. Haar, and E. Fabre. Markov Nets: probabilistic Models for Distributed and Concurrent Systems. INRIA *Report* **4235**, 2001; available electronically at http://www.inria.fr/rrrt/rr-4754.html.
5. B. Caillaud, E. Badouel and Ph. Darondeau. Distributing Finite Automata through Petri Net Synthesis. *Journal on Formal Aspects of Computing,* 13, 447–470, 2002.
6. C. Cassandras and S. Lafortune. *Introduction to discrete event systems.* Kluwer Academic Publishers, 1999.
7. J.-M. Couvreur, S. Grivet, D. Poitrenaud, Unfolding of Products of Symmetrical Petri Nets, 22nd International Conference on Applications and Theory of Petri Nets (ICATPN 2001), Newcastle upon Tyne, UK, June 2001, LNCS 2075, pp. 121–143.
8. J. Desel, and J. Esparza. *Free Choice Petri Nets.* Cambridge University Press, 1995.
9. J. Engelfriet. *Branching Processes of Petri Nets.* Acta Informatica 28, 1991, pp. 575–591.

10. J. Esparza, S. Römer. An unfolding algorithm for synchronous products of transition systems. in proc. of CONCUR'99, LNCS Vol. 1664, Springer Verlag, 1999.
11. E. Fabre, A. Benveniste, C. Jard. Distributed diagnosis for large discrete event dynamic systems. In *Proc of the IFAC congress,* Jul. 2002.
12. E. Fabre. Compositional models of distributed and asynchronous dynamical systems. In *Proc of the 2002 IEEE Conf. on Decision and Control,* 1–6, Dec. 2002, Las Vegas, 2002.
13. E. Fabre. Monitoring distributed systems with distributed algorithms. In *Proc of the 2002 IEEE Conf. on Decision and Control,* 411–416, Dec. 2002, Las Vegas, 2002.
14. E. Fabre. Distributed diagnosis for large discrete event dynamic systems. In preparation.
15. E. Fabre. Convergence of the turbo algorithm for systems defined by local constraints. IRISA Res. Rep. 1510, 2003.
16. G.C. Goodwin and K.S. Sin. *Adaptive Filtrering, Prediction, and Control.* Prentice-Hall, Upper Sadle River, N.J. 1984.
17. S. Haar. Probabilistic Cluster Unfoldings. *Fundamenta Informaticae* 53(3–4), 281–314, 2002.
18. L. Lamport and N. Lynch. Distributed Computing: Models and Methods. in *Handbook of Theoretical Computer Science, Volume B: Formal Models and Semantics,* Jan van Leeuwen, editor, Elsevier (1990), 1157–1199.
19. S.L. Lauritzen. *Graphical Models,* Oxford Statistical Science Series 17, Oxford University Press, 1996.
20. S.L. Lauritzen and D.J. Spiegelhalter. Local computations with probabilities on graphical structures and their application to expert systems. *J. Royal Statistical Society, Series B,* 50(2), 157–224, 1988.
21. R.J. McEliece, D.J.C. MacKay, J.-F. Cheng, Turbo Decoding as an Instance of Pearl's Belief Propagation Algorithm. *IEEE Transactions on Selected Areas in Communication,* 16(2), 140–152, Feb. 1998.
22. M. Nielsen and G. Plotkin and G. Winskel. Petri nets, event structures, and domains. Part I. *Theoretical Computer Science* **13**:85–108, 1981.
23. J. Pearl. Fusion, Propagation, and Structuring in Belief Networks. *Artificial Intelligence,* 29, 241–288, 1986.
24. L.R. Rabiner and B.H. Juang. An introduction to Hidden Markov Models. *IEEE ASSP magazine* **3**, 4–16, 1986.
25. M. Raynal. *Distributed Algorithms and Protocols.* Wiley & Sons, 1988.
26. W. Reisig. *Petri nets.* Springer Verlag, 1985.
27. M. Sampath, R. Sengupta, K. Sinnamohideen, S. Lafortune, and D. Teneketzis. Failure diagnosis using discrete event models. *IEEE Trans. on Systems Technology,* 4(2), 105–124, March 1996.
28. Y. Weiss, W.T. Freeman, On the Optimality of Solutions of the Max-Product Belief-Propagation Algorithm in Arbitrary Graphs. *IEEE Trans. on Information Theory,* 47(2), 723–735, Feb. 2001.

Synthesis of Distributed Algorithms Using Asynchronous Automata

Alin Ştefănescu[1], Javier Esparza[1], and Anca Muscholl[2]

[1] Institut für Formale Methoden der Informatik
Universitätsstr. 38, 70569 Stuttgart, Germany
[2] LIAFA, Université Paris VII
2, place Jussieu, case 7014, F-75251 Paris Cedex 05

Abstract. We apply the theory of asynchronous automata to the synthesis problem of closed distributed systems. We use safe asynchronous automata as implementation model, and characterise the languages they accept. We analyze the complexity of the synthesis problem in our framework. Theorems by Zielonka and Morin are then used to develop and implement a synthesis algorithm. Finally, we apply the developed algorithms to the classic problem of mutual exclusion.

1 Introduction

We address the problem of automatically synthesising a finite-state, closed distributed system from a given specification. Seminal papers in this area are [EC82, MW84], where synthesis algorithms from temporal logic specifications are developed. The algorithms are based on tableau procedures for the satisfiability problem of CTL and LTL.

These approaches suffer from the limitation that the synthesis algorithms produce a sequential process P, and not a distributed implementation, i.e., a tuple $(P_1, \ldots, P_n)$ of communicating processes. The solution suggested in these works is to first synthesise the sequential solution, and then decompose it. However, since distribution aspects like concurrency and independency of events are not part of the CTL or LTL specification (and cannot be, since they are not bisimulation invariant), the solution may be impossible to distribute while keeping the intended concurrency. (This is in fact what happens with the solutions of [EC82,MW84] to the mutual exclusion problem)

A better approach to the problem consists in formally specifying not only the properties the system should satisfy, but also its architecture (how many components, and how they communicate). This approach was studied in [PR89] for *open* systems, in which the environment is an *adversary* of the system components, and the question is whether the system has a strategy that guarantees the specification against all possible behaviours of the environment. The realization problem (given the properties and the architecture, decide if there exists an implementation) was shown to be undecidable for arbitrary architectures, and decidable but non-elementary for hierarchical architectures vs. LTL specifications. Recent work [KV01] extends the decidability result (and the upper bound)

R. Amadio, D. Lugiez (Eds.): CONCUR 2003, LNCS 2761, pp. 27–41, 2003.

to CTL* specifications and linear architectures. To the best of our knowledge the synthesis procedures have not been implemented or tested on small examples.

In this paper we study the realization problem for the simpler case of *closed* systems, the original class of systems considered in [EC82,MW84]. This problem has been studied with unlabelled Petri nets (see e.g. [BD98]) and product transition systems (see [CMT99] and the references therein) as notions of implementation. In this paper, we attack the problem using *asynchronous automata* [Zie87,DR95]. Asynchronous automata can be seen as a tuple of concurrent processes communicating in a certain way (or as 1-safe *labelled* Petri nets). In our approach, a specification consists of two parts: a regular language L over an alphabet Σ of actions, containing all the finite executions that the synthesised system should be able to execute, and a tuple $(\Sigma_1, \ldots, \Sigma_n)$ of *local alphabets* indicating the actions in which the processes to be synthesised are involved; an action can be executed only if all processes involved in it are willing to execute it. The synthesis problem consists of producing a so-called *safe* asynchronous automaton whose associated processes have $(\Sigma_1, \ldots, \Sigma_n)$ as alphabets, and whose language is included in L (together with some other conditions to remove trivial solutions). The main advantage of our approach with respect to those of [BD98, CMT99] is its generality: Unlabelled Petri nets and product transition systems can be seen as strict subclasses of safe asynchronous automata.

The first two contributions of the paper are of theoretical nature. The first one is a refinement of Zielonka's theorem [Zie87], a celebrated central result of the theory of Mazurkiewicz traces. The refinement characterises the languages recognised by safe asynchronous automata, which we call *implementable* languages. (This result was also announced in [Muk02] without proof.) This result allows to divide the synthesis problem into two parts: (1) given a specification L, $(\Sigma_1, \ldots, \Sigma_n)$, decide if there exists an implementable language $L' \subseteq L$, and (2) given a such L', obtain a safe asynchronous automaton with L' as language. In the second contribution, we find that part (1) is undecidable, therefore we restrict our attention to an NP-complete subclass of solutions for which reasonable heuristics can be developed.

The third and main contribution of the paper is the development of heuristics to solve (1) and (2) in practice, their application to the mutual exclusion problem, and the evaluation of the results. The heuristic for (2) uses a result by Morin [Mor98] to speed up the synthesis procedure given by Zielonka in [Zie87]. Our heuristics synthesise two (maybe not 'elegant' but) new and far more realistic shared-variables solutions to the mutex problem than those of [EC82,MW84] (the results of [PR89], being for open systems and very generally applicable, did not provide any better automatically generated solution to the mutual exclusion problem). We make good use of the larger expressivity of asynchronous automata compared to unlabelled Petri nets and product transition systems: The first solution cannot be synthesised using Petri nets, and the second – the most realistic – cannot be synthesised using Petri nets or product transition systems.

The paper is structured as follows. Section 2 introduces asynchronous automata and Zielonka's theorem. In Sect. 3 we present the characterisation of implementable languages. Section 4 describes the synthesis problem together with heuristics for the construction of a solution and discusses complexity issues. Section 5 shows the synthesis procedure at work on the mutual exclusion problem. All the proofs can be found in [SEM03].

2 Preliminaries

We start with some definitions and notations about automata and regular languages. A *finite automaton* is a five tuple $\mathcal{A} = (Q, \Sigma, \delta, I, F)$ where Q is a finite set of *states*, Σ is a finite alphabet of *actions*, $I, F \subseteq Q$ are sets of *initial* and *final* states, respectively, and $\delta \subseteq Q \times \Sigma \times Q$ is the transition relation. We write $q \xrightarrow{a} q'$ to denote $(q, a, q') \in \delta$. The language recognised by $\mathcal{A}$ is defined as usual. A language is *regular* if it is recognised by some finite automaton. Given a language L, its *prefix closure* is the language containing all words of L together with all their prefixes. A language L is *prefix-closed* if it is equal to its prefix closure. Given two languages $L_1, L_2 \subseteq \Sigma^*$, we define their *shuffle* as $\text{shuffle}(L_1, L_2) := \{u_1 v_1 u_2 v_2 \ldots u_k v_k \mid k \geq 1, u_1 \ldots u_k \in L_1, v_1 \ldots v_k \in L_2 \text{ and } u_i, v_i \in \Sigma^*\}$.

We recall that regular languages are closed under boolean operations, that the prefix-closure of a regular language is regular, and that the shuffle of two regular languages is regular.

2.1 Asynchronous Automata

Let Σ be a nonempty, finite alphabet of *actions*, and let *Proc* be a nonempty, finite set of *process labels*. A *distribution* of Σ over *Proc* is a function $\Delta\colon Proc \to 2^{\Sigma} \backslash \emptyset$. Intuitively, Δ assigns to each process the set of actions it is involved in, which are the actions that cannot be executed without the process participating in it. It is often more convenient to represent a distribution by the *domain* function $dom : \Sigma \to 2^{Proc} \backslash \emptyset$ that assigns to each action the processes that execute it. We call the pair (Σ, dom) a *distributed alphabet*. A distribution induces an *independence relation* $\|\colon \Sigma \times \Sigma$ as follows: $\forall a, b \in \Sigma :\ a \| b \Leftrightarrow dom(a) \cap dom(b) = \emptyset$. I.e., two actions are independent if no process is involved in both. The intuition is that independent actions may occur concurrently.

An asynchronous automaton over a distributed alphabet is a finite automaton that can be distributed into communicating local automata. The states of the automaton are tuples of local states of the local automata.

Definition 1. An *asynchronous automaton* $\mathcal{AA}$ over a distributed alphabet (Σ, dom) is a finite automaton $(Q, \Sigma, \delta, I, F)$ such that there exist

- a family of sets of *local states* $(Q_k)_{k \in Proc}$, and
- a relation $\delta_a \subseteq \prod_{k \in dom(a)} Q_k \times \prod_{k \in dom(a)} Q_k$ for each action $a \in \Sigma$

satisfying the following properties:

- $Q \subseteq \prod_{k \in Proc} Q_k$, with $I, F \subseteq Q$ initial and final states, and

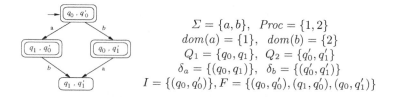

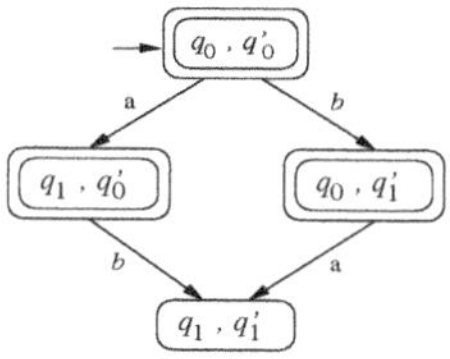

$$\Sigma = \{a, b\}, \ \ Proc = \{1, 2\}$$
$$dom(a) = \{1\}, \ \ dom(b) = \{2\}$$
$$Q_1 = \{q_0, q_1\}, \ \ Q_2 = \{q_0', q_1'\}$$
$$\delta_a = \{(q_0, q_1)\}, \ \ \delta_b = \{(q_0', q_1')\}$$
$$I = \{(q_0, q_0')\}, F = \{(q_0, q_0'), (q_1, q_0'), (q_0, q_1')\}$$

Fig. 1. An asynchronous automaton together with its formal description

- $(q, a, q') \in \delta \Leftrightarrow \begin{cases} \forall k \notin dom(a) : q_k = q'_k \\ \big((q_k)_{k \in dom(a)}, (q'_k)_{k \in dom(a)}\big) \in \delta_a \end{cases}$
  where q_k denotes the k-th component of q, and $(q_k)_{k \in dom(a)}$ denotes the projection of q onto $dom(a)$.

The language recognised by an asynchronous automaton is the language it recognises as a finite automaton. If all δ_a's are functions and I contains only one element, then $\mathcal{AA}$ is called *deterministic.*

Figure 1 shows an asynchronous automaton. Intuitively, each set Q_k represents the set of states of a sequential component. Whether there is an a-transition between two states depends only on the projections of the states onto $dom(a)$, the local states of the other components are irrelevant; moreover the execution of a only changes the local state of the processes in $dom(a)$. In particular, if there is an a-transition between two global states q_1, q_2, then there must also be a-transitions between any states q'_1, q'_2 such that the projections of q_1, q'_1 and q_2, q'_2 on $dom(a)$ coincide. It is easy to see that, as a consequence, every asynchronous automaton satisfies the *independent* and *forward diamond rules*:

- **ID** : $q_1 \overset{a}{\to} q_2 \overset{b}{\to} q_4 \wedge a \| b \Rightarrow \exists q_3 : q_1 \overset{b}{\to} q_3 \overset{a}{\to} q_4$
- **FD** : $q_1 \overset{a}{\to} q_2 \wedge q_1 \overset{b}{\to} q_3 \wedge a \| b \Rightarrow \exists q_4 : q_2 \overset{b}{\to} q_4 \wedge q_3 \overset{a}{\to} q_4$.

Finally, observe that the accepting conditions of asynchronous automata are *global*: We need to know the local states of *all* the processes in order to determine if the tuple of local states is a final state.

2.2 Zielonka's Theorem

Zielonka's theorem characterises the languages accepted by asynchronous automata. Given a distributed alphabet (Σ, dom), we say that $L \subseteq \Sigma^*$ is a *trace language* if L is closed under the independence relation $\|$ associated to *dom*:

$$\forall a, b \in \Sigma \text{ and } \forall w, w' \in \Sigma^* : wabw' \in L \wedge a \| b \Rightarrow wbaw' \in L.$$

Theorem 1. *[Zie87] Let (Σ, dom) be a distributed alphabet, and let $L \subseteq \Sigma^*$. L is recognised by a finite asynchronous automaton with distribution dom if and only if it is a regular trace language. Moreover, if L is recognised by an asynchronous automaton, then it is also recognised by a deterministic asynchronous automaton.*

The proof of the theorem is constructive. Zielonka defines an effectively computable equivalence relation $\approx_Z \subseteq \Sigma^* \times \Sigma^*$ of finite index. The definition of $\approx_Z$ can be found in [Zie87]. Now, let T_L be the infinite automaton having L as set of states, and $w \xrightarrow{a} wa$ as transitions. The asynchronous automaton of Theorem 1 is the quotient of T_L under $\approx_Z$. The size of the automaton is single exponential in the size of the minimal deterministic automaton recognising L, and double exponential in the size of *Proc*.

The following shows that in order to decide if a language is a regular trace language, it suffices to compute the minimal automaton recognising it, and check if it satisfies ID.

Proposition 1. *Let (Σ, dom) be a distributed alphabet, and let $L \subseteq \Sigma^*$ regular. The following conditions are equivalent:*

1. *L is a regular trace language;*
2. *the minimal deterministic finite automaton recognising L satisfies ID.*

3 Implementable Specifications

As mentioned in the introduction, we use regular languages as specification of the set of global *behaviours* of a distributed system, where a behaviour is a finite sequence of actions. In this setting, asynchronous automata are not a realistic implementation model. The reason is best explained by means of an example. Let $\Sigma = \{a, b\}$ and $dom(a) = \{1\}$, $dom(b) = \{2\}$, and consider the language $L = \{\epsilon, a, b\}$. Intuitively, (L, dom) cannot be implemented: Since L contains both a and b, and a and b are executed *independently* of each other, nothing can prevent an implementation from executing ab and ba as well, which do not belong to L. However, the asynchronous automaton of Fig. 1 recognises L. The reason is that we can choose the global final states as $\{(0,0), (1,0), (0,1)\}$, excluding $(1,1)$. (Notice that if we remove $(1,1)$ from the set of states the automaton is no longer asynchronous, because it does not satisfy FD.) In our context, in which runs of the automaton should represent behaviours of a distributed system, this is not acceptable: We cannot declare *a posteriori* that a sequence of actions we have observed is not a behaviour because the state it reaches as non-final.

This example shows that we have to restrict our attention to asynchronous automata in which all states reachable from the initial states are final. We call such automata *safe*[1].

As we mentioned in the introduction, the synthesis of closed distributed systems has been studied before using unlabelled Petri nets [BD98] and product transition systems [Mor98,CMT99] as implementation models. Both models can be seen as subclasses of safe asynchronous automata in which, for each action a, the relation δ_a satisfies an additional condition. In the case of Petri nets, δ_a

[1] Safe (asynchronous) automata were studied by Zielonka in [Zie89]. *Safe* there means something weaker: All reachable states are co-reachable (i.e. there is a path from that state to a final one). However, the difference between the two definitions of *safe* vanishes when the recognised language is prefix-closed.

must contain at most one element. In the case of product transition systems, δ_a must have a product form: There must be a family of relations $\delta_a^k \subseteq Q_k \times Q_k$ such that $\delta_a = \prod_{k \in Proc} \delta_a^k$.

In the rest of this section we obtain the equivalent of Theorem 1 and Proposition 1 for safe asynchronous automata.

Definition 2. A regular trace language $L \subseteq \Sigma^*$ is called *implementable* if it satisfies:

- *prefix-closedness*: $\forall w, w', w'' \in \Sigma^* : w = w'w'' \in L \Rightarrow w' \in L$
- *safe-branching*[2]: $\forall w \in \Sigma^* : wa \in L \wedge wb \in L \wedge a \| b \Rightarrow wab \in L$.

Theorem 2. *Let (Σ, dom) be a distributed alphabet, and let $L \subseteq \Sigma^*$. L is recognised by a finite safe asynchronous automaton with distribution dom if and only if it is an implementable trace language. Moreover, if L is recognised by a safe asynchronous automaton, then it is also recognised by a safe deterministic asynchronous automaton.*

A proof is given in [SEM03] and a constructive one follows from Proposition 3.

Proposition 2. *Let (Σ, dom) be a distributed alphabet, and let $L \subseteq \Sigma^*$ regular. The following conditions are equivalent:*

1. *L is an implementable language;*
2. *the minimal deterministic finite automaton recognising L is safe and satisfies ID and FD.*

This result provides an inexpensive test to check if a specification (L, dom) is implementable: Compute the minimal automaton recognising L and check if it satisfies ID and FD and if all its states are final. These checks have linear time complexity in the size of the minimal automaton and in the size of the independence relation generated by *dom*.

Remark 1. Testing whether a specification is implementable is PSPACE-complete, when the input is a regular expression or a non-deterministic automaton (it can be easily shown that both checking ID and FD are PSPACE-complete).

It is not difficult to show that implementable languages are a proper superset of the Petri net languages and the languages of product transition systems (see [Zie87]). As we will see in Sect. 5.1, this is the fact that will allow us to derive implementations in our case studies.

4 The Synthesis Problem

In our setting, the synthesis problem is: Given a distributed alphabet (Σ, dom) and a regular language L_{Spec}, represented as a deterministic finite automaton $\mathcal{A}_{Spec}$, is there a safe asynchronous automaton $\mathcal{AA}$ such that $L(\mathcal{AA}) \subseteq$

2 In [Maz87] the property of *safe-branching* is called *properness*.

$L(\mathcal{A}_{Spec})$? In addition, we require that all the actions in Σ appear in $L(\mathcal{AA})$, because we are not interested in trivial solutions like $L(\mathcal{AA}) = \emptyset$ or partial solutions in which only some of the processes are executing actions. By definition, for a given language L, let $\Sigma(L) := \{a \in \Sigma \mid \exists u, v \in \Sigma^* \text{ with } uav \in L\}$ denote the actions appearing in L. Then, the set of actions *appearing* in an (asynchronous) automaton $\mathcal{A}$ is just $\Sigma(\mathcal{A}) := \Sigma(L(\mathcal{A}))$. We are now able to formulate:

Problem 1. (Synthesis problem) Given a distributed alphabet (Σ, dom) and a deterministic finite automaton $\mathcal{A}_{Spec}$ such that $\Sigma(\mathcal{A}_{Spec}) = \Sigma$, is there a safe asynchronous automaton $\mathcal{AA}$ such that $L(\mathcal{AA}) \subseteq L(\mathcal{A}_{Spec})$ and $\Sigma(\mathcal{AA}) = \Sigma$?

Theorem 3. *The synthesis problem is undecidable.*

Because of the undecidability of synthesis problem stated in Theorem 3, we attack a more modest but 'only' NP-complete problem, for which, as we can see, we can develop reasonable heuristics. This requires to introduce the notion of a subautomaton. We say that $\mathcal{A}' = (Q', \Sigma', \delta', I', F')$ is a *subautomaton* of $\mathcal{A} = (Q, \Sigma, \delta, I, F)$ if $Q' \subseteq Q$, $\Sigma' \subseteq \Sigma$, $I' \subseteq I$, $F' \subseteq F$ and $\delta' \subseteq \delta$.

In the *subautomata synthesis problem* we search for the language of $\mathcal{AA}$ only among the languages of the subautomata of $\mathcal{A}_{Spec}$. More precisely, we examine the languages of the subautomata, which are obviously included in $L(\mathcal{A}_{Spec})$, and determine if some of them is the language of an asynchronous automaton. Since the languages of safe asynchronous automata are those implementable, what we in fact do is to consider the following problem:

Problem 2. (Subautomata synthesis) Given a distributed alphabet (Σ, dom) and a deterministic finite automaton $\mathcal{A}_{Spec}$ such that $\Sigma(\mathcal{A}_{Spec}) = \Sigma$, is there a safe subautomaton $\mathcal{A}'$ of $\mathcal{A}_{Spec}$ with $\Sigma(\mathcal{A}') = \Sigma$ satisfying ID and FD?

A positive solution to an instance of this problem implies a positive solution to the same instance of the synthesis problem.

Theorem 4. *The subautomata synthesis problem is NP-complete.*

Let us now summarize our approach:

1. Choose the set of actions Σ of the system and a distribution Δ.
2. Describe the 'good' behaviours of the system as a regular language L_{Spec}. Usually, we give L_{Spec} as a base language (e.g. a shuffle of local behaviours), from which we filter out undesired behaviours (e.g. behaviours that lead to two processes in a critical section).
3. Construct $\mathcal{A}$ (usually, the minimal deterministic finite automaton) satisfying $L(\mathcal{A}) = L_{Spec}$.
4. Find a safe subautomaton $\mathcal{A}'$ of $\mathcal{A}$ with $\Sigma(\mathcal{A}') = \Sigma$ satisfying ID and FD. (see Sect. 4.1)
5. Apply Theorem 2 to obtain a safe asynchronous automaton $\mathcal{AA}$ satisfying $L(\mathcal{AA}) = L(\mathcal{A}')$. [3] (see Sect. 4.2)

[3] Note that we can apply Theorem 2, because the language of a safe automaton satisfying ID and FD is implementable.

4.1 Constructing a Subautomaton

Given an automaton $\mathcal{A}$, finding a safe subautomaton $\mathcal{A}'$ satisfying $\Sigma(\mathcal{A}') = \Sigma$, ID, and FD is NP-complete, so in the worst case this is exponentially expensive. In our experiments, we found two natural heuristics helpful in this problem:

1. [*destructive*] Starting with the initial automaton $\mathcal{A}$, we remove states and transitions that prevent the properties of safety, ID and FD to hold. So, if we have non-final states, we remove them; if we have a conflict w.r.t. FD (e.g., $\exists q_1 \stackrel{a}{\to} q_2$ and $\exists q_1 \stackrel{b}{\to} q_3$ with $a \| b$, but there exists no state q_4 such that $\exists q_2 \stackrel{b}{\to} q_4$ and $\exists q_3 \stackrel{a}{\to} q_4$), we remove one of the transition involved in the conflict (e.g., removing $q_1 \stackrel{b}{\to} q_3$ will solve the conflict); something similar for ID. In the process of removal we want to preserve $\Sigma(\mathcal{A}') = \Sigma$.
2. [*constructive*] Starting with the empty subautomaton, we add states and transitions until we find a safe subautomaton $\mathcal{A}'$ satisfying ID and FD. We apply a breadth-first traversal together with a 'greedy strategy' which selects transitions labelled by new action names and we do not add transitions violating the ID and FD rules and we do not add non-final states.

In both of the above strategies, we stop when we find a subautomaton satisfying our properties. Therefore, in general, the first heuristic will produce a larger solution than the second one. Larger solutions represent more behaviours, so better implementations for our synthesis problem. Unfortunately, this large subautomaton will serve as an input for Zielonka's procedure and this may blow-up the state space of the solution. That is why the second heuristic is usually preferred and the experimental results in Sect. 5.2 witness this fact.

4.2 Constructing an Asynchronous Automaton

The proof of Zielonka's theorem provides an algorithm to automatically derive an asynchronous automaton from an implementable language L (obtained as in the previous subsection). We start by giving here a version of the algorithm. The version is tailored so that we can easily add a heuristic that we describe in the second half of the section. Loosely speaking, the algorithm proceeds by *unfolding the minimal deterministic automaton recognising L until an asynchronous automaton is obtained.*

Data structure The algorithm maintains a deterministic reachable automaton $\mathcal{A}$ in which all states are final. The transitions of $\mathcal{A}$ are coloured *green*, *red*, or *black*. The algorithm keeps the following invariants:

1. The automaton $\mathcal{A}$ is deterministic and recognises L.
2. *Green* transitions form a directed spanning-tree of $\mathcal{A}$, i.e., a directed tree with the initial state q_0 as root and containing all states of $\mathcal{A}$.
3. Let $W(q)$ be the unique word w such that there is a path $q_0 \stackrel{w}{\to} q$ in the spanning-tree. For any $q \neq q'$ we have $W(q) \not\approx_Z W(q')$. (Notice that if a transition $q \stackrel{a}{\to} q'$ is green, then $W(q) \cdot a = W(q')$.)
4. A transition $q \stackrel{a}{\to} q'$ is *red* if $W(q) \cdot a \not\approx_Z W(q')$.

5. All other transitions are *black*.

Initially, $\mathcal{A}$ is the minimal deterministic finite automaton $\mathcal{A}_0 = (Q_0, \Sigma, \delta_0, q_0, Q_0)$ recognising the implementable language L. The set of green transitions can be computed by means of well-known algorithms. The other colours are computed to satisfy the invariants.

Algorithm If the current automaton has no red transitions, then the algorithm stops. Otherwise, it chooses a red transition $q \stackrel{a}{\to} q'$, and proceeds as follows:

a. Deletes the transition $q \stackrel{a}{\to} q'$.

b. If there is a state q'' such that $W(q) \cdot a \approx_Z W(q'')$ then the algorithm adds a black transition $q \stackrel{a}{\to} q''$.

c. Otherwise, the algorithm
  - *c*1. creates a new (final) state r,
  - *c*2. adds a new green transition $q \stackrel{a}{\to} r$ (and so $W(r) := W(q) \cdot a$), and
  - *c*3. for every transition $q' \stackrel{b}{\to} q''$, adds a new transition $r \stackrel{b}{\to} s$ with $s := \delta_0(q_0, W(r) \cdot b)$. The new transition is coloured red if $W(r) \cdot b \not\approx_Z W(s)$ and black otherwise.

Proposition 3. *The algorithm described above always terminates and its output is a safe deterministic finite asynchronous automaton recognising the implementable language L.*

Unfortunately, as we will see later in one of our case studies, the algorithm can produce automata with many more states than necessary. We have implemented a heuristic that allows to 'stop early' if the automaton synthesised so far happens to already be a solution, and otherwise guides the algorithm in the choice of the next red transition.

For this, we need a test that, given a distributed alphabet (Σ, dom) and an automaton $\mathcal{A}$, checks if $\mathcal{A}$ is an asynchronous automaton with respect to dom (i.e., checks the existence of the sets Q_k and the relations δ_a). Moreover, if $\mathcal{A}$ is not asynchronous, the test should produce a "witness" transition of this fact. Fortunately, Morin provides in [Mor98] precisely such a test:

Theorem 5. *[Mor98] Let $\mathcal{A}$ be a deterministic automaton and dom be a distribution. There is the least family of equivalences $(\equiv_k)_{k \in Proc}$ over the states of $\mathcal{A}$ such that (below we denote by $q \equiv_{dom(a)} q'$ if $\forall k \in dom(a) : q \equiv_k q'$)*

DE$_1$: $q \stackrel{a}{\to} q' \wedge k \notin dom(a) \Rightarrow q \equiv_k q'$

DE$_2$: $q_1 \stackrel{a}{\to} q_1' \wedge q_2 \stackrel{a}{\to} q_2' \wedge q_1 \equiv_{dom(a)} q_2 \Rightarrow q_1' \equiv_{dom(a)} q_2'$

Moreover $\mathcal{A}$ is an asynchronous automaton over dom iff the next two conditions hold for any states q, q' and any action a:

DS$_1$: $(\forall k \in Proc : q_1 \equiv_k q_2) \Rightarrow q_1 = q_2$

DS$_2$: $(\exists q_1' : q_1 \stackrel{a}{\to} q_1' \wedge q_1 \equiv_{dom(a)} q_2) \Rightarrow \exists q_2' : q_2 \stackrel{a}{\to} q_2'$

It is not difficult to show that the least equivalences $(\equiv_k)_{k \in Proc}$ can be computed in polynomial time by means of a fixpoint algorithm, and so Theorem 5 provides a polynomial test to check if $\mathcal{A}$ is asynchronous.

Because we are interested only in an asynchronous automaton accepting the same language as the initial automaton, a weaker version of Theorem 5 suffices: If the least family of equivalences satisfying DE_1 and DE_2 also satisfies DS_2 (but not necessarily DS_1), there exists an asynchronous automaton recognising the same language as $\mathcal{A}$.

Notice that if $\mathcal{A}$ passes the test then we can easily derive the sets Q_k of local states and the δ_a's functions for every action a: Q_k contains the equivalence classes of $\equiv_k$; given two classes q, q', we have $(q, q') \in \delta_a$ iff $\mathcal{A}$ contains an a-transition between some representatives of q, q'. We remark for later use in our case studies that the proof of [Mor98] proves in fact something stronger than Theorem 5: Any equivalence satisfying DE_1, DE_2, and DS_2 can be used to obtain an asynchronous automaton language-equivalent with $\mathcal{A}$. The least family is easy to compute, but it yields an implementation in which the sets Q_k are too large.

If $\mathcal{A}$ does not pass the test (this implies a red transition involved in the failure), the heuristic will propose a red transition to be processed by the algorithm. We find this transition by applying Morin's test to the subautomaton $\mathcal{A}_{g\&b}$ containing only the green and black transitions of $\mathcal{A}$. There are two cases: (1) the test fails and then we can prove that there is a red edge involved in the failure of DS_2 on $\mathcal{A}_{g\&b}$: $\exists q_1 \xrightarrow{a} q_1'$ green or black and $q_1 \equiv^{g\&b}_{dom(a)} q_2$ and $\exists q_2 \xrightarrow{a} q_2'$ red or (2) the test is successful and then we iteratively add red transitions to the subautomaton $\mathcal{A}_{g\&b}$ until DS_2 is violated. In either case, we find a red transition as a candidate for the unfolding algorithm.

5 Case Study: Mutual Exclusion

A *mutual exclusion* (*mutex* for short) situation appears when two or more processes are trying to access for 'private' use a common resource. A distributed solution to the mutex problem is a collection of programs, one for each process, such that their concurrent execution satisfies three properties: *mutual exclusion* (it is never the case that two processes have simultaneous access to the resource), *absence of starvation* (if a process requests access to the resource, the request is eventually granted), and *deadlock freedom.*

We consider first the problem for two processes. Let the actions be

$$\Sigma := \{req_1, enter_1, exit_1, req_2, enter_2, exit_2\}$$

with the intended meanings: request access to, enter and exit the critical section giving access to the resource. The indices 1 and 2 specify the process that executes the action.

We fix now a distribution. Obviously, we wish to have two processes P_1, P_2 such that $\Delta(P_1) = \{req_1, enter_1, exit_1\}$ and $\Delta(P_2) = \{req_2, enter_2, exit_2\}$. We also want $req_1 \| req_2$ so we need at least two extra processes V_1 and V_2, such that $\Delta(V_1)$ contains req_1 but not req_2, and $\Delta(V_2)$ contains req_2 but not req_1. So let:

$\Delta(V_1) = \{req_1, enter_1, exit_1, enter_2\}$ and $\Delta(V_2) = \{req_2, enter_2, exit_2, enter_1\}$.[4]

[4] We could also add $exit_1$ to $\Delta(V_1)$ and $exit_2$ to $\Delta(V_2)$; the solution does not change.

Next, we define a regular language, $Mutex_1$, specifying the desired behaviours of the system. We want $Mutex_1$ to be the maximal language satisfying the following conditions:

1. $Mutex_1$ is included in the shuffle of prefix-closures of $(req_1 enter_1 exit_1)^*$ and $(req_2 enter_2 exit_2)^*$.
   I.e., the processes execute $req_i enter_i exit_i$ in cyclic order.
2. $Mutex_1 \subseteq \Sigma^* \setminus [\Sigma^* enter_1 (\Sigma \setminus exit_1)^* enter_2 \Sigma^*]$ and its dual version.
   I.e., a process must exit before the other one can enter. This guarantees *mutual exclusion.*
3. $Mutex_1 \subseteq \Sigma^* \setminus [\Sigma^* req_1 (\Sigma \setminus enter_1)^* enter_2 (\Sigma \setminus enter_1)^* enter_2 \Sigma^*]$ and dual.
   I.e., after a request by one process the other process can enter the critical section at most once. This guarantees *absence of starvation.*
4. For any $w \in Mutex_1$, there exists an action $a \in \Sigma$ such that $wa \in Mutex_1$.
   This guarantees *deadlock freedom.*

Condition 3 needs to be discussed. In our current framework we cannot deal with 'proper' liveness properties, like: If a process requests access to the critical section, then the access will eventually be granted. This is certainly a shortcoming of our current framework. In this example, we enforce absence of starvation by putting a concrete bound on the number of times a process can enter the critical section after a request by the other process.

The largest language satisfying conditions 1-3 is regular because of the closure properties of regular languages, and a minimal automaton recognising it can be easily computed. Since it is deadlock-free, it recognises the largest language satisfying conditions 1-4. [5]

It turns out that the minimal automaton $\mathcal{A}_1$ for $Mutex_1$ is safe, satisfies ID, FD, and $\Sigma(\mathcal{A}_1) = \Sigma$. Using Proposition 2 the recognised language is implementable. This allows us to apply Zielonka's construction, that yields a safe asynchronous automaton with **34** states. Applying our heuristic based on Morin's test we obtain that the minimal automaton recognising $Mutex_1$ and having **14** states, is already an asynchronous automaton. Families of local states and transitions can be constructed using Morin's theorem. The processes P_1 and P_2 have three local states each, while the processes V_1 and V_2 have 7 states.

We can now ask if the solution can be simplified, i.e., if there is a smaller family of local states making the minimal automaton asynchronous. This amounts to finding a larger family $(\equiv_k)_{(k \in Proc)}$ of equivalences satisfying the properties of Theorem 5. This can be done by merging equivalence classes, and checking if the resulting equivalences still satisfy the properties. We have implemented this procedure and it turns out that there exists another solution in which V_1 and V_2 have only 4 states. Figure 2 (top) shows the resulting asynchronous automaton, translated into pseudocode for legibility. There $\langle com \rangle$ denotes that the command *com* is executed in one single atomic step. We have represented the processes

[5] If this had not been the case, the largest automaton would have been obtained by removing all states not contained in any infinite path.

```
Initialization: v1 := 0; v2 := 0
                         Process 1 || Process 2
repeat forever                     || repeat forever
  [NCS1];                          ||   [NCS2];
  v1 := 1;                         ||   v2 := 1;
  ⟨ await (v1 ∈ {1,3} and v2 ∈ {0,1}) then
                                   ||   ⟨ await (v1 ∈ {0,1} and v2 ∈ {1,3}) then
    (v1 := 2 and (if v2 = 1 then v2 := 3)) ⟩;
                                   ||     ((if v1 = 1 then v1 := 3) and v2 := 2) ⟩;
  [CS1];                           ||   [CS2];
  v1 := 0                          ||   v2 := 0
end repeat                         || end repeat
```

```
Initialization: v1 := 0; v2 := 0
                         Process 1 || Process 2
ncs1: [NCS1];                      || ncs2: [NCS2];
  ⟨ case (v1 = 0): v1 := 1; goto e1 ||   ⟨ case (v2 = 0): v2 := 1; goto e2
    case (v1 = 2): v1 := 1; goto e1' ||     case (v2 = 2): v2 := 3; goto e2
    case (v1 = 3): v1 := 4; goto e1' ⟩ || e2: ⟨ await v1 ∈ {0,2,3,4} then
e1: ⟨ await v2 ∈ {0,1} then        ||     case (v1 = 0): v1 := 2; goto cs2
    case (v2 = 0): goto cs1        ||     case (v1 = 2): v1 := 0; goto cs2
    case (v2 = 1): goto cs1' ⟩     ||     case (v1 = 3): v1 := 2; goto cs2
e1': ⟨ await v2 ∈ {2,3} then       ||     case (v1 = 4): v1 := 1; goto cs2 ⟩
    case (v2 = 2): v2 := 0; goto cs1 || cs2: [CS2];
    case (v2 = 3): v2 := 1; goto cs1' ⟩ ||   case (v2 = 1): v2 := 2; goto ncs2
cs1: [CS1]; v1 := 0; goto ncs1     ||     case (v2 = 3): v2 := 0; goto ncs2
cs1': [CS1]; v1 := 3; goto ncs1    ||
```

Fig. 2. The two synthesised solutions for Mutex (N=2)

V_1 and V_2 as two variables with range $[0, 1, 2, 3]$. [6] By construction, the algorithm satisfies mutual exclusion, absence of starvation, and deadlock freedom. Moreover, the two processes can make requests independently of each other.

Using the results of [BD98] it is easy to show that $Mutex_1$ is not a Petri net language. However, it is a product language in the sense of [CMT99]. The results of [CMT99] also allow to derive the solution of Fig. 2. In this case, asynchronous automata do not have an advantage.

5.1 Mutual Exclusion Revisited

The mutex algorithm of the previous section requires to update the variables v_1 and v_2 before entering the critical section in one single atomic action, which is difficult to implement. Is it possible to obtain a solution that avoids this problem? We observe that the problem lies in the distribution we have chosen. We have $\Delta(V_1) \cap \Delta(V_2) = \{enter_1, enter_2\}$, and so both V_1 and V_2 are involved in the *enter* actions, which means that the implementation of both $enter_1$ and $enter_2$ requires to update both of v_1 and v_2 in a single atomic action. So we look for a different distribution in which $\Delta(V_1) \cap \Delta(V_2) = \emptyset$. We take:

$$\Delta(V_1) = \{req_1, enter_2, exit_1\} \text{ and } \Delta(V_2) = \{req_2, enter_1, exit_2\}.$$

[6] The pseudocode was derived by hand, but it would be not difficult to automatise the process.

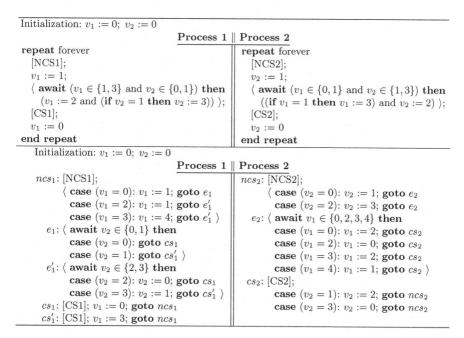

Unfortunately, $Mutex_1$ is not implementable anymore under this new distribution. The minimal automaton fails to satisfy FD: There is a state in which both $enter_1$ and $enter_2$ are enabled (and $enter_1 \| enter_2$), but there is no converging state to close the diamond. We then apply first heuristic from Sect. 4.1 and we indeed find a subautomaton satisfying ID and FD, deadlock-free and containing all the actions.

Zielonka's construction yields a safe asynchronous automaton with **4799** states. Fortunately, our heuristic yields an asynchronous automaton with only **20** states (see [SEM03]). Once distributed over the four processes of the specification (and merging local states if possible), we obtain the pseudocode shown in Fig. 2 (bottom). The variables v_1 and v_2 range over $[0, 1, 2, 3, 4]$ and $[0, 1, 2, 3]$ respectively. The labels associated with the commands suggest their type, for example r_1 means a request of the first process and x_2 means an exit from the critical section of the second process. Notice that the command corresponding to a label is executed atomically and that the program pointers for the two components advance only as a result of a **goto** command.

The components are now asymmetric, due to the fact that the first heuristic 'solved' the FD conflict by removing an $enter_2$ transition. Yet the algorithm is starvation-free: If the second process request access to the critical section, it will receive it as soon as possible.

The language $Mutex_2$ is neither a Petri net language nor the language of a product of transition systems, and so the procedures of [BD98,CMT99] cannot be applied.

5.2 More Processes

When we consider the mutual exclusion problem for an arbitrary number of processes $N \geq 2$, we choose the alphabet $\Sigma = \cup_{1 \leq i \leq N}\{req_i, enter_i, exit_i\}$. There exist several distributions of the actions. We choose generalizations of the two distribution used for $N = 2$. For $1 \leq i \leq N$:

- $\Delta_1(P_i) := \{req_i, exit_i, enter_1, \ldots, enter_N\}$
- $\Delta_2(P_i) := \{req_i, exit_i, enter_i\}$, $\Delta_2(V_i) = \Delta_1(P_i) \setminus enter_i$

We also generalize the regular specification of the problem. E.g., the mutual exclusion property is specified as: $\Sigma^* \setminus \bigcup_{i \neq j}[\Sigma^* enter_i (\Sigma \setminus exit_i)^* enter_j \Sigma^*]$.

The experiments for $N = 2, 3, 4, 5$ are presented in Table 1. In the first column, we give the parameters of the problem. In the second column, we give the size of the minimal automaton accepting the regular specification together with the number of the processors in the distribution. (The tool AMoRE [Amo] was used to construct the minimal automata recognising the regular specification.) In each of the following columns *size* represents the global state space of the solution (the asynchronous automaton) and *time* is the computation time given in seconds. A dash symbol '–' represents the fact that the system run out of memory without finding a solution. The third and fourth columns give the results after applying the first and respectively second heuristic in Sect. 4.1, followed by Zielonka's procedure. The fifth and sixth columns give the results after applying

Table 1. Experimental results

Problem	Input		Zielonka 1		Zielonka 2		Heuristic 1		Heuristic 2	
	$\|\mathcal{A}\|$	$\|\Delta\|$	size	time	size	time	size	time	size	time
Mutex(2,Δ_1)	14	4	34	<0.01	23	<0.01	14	<0.01	10	<0.01
Mutex(2,Δ_2)	14	4	4799	5.30	2834	2.66	17	<0.01	16	<0.01
Mutex(3,Δ_1)	107	6	–	–	–	–	107	<0.01	30	<0.01
Mutex(3,Δ_2)	107	6	–	–	–	–	–	–	58	0.11
Mutex(4,Δ_1)	1340	8	–	–	–	–	1340	0.31	62	1.25
Mutex(4,Δ_2)	1340	8	–	–	–	–	–	–	157	3.83
Mutex(5,Δ_1)	25338	10	–	–	–	–	25338	170.95	147	1000.76
Mutex(5,Δ_2)	25338	10	–	–	–	–	–	–	387	1053.79

the first and respectively second heuristic in Sect. 4.1, followed by the heuristic in Sect. 4.2. (The experiments were performed on a machine with 2.4 GHz CPU and 1 GB RAM.)

6 Further Remarks

We have proposed to apply the theory of asynchronous automata to the problem of synthesising closed distributed algorithms. We have observed that the right implementation model are safe asynchronous automata, and we have characterised their languages. We defined the synthesis problem in our framework and proved that it is undecidable, therefore we focused our attention on an NP-complete subclass of solutions. We have implemented Zielonka's algorithm, and observed that it leads to large implementations even for natural and relevant case studies where much smaller implementations exist. We have derived heuristics to make the synthesis problem more feasible in practice. We have used the heuristics to automatically generate mutual exclusion algorithms.

Obtaining 'Elegant' Solutions: Our solutions to the mutex problem are not 'elegant': They use variables with larger domains than those appearing in the literature, and a human finds it difficult to understand why they are correct. Notice, however, that this is the case with virtually all computer generated outputs, whether they are HTML text, program code, or a computer generated proof of a formula in a logic. Our solutions are correct and relatively small.

Specifying with Temporal Logic: Notice that our approach is compatible with giving specifications as LTL temporal logic formulas over finite strings, since the language of finite words satisfying a formula is known to be regular, and an automaton recognising this language can be effectively computed.

Dealing with Liveness Properties: Currently our approach cannot deal with liveness properties. Loosely speaking, 'eventually' properties have to be transformed into properties of the form 'before this or that happens'. Dealing with liveness properties requires to consider the theory of asynchronous automata on infinite words, for which not much is known yet (see Chap. 11 of [DR95]). The approaches of [BD98,CMT99] take a transition system as specification, and so do not consider liveness properties either. The approach of [MT02] can deal with liveness properties, but it can only synthesise controllers satisfying certain conditions (clocked controllers). These conditions would not appear in a reformulation of our results in a distributed controllers synthesis.

Acknowledgments. We thank Volker Diekert, Rémi Morin, Madhavan Mukund, Holger Petersen, and several anonymous referees for useful comments. This work was partially supported by EPSRC grant GR64322/01.

References

[Amo] AMoRE. http://www-i7.informatik.rwth-aachen.de/d/research/amore.html

[BD98] E. Badouel and Ph. Darondeau. Theory of regions. In *Lectures on Petri Nets I (Basic Models)*, LNCS 1491 (1998) 529–588.

[CMT99] I. Castellani, M. Mukund, and P.S. Thiagarajan. Synthesizing distributed transition systems from global specifications. In *Proc. FSTTCS* 19, LNCS 1739 (1999) 219–231.

[DR95] V. Diekert and G. Rozenberg (Eds.) *The Book of Traces*. World Scientific, 1995.

[EC82] E.A. Emerson and E.M. Clarke. Using branching time temporal logic to synthesize synchronization skeletons. *Science of Computer Programming* 2 (1982) 241–266.

[KV01] O. Kupferman and M.Y. Vardi. Synthesizing distributed systems. In *Proc. 16th IEEE Symp. on Logic in Computer Science*, (2001).

[Maz87] A. Mazurkiewicz. Trace theory. In LNCS 255 (1987) 279–324.

[Mor98] R. Morin. Decompositions of asynchronous systems. In *CONCUR'98*, LNCS 1466 (1998) 549–564.

[MT02] P. Madhusudan and P.S. Thiagarajan. A decidable class of asynchronous distributed controllers. In *Proc. CONCUR'02*, LNCS 2421 (2002) 145–160.

[Muk02] M. Mukund. From global specifications to distributed implementations. In *Synthesis and Control of Discrete Event Systems*, B. Caillaud, P. Darondeau, L. Lavagno (Eds.), Kluwer (2002) 19–34.

[MW84] Z. Manna and P. Wolper. Synthesis of communicating processes from temporal logic. *ACM TOPLAS* 6(1) (1984) 68–93.

[PR89] A. Pnueli and R. Rosner. On the synthesis of an asynchronous reactive module. In *Proc. 16th ICALP*, LNCS 372 (1989) 652–671.

[SEM03] A. Ştefănescu, J. Esparza, and A. Muscholl. Synthesis of distributed algorithms using asynchronous automata. Available at: http://www.fmi.uni-stuttgart.de/szs/publications/stefanan/concur03_full.ps

[Zie87] W. Zielonka. Notes on finite asynchronous automata. *R.A.I.R.O. Inform. Théor. Appl.* 21 (1987) 99–135.

[Zie89] W. Zielonka. Safe executions of recognizable trace languages by asynchronous automata. In LNCS 363 (1989) 278–289.

Compression of Partially Ordered Strings

Rajeev Alur[1], Swarat Chaudhuri[1], Kousha Etessami[2], Sudipto Guha[1], and Mihalis Yannakakis[3]

• Department of Computer and Information Science, University of Pennsylvania
• School of Informatics, University of Edinburgh
• Department of Computer Science, Stanford University

Abstract. We introduce the problem of compressing partially ordered strings: given string $\sigma \in \Sigma^*$ and a binary independence relation I over Σ, how can we compactly represent an input if the decompressor is allowed to reconstruct any string that can be obtained from σ by repeatedly swapping adjacent independent symbols? Such partially ordered strings are also known as Mazurkiewicz traces, and naturally model executions of concurrent programs. Compression techniques have been applied with much success to sequential program traces not only to store them compactly but to discover important profiling patterns within them. For compression to achieve similar aims for concurrent program traces we should exploit the extra freedom provided by the independence relation. Many popular string compression schemes are grammar-based schemes that produce a small context-free grammar generating uniquely the given string. We consider three classes of strategies for compression of partially-ordered strings: (*i*) we adapt grammar-based schemes by *rewriting* the input string σ into an "equivalent" one before applying grammar-based string compression, (*ii*) we represent the input by a collection of *projections* before applying (*i*) to each projection, and (*iii*) we combine (*i*) and (*ii*) with *relabeling* of symbols. We present some natural algorithms for each of these strategies, and present some experimental evidence that the extra freedom does enable extra compression. We also prove that a strategy of projecting the string onto each pair of dependent symbols can indeed lead to exponentially more succinct representations compared with only rewriting, and is within a factor of $|\Sigma|^{\cdot}$ of the optimal strategy for combining projections with rewriting.

1 Introduction

Algorithms for text compression view the input as a linearly ordered sequence of symbols and try to discover repeating patterns so that the input can be represented more compactly. In this paper, we initiate the study of compression of partially ordered strings. Given an independence relation over an alphabet, two strings are said to be equivalent if one can be obtained from the other by repeatedly commuting adjacent independent symbols. An equivalence class of such a type is known as a Mazurkiewicz trace in concurrency theory [Maz87,

R. Amadio, D. Lugiez (Eds.): CONCUR 2003, LNCS 2761, pp. 42–56, 2003.

DM97]. The new compression problem is then to compactly represent an input string if the decompressor is allowed to output *any* string that is equivalent to the original string. For instance, if all the symbols are pair-wise independent of each other, then a string can simply be represented by listing the number of occurences of each occuring symbol of the alphabet in the string. In this case, the original string may be uncompressible, but the extra freedom afforded by independence allows a representation that is logarithmic in the original size.

Many popular algorithms for string compression, such as the Lempel-Ziv algorithms [ZL77,ZL78] and SEQUITUR [NW97], are variant of grammar-based schemes, which work by essentially computing a small context-free grammar that generates the input string uniquely (see [KY00,CLL+02]). Such grammars are deterministic and contain no cycles, and hence can be viewed simply as hierarchical representations of the string. Larus ([Lar99]), using the SEQUITUR scheme, has shown that such compact hierarchical representations of sequential program traces can be used profitably to extract a variety of useful profiling information, such as detection of hotspots and hot subpaths, for analyzing and optimizing a program's dynamic behavior ([Lar99,BL00]).

While executions of sequential programs can be described naturally by strings of events, the behavior of a concurrent system is more appropriately modeled as a partially-ordered sequence of events [Lam78,Pra86,Maz87], reflecting the fact that if events occurring on distinct processes are not causally related their actual order of occurrence may be irrelevant. Message sequence charts (MSCs) offer a visual depiction of message exchanges in a concurrent system, and are used, e.g., for describing high-level requirements in the Unified Modeling Language [BJR97]. MSCs are also best formalized as partially ordered strings. Model checking tools like SPIN [Hol97] generate MSCs as outputs for simulation runs and counterexample traces. Hierarchical representations of MSCs can be used to improved comprehension and visualization of such outputs which are often large. All this suggests that compression of partially ordered strings should be used for concurrent program traces to achieve similar aims as string compression achieves for sequential program executions. In doing so, however, we should exploit the extra freedom provided by the independence relation to find patterns that are not available in a fixed sequential view of a partially ordered trace.

While compression has been studied for decades from both theoretical and practical viewpoints, we are not aware of any research that explicitly addresses compression of partially ordered strings.[1]

Our first class of algorithms involves adaptation of grammar-based schemes directly to partial-order strings. For strings it is NP-hard to find an optimal grammar ([SS82]) but such a grammar is approximable to within a log factor in polynomial time [CLL+02]. We present two algorithms for finding potentially smaller grammar representations by exploiting the independence relation. Our first algorithm is a modification of SEQUITUR ([NW97]) that greedily chooses the next symbol to be processed from the minimal elements of the remaining

[1] Based on the work we have initiated here, S. Savari has begun an information-theoretic study of such structures based on entropy considerations [Sav03a,Sav03b].

partial order by giving preference to the one that would lead to an already encountered pattern. Second is an offline algorithm that repeatedly replaces the most frequently occurring pair of dependent *or independent* symbols by a new nonterminal. As such, it does not strictly speaking produce a string grammar, but rather a limited form of more general graph grammars ([Eng97]). We report on a prototype implementation of these algorithms, and experimental results that indicate improvements in compression.

Our second class of algorithms consists of representing a string by an adequate collection of projections onto subsets of the alphabet, and then compressing each projection by a grammar-based string compression algorithm or by one of the algorithms of the first class. A necessary and sufficient condition for being able to reconstruct the original string up to equivalence is that each pair of dependent symbols must occur in one of the projections. A natural strategy for projection is to project the string onto every pair of dependent symbols. Surprisingly, this strategy can be exponentially more succinct than the optimal representation using just rewriting. In fact, this exponential gap holds even for ordinary strings (that is, when the independence relation is empty). Furthermore, the strategy of projecting onto dependent pairs produces output within a factor of d of that of the optimal algorithm in this class, where d is the number of dependent pairs, and this factor is tight. When the alphabet is partitioned into k sets such that symbols are dependent iff they belong to the same partition, then the natural strategy is to project the input string onto each of the partitions. Compared to compressing the original string, this can be exponentially better in the best case, and it is always within a factor of k compared to the optimal algorithm using just rewriting.

Finally, the third class of algorithms allows collapsing of symbols using relabeling in addition to the projections and rewriting. One strategy in this class is the following. For every symbol a, we project the string onto a and all the symbols dependent on a, then collapse all these dependent symbols to a single symbol b. This leads to $|\Sigma|$ strings, each over a two-letter alphabet, and can be compressed separately. We show how to reconstruct the original string, up to equivalence, from these projections.

2 Grammar-Based Compression up to Equivalence

2.1 Equivalence Classes of Strings and Labeled Partial Orders

Our model consists of a set Σ of terminals and an irreflexive symmetric *independence relation* $I \subseteq \Sigma \times \Sigma$. Two terminals a,b are said to be independent if $(a,b) \in I$. Intuitively, two strings are equivalent if one can be obtained from the other by a sequence of swaps of adjacent independent symbols. Formally, $\equiv_I$ is the smallest binary equivalence relation on Σ^* satisfying $\sigma ab\tau \equiv_I \sigma ba\tau$, for all $(a,b) \in I$ and for all strings $\sigma, \tau \in \Sigma^*$. We shall represent the equivalence class corresponding to a string σ by $[\sigma]_{\equiv_I}$. Such equivalence classes are called Mazurkiewicz traces in the concurrency literature [Maz87].

Equivalence classes induced by $\equiv_I$ correspond to *labeled partial orders* of a particular form. A labeled partial order respecting I is a structure $P = (V, E, \lambda)$, where V is a finite set of nodes, E is a set of edges over V such that the reflexive-transitive closure E^* is a partial order over V, and $\lambda : V \to \Sigma$ is a labeling of nodes by terminals such that for all $u, v \in V$,

1. if $(u, v) \in E$, then $(\lambda(u), \lambda(v)) \notin I$,
2. if $(\lambda(u), \lambda(v)) \notin I$, then either $(u, v) \in E^*$ or $(v, u) \in E^*$.

A *linearization* σ of the labeled partial order $P = (V, E, \lambda)$ is a string $\sigma_1\sigma_2 \cdots \sigma_{|V|}$ over Σ such that there exists an ordering $v_1 v_2 \cdots v_{|V|}$ of the nodes in V satisfying (1) $\sigma_i = \lambda(v_i)$ for $1 \le i \le |V|$, and (2) for all $(v_i, v_j) \in E$, $i < j$. We can define a correspondence between equivalence classes of strings and labeled partial orders. Namely, given a string σ and an independence relation I, there is an algorithm to construct the labeled partial order $P_{\sigma,I}$ with $|\sigma|$ vertices whose linearizations are the strings in $[\sigma]_{\equiv_I}$. The details of the algorithm to construct $P_{\sigma,I}$ are standard, and omitted from this abstract.

2.2 Grammar-Based Compression

In grammar-based compression algorithms for strings, given an input string σ, the algorithm computes a context-free grammar G that generates the singleton language $\{\sigma\}$. The grammar G then serves as a succinct hierarchical representation of σ. From now on, we shall refer to such a grammar as a *grammar for* σ. Over the years, several interesting grammar-based string compression algorithms have been proposed. Of them, the algorithm Sequitur [NW97] has been used for compression as well as to gather profiling information from program executions [Lar99,BL00,GRM03], and is of particular interest to us. Sequitur is an online algorithm that greedily constructs a hierarchy out of an input string. It scans the input from left to right, identifies repeated pairs of adjacent symbols (digrams) in the representation of the input that it has processed so far, and replaces them by nonterminals. A grammar rule maps every nonterminal to the digram it represents.

A good measure of the performance of a grammar-based compression algorithm is the size of the grammar, where the size of a grammar G is defined to be the sum of the lengths of the right-hand sides of all the rules in G. The optimal grammar-based compression algorithm needs to find the smallest grammar for the given input string. Unfortunately, this problem is NP-complete [SS82]. However, some recent research is aimed at finding approximation algorithms for this problem: Lehman et al [LS02] find approximation ratios for some previously proposed grammar-based compression algorithms (e.g., the well known LZ78 has an approximation ratio $O((n/\log\, n)^{2/3})$), and prove the hardness of approximating the smallest grammar beyond a certain constant factor; Charikar et al [CLL$^+$02] present an algorithm with an approximation ratio $O(\log(n/g^*))$, where g^* is the size of the smallest grammar.

2.3 Compression up to Equivalence

In this paper, we are interested in generating a small grammar-based representation of a given string *up to the equivalence induced by an independence relation.* We propose three different methodologies for achieving this, and pose three different optimization problems that these methods correspond to.

Finding Optimal Equivalent Strings. In our first approach, we find a string that is equivalent to the input string and can be represented by a small grammar. The output is the grammar for this string. For example, suppose $\Sigma = \{a, b, c\}$ and b and c are independent of each other. Then, the strategy of clustering all the b's (and c's) together between every pair of a's is a good heuristic to increase compressibility. For instance, *abccbacbbc* will be rewritten to $(ab^2c^2)^2$ to reduce the size of the grammar-based representation. The corresponding optimization problem is as follows. Let $C(\sigma)$ represent the size of the smallest grammar for a given string σ. Then, given a string σ and an independence relation I, the problem is to find $\tau \in [\sigma]_{\equiv_I}$ such that $C(\tau)$ is the minimum of the set $\{C(\sigma') \mid \sigma' \in [\sigma]_{\equiv_I}\}$. From now on, we refer to this optimal value $C(\tau)$ as $C_I(\sigma)$.

Projections and Compression. In our second approach, we consider the compression algorithms that project the input string onto a sequence of subsets of Σ such that the original string (up to equivalence) can be recovered from these projections, and compress the projections separately. In the example with $\Sigma = \{a, b, c\}$ with b and c independent, we can represent σ by two projections, one onto $\{a, b\}$ and one onto $\{a, c\}$, and compress the two separately (e.g., *abcccbacbbc* will be replaced by the pair (*abbabb*, *acccacc*)).

The projection of a string σ on a subalphabet $\Sigma' \subseteq \Sigma$ is obtained by erasing all symbols in σ that are not in Σ', and is represented by $\sigma \uparrow \Sigma'$. Subalphabets $\Sigma_1, \Sigma_2, \ldots, \Sigma_m \subseteq \Sigma$ *cover* an independence relation I, if there is a *reconstruction algorithm* A such that, for all strings σ, given the projections $\sigma \uparrow \Sigma_i$, A outputs some $\sigma' \in [\sigma]_{\equiv_I}$.

In this case, the compression methodology is as follows. We first project the input string σ on a set of covering subalphabets. Then we find grammars for these projections using an approximation algorithm for string compression. The compressed representation of the string (and the equivalence class) is the collection of all these grammars. In order to uncompress, we regenerate the projections from their grammars and use a reconstruction algorithm to generate a string equivalent to σ. Formally, the optimization problem is as follows. Given a σ and an independence relation I, find a cover $\Sigma_1, \Sigma_2, \ldots, \Sigma_m$ for I such that $\sum_{i=1}^{m} C_I(\sigma \uparrow \Sigma_i)$ is minimized. Let us denote $\sum_{i=1}^{m} C_I(\sigma \uparrow \Sigma_i)$ for the optimal cover by $C_I^p(\sigma)$.

Relabeling, Projections, and Compression. In our third approach, we allow relabeling of symbols during projections as long as the original string can be recovered up to equivalence. Going back to our example with independent b's and c's, we can represent a string by a pair, where the first one is a projection onto $\{a, b\}$ and the second one is obtained by renaming b to c. For instance, $abcbccacbcbb$ can be represented by $(abbabbb, ac^5ac^5)$. In this example, it is clear that the original string can be reconstructed up to equivalence, and relabeling can be exploited to minimize grammar sizes.

A *relabeling* γ is a function from Σ to Σ, and we use $\gamma(\sigma)$ to denote the string obtained from σ by replacing each symbol s in σ by the corresponding symbol $\gamma(s)$. A sequence of subalphabets $\Sigma_1, \Sigma_2, \dots, \Sigma_m \subseteq \Sigma$ and a corresponding sequence of relabelings $\gamma_1, \gamma_2, \dots \gamma_m$ are said to *cover* an independence relation I if there is a *reconstruction algorithm* A such that, for all strings σ, given renamed projections $\gamma_i(\sigma \uparrow \Sigma_i)$, outputs some $\sigma' \in [\sigma]_{\equiv_I}$. The optimization problem is defined as in the previous case. Given a string σ and an independence relation I, find a set of subalphabets $\Sigma_1, \Sigma_2, \dots, \Sigma_m$ together with relabeling functions $\gamma_1, \gamma_2, \dots \gamma_m$ such that the two sequences cover I, and $\sum_{i=1}^{m} C_I(\gamma_i(\sigma \uparrow \Sigma_i))$ is minimized. Let us denote the optimal sum by $C_I^{pr}(\sigma)$. Note that, by definition,

$$C_I^{pr}(\sigma) \leq C_I^{p}(\sigma) \leq C_I(\sigma) \leq C(\sigma).$$

3 Compression Algorithms

3.1 Locally Greedy Algorithm for Finding Good Linearizations

We first describe an algorithm that takes labeled partial orders as inputs, and outputs grammars for certain "good" linearizations. Given a string σ and independence relation I, we can first construct the partial order $P_{\sigma,I}$. Algorithm of Fig. 1 is an online algorithm inspired by Sequitur [NW97] and traverses the input partial order P from top to bottom. At each step, one of the *minimal nodes* (nodes without any incoming edges from unprocessed nodes) is chosen and removed from P. The choice is made greedily by giving preference to a node that will create a digram that has already appeared. Its label a is appended to a list L representing the part of the input already seen. Following Sequitur, we enforce digram uniqueness on L; that is, if a digram xy occurs at two separate locations on L, they are to be replaced by a nonterminal. If this digram has not been seen in the input processed so far, we add a rule $A \rightarrow xy$, for some new nonterminal A, to the grammar.

In our implementation of this algorithm, we maintain a map from digrams to positions in L. This map is maintained as a hashtable, so that we are able to match rules in constant time. Changes to the list L – required when a digram is replaced by a nonterminal – are implemented through low-level pointer operations. At each step we contract one edge of the partial order; we terminate when there are no edges left to explore. Since the edge relation is the covering relation of the partial order, there are at most a linear number of edges. If n is the length of the input string, and k is the width of the partial order $P_{\sigma,I}$ (that is, the maximum number of pair-wise unordered symbols), then the algorithm runs in time $O(k \cdot n)$.

Consider the labeled partial order P corresponding to the string $cabcbac$ with a and b independent. Let us follow a run of this algorithm on P. The stages of

input : Labeled partial order $P = (V, E, \lambda)$.
output : Grammar G for some linearization of P.
begin
  $G := \emptyset$.
  List of symbols $L := [\lambda(w)]$ for some minimal element w of P. Remove w from P.
  Hashtable of digrams $D := \emptyset$.
  repeat
    Min := Set of minimal elements of P.
    p:= last element appended to L.
    if *there is* $v \in Min$ *and digram* $A \to u\lambda(v)$ *in* D **then**
      Remove v from P. Append $q = \lambda(v)$ to L
      Replace the pair pq at the end of L by nonterminal A
      If the rule $A \to pq$ is not already in G, then add it. In this case there is a previous unreplaced occurrence of pq pointed to by δ in the hashtable. Replace that as well.
      Update D with digrams generated by these changes. If the digram uniqueness property is found to be violated, repeatedly replace the violating digrams by nonterminals till there is no repetition.
    else
      Choose some arbitrary $v \in Min$. Remove v from P.
      Add a digram $A \to p\lambda(v)$ to D for some new nonterminal A. Make it "point" to the current last position in L.
      Append $\lambda(v)$ to L.
    end
  until $Min = \emptyset$.
  $G := G \cup \{S \to L\}$, where S is a new *starting nonterminal.*
  Output G.
end

Algorithm 1: Top-to-bottom

Step	List L	Comments
1	c	Only one choice.
2	ca	Symbol a chosen arbitrarily.
3	cab	No other choice.
4	$cabc$	No other choice.
5	$cabca$	Choice made to repeat digram ca. Rule $A \to ca$ added.
6	$AbAb$	Symbol b appended. Digram Ab repeated. Add rule $B \to Ab$.
7	BBc	End of partial order reached. Add rule $S \to BBc$.

Fig. 1. Sample run of the Top-to-bottom Algorithm

the algorithm are described in the table in Fig. 1. The key step is step 5, where a is preferred over b as it causes a repeating digram.

3.2 Replace Most Frequent Pair

Our next algorithm is a greedy offline algorithm that chooses the most frequently occurring pair of dependent or independent symbols, and replaces this digram by a nonterminal. Consider a labeled partial order $P = (V, E, \lambda)$. The *frequency* of a pair of dependent symbols (p, q) is the maximum number of edges of the form (u, v) with $\lambda(u) = p$ and $\lambda(v) = q$ such that no two edges share an end-point (note that sharing of end-points can happen when $p = q$); while the frequency of a pair of independent symbols (p, q) is the maximum number of pair-wise disjoint sets of nodes of the form $\{u, v\}$ such that $\lambda(u) = p$, $\lambda(v) = q$, and neither uE^*v nor vE^*u. The *contraction* of $(u, v) \in E$ by a node w is the following operation on P: remove u, v from V; add w to V; replace $(s, t) \in E$, where $t \in \{u, v\}$ and $s \neq u$, by (s, w); replace $(s, t) \in E$, where $s \in \{u, v\}$ and $t \neq v$, by (w, t); and remove (u, v). For a pair (u, v) of unrelated nodes, the contraction by a node w is defined similarly: remove u, v from V, add w to V; replace $(s, t) \in E$ with $t \in \{u, v\}$ by (s, w); and replace $(s, t) \in E$ with $s \in \{u, v\}$ by (w, t). Finally, we will modify our definition of the labeling function λ a bit so that a labeled partial order can also have nodes labeled with arbitrary nonterminals. The definitions of frequency and contraction apply to such nodes also. If such a new node w is labeled with a new nonterminal A, then A is declared to be dependent on all the symbols that are dependent on p as well as the symbols dependent on q.

At each step of this algorithm, we identify a pair of symbols (p, q) with the maximum frequency. Then we add a rule $A \to pq$, for some new nonterminal A, and contract a disjoint collection of node pairs labeled (p, q) by a node labeled A. Computing the frequency of dependent pairs is straightforward, we simply need to scan all the edges and maintain a count for every pair of symbols. Computing the frequency of independent symbols requires more care, we need to make sure that if a node labeled p is unrelated to two nodes labeled q, then only one pair gets counted to the frequency of (p, q). In this case, matching the p-labeled node with the first possible q-labeled node that is a potential match, is a safe strategy to maximize the count of disjoint pairs. Note that the resulting grammar is not, strictly speaking, a string grammar because we are also allowed to introduce new nonterminals for pairs of independent symbols. Rather, it can be viewed as a limited form of more general graph grammars ([Eng97]), and hence as a generalization of the grammar-based string compression approach to a graph grammar-based approach for compression of partial orders.

Consider again the labeled partial order P corresponding to the string *cabcbac* with a and b independent. At the first step, we have to choose a set of disjoint edges labeled by the same symbol-pairs. We arbitrarily choose the symbol-pair (a, c) (we could also have chosen (b, c), (c, a), (a, b) or (c, b), all have frequency 2), add the rule $A \to ac$, and contract. The partial order now becomes the one corresponding to the string $cbAbA$. At the next step, we contract the two edges labeled (b, A) and add a rule $B \to bA$. The partial order now becomes a chain cBB. There is no way to contract further.

```
input  : Projections σ_i, 1 ≤ i ≤ m, with σ_i = σ↑Σ_i. The following condition
         is satisfied: for all (a,b) ∉ I, there is an i such that a,b ∈ Σ_i.
output : A string σ' satisfying σ' ≡_I σ.

begin
    p_i ⟵ 1 for each 1 ≤ i ≤ m
    Proj_a ⟵ {i : a ∈ Σ_i} for each a ∈ Σ
    j ⟵ 0
    repeat
        Select a ∈ Σ such that for all i ∈ Proj_a, we have p_i ≤ |σ_i| and σ_i(p_i) = a
        p_i := p_i + 1 for all i ∈ Proj_a
        σ'(j) := a; j := j + 1
    until no such a can be selected.
end
```

Algorithm 2: A reconstruction algorithm

3.3 Algorithms Using Projections

The first step in the algorithms that employ projection is to compute a cover for the given independence relation. The next theorem identifies a key property of the cover. (We have been informed that [CP85] contains this result. We provide a proof here for completeness and because our proof provides an efficient reconstruction algorithm which we use.)

Theorem 1. *([CP85]) Subalphabets $\Sigma_1, \Sigma_2, \ldots, \Sigma_m$ cover an independence relation I iff for all $(a,b) \notin I$, there is an i such that $a, b \in \Sigma_i$.*

Proof: ($\Rightarrow$) Suppose there is a pair of symbols $(a,b) \notin I$ such that there is no i with $a, b \in \Sigma_i$. Then, the projections of the non-equivalent strings ab and ba will be identical, and hence, reconstruction is impossible.

($\Leftarrow$) For this direction, we give a reconstruction algorithm for any set of subalphabets satisfying the above condition. In Algorithm 2, $\sigma(i)$ represents the i-th symbol of σ. The algorithm keeps a *current pointer* $1 \leq p_i \leq |\sigma_i|$ for each projection $1 \leq i \leq m$. For each projection we advance this pointer from the beginning to the end. The correctness proof is omitted due to lack of space. ∎

The reconstruction algorithm, with appropriate book-keeping, can be made to run in time linear in the size of its input (that is, the sum of the sizes of the projections). In the context of this theorem, a reasonable strategy is to project on a set of maximal cliques covering all the edges in the graph for the complement D of the independence relation. We present two special cases of this methodology.

- The algorithm **edge-cover** projects on the set of subalphabets $\{a, b\}$ for every pair (a,b) in the complement D of the independence relation I. This algorithm can be optimized by considering only the pairs (a,b) such that the dependency is realized within the input string σ, that is, the partial order $P_{\sigma,I}$ contains an edge whose endpoints are labeled with a and b.

- An interesting special case is when the independence relation is a k-partite graph: the alphabet Σ is partitioned into sets $\Sigma_1, \ldots \Sigma_k$ such that two symbols are independent iff they belong to separate partitions. In this case, this partition makes a natural choice for the clique cover.

3.4 Algorithms with Relabeling and Projection

If the subalphabets Σ_j and relabelings γ_j cover an independence relation I, then for $(a,b) \notin I$, there must be an index j such that $a, b \in \Sigma_j$ and $\gamma_j(a) \neq \gamma_j(b)$. That is, a necessary condition for reconstruction is that every pair of dependent symbols must belong to a projection whose relabeling does not collapse them.

Now, we present an alternative to the covering the dependency graph using cliques. Given an independence relation I, for every symbol $a \in \Sigma$, let $\Sigma_a = \{b \in \Sigma \mid (a,b) \notin I\}$ be the set of symbols dependent on a. Let # be a special symbol that is dependent on every symbol, and let γ_a be the relabeling that maps a to a and renames all other symbols to #. The strategy **star-cover** is to project the input string onto Σ_a, and apply the relabeling γ_a before applying the standard string compression. Note that, like the edge-cover algorithm of the previous section, this strategy also leads to a collection of 2-symbol strings, but now, we are guaranteed that we have only $|\Sigma|$ projections, one per symbol in Σ.

Theorem 2. *The subalphabets Σ_a and relabelings γ_a, for each $a \in \Sigma$, cover the independence relation I.*

Proof: The reconstruction algorithm is similar to Algorithm 2. Let $\sigma_a = \gamma_a(\sigma \uparrow \Sigma_a)$. As before, we maintain a pointer p_a for each projection σ_a. At every step, we try to select a symbol $a \in \Sigma$ such that $\sigma_a(p_a) = a$ and for each $b \in \Sigma_a$ with $b \neq a$, $\sigma_b(p_b) = \#$. If such a symbol a is found, the algorithm outputs a, and increments all the pointers p_b for $b \in \Sigma_a$. If there are two such symbols, then they must be independent, and the choice does not matter. ■

3.5 Experiments

In this section, we discuss preliminary experimental results for the top-to-bottom and replace-most-frequent compression algorithms presented earlier. We experimented with two distributed programs shipped as demos with the popular SPIN verification toolkit [Hol97]. One of them (`mobile1`) is a model of a cellphone hand-off strategy, the other (`pftp`) is a flow control protocol. These models consist of a number of processes communicating through message channels. Now consider the natural alphabet of *send* and *receive* events. There is a natural dependence relation on this alphabet: any send is dependent on the corresponding receive. Also, the local clock for every process defines a dependence between any two sends or receives that it participates in. Such an independence relation induces a special subclass of labeled partial orders called *message sequence charts*. We made SPIN perform random simulations of `pftp` and `mobile1` and produce message sequence charts (MSCs) of different lengths. These MSCs were fed as inputs

MSC size	Sequitur (random linearization)	Top-to-bottom	Replace-most-frequent
20000	13800	5612	4203
40000	24945	9679	7123
60000	35490	13441	12226
80000	45617	16641	22157
100000	55228	19759	-

Fig. 2. Grammar representations constructed by different algorithms: `mobile1`

MSC size	Sequitur (random linearization)	Top-to-bottom	Replace-most-frequent
20000	7048	4474	3457
40000	12470	7571	5128
60000	17433	10700	12461
80000	22026	13453	15233
100000	27081	15456	-

Fig. 3. Grammar representations constructed by different algorithms: `pftp`

MSC size	Size of Sequitur output on different linearizations
20000	5612, 7381, 8909, 11584, 13800
40000	9679, 12303, 18911, 21526, 24945
60000	13441, 20121, 27212, 31443, 35490
80000	16641, 23117, 30235, 39318, 45617
100000	19759, 30257, 38221, 47116, 55228

Fig. 4. Impact of the choice of linearization on Sequitur

to implementations of algorithms Top-to-bottom and Replace-most-frequent. We also fed random linearizations of these charts to the string compression algorithm Sequitur. A performance comparison is described in Figs. 2 and 3. The tables compare average sizes of grammar representations of MSCs of given lengths. The quadratic-time algorithm Replace-most-frequent did not terminate within a reasonable time for the longest input.

The above results suggest there is a practical advantage in choosing a linearization judiciously (as opposed to randomly). We experimented with this separation more, by studying the performance of the Sequitur algorithm on different linearizations of an MSC outputted by `mobile1` (Fig. 4). These linearizations are chosen with various "degrees" of arbitrariness. More precisely, while generating a linearization, we proceed along the lines of the Top-to-bottom algorithm, but make random choices at *some* of the steps. Of course, we cannot hope to generate the entire spectrum of linearizations of a large MSC this way; however, we do seem to get linearizations nicely covering the space between random and greedily chosen linearizations. Note that it is very possible that linearizations with much smaller grammars exist; it is just not easy to find them.

4 Bounds on Performance

In this section, we provide some theoretical bounds for the compression problem and strategies mentioned in the previous section.

We first demonstrate an exponential separation between the optimals C_I and C_I^p. We encode the sequence $\langle 0, 1, 2, \ldots, 2^k - 1\rangle$ in binary as follows. Our alphabet is $\Sigma = \{\#\} \cup (\cup_{i=1,\ldots,k}\{b_i^0, b_i^1\})$. The special symbol # will separate two successive numbers in the sequence. The encoding of a number that has 0 or 1 as its j-th bit will have, respectively, b_j^0 or b_j^1 as its j-th bit. That is, we consider the string

$$\sigma = \#b_1^0 b_2^0 ... b_k^0 \# b_1^0 b_2^0 \ldots b_k^1 \# b_1^0 \ldots b_{k-1}^1 b_k^0 \# \ldots .$$

Our independence relation is $I = \{(b_i^p, b_j^q) : b_i^p \neq b_j^q\}$. In other words, distinct b_i^p-s are independent of each other and are all dependent on #. In any string that is equivalent to σ, the set of symbols between every pair of #'s encodes a distinct number $0 \leq n < 2^k$. This makes such a string incompressible using grammar-based algorithms; intuitively, every interval between successive #'s contributes at least one symbol to the grammar. Formally, we show that the application of the Lempel-Ziv compression algorithm [ZL77] to σ compresses it at most by a factor k. Then we will use a relation between $C_I(\sigma)$ and this compressed form proved by Charikar et al. [CLL$^+$02] to show that $C_I(\sigma)$ is smaller than σ by at most a factor of k. Finally, we show that $C_I^p(\sigma)$ is logarithmic in $|\sigma|$.

Lemma 1. *If $\tau \equiv_I \sigma$, then LZ77-encoding of τ is $\Omega(2^k)$.*

Proof: The LZ77 algorithm describes a string w using a sequence $s_1 s_2 \ldots s_d$ of *widgets*. Each widget s_i is either a symbol of the alphabet of w or of the form $s_i = (j, r)$. Intuitively, the latter means "start at the position j of the string encoded by $s_1 \ldots s_{i-1}$ and read the next r symbols." More precisely, a widget (j, r) represents the substring $w(j)w(j+1)\ldots w(j+r-1)$, assuming the length of the string represented by $s_1 \ldots s_{i-1}$ is at least j.

We show there is no way to encode any consistent ordering of σ with fewer than $c2^k$ widgets, for some c. Assume $S = s_1 \ldots s_d$ is the LZ77 form for some ordering τ of σ. Then no widget of the form $s_i = (j, r)$ in S can encode a substring containing two or more occurrences of symbol #. This is because the set of b's that occurs between each pair of #'s in τ is unique, and thus there is no part of $s_1 \ldots s_{i-1}$ that one can refer to obtain the same set, irrespective of how b's are ordered. Consequently, we can have at most two # in the string denoted by (j, r), and thus $r < 2k$. Then the claim holds for $c = 1/2$. ■

The result of [CLL$^+$02] shows that if λ is the length of the LZ77-encoding of a string σ, then $\lambda \leq C(\sigma) \log|\sigma|$. It follows that $C_I(\sigma) = \Omega(2^k)$. Suppose the edge-cover algorithm will project σ onto subalphabets $\Sigma_{i,d} = \{b_i^d, \#\}$, where $i \in \{1, \ldots, k\}$ and $d \in \{0, 1\}$. There are $2k$ such subalphabets. It can be shown that each of these projections has a periodic nature and, as a result, a grammar of size $O(k)$. For instance, the projection on b_k^0 and # is $(\#b_k^0\#)^{2^{k-1}}$. This shows that $C_I^p(\sigma) = O(k^2)$. Note that the choice of k is arbitrary in the above. Consequently, we conclude the following theorem:

Theorem 3. *For each n, there is an alphabet Σ, an independence relation I and a string σ such that $|\sigma| \geq n$ and $C_I(\sigma) = \Omega(2^{|\sigma|} C_I^p(\sigma))$.*

It is worth noting the exponential separation holds even when the independence relation is empty, that is, even for compressing ordinary strings. Consider the string in the above proof. Clearly, σ itself cannot be compressed. Now, all symbols are pair-wise dependent, and there are $O(k^2)$ projections. It is easy to verify that projections onto each pair $\{b_i^p, b_j^q\}$ is periodic and has a grammar of size $O(k)$. Thus, $C^p(\sigma) = O(k^3)$.

Now we proceed to give an upper bound for the edge-cover algorithm which, given a string σ, projects it onto each edge (a, b) in the complement D of the independence relation I. Let these projections be called $\pi_1, \pi_2, \ldots, \pi_k$. Let $C_I^e(\sigma)$ be the sum of $C(\pi_i)$.

Theorem 4. *For all strings σ and independence relations I, $C_I^e(\sigma) \leq |\Sigma|^2 C_I^p(\sigma)$.*

Proof: Consider the projection π of the string σ on a pair (a, b) of dependent symbols. In the optimal projection based algorithm, one of the covering sub-alphabets Σ_j must include the pair (a, b). Let τ be a string that is equivalent to $\sigma \uparrow \Sigma_j$. Consider a grammar G for τ. Note that $\tau \uparrow \{a, b\}$ equals π. We can remove all other terminals from each rule of G to get a grammar for π without increasing the size of G. Therefore, $C(\pi) \leq C(\tau)$. Hence, $C(\pi) \leq C_I(\sigma \uparrow \Sigma_j)$, and $C(\pi) \leq C_I^p(\sigma)$. There are at most $|\Sigma|^2$ edges in D, and the result follows. ∎

To compress projections of σ onto single pairs of dependent symbols, we can use any grammar based algorithm, in particular, the algorithm by Charikar et al [CLL+02], thereby approximating $C_I^p(\sigma)$ up to factor $|\Sigma|^2 \log(|\sigma|/g^*)$, where g^* is the size of the optimal grammar. The bound for the edge-cover strategy is tight. Suppose $\Sigma = \{a_1, \ldots a_k\}$ such that all symbols are dependent. Consider the string $\sigma = (a_1 \cdots a_k)^n$. The grammar of σ is of size $k + \log\ n$, while the grammar for each $\sigma \uparrow \{a_i, a_j\}$ is of size $\log\ n$, and thus, $C_I^e(\sigma)$ is $k^2 \log\ n$.

An interesting special case for the clique-cover is when the alphabet is union of disjoint alphabets $\Sigma_1, \ldots \Sigma_k$ and two symbols are dependent iff they belong to the same partition Σ_i. In this case, a natural choice for cover is the partition $\Sigma_1, \ldots \Sigma_k$. Let $C_I^c(\sigma)$ denote the sum of $C(\sigma \uparrow \Sigma_i)$. This strategy can be quite beneficial over compressing the original string. For example, if there are two independent symbols a and b, then a random string σ won't compress well, while the two projections onto individual symbols carry the minimal information, namely, the number of a's and number of b's. As the next theorem shows, projecting onto cliques can be worse than compressing the original string, or even an equivalent linearization, by at most by a factor of the number of cliques.

Theorem 5. *If the alphabet is a disjoint union of k cliques, then for any string σ, $C_I^c(\sigma) \leq k C_I(\sigma)$.*

Proof: The proof is similar to the proof of Theorem 4. Consider any clique Σ_i in D, let $\pi_i = \sigma \uparrow \Sigma_i$, and let τ be any string equivalent to σ. Since all the terminals in Σ_i depend on each other, one can show that the size of the optimal grammar for π_i is bounded by the size of the optimal grammar for τ. ∎

Again, this bound is tight. Suppose $\Sigma = \{a_1, \ldots a_k\}$ such that all symbols are independent (that is, there are k singleton cliques). Consider the string $\sigma = (a_1 \cdots a_k)^n$. The grammar of σ is of size $k + \log\ n$, while the grammar for each $\sigma \uparrow \{a_i\}$ is of size $\log\ n$, and thus, $C_I^c(\sigma)$ is $k \log\ n$.

Finally, for the star-cover algorithm that uses both projections and relabeling, we can show that every relabeled projection $\gamma_a(\sigma \uparrow \Sigma_a)$ has a grammar of size at most that of the smallest grammar for any string equivalent to the original.

Theorem 6. *For all strings σ and independence relations I, for each $a \in \Sigma$, $C(\gamma_a(\sigma \uparrow \Sigma_a)) \leq C_I(\sigma)$.*

5 Conclusions

In this paper, we have formulated and initiated the study of the compression problem of partially ordered strings. It is worth noting that even for compression of ordinary strings, the use of projections and relabeling, and the resulting succinctness of the representation, has not been studied earlier. While we have shown that projection can lead to exponential succinctness for a class of strings, it remains to be seen if projections, possibly augmented with relabeling, can be engineered to lead to practical general compression techniques.

There are many directions for future research. The application to profiling of executions of concurrent programs, and for visualization large MSCs generated by tools like SPIN in compact form, both seem promising. A recent paper applies standard string compression techniques to parallel program executions [GRM03], and our techniques can potentially improve their results. Compression of partially ordered strings can be studied from an information theoretic perspective. Based on the work we have initiated here, Savari has begun a study of the graph entropy of such structures and of rewriting strings to normal forms [Sav03a,Sav03b]. We would also like to sharpen the approximability of the optimization measures introduced in this paper. In particular, approximability bounds for the measure C_I, and improving the $|\Sigma|^2$ bound for the measure C_I^p, are open problems. Finally, it would be interesting to study compression of labeled partial orders based on more general classes of graph grammars ([Eng97]) than those implicit in Algorithm 1.

Acknowledgements. Thanks to Serap Savari for discussions and comments.

References

[BJR97] G. Booch, I. Jacobson, and J. Rumbaugh. *Unified Modeling Language User Guide.* Addison Wesley, 1997.

[BL00] T. Ball and J. Larus. Using paths to measure, explain, and enhance program behavior. *IEEE Computer*, 33(7):57–65, 2000.

[CLL⁺02] M. Charikar, E. Lehman, D. Liu, R. Panigrahy, M. Prabhakaran, A. Rasala, A. Sahai, and A. Shelat. Approximating the smallest grammar: Kolmogorov complexity in natural models. In *Proceedings of the 34th ACM Symposium on Theory of Computing*, pages 792–801, 2002.

[CP85] R. Cori and D. Perrin. Sur la Reconnaissabilite dans les monoides partiellement commutatifs libres. R.A.I.R.O.-Informatique Thorique et Applications, 19:21-32, 1985.

[DM97] V. Diekert and Y. Metivier. Partial commutation and traces. In *Handbook of Formal Languages: Beyond Words*, pages 457–534. Springer, 1997.

[Eng97] J. Engelfriet. Context-free graph grammars. In *Handbook of Formal Languages, vol. 3*, ed. G. Rozenberg and A. Salomaa, Springer-Verlag, 1997.

[GRM03] A. Goel, A. Roychoudhury, and T. Mitra. Compactly representing parallel program executions. In *Proceedings of the ACM Symposium on Principles and Practice of Parallel Programming*, 2003.

[Hol97] G.J. Holzmann. The model checker SPIN. *IEEE Transactions on Software Engineering*, 23(5):279–295, 1997.

[KY00] J. Kieffer and E. Yang. Grammar-based codes: a new class of universal lossless source codes. *IEEE Transactions on Information Theory*, 46:737–754, 2000.

[Lam78] L. Lamport. Time, clocks, and the ordering of events in a distributed system. *Communications of the ACM*, 21:558–565, 1978.

[Lar99] J. Larus. Whole program paths. In *Proceedings of the ACM Conference on Programming Languages Design and Implementation*, pages 259–269, 1999.

[LS02] E. Lehman and A. Shelat. Approximation algorithms for grammar-based compression. In *Proceedings of the 13th ACM-SIAM Symposium on Discrete Algorithms*, pages 205–212, 2002.

[Maz87] A. Mazurkiewicz. Trace theory. In *Advances in Petri nets: Proceedings of an advanced course*, LNCS 255, pages 279–324. Springer-Verlag, 1987.

[NW97] C. Nevill-Manning and I. Witten. Identifying hierarchical structure in sequences: A linear-time algorithm. *Journal of Artificial Intelligence Research*, 7:67–82, 1997.

[Pra86] V.R. Pratt. Modeling concurrency with partial orders. *International Journal of Parallel Programming*, 15(1), 1986.

[Sav03a] S. Savari. Concurrent processes and interchange entropy. In *IEEE Int. Symp. on Information Theory*, 2003. To appear.

[Sav03b] S. Savari. On compressing interchange classes of events in a concurrent system. IEEE Data Compression Conference, 2003.

[SS82] J. Storer and T.G. Szymanski. Data compression via textual substitution. *Journal of the ACM*, 29:928–951, 1982.

[ZL77] J. Ziv and A. Lempel. A universal algorithm for sequential data compression. *IEEE Transactions on Information Theory*, 23(3):337–343, 1977.

[ZL78] J. Ziv and A. Lempel. Compression of individual sequences via variable-rate coding. *IEEE Transactions on Information Theory*, 24(5):530–536, 1978.

Bundle Event Structures and CCSP

Rob van Glabbeek[1,⋆] and Frits Vaandrager[2]

[1] National ICT Australia
School of Computer Science & Engineering, Univ. of New South Wales, Sydney 2052, Australia
rvg@cs.stanford.edu
[2] Nijmegen Institute for Computing and Information Sciences, University of Nijmegen
Postbus 9010, 6500 GL Nijmegen, The Netherlands
fvaan@cs.kun.nl

Abstract. We investigate which event structures can be denoted by means of closed CCS ∪ CSP expressions. Working up to isomorphism we find that

- all denotable event structures are bundle event structures,
- upon adding an infinitary parallel composition all bundle event structures are denotable,
- without it every finite bundle event structure can be denoted,
- as well as every countable prime event structure with binary conflict.

Up to hereditary history preserving bisimulation equivalence finitary conflict can be expressed in terms of binary conflict. In this setting all countable stable event structures are denotable.

Introduction

In concurrency theory many languages for the representation of concurrent systems have been proposed, including CCS, SCCS, CSP, MEIJE, ACP, COSY and LOTOS, all in several variations. Although most of these languages were originally equipped with an interleaving semantics, concurrency respecting interpretations have been proposed by various authors, using semantical models like Petri nets, event structures, transition systems – optionally with additional structure to represent causal independence –, causal trees, families of posets, etc. In recent years it has been established that there are canonical translations between most of these models, thereby making them into different representations of one and the same semantic concept [13,17,2,19,6]. In addition, the languages mentioned above are to a large extent intertranslatable, and can be regarded as dialects of one and the same system specification language.

This paper deals with the question which of these unified semantic objects can be denoted by closed expressions in this unified language. As a representative semantic model we take the *event structures* from WINSKEL [17]. Our findings can then be transmitted to other models by means of the canonical translations found in the literature. As a representative language we combine some operators from CCS [12] and CSP [3, 9]. Following a suggestion of Mogens Nielsen, such a combination is called CCSP. Our version of CCSP is sufficiently expressive to emulate most constructions from other

⋆ This work was performed in part while the first author was employed at Stanford University. It was supported by ONR under grant number N00014-92-J-1974.

R. Amadio, D. Lugiez (Eds.): CONCUR 2003, LNCS 2761, pp. 57–71, 2003.

languages found in the literature, including the ones provided with an event structure semantics in [17]. The chosen combination of operators appears to be optimal for carrying out the constructions in this paper. However, many other combinations would lead to the same results.

In [17] the subclass of *stable event structures* is defined, as well as the further subclass of *prime event structures*. In [18] a subclass of event structures with a *binary conflict relation* is proposed (see Fig. 1 below). The prime event structures with binary conflict are exactly the (finitary) event structures originally introduced in [13]. It is well known that unstable event structures cannot be represented in CCSP-like languages in a causality respecting way. It is an interesting quest to extend such languages with novel operators that make this possible. This quest is not pursued here; we will be happy to just find out which of the stable event structures are denotable.

It is unreasonable to expect to find a CCSP expression denoting a given event structure exactly. Hence we will try to find for any given stable event structure a CCSP expression whose denotation as event structure is semantically equivalent. This makes our quest parametrised by the choice of a suitable semantic equivalence. We consider three choices for this parameter: *isomorphism, history preserving bisimulation equivalence* and (in this introduction only) *ST-bisimulation equivalence*.

Denotability up to Isomorphism. Up to isomorphism we characterise the denotable event structures as the *bundle event structures* proposed in LANGERAK [11]. As we will recall in Sect. 1, these include all prime event structures with binary conflict, and are included in the stable event structures with binary conflict (cf. Fig. 1). In [11] examples can be found showing that these inclusions are strict.

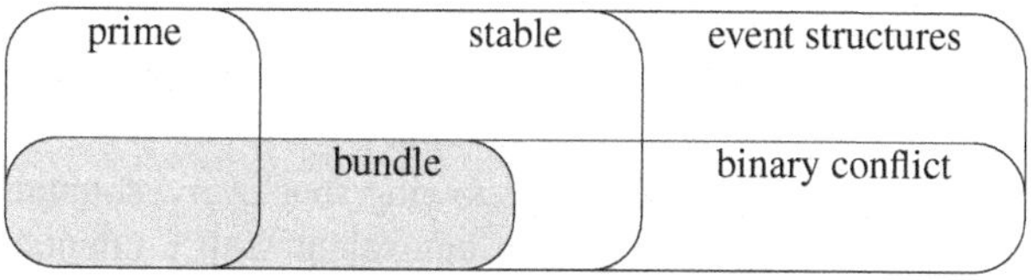

Fig. 1. *Several classes of event structures*

Our characterisation of the bundle event structures as the event structures that can be expressed by CCSP expressions is exact when dealing with the original finite bundle event structures and recursion-free CCSP. Our characterisation is also exact when dealing with arbitrary infinite bundle event structures and a version of CCSP with an infinite parallel composition operator. However, when dealing with countable event structures and CCSP expressions with arbitrary systems of recursion equations, or with recursive enumerable event structures and a recursive enumerable version of CCSP, all we can show is that the denotable event structures are a subclass of the bundle event structures that include all prime event structures with binary conflict.

In Sect. 2 a denotational semantics of CCSP is given in terms of event structures with binary conflict. This semantics follows the standard lines of [17,18]. In the same section we show that the class of bundle event structures is closed under the CCSP operators, thereby establishing that CCSP expressions can denote bundle event structures only.

Along the same lines one can show that the bundle event structures are closed under *action refinement* [5], the choice operators □ and ⊓ of CSP, and many other operators found in the literature. We are not aware of any operator interpreted on event structures for which the class of bundle event structures is not closed.

In Sect. 3 we show that up to isomorphism

- every *finite* bundle event structure can be denoted by a recursion-free CCSP expression,
- every *countable* prime event structure with binary conflict can be denoted by a CCSP expression,
- and *every* bundle event structure can be denoted by a CCSP expression with an infinitary parallel composition.

We also provide a recursive enumerable version of the second result. The same results can be obtained for the language of [17,18].

Denotability up to Hereditary History Preserving Bisimulation Equivalence. The concept of *history preserving bisimulation equivalence* stems from RABINOVICH & TRAKHTENBROT [15] and was adapted to event structures in VAN GLABBEEK & GOLTZ [5]. There it was suggested that the notion could be regarded as the coarsest equivalence that takes the interplay of causality and branching time completely into account. This makes the equivalence a semantically interesting choice of parameter to instantiate our quest with. We arrive at the positive conclusion that up to history preserving bisimulation *every* countable stable event structure can be denoted by a CCSP expression. This result is obtained in three steps, the first of which is the aforementioned denotability by CCSP expressions of countable prime event structures with binary conflict. In Sect. 4 we extend this to countable stable event structures, by observing that that every countable stable event structure with binary conflict is history preserving bisimulation equivalent with a countable prime event structure with binary conflict.

In Sect. 5 we complete the proof by showing that every countable stable event structure is history preserving bisimulation equivalent with a countable stable event structure with binary conflict. This result was first claimed by us in [8] for finite prime event structures. The claim was strengthened in [6] to include infinite ones. The first published proof (for prime event structures) appears in NIELSEN & WINSKEL [14], who discovered the result independently. Their proof is somewhat nonconstructive however, in the sense that there is no construction giving a specific countable event structure with binary conflict for any given countable stable event structure with arbitrary conflict. Our proof offers such a construction and is somewhat shorter as well.

The results above hold even when merely working up to *hereditary history preserving bisimulation equivalence*, which is a finer variant of history preserving bisimulation equivalence, proposed by BEDNARCZYK [1].

Denotability up to ST-Bisimulation Equivalence. The coarser *ST-bisimulation equivalence*, proposed in VAN GLABBEEK & VAANDRAGER [7], respects branching time and the possibility of actions to overlap in time, but abstracts from the faithful modelling of causality. By Theorem 1 in VAN GLABBEEK & PLOTKIN [6], every event structure, stable or not, is ST-bisimulation equivalent to a prime event structure. This result keeps being valid when assuming and requiring countability. It follows that up to ST-bisimulation equivalence every event structure can be denoted by a CCSP expression.

1 Bundle Event Structures

Bundle event structures are introduced in LANGERAK [11]. Here we add the alphabets for *typed* bundle event structures and generalise the notion to structures with infinite sets of events.

Definition 1. A *(typed) bundle event structure* is a 5-tuple $\mathrm{E} = (E, \#, \mapsto, A, l)$ where

- E is a set of *events*,
- $\# \subseteq E \times E$ is an irreflexive and symmetric relation, the *conflict relation*,
- $\mapsto \subseteq 2^E \times E$ is the *bundle set*, satisfying $X \mapsto e \Rightarrow \forall e_1, e_2 \in X.\, e_1 \neq e_2 \Rightarrow e_1 \# e_2$,
- A is a set of *actions*, the *alphabet* of E,
- and $l : E \to A$ is the *labelling function*.

A bundle event structure represents a concurrent system in the following way: action names $a \in A$ represent actions the system might perform and an event $e \in E$ labelled with a represents an occurrence of a during a possible run of the system. In order for e to happen it is necessary that for every bundle $X \mapsto e$, one of the elements of X occurred previously. The conflict $d \# e$ means that the events d and e cannot happen both in the same run.

The components of a bundle event structure E will be denoted by $E_{\mathrm{E}}, \#_{\mathrm{E}}, \mapsto_{\mathrm{E}}, A_{\mathrm{E}}$ and l_{E}; a convention that will also apply to other structures given as tuples.

The behaviour of a bundle event structure is described by explaining which subsets of events constitute possible (partial) runs of the represented system (thus formalising the interpretation of the bundle sets and the conflict relation). These subsets are called *configurations*. The causal relationships between events in a configuration x can be represented by a partial order $<_x$.

Definition 2. The set $C(\mathrm{E})$ of (finite) *configurations* of a bundle event structure $\mathrm{E} = (E, \#, \mapsto, A, l)$ consists of those finite $x \subseteq E$ which are

- *conflict-free*: $\# \cap (x \times x) = \emptyset$,
- and *secured*:
  $\exists e_1, \dots, e_n\ (n \geq 0) \colon x = \{e_1, \dots, e_n\} \wedge \forall i < n\ (Y \mapsto e_{i+1} \Rightarrow \{e_1, \dots, e_i\} \cap Y \neq \emptyset)$.

The *causality relation* $<_x$ on $x \in C(\mathrm{E})$ is $\{(d, e) \in x \times x \mid \exists Y : d \in Y \mapsto e\}^+$. Here R^+ denotes the transitive closure of a relation R.

Following [5], we only consider finite configurations here; since the infinite configurations which are usually considered are completely determined by the finite ones, this causes no loss of generality. Note that if $e \in x \in C(\mathrm{E})$ and $Y \mapsto_{\mathrm{E}} e$ then $x \cap Y$ has exactly one element. Hence $<_x$ is always a partial order.

We now define the *prime* and the *stable* event structures with binary conflict, stemming from WINSKEL [18], and show that the bundle event structures can be regarded as a generalisation of the former and a special case of the latter.

Definition 3. A *(typed) prime event structure with binary conflict* is a 5-tuple $\mathrm{E} = (E, <, \#, A, l)$ where

- E, A, and l are as above,
- $< \subseteq E \times E$ is a partial order such that $\forall e \in E : \{d \in E \mid d < e\}$ is finite,
- and $\# \subseteq E \times E$ is an irreflexive, symmetric relation satisfying

$$\forall d, e, f \in E : d < e \wedge d\#f \Rightarrow e\#f.$$

Here $d < e$ means that d is a prerequisite for e. Prime event structures with binary conflict can be regarded as special bundle event structures, by defining

$$X \mapsto e \;\Leftrightarrow\; X = \{d\} \wedge d < e.$$

The definition of configurations given above is then consistent with the one in [18].

Definition 4. A *(typed) event structure with binary conflict* is a 5-tuple $\mathrm{E} = (E, \#, \vdash, A, l)$ where

- E, $\#$, A, and l are as for bundle event structures, $Con = \{X \subseteq E \mid X \text{ finite and } d\#e \text{ for no } d, e \in X\}$ is the *consistency predicate*,
- and $\vdash \subseteq Con \times E$ is the *enabling relation*, satisfying

$$X \vdash e \wedge X \subseteq Y \in Con \Rightarrow Y \vdash e.$$

E is *stable* if $Y \vdash e$ implies that there is a least subset X of Y with $X \vdash e$.

$X \vdash e$ means that X is a *possible cause* of e in the sense that e can occur only if for certain Y with $Y \vdash e$ all events in Y have occurred before.

Definition 5. The set $C(\mathrm{E})$ of (finite) *configurations* of an event structure with binary conflict $\mathrm{E} = (E, \#, \vdash, A, l)$ consists of those finite $x \subseteq E$ which are

- *conflict-free*: $\# \cap (x \times x) = \emptyset$, i.e. $x \in Con$,
- and *secured*: $\exists e_1, ..., e_n \; (n \geq 0) : x = \{e_1, ..., e_n\} \wedge \forall i < n \; \{e_1, ..., e_i\} \vdash e_{i+1}$.

The *causality relation* $<_x$ on x is $\{(d, e) \in x \times x \mid \forall Y : Y \vdash e \Rightarrow d \in Y\}^+$.

The causality relation gives a faithful description of the causal relations in a configuration only if E is stable. As shown in [17,18], unstable event structures can model causal relationships that cannot be captured in terms of partial orders. The following shows how bundle event structures can be regarded as special stable event structures with binary conflict.

Definition 6. Given a bundle event structure $\mathrm{E} = (E, \#, \mapsto, A, l)$, the *associated* event structure with binary conflict $\mathcal{E}(\mathrm{E}) = (E, \#, \vdash, A, l)$ is given by

$$X \vdash e \;\Leftrightarrow\; X \in Con \wedge \forall Y \colon (Y \mapsto e \Rightarrow X \cap Y \neq \emptyset).$$

Proposition 1. *$\mathcal{E}(\mathrm{E})$ is always stable. Moreover, the translation $\mathcal{E}$ preserves configurations and the causality relations $<_x$ on them.*

Proof. Straightforward.

2 A Denotational Event Structure Semantics of CCSP

CCSP is parametrised by the choice of an infinite set *Act* of actions, that we will assume to be fixed for this paper. We also assume an infinite set V of *variable names*. A *variable* is a pair X_A with $X \in V$ and $A \subseteq Act$. The syntax of CCSP is given by

$$P ::= 0_A \mid aP \mid P + P \mid P \| P \mid R(P) \mid X_A \mid \langle X_A | S \rangle \text{ (with } X_A \in V_S)$$

with $A \subseteq Act$, $a \in Act$, $R \subseteq Act \times Act$, $X \in V$ and S a *recursive specification*: a set of equations $\{X_A = P_{X_A} \mid X_A \in V_S\}$ with $V_S \subseteq V \times Act$ (the *bound variables* of S) and $\alpha(P_{X_A}) = A$ for all $X_A \in V_S$ (were $\alpha(P_{X_A})$ is defined below). The constant 0_A represents a process that is unable to perform any action. The process aP first performs the action a and then proceeds as P. The process $P + Q$ will behave as either P or Q, $\|$ is a partially synchronous parallel composition operator, R a renaming, and $\langle X_A | S \rangle$ represents the X_A-component of a solution of the system of recursive equations S. A CCSP expression P is *closed* if every occurrence of a variable X_A occurs in a subexpression $\langle Y_B | S \rangle$ of P with $X_A \in V_S$. An expression $a0_\emptyset$ is abbreviated a.

Just like the version of CSP from HOARE [9], the version of CCSP used here is a typed language, in the sense that with every process P an explicit alphabet $\alpha(P) \subseteq Act$ is associated, which is a superset of the set of all actions the process could possibly perform. This alphabet is exploited in the definition of $P\|Q$: actions in the intersection of the alphabets of P and Q are required to synchronise, whereas all other actions of P and Q happen independently. Because of this, processes with different alphabets may never be identified, even if they can perform the same set of actions and are alike in all other aspects. It is for this reason that we interpret CCSP in terms of *typed* event structures. The constant 0 and the variables are indexed with an alphabet. The alphabet of an arbitrary CCSP expression is given by:

- $\alpha(0_A) = \alpha(X_A) = \alpha(\langle X_A | S \rangle) = A$
- $\alpha(aP) = \{a\} \cup \alpha(P)$
- $\alpha(P + Q) = \alpha(P\|Q) = \alpha(P) \cup \alpha(Q)$
- $\alpha(R(P)) = \{b \mid \exists a \in \alpha(P) : (a, b) \in R\}$.

Substitutions of expressions for variables are allowed only if the alphabets match. For this reason a recursive specification S is declared syntactically incorrect if $\alpha(P_{X_A}) \neq A$ for some $X_A \in V_S$.

Below we define the CCSP operators formally on the domains of (typed) bundle and stable event structures with binary conflict. As our bundle and stable interpretations agree on the components E, $\#$, A and l, they will be given as 6-tuples $(E, \#, \mapsto, \vdash, A, l)$, so that the bundle interpretation is found by dropping $\vdash$, and the stable interpretation by dropping $\mapsto$. When E is an event structure representing a CCSP expression P then $A_{\mathrm{E}} = \alpha(P)$. Hence we can abstain from explicitly mentioning the A-component in the forthcoming constructions.

Definition 7. The operators of CCSP are defined on event structures as follows:
Inaction: $\mathrm{E}_{0_A} = (\emptyset, \emptyset, \emptyset, \emptyset, A, \emptyset)$.

Action prefix (for $a \in Act$):

- $E_{a\mathrm{E}} = \{a\} \cup \{\hat{a}e \mid e \in E_{\mathrm{E}}\}$
- $l_{a\mathrm{E}}(a) = a$ and $l_{a\mathrm{E}}(\hat{a}e) = l_{\mathrm{E}}(e)$
- $\#_{a\mathrm{E}} = \{(\hat{a}e, \hat{a}e') \mid e\#_{\mathrm{E}}e'\}$
- $\mapsto_{a\mathrm{E}} = \{(\hat{a}X, \hat{a}e) \mid X \mapsto_{\mathrm{E}} e\} \cup \{(\{a\}, \hat{a}e) \mid e \in E_{\mathrm{E}}\}$ in which $\hat{a}X = \{\hat{a}e \mid e \in X\}$
- $\vdash_{a\mathrm{E}} = \{(X, a) \mid X \in Con_{a\mathrm{E}}\} \cup \{(\hat{a}X \cup \{a\}, \hat{a}e) \mid X \vdash_{\mathrm{E}} e\}$.

Alternative composition:

- $E_{\mathrm{E}_1+\mathrm{E}_2} = \{+_1 e \mid e \in E_{\mathrm{E}_1}\} \cup \{+_2 e \mid e \in E_{\mathrm{E}_2}\}$
- $l_{\mathrm{E}_1+\mathrm{E}_2}(+_i e) = l_{\mathrm{E}_i}(e)$ $(i = 1, 2)$
- $\#_{\mathrm{E}_1+\mathrm{E}_2} = \{(+_i e, +_i e') \mid e\#_{\mathrm{E}_i}e',\ i = 1, 2\} \cup \{(+_i e, +_j f) \mid e \in E_{\mathrm{E}_i},\ f \in E_{\mathrm{E}_j},\ i \neq j\}$
- $\mapsto_{\mathrm{E}_1+\mathrm{E}_2} = \{(+_i X, +_i e) \mid X \mapsto_{\mathrm{E}_i} e,\ i = 1, 2\}$ in which $+_i X = \{+_i e \mid e \in X\}$
- $\vdash_{\mathrm{E}_1+\mathrm{E}_2} = \{(+_i X, +_i e) \mid X \vdash_{\mathrm{E}_i} e,\ i = 1, 2\}$.

Parallel composition:

- $E_{\mathrm{E}\|\mathrm{F}} = \{(e\|*) \mid l_{\mathrm{E}}(e) \notin A_{\mathrm{F}}\} \cup \{(*\|f) \mid l_{\mathrm{F}}(f) \notin A_{\mathrm{E}}\} \cup \{(e\|f) \mid l_{\mathrm{E}}(e) = l_{\mathrm{F}}(f) \in A_{\mathrm{E}} \cap A_{\mathrm{F}}\}$
- $l_{\mathrm{E}\|\mathrm{F}}(e\|*) = l_{\mathrm{E}}(e)$, $l_{\mathrm{E}\|\mathrm{F}}(*\|f) = l_{\mathrm{F}}(f)$ and $l_{\mathrm{E}\|\mathrm{F}}(e\|f) = l_{\mathrm{E}}(e) = l_{\mathrm{F}}(f)$
- $\#_{\mathrm{E}\|\mathrm{F}} = \{(e\|f, e'\|f') \mid (e\|f \neq e'\|f') \wedge (e\#_{\mathrm{E}}e' \vee e = e' \neq * \vee f\#_{\mathrm{F}}f' \vee f = f' \neq *)\}$
- $\mapsto_{\mathrm{E}\|\mathrm{F}} = \{(X\|\mathrm{F}, e\|f) \mid X \mapsto_{\mathrm{E}} e\} \cup \{(\mathrm{E}\|Y, e\|f) \mid Y \mapsto_{\mathrm{F}} f\}$
  in which $X\|\mathrm{F} = \{(e\|f) \in E_{\mathrm{E}\|\mathrm{F}} \mid e \in X\}$ and $\mathrm{E}\|Y = \{(e\|f) \in E_{\mathrm{E}\|\mathrm{F}} \mid f \in Y\}$
- $\vdash_{\mathrm{E}\|\mathrm{F}} = \{(X, e\|f) \mid (e = * \vee \pi_1(X) \vdash_{\mathrm{E}} e) \wedge (f = * \vee \pi_2(X) \vdash_{\mathrm{F}} f)\}$
  in which $\pi_1(X) = \{e \in E_{\mathrm{E}} \mid \exists f \in E_{\mathrm{F}} \cup \{*\} : e\|f \in X\}$
  and $\pi_2(X) = \{f \in E_{\mathrm{F}} \mid \exists e \in E_{\mathrm{E}} \cup \{*\} : e\|f \in X\}$.

Relational renaming (for $R \subseteq Act \times Act$):

- $E_{R(\mathrm{E})} = \{R_b e \mid e \in E_{\mathrm{E}},\ (l_{\mathrm{E}}(e), b) \in R\}$
- $l_{R(\mathrm{E})}(R_b e) = b$
- $\#_{R(\mathrm{E})} = \{(R_b e, R_c e') \mid e\#_{\mathrm{E}}e' \vee (e = e' \wedge b \neq c)\}$
- $\mapsto_{R(\mathrm{E})} = \{(R(X), R_b e) \mid X \mapsto_{\mathrm{E}} e\}$
  in which $R(X) = \{R_b e \mid e \in X \wedge (l_{\mathrm{E}}(e), b) \in R\}$
- $\vdash_{R(\mathrm{E})} = \{(X, R_b e) \mid X \in Con \wedge R^{-1}(X) \vdash e\}$
  in which $R^{-1}(X) = \{e \in E_{\mathrm{E}} \mid \exists b \in A_{R(\mathrm{E})} : R_b e \in X\}$.

The semantics for 0, $a\mathrm{E}$, $\mathrm{E} + \mathrm{F}$ and $\mathrm{E}\|\mathrm{F}$ follows the lines of [17,18,2,11]. *Relational renaming* appears in [16] and [4]. For every relation $R \subseteq Act \times Act$ there is an operator R that replaces each occurrence of an action a by fresh occurrences of the actions $\{b \mid aRb\}$. These occurrences are pairwise in conflict, and inherit their causal relationships from their source. The *relabelling* operators of CCS [12], CSP [3] and WINSKEL [18] are special cases where R is a function; the *inverse image operator* of CSP [3] is the special case where R is the inverse of a function. In case $R(a) = \emptyset$, the definition implies that the events labelled a are removed; thus also the *restriction* operators of CCS and [18] constitute special cases of relational renaming. Relational renaming in turn is a special case of *action refinement*, as studied for instance in [5]. Also note that every relational renaming operator can be written as the composition of an inverse image operator and a functional renaming operator.

The meaning of the recursion constructs $\langle X_A|S\rangle$ can be given by means of least fixed point techniques, see e.g. [17,18]. The fact that we allow recursive specifications of arbitrary size (in [17,18] they are of size 1) does not create complications; we will not repeat the definitions here. Following the standard denotational approach this yields a bundle event structure $[\![P]\!]$ and a general event structure $[\![P]\!]_{\mathcal{E}}$ for every closed CCSP expression P. For open CCSP expressions $[\![P]\!]$ and $[\![P]\!]_{\mathcal{E}}$ are functions from valuations of the variables to event structures.

Proposition 2. *For every CCSP expression P we have $\mathcal{E}([\![P]\!]) = [\![P]\!]_{\mathcal{E}}$. Hence the bundle event structures, seen as a subclass of the stable event structures, are closed under the operators of CCSP.*

Proof. Straightforward with Definition 7.

3 Denoting Bundle Event Structures in CCSP

In this section we address the question which event structures can be denoted by closed CCSP expressions of various kinds. As the events in structures $[\![P]\!]$ with P a closed CCSP expression have very particular names, whose choice seems to carry little semantic relevance, it is for this purpose most appropriate to study event structures up to isomorphism. Here two event structures E and F are *isomorphic* ($\mathrm{E} \cong \mathrm{F}$) iff $A_{\mathrm{E}} = A_{\mathrm{F}}$ and there exists a bijection between their sets of events preserving $\mapsto$ (resp. $\vdash$), $\#$ and labelling. Later we will see if the class of denotable event structures increases when considering a coarser equivalence.

The following proposition allows us to exchange, within a recursion-free CCSP expression, a closed subexpression by another expression denoting an isomorphic event structure.

Proposition 3. (Congruence) *Let* E, E' *and* F *be event structures with* $\mathrm{E} \cong \mathrm{E}'$. *Then* $a\mathrm{E} \cong a\mathrm{E}'$ *for* $a \in Act$, $\mathrm{E}+\mathrm{F} \cong \mathrm{E}'+\mathrm{F}$, $\mathrm{F}+\mathrm{E} \cong \mathrm{F}+\mathrm{E}'$, $\mathrm{E}\|\mathrm{F} \cong \mathrm{E}'\|\mathrm{F}$, $\mathrm{F}\|\mathrm{E} \cong \mathrm{F}\|\mathrm{E}'$ *and* $R(\mathrm{E}) \cong R(\mathrm{E}')$ *for* $R : Act \to Act$.

Proof. Immediate from the definitions.

The next one, essentially due to HOARE [9], allows us to drop brackets and abstract from the order of components in nested parallel compositions.

Proposition 4. *Let* E, F *and* G *be event structures. Then* $\mathrm{E}\|(\mathrm{F}\|\mathrm{G}) \cong (\mathrm{E}\|\mathrm{F})\|\mathrm{G}$ *and* $\mathrm{E}\|\mathrm{F} \cong \mathrm{F}\|\mathrm{E}$.

Proof. Straightforward.

Now we are ready to state the main theorems. We start with the simplest case of finite prime event structures with binary conflict.

Theorem 1. *For every finite prime event structure with binary conflict* E *there is a closed recursion-free CCSP expression P such that* $[\![P]\!] \cong$ E.

Proof. As we are interested in E only up to isomorphism, w.l.o.g. we may assume that $E_{\mathrm{E}} \subset Act$, i.e. the names of the events can also be used as names of CCSP actions. Let E′ be the variant of E in which every event is labelled by itself. We first build a CCSP expression denoting E′ (up to isomorphism), by encoding all events and all elements of the conflict and causality relation of E′ in terms of CCSP constructions. Subsequently, an expression for E is obtained by applying a renaming operator. Let $E_{\mathrm{E}} = \{a_1, ..., a_n\}$, $\#_{\mathrm{E}} = \{(b_1, c_1), ..., (b_m, c_m)\}$ and $<_{\mathrm{E}} = \{(d_1, e_1), ..., (d_k, e_k)\}$. Then

$$P = l_{\mathrm{E}}\left[a_1 \| \cdots \| a_n \| (b_1 + c_1) \| \cdots \| (b_m + c_m) \| (d_1 e_1) \| \cdots \| (d_k e_k)\right] + 0_{A_{\mathrm{E}}}.$$

Here l_{E} is not only the labelling function of E, but also one of the renaming operators of CCSP. Note that the actions b_i and c_i $(i = 1, ..., m)$, as well as d_j and e_j $(j = 1, ..., k)$ are among the actions $a_1, ..., a_n$. We have that $[\![a \| (a + b)]\!] \cong [\![a + b]\!]$ and $[\![a \| (ab)]\!] \cong [\![b \| (ab)]\!] \cong [\![ab]\!]$. Hence it would suffice to list as $a_1, ..., a_n$ only those events not in conflict or in any causal relationship with another event. It is routine to check that the constructed expression denotes E (up to isomorphism). The term $0_{A_{\mathrm{E}}}$ is added in case the alphabet of E contains actions that do not arise as the label of any event.

It is interesting to observe that the relational renaming operator is not needed in this proof; functional renaming would suffice. The proof above holds for the syntax of CSP – as in [3] – as well. The same cannot be done in CCS [12], because there only handshaking communication is available.

Now we pass to the case of finite bundle event structures.

Theorem 2. *For every finite bundle event structure* E *there is a closed recursion-free CCSP expression P such that* $[\![P]\!] \cong$ E.

Proof. The proof goes along the same lines as the previous one, except that instead of causal links $d <_{\mathrm{E}} e$ we now have to encode bundles $X \mapsto_{\mathrm{E}} e$. Let $X = \{d_1, ..., d_h\}$. Then the bundle $X \mapsto_{\mathrm{E}} e$ is represented by $R(de)$ where d is a fresh action and R is the relational renaming $\{(d, d_1), (d, d_2), ..., (d, d_h)(e, e)\}$.

One may wonder to what extent relational renaming is really needed here. For the language CCSP as given here it is, because with a straightforward structural induction one can check that all bundle event structures that can be denoted by recursion-free CCSP expressions with merely functional renaming only have bundles $X \mapsto e$ in which all events in X have the same label. However, there are other process algebraic operators that can take over the rôle of relational renaming.

Theorem 3. *For every finite bundle event structure* E *there is a closed recursion-free expression P in the language from* WINSKEL [18] *such that* $[\![P]\!] \cong$ E.

Proof. Winskel's language does not have relational renaming, but only functional renaming and a restriction operator $\lceil$. The restriction E$\lceil A$ behaves like E but with its events restricted to those with labels which lie in the set A. The parallel composition

$\times$ in Winskel's language allows every pair of events to synchronise; if e_0 is labelled a_0 and e_1 is labelled a_1 the synchronisation event is then labelled (a_0, a_1). Events need not synchronise however; an event e_0 in the first component that does not synchronise with any event of the second will be labelled by $(a_0, *)$, where a_0 is the label of e_0. For the rest Winskel's language is the same as CCSP, but untyped. The parallel operator of CCSP can be defined in terms of the operators $\times$, functional renaming and restriction. Although in a setting without recursion it is not possible to define CCSP's relational renaming operation in terms of the operations of Winskel's language, we can, for any finite bundle event structure E in which all actions have a different label and any image finite relational renaming R, define a context $C_{R,\mathrm{E}}[.]$ in Winskel's language that behaves like R. In the definition of this context, we use as a derived construct the *interleaving* operator $|||$ that is given by

$$P|||Q = f((P \times Q)\lceil\{(a,*) \mid a \in \alpha(P)\} \cup \{(*,b) \mid b \in \alpha(Q)\}),$$

where f is a functional renaming that renames each action $(a,*)$ into a, and each action $(*,b)$ into b. Let E and R be as stipulated. Now the context $C_{R,\mathrm{E}}[.]$ can be defined as

$$C_{R,\mathrm{E}}[.] = g(([.] \times (a_1||| \cdots |||a_n))\lceil R'),$$

where $\{a_b \mid \exists e \in E_{\mathrm{E}} \wedge l_{\mathrm{E}}(e) = a \wedge (a,b) \in R\} = \{a_1, \ldots a_n\}$, $R' = \{(a, a_b) \mid (a,b) \in R\}$ and g is the functional renaming that renames each action (a, a_b) into b. We claim that $[\![R(\mathrm{E})]\!] = [\![C_{R,\mathrm{E}}[\mathrm{E}]]\!]$ for each bundle event structure E in which all actions have a different label and any image finite relational renaming R. Using $C_{R,de}(de)$ instead of $R(de)$ in the proof of Theorem 2 now yields the required result.

In the presence of sequential composition, such as the operator ; in CSP, a bundle $\{d_1, \ldots, d_h\} \mapsto_{\mathrm{E}} e$ can also be represented as $(d_1 + \cdots + d_h); e$. However, a semantics of ; requires the introduction of a special event-label $\surd$, or some other additional structure, that helps to distinguish deadlock from successful termination. Arbitrary bundle event structures with this additional structure can in general not be represented by CSP expressions, at least not with the method employed here. Nevertheless, the construction above would work when taking ; to be a *sequencing* operator [5] that starts its second argument as soon as its first argument can perform no further actions.

Next we turn to infinite bundle event structures. Obviously *any* bundle event structure can be denoted, up to isomorphism, in a variant of CCSP with a suitable infinitary parallel composition $\|_{i \in I} P_i$. If we stick to the binary versions of $\|$ and $+$ it is straightforward to check that only countable event structures can be denoted (event structures with countably many events and only countably many bundles), even in the presence of arbitrary large recursive specifications. Thus, the best we can hope for is that every countable bundle event structure can be denoted by a CCSP expression. We are not sure if this is true; however, it can be established for prime event structures with binary conflict.

Theorem 4. *For every countable prime event structure with binary conflict* E *there is a closed CCSP expression P such that* $[\![P]\!] \cong \mathrm{E}$.

Proof. Although only the binary parallel composition exists in the syntax, a countable parallel composition $P_0 \| P_1 \| \cdots$ can be created with infinite unguarded recursion, namely as $\langle X^0_{A_0} | S \rangle$ where S contains the equations $X^i_{A_i} = P_i \| X^{i+1}_{A_{i+1}}$ where $A_i = \bigcup_{j \geq i} \alpha(P_j)$, for $i \in \mathbb{N}$. However, the denotational interpretation of such a system of equations contains only events whose existence can be proved by unwinding the recursion a finite number of times. There are for instance no events in $[\![\langle X_A | (X_A = a \| X_A) \rangle]\!]$. Thus, for the generated parallel composition to be useful, we need to require that for each $a \in A_0$ there is an i with $a \notin A_i$, i.e. $\bigcap_{i \in \mathbb{N}} A_i = \emptyset$. (*)

Now let $E_{\mathrm{E}} = \{a_i \mid i \in \mathbb{N}\}$, where the numbering is chosen in such a way that $a_i <_{\mathrm{E}} a_j \Rightarrow i < j$. As $<_{\mathrm{E}}$ is a partial order in which $\{d \in E \mid d <_{\mathrm{E}} e\}$ is finite for all $e \in E_{\mathrm{E}}$, this is always possible. Then the (possibly infinite) parallel composition $\|\{a_j \mid a_i <_{\mathrm{E}} a_j\}$ contains all events that have a_i as a causal predecessor, executed in parallel. Hence $[\![a_i(\|\{a_j \mid a_i <_{\mathrm{E}} a_j\})]\!]$ is the fragment of the desired event structure that contains all causal links starting in a_i. Its alphabet is contained in $\{a_j \mid j \geq i\}$. The parallel composition of all such event structures for $i \in \mathbb{N}$ therefore contains all causal links of E, and satisfies (*). This structure is to be put in parallel with one containing all conflicts, constructed in a similar way. As $\#_{\mathrm{E}}$ is irreflexive and symmetric, we only need to implement the conflicts $a_i \#_{\mathrm{E}} a_j$ with $i < j$. We find that E is denoted by

$$P = l_{\mathrm{E}} \left[\|_{i=0}^{\infty} (a_i(\|\{a_j \mid a_i <_{\mathrm{E}} a_j\})) \| \; \|_{i=0}^{\infty} (a_i + \|\{a_j \mid j > i \wedge a_i \#_{\mathrm{E}} a_j\}) \right] + 0_{A_E}$$

We leave it as an open problem whether the same can be achieved using only finite recursive specifications.

Due to the presence of uncountably many renaming operators, the signature of CCSP is undecidable. This can be changed by only allowing recursive enumerable renaming operators, i.e. operators $R \subseteq Act \times Act$ for which there exists a Turing machine enumerating all pairs $(a, b) \in R$. Such renaming operators can be represented by the source code describing the generating Turing machine. Codes are finite objects, and it is decidable whether a piece of text is the source code describing such a Turing machine. Now define a recursive enumerable version of CCSP, call it $\mathrm{CCSP}^{\mathrm{r.e.}}$, by requiring

- that Act is a r.e. set and all renaming operators are r.e.,
- that only r.e. subsets of Act are allowed as indices of 0 and the variables.
- and that recursive specifications S, seen as functions from V_S to the CCSP expressions, should be *primitive recursive*, with V_S a primitive decidable set.

This makes the signature of the language decidable. The primitive recursive requirement on S even makes it decidable whether a variable in a $\mathrm{CCSP}^{\mathrm{r.e.}}$ expression is free [4]. Now we have the following recursive enumerable version of Theorem 4:

Theorem 5. *Let* E *be a prime event structure with binary conflict such that* E_{E}, $\#_{\mathrm{E}}$, $<_{\mathrm{E}}$, A_{E} *are recursive enumerable sets,* l_{E} *is a recursive function, and there is an algorithm that for every event returns the finite set of its causal predecessors. Then there is a closed* $\mathrm{CCSP}^{\mathrm{r.e.}}$ *expression* P *such that* $[\![P]\!] \cong \mathrm{E}$.

4 Denoting Stable Event Structures with Binary Conflict in CCSP

In this section we infer from Theorem 4 that up to hereditary history preserving bisimulation equivalence any countable stable event structure with binary conflict can be denoted by a closed CCSP expression.

Definition 8. Two stable event structures E and F are *history preserving bisimulation equivalent* ($\mathrm{E} \leftrightarrow_h \mathrm{F}$) iff $A_{\mathrm{E}} = A_{\mathrm{F}}$ and there exists a relation $R \subseteq C(\mathrm{E}) \times C(\mathrm{F}) \times \mathcal{P}(E_{\mathrm{E}} \times E_{\mathrm{F}})$ – called a *history preserving bisimulation* – such that $(\emptyset, \emptyset, \emptyset) \in R$ and whenever $(x, y, f) \in R$ then

- f is an isomorphism between $(x, <_x, l_{\mathrm{E}} \upharpoonright x)$ and $(y, <_y, l_{\mathrm{F}} \upharpoonright y)$,
- $x \subseteq x' \in C(\mathrm{E}) \Rightarrow \exists y', f'$ with $y \subseteq y' \in C(\mathrm{F})$, $(x', y', f') \in R$ and $f' \upharpoonright x = f$,
- $y \subseteq y' \in C(\mathrm{F}) \Rightarrow \exists x', f'$ with $x \subseteq x' \in C(\mathrm{E})$, $(x', y', f') \in R$ and $f' \upharpoonright x = f$.

The bisimulation and the equivalence are *hereditary* ($\mathrm{E} \leftrightarrow_{hh} \mathrm{F}$) if moreover

- $x \supseteq x' \in C(\mathrm{E}) \Rightarrow \exists y', f'$ with $y \supseteq y' \in C(\mathrm{F})$, $(x', y', f') \in R$ and $f' = f \upharpoonright x'$,
- $y \supseteq y' \in C(\mathrm{F}) \Rightarrow \exists x', f'$ with $x \supseteq x' \in C(\mathrm{E})$, $(x', y', f') \in R$ and $f' = f \upharpoonright x'$.

R is *functional* if $R = \{(x, f(x), f \upharpoonright x) \mid x \in C(\mathrm{E})\}$ for a function $f \colon E_{\mathrm{E}} \to E_{\mathrm{F}}$.

Note that a functional bisimulation is always hereditary. Moreover, when checking that a function $f : E_{\mathrm{E}} \to E_{\mathrm{F}}$ induces an history preserving bisimulation, the second requirement is trivially fulfilled. JOYAL, NIELSEN & WINSKEL [10] characterised a functional history preserving bisimulation as a categorical construction called *open map*.

Definition 8 also applies when E and F are prime or bundle event structures, or when one of them is prime and the other is stable. We now show that every (countable) stable event structure with binary conflict is hereditary history preserving bisimulation equivalent with a (countable) prime event structure with binary conflict.

Definition 9. *Given a stable event structure* $\mathrm{E} = (E, \#, \vdash, A, l)$ *with binary conflict, the associated prime event structure* $\mathrm{E}' = (E', <, \#', A, l')$ *is given by*

- $E' = \{e_x \mid e \in x \in C(\mathrm{E})$ *and x is a minimal configuration containing* $e\}$,
- $d_x < e_y$ *iff* $x \subset y$,
- $d_x \#' e_y$ *iff* E *has no configuration containing both x and y,*
- $l'(e_X) = l(e)$.

E' is obviously a prime event structure with binary conflict, and if E is countable, then so is E'. Moreover, it is not too hard to check that the function $f : E' \to E$ given by $f(e_X) = e$ for $e \in E'$ induces a history preserving bisimulation. Therefore, any (countable) stable event structure with binary conflict is hereditary history preserving bisimulation equivalent with a (countable) prime event structure with binary conflict. This result also follows from the category theoretic results in [10]. In view of this, Theorem 4 implies

Theorem 6. *For every countable stable event structure with binary conflict* E *there is a closed CCSP expression P such that* $[\![P]\!] \leftrightarrow_{hh} \mathrm{E}$.

5 Arbitrary Conflict Reduces to Binary Conflict

In [17] event structures of the form $(E, Con, \vdash, A, l)$ appear, in which the predicate Con of Definition 4 is explicitly given rather then generated by a binary conflict relation. It is postulated that Con is a downwards closed nonempty set of finite sets of events. The configurations of such event structures and the causality relations on them are determined exactly as in Definition 5. Note that Con can equivalently be represented by its complement: an upwards closed set CONFL of finite nonempty sets of events. Another equivalent representation is in terms of the minimal members of CONFL: a collection $\#$ of finite nonempty sets of events, such that there are no two different $\gamma, \gamma' \in \#$ with $\gamma \subset \gamma'$. Now a finite set x is *consistent* or *conflict-free* if $\gamma \subseteq x$ for no $\gamma \in \#$. In this representation event structures with a binary conflict relation are literally a special case of the ones with arbitrary conflict relations. Statement $\gamma \in \#$ means that the events in γ cannot *all* happen in the same run. It does not place a restriction on proper subsets of γ.

In this section we show that every (countable) stable event structure is history preserving bisimulation equivalent to a (countable) prime event structure with binary conflict. For finite prime event structures this theorem was claimed by us in [8]. The generalisation to infinite event structures was reported in [6]. The same theorem has been discovered independently by NIELSEN & WINSKEL [14], where the first published proof can be found. Although our proof is based on the same idea as the one of [14], it is somewhat shorter and more constructive.

Definition 10. Let E be a countable event structure with arbitrary conflict. For $e \in E_{\mathrm{E}}$ let $\#_e$ be the set of conflicts involving e: $\#_e = \{\gamma \in \#_{\mathrm{E}} \mid e \in \gamma\}$. Define the event structure $2(\mathrm{E})$ by

- $E_{2(\mathrm{E})} = \{(e,t) \mid e \in E_{\mathrm{E}},\ t\colon \#_e \to \mathbb{N}$ with t recursive and $\forall \gamma \in \#_e\colon t(\gamma) < |\gamma| - 1\}$
- $A_{2(\mathrm{E})} = A_{\mathrm{E}}$ and $l_{2(\mathrm{E})}(e,t) = l_{\mathrm{E}}(e)$
- $\#_{2(\mathrm{E})} = \{((e,t),(e',t')) \mid (e = e' \wedge t \neq t') \vee (e \neq e' \wedge \exists \gamma \in \#_e : t(\gamma) = t'(\gamma))\}$
- $\mapsto_{2(\mathrm{E})} = \{(2(X),(e,t)) \mid X \mapsto_{\mathrm{E}} e\}$ in which $2(X) = \{(e,t) \mid e \in X\}$
- $\vdash_{2(\mathrm{E})} = \{(X,(e,t)) \mid \pi_1(X) \vdash e\}$ in which $\pi_1(X) = \{e \in E_{\mathrm{E}} \mid \exists t : (e,t) \in X\}$.

The idea behind this definition is the following: every member γ of the conflict relation on E has $|\gamma|$ elements, of which only $|\gamma| - 1$ can be executed. This can be modelled as an allocation of $|\gamma| - 1$ seats to $|\gamma|$ events. Let us number these seats from 0 to $|\gamma| - 2$. The event that is last in grabbing a seat can not happen. In general an event can occur in many elements γ of $\#_{\mathrm{E}}$, namely the ones in $\#_e$. In order to happen it has to grab a seat for each of these γ's. Now $2(\mathrm{E})$ is an event structure where this abstract notion of conflict has been implemented on a more down-to-earth level. The new events are allocations of old events to seats. To be precise, they are pairs (e,t), where e is the name of the source event and t a function that for every competition $\gamma \in \#_e$ in which e participates selects a seat $t(\gamma) < |\gamma| - 1$. In [14] a pair (e,t) is called an event with a *twist*; hence the choice of the letter t. Now the new events, which are old events allocated to seats, inherit their labelling and their causal dependencies from their source events. The causal dependencies are implemented by $\mapsto_{2\mathrm{E}}$ or $\vdash_{2\mathrm{E}}$, depending on whether the original event structure was a bundle, or a general one. Compare these definitions with the relational renaming operator in Sect. 2. The conflict relation on 2(E) is binary. The first set of conflicts ensures that an event can occur with only one allocation to seats in the various conflicts. The second set, that no two events are assigned the same seat in any particular conflict. This implements the abstract notion of conflict in E.

When an event e' occurs is does not really matter which seats it chooses in the various conflicts it participates in, as long as these seats are not yet taken by other events. For each event e that happened already, the chosen seats are given by the function t it happened with. Now e' has to choose an allocation function t' that is different from t in each conflict that involves both e' and e. In order to make such a choice in a computationally respectful way, we assume that all allocation functions of events that happened previously are recursive. When e is about to happen, it can then calculate which seats are still free and choose a function that is recursive as well. (A function $t : \#_e \rightarrow \mathbb{N}$ is *recursive* if there is a partial recursive function $t' : \mathcal{P}_{fin}(E_\mathrm{E}) \rightarrow \mathbb{N}$ with $t = t' \lceil \#_e$ a total function. There is no need to assume that $\#_e$ is a decidable set.) The resulting requirement in Definition 10 that all functions t should be recursive, ensures that 2(E) is still countable. Without the recursiveness requirement this would not be the case.

Theorem 7. *Let* E *be a countable stable event structure. Then* 2(E) *is a countable stable event structure with binary conflict and the function* $f : 2(\mathrm{E}) \rightarrow \mathrm{E}$ *given by* $f(e,t) = e$ *induces a history preserving bisimulation. Hence* $2(\mathrm{E}) \leftrightarrow_{hh} \mathrm{E}$.

Proof. As $(e,t)\#_{2(\mathrm{E})}(e,t')$ for $t \neq t'$, it follows immediately that $f \lceil x$ is injective for every configuration x. Now suppose $f(x)$ contains a conflict $\gamma \in \#_\mathrm{E}$. Then in x there must be events $\{(e_i,t_i) \mid e_i \in \gamma\}$ with $t(e_i) < |\gamma| - 1$. Hence two of these events must be in conflict, contradicting that x is a configuration. If follows that $f(x)$ is conflict-free. It is immediate from the definition of $\mapsto_{2(\mathrm{E})}$ resp. $\vdash_{2\mathrm{E}}$ that if $(e_1,t_1),...,(e_n,t_n)$ secures x in 2(E) then $e_1,...,e_n$ secures $f(x)$. Hence x is a configuration of E. It is also immediate from the definition of $\mapsto_{2(\mathrm{E})}$ resp. $\vdash_{2\mathrm{E}}$ that f preserves $<_x$ and labelling.

Now suppose $x \in C(2(\mathrm{E}))$ and $f(x) \subseteq y' \in C(\mathrm{E})$. We need to show that there is an $x' \in C(2(\mathrm{E}))$ with $x \subseteq x'$ and $f(x') = y'$. By induction on $|y'|$ it suffices to restrict attention to the case that there is exactly one event in $y' - f(x)$, call it e. As y' is conflict-free, for every $\gamma \in \#_e$ we have that $|\gamma \cap f(x)| \leq |\gamma| - 2$. Hence there exists a recursive $t : \#_e \rightarrow \mathbb{N}$ satisfying, for all $\gamma \in \#_e$, $t(\gamma) < |\gamma| - 1$ and for no $(e',t') \in x$: $t'(\gamma) = t(\gamma)$. It follows that $x' \stackrel{\mathrm{def}}{=} x \cup \{(e,t)\}$ is conflict-free. Moreover, any securing $(e_1,t_1),...,(e_n,t_n)$ of x can be extended with (e,t) into a securing of x'. This follows because $e_1,...,e_n,e$ is a securing of y', using the definition of $\mapsto_{2(\mathrm{E})}$ resp. $\vdash_{2\mathrm{E}}$. Thus $x' \in C(2(\mathrm{E}))$, which had to be proved. The other requirement for f inducing a history-preserving bisimulation is trivial.

By combining this insight with Theorem 6 it follows immediately that up to hereditary history preserving bisimulation equivalence all countable stable event structures with arbitrary conflicts are expressible in CCSP.

Acknowledgement. Thanks to Rom Langerak for valuable feedback.

References

1. M. Bednarczyk (1991): *Hereditary history preserving bisimulation, or what is the power of the future perfect in program logics.* Technical report, Polish Academy of Sciences, Gdańsk. Available at `ftp://ftp.ipipan.gda.pl/marek/historie.ps.gz`.
2. G. Boudol & I. Castellani (1994): *Flow models of distributed computations: Three equivalent semantics for CCS. Information and Computation* 114(2), pp. 247–314.
3. S.D. Brookes, C.A.R. Hoare & A.W. Roscoe (1984): *A theory of communicating sequential processes. Journal of the ACM* 31(3), pp. 560–599.
4. R.J. van Glabbeek (1994): *On the expressiveness of ACP (extended abstract).* In A. Ponse, C. Verhoef & S.F.M. van Vlijmen, editors: Proceedings First Workshop on the *Algebra of Communicating Processes,* ACP94, Utrecht, Workshops in Computing, Springer, pp. 188–217. Available at `http://boole.stanford.edu/pub/acp.ps.gz`.
5. R.J. van Glabbeek & U. Goltz (2001): *Refinement of actions and equivalence notions for concurrent systems. Acta Informatica* 37, pp. 229–327.
6. R.J. van Glabbeek & G.D. Plotkin (1995): *Configuration structures (extended abstract).* In D. Kozen, editor: Proceedings 10th Annual IEEE Symposium on *Logic in Computer Science* (LICS95), San Diego, USA, IEEE Computer Society Press, pp. 199–209.
7. R.J. van Glabbeek & F.W. Vaandrager (1987): *Petri net models for algebraic theories of concurrency (extended abstract).* In J.W. de Bakker, A.J. Nijman & P.C. Treleaven, editors: Proceedings *PARLE, Parallel Architectures and Languages Europe,* Eindhoven, The Netherlands, June 1987, Vol. II: Parallel Languages, LNCS 259, Springer, pp. 224–242.
8. R.J. van Glabbeek & F.W. Vaandrager (1991): *The difference between splitting in n and $n+1$ (abstract).* In E. Best & G. Rozenberg, eds.: *Proc. 3rd Workshop on Concurrency and Compositionality, Goslar, GMD-Studien Nr. 191,* Universität Hildesheim, pp. 117–121. Full version in *Information and Computation* 136(2), 1997, pp. 109–142.
9. C.A.R. Hoare (1985): *Communicating Sequential Processes.* Prentice Hall.
10. A. Joyal, M. Nielsen & G. Winskel (1996): *Bisimulation from open maps. Information and Computation* 127(2), pp. 164–185.
11. R. Langerak (1992): *Transformations and Semantics for LOTOS.* PhD thesis, Department of Computer Science, University of Twente.
12. R. Milner (1989): *Communication and Concurrency.* Prentice Hall, Englewood Cliffs.
13. M. Nielsen, G.D. Plotkin & G. Winskel (1981): *Petri nets, event structures and domains, part I. Theoretical Computer Science* 13(1), pp. 85–108.
14. M. Nielsen & G. Winskel (1996): *Petri nets and bisimulation. Theoretical Computer Science* 153, pp. 211–244.
15. A. Rabinovich & B.A. Trakhtenbrot (1988): *Behavior structures and nets. Fundamenta Informaticae* 11(4), pp. 357–404.
16. F.W. Vaandrager (1993): *Expressiveness results for process algebras.* In J.W. de Bakker, W.P. de Roever & G. Rozenberg, eds.: *Proc. REX Workshop on Semantics: Foundations and Applications,* Beekbergen, The Netherlands, June 1992, LNCS 666, Springer, pp. 609–638.
17. G. Winskel (1987): *Event structures.* In W. Brauer, W. Reisig & G. Rozenberg, editors: *Petri Nets: Applications and Relationships to Other Models of Concurrency, Advances in Petri Nets 1986, Part II; Proceedings of an Advanced Course,* Bad Honnef, September 1986, LNCS 255, Springer, pp. 325–392.
18. G. Winskel (1989): *An introduction to event structures.* In J.W. de Bakker, W.P. de Roever & G. Rozenberg, editors: *REX School and Workshop on Linear Time, Branching Time and Partial Order in Logics and Models for Concurrency,* Noordwijkerhout, The Netherlands, May/June 1988, LNCS 354, Springer, pp. 364–397.
19. G. Winskel & M. Nielsen (1995): *Models for concurrency.* In S. Abramsky, D. Gabbay & T. Maibaum, editors: *Handbook of Logic in Computer Science,* Oxford University Press, pp. 1–148.

Syntactic Formats for Free
An Abstract Approach to Process Equivalence

Bartek Klin and Paweł Sobociński

BRICS*
University of Aarhus, Denmark

Abstract. A framework of Plotkin and Turi's, originally aimed at providing an abstract notion of bisimulation, is modified to cover other operational equivalences and preorders. Combined with bialgebraic methods, it yields a technique for the derivation of syntactic formats for transition system specifications which guarantee operational preorders to be precongruences. The technique is applied to the trace preorder, the completed trace preorder and the failures preorder. In the latter two cases, new syntactic formats ensuring precongruence properties are introduced.

1 Introduction

Structural operational semantics [18,2] is one of the most fundamental frameworks for providing a precise interpretation of programming and specification languages. It is usually presented as a labelled transition system (LTS), in which states (sometimes called processes) are closed terms over some syntactic signature, and transitions are labelled with elements of some fixed set of actions. The transition relation is in turn specified by a set of derivation rules.

Many operational equivalences and preorders have been defined on processes. Among these are: bisimulation equivalence [17], simulation preorder, trace preorder, completed trace preorder, failures preorder [13,21] and many others (for a comprehensive list see [10]). In the case of processes without internal actions, all of the above have been given modal characterisations [10].

Reasoning about operational equivalences and preorders is significantly easier when they are congruences (resp. precongruences). This facilitates compositional reasoning and full substitutivity. In general, operational equivalences (preorders) are not necessarily congruences (resp. precongruences) on processes defined by operational rules. Proofs of such congruence results for given transition system specifications can be quite demanding.

One way to ensure congruential properties is to impose syntactic restrictions (syntactic formats) on operational rules. Many such formats have been developed. For bisimulation equivalence, the examples are: de Simone format [23], GSOS [8], and ntyft/ntyxt [11]. For trace equivalence, examples include [27,5],

* Basic Research in Computer Science (`www.brics.dk`), funded by the Danish National Research Foundation.

R. Amadio, D. Lugiez (Eds.): CONCUR 2003, LNCS 2761, pp. 72–86, 2003.

while decorated trace preorders have been provided with formats in [6]. For an overview of the subject see [2].

The search for an abstract theory of bisimulation and 'well-behaved' operational semantics has led to development of final coalgebra semantics [22], and bialgebraic semantics [25,26] of processes. In these frameworks, the notion of a transition system is parametrised by a notion of behaviour. Bisimulation is modelled abstractly as a span of coalgebra morphisms.

In [25,26] it was shown how to define operational rules on an abstract level, guaranteeing bisimulation equivalence (defined abstractly, using spans of coalgebra morphisms) to be a congruence. At the core of this so-called abstract GSOS is the modelling of a transition system specification as a natural transformation

$$\lambda : \Sigma(\mathrm{id} \times B) \to BT$$

where Σ is the syntactic endofunctor, T is the monad freely generated from Σ, and B is some behaviour endofunctor. In the special case of the behaviour endofunctor $\mathcal{P}_{\mathrm{f}}(A \times -)$, the abstract operational rules specialise to GSOS rules.

The abstract framework which defines bisimulation as a span of coalgebra morphisms is not sufficient for certain purposes [19]. Recently, another abstract notion of bisimulation, based on topologies (or complete boolean algebras) of tests, has been proposed [20,24].

In this paper we show that the latter abstract definition of bisimulation can be modified in a structured manner, to yield other known equivalences and preorders. We illustrate this approach on trace, completed trace and failures preorders. This constitutes a systematic approach to operational preorders, such as those based on testing scenarios [10], modal logics [10], and quantales [1].

Although the framework is general, in this paper we shall concentrate on the category of sets and functions, **Set**. We define the *test-suite fibration* with total category **Set*** having as objects pairs consisting of a set X and a test suite (a subset of $\mathcal{P}X$) over X. We lift the abstract-GSOS framework to **Set*** by describing how to lift the syntactic functor Σ and the behaviour functor B. By changing how B lifts to **Set*** we alter the specialisation preorder of certain test suites in **Set***. In particular, taking liftings which strongly resemble fragments of the Hennessy-Milner logic [12] causes the specialisation preorder to vary between known operational preorders. The abstract framework guarantees precongruence properties. The only hurdle is proving that a particular transition system specification (natural transformation) λ lifts to a natural transformation in **Set***:

$$\lambda : \Sigma^*(\mathrm{id} \times B^*) \to B^* T^*.$$

The consideration of which properties λ must satisfy in order to lift provides us with syntactic sub-formats of GSOS which guarantee precongruence properties for various operational preorders.

In this paper, we illustrate this approach by presenting precongruence formats for the trace preorder, the completed trace preorder and the failures preorder. The format derived for the trace preorder coincides with the well known de

Simone format [27]. The format derived for the completed trace preorder is, to the best of our knowledge, the first such format published. The format derived for the failures preorder is incomparable with the analogous format given in [6].

The structure of the paper is as follows. After §2 of preliminaries, we present the three obtained syntactic formats in §3, together with some examples and counterexamples from literature. The remaining sections are devoted to proving that the presented formats are indeed precongruence formats w.r.t. their respective preorders, and at the same time to illustrating the method of deriving such formats from a given operational preorder. In §4, we recall the basics of bialgebraic semantics. In §5, we present an abstract approach to operational preorders based on the notion of a test suite. In §6, this approach is merged with the bialgebraic framework to yield a general way of checking whether a given operational preorder is a congruence for a given transition system specification. Finally, in §7, we show that the formats presented in §3 ensure the respective precongruence results. We conclude in §8 by showing possible directions of future work. Due to lack of space, most proofs are left to the full version of this paper [15].

Acknowledgements. Most of the contents of §5 and §6 is a modified version of the framework developed (and, unfortunately, not yet published) by Gordon Plotkin [20] and Daniele Turi [24]. Thanks also goes to Mikkel Nygaard for reading the paper and providing us with many valuable comments. The first author is also grateful to Daniele Turi for introducing him to the subject and for inspiration, and to Gordon Plotkin for discussions and encouragement.

2 Preliminaries

A *labelled transition system* (LTS) is a set P of *processes*, a set A of *actions*, and a *transition relation* $\longrightarrow \subseteq P \times A \times P$. As usual, we write $p \xrightarrow{a} p'$ instead of $\langle p, a, p' \rangle \in \longrightarrow$. An LTS is *finitely branching*, if for every process p there are only finitely many transitions $p \xrightarrow{a} p'$.

Given a set of actions A, three sets of *modal formulae* $\mathcal{F}_{\mathsf{Tr}}$, $\mathcal{F}_{\mathsf{CTr}}$, and $\mathcal{F}_{\mathsf{Fl}}$ are given by the following BNF grammars:

$\mathcal{F}_{\mathsf{Tr}}$	$\mathcal{F}_{\mathsf{CTr}}$	$\mathcal{F}_{\mathsf{Fl}}$
$\phi ::= \top \mid \langle a \rangle \phi$	$\phi ::= \top \mid \langle a \rangle \phi \mid \tilde{A}$	$\phi ::= \top \mid \langle a \rangle \phi \mid \tilde{Q}$

where a ranges over A, and Q ranges over subsets of A. Formulae in $\mathcal{F}_{\mathsf{Tr}}$ are called *traces* over A. Formulae in $\mathcal{F}_{\mathsf{CTr}}$ ended with $\tilde{A}$ are called *completed traces*, and formulae in $\mathcal{F}_{\mathsf{Fl}}$ — *failures*.

Given an LTS, the satisfaction relation $\models$ between processes and modal formulae is defined inductively as follows:

$$
\begin{aligned}
&p \models \top \\
&p \models \langle a \rangle \phi && \iff p' \models \phi \text{ for some } p' \text{ such that } p \xrightarrow{a} p' \\
&p \models \tilde{Q} && \iff \text{there is no } a \in Q,\ p' \in P \text{ such that } p \xrightarrow{a} p'
\end{aligned}
$$

Then three operational preorders on the set of processes are defined: the *trace preorder* $\sqsubseteq_{\mathsf{Tr}}$, the *completed trace preorder* $\sqsubseteq_{\mathsf{CTr}}$, and the *failures preorder* $\sqsubseteq_{\mathsf{Fl}}$:

$$p \sqsubseteq_W p' \iff (\forall \phi \in \mathcal{F}_W . p \models \phi \Longrightarrow p' \models \phi)$$

where $W \in \{\mathsf{Tr}, \mathsf{CTr}, \mathsf{Fl}\}$.

In the context of structural operational semantics, processes are usually closed terms over some signature. A *signature* Σ is a set (also denoted Σ) of *language constructs*, together with an *arity function* $ar : \Sigma \to \mathbb{N}$. For a given set X of *variables*, ΣX is the set of expressions of the form $\mathtt{f}(x_1, \ldots, x_{ar(\mathtt{f})})$, where $\mathtt{f} \in \Sigma$ and $x_i \in X$. Given a signature Σ and a set X, the set $T_\Sigma X$ of *terms* over Σ with variables X is (isomorphic to) the least fixpoint of the operator

$$\Phi Y = X + \Sigma Y$$

where $+$ denotes disjoint union of sets. When describing terms from $T_\Sigma X$ the injections $\iota_1 : X \to T_\Sigma X$ and $\iota_2 : \Sigma T_\Sigma X \to T_\Sigma X$ will often be omitted, i.e., we will write $\mathtt{f}(x, y)$ rather than $\iota_2(\mathtt{f}(\iota_1(x), \iota_1(y)))$. Also the subscript in $T_\Sigma X$ will be omitted if Σ is irrelevant or clear from the context. Elements of $T\emptyset$ are called *closed terms* over Σ.

For a term $t \in TX$ and a function $\sigma : X \to Y$, $t[\sigma]$ will denote the term in TY resulting from t by simultaneously replacing every $x \in X$ with $\sigma(x)$.

In the following, we assume a fixed, infinite set of variables Ξ, ranged over by $\mathtt{x}_1, \mathtt{x}_2, \ldots, \mathtt{y}_1, \mathtt{y}_2, \ldots$. Terms with variables from Ξ will be typeset $\mathtt{t}$, $\mathtt{t}'$, etc.

Let us fix an arbitrary set of labels A. For a signature Σ, a *positive Σ-literal* is an expression $\mathtt{t} \xrightarrow{a} \mathtt{t}'$, and a *negative Σ-literal* is an expression $\mathtt{t} \stackrel{a}{\not\rightarrow}$, where $\mathtt{t}, \mathtt{t}' \in T\Xi$ and $a \in A$. A *transition rule* ρ over Σ is an expression $\frac{H}{\alpha}$, where H is a set of Σ-literals and α is a positive Σ-literal. Elements of H are called *premises* of ρ, and α — the *conclusion* of ρ. The left-hand side and the right-hand side of the conclusion of ρ are called the *source* and the *target* of ρ, respectively. A *transition system specification* over Σ is a set of transition rules over Σ.

Similarly, a *Σ-semiliteral* is either a negative Σ-literal, or an expression $\mathtt{t} \xrightarrow{a}$, where $\mathtt{t} \in T\Xi$ and $a \in A$. A positive literal $\mathtt{t} \xrightarrow{a} \mathtt{t}'$ *completes* the semiliteral $\mathtt{t} \xrightarrow{a}$, and we say that a negative literal completes itself.

In the following definition assume a fixed signature Σ, and a *finite* set A.

Format 1 (GSOS). A transition system specification R is in GSOS [8] format if every rule $\rho \in R$ is of the form

$$\frac{\left\{ \mathtt{x}_i \xrightarrow{a_{ij}} \mathtt{y}_{ij} \; : \; i \leq n, \; j \leq m_i \right\} \cup \left\{ \mathtt{x}_i \stackrel{b_{ik}}{\not\rightarrow} \; : \; i \leq n, k \leq n_i \right\}}{\mathtt{f}(\mathtt{x}_1, \ldots, \mathtt{x}_n) \xrightarrow{c} \mathtt{t}}$$

with $\mathtt{f} \in \Sigma$ and $n = ar(\mathtt{f})$, such that $\mathtt{x}_i \in \Xi$ and $\mathtt{y}_{ij} \in \Xi$ are all distinct and are the only variables that occur in ρ. In the following, we will consider only *image finite* GSOS specifications, i.e. those with finitely many rules for each construct $\mathtt{f} \in \Sigma$ and action $c \in A$.

Given a transition system specification R in GSOS format, one defines a notion of a provable positive literal in a straightforward way. The set of all provable literals forms a finitely branching LTS with closed terms over Σ as processes, and with positive closed literals as transitions (for details, see [2]).

An operational preorder $\sqsubseteq$ is a *precongruence* with respect to a transition system specification R, if in the LTS induced by R, for each $\mathtt{f} \in \Sigma$ with arity n, if $t_1 \sqsubseteq t'_1, \ldots, t_n \sqsubseteq t'_n$, then $\mathtt{f}(t_1, \ldots, t_n) \sqsubseteq \mathtt{f}(t'_1, \ldots, t'_n)$.

The examples in §3 are based on basic process algebra **BPA**. Assuming a finite set A of actions, its syntax Σ is defined by the BNF grammar $t ::= \mathtt{0} \mid \alpha t \mid t{+}t$ and the transition system specification **BPA** over Σ is a collection of rules

$$\frac{}{\alpha\mathtt{x} \xrightarrow{\alpha} \mathtt{x}} \qquad \frac{\mathtt{x} \xrightarrow{\alpha} \mathtt{x}'}{\mathtt{x}+\mathtt{y} \xrightarrow{\alpha} \mathtt{x}'} \qquad \frac{\mathtt{y} \xrightarrow{\alpha} \mathtt{y}'}{\mathtt{x}+\mathtt{y} \xrightarrow{\alpha} \mathtt{y}'}$$

where α ranges over A. When presenting terms over the above syntax, the trailing 0's will be omitted. It is easy to see that **BPA** is in the GSOS format.

3 Precongruence Formats

In this section we introduce the syntactic formats derived using the framework described in the latter parts of the paper. The precongruence properties of these formats are formally stated in §7.

Format 2 (Tr-format). A set of GSOS rules R is in Tr-*format*, if for each $\rho \in R$, all premises of ρ are positive, and no variable occurs more than once in the left-hand sides of premises and in the target of ρ.

It is easy to see that this format coincides with the well-known de Simone format [23]. The fact that this syntactic format ensures the trace preorder to be a precongruence was first proved in [27].

We proceed to define an analogous syntactic format for the completed trace preorder.

Definition 1 (CTr-testing set). A CTr-*testing set* P over a set of variables $\{\mathtt{x}_1, \ldots, \mathtt{x}_n\}$ is a set of semiliterals of the form

$$P = \{\, \mathtt{x}_i \stackrel{a_i}{\not\longrightarrow} \; : i \in I \,\} \cup \{\, \mathtt{x}_i \xrightarrow{a} \; : i \in J, a \in A \,\}$$

where $I, J \subseteq \{1, \ldots, n\}$.

Format 3 (CTr-format). A set of GSOS rules R is in CTr-*format*, if

1. For each rule $\rho \in R$:
   - if ρ has a negative premise $\mathtt{x} \stackrel{a}{\not\longrightarrow}$, than for every label $b \in A$, ρ has also the negative premise $\mathtt{x} \stackrel{b}{\not\longrightarrow}$,
   - no variable occurs more than once in the target of ρ,

  - no variable occurs simultaneously in the left-hand side of a premise and in the target of ρ,
  - no variable occurs simultaneously in the left-hand side of a positive premise and in the left-hand side of any other premise of ρ.
2. For each construct $\mathtt{f}(\mathtt{x}_1, \ldots, \mathtt{x}_n)$ of the language, there exists a sequence $P_1, \ldots, P_k$ of CTr-testing sets over $\{\mathtt{x}_1, \ldots, \mathtt{x}_n\}$, such that
  - For every (possibly renamed) rule $\rho \in R$ with source $\mathtt{f}(\mathtt{x}_1, \ldots, \mathtt{x}_n)$ there exists a sequence $p_1, \ldots p_k$ of semiliterals from $P_1, \ldots, P_k$ respectively, such that for every $i \in \{1, \ldots, k\}$ there exists a premise r of ρ such that r completes p_i.
  - For every sequence $p_1, \ldots, p_k$ of semiliterals from $P_1, \ldots, P_k$ respectively, there exists a (possibly renamed) rule $\rho \in R$ with source $\mathtt{f}(\mathtt{x}_1, \ldots, \mathtt{x}_n)$ such that for each premise r of ρ there exists an $i \in \{1, \ldots, k\}$ such that r completes p_i.

Proposition 2. BPA is in CTr-format. □

The following example is taken from [2]. Assume $A = \{a, b\}$, and extend **BPA** with an operational rule for the so-called *encapsulation operator* $\partial_{\{b\}}$:

$$\frac{\mathtt{x} \xrightarrow{a} \mathtt{y}}{\partial_{\{b\}}(\mathtt{x}) \xrightarrow{a} \partial_{\{b\}}(\mathtt{y})}$$

Then it is easy to check that $aa + ab \sim_{\mathsf{CTr}} a(a+b)$ but that $\partial_{\{b\}}(aa+ab) \not\sqsubseteq_{\mathsf{CTr}} \partial_{\{b\}}(a(a+b))$.

Another example of an operational construct that is not well behaved with respect to completed traces is the *synchronous composition*, as shown in [27]. Here, we add the rules

$$\frac{\mathtt{x} \xrightarrow{\alpha} \mathtt{x}' \quad \mathtt{y} \xrightarrow{\alpha} \mathtt{y}'}{\mathtt{x} \times \mathtt{y} \xrightarrow{\alpha} \mathtt{x}' \times \mathtt{y}'}$$

where α ranges over $A = \{a, b\}$. Here it is easy to see that $aa \times (aa + ab) \not\sqsubseteq_{\mathsf{CTr}} aa \times a(a+b)$.

These two examples have led the authors of [2] to speculate that one cannot hope for a general syntactic congruence format for completed trace equivalence.

Proposition 3. The semantics for the encapsulation operator ∂ and the synchronous composition $\times$ are not in CTr-format. □

For a non-trivial example of a transition system specification in CTr-format, extend **BPA** with *sequential composition*, defined by rules

$$\frac{\mathtt{x} \xrightarrow{\alpha} \mathtt{x}'}{\mathtt{x};\mathtt{y} \xrightarrow{\alpha} \mathtt{x}';\mathtt{y}} \qquad \frac{\mathtt{x} \stackrel{a}{\not\rightarrow} \text{ for all } a \in A \quad \mathtt{y} \xrightarrow{\alpha} \mathtt{y}'}{\mathtt{x};\mathtt{y} \xrightarrow{\alpha} \mathtt{y}'}$$

where α ranges over A.

Proposition 4. **BPA** extended with sequential composition is in CTr-format. □

We proceed to define a precongruence syntactic format for the failures preorder.

Definition 5 (Fl-testing set). An Fl-*testing set* P over a set of variables $\{\mathtt{x}_1, \dots, \mathtt{x}_n\}$ is a set of semiliterals of the form

$$P = \left\{ \mathtt{x}_i \overset{a_i}{\not\longrightarrow} \; : i \in I \right\} \cup \left\{ \mathtt{x}_i \overset{b_{ij}}{\longrightarrow} \; : 1 \leq i \leq n, 1 \leq j \leq m_i \right\}$$

(where $I \subseteq \{1, \dots, n\}$, $m_i \in \mathbb{N}$), such that for any labels $a, b \in A$, if $\mathtt{x}_i \overset{a}{\longrightarrow} \in P$ and $\mathtt{x}_i \overset{b}{\not\longrightarrow} \in P$ then $\mathtt{x}_i \overset{b}{\longrightarrow} \in P$.

Format 4 (Failures Format). A set of GSOS rules R is in Fl-*format*, if

1. For each rule $\rho \in R$:
   - no variable occurs more than once in the target of ρ,
   - no variable occurs simultaneously in the left-hand side of a premise and in the target of ρ,
   - no variable occurs simultaneously in the left-hand side of a positive premise and in the left-hand side of any other premise of ρ.
2. For each construct $\mathtt{f}(\mathtt{x}_1, \dots, \mathtt{x}_n)$ of the language, and for each set of labels $Q \subseteq A$, there exists a sequence $P_1, \dots, P_k$ of Fl-testing sets over $\{\mathtt{x}_1, \dots, \mathtt{x}_n\}$, such that
   - For every (possibly renamed) rule $\rho \in R$ with the conclusion of the form $\mathtt{f}(\mathtt{x}_1, \dots, \mathtt{x}_n) \overset{a}{\longrightarrow} \mathtt{t}$ with $a \in Q$ and an arbitrary $\mathtt{t}$, there exists a sequence $p_1, \dots, p_k$ of semiliterals from $P_1, \dots, P_k$ respectively, such that for every $i \in \{1, \dots, k\}$ there exists a premise r of ρ such that r completes p_i.
   - For every sequence $p_1, \dots, p_k$ of semiliterals from $P_1, \dots, P_k$ respectively, there exist a label $a \in Q$, a term $\mathtt{t}$, and a (possibly renamed) rule $\rho \in R$ with the conclusion $\mathtt{f}(\mathtt{x}_1, \dots, \mathtt{x}_n) \overset{a}{\longrightarrow} \mathtt{t}$ such that for each premise r of ρ there exists an $i \in \{1, \dots, k\}$ such that r completes p_i.

Proposition 6. **BPA** is in Fl-format. □

In [7] it was shown that the failures preorder is not a precongruence for **BPA** extended with sequential composition.

Proposition 7. If A contains at least two different labels a, b, then **BPA** extended with sequential composition is not in Fl-format. □

The Fl-format excludes many examples of transition system specifications that behave well with respect to the failures preorder. Many of these examples are covered by the 'failure trace format' introduced in [6]. However, the latter format excludes also some examples covered by FI-format. Indeed, assume $a, b \in A$ and extend **BPA** with two unary constructs $\mathtt{g}, \mathtt{h}$ and rules (where α ranges over A)

$$\frac{\mathtt{x} \xrightarrow{\alpha} \mathtt{x}'}{\mathtt{g}(\mathtt{x}) \xrightarrow{\alpha} \mathtt{h}(\mathtt{x}')} \qquad \frac{\mathtt{x} \overset{a}{\not\longrightarrow}}{\mathtt{h}(\mathtt{x}) \xrightarrow{b} 0}$$

Proposition 8. **BPA** extended with g and h as above, is in FI-format. □

However, the rules above are not in the 'failure trace format' proposed in [6]. This means that FI-format is incomparable with that format.

4 An Abstract Approach

In this section we shall recall the foundations needed for the framework described in §5 and §6. First, we briefly recall how LTS can be described as coalgebras for a specific behaviour endofunctor and briefly recall final coalgebra semantics. We then proceed to recall several notions from the abstract approach to operational semantics of Plotkin and Turi [26].

In the following, $\mathcal{P} : \mathbf{Set} \to \mathbf{Set}$ will denote the (covariant) powerset functor. The (covariant) finite powerset functor $\mathcal{P}_{\mathrm{f}} : \mathbf{Set} \to \mathbf{Set}$ takes a set to the set of its *finite* subsets. The reader is referred to [16] for any unexplained categorical notation used henceforward.

There is a bijection between the set of finitely branching LTS over a fixed set of actions A and the coalgebras of the functor $\mathcal{P}_{\mathrm{f}}(A \times -)$. Indeed, given an LTS $\langle P, A, \longrightarrow \rangle$ let

$$h : P \to \mathcal{P}_{\mathrm{f}}(A \times P)$$

be defined by $h(p) = \{ \langle a, p' \rangle \, : \, p \xrightarrow{a} p' \}$.

The functor $\mathcal{P}_{\mathrm{f}}(A \times -)$ has a final coalgebra $\varphi : S \to \mathcal{P}_{\mathrm{f}}(A \times S)$. The carrier S of this coalgebra may be described as the set of synchronisation trees with edges having labels from A, quotiented by bisimulation [4,25].

In the following we specialise the framework of [26] to the category **Set** and behaviour functor $\mathcal{P}_{\mathrm{f}}(A \times -)$. Any syntactic signature Σ determines a so-called syntactic endofunctor $\Sigma : \mathbf{Set} \to \mathbf{Set}$ which acts on sets by sending

$$\Sigma X = \coprod_{\mathtt{f} \in \Sigma} X^{ar(\mathtt{f})} \tag{1}$$

and the action on functions is the obvious one. The functor Σ freely generates a monad $\langle T, \mu, \eta \rangle : \mathbf{Set} \to \mathbf{Set}$. It turns out that TX is (isomorphic to) the set of all terms over Σ with variables from X.

Theorem 9 ([26]). There is a correspondence between sets of rules in the GSOS format (Format 1) and natural transformations

$$\lambda : \Sigma(\mathrm{id} \times \mathcal{P}_{\mathrm{f}}(A \times -)) \to \mathcal{P}_{\mathrm{f}}(A \times T-)$$

Moreover, the correspondence is 1-1 up to equivalence of sets of rules. □

Assume a natural transformation $\lambda : \Sigma(\mathrm{id} \times B) \to BT$. A λ-*model* is a pair

$$\Sigma X \xrightarrow{h} X \xrightarrow{g} BX$$

such that $g \circ h = Bh^{\sharp} \circ \lambda_X \circ \Sigma \langle \mathrm{id}, g \rangle$, ($h^{\sharp} : TX \to X$ is the inductive extension of h). A λ-model morphism between $\Sigma X \xrightarrow{h} X \xrightarrow{g} BX$ and $\Sigma X' \xrightarrow{h'} X' \xrightarrow{g'} BX'$ is a morphism $f : X \to X'$ which is simultaneously a Σ-algebra morphism and a B-coalgebra morphism, ie. $h' \circ \Sigma f = g \circ h$ and $g' \circ f = Bf \circ g$. Let λ-**Mod** denote the category of λ-models and λ-model morphisms.

Theorem 10 ([26]). Suppose that $\mathbf{C}$ is a category, Σ is an endofunctor which freely generates a monad T and B is an endofunctor which cofreely generates a comonad D. Then the following hold:

1. λ-**Mod** has an initial and final object,
2. the carrier and algebra part of the initial λ-model is the initial Σ-algebra,
3. the carrier and coalgebra part of the final λ-model is the final B-coalgebra,
4. the coalgebra part of the initial λ-model is the so-called *intended operational model* of λ. □

In particular, if $\mathbf{C} = \mathbf{Set}$ and $B = \mathcal{P}_{\mathrm{f}}(A \times -)$, then the intended operational model of λ is the LTS generated by the GSOS rules associated to λ.

5 Process Equivalences from Fibred Functors

In this section, we introduce the central concept of a test suite fibration. This is a modification of the yet unpublished framework [20,24] due to Plotkin and Turi. In that approach, the test suites (Definition 11) are necessarily *topologies*, that is, they satisfy certain closure properties. We relax this definition and require only that a test suite contains the largest test. This modification allows us to consider operational preorders and equivalences different from bisimulation. Also, the original framework was developed largely for **Cppo**-enriched categories, here we deal primarily with **Set**.

We define $2 = \{\mathtt{tt}, \mathtt{ff}\}$. Given a function $f : X \to Y$ and subsets $V \subseteq X$, $V' \subseteq Y$, we shall use $f(V)$ to denote the set $\{ y \in Y : \exists x \in V. fx = y \}$ and similarly $f^{-1}(V')$ to denote $\{ x \in X : fx \in V' \}$. Given a set $\tau \subseteq \mathcal{P}X$, the *specialisation preorder* of τ is defined by

$$x \leq_\tau x' \quad \text{iff} \quad \forall V \in \tau. x \in V \Rightarrow x' \in V$$

For an introduction to fibrations and related terminology, the reader is referred to the first chapter of [14].

Definition 11 (Test suite). A *test* on a set X is a function $V : X \to 2$. We say that an element x *passes a test* V iff $Vx = \mathtt{tt}$. A *test suite* on X is a collection of tests on X which includes the maximal test, that is, the function constant at $\mathtt{tt}$. Let X^* denote the poset of test suites on X ordered by inclusion.

We can define a functor $(-)^* : \mathbf{Set}^{\mathrm{op}} \to \mathbf{Pos}$ which sends a set to the poset of test suites X^* and sends a function $f : X \to Y$ to $f^* : Y^* \to X^*$ defined by

$$f^*\tau' = \{\, V' \circ f \, : \, V' \in \tau' \,\}\,.$$

Intuitively, we think of tests on X as subsets of X. Then f^* is the pre-image operation, taking each test on Y to the test on X which maps to Y via f.

Definition 12 (Test suite fibration). A *fibration of test suites* for $(-)^*$ is the fibration obtained using the Grothendieck construction, ie. the total category $\mathbf{Set}^*$ has

- objects: pairs $\langle X, \tau \rangle$ where $X \in \mathbf{Set}$ and $\tau \in X^*$, τ is a test suite.
- arrows: $\langle X, \tau \rangle \xrightarrow{f} \langle X', \tau' \rangle$ iff $f : X \to X'$ and $f^*\tau' \subseteq \tau$.

It is then standard that the obvious forgetful functor $U : \mathbf{Set}^* \to \mathbf{Set}$ taking $\langle X, \tau \rangle$ to X is a fibration.

It will be useful to define various operations on test suites τ. Letting $\nabla : 2 + 2 \to 2$ be the codiagonal and $\wedge : 2 \times 2 \to 2$ be logical-and, we let

$$\begin{aligned} \tau \oplus \tau' &= \{\, \nabla \circ (V + V') \, : \, V \in \tau, V' \in \tau' \,\} \\ \tau \otimes \tau' &= \{\, \wedge \circ (V \times V') \, : \, V \in \tau, V' \in \tau' \,\} \\ \tau \bowtie \tau' &= \{\, V \circ \pi_1 \, : \, V \in \tau \,\} \cup \{\, V' \circ \pi_2 \, : \, V' \in \tau' \,\}\,. \end{aligned}$$

It is easy to check that given two test suites, families $\tau \oplus \tau'$, $\tau \otimes \tau'$ and $\tau \bowtie \tau'$ are test suites. Intuitively, given test suites τ and τ' on X and Y, $\tau \oplus \tau'$ is the test suite on $X + Y$ obtained by taking (disjoint) unions of tests from τ on X and τ' on Y, $\tau \otimes \tau'$ is the test suite on $X \times Y$ consisting of tests built by performing a test from τ on X *and simultaneously* performing a test from τ' on Y and accepting when both tests accept; finally, $\tau \bowtie \tau'$ is the test on $X \times Y$ which consists of *either* a test from τ on X *or* a test from τ' on Y.

Proposition 13. The category $\mathbf{Set}^*$ has coproducts and products:

$$\begin{aligned} \langle X, \tau \rangle + \langle Y, \tau' \rangle &= \langle X + Y, \tau \oplus \tau' \rangle \\ \langle X, \tau \rangle \times \langle Y, \tau' \rangle &= \langle X \times Y, \tau \bowtie \tau' \rangle \end{aligned}$$

Let $B : \mathbf{Set} \to \mathbf{Set}$ be some behaviour endofunctor. A *lifting* of B to $\mathbf{Set}^*$ is an endofunctor $B^* : \mathbf{Set}^* \to \mathbf{Set}^*$ such that, for some $B_X : X^* \to (BX)^*$ we have $B^* \langle X, \tau \rangle = \langle BX, B_X\tau \rangle$ and $B^*f = Bf$. It turns out that there are many possible choices for B_X giving different liftings of B to $\mathbf{Set}^*$. One systematic way to construct such liftings is via families of functions from $B2$ to 2. Intuitively, such functions correspond to modalities like those in the Hennessy-Milner logic. In the original framework due to Plotkin and Turi [20,24] the canonical choice of *all* functions from $B2$ to 2 is taken.

For any X, let $\mathrm{Cl}_X : \mathcal{P}\mathcal{P}X \to X^*$ denote a closure operator. We shall only demand that for all $f : X \to Y$ and $Z \subseteq \mathcal{P}Y$ we have $\mathrm{Cl}_X\, f^*Z = f^*\,\mathrm{Cl}_Y\, Z$

(with the obvious extension of the domain of f^* from Y^* to $\mathcal{PP}Y$). Intuitively, a closure operator corresponds to a set of propositional connectives.

Given an arbitrary family W of functions $B2 \to 2$, we define an operator $B^W_X : X^* \to (BX)^*$ as follows:

$$B^W_X(\tau) = \mathrm{Cl}_{BX}\{\, w \circ BV \,:\, w \in W \text{ and } V \in \tau \,\}.$$

We are now in a position to construct a lifting of B to $\mathbf{Set}^*$. Indeed, we let $B^W \langle X, \tau\rangle = \langle BX, B^W_X \tau\rangle$ and $B^W f = Bf$. It turns out that this defines an endofunctor B^W on $\mathbf{Set}^*$.

Theorem 14. Suppose that $B : \mathbf{Set} \to \mathbf{Set}$ has a final coalgebra $\varphi : S \to BS$. Then $\varphi : \langle S, M^W\rangle \to \langle BS, B^W_S M^W\rangle$ is a final B^W coalgebra where M^W is the least fixpoint of the operator $\Phi(\tau) = \varphi^* B^W_S \tau$ on S^*. □

Suppose that $B : \mathbf{Set} \to \mathbf{Set}$ lifts to a functor $B^W : \mathbf{Set}^* \to \mathbf{Set}^*$ with B^W defined as before.

Theorem 15. Take any coalgebra $h : X \to BX$, and let $k : X \to S$ be the unique coalgebra morphism from h to the final B-coalgebra. Then k^*M (where $\langle S, M\rangle$ is the carrier of the final B^W-coalgebra) is the least test suite τ on X such that $h : \langle X, \tau\rangle \to \langle BX, B^W_X \tau\rangle$ is a morphism in $\mathbf{Set}^*$. □

From now on we shall assume a finite set of labels A and confine our attention to the endofunctor $BX = \mathcal{P}_\mathrm{f}(A \times X)$ on $\mathbf{Set}$.

Assuming $a \in A$ and $Q \subseteq A$, let $w_{\langle a\rangle}, w_{rQ} : B2 \to 2$ denote the functions

$$w_{\langle a\rangle}X = \begin{cases} \mathtt{tt} & \text{if } \langle a, \mathtt{tt}\rangle \in X \\ \mathtt{ff} & \text{otherwise,} \end{cases} \qquad w_{rQ}X = \begin{cases} \mathtt{tt} & \text{if } \forall a \in Q\, \forall v \in 2.\, \langle a, v\rangle \notin X \\ \mathtt{ff} & \text{otherwise.} \end{cases}$$

We shall now define three subsets of maps $B2 \to 2$:

$$\mathsf{Tr} = \{\, w_{\langle a\rangle} \,:\, a \in A \,\} \qquad \mathsf{CTr} = \mathsf{Tr} \cup \{w_{rA}\} \qquad \mathsf{Fl} = \mathsf{Tr} \cup \{\, w_{rQ} \,:\, Q \subseteq A \,\}$$

The set Tr together with the closure operator $\mathrm{Cl}^\top_X(\tau) = \tau \cup \{X\}$, determines B^{Tr}_X for any X and therefore determines a lifting of B to $B^{\mathsf{Tr}} : \mathbf{Set}^* \to \mathbf{Set}^*$. Similarly, CTr with $\mathrm{Cl}^\top$ and and Fl with $\mathrm{Cl}^\top$ determine liftings B^{CTr} and B^{Fl} respectively.

The following Theorems 16–18 show that the specialisation preorders in the final B^{Tr}, B^{CTr} and B^{Fl}-coalgebras coincide with the trace, the completed trace and the failures preorders. We use these facts to prove Theorem 19 which states that given any $h : X \to \mathcal{P}_\mathrm{f}(A \times X)$, the specialisation preorders on certain test suites on X coincide with these operational preorders.

Theorems 16–18. In the final B^W-coalgebra, the specialization preorder coincides with $\sqsubseteq_W$, where $W \in \{\mathsf{Tr}, \mathsf{CTr}, \mathsf{Fl}\}$.

Theorem 19. Suppose that $h : X \to \mathcal{P}_\mathrm{f}(A \times X)$ is a coalgebra (LTS), $\varphi : S \to BS$ is the final B-coalgebra and that $k : X \to S$ is the unique morphism given by finality. Then $x \leq_{k^*M^W} x'$ if and only if $x \sqsubseteq_W x'$, where $W \in \{\mathsf{Tr}, \mathsf{CTr}, \mathsf{Fl}\}$ and $\langle S, M^W\rangle$ is the carrier of the final B^W-coalgebra. □

6 Application: Congruence Formats from Bialgebras

To lift the bialgebraic framework to the total category **Set***, we need a way to lift the syntactic and the behaviour functors together with the natural transformation λ. Various ways to lift the behaviour B were shown in the previous section, now we proceed to show a lifting of the syntactic functor.

Given a syntactic endofunctor Σ on **Set** defined as in Eq. (1), define an endofunctor Σ^* on **Set***: $\Sigma^* \langle X, \tau \rangle = \langle \Sigma X, \Sigma_X \tau \rangle$, where

$$\Sigma_X \tau = \mathrm{Cl}^{\cup} \left(\bigoplus_{\mathtt{f} \in \Sigma} \tau^{\otimes ar(\mathtt{f})} \right)$$

where $\mathrm{Cl}^{\cup}$ is closure under arbitrary unions, and $\tau^{\otimes n}$ denotes $\overbrace{\tau \otimes \tau \otimes \cdots \otimes \tau}^{n \text{ times}}$. On arrows, given $f : \langle X, \tau \rangle \to \langle X', \tau' \rangle$, we define simply $\Sigma^* f = \Sigma f$. It turns out that Σ^* defined this way is an endofunctor on **Set***.

Theorem 20. Suppose that an endofunctor F lifts to a endofunctor F^*, and has an initial algebra $\psi : FN \to N$. Then $\psi : \langle FN, F_N P \rangle \to \langle N, P \rangle$ is the initial F^* algebra where P is the greatest fixpoint of the operator $\Psi(\tau) = (\psi^{-1})^* F_N \tau$.

Corollary 21. For any syntactic endofunctor Σ, the functor Σ^* freely generates a monad T^* that lifts the monad T freely generated by Σ. □

A similar corollary about a behaviour B^W cofreely generating a comonad D^W can be drawn from Theorem 14. These two corollaries allow us to apply Theorem 10 for the category **Set*** and endofunctors Σ^* and B^W.

The following theorem is a crucial property of the endofunctor Σ^*. Indeed, varying the definition of Σ^* in our framework would lead to definition of various precongruence formats, but only as long as the following property holds.

Theorem 22. For any Σ^*-algebra $h : \langle \Sigma X, \Sigma_X \tau \rangle \to \langle X, \tau \rangle$, the specialisation preorder $\leq_\tau$ is a precongruence on $h : \Sigma X \to X$. □

We now have the technology needed to prove the main result of this section.

Consider a natural transformation $\lambda : \Sigma(\mathrm{id} \times B) \to BT$. By Theorem 10, the coalgebraic part of the initial λ-model has $N = T\emptyset$ as its carrier, and it is the intended operational model of λ. If $B = \mathcal{P}_{\mathrm{f}}(A \times -)$, then the intended operational model is the LTS generated by GSOS rules associated to λ. Let $k : N \to S$ be the final coalgebra morphism from the intended operational model to the final B-coalgebra. Assume B lifts to some B^* as before, and let $\langle S, M \rangle$ be the carrier of the final B^*-coalgebra.

Theorem 23. If λ lifts to a natural transformation in the total category:

$$\lambda : \Sigma^*(\mathrm{id} \times B^*) \to B^* T^*.$$

then the specialisation preorder on $k^* M$ is a precongruence on N.

Proof. (Sketch) In diagram *(i)* below, the left column is the initial λ-model while the right column is the final λ-model; the λ-model morphism k is the unique morphism making both squares commutative.

$$
\begin{array}{ccc}
\Sigma^* \langle N, P\rangle & \xrightarrow{\Sigma^* k} & \Sigma^* \langle S, M\rangle \\
\downarrow{\scriptstyle \psi} & & \downarrow{\scriptstyle \delta} \\
\langle N, P\rangle & \xrightarrow{k} & \langle S, M\rangle \\
\downarrow{\scriptstyle \epsilon} & & \downarrow{\scriptstyle \varphi} \\
B^* \langle N, P\rangle & \xrightarrow[B^* k]{} & B^* \langle S, M\rangle
\end{array}
\qquad
\begin{array}{ccc}
\Sigma^* \langle N, k^*M\rangle & \xrightarrow{\Sigma^* k} & \Sigma^* \langle S, M\rangle \\
\downarrow{\scriptstyle \psi} & & \downarrow{\scriptstyle \delta} \\
\langle N, k^*M\rangle & \xrightarrow{k} & \langle S, M\rangle \\
\downarrow{\scriptstyle \epsilon} & & \downarrow{\scriptstyle \varphi} \\
B^* \langle N, k^*M\rangle & \xrightarrow[B^* k]{} & B^* \langle S, M\rangle
\end{array}
$$

(i) *(ii)*

Our goal is to show that *(ii)* above a diagram in **Set***. If all the morphisms are defined then its commutativity follows from the commutativity of *(i)*. By Theorem 15, $\epsilon : \langle N, k^*M\rangle \to B^* \langle N, k^*M\rangle$ is a B^*-coalgebra.

Now $\psi^*k^*M = (\Sigma k)^*\delta^*M \subseteq (\Sigma k)^*(\Sigma_S M) \subseteq \Sigma_X(k^*M)$ where we use the fact that δ is a morphism in **Set*** and the fact that Σ^* is a functor. Thus $\psi : \langle N, k^*M\rangle \to B^* \langle N, k^*M\rangle$ is a Σ^*-algebra and by Theorem 22 the specialisation preorder of k^*M is a precongruence. □

7 Precongruence Formats for (Almost) Free

In this section we consider a syntactic endofunctor Σ with a freely generated monad T, the behaviour functor $BX = \mathcal{P}_{\mathrm{f}}(A \times X)$, and a set R of GSOS rules with the corresponding natural transformation $\lambda : \Sigma(\mathrm{id} \times B) \to BT$. The purpose is to describe syntactic conditions on R that would ensure that λ lifts to a natural transformation $\lambda : \Sigma^*(\mathrm{id} \times B^W) \to B^W T^*$, where $W \in \{\mathsf{Tr}, \mathsf{CTr}, \mathsf{Fl}\}$. As a consequence of Theorem 23, such syntactic conditions ensure that the respective operational preorders are precongruences.

Theorems 24–26. For $W \in \{\mathsf{Tr}, \mathsf{CTr}, \mathsf{Fl}\}$, if R is in W-format (see Form. 2-4), then $\lambda : \Sigma^*(Id \times B^W) \to B^W T^*$ is a natural transformation in **Set***. □

8 Conclusions

We have presented an abstract coalgebraic approach to the description of various operational preorders, via a fibration of test suites. In Theorems 16–18 we illustrated this approach on the trace preorder, the completed trace preorder and the failures preorder. Combined with bialgebraic methods, this framework allows the derivation of syntactic subformats of GSOS which guarantee that the above operational preorders are precongruences. Theorem 23 is a guideline in the search for such formats, and Theorems 24–26 are applications of the framework.

The generality and abstractness of Theorem 23 prompted us to coin the expression 'precongruence format for free'. However, it must be stressed that to derive a format for a given operational preorder remains a non-trivial task. Indeed, the proofs of Theorems 24–26 are quite long and technical. The expression 'for free' reflects the fact that Theorem 23 lets us prove precongruence properties without considering the global behaviour (e.g. traces) of processes. Instead, one considers only simple test constructions, corresponding intuitively to single modalities.

Related abstract approaches to operational preorders and equivalences include those based on modal characterisations [10] and quantales [1]. In the latter framework, no syntactic issues have been addressed. In the former, some general precongruence formats have been obtained by attempting to decompose modal formulae according to given operational rules [7]. This technique bears some resemblance to our approach, and the precise connections have to be investigated.

There are several possible directions of future work. Firstly, the approach presented here can be extended to deal with other operational preorders and equivalences described in literature. Secondly, one can move from the GSOS format (and its subformats) to the more general (safe) ntree format [9], which can also be formalised in the bialgebraic framework [26]. Thirdly, the abstract framework of test suites seems to be general enough to cover other notions of process behaviour (e.g. involving store), or even other underlying categories (e.g. complete partial orders instead of sets). It may prove interesting to formalise various operational preorders in such cases and to find precongruence formats for them.

References

1. S. Abramsky and S. Vickers. Quantales, observational logic and process semantics. *Math. Struct. in Comp. Sci.*, 3:161–227, 1993.
2. L. Aceto, W. Fokkink, and C. Verhoef. Structural operational semantics. In J. Bergstra, A. Ponse, and S. Smolka, editors, *Handbook of Process Algebra*. Elsevier, 1999.
3. P. Aczel and N. Mendler. A final coalgebra theorem. In D. H. Pitt et al., editors, *Proc. CTCS'89*, volume 389 of *LNCS*, pages 357–365, 1989.
4. M. Barr. Terminal coalgebras in well-founded set theory. *Theoretical Computer Science*, 114:299–315, 1993.
5. B. Bloom. When is partial trace equivalence adequate? *Formal Aspects of Computing*, 6:25–68, 1994.
6. B. Bloom, W. Fokkink, and R. J. van Glabbeek. Precongruence formats for decorated trace preorders. In *Logic in Computer Science*, pages 107–118, 2000.
7. B. Bloom, W.J. Fokkink, and R.J. van Glabbeek. Precongruence formats for decorated trace semantics. *ACM Transactions on Computational Logic*. To appear.
8. B. Bloom, S. Istrail, and A. Meyer. Bisimulation can't be traced. *Journal of the ACM*, 42:232–268, 1995.
9. W. Fokkink and R. van Glabbeek. Ntyft/ntyxt rules reduce to ntree rules. *Information and Computation*, 126:1–10, 1996.
10. R.J. van Glabbeek. The linear time-branching time spectrum I. In J. Bergstra, A. Ponse, and S. Smolka, editors, *Handbook of Process Algebra*. Elsevier, 1999.
11. J.F. Groote. Transition system specifications with negative premises. *Theoret. Comput. Sci.*, 118:263–299, 1993.

12. M. Hennessy and R. Milner. Algebraic laws for nondeterminism and concurrency. *Journal of the ACM*, 32:137–161, 1985.
13. C.A.R. Hoare. *Communicating Sequential Processes*. Prentice Hall, 1985.
14. B. Jacobs. *Categorical Logic and Type Theory*, volume 141 of *Studies in Logic and the Foundations of Mathematics*. North Holland, Elsevier, 1999.
15. B. Klin and P. Sobociński. Syntactic formats for free: An abstract approach to process equivalence. BRICS Report RS-03-18, Aarhus University, 2003. Available from `http://www.brics.dk/RS/03/18/BRICS-RS-03-18.pdf`.
16. S. Mac Lane. *Categories for the Working Matematician*. Springer-Verlag, 1998.
17. D.M. Park. Concurrency on automata and infinite sequences. In P. Deussen, editor, *Conf. on Theoretical Computer Science*, volume 104 of *Lecture Notes in Computer Science*. Springer Verlag, 1981.
18. G. Plotkin. A structural approach to operational semantics. DAIMI Report FN-19, Computer Science Department, Aarhus University, 1981.
19. G. Plotkin. Bialgebraic semantics and recursion (extended abstract). In A. Corradini, M. Lenisa, and U. Montanari, editors, *Electronic Notes in Theoretical Computer Science*, volume 44. Elsevier Science Publishers, 2001.
20. G. Plotkin. Bialgebraic semantics and recursion. Invited talk, Workshop on Coalgebraic Methods in Computer Science, Genova, 2001.
21. A. W. Roscoe. *The Theory and Practice of Concurrency*. Prentice Hall, 1997.
22. J. Rutten and D. Turi. Initial algebra and final coalgebra semantics for concurrency. In J. de Bakker et al., editors, *Proc. of the REX workshop A Decade of Concurrency – Reflections and Perspectives*, *LNCS* vol. 803, pp. 530–582. Springer-Verlag, 1994.
23. R. de Simone. Higher-level synchronising devices in Meije-SCCS. *Theoret. Comput. Sci.*, 37:245–267, 1985.
24. D. Turi. Fibrations and bisimulation. Unpublished notes.
25. D. Turi. *Functorial Operational Semantics and its Denotational Dual*. PhD thesis, Vrije Universiteit, Amsterdam, 1996.
26. D. Turi and G. Plotkin. Towards a mathematical operational semantics. In *Proceedings 12th Ann. IEEE Symp. on Logic in Computer Science, LICS'97, Warsaw, Poland, 29 June – 2 July 1997*, pages 280–291. IEEE Computer Society Press, 1997.
27. Frits W. Vaandrager. On the relationship between process algebra and input/output automata. In *Logic in Computer Science*, pages 387–398, 1991.

Priority Rewrite Systems for OSOS Process Languages

Irek Ulidowski

Department of Mathematics and Computer Science
University of Leicester, University Road, Leicester LE1 7RH, UK
I.Ulidowski@mcs.le.ac.uk

Abstract. We propose a procedure for generating a Priority Rewrite System (PRS) for an arbitrary process language in the OSOS format. Rewriting of process terms is sound for bisimulation and head normalising within the produced PRSs. For a subclass of process languages representing finite behaviours the generated PRSs are strongly normalising (terminating), confluent and complete for bisimulation for closed terms modulo associativity and commutativity of the choice operator. We illustrate the usefulness of our procedure with several examples.

1 Introduction

Structural Operational Semantics (SOS) [20,3] is a method for assigning operational meaning to operators of process languages. The main components of SOS are transition rules, or simply SOS rules, which describe how the behaviour of a composite process depends on the behaviour of its component processes. A general syntactic form of transition rules is called a *format*. A process operator is in a format if all its SOS rules are in the format, and a process language, often abbreviated by PL, is in a format if all its operators are in the format. Many general formats have been proposed and a wealth of important results and specification and verification methods for PLs in these formats have been developed: see [3] for an overview.

In order to realise the potential of general PLs supporting software tools need to be developed. Such tools would accept general PLs as input languages and perform tasks such as simulation, model checking and equivalence checking, refinement and testing. Several such tools already exist. For example, we can use *Process Algebra Compiler* [22] to change the input PL to the *Concurrency Workbench of New Century* [12].

Alternatively, we can utilise the existing term rewrite and theorem prover software tools to analyse properties of processes of general PLs. To this end several procedures for automatic derivation of axioms systems and term rewrite systems for PLs in several formats were proposed [2,1,11,24,8]. The present paper continues the research of Aceto, Bloom and Vaandrager [2] and Bosscher [11], and extends and generalises it further. A new procedure for deriving *Priority Rewrite Systems* for bisimulation is presented. Our work delivers the following improvements: (a) priority rewrite rules are often simpler than those in [2,1,11], (b) the number of rewrite rules for typical operators is smaller (Remark 1 and 2), and (c) the priority order increases the effectiveness of term rewriting by reducing the nondeterminism inherent in rewriting [25]. We work with *Ordered* SOS PLs [26] (OSOS) instead of the GSOS PLs [10]. The proposed procedure generates term rewrite systems with a priority order on rewrite rules instead of axiom systems or

R. Amadio, D. Lugiez (Eds.): CONCUR 2003, LNCS 2761, pp. 87–102, 2003.

ordinary term rewrite systems as in [2,11]. We illustrate this with an example. Consider the priority operator 'θ' [6]. For a given irreflexive partial order $\gg$ on actions process $\theta(p)$ is a restriction of p such that, in any state of p, action a can happen only if no action b with $b \gg a$ is possible in that state. If $B_a = \{b \mid b \gg a\}$, then θ is defined by the following GSOS rules, one for each action a, where expressions of the form $X \stackrel{b}{\nrightarrow}$ in the premises are called *negative premises*.

$$\frac{X \stackrel{a}{\rightarrow} X' \quad \{X \stackrel{b}{\nrightarrow}\}_{b \in B_a}}{\theta(X) \stackrel{a}{\rightarrow} \theta(X')} \qquad \frac{X \stackrel{a}{\rightarrow} X' \quad \{Y \stackrel{b}{\nrightarrow}\}_{b \in B_a}}{X \triangle Y \stackrel{a}{\rightarrow} \theta(X')}$$

The second procedure in [2], also described in [1], produces the following axioms (laws) for θ, where the basic operators of CCS, namely '+', prefixing and '$\mathbf{0}$', are used. Since rules for θ have several copies of X in the premises an auxiliary binary operator $\triangle$, defined above, is used by the second procedure in [2]. The axioms for θ consist of the axiom that makes copies of X, and the axioms for $\triangle$ consisting of the distributivity axiom, peeling axioms, and action and inaction axioms:

$$\begin{array}{rlrl} \theta(X) = X \triangle X & & a.X \triangle (b.Y + Z) = a.X \triangle Z & \text{if } \neg(b > a) \\ (X + Y)\triangle Z = X \triangle Z + Y \triangle Z & & a.X \triangle (b.Y + Z) = \mathbf{0} & \text{if } b > a \\ \mathbf{0} \triangle X = \mathbf{0} & & a.X \triangle \mathbf{0} = a.\theta(X) & \end{array}$$

The priority operator can be defined equivalently, and perhaps more intuitively, by *positive* GSOS rules equipped with an *ordering* that represents the priority order on actions: the ordering has the same effect as negative premises in rules. This is the idea behind the *Ordered* SOS format [26]. The OSOS rules for θ are, one for each action a,

$$\frac{X \stackrel{a}{\rightarrow} X'}{\theta(X) \stackrel{a}{\rightarrow} \theta(X')}\ r_a$$

and the ordering $>$ is such that $r_b > r_a$ whenever $b \gg a$. The ordering prescribes that rule r_a can be applied to derive transitions of $\theta(p)$ if no higher priority rule, e.g. r_b, can be applied to $\theta(p)$. This suggests an axiomatisation algorithm: derive axioms from the OSOS rules and then "order" them appropriately according to the ordering on the SOS rules. We orient the axioms from left to right to obtain rewrite rules and define a *priority* order, an irreflexive partial order (irreflexive and transitive), on these rewrite rules. Thus, we obtain a *Priority Rewrite System* (PRS) originated by Baeten, Bergstra, Klop and Weijland [7]. The derived PRS for θ is listed below and contains one rewrite rule θ_{pr} for each pair of a and b such that $b \gg a$, and one θ_{act} rule for each a.

$$\begin{array}{ll} \theta_{pr} : \theta(a.X + b.Y + Z) \to \theta(b.Y + Z) & \\ \theta_{dn} : \quad \theta(X + \mathbf{0}) \to \theta(X) & \\ \theta_{ds} : \quad \theta(X + Y) \to \theta(X) + \theta(Y) & \theta_{act} : \theta(a.X) \to a.\theta(X) \\ & \theta_{nil} : \quad \theta(X) \to \mathbf{0} \end{array}$$

The priority order $\succ$ on the above rewrite rules is defined by $\theta_{pr} \succ \theta_{dn} \succ \theta_{ds}$ and $\{\theta_{ds}, \theta_{act}\} \succ \theta_{nil}$. Note, that we have fewer rewrite rules (schemes) than the axioms (schemes) above, and no need for the auxiliary $\triangle$. Our PRSs are sound for bisimulation, and head normalising. For OSOS PLs generating finite behaviours (linear and syntactically well-founded OSOS PLs) the generated PRSs are also strongly normalising and confluent. Finally, for the mentioned subclass of OSOS PLs, the generated PRSs are complete for bisimulation.

2 Preliminaries

This section recalls the notions of bisimulation and the GSOS and OSOS formats. OSOS and GSOS have the same expressive power [26]: OSOS is a reformulation of GSOS where rules of the form (2) together with orderings on rules have the same capability to define operational semantics as rules of the form (1).

2.1 Bisimulation

A labelled transition system (LTS) is a structure $(\mathcal{P}, A, \rightarrow)$, where $\mathcal{P}$ is the set of processes, A is the set of actions and $\rightarrow \subseteq \mathcal{P} \times A \times \mathcal{P}$ is a *transition relation*. $\mathcal{P}$ is ranged over by $p, q, p', q', \ldots$. The set Act is a finite set of actions and it is ranged over by a, b, c and their subscripted versions. The action τ is the silent action but we do not treat it differently from other actions. We permit Act to have a structure: for example Act may consist of action labels and co-labels as in CCS [18]. We write $p \xrightarrow{a} q$ for $(p, a, q) \in \rightarrow$ and read it as process p performs a and in doing so becomes process q. Expressions of the form $p \xrightarrow{a} q$ will be called *transitions*. We write $p \xrightarrow{a}$ when there is q such that $p \xrightarrow{a} q$, and $p \not\xrightarrow{a}$ otherwise. We recall the definition of bisimulation [18]:

Definition 1. Given $(\mathcal{P}, \mathsf{Act}, \rightarrow)$, a relation $R \subseteq \mathcal{P} \times \mathcal{P}$ is a *bisimulation* if, for all $(p, q) \in R$ and all $a \in \mathsf{Act}$, $p \xrightarrow{a} p'$ implies $\exists q'.(q \xrightarrow{a} q'$ and $p'Rq')$ and $q \xrightarrow{a} q'$ implies $\exists p'.(p \xrightarrow{a} p'$ and $p'Rq')$ hold. We write $p \sim q$ if there $\exists$ a bisimulation R such that pRq.

2.2 GSOS and OSOS Formats

The OSOS format [26] is an alternative to the GSOS format [10]. The reader can find in [26] the motivation for the OSOS format and many examples of its application. Before we recall the definitions of the formats we introduce several notions and notations.

Var is a countable set of variables ranged over by $X, X_i, Y, Y_i, \ldots$. Σ_n is a set of operators with arity n. A signature Σ is a collection of all Σ_n and it is ranged over by $f, g, \ldots$. The members of Σ_0 are called *constants*; $\mathbf{0} \in \Sigma_0$ is the deadlocked process operator. The set of *open terms* over Σ with variables in $V \subseteq$ Var, denoted by $\mathbb{T}(\Sigma, V)$, is ranged over by t, t', s, s'. $Var(t) \subseteq$ Var is the set of variables in a term t. The set of *closed terms*, written as $\mathrm{T}(\Sigma)$, is ranged over by $p, q, \ldots$. In the setting of process languages these terms will often be called process terms. A Σ context with n holes $C[X_1, \ldots, X_n]$ is a member of $\mathbb{T}(\Sigma, \{X_1, \ldots, X_n\})$, where all X_i are distinct. If $t_1, \ldots, t_n$ are Σ terms, then $C[t_1, \ldots, t_n]$ is the term obtained by substituting t_i for X_i for $1 \leq i \leq n$.

We will use bold italic font to abbreviate the notation for sequences. For example, a sequence of process terms $p_1, \ldots, p_n$, for any $n \in \mathbb{N}$, will often be written as $\boldsymbol{p}$ when the length is understood from the context, for example $f(\boldsymbol{p})$.

A *closed substitution* is a mapping $\mathsf{Var} \to \mathrm{T}(\Sigma)$. Closed substitutions are ranged over by σ; they extend to $\mathbb{T}(\Sigma, \mathsf{Var}) \to \mathrm{T}(\Sigma)$ mappings in a standard way.

Definition 2. [10] A GSOS rule is an expression of the form

$$\frac{\{\, X_i \stackrel{a_{ij}}{\to} Y_{ij} \,\}_{i \in I, j \in J_i} \quad \{\, X_k \stackrel{b_{kl}}{\not\to} \,\}_{k \in K, l \in L_k}}{f(\boldsymbol{X}) \stackrel{a}{\to} C[\boldsymbol{X}, \boldsymbol{Y}]}, \tag{1}$$

where $\boldsymbol{X}$ is the sequence $X_1, \ldots, X_n$ and $\boldsymbol{Y}$ is the sequence of all Y_{ij}, and all process variables in $\boldsymbol{X}$ and $\boldsymbol{Y}$ are distinct. Variables in $\boldsymbol{X}$ are the *arguments* of f. Moreover, I and K are subsets of $\{1, \ldots, n\}$ and all J_i and L_k, for $i \in I$ and $k \in K$, are finite subsets of $\mathbb{N}$, and $C[\boldsymbol{X}, \boldsymbol{Y}]$ is a context.

Let r be the rule of the form (1). Operator f is the *operator* of r and $rules(f)$ is the set of rules with the operator f. Expressions $t \stackrel{a}{\to} t'$ and $t \stackrel{a}{\not\to}$, where $t, t' \in \mathbb{T}(\Sigma, V)$, are called *transitions* and *negative transitions* respectively. Transitions are ranged over by T and T'. If transition T is $X \stackrel{a}{\to} X'$, we will sometime write $\neg T$ to denote $X \stackrel{a}{\not\to}$. A (negative) transition which involves only closed terms is called a *closed* (negative) transition. The set of transitions and negative transitions above the horizontal bar in r is called the *premises* of r, written $pre(r)$. The transition below the bar in r is the *conclusion*. Action a in the conclusion of r is the *action* of r, and $f(\boldsymbol{X})$ and $C[\boldsymbol{X}, \boldsymbol{Y}]$ are the *source* and *target* of r, respectively. The i-th argument X_i is *active* in r if $X_i \stackrel{a_{ij}}{\to} Y_{ij}$ or $X_i \stackrel{b_{kl}}{\not\to}$ is a premise of r for some a_{ij} and b_{kl}. The set of all i such that X_i is active in r is denoted by $active(r)$. Moreover, the i-th argument of f is *active* if $i \in active(r')$ for some rule r' for f.

Definition 3. A *positive* GSOS rule (transition rule, or OSOS rule, or simply a rule) is a GSOS rule with $K = \emptyset$. With the notation as in Definition 2, it has the form

$$\frac{\{\, X_i \stackrel{a_{ij}}{\to} Y_{ij} \,\}_{i \in I, j \in J_i}}{f(\boldsymbol{X}) \stackrel{a}{\to} C[\boldsymbol{X}, \boldsymbol{Y}]}. \tag{2}$$

Next, we recall the notion of *ordering* on rules [26]. An ordering on OSOS rules for operator f, $>_f$, is a binary relation over the rules for f. For the purpose of this paper we assume without loss of generality that orderings are *irreflexive* (i.e. $r > r$ never holds) and *transitive*. In general, there are situations where non-transitive or not irreflexive relations are useful orderings on rules [26]. Expression $r >_f r'$ is interpreted as r having higher priority than r' when deriving transitions of terms with f as the outermost operator. Given Σ, $>_\Sigma$, or simply $>$ if Σ is known from the context, is $\bigcup_{f \in \Sigma} >_f$.

Definition 4. A GSOS PL is a tuple (Σ, A, R), where Σ is a finite set of operators, $A \subseteq \mathsf{Act}$, R is a finite set of GSOS rules for operators in Σ such that all actions mentioned in the rules belong to A. An operator of a GSOS PL is called a GSOS operator.

An *Ordered SOS* (OSOS) PL is a tuple $(\Sigma, A, R, >)$, where Σ is a finite set of operators, $A \subseteq \mathsf{Act}$, R is a finite set of rules for operators in Σ, written as $rules(\Sigma)$, such that all actions mentioned in the rules belong to A, and $>$ is an ordering on $rules(\Sigma)$. An operator of an OSOS PL is called an OSOS operator.

Given an OSOS process language $G = (\Sigma, A, R, >)$, we associate a unique transition relation $\rightarrow$ with G. The details are given in [26]. Having the transition relation for G, we easily construct $(\mathrm{T}(\Sigma), A, \rightarrow)$, the LTS for G. Bisimulation is defined over this LTS as in Definition 1. Since GSOS and OSOS are equally expressive, namely every GSOS process language can be equivalently given as an OSOS process language and vice versa [26], bisimulation is a congruence for all OSOS PLs.

An OSOS PL H is a *disjoint extension* of an OSOS PL G, written as $G \leq H$, if the signature, the rules and the orderings of H include those of G, and H introduces no new rules and orderings for the operators in G.

2.3 GSOS = OSOS

The expressiveness results in [26] show that in general OSOS rules with orderings have the same effect as GSOS rules with negative premises.

Example 1. Consider the OSOS and GSOS definitions of the sequential composition

$$\frac{X \xrightarrow{a} X'}{X;Y \xrightarrow{a} X';Y}\, r_{a*} \quad > \quad \frac{Y \xrightarrow{c} Y'}{X;Y \xrightarrow{c} Y'}\, r_{*c} \qquad\qquad \frac{\{X \stackrel{a}{\not\rightarrow}\}_{a \in \mathsf{Act}}\; Y \xrightarrow{c} Y'}{X;Y \xrightarrow{c} Y'}\, r_{nc}$$

operator ';'. Rules r_{a*} and r_{*c}, for all actions a and c, together with $>$ defined by $r_{a*} > r_{*c}$, for all a and c, comprise the OSOS formulation. Rules r_{a*} and r_{nc}, for all a and c, form the GSOS definition.

Most of the typical process operators that are (or may be) defined by GSOS rules with negative premises have natural and efficient OSOS formulations. We mention just a few: priority choice (Sect. 4), hide operator of ET-LOTOS [16] (Example 9) and several delay operators [19,13,5]. There are operators whose OSOS definitions are more efficient than GSOS definitions: timed extensions of traditional operators such as parallel composition of *TPL* (Example 2), and others [25]. However, we cannot think of a popular operator whose GSOS formulation is more efficient than its OSOS formulation. Further discussion can be found in [25].

Example 2. Consider Hennessy and Regan's *Temporal Process Language* (TPL) [13]. It has a delay operator '$\lfloor\,\rfloor(\,)$' defined by the following GSOS rules, where a is any action except τ and σ, and the action σ denotes the passage of one time unit.

$$\frac{X \xrightarrow{a} X'}{\lfloor X \rfloor(Y) \xrightarrow{a} X'}\, r_a \qquad \frac{X \xrightarrow{\tau} X'}{\lfloor X \rfloor(Y) \xrightarrow{\tau} X'}\, r_\tau \qquad \frac{X \stackrel{\tau}{\not\rightarrow}}{\lfloor X \rfloor(Y) \xrightarrow{\sigma} Y}\, r_\sigma$$

The OSOS formulation of $\lfloor\,\rfloor(\,)$ uses the first two rules above and the rule $\sigma_\emptyset$: $\lfloor X \rfloor(Y) \xrightarrow{\sigma} Y$. The ordering is $r_\tau > \sigma_\emptyset$. The parallel composition '$\|$' of TPL is a timed extension of the CCS parallel with the following non-GSOS rule r.

$$\frac{X \xrightarrow{\sigma} X' \quad Y \xrightarrow{\sigma} Y' \quad X \parallel Y \stackrel{\tau}{\nrightarrow}}{X \parallel Y \xrightarrow{\sigma} X' \parallel Y'}\,r \qquad \frac{X \xrightarrow{\sigma} X' \quad Y \xrightarrow{\sigma} Y'}{X \parallel Y \xrightarrow{\sigma} X' \parallel Y'}\,r_\sigma$$

The rule requires that $p \parallel q$ can pass time if both p and q can pass time, are stable and cannot communicate. The OSOS formulation has the standard CCS rules (were communication rules are denoted by $r_{a\bar{a}}$) and the above timed rule r_σ, which is placed below all the rules with the action τ, namely the two τ-rules and all communication rules $r_{a\bar{a}}$. The GSOS formulation of $\parallel$ [25] contains the rules

$$\frac{X \xrightarrow{\sigma} X' \quad Y \xrightarrow{\sigma} Y' \qquad X \stackrel{\tau}{\nrightarrow} \quad Y \stackrel{\tau}{\nrightarrow} \quad \{\neg T_a \mid a \in \mathsf{Act} \setminus \{\tau, \sigma\} \wedge T \in pre(r_{a\bar{a}})\}}{X \parallel Y \xrightarrow{\sigma} X' \parallel Y'}$$

where the premise $\neg T$ for a fixed a is either $X \stackrel{a}{\nrightarrow}$ or $Y \stackrel{\bar{a}}{\nrightarrow}$. The OSOS formulation uses one timed rule r_σ and the ordering which places r_σ below all the rules for $\parallel$ with the action τ, whereas the GSOS formulation uses 2^k timed rules, where $k = \mid \mathsf{Act} \setminus \{\tau, \sigma\} \mid$.

2.4 Classes of GSOS and OSOS Operators

The axiomatisation algorithms in [2] produce several types of axioms (laws) for GSOS operators depending on the form of operators' definitions. Three types of definitions, and hence three classes of operators, are defined: *smooth*, *distinctive* and *discarding*. Our PRS algorithm relies on partitioning OSOS operators into *free of implicit copies* operators and *simply distinctive* operators.

A GSOS rule is smooth [2] if it has the form

$$\frac{\{\, X_i \xrightarrow{a_i} Y_i \,\}_{i \in I} \quad \{\, X_k \stackrel{b_{kl}}{\nrightarrow} \,\}_{k \in K, l \in L_k}}{f(X_1, \ldots, X_n) \xrightarrow{a} C[\boldsymbol{X}, \boldsymbol{Y}]}\,,$$

where I and K are distinct sets and $I \cup K = \{1, \ldots, n\}$, and no X_i appears in $C[\boldsymbol{X}, \boldsymbol{Y}]$ when $i \in I$. A GSOS operator is smooth if all its defining rules are smooth.

Multiple occurrences of process variables in rules are called *copies*. They are either *explicit* or *implicit* copies [23,26]. Given a rule r as in Definition 3, explicit copies are the multiple occurrences of variables Y_{ij} and X_i, for $i \notin I$, in the target $C[\boldsymbol{X}, \boldsymbol{Y}]$. The implicit copies are the multiple occurrences of X_i in the premises of r and the occurrences, not necessarily multiple, of variables X_i in $C[\boldsymbol{X}, \boldsymbol{Y}]$ when $i \in I$. Consider

$$\frac{X_1 \xrightarrow{a_{11}} Y_{11} \quad X_1 \xrightarrow{a_{12}} Y_{12} \quad X_2 \xrightarrow{a_{21}} Y_{21}}{h(X_1, X_2, X_3, X_4) \xrightarrow{a} g(X_2, X_3, X_3, X_4, Y_{11}, Y_{11})}\,.$$

There are implicit copies of X_1 in the premises and implicit copies of X_2 in the target (just one occurrence). The copies of X_3 and Y_{11} are explicit. There are no copies of X_4.

A rule with no implicit copies is free of implicit copies, and an OSOS operator is free of implicit copies if its rules are free of implicit copies. We notice that smooth GSOS rules can be defined using the notion of implicit copies: A GSOS rule of the form (1) is smooth if it has no implicit copies, I and K are distinct sets and $I \cup K = \{1, \ldots, n\}$. Hence, if a GSOS operator is smooth, then its OSOS formulation is free of implicit copies [25]. The converse is not valid as there are non-smooth GSOS operators whose OSOS formulations are free of implicit copies. The examples are the priority operator θ and the timed versions of the parallel operator of TPL in Example 2 and the hiding operator of ET-LOTOS [16] given in [27] and recalled in Example 9.

A GSOS operator f is distinctive [2] if it is smooth and satisfies the *distinctiveness* property: for each argument i, the argument either appears in positive premises of all rules for f or in none of them, and also, for each pair of different rules for f, there is an argument for which both rules have the same positive premise but with a different action. We use a similar notion. An OSOS operator f is simply distinctive if it is free of implicit copies and satisfies distinctiveness. It is not the case that if an OSOS formulation of a GSOS operator is simply distinctive, then the GSOS operator is distinctive: consider θ. The converse is not valid [25]. However, we cannot think of a popular distinctive GSOS operator which has no simply distinctive OSOS formulation.

3 Term Rewrite Systems

We recall the basic notions of term rewriting [15,4]. A Term Rewriting System (TRS) $\mathcal{R}$ is a pair (Σ, R) where Σ is a signature and R is a set of *reduction rules* or *rewrite* rules. We associate a countable set of variable $V \subseteq \mathit{Var}$ with each TRS. A reduction rule is a pair of terms (t, s) over $\mathbb{T}(\Sigma, V)$ and it is written as $t \to s$. Two conditions are imposed on the terms of reduction pairs: t is not a variable, and $\mathit{Var}(s) \subseteq \mathit{Var}(t)$. Often a reduction rule has a name, for example r, and thus we write $r : t \to s$.

A reduction rule $r : t \to s$ can be seen as a prescription for deriving *rewrites* $\sigma t \to \sigma s$ for all substitutions σ, where a rewrite is a closed instance of a reduction rule. The left-hand side σt is called a *redex*, more precisely r-redex. The right-hand side σs is called a *contractum*. A σt redex may be replaced by its contractum σs in an arbitrary context $C[\,]$ giving rise to a *reduction step* (one-step rewriting): $C[\sigma t] \to_r C[\sigma s]$. We call $\to_r$ the *one-step reduction relation* generated by r. The one-step reduction relation of a TRS $\mathcal{R}$, denoted by $\to_{\mathcal{R}}$ or simply by $\to$, is defined as the union of $\to_r$ for all $r \in \mathrm{R}$. Let R be a set of rewrites. The closure of R under closed contexts is denoted by $\to_R$. The transitive and reflexive closure of $\to$ $(\to_R)$ is called *reduction* (*R-reduction*) and is written as $\twoheadrightarrow$ $(\twoheadrightarrow_R)$. A reduction of term $f(t_1, \ldots, t_n)$ is *internal* if it occurs solely in the subterms $t_1, \ldots, t_n$ leaving the head operator f unaffected. A term t is *strongly normalising* (*terminating*) if it has no infinite reductions; and t is called *confluent* if any two reducts of t are *convergent* (or *joinable*), namely have a common reduct. A TRS is strongly normalising and confluent if all its terms have these properties.

The notions that are crucial in proving confluence are *overlap* and *critical pair* [15, 4]. For confluence we use the result due Knuth and Bendix [15] that states that if a TRS is strongly normalising, then it is confluent iff all its critical pairs are convergent.

3.1 Rewriting modulo AC

The application of term rewriting in concurrency is somewhat complicated by the need to preserve the commutativity and associativity of the nondeterministic choice operator $+$: namely $X + Y = Y + X$ and $X + (Y + Z) = (X + Y) + Z$, respectively. These properties are denoted by e_1 and e_2. The equations cannot be oriented without losing the normalising property. For example, if we turn e_1 into $X + Y \to Y + X$, then $t + s \to s + t \to t + s \to \cdots$. Therefore, we shall use term rewriting modulo the commutativity and the associativity of $+$. We denote the axioms e_1 and e_2 by AC and the equivalence class of terms t under AC by $[t]_{AC}$. For terms t, t' and s such that $t \in [t']_{AC}$ if $t' \to s$, then we shall write $t \to_{AC} s$ and $[t]_{AC} \to_{AC} [s]_{AC}$. We define $t \twoheadrightarrow_{AC} s$ and $[t]_{AC} \twoheadrightarrow_{AC} [s]_{AC}$ as the appropriate transitive reflexive closures of $\to_{AC}$. The internal reductions of $t \twoheadrightarrow_{AC} s$ and $[t]_{AC} \twoheadrightarrow_{AC} [s]_{AC}$ are defined in the corresponding way to the internal reductions of $\twoheadrightarrow$. Henceforth, we drop all subscripts AC.

Example 3. Consider $\Sigma = \{(\mathbf{0}, 0), (+, 2)\} \cup \{(a., 1) \mid a \in \mathsf{Act}\}$, where $\mathbf{0}$ is the deadlocked process operator, '$a.$' are the prefixing with actions a operators, for all $a \in \mathsf{Act}$, and $+$ is the CCS choice operator. The transition rules for Σ are given in Definition 7. Let (Σ, R) be a TRS with the following set R of reduction rules: $r_1 : \ X + \mathbf{0} \to X$ and $r_2 : \ X + X \to X$. Term $a.X + (a.X + \mathbf{0})$ reduces to $a.X$ as follows: $a.X + (a.X + \mathbf{0}) \to_{r_1} a.X + a.X \to_{r_2} a.X$. There is another reduction modulo AC to $a.X$: $a.X + (a.X + \mathbf{0}) = (a.X + a.X) + \mathbf{0} \to_{r_2} a.X + \mathbf{0} \to_{r_1} a.X$. Hence, $[a.X + (a.X + \mathbf{0})] \twoheadrightarrow [a.X]$.

(Σ, R) is strongly normalising. Interpret $\mathbf{0}$, $a.X$ and $X + Y$ as polynomials 2, $2X$ and $X + Y$ to obtain polynomial termination modulo AC. The TRS is confluent modulo AC. Reduction rules r_1 and r_2 have a simple overlap which replaces X with $\mathbf{0}$. We have $\mathbf{0} + \mathbf{0} \to_{r_1} \mathbf{0}$ and $\mathbf{0} + \mathbf{0} \to_{r_2} \mathbf{0}$. Hence, the only critical pair is $([\mathbf{0}], [\mathbf{0}])$, and it is joinable.

3.2 Priority Rewriting

As transition rules for process operators can be equipped with orderings that indicate which transition rules to apply first, reduction rules can also have an ordering associated with them. This ordering, called *priority* order, specifies the order in which rewrite rules are to be used to rewrite a term. This is illustrated by the following simple example.

Example 4. The TRS from Example 3 is now equipped with a priority order $\succ$ defined by $r_1 \succ r_2$. As before, $a.X + (a.X + \mathbf{0}) \twoheadrightarrow [a.X]$ because $a.X + (a.X + \mathbf{0}) \to_{r_1} a.X + a.X$, and since $a.X$ cannot be reduced to $\mathbf{0}$, $a.X + a.X$ then reduces to $a.X$ by rule r_2. However, the second reduction from Example 3 is not correct (intended) in this new setting. After $a.X + (a.X + \mathbf{0}) = (a.X + a.X) + \mathbf{0}$, both r_1 and r_2 can be used; but since r_1 has priority over r_2 we must apply r_1: $(a.X + a.X) + \mathbf{0} \to_{r_1} a.X + a.X$. Now, only r_2 can be used.

Next, consider term $t \equiv (a.X + \mathbf{0}) + (a.X + \mathbf{0})$. The term is an r_2-redex, it is not an r_1-redex although it contains r_1-redexes. We may wish to reduce the term with r_2 ahead of r_1. This is not intended in the new setting: we must either use higher priority r_1 to reduce subterms $a.X + \mathbf{0}$ to $a.X$ first, or use AC to convert t to r_1-redex $((a.X + \mathbf{0}) + a.X) + \mathbf{0}$ and reduce it as $[((a.X + \mathbf{0}) + a.X) + \mathbf{0}] \to_{r_1} [(a.X + a.X) + \mathbf{0}] \to_{r_1} [a.X + a.X] \to_{r_2} [a.X]$.

In general, a rewrite rule r_2 with a lower priority than r_1 can be applied to term t in favour of r_1, if no **internal** reduction (reduction sequence leaving head operator unaffected) modulo AC of t can produce a contractum that is an r_1-redex. We recall the basic notions of term rewriting with priority [7,21,28].

Definition 5. A Priority Rewrite System, or PRS for short, is a tuple $(\Sigma, \mathrm{T}, \succ)$, where (Σ, T) is a TRS and $\succ$ is a partial order on T called *priority order*. Let $\mathcal{P} = (\Sigma, \mathrm{T}, \succ)$ be a PRS, and let R be a set of rewrites for $\mathcal{P}$, namely closed substitutions of reduction rules of $\mathcal{P}$. The rewrite $r : t \twoheadrightarrow s$ is *correct* with respect to R (modulo AC) if there is no internal reduction $[t] \twoheadrightarrow_R [t']$ such that $r' : t' \twoheadrightarrow s' \in R$ with $r' \succ r$. R is *sound* if all its rewrites are correct w.r.t. R. R is *complete* if it contains all rewrites of $\mathcal{P}$ which are correct w.r.t. R. $\mathcal{P}$ is *well-defined* if it has a unique sound and complete rewrite set; this set is called the *semantics* of $\mathcal{P}$.

If the underlying TRS of a PRS is strongly normalising modulo AC, then the PRS is well-defined and strongly normalising modulo AC. Hence, the PRS from Example 4 is well-defined and strongly normalising modulo AC; it is also confluent.

Definition 6. Let $G = (\Sigma, A, S, >)$ be an OSOS process language. Let $\mathcal{P} = (\Sigma, \mathrm{T}, \succ)$ be a well-defined PRS with its unique sound and complete rewrite set R. A rewrite $t \twoheadrightarrow s$ of R, where t and s are closed Σ terms, is *sound for bisimulation* if $t \sim s$. A rewrite rule $\mathrm{T} \ni r_0 : l \twoheadrightarrow r$ is sound for bisimulation if every r_0-rewrite, which is correct with respect to the semantics of $\mathcal{P}$, is sound for bisimulation. $\mathcal{P}$ is sound for bisimulation if all its rewrite rules are. The set R is *complete for bisimulation* if whenever $t \sim s$, then t R-reduces modulo AC to normal form t', s R-reduces modulo AC to normal form s' and $[t'] = [s']$. $\mathcal{P}$ is complete for strong bisimulation, if its rewrite set R is.

4 Basic Language

Our language contains a new operator, called *priority choice*, which is denoted by '$\rhd$'.

Definition 7. The basic language, B, is an OSOS process language $(\Sigma_{\mathsf{B}}, A, R, >)$, where $\Sigma_{\mathsf{B}} = \Sigma_0 \cup \Sigma_1 \cup \Sigma_2$ with $\Sigma_0 = \{(\mathbf{0}, 0)\}$, $\Sigma_1 = \{(a., 1) \mid a \in A\}$ and $\Sigma_2 = \{(+, 2), (\rhd, 2)\}$, $A \subset_{fin}$ Act, and R and $>$ are the set of rules and the ordering as follows, where every rule for $\rhd$ with X in the premises is above every rule for $\rhd$ with Y in the premises:

$$a.X \xrightarrow{a} X \qquad \frac{X \xrightarrow{a} X'}{X + Y \xrightarrow{a} X'} \qquad \frac{Y \xrightarrow{a} Y'}{X + Y \xrightarrow{a} Y'} \qquad \frac{X \xrightarrow{a} X'}{X \rhd Y \xrightarrow{a} X'} > \frac{Y \xrightarrow{a} Y'}{X \rhd Y \xrightarrow{a} Y'}$$

The prefixing $a.$ binds stronger than $\rhd$, which in turn binds stronger than $+$.

B generates the LTS $B = (\mathrm{T}(\Sigma_{\mathsf{B}}), A, \rightarrow)$. Bisimulation over B is defined accordingly. Let $\mathcal{B}$ be the PRS for B defined in Table 1. Notice, that reduction rules $+_{nil}$, $+_{ice}$ and $\rhd_{act}$ on their own are sound for bisimulation, but $\rhd_{ds1}$ is not sound on its own. Let $\sigma X = \mathbf{0}$, $\sigma Y = a.\mathbf{0}$ and $\sigma Z = b.\mathbf{0}$; then $\sigma((X + Y) \rhd Z) \sim a.\mathbf{0}$ and $\sigma(X \rhd Z + Y \rhd Z) \sim a.\mathbf{0} + b.\mathbf{0}$. However, putting $\rhd_{ds1}$ below $\rhd_{dn1}$ solves this problem as $\rhd_{ds1}$ can only be applied when neither σX nor σY reduces to $\mathbf{0}$.

Theorem 1. $\mathcal{B}$ is strongly normalising and confluent modulo AC, and sound and complete for bisimulation.

Table 1. Reduction rules and the priority order for B

$+_{nil}$:	$X + \mathbf{0} \to X$	$\rhd_{dn1}$:	$(X + \mathbf{0}) \rhd Z \to X \rhd Z$
$+_{ice}$:	$X + X \to X$	$\rhd_{ds1}$:	$(X + Y) \rhd Z \to X \rhd Z + Y \rhd Z$
		$\rhd_{act}$:	$a.X \rhd Y \to a.X$
		$\rhd_{nil}$:	$X \rhd Y \to Y$

$$+_{nil} \succ +_{ice} \text{ and } \rhd_{dn1} \succ \rhd_{ds1} \text{ and } \{\rhd_{ds1}, \rhd_{act}\} \succ \rhd_{nil}$$

5 Rewrite Rules for Arbitrary OSOS Operators

Arbitrary OSOS operators can be grouped into three disjoint sets: operators that are not free of implicit copies, operators that are free of implicit copies and are not simply distinctive, and simply distinctive operators. For each type of operators (and auxiliary operators) Lemmas 1, 2 and 3 specify the type of rewrite rules and the associated with them priority orders.

If an OSOS operator (f, n) is not free of implicit copies, then we can construct a free of implicit copies OSOS operator (f^c, m), with $m > n$, that does the job of f. This is achieved in the same way as for GSOS operators which are not smooth due to implicit copies [2] (Lemmas 5.1 and 5.2) and, therefore, is left without a proof.

Lemma 1. Let G be an OSOS PL with signature Σ. Let $\mathcal{P} = (\Sigma, \mathrm{R}, \succ)$ be a well-defined PRS that is sound for bisimulation. Suppose $(f, n) \in \Sigma$ is an operator not free of implicit copies. Then, there is a disjoint extension of G' of G with a free of implicit copies operator (f^c, m) such that $m > n$, and a PRS $\mathcal{P}'$ is $(\Sigma \cup \{f^c\}, \mathrm{R} \cup \{f_{copy}\}, \succ)$, where is f_{copy} : $f(\boldsymbol{X}) \to f^c(\boldsymbol{Y})$ is the **copying** rewrite rule, for some vector $\boldsymbol{X}$ of n distinct variables and a vector $\boldsymbol{Y}$ of m variables from $\boldsymbol{X}$, such that $\mathcal{P}'$ is sound for bisimulation.

As an example consider operator $(h, 4)$ from Sect. 2.4. The operator has implicit copies of its first two arguments and the operator h^c, the free of implicit copies version of h required by Lemma 1, uses two extra arguments as follows:

$$\frac{X_1^1 \stackrel{a_{11}}{\to} Y_{11} \quad X_1^2 \stackrel{a_{12}}{\to} Y_{12} \quad X_2^1 \stackrel{a_{21}}{\to} Y_{21}}{h^c(X_1^1, X_1^2, X_2^1, X_2^2, X_3, X_4) \stackrel{a}{\to} g(X_2^2, X_3, X_3, X_4, Y_{11}, Y_{11})}$$

The copying rewrite rule for h is $h(X_1, X_2, X_3, X_4) \to h^c(X_1, X_1, X_2, X_2, X_3, X_4)$.

Remark 1. Non-smooth operators that have no implicit copies but test some arguments both positively and negatively, for example θ and the parallel in Example 2, require auxiliary operators that copy the relevant arguments (Lemma 5.2 in [2]). As we use orderings on rules, such auxiliary copying operators are not needed in our setting.

If an operator (f, n) is free of implicit copies and is not simply distinctive, then $rules(f)$ and the ordering can be partitioned into a number of sets of simply distinctive rules with the orderings inherited from the original ordering, thus leading to a rewrite rule corresponding to the distinctifying law in [2]. We need the following notation. Assume an OSOS PL with signature Σ that contains operators $+$ and $\rhd$. The sets of *auxiliary terms* and sum terms are defined inductively as follows: (a) $f(\boldsymbol{X})$ is a *sum* term for each $f \in \Sigma \setminus \{+, \rhd\}$; (b) if s and t are sum terms, then $t + s$ is a sum term; (c) if s and t are sum terms, then $s \rhd t$ is an auxiliary term; (d) if s is a sum term and t is an auxiliary term, then $s + t, t + s$ and $s \rhd t$ are auxiliary terms ($t \rhd s$, for example $(f \rhd f' + f'') \rhd g$, is not an auxiliary term).

Lemma 2. Let G be an OSOS PL with signature Σ such that $\mathsf{B} \leq G$. Let $\mathcal{P} = (\Sigma, \mathrm{R}, \succ)$ be a well-defined PRS that is sound for bisimulation. Suppose $(f, n) \in \Sigma$ is free of implicit copies operator which is not simply distinctive. Then, there is

- G' such that $G \leq G'$ with l simply distinctive operators (f_i, n), thus creating new Σ',
- an auxiliary term $AuxiliaryTerm[f_1(\boldsymbol{X}), \ldots, f_l(\boldsymbol{X})]$ constructed from all operators (f_i, n) and involving only the operators (f_i, n), and
- PRS $\mathcal{P}' = (\Sigma', \mathrm{R} \cup \{f_{aux}\}, \succ)$, where $f_{aux} : f(\boldsymbol{X}) \to AuxiliaryTerm[f_1(\boldsymbol{X}), \ldots, f_l(\boldsymbol{X})]$ is the **auxiliary** rewrite rule (note, f_{aux} is "unordered" w.r.t. rewrite rules in R),

and the PRS $\mathcal{P}'$ is sound for bisimulation.

We find the auxiliary operators f_is by partitioning $rules(f)$ into sets R_i such that (a) each R_i defines a simply distinctive operator and (b) for every pair of R_i, R_j, where $i, j \in \{1, \ldots, l\}$, either $R_i > R_j$, or $R_j > R_i$ or $R_i \natural R_j$. In other words, either R_i is wholly above R_j, or vice versa, or the ordering between the rules in R_i and R_j is empty. For each such set we change the operator of each SOS rule from f to the relevant f_i leaving the rest unchanged. Thus, we obtain l distinctive operators (f_i, n). As for constructing auxiliary terms required by Lemma 2, a general procedure can be found in [25].

Example 5. Consider B extended with the CCS parallel composition operator '$\|$' which is not simply distinctive but free of implicit copies. Assume that, for each $a \in A \subseteq \mathsf{Act} \setminus \{\tau\}$, we have $\bar{a} \in \mathsf{Act}$ and $\bar{\bar{a}} = a$. Lemma 2 requires three auxiliary simply distinctive operators for $\|$. These operators are the left-merge, written as '$\lfloor\!\lfloor$', the right-merge, written as '$\rfloor\!\rfloor$', and the communication merge, written as '$|$', with standard definitions as in [9, 2]. Since there is no ordering on rules the priority term does not involve $\rhd$, and the auxiliary rewrite rule is an instance of the distinctifying law and rewrite rule [2,11]: $X \| Y \to (X \mid Y) + (X \lfloor\!\lfloor Y) + (X \rfloor\!\rfloor Y)$.

Example 6. The operator ; from Example 1 is not simply distinctive; it is, however, free of implicit copies. The operator has two active arguments and its rules can be easily partitioned into two sets defining simply distinctive operators. The set of rules r_{a*} is wholly above the set of rules r_{*c}. The auxiliary operators $;_1$ and $;_2$ required by Lemma 2 are defined below, and the auxiliary rewrite rule is $X;Y \to (X;_1 Y) \rhd (X;_2 Y)$.

$$\frac{X \xrightarrow{a} X'}{X;_1 Y \xrightarrow{a} X';Y} \qquad \frac{Y \xrightarrow{c} Y'}{X;_2 Y \xrightarrow{c} Y'}$$

Example 7. Consider a version of the CCS parallel that gives priority to communication over concurrency. The operator is defined simply by putting each and every communication rule above all the concurrency rules. The GSOS definition of this operator is quite awkward (see a similar operator in Example 2). As in Example 5, we need the three auxiliary operators $\lfloor\!\lfloor$, $\rfloor\!\rfloor$ and $|$, and, because of the ordering on rules, the auxiliary rewrite rule uses $\rhd$ as well as $+$: $X \parallel Y \rightarrow (X \mid Y) \rhd (X \lfloor\!\lfloor Y + X \rfloor\!\rfloor Y)$.

Example 8. The auxiliary rewrite rule for $\lfloor\ \rfloor(\ )$ in Example 2 is $\lfloor X\rfloor(Y) \rightarrow \lfloor X\rfloor(Y)_a + (\lfloor X\rfloor(Y)_\tau \rhd \lfloor X\rfloor(Y)^\sigma)$, whereas the GSOS version of the operator has the rule $\lfloor X\rfloor(Y) \rightarrow \lfloor X\rfloor(Y)_\alpha + \lfloor X\rfloor(Y)_\sigma$. Here, the auxiliary operators $\lfloor\ \rfloor(\ )_x$ are defined by the appropriate rules based on rules r_x from Example 2, and $\lfloor\ \rfloor(\ )^\sigma$ is defined by $\lfloor X\rfloor(Y)^\sigma \stackrel{\sigma}{\rightarrow} Y$.

Finally, we consider simply distinctive operators. For such operators there are priority resolving, distributivity, action and deadlock rewrite rules. First, we introduce some useful notations. When an OSOS rule r has no implicit copies (has the form (3)), the *trigger* of r is an n-tuple $(a_1,\ldots,a_n)$, where $a_i = *$ if $i \notin I$. We often write $\boldsymbol{a}$ for $(a_1,\ldots,a_n)$, and $\boldsymbol{a}.\boldsymbol{X}$ denotes the vector $a_1.X_1,\ldots,a_n.X_n$, where if $a_i = *$, then $a_i.X_i$ is simply X_i. When we write $\boldsymbol{a}.\boldsymbol{Y} + \boldsymbol{b}.\boldsymbol{Z}$, we assume that the summand vectors, $\boldsymbol{a}.\boldsymbol{Y}$ and $\boldsymbol{b}.\boldsymbol{Z}$, have the same length and are summed elementwise. Correspondingly, for $\boldsymbol{X} + \boldsymbol{a}.\boldsymbol{Y}$.

Lemma 3. Let G be an OSOS PL with Σ such that $\mathsf{B} \leq G$. Let $\mathcal{P} = (\Sigma, \mathrm{R}, \succ)$ be a well-defined PRS that is sound for bisimulation. Let $(f,n) \in \Sigma$ be a simply distinctive operator defined by rules of the following form, where $Y_i = X_i'$ if $i \in I$, and $Y_i = X_i$ otherwise.

$$\frac{\{\, X_i \stackrel{a_i}{\rightarrow} X_i' \,\}_{i\in I}}{f(X_1,\ldots,X_n) \stackrel{a}{\rightarrow} C[\boldsymbol{Y}]} \qquad (3)$$

1. For each pair of rules r and r' of the form (3) with non-empty premises such that $r > r'$, and for triggers $\boldsymbol{a}.\boldsymbol{Y}$ and $\boldsymbol{b}.\boldsymbol{Z}$ of r and r' respectively, the **priority resolving** rewrite rule is $f_{pr} :\ f(\boldsymbol{X} + \boldsymbol{a}.\boldsymbol{Y} + \boldsymbol{b}.\boldsymbol{Z}) \rightarrow f(\boldsymbol{X} + \boldsymbol{a}.\boldsymbol{Y})$.
2. For each active argument i of f the following are the **distributivity** rewrite rules for f and i: $f_{dn(i)} :\ f(\ldots,X_i + \mathbf{0},\ldots) \rightarrow f(\ldots,X_i,\ldots)$ and $f_{ds(i)} :\ f(\ldots,X_i + Y_i,\ldots) \rightarrow f(\ldots,X_i,\ldots) + f(\ldots,Y_i,\ldots)$. The priority order satisfies $f_{pr} \succ f_{dn(i)}$ for each rule f_{pr} and each active i, and $f_{dn(i)} \succ f_{ds(i)}$ for each active i.
3. For each rule for f with the form (3) and trigger $\boldsymbol{a}_i.\boldsymbol{X}$ the **action** rewrite rule is $f_{act} :\ f(\boldsymbol{a}_i.\boldsymbol{X}) \rightarrow a.C[\boldsymbol{X}]$. If f has no active arguments, then f_{act} is $f(\boldsymbol{X}) \rightarrow a.C[\boldsymbol{X}]$.
4. The **deadlock** rewrite rule is $f_{nil} :\ f(\boldsymbol{X}) \rightarrow \mathbf{0}$, and the priority order satisfies $\{f_{ds(i)}, f_{act}\} \succ f_{nil}$ for all rewrite rules $f_{ds(i)}$ and f_{act}.

Then, there is a PRS $\mathcal{P}' = (\Sigma, \mathrm{R}', \succ')$, where R' is R extended with all the listed above priority resolving, distributivity, action and deadlock rules for f, and $\succ'$ is $\succ$ extended with the listed above orderings. Moreover, $\mathcal{P}'$ is sound for bisimulation.

Remark 2. There are fewer rules for typical simply distinctive OSOS operators than for their GSOS formulations mainly due to a large number of inaction rules for GSOS operators. Overall, a typical OSOS operator has fewer rewrite rules than its GSOS formulation, even though it may use more auxiliary operators (Example 8). Our algorithm produces $k + 10$ rules for the OSOS $\lfloor\ \rfloor()$ compared with $2k + 6$ rules for the GSOS version of $\lfloor\ \rfloor()$ generated by the second procedure in [2], where $k = \mid \mathsf{Act} \setminus \{\tau, \sigma\} \mid$.

If the ordering on $rules(f)$ is empty, then there would be no priority resolving rewrite rules for f. Out of the operators that we have discussed so far only the priority operator θ is simply distinctive with a non-trivial ordering on its rules. All the rewrite rules and the priority order for θ required by Lemma 3 have been given in the Introduction. Another example of a simply distinctive operator with a non-trivial ordering on its SOS rules is the **hide** operator of ET-LOTOS [16]:

Example 9. Our definition of **hide** employs simple OSOS rules instead of rules with negative premises and a *lookahead* as in [16]. The defining rules r^c and r_a are

$$\frac{X \xrightarrow{c} X'}{\textbf{hide } A \textbf{ in } X \xrightarrow{c} \textbf{hide } A \textbf{ in } X'}\ c \notin A \qquad\qquad \frac{X \xrightarrow{a} X'}{\textbf{hide } A \textbf{ in } X \xrightarrow{\tau} \textbf{hide } A \textbf{ in } X'}\ a \in A$$

respectively, where $\sigma \notin A$, r^σ is a timed rule and the ordering is $r_a > r^\sigma$ for all $a \in A$. The rewrite rules required by Lemma 3 are given below.

$$\begin{array}{ll}
\textbf{hide}_{pr}: & \textbf{hide } A \textbf{ in } (a.X + \sigma.Y + Z) \to \textbf{hide } A \textbf{ in } (a.X + Z) \\
\textbf{hide}_{dn}: & \textbf{hide } A \textbf{ in } (X + \mathbf{0}) \to \textbf{hide } A \textbf{ in } (X) \\
\textbf{hide}_{ds}: & \textbf{hide } A \textbf{ in } (X + Y) \to \textbf{hide } A \textbf{ in } X\ +\ \textbf{hide } A \textbf{ in } Y \\
\textbf{hide}^{a}_{act}: & \textbf{hide } A \textbf{ in } (a.X) \to a.(\textbf{hide } A \textbf{ in } X) \\
\textbf{hide}^{\tau}_{act}: & \textbf{hide } A \textbf{ in } (a.X) \to \tau.(\textbf{hide } A \textbf{ in } X) \\
\textbf{hide}^{\sigma}_{act}: & \textbf{hide } A \textbf{ in } (\sigma.X) \to \sigma.(\textbf{hide } A \textbf{ in } X) \\
\textbf{hide}_{nil}: & \textbf{hide } A \textbf{ in } X \to \mathbf{0}
\end{array}$$

We have one $\textbf{hide}_{pr}$ rule for every $a \in A$, one $\textbf{hide}^{a}_{act}$ rule for every $a \notin A \cup \{\sigma\}$, and one $\textbf{hide}^{\tau}_{act}$ for every $a \in A$. The priority order $\succ$ satisfies $\textbf{hide}_{pr} \succ \textbf{hide}_{dn} \succ \textbf{hide}_{ds}$ for all priority resolving rules $\textbf{hide}_{pr}$. Moreover, $\{\textbf{hide}_{ds}, \textbf{hide}^{a}_{act}, \textbf{hide}^{\tau}_{act}, \textbf{hide}^{\sigma}_{act}\} \succ \textbf{hide}_{nil}$.

Theorem 2. Let G be an OSOS process language, and let G' and $\mathcal{P}$ be the OSOS process language and the PRS respectively that are produced by the algorithm in Fig. 1. Then, $\mathcal{P}$ is head normalising and sound for bisimulation.

6 Termination, Confluence, and Completeness for Bisimulation

Any practically useful process language must contain a mechanism for representing infinite behaviour. Most often this is done by means of process constants (or variables)

Input: OSOS process language $G = (\Sigma, A, R, >)$ and PRS $\mathcal{P} = (\Sigma, \emptyset, \emptyset)$.

1. If G is not a disjoint extension of B, then add to G disjointly B. Call the resulting language G'''. $\mathcal{P}$ becomes $(\Sigma_{G'''}, \mathrm{R}''', \succ''')$, where R''' and $\succ'''$ are as in Table 1.
2. For each operator $f \in G'''$ which is not free of implicit copies apply the construction of Lemma 1 to obtain a free of implicit copies operator f^c. G''' extended disjointly with all f^c, for all not free of implicit copies operators f of G''', is denoted by G''. $\mathcal{P}$ becomes $(\Sigma_{G''}, \mathrm{R}''' \cup \mathrm{R}_{copy}, \succ''')$, where R_{copy} is the set of copy rewrite rules produced by Lemma 1.
3. For each free of implicit copies operator $f \in \Sigma_{G''} \setminus \Sigma_{\mathsf{B}}$ which is not simply distinctive apply the construction of Lemma 2 to produce simply distinctive auxiliary operators $f_1, \ldots, f_l$. A language G' is the result of extending disjointly G'' with all auxiliary operators for all free of implicit copies and not simply distinctive operators of G''. $\mathcal{P}$ becomes $(\Sigma_{G'}, \mathrm{R}'', \succ'')$, where R'' is $\mathrm{R}''' \cup \mathrm{R}_{copy}$ extended with all the auxiliary rewrite rules required by Lemma 2.
4. For each simply distinctive f in $\Sigma_{G'} \setminus \Sigma_{\mathsf{B}}$ extend R'' and $\succ''$ with all the priority resolving rewrite rules as in Lemma 3 to obtain the PRS $\mathcal{P}'' = (\Sigma_{G'}, \mathrm{R}', \succ')$.
5. For each simply distinctive f in $\Sigma_{G'} \setminus \Sigma_{\mathsf{B}}$ extend R' and $\succ'$ with all the remaining rewrite rules and the priority orders from Lemma 3. The resulting PRS is $\mathcal{P} = (\Sigma_{G'}, \mathrm{R}, \succ)$.

Output: G' such that $G \leq G'$, and a sound for bisimulation and head normalising PRS $\mathcal{P}$.

Fig. 1. The PRS construction algorithm for OSOS process languages

that are defined by *mutual recursion*. A simple semaphore can be modelled by Sem and Sem' which are defined by $Sem \stackrel{up}{\to} Sem'$ and $Sem' \stackrel{down}{\to} Sem$, respectively. Sem and Sem' are simply distinctive OSOS operators. The PRS for Sem and Sem' is $Sem \to up.Sem' \;\succ\; Sem \to \mathbf{0}$ and $Sem' \to down.Sem \;\succ\; Sem' \to \mathbf{0}$. Not surprisingly, processes such as Sem have non-terminating reductions: $Sem \to up.Sem' \to up.down.Sem \to up.down.up.Sem' \to \cdots$. The properties of PRSs with operators such as Sem are the subject of *infinitary rewriting* [14], and will be investigated in future. However, there is a subclass of OSOS PLs containing processes with finite behaviour [2]: Let G be an OSOS PL. A term $p \in \mathrm{T}(\Sigma_G)$ is *well-founded* if there exists no infinite sequence $p_0, a_0, p_1, a_1, \ldots$ with $p \equiv p_0$ and $p_i \stackrel{a_i}{\to} p_{i+1}$ for all $i \geq 1$. G is well-founded if all its terms are well-founded. Well-foundedness of OSOS PLs is not decidable, but *syntactical well-foundedness* is. If a PL is *linear* and syntactically well-founded, then it is well-founded.

Theorem 3. Let G be a syntactically well-founded and linear OSOS process language, and let G' and $\mathcal{P}$ be the OSOS process language and the PRS respectively that are produced by the algorithm in Fig. 1. Then, $\mathcal{P}$ is strongly normalising modulo AC, confluent and complete for bisimulation on closed terms over G'.

7 Conclusion and Future Work

We have described how to produce, for an arbitrary OSOS PL, a PRS that is head normalising and sound for bisimulation. When a PL in question is syntactically well-founded and linear, then its PRS is strongly normalising and confluent, and two processes are bisimilar iff they can be reduced to the same normal form modulo AC. We are planning to adapt our algorithm to other process equivalences.

There are well developed techniques for equational reasoning about processes that are not well-founded processes, for example, *regular* processes [17] and reasoning about such processes w.r.t. bisimulation. We can prove equalities between such processes by using (a) the standard axioms to "unwind" recursive processes to hnf and (b) the *Recursive Specification Principle* (RSP) [9]. It would be interesting to research a class of OSOS PLs corresponding to Aceto's class of *regular infinitary* GSOS PLs [1], and investigate rewriting/equational reasoning in PLs within the class using our PRSs and the RSP.

References

1. L. Aceto. Deriving complete inference systems for a class of GSOS languages generating regular behaviours. In *CONCUR'94*, volume 789 of *LNCS*. Springer, 1994.
2. L. Aceto, B. Bloom, and F.W. Vaandrager. Turning SOS rules into equations. *Information and Computation*, 111:1–52, 1994.
3. L. Aceto, W. Fokkink, and C. Verhoef. Structured operational semantics. In J.A. Bergstra, A. Ponse, and S.A. Smolka, editors, *Handbook of Process Algebra*, pages 197–292. Elsevier Science, 2001.
4. F. Baader and T. Nipkow. *Term Rewriting and All That*. Cambridge University Press, 1998.
5. J.C.M. Baeten. Embedding untimed into timed process algebra: the case for explicit termination. In L. Aceto and B. Victor, editors, *EXPRESS'00*. BRICS, 2000.
6. J.C.M. Baeten, J.A. Bergstra, and J.W. Klop. Syntax and defining equations for an interrupt mechanism in process algebra. *Fundamenta Informaticae*, XI(2):127–168, 1986.
7. J.C.M. Baeten, J.A. Bergstra, J.W. Klop, and W.P. Weijland. Term-rewriting systems with rule priorities. *Theoretical Computer Science*, 67:283–301, 1989.
8. J.C.M. Baeten and E.P. de Vink. Axiomatizing GSOS with termination. In H. Alt and A. Ferreira, editors, *Proceedings of STACS 2002*, volume 2285 of *LNCS*. Springer, 2002.
9. J.C.M. Baeten and W.P. Weijland. *Process Algebra*, volume 18 of *Cambridge Tracts in Theoretical Computer Science*. Cambridge University Press, 1990.
10. B. Bloom, S. Istrail, and A.R. Meyer. Bisimulation can't be traced. *Journal of the ACM*, 42(1):232–268, 1995.
11. D.J.B. Bosscher. Term rewriting properties of SOS axiomatisations. In *Proceedings of TACS'94*, volume 1000 of *LNCS*. Springer, 1994.
12. R. Cleaveland and S. Sims. The Concurrency Workbench of New Century. http://www.cs.sunysb.edu/~cwb/.
13. M. Hennessy and T. Regan. A process algebra for timed systems. *Information and Computation*, 117:221–239, 1995.
14. J.R. Kennaway and F.J. de Vries. Infinitary rewriting. In J.W. Terese, editor, *Term Rewriting Systems*. Cambridge University Press, 2002.
15. J.W. Klop. Term rewriting systems. In S. Abramsky, D. Gabbay, and T. Maibaum, editors, *Handbook of Logic in Computer Science*, pages 1–116. Oxford University Press, 1992.
16. L. Léonard and G. Leduc. A formal definition of time in LOTOS. *Formal Aspects of Computing*, 10:248–266, 1998.
17. R. Milner. A complete inference system for a class of regular behaviours. *Journal of Computer System Sciences*, 28:439–466, 1984.
18. R. Milner. *Communication and Concurrency*. Prentice Hall, 1989.

19. X. Nicollin and J. Sifakis. The algebra of timed processes, ATP: theory and application. *Information and Computation*, 114:131–178, 1994.
20. G. Plotkin. A structural approach to operational semantics. Technical Report DAIMI FN-19, Aarhus University, 1981.
21. M. Sakai and Y. Toyama. Semantics and strong sequentiality of priority term rewriting systems. *Theoretical Computer Science*, 208:87–110, 1998.
22. S. Sims. The Process Algebra Compiler. http://www.reactive-systems.com/pac/.
23. I. Ulidowski. *Local Testing and Implementable Concurrent Processes*. PhD thesis, Imperial College, University of London, 1994.
24. I. Ulidowski. Finite axiom systems for testing preorder and De Simone process languages. *Theoretical Computer Science*, 239(1):97–139, 2000.
25. I. Ulidowski. Priority rewrite systems for OSOS process languages. Technical Report 2002/30, Department of Mathematics and Computer Science, Leicester University, 2002. Updated version at http://www.mcs.le.ac.uk/~iulidowski/PRS.html.
26. I. Ulidowski and I.C.C. Phillips. Ordered SOS rules and process languages for branching and eager bisimulations. *Information and Computation*, 178(1):180–213, 2002.
27. I. Ulidowski and S. Yuen. General process languages with time. Technical Report 2000/41, Department of Mathematics and Computer Science, Leicester University, 2000.
28. J. van de Pol. Operational semantics of rewriting with priorities. *Theoretical Computer Science*, 200:289–312, 1998.

Quantitative Verification and Control via the Mu-Calculus*

Luca de Alfaro

Department of Computer Engineering, UC Santa Cruz, USA

Abstract. Linear-time properties and symbolic algorithms provide a widely used framework for system specification and verification. In this framework, the verification and control questions are phrased as *boolean* questions: a system either satisfies (or can be made to satisfy) a property, or it does not. These questions can be answered by symbolic algorithms expressed in the μ-calculus. We illustrate how the μ-calculus also provides the basis for two quantitative extensions of this approach: a *probabilistic* extension, where the verification and control problems are answered in terms of the probability with which the specification holds, and a *discounted* extension, in which events in the near future are weighted more heavily than events in the far away future.

1 Introduction

Linear-time properties and symbolic algorithms provide a widely adopted framework for the specification and verification of systems. In this framework, a *property* is a set of linear sequences of system states. Common choices for the specification of system properties are *temporal logic* [MP91] and *ω-regular automata* [BL69,Tho90]. The *verification question* asks whether a system satisfies a property, that is, whether all the sequences of states that can be produced during the activity of the system belong to the property. Similarly, the *control question* asks whether it is possible to choose (a subset of) the inputs to the system to ensure that the system satisfies a property. These questions can be answered by algorithms that operate on sets of states, and that correspond to the iterative evaluation of μ-calculus fixpoint formulas [Koz83b,EL86,BC96]. This approach is often called the *symbolic* approach to verification and control, since the algorithms are often able to take advantage of compact representations for sets of states, thus providing an efficient way to answer the verification and control questions on systems with large (and, under some conditions [HM00,dAHM01b], infinite) state spaces. The approach is completed by property-preserving equivalence relations, such as bisimulation [Mil90] (for verification) and alternating bisimulation [AHKV98] (for control).

We refer to this approach as the *boolean* setting for verification and control. Indeed, the verification question is answered in a boolean fashion (either a system

* This research was supported in part by the NSF CAREER award CCR-0132780, the NSF grant CCR-0234690, and the ONR grant N00014-02-1-0671.

R. Amadio, D. Lugiez (Eds.): CONCUR 2003, LNCS 2761, pp. 103–127, 2003.

satisfies a property, or it does not). Correspondingly, the symbolic verification algorithms are boolean in nature: the subsets of states on which they operate can be (and very often are [Bry86]) represented by their characteristic functions, that are mappings from states to $\{0,1\}$. Bisimulation itself can be seen as a binary distance function, associating distance 0 to two states if they are bisimilar, and distance 1 if they are not. In this paper, we illustrate how all the elements of this approach, namely, linear properties, symbolic algorithms, and equivalence relations, can be extended to a *quantitative* settings, where the control and verification questions are given quantitative answers, where the algorithms operate on mappings from states to real numbers, and where the equivalence relations correspond to real-valued distances [HK97,DGJP99,vBW01b,DEP02]. We consider two such quantitative settings: a *probabilistic* setting, where the verification and control questions are answered in terms of the probability that the system exhibits the desired property, and a *discounted* setting, where events in the near future are weighted more than those in the distant future. Our extensions rely on *quantitative* versions of the μ-calculus for solving the verification and control problems and, in the discounted setting, even for expressing the linear-time (discounted) specifications.

1.1 Games

We develop the theory for the case of two-player stochastic games [Sha53,Eve57, FV97], also called *concurrent probabilistic games* [dAH00], and for control goals. A *stochastic game* is played over a state space. At each state, player 1 selects a move, and simultaneously and independently, player 2 selects a move; the game then proceeds to a successor state according to a transition probability determined by the current state and by the selected moves. An outcome of a game, called *trace,* consists in the infinite sequence of states that are visited in the course of the game. We say that a linear property holds for a trace if the trace belongs to the property. A simple example of game is the game MATCHBIT. The game MATCHBIT can be in one of two states, s_{try} or s_{goal}. In state s_{try}, player 1 chooses a bit $b_1 \in \{0,1\}$, and player 2 chooses a bit $b_2 \in \{0,1\}$. If $b_1 = b_2$, the game proceeds to state s_{goal}; otherwise, the game stays in state s_{try}. The state s_{goal} is *absorbing:* once entered it, the game never leaves it.

Games are a standard model for control problems: the moves of player 1 model the inputs from the controller, while the moves of player 2 model the remaining inputs along with the internal nondeterminism of the system. Stochastic games generalize transition systems, Markov chains, Markov decision processes [Ber95], and turn-based games.[1] The verification setting can be recovered as a special case of the control setting, corresponding to games where only one player has a choice of moves.

[1] For many of these special classes of systems, there are algorithms for solving verification and control problems that have better worst-case complexity than those that can be obtained by specializing the algorithms for stochastic games. A review of the most efficient known algorithms for these structures is beyond the scope of this paper.

1.2 Probabilistic Verification and Control

In systems with probabilistic transitions, such as Markov chains, it is possible that a linear property does not hold for all traces, but nevertheless holds with positive probability. Likewise, even in games with deterministic transitions, player 1 may not be able to ensure that a property holds on all traces, but may nevertheless be able to ensure that it holds with some positive probability [dAHK98]. For example, consider again the game MATCHBIT, together with the property of reaching s_{goal} (consisting of all traces that contain s_{goal}). Starting from s_{try}, player 1 is not able to ensure that all traces reach s_{goal}: whatever sequence of bits player 1 chooses, there is always the possibility that player 2 chooses the complementary sequence, confining the game to s_{try}. Nevertheless, if player 1 chooses each bit 0 and 1 with equal probability, in each round he will proceed to s_{goal} with probability 1/2, so that s_{goal} is reached with probability 1. Another example is provided by the game MATCHONE, a variant of MATCHBIT where the bits can be chosen once only. The game MATCHONE can be in one of three states s_{try}, s_{goal}, and s_{fail}. At s_{try}, players 1 and 2 choose bits b_1 and b_2; if $b_1 = b_2$, the game proceeds to s_{goal}, otherwise it proceeds to s_{fail}. Both s_{goal} and s_{fail} are absorbing states. In the game MATCHONE, the maximal probability with which player 1 can ensure reaching s_{goal} is 1/2.

Hence, it is often of interest to consider a probabilistic version of verification and control problems, that ask the maximal probability with which a property can be guaranteed to hold. We are thus led to the problem of computing the maximal probability with which player 1 can ensure that an ω-regular property holds in a stochastic game. This problem can be solved with quantitative μ-calculus formulas that are directly derived from their boolean counterparts used to solve boolean control problems.

Specifically, [EJ91] showed that for turn-based games with deterministic transitions, the set of states from which player 1 can ensure that an ω-regular specification holds can be computed in a μ-calculus based on the set-theoretic operators $\cup, \cap$ and on the *controllable predecessor* operator *Cpre*. For a set T of states, the set $Cpre(T)$ consists of the states from which player 1 can ensure a transition to T in one step. As an example, consider the *reachability* property $\Diamond T$, consisting of all the traces that contain a state in T. The set of states from which player 1 can ensure that all traces are in $\Diamond T$ can be computed by letting $R_0 = T$, and for $k = 0, 1, 2, \ldots$, by letting $R_{k+1} = T \cup Cpre(R_k)$. The set R_k consists of the states from which player 1 can force the game to T in at most k steps; in a finite game, the solution is thus given by $\lim_{k\to\infty} R_k$. Computing the sequence $R_0, R_1, R_2, \ldots$ of states corresponds to evaluating by iteration the least fixpoint of $R = T \cup Cpre(R)$, which is denoted in μ-calculus as $\mu x.(T \cup Cpre(R))$: this formula is thus a μ-calculus solution formula for reachability. Solution formulas are known for general parity conditions [EJ91], and this suffices for solving games with respect to arbitrary ω-regular properties.

The solution formulas for the probabilistic setting can be obtained simply by giving a *quantitative* interpretation to the solution formulas of [EJ91]. In this quantitative interpretation, subsets of states are replaced by *state valuations* that associate with each state a real number in the interval $[0, 1]$; the set operators $\cup$, $\cap$ are replaced by the pointwise maximum and minimum operators $\sqcup$, $\sqcap$ [Rei80, FH82,Koz83a,Fel83]. The operator *Cpre* is replaced by an operator *Qpre* that, given a state valuation f, gives the state valuation $Qpre(f)$ associating with each state the maximal expectation of f that player 1 can achieve in one step.

As an example, consider again the goal $\Diamond T$. Denote by $\chi(T)$ the characteristic function of T, that assigns value 1 to states in T and value 0 to states outside T. We can compute the maximal probability of reaching T by letting $f_0 = \chi(T)$ and, for $k = 0, 1, 2, \ldots$, by letting $f_{k+1} = \chi(T) \sqcup Qpre(f_k)$. It is not difficult to see that $f_k(s)$ is the maximal probability with which player 1 can reach T from state s. The limit of f_k for $k \to \infty$, which corresponds to the least fixpoint $\mu x.(\chi(T) \sqcup Qpre(x))$, associates with each state the maximal probability with which player 1 can ensure $\Diamond T$. As an example, in the game MATCHBIT we have $f_0(s_{try}) = 0$ and, for $k \geq 0$, $f_k(s_{try}) = 1 - 2^{-k}$; the limit $\lim_{k\to\infty} 1 - 2^{-k} = 1$ is indeed the probability with which player 1 can ensure reaching s_{goal} from s_{try}. We note that the case for reachability games is in fact well known from classical game theory (see, e.g., [FV97]). However, this quantitative interpretation of the μ-calculus yields solution formulas for the complete set of ω-regular properties [dAM01]. Moreover, even for reachability games, the μ-calculus approach leads to simpler correctness arguments for the solution formula, since it is possible to exploit the complementation of μ-calculus and the connection between μ-calculus formulas and winning strategies in the construction of the arguments [dAM01].

1.3 Discounted Verification and Control

The probabilistic setting is quantitative with respect to states, but not with respect to traces: while state valuations are quantitative, each trace is still evaluated in a boolean way: either it is in the property, or it is not. This boolean evaluation of traces does not enable us to specify "how well" a specification is met. For instance, a trace satisfies $\Diamond T$ as long as the set T of target states is reached, no matter how long it takes to reach it: no prize is placed on reaching T sooner than later, and even if T is reached in a much longer time than the reasonable life expectancy of the system, the property nevertheless holds. Furthermore, if a property *does not* hold, the boolean evaluation of traces does not provide a notion of property approximation. For example, the safety property $\Box T$ is violated if a state outside T is ever reached: no prize is placed on staying in T as long as possible, and the property fails even if the system stays in T for an expected time much larger than the system's own expected life time. As these examples illustrate, the boolean evaluation of traces is sensitive to changes in behavior that occur arbitrarily late: in technical terms, ω-regular properties are not continuous in the Cantor topology, which assigns distance 2^{-k} to traces that are identical up to position $k - 1$, and differ at k. Discounted control and verification proposes to remedy this situation by weighting events that occur in the near future more heavily than events that occur in the far-away future [dAHM03].

Discounting reachability and safety properties is easy. For reachability, we assign the value α^k to traces that reach the target set after k steps, for $\alpha \in [0,1]$; for safety, we assign the value $1-\alpha^k$ to traces that stay in the safe set of states for k steps. For more complex temporal-logic properties, however, many discounting schemes are possible. For example, a Büchi property $\Box\Diamond B$ consists of all the traces that visit a subset B of states infinitely many times [MP91]. We can discount the property $\Box\Diamond B$ in several ways: on the basis of the time required to reach B, or on the basis of the number of visits to B, or on the basis of some more complex criterion (for instance, the time required to visit B twice). On the other hand, the predecessor operators of the μ-calculus provide a natural locus for discounting the next-step future. Discounted μ-calculus replaces *Qpre* with two discounted versions, $\alpha\, Qpre$ and $(1-\alpha) + \alpha\, Qpre$, where $\alpha \in [0,1]$ is a discount factor. Using these operators, we can write the solution to discounted reachability games as $\phi_{\alpha\text{-}reach} = \mu x.(\chi(T) \sqcup \alpha\, Qpre(x))$, and the solution to discounted safety games as $\phi_{\alpha\text{-}safety} = \nu x.(\chi(T) \sqcap (1-\alpha) + \alpha\, Qpre(x))$.

We propose to use discounted μ-calculus as the common basis for the specification, verification, and control of discounted properties. We define *discounted properties* as the *linear semantics* of formulas of the discounted μ-calculus. The resulting setting is continuous in the Cantor topology, and provides notions of satisfaction quality and approximation for linear properties.

1.4 Linear and Branching Semantics for the μ-Calculus

Given a formula ϕ of the μ-calculus, we can associate a *linear semantics* to ϕ by evaluating it on linear traces, and by taking the value on the first state. This linear semantics is often, but not always, related to the evaluation of the formula on the game, which we call the *branching semantics*. As an example, if we evaluate the fixpoint $\phi_{reach} = \mu x.(T \cup Cpre(R))$ on a trace $s_0, s_1, s_2, \ldots$, we have that $s_0 \in \phi_{reach}$ if there is $k \in \mathbb{N}$ such that $s_k \in T$. Hence, the linear semantics of ϕ_{reach}, denoted $[\phi_{reach}]^{\mathrm{blin}}$, coincides with $\Diamond T$, In this case, we have that the formula ϕ_{reach}, evaluated on a game, returns exactly the states from which player 1 can ensure $[\phi_{reach}]^{\mathrm{blin}}$. This connection does not hold for all formulas. For example, consider the formula $\psi = \mu x.(Cpre(x) \cup \nu y.(T \cap Cpre(y)))$. If we evaluate this formula on a trace $s_0, s_1, s_2, \ldots$, we can show that $s_k \in \nu y.(T \cap Cpre(y))$ iff we have $s_j \in T$ for all $j \geq k$. Hence, we have $s_0 \in \psi$ iff there is $k \in \mathbb{N}$ such that $s_j \in T$ for all $j \geq k$: in other words, the linear semantics $[\psi]^{\mathrm{blin}}$ coincides with the co-Büchi property $\Diamond\Box T$ [Tho90]. On the other hand, the formula ϕ, evaluated on a game, does *not* correspond to the states from which player 1 can ensure $\Diamond\Box T$ (see Example 1 in Sect. 3).

In the boolean setting, the linear and branching semantics are related for all *strongly deterministic* formulas [dAHM01a], a set of formulas that includes the solution formulas for games with respect to ω-regular properties [EJ91]. We show that this correspondence carries through to the probabilistic and discounted settings. Indeed, in both the probabilistic and the discounted settings, we show that the values computed by strongly deterministic formulas is equal to the maximal expectation of their linear semantics that player 1 can ensure.

In the discounted setting, the linear semantics of discounted μ-calculus provides the specification language, and the branching semantics provides the verification algorithms. For example, the value of the discounted formula $\phi_{\alpha\text{-}reach}$ on the first state of a trace $s_0, s_1, s_2, \ldots$ is α^k, where $k = \min\{j \in \mathbb{N} \mid s_j \in T\}$. Hence, the linear semantics $[\phi_{\alpha\text{-}reach}]^{\text{blin}_\alpha}$ associates the value α^k to traces that reach T in k steps. The same formula $\phi_{\alpha\text{-}reach}$, evaluated on a game, yields the maximum value of $[\phi_{\alpha\text{-}reach}]^{\text{blin}_\alpha}$ that player 1 can achieve. Similarly, the value of $\phi_{\alpha\text{-}safety}$ on the first state of a trace $s_0, s_1, s_2, \ldots$ is $1 - \alpha^k$, where $k = \min\{j \in \mathbb{N} \mid s_j \notin T\}$. Hence, the linear semantics $[\phi_{\alpha\text{-}safety}]^{\text{blin}_\alpha}$ associates the value $1 - \alpha^k$ to traces that stay in T for k steps. The same formula $\phi_{\alpha\text{-}safety}$, evaluated on a game, yields the maximum value of $[\phi_{\alpha\text{-}safety}]^{\text{blin}_\alpha}$ that player 1 can achieve. Again, this correspondence holds for a set of formulas that includes the solution formulas of games with parity conditions.

1.5 Quantitative Equivalence Relations

The frameworks for probabilistic and discounted verification are complemented by quantitative equivalence relations [HK97,DGJP99,vBW01b]. We show that, just as CTL and CTL* characterize ordinary bisimulation [Mil90], so probabilistic and discounted μ-calculus characterize probabilistic and discounted bisimulation [dAHM03].

Credits. This paper is based on joint work with Thomas A. Henzinger and Rupak Majumdar on the connection between games, μ-calculus, and linear properties [dAHM01a,dAM01,dAHM03]. I would like to thank Marco Faella, Rupak Majumdar, Mariëlle Stoelinga, and an anonymous reviewer for reading a preliminary version of this work and for providing many helpful comments and suggestions.

2 Preliminaries

2.1 The μ-Calculus

Syntax. Let $\mathcal{P}$ be a set of predicate symbols, $\mathcal{V}$ be a set of variables, and $\mathcal{F}$ be a set of function symbols. The formulas of μ-calculus are generated by the grammar

$$\phi ::= p \mid x \mid \neg\phi \mid \phi \vee \phi \mid \phi \wedge \phi \mid f(\phi) \mid \mu x.\phi \mid \nu x.\phi, \tag{1}$$

for predicates $p \in \mathcal{P}$, variables x, and functions $f \in \mathcal{F}$. In the two quantifications $\mu x.\phi$ and $\nu x.\phi$, we require that all occurrences of x in ϕ have even polarity, that is, they occur in the scope of an even number of negations ($\neg$). We assume that for each function $f \in \mathcal{F}$ there is *dual* function $\text{Dual}(f) \in \mathcal{F}$, with $\text{Dual}(\text{Dual}(f)) = f$. Given a closed formula ϕ of μ-calculus, the following transformations enable us to push all negations to the predicates:

$$\neg(\phi_1 \vee \phi_2) \rightsquigarrow (\neg\phi_1) \wedge (\neg\phi_2) \qquad \neg(\mu x.\phi) \rightsquigarrow \nu x.\neg\phi[\neg x/x] \tag{2}$$

$$\neg(\phi_1 \wedge \phi_2) \rightsquigarrow (\neg\phi_1) \vee (\neg\phi_2) \qquad \neg(\nu x.\phi) \rightsquigarrow \mu x.\neg\phi[\neg x/x] \tag{3}$$

$$\neg f(\phi) \rightsquigarrow \text{Dual}(f)(\neg\phi), \tag{4}$$

where $\phi[\neg x/x]$ denotes the formula in which all free occurrences of x are replaced by $\neg x$. We will be particularly interested in the of formulas in *EJ-form*. These formulas take their name from the authors of [EJ91], where it was shown that they suffice for solving turn-based games; as we will see, these formulas can be uniformly used for solving boolean, probabilistic, and discounted control problems with respect to parity conditions. For $f \in \mathcal{F}$, a formula ϕ is in *EJ-form* if it can be written as

$$\phi ::= \gamma_n x_n.\gamma_{n-1} x_{n-1} \cdots \gamma_0 x_0. \bigvee_{i=0}^{n} (\chi_i \wedge f(x_i)),$$

$$\chi ::= p \mid \neg\chi \mid \chi \vee \chi \mid \chi \wedge \chi,$$

where for $0 \leq i \leq n$, we have $x_i \in \mathcal{V}$, and where $f \in \mathcal{F}$ and $p \in \mathcal{P}$. For $0 \leq i \leq n$, the fixpoint quantifier γ_i is ν if i is even, and is μ if i is odd. A fixpoint formula ϕ is in *strongly deterministic form* [dAHM01a] iff ϕ consists of a string of fixpoint quantifiers followed by a quantifier-free part ψ generated by the following grammar:

$$\psi ::= p \mid \neg p \mid \psi \vee \psi \mid p \wedge \psi \mid \neg p \wedge \psi \mid f(\chi),$$

$$\chi ::= x \mid \chi \vee \chi.$$

Semantics. The semantics of μ-calculus is defined in terms of lattices. A *lattice* $\mathbb{L} = (E, \preceq)$ consists of a set E of elements and of a partial order $\preceq$ over E, such that every pair of elements $v_1, v_2 \in E$ has a unique greatest lower bound $v_1 \sqcap v_2$ and least upper bound $v_1 \sqcup v_2$. A lattice is *complete* if every (not necessarily finite) non-empty subset of E has a greatest lower bound and a least upper bound in E. A *value lattice* is a complete lattice together with a *negation* operator $\sim$, satisfying $\sim\sim v = v$ for all $v \in E$, and $\sim\sqcap E' = \bigsqcup\{\sim v \mid v \in E'\}$ for all $E' \subseteq E$ [Ros90, chapter 6]. A μ-calculus *interpretation* $(\mathbb{L}, [\![\cdot]\!])$ consists of a value lattice $\mathbb{L} = (E, \preceq)$ and of an interpretation $[\![\cdot]\!]$ that maps every predicate $p \in \mathcal{P}$ to a lattice element $[\![p]\!] \in E$, and that maps every function $f \in \mathcal{F}$ to a function $[\![f]\!] \in (E \mapsto E)$. We require that for all $f \in \mathcal{F}$ and all $v \in E$, we have $\sim[\![f]\!](v) = [\![\mathrm{Dual}(f)]\!](\sim v)$. A *variable environment* is a function $e : \mathcal{V} \mapsto E$ that associates a lattice element $e(x) \in E$ to each variable $x \in \mathcal{V}$. For $x \in \mathcal{V}$, $v \in E$, and a variable environment e, we denote by $e[x := v]$ the variable environment defined by $e[x := v](x) = v$, and $e[x := v](y) = e(y)$ for $x \neq y$. Given an interpretation $\mathcal{I} = (\mathbb{L}, [\![\cdot]\!])$, and a variable environment e, every μ-calculus formula ϕ specifies a lattice element $[\![\phi]\!]_e^{\mathcal{I}} \in E$, defined inductively as follows, for $p \in \mathcal{P}$, $f \in \mathcal{F}$, and $x \in \mathcal{V}$:

$$[\![p]\!]_e^{\mathcal{I}} = [\![p]\!] \qquad [\![\phi_1 \vee \phi_2]\!]_e^{\mathcal{I}} = [\![\phi_1]\!]_e^{\mathcal{I}} \sqcup [\![\phi_2]\!]_e^{\mathcal{I}}$$

$$[\![\neg p]\!]_e^{\mathcal{I}} = \sim[\![p]\!]_e^{\mathcal{I}} \qquad [\![\phi_1 \wedge \phi_2]\!]_e^{\mathcal{I}} = [\![\phi_1]\!]_e^{\mathcal{I}} \sqcap [\![\phi_2]\!]_e^{\mathcal{I}}$$

$$[\![x]\!]_e^{\mathcal{I}} = e(x) \qquad [\![\mu x.\phi]\!]_e^{\mathcal{I}} = \sqcap\{v \in E \mid v = [\![\phi]\!]_{e[x:=v]}^{\mathcal{I}}\}$$

$$[\![f(\phi)]\!]_e^{\mathcal{I}} = [\![f]\!]([\![\phi]\!]_e^{\mathcal{I}}) \qquad [\![\nu x.\phi]\!]_e^{\mathcal{I}} = \sqcup\{v \in E \mid v = [\![\phi]\!]_{e[x:=v]}^{\mathcal{I}}\}.$$

All right-hand-side (semantic) operations are performed over the value lattice $\mathbb{L}$. It is easy to show that if $\phi \rightsquigarrow \phi'$ by (2)–(4), then $[\![\phi]\!]_e^{\mathcal{I}} = [\![\phi']\!]_e^{\mathcal{I}}$. A μ-calculus formula ϕ is *closed* if all its variables are bound by one of the μ or ν fixpoint quantifiers. If ϕ is closed, then the value $[\![\phi]\!]_e^{\mathcal{I}}$ does not depend on e, and we write simply $[\![\phi]\!]^{\mathcal{I}}$.

2.2 Game Structures

We develop the theory for stochastic game structures. For a finite set A, we denote by $\text{Distr}(A)$ the set of probability distributions over A. A (two-player stochastic) *game structure* $\mathcal{G} = \langle S, \mathcal{M}, \Gamma_1, \Gamma_2, \delta \rangle$ consists of the following components [AHK02,dAHK98]:

- A finite set S of states.
- A finite set $\mathcal{M}$ of moves.
- Two move assignments $\Gamma_1, \Gamma_2 \colon S \mapsto 2^{\mathcal{M}} \setminus \emptyset$. For $i \in \{1, 2\}$, the assignment Γ_i associates with each state $s \in S$ the nonempty set $\Gamma_i(s) \subseteq \mathcal{M}$ of moves available to player i at state s.
- A probabilistic transition function $\delta \colon S \times \mathcal{M}^2 \mapsto \text{Distr}(S)$, that gives the probability $\delta(s, a_1, a_2)(t)$ of a transition from s to t when player 1 plays move a_1 and player 2 plays move a_2.

At every state $s \in S$, player 1 chooses a move $a_1 \in \Gamma_1(s)$, and simultaneously and independently player 2 chooses a move $a_2 \in \Gamma_2(s)$. The game then proceeds to the successor state $t \in S$ with probability $\delta(s, a_1, a_2)(t)$. We denote by $\tau(s, a_1, a_2) = \{t \in S \mid \delta(s, a_1, a_2)(t) > 0\}$ the set of *destination states* when actions a_1, a_2 are chosen at s. In general, the players can randomize their choice of moves at a state. We denote by $\mathcal{D}_i(s) \subseteq \text{Distr}(\mathcal{M})$ the set of move distributions available to player $i \in \{1, 2\}$ at $s \in S$, defined by

$$\mathcal{D}_i(s) = \{\zeta \in \text{Distr}(\mathcal{M}) \mid \zeta(a) > 0 \textit{ implies } a \in \Gamma_i(s)\}.$$

For $s \in S$ and $\zeta_1 \in \mathcal{D}_1(s)$, $\zeta_2 \in \mathcal{D}_2(s)$, we denote by $\hat{\delta}(s, \zeta_1, \zeta_2)$ the next-state probability distribution, defined for all $t \in S$ by

$$\hat{\delta}(s, \zeta_1, \zeta_2)(t) = \sum_{a_1 \in \Gamma_1(s)} \sum_{a_2 \in \Gamma_2(s)} \delta(s, a_1, a_2)(t)\, \zeta_1(a_1)\, \zeta_2(a_2).$$

A (randomized) *strategy* π_i for player $i \in \{1, 2\}$ is a mapping $\pi_i : S^+ \mapsto \text{Distr}(\mathcal{M})$ that associates with every sequence of states $\overline{s} \in S^+$ the move distribution $\pi_i(\overline{s})$ used by player i when the past history of the game is $\overline{s}$; we require that $\pi_i(\overline{s}s) \in \mathcal{D}_i(s)$ for all $\overline{s} \in S^*$ and $s \in S$. We indicate with Π_i the set of all strategies for player $i \in \{1, 2\}$.

Given an initial state $s \in S$ and two strategies $\pi_1 \in \Pi_1$ and $\pi_2 \in \Pi_2$, we define the set $\mathit{Outcomes}(s, \pi_1, \pi_2) \subseteq S^\omega$ to consist of all the sequences of states $s_0, s_1, s_2, \ldots$ such that $s_0 = s$ and such that for all $k \geq 0$ there are moves $a_1^k, a_2^k \in \mathcal{M}$ such that $\pi_1(s_0, \ldots, s_k)(a_1^k) > 0$, $\pi_2(s_0, \ldots, s_k)(a_2^k) > 0$, and

$s_{k+1} \in \tau(s_k, a_1^k, a_2^k)$. Given a trace $\sigma = s_0, s_1, s_2, \ldots \in S^\omega$, we denote by σ_k its k-th state s_k.

An initial state $s \in S$ and two strategies $\pi_1 \in \Pi_1$ and $\pi_2 \in \Pi_2$ uniquely determine a stochastic process $(S^\omega, \Omega, \mathrm{Pr}_s^{\pi_1,\pi_2})$ where $\Omega \subseteq 2^{S^\omega}$ is the set of measurable sets, and where $\mathrm{Pr}_s^{\pi_1,\pi_2} : \Omega \mapsto [0,1]$ assigns a probability to each measurable set [KSK66,FV97]. In particular, for a measurable set of traces $A \in \Omega$, we denote by $\mathrm{Pr}_s^{\pi_1,\pi_2}(A)$ the probability that the game follows a trace in A, and given a measurable function $f : S^\omega \mapsto \mathbb{R}$, we denote by $\mathrm{E}_s^{\pi_1,\pi_2}(f)$ the expectation of f in $(S^\omega, \Omega, \mathrm{Pr}_s^{\pi_1,\pi_2})$.

We denote by $S_{\mathcal{G}}$, $\mathcal{M}_{\mathcal{G}}$, $\Gamma_1^{\mathcal{G}}$, $\Gamma_2^{\mathcal{G}}$, and $\delta_{\mathcal{G}}$ the individual components of a game structure $\mathcal{G}$.

Special Classes of Game Structures. Transition systems, turn-based games, and Markov decision processes are special cases of deterministic game structures. A game structure $\mathcal{G}$ is *deterministic* if for all states $s, t \in S_{\mathcal{G}}$ and all moves $a_1, a_2 \in \mathcal{M}_{\mathcal{G}}$ we have $\delta_{\mathcal{G}}(s, a_1, a_2)(t) \in \{0,1\}$. The structure $\mathcal{G}$ is *turn-based* if at every state at most one player can choose among multiple moves; that is, if for all states $s \in S_{\mathcal{G}}$, there exists at most one $i \in \{1,2\}$ with $|\Gamma_i^{\mathcal{G}}(s)| > 1$. The turn-based deterministic game structures coincide with the games of [BL69, Con92,Tho95]. For $i \in \{1,2\}$, the structure $\mathcal{G}$ is *player-i* if at every state only player i can choose among multiple moves; that is, if $|\Gamma_{3-i}^{\mathcal{G}}(s)| = 1$ for all states $s \in S$. Player-1 and player-2 structures (called collectively *one-player* structures) coincide with Markov decision processes [Der70]. The player-i deterministic game structures coincide with transition systems: in every state, each available move of player i determines a unique successor state.

3 Boolean Verification and Control

Given a game structure $\mathcal{G} = \langle S, \mathcal{M}, \Gamma_1, \Gamma_2, \delta \rangle$, a *linear property* of $\mathcal{G}$ is a subset $\Phi \subseteq S^\omega$ of its state sequences. Given a linear property $\Phi \subseteq S^\omega$, we let

$$\langle 1 \rangle_{\mathcal{G}}^{\mathrm{b}} \Phi = \{s \in S \mid \exists \pi_1 \in \Pi_1 . \forall \pi_2 \in \Pi_2 . \mathit{Outcomes}(s, \pi_1, \pi_2) \subseteq \Phi\} \tag{5}$$

$$\langle 2 \rangle_{\mathcal{G}}^{\mathrm{b}} \Phi = \{s \in S \mid \exists \pi_2 \in \Pi_2 . \forall \pi_1 \in \Pi_1 . \mathit{Outcomes}(s, \pi_1, \pi_2) \subseteq \Phi\} . \tag{6}$$

The set $\langle 1 \rangle_{\mathcal{G}}^{\mathrm{b}} \Phi$ is the set of states from which player 1 can ensure that the game outcome is in Φ; the set $\langle 2 \rangle_{\mathcal{G}}^{\mathrm{b}} \Phi$ is the symmetrically defined set for player 2. We consider the control problems of computing the sets (5) and (6). We note that for player-1 deterministic game structures, computing (5) corresponds to solving the *existential* verification problem "is there a trace in Φ?", and for player-2 game structures computing (5) corresponds to solving the *universal* verification problem "are all traces in Φ?". We review the well-known solution of these control problems for the case in which Φ is a reachability property, a safety property, and a parity property. For a subset $T \subseteq S$ of states, the *safety* property $\Box T = \{s_0, s_1, s_2, \ldots \in S^\omega \mid \forall k . s_k \in T\}$ consists of all traces that stay always in T, and the *reachability* property $\Diamond T = \{s_0, s_1, s_2, \ldots \in S^\omega \mid \exists k . s_k \in T\}$ consists of all traces that contain a state in T. Consider any tuple $\mathcal{A} = \langle T_1, T_2, \ldots, T_m \rangle$ such that $T_1, T_2, \ldots, T_m$ form a partition of S into $m > 0$ mutually disjoint subsets. Given a trace $\sigma = s_0, s_1, s_2, \ldots \in S^\omega$, we denote by $\mathit{Index}(\sigma, \mathcal{A})$ the largest $i \in \{1, \ldots, m\}$ such that $s_k \in T_i$ for infinitely many $k \in \mathbb{N}$. Then, the *parity* property $\mathit{Parity}(\mathcal{A})$ is defined as $\mathit{Parity}(\mathcal{A}) = \{\sigma \in S^\omega \mid \mathit{Index}(\sigma, \mathcal{A}) \text{ is even}\}$. The relevance of parity properties is due to the fact that any ω-regular property can be specified with a deterministic automaton with a parity accepting condition [Tho90]. Hence, we can transform any verification problem with respect to an ω-regular condition into a verification problem with respect to a parity condition by means of a simple automaton product construction (see for instance [dAHM01a]).

3.1 Boolean μ-Calculus

For all three classes of properties (safety, reachability, and parity), the solution of the boolean control problems (5)–(6) can be written in μ-calculus interpreted over the lattice of subsets of states. Precisely, given a set S of states, the set $\mathcal{BMC}_S$ of *boolean μ-calculus* formulas consists of all μ-calculus formulas defined over the set of predicates $\mathcal{P}_S = 2^S$ and the set of functions $\mathcal{F}_b = \{\mathit{pre}_1, \mathit{dpre}_1, \mathit{pre}_2, \mathit{dpre}_2\}$, where $\mathrm{Dual}(\mathit{pre}_1) = \mathit{dpre}_1$ and $\mathrm{Dual}(\mathit{pre}_2) = \mathit{dpre}_2$. Given a game structure $\mathcal{G} = \langle S, \mathcal{M}, \Gamma_1, \Gamma_2, \delta \rangle$, we interpret the formulas of $\mathcal{BMC}_S$ over the lattice $\mathbb{L}(2^S, \subseteq)$ of subsets of S, ordered according to set inclusion. Negation is set complementation: for all $T \subseteq S$, we let $\sim T = S \setminus T$. The predicates are interpreted as themselves: for all $p \in \mathcal{P}$, we let $[\![p]\!]^{\mathrm{b}}_{\mathcal{G}} = p$. The functions pre_1, dpre_1, pre_2, and dpre_2 are called *predecessor operators,* and they are interpreted as follows:

$$[\![\mathit{pre}_1]\!]^{\mathrm{b}}_{\mathcal{G}}(X) = \{s \in S \mid \exists a_1 \in \Gamma_1(s). \forall a_2 \in \Gamma_2(s). \tau(s, a_1, a_2) \subseteq X\} \tag{7}$$

$$[\![\mathit{dpre}_1]\!]^{\mathrm{b}}_{\mathcal{G}}(X) = \{s \in S \mid \forall a_1 \in \Gamma_1(s). \exists a_2 \in \Gamma_2(s). \tau(s, a_1, a_2) \cap X \neq \emptyset\} \tag{8}$$

$$[\![\mathit{pre}_2]\!]^{\mathrm{b}}_{\mathcal{G}}(X) = \{s \in S \mid \exists a_2 \in \Gamma_2(s). \forall a_1 \in \Gamma_1(s). \tau(s, a_1, a_2) \subseteq X\} \tag{9}$$

$$[\![\mathit{dpre}_2]\!]^{\mathrm{b}}_{\mathcal{G}}(X) = \{s \in S \mid \forall a_2 \in \Gamma_2(s). \exists a_1 \in \Gamma_1(s). \tau(s, a_1, a_2) \cap X \neq \emptyset\}. \tag{10}$$

Intuitively, $[\![\mathit{pre}_1]\!]^{\mathrm{b}}_{\mathcal{G}}(X)$ is the set of states from which player 1 can force a transition to X in $\mathcal{G}$, and $[\![\mathit{dpre}_1]\!]^{\mathrm{b}}_{\mathcal{G}}(X)$ is the set of states from which player 1 is unable to avoid a transition to X in $\mathcal{G}$. The functions pre_2 and dpre_2 are interpreted symmetrically. We denote by $\mathit{bin}(\mathcal{G}) = (\mathbb{L}(2^S, \subseteq), [\![\cdot]\!]^{\mathrm{b}}_{\mathcal{G}})$ the resulting interpretation for μ-calculus. For a closed formula $\phi \in \mathcal{BMC}_S$, we write $[\![\phi]\!]^{\mathrm{b}}_{\mathcal{G}}$ rather than $[\![\phi]\!]^{\mathit{bin}(\mathcal{G})}$, and we omit $\mathcal{G}$ when clear from the context. For a game $\mathcal{G}$, a subset $T \subseteq S_{\mathcal{G}}$ of states and player $i \in \{1, 2\}$, we have then

$$\langle i \rangle^{\mathrm{b}}_{\mathcal{G}} \Diamond T = [\![\mu x.(T \vee \mathit{pre}_i(x))]\!]^{\mathrm{b}}_{\mathcal{G}}. \tag{11}$$

This formula can be understood by considering its iterative computation: we have that $[\![\mu x.(T \vee \mathit{pre}_i(x))]\!]^{\mathrm{b}}_{\mathcal{G}} = \lim_{k \to \infty} X_k$, where $X_0 = \emptyset$ and, for $k \in \mathbb{N}$, where $X_{k+1} = T \cup [\![\mathit{pre}_i]\!]^{\mathrm{b}}_{\mathcal{G}}(X_k)$: it is easy to show by induction that the set that

X_k consists of the states of S from which player i can force the game to T in at most k steps. The equation (11) then follows by taking the limit $k \to \infty$. Similarly, for safety properties we have, for $i \in \{1,2\}$:

$$\langle i \rangle^{\mathrm{b}}_{\mathcal{G}} \Box T = [\![\nu x.(T \wedge pre_i(x))]\!]^{\mathrm{b}}_{\mathcal{G}}. \tag{12}$$

Again, to understand this formula it helps to consider its iterative computation. We have $[\![\nu x.(T \wedge pre_1(x))]\!]^{\mathrm{b}}_{\mathcal{G}} = \lim_{k\to\infty} X_k$, where $X_0 = S$ and, for $k \in \mathbb{N}$, where $X_{k+1} = T \cap [\![pre_1]\!]^{\mathrm{b}}_{\mathcal{G}}(X_k)$; it is easy to see that X_k consists of the states of $\mathcal{G}$ from which player 1 can guarantee that the game stays in T for at least k steps. The equation (12) then follows by taking the limit $k \to \infty$. The solution of control and verification problems for parity properties is given by the following result.

Theorem 1. [EJ91] *For all game structures $\mathcal{G}$, all partitions $\langle T_1, T_2, \dots, T_m \rangle$ of $S_{\mathcal{G}}$, and all $i \in \{1,2\}$, we have*

$$\langle i \rangle^{\mathrm{b}}_{\mathcal{G}} Parity(\langle T_1, \dots, T_m \rangle) = [\![\gamma_m x_m \cdots \gamma_1 x_1 . \bigvee_{j=1}^{m} (T_j \wedge pre_i(x_j))]\!]^{\mathrm{b}}_{\mathcal{G}} \tag{13}$$

Given an EJ-form μ-calculus formula $\phi = \gamma_m x_m \cdots \gamma_1 x_1 . \bigvee_{j=1}^{m} (T_j \wedge pre_1(x_j))$, we define the parity property $PtyOf(\phi)$ corresponding to ϕ by $PtyOf(\phi) = Parity(\langle T_1, \dots, T_m \rangle)$. With this notation, we can restate (13) as follows.

Corollary 1. *For all game structures $\mathcal{G}$, all $i \in \{1,2\}$ and all closed EJ-form μ-calculus formulas $\phi \in \mathcal{BMC}_{S_{\mathcal{G}}}$ containing only the function symbol pre_i, we have $\langle i \rangle^{\mathrm{b}}_{\mathcal{G}} PtyOf(\phi) = [\![\phi]\!]^{\mathrm{b}}_{\mathcal{G}}$.*

Lack of Determinacy. In boolean μ-calculus, the operators pre_1 and pre_2 are not the dual one of the other. This implies that boolean control problems are not determined: for $\Phi \subseteq S^\omega$, the equality $S \setminus \langle 1 \rangle^{\mathrm{b}}_{\mathcal{G}} \Phi = \langle 2 \rangle^{\mathrm{b}}_{\mathcal{G}} (S^\omega \setminus \Phi)$ does *not* hold for all game structures $\mathcal{G}$ and all properties Φ. Intuitively, the fact that player 1 is unable to ensure the control goal Φ does not entail that player 2 is able to ensure the control goal $\neg\Phi$. For example, there are game structures $\mathcal{G}$ where for some $T \subseteq S_{\mathcal{G}}$ we have

$$S_{\mathcal{G}} \setminus (\langle 1 \rangle^{\mathrm{b}}_{\mathcal{G}} \Diamond T) = [\![\neg \mu x.(T \vee pre_1(x))]\!]^{\mathrm{b}}_{\mathcal{G}} = [\![\nu x.(\neg T \wedge dpre_1(x))]\!]^{\mathrm{b}}_{\mathcal{G}}$$
$$\neq [\![\nu x.(\neg T \wedge pre_2(x))]\!]^{\mathrm{b}}_{\mathcal{G}} = \langle 2 \rangle^{\mathrm{b}}_{\mathcal{G}} \Box (\neg T).$$

An example is the game structure MATCHBIT: as explained in the introduction we have $s_{try} \notin \langle 1 \rangle^{\mathrm{b}} \Diamond \{s_{goal}\}$; on the other hand, it can be easily seen that $s_{try} \notin \langle 2 \rangle^{\mathrm{b}} \Box \{s_{try}\}$.

3.2 The Linear Semantics of Boolean μ-Calculus

Theorem 1 establishes a basic connection between linear parity properties and verification algorithms expressed in μ-calculus. Here, we shall develop a connection between linear properties *expressed in μ-calculus,* and their verification algorithms also expressed in μ-calculus. To this end, we provide a linear semantics for μ-calculus, obtained by evaluating μ-calculus on linear traces.

A trace $\sigma \in S^\omega$ gives rise to an interpretation $\mathcal{I}^b_\sigma = (\mathbb{L}(2^{\mathbb{N}}, \subseteq), [\![\cdot]\!]^{\mathrm{b}}_\sigma)$ for μ-calculus, where $\mathbb{L}(2^{\mathbb{N}}, \subseteq)$ is the lattice of subsets of natural numbers ordered according to set inclusion, and where all predicates $p \in 2^S$ are interpreted as the sets of indices of states in p, i.e., $[\![p]\!]^{\mathrm{b}}_\sigma = \{k \in \mathbb{N} \mid \sigma_k \in p\}$. The definitions (7)–(10) can be simplified, since every location of the trace has a single successor: for all $i \in \{1,2\}$ and all $X \subseteq \mathbb{N}$ we let $[\![pre_i]\!]^{\mathrm{b}}_\sigma(X) = [\![dpre_i]\!]^{\mathrm{b}}_\sigma(X) = \{k \in \mathbb{N} \mid k+1 \in X\}$. We define the *boolean linear semantics* $[\phi]^{\mathrm{blin}_S}$ over the set of states S of a closed μ-calculus formula $\phi \in \mathcal{BMC}_S$ to consist of all traces whose first state is in the semantics of ϕ: specifically, $[\phi]^{\mathrm{blin}_S} = \{\sigma \in S^\omega \mid \sigma_0 \in [\![\phi]\!]^{\mathcal{I}^b_\sigma}\}$. In contrast, we call the semantics $[\![\phi]\!]^{\mathrm{b}}_{\mathcal{G}}$ defined over a game structure $\mathcal{G}$ the *branching* semantics of the μ-calculus formula ϕ. The following lemma states that for formulas in EJ-form, the parity property corresponding to the formula coincides with the linear semantics of the formula.

Lemma 1. *For all sets of states S, all $i \in \{1,2\}$ and all closed EJ-form μ-calculus formulas $\phi \in \mathcal{BMC}_S$ containing only the function symbol pre_i, we have $PtyOf(\phi) = [\phi]^{\mathrm{blin}_S}$.*

This leads easily to the following result, which relates the linear and branching semantics of EJ-form formulas.

Corollary 2. *For all game structures $\mathcal{G}$, all $i \in \{1,2\}$ and all closed EJ-form μ-calculus formulas $\phi \in \mathcal{BMC}_{S_\mathcal{G}}$ containing only the function symbol pre_i, we have $[\![\phi]\!]^{\mathrm{b}}_{\mathcal{G}} = \langle i \rangle^{\mathrm{b}}_{\mathcal{G}} PtyOf(\phi) = \langle i \rangle^{\mathrm{b}}_{\mathcal{G}} [\phi]^{\mathrm{blin}_{S_\mathcal{G}}}$.*

In fact, the relationship between the linear and branching semantics holds for all μ-calculus formulas in strongly deterministic form.

Theorem 2. [dAHM01a] *For all game structures $\mathcal{G}$, all closed μ-calculus formulas $\phi \in \mathcal{BMC}_{S_\mathcal{G}}$, and all players $i \in \{1,2\}$, if ϕ is in strongly deterministic form and contains only the function symbol pre_i, then $[\![\phi]\!]^{\mathrm{b}}_{\mathcal{G}} = \langle i \rangle^{\mathrm{b}}_{\mathcal{G}} [\phi]^{\mathrm{blin}_{S_\mathcal{G}}}$.*

We will see that the linear and branching semantics of strongly deterministic (and in particular, EJ-form) formulas are related in all the settings considered in this paper, namely, in the boolean, probabilistic, and discounted settings. The linear and branching semantics of formulas are not always related, as the following example demonstrates.

Example 1. [dAHM01a] Consider the formula $\phi = \mu x.(pre_1(x) \vee \nu y.(B \wedge pre_1(y)))$, where $B \subseteq S$ is a set of states. The linear semantics $[\phi]^{\mathrm{blin}_S}$ consists of all the traces $\sigma = s_0, s_1, s_2, \ldots$ for which there is a $k \in \mathbb{N}$ such that $s_i \in B$ for all $i \geq k$, that is, of all the traces that eventually enter B, and never leave it again; using temporal-logic notation, we indicate this set of traces by $[\Diamond\Box B]_S$. In fact, we have $s_k \in [\![\nu y.(B \wedge pre_1(y))]\!]^{\mathcal{I}^b_\sigma}$ only if $s_i \in B$ for all $i \geq k$, and we have that $s_0 \in [\![\phi]\!]^{\mathcal{I}^b_\sigma}$ iff there is $k \in \mathbb{N}$ such that $s_k \in [\![\nu y.(B \wedge pre_1(y))]\!]^{\mathcal{I}^b_\sigma}$. However, consider a deterministic player-2 structure $\mathcal{G} = \langle S, \mathcal{M}, \Gamma_1, \Gamma_2, \delta \rangle$ with $S = \{s, t, u\}$, $\mathcal{M} = \{a, b, \bullet\}$, and $\Gamma_2(s) = \{a, b\}$, $\Gamma_2(t) = \Gamma_2(u) = \{a\}$, and transition relation given by $\tau(s, \bullet, a) = \{s\}$, $\tau(s, \bullet, b) = \{t\}$, $\tau(t, \bullet, a) = \{u\}$, $\tau(u, \bullet, a) = \{u\}$, where

• is the single move available to player 1. For $B = \{s, u\}$ it is easy to see that $\langle 1\rangle^{\rm b}_{\mathcal{G}}[\phi]^{\rm blins} = \langle 1\rangle^{\rm b}_{\mathcal{G}}[\Diamond\Box B]^{\rm blins} = \{s, t, u\}$, while $[\![\phi]\!]^{\rm b}_{\mathcal{G}} = \{t, u\}$. ∎

In [dAHM01a], it is shown that in general the linear and branching semantics of μ-calculus formulas are related on all game structures iff they are related on all player-1 and all player-2 game structures.

4 Probabilistic Verification and Control

The boolean control problem asks whether a player can guarantee that all outcomes are in a desired linear property. The *probabilistic* control problem asks what is the *maximal probability* with which a player can guarantee that the outcome of the game belongs to the desired linear property. Given a game structure $\mathcal{G} = \langle S, \mathcal{M}, \Gamma_1, \Gamma_2, \delta\rangle$ and a property $\Phi \subseteq S^\omega$, we consider the two *probabilistic control problems* consisting in computing the functions $\langle 1\rangle^{\rm P}_{\mathcal{G}}\Phi, \langle 2\rangle^{\rm P}_{\mathcal{G}}\Phi : S \mapsto [0, 1]$ defined by:

$$\langle 1\rangle^{\rm P}_{\mathcal{G}}\Phi = \lambda s \in S.\ \sup_{\pi_1\in\Pi_1}\ \inf_{\pi_2\in\Pi_2}\ \Pr_s^{\pi_1,\pi_2}(\Phi) \tag{14}$$

$$\langle 2\rangle^{\rm P}_{\mathcal{G}}\Phi = \lambda s \in S.\ \sup_{\pi_2\in\Pi_2}\ \inf_{\pi_1\in\Pi_1}\ \Pr_s^{\pi_1,\pi_2}(\Phi). \tag{15}$$

where $\lambda s \in S.f(s)$ is the usual λ-calculus notation for a function that maps each $s \in S$ into $f(s)$.

4.1 Probabilistic μ-Calculus

For the case in which Φ is a reachability, safety, or parity property, we can compute the functions (14), (15) using a *probabilistic* interpretation of μ-calculus [dAM01]. Precisely, given a set S of states, the set $\mathcal{PMC}_S$ of *probabilistic μ-calculus* formulas consists of all μ-calculus formulas defined over the set of predicates $\mathcal{P}_S = 2^S$ and the set of functions $\mathcal{F}_q = \{pre_1, pre_2\}$, where $\mathrm{Dual}(pre_1) = pre_2$. Given a game structure $\mathcal{G} = \langle S, \mathcal{M}, \Gamma_1, \Gamma_2, \delta\rangle$, we interpret these formulas over the lattice $\mathbb{L}(S \mapsto [0, 1], \leq)$ of functions $S \mapsto [0, 1]$, ordered pointwise: for $f, g : S \mapsto [0, 1]$ and $s \in S$, we have $(f \sqcup g)(s) = \max\{f(s), g(s)\}$ and $(f \sqcap g)(s) = \min\{f(s), g(s)\}$. Negation is defined by $\sim f = \lambda s \in S.1 - f(s)$. The predicates are interpreted as *characteristic functions:* for all $p \in \mathcal{P}_S$, we let $[\![p]\!]^{\rm P}_{\mathcal{G}} = \chi(p)$, where $\chi(p)$ is defined for all $s \in S$ by $\chi(p)(s) = 1$ if $s \in p$, and $\chi(p)(s) = 0$ otherwise. The interpretations of pre_1 and pre_2 are defined as follows, for $X : S \mapsto [0, 1]$:

$$[\![pre_1]\!]^{\rm P}_{\mathcal{G}}(X) = \lambda s \in S.\ \sup_{\zeta_1\in\mathcal{D}_1(s)}\ \inf_{\zeta_2\in\mathcal{D}_2(s)}\ \mathrm{E}_\circ(X \mid s, \zeta_1, \zeta_2) \tag{16}$$

$$[\![pre_2]\!]^{\rm P}_{\mathcal{G}}(X) = \lambda s \in S.\ \sup_{\zeta_2\in\mathcal{D}_2(s)}\ \inf_{\zeta_1\in\mathcal{D}_1(s)}\ \mathrm{E}_\circ(X \mid s, \zeta_1, \zeta_2) \tag{17}$$

where $\mathrm{E}_\circ(X \mid s, \zeta_1, \zeta_2) = \sum_{t \in S} \hat{\delta}(s, \zeta_1, \zeta_2)(t)\, X(t)$ is the next-step expectation of X, given that player 1 and player 2 choose their moves according to distributions ζ_1 and ζ_2, respectively. Intuitively, $[\![pre_1]\!]^{\mathrm{P}}_{\mathcal{G}}(X)$ is the function that associates with each $s \in X$ the maximal expectation of X that player 1 can achieve in one step. In particular, for $T \subseteq S$, $[\![pre_i]\!]^{\mathrm{P}}_{\mathcal{G}}(\chi(T))$ is the maximal probability with which player i can force a transition to T in one step. We note that, unlike in the boolean case, in probabilistic μ-calculus the operators pre_1 and pre_2 are dual, so that the calculus requires only two predecessor operators, rather than four. The duality follows directly from the minimax theorem [vN28]: for all $X : S \mapsto [0,1]$ and all $s \in S$, we have

$$\begin{aligned}
1 - [\![pre_1]\!]^{\mathrm{P}}_{\mathcal{G}}(X)(s) &= 1 - \sup\nolimits_{\zeta_1 \in \mathcal{D}_1(s)} \inf\nolimits_{\zeta_2 \in \mathcal{D}_2(s)} \mathrm{E}_\circ(X \mid s, \zeta_1, \zeta_2) \\
&= \inf\nolimits_{\zeta_1 \in \mathcal{D}_1(s)} \sup\nolimits_{\zeta_2 \in \mathcal{D}_2(s)} 1 - \mathrm{E}_\circ(X \mid s, \zeta_1, \zeta_2) \\
&= \sup\nolimits_{\zeta_2 \in \mathcal{D}_2(s)} \inf\nolimits_{\zeta_1 \in \mathcal{D}_1(s)} \mathrm{E}_\circ(\sim X \mid s, \zeta_1, \zeta_2) \\
&= [\![pre_2]\!]^{\mathrm{P}}_{\mathcal{G}}(\sim X)(s).
\end{aligned}$$

We denote by $prb(\mathcal{G}) = (\mathbb{L}(S \mapsto [0,1], \leq), [\![\cdot]\!]^{\mathrm{P}}_{\mathcal{G}})$ the resulting interpretation for μ-calculus. For a closed formula $\phi \in \mathcal{BMC}_S$, we write $[\![\phi]\!]^{\mathrm{P}}_{\mathcal{G}}$ rather than $[\![\phi]\!]^{prb(\mathcal{G})}$, and we omit $\mathcal{G}$ when clear from the context. The solutions to probabilistic control problems with respect to reachability, safety, and parity properties can then be written in μ-calculus as stated by the following theorem.

Theorem 3. [dAM01] *For all game structures $\mathcal{G}$, all $i \in \{1,2\}$, all $T \subseteq S_{\mathcal{G}}$ and all partitions $\mathcal{A} = \langle T_1, T_2, \ldots, T_m \rangle$ of $S_{\mathcal{G}}$, we have:*

$$\langle i \rangle^{\mathrm{P}}_{\mathcal{G}} \Diamond T = [\![\mu x.(T \vee pre_i(x))]\!]^{\mathrm{P}}_{\mathcal{G}} \tag{18}$$

$$\langle i \rangle^{\mathrm{P}}_{\mathcal{G}} \Box T = [\![\nu x.(T \wedge pre_i(x))]\!]^{\mathrm{P}}_{\mathcal{G}} \tag{19}$$

$$\langle i \rangle^{\mathrm{P}}_{\mathcal{G}} Parity(\langle T_1, \ldots, T_m \rangle) = [\![\gamma_m x_m \cdots \gamma_1 x_1. \bigvee_{i=1}^{m} (T_i \wedge pre_i(x_i))]\!]^{\mathrm{P}}_{\mathcal{G}}. \tag{20}$$

The above solution formulas are the analogous to (11), (12), and (13), even though the proof of their correctness requires different arguments. The argument for reachability games is as follows. The fixpoint (18) can be computed iteratively by $[\![\mu x.(T \vee pre_i(x))]\!]^{\mathrm{P}}_{\mathcal{G}} = \lim_{k \to \infty} X_k$, where $X_0 = \lambda s.0$ and, for $k \in \mathbb{N}$, where $X_{k+1} = \chi(T) \sqcup [\![pre_i]\!]^{\mathrm{P}}_{\mathcal{G}}(X_k)$. It is then easy to show by induction that $X_k(s)$ is the maximal probability with which player i can reach T from $s \in S$ in at most k steps. In fact, (18) is simply a restatement in μ-calculus of the well-known fixpoint characterization of the solution of positive stochastic games (see, e.g., [FV97]). The solution (19) can also be understood in terms of the iterative evaluation of the fixpoint. We have $[\![\nu x.(T \wedge pre_i(x))]\!]^{\mathrm{P}}_{\mathcal{G}} = \lim_{k \to \infty} X_k$, where $X_0 = \lambda s.1$, and for $x \in \mathbb{N}$, where $X_{k+1} = \chi(T) \sqcap [\![pre_i]\!]^{\mathrm{P}}_{\mathcal{G}}(X_k)$. It can be shown by induction that $X_k(s)$ is equal to the maximal probability of staying in T for at least k steps that player i can achieve from $s \in S$. The detailed arguments can be found in [dAM01].

We note that on deterministic turn-based structures (and their special cases, such as transition systems), the boolean and probabilistic control problems are equivalent, as are the boolean and probabilistic μ-calculi. Indeed, for all deterministic turn-based game structures $\mathcal{G}$ with set of states S, and for all properties $\Phi \subseteq S^\omega$, all $i \in \{1,2\}$, and all closed μ-calculus formulas ϕ containing only functions pre_1 and pre_2, we have that $\chi(\langle i\rangle^{\mathrm{b}}_{\mathcal{G}}\Phi) = \langle i\rangle^{\mathrm{p}}_{\mathcal{G}}\Phi$ and $\chi([\![\phi]\!]^{\mathrm{b}}_{\mathcal{G}}) = [\![\phi]\!]^{\mathrm{p}}_{\mathcal{G}}$.

Determinacy. As a consequence of the duality between the pre_1 and pre_2 operators, probabilistic control problems are determined, unlike their boolean counterparts: in particular, [Mar98] proves that for all games $\mathcal{G}$, all sets $\Phi \subseteq S^\omega_{\mathcal{G}}$ in the Borel hierarchy, and all $s \in S_{\mathcal{G}}$, we have $1 - \langle 1\rangle^{\mathrm{p}}_{\mathcal{G}}\Phi(s) = \langle 2\rangle^{\mathrm{p}}_{\mathcal{G}}(S^\omega \setminus \Phi)(s)$. While the proof of this result requires advanced arguments, the case in which Φ is a parity property follows elementarily from our μ-calculus solution formula (20), and from the duality of pre_1 and pre_2. In fact, consider a partition $\mathcal{A} = \langle T_1, T_2, \ldots, T_m\rangle$ of S. Letting $U_1 = \emptyset$ and $U_{i+1} = T_i$ for $1 \le i \le m$, we have:

$$\begin{aligned}
1 - \langle 1\rangle^{\mathrm{p}}_{\mathcal{G}} Parity(\langle T_1, \ldots, T_m\rangle) &= 1 - [\![\gamma_m x_m \cdots \gamma_1 x_1 . \textstyle\bigvee_{j=1}^m (T_j \wedge pre_1(x_j))]\!]^{\mathrm{p}}_{\mathcal{G}} \\
&= [\![\gamma_{m+1} x_m \cdots \gamma_2 x_1 . \textstyle\bigvee_{j=1}^m (T_j \wedge pre_2(x_j))]\!]^{\mathrm{p}}_{\mathcal{G}} \\
&= \langle 2\rangle^{\mathrm{p}}_{\mathcal{G}} Parity(\langle U_1, \ldots, U_{m+1}\rangle) \\
&= \langle 2\rangle^{\mathrm{p}}_{\mathcal{G}}(S^\omega \setminus Parity(\langle T_1, \ldots, T_m\rangle)).
\end{aligned}$$

4.2 The Linear Semantics of Probabilistic μ-Calculus

The solution (20) of parity control problems can be restated as follows. For player $i \in \{1,2\}$, all game structures $\mathcal{G}$, and all EJ-form formulas ϕ containing only the function symbol pre_i, we have $[\![\phi]\!]^{\mathrm{p}}_{\mathcal{G}} = \langle i\rangle^{\mathrm{p}}_{\mathcal{G}} PtyOf(\phi)$. Using Lemma 1, we can therefore relate the linear and branching semantics of ϕ as follows.

Theorem 4. *For all game structures $\mathcal{G}$, all $i \in \{1,2\}$, and all closed μ-calculus formulas $\phi \in \mathcal{PMC}_{S_\mathcal{G}}$ in EJ-form containing only the function symbol pre_i, we have that $[\![\phi]\!]^{\mathrm{p}}_{\mathcal{G}} = \langle i\rangle^{\mathrm{p}}_{\mathcal{G}} [\phi]^{\mathrm{blins}_{S_\mathcal{G}}}$.*

This theorem relates the branching semantics $[\![\phi]\!]^{\mathrm{p}}_{\mathcal{G}}$ of *probabilistic* μ-calculus with the linear semantics $[\phi]^{\mathrm{blins}_{S_\mathcal{G}}}$ of *boolean* μ-calculus. In order to relate branching and linear semantics of *probabilistic* μ-calculus, we define a probabilistic linear semantics $[\cdot]^{\mathrm{plins}}$ of probabilistic μ-calculus.

A trace $\sigma \in S^\omega$ gives rise to an interpretation $\mathcal{I}^p_\sigma = (\mathbb{L}(N \mapsto [0,1], \le), [\![\cdot]\!]^{\mathrm{p}}_\sigma)$ for μ-calculus, where $(\mathbb{L}(N \mapsto [0,1], \le)$ is the lattice of functions $\mathbb{N} \mapsto \{0,1\}$ ordered pointwise, where a predicate $p \in 2^S$ is interpreted as its characteristic function, i.e., for all $k \ge 0$ we have $[\![p]\!]^{\mathrm{p}}_\sigma(k) = 1$ if $\sigma_k \in p$, and $[\![p]\!]^{\mathrm{p}}_\sigma(k) = 0$ if $\sigma_k \notin p$. Similarly to the boolean case, the definitions (16)–(17) can be simplified, since every state of the trace has a single successor. For all $X : \mathbb{N} \mapsto [0,1]$ and $i \in \{1,2\}$ we let $[\![pre_i]\!]^{\mathrm{p}}_\sigma(X) = \lambda k. X(k+1)$. Given a closed μ-calculus formula

$\phi \in \mathcal{PMC}_S$, we define the *probabilistic linear semantics* $[\phi]^{\text{plins}} : S^\omega \mapsto [0,1]$ of ϕ over the set of states S by taking the value of ϕ over the first state of the trace: specifically, we let $[\phi]^{\text{plins}}(\sigma) = [\![\phi]\!]^{\mathcal{I}^p_\sigma}(0)$.

In the definitions (14), (15) of the probabilistic control and verification problems, the property Φ is a set of traces. To complete our connection with the probabilistic linear semantics $[\cdot]^{\text{plins}}$, we need to define a probabilistic version of these problems. Let $h : S^\omega \mapsto [0,1]$ be a function that is measurable in the probability space $(S^\omega, \Omega, \text{Pr}_s^{\pi_1,\pi_2})$, for all $\pi_1 \in \Pi_1$ and $\pi_2 \in \Pi_2$. We define:

$$\langle 1 \rangle_{\mathcal{G}}^{\text{q}} h = \lambda s \in S \, . \, \sup_{\pi_1 \in \Pi_1} \inf_{\pi_2 \in \Pi_2} \text{E}_s^{\pi_1,\pi_2}(h)$$

$$\langle 2 \rangle_{\mathcal{G}}^{\text{q}} h = \lambda s \in S \, . \, \sup_{\pi_2 \in \Pi_2} \inf_{\pi_1 \in \Pi_1} \text{E}_s^{\pi_1,\pi_2}(h).$$

The relationship between the probabilistic linear and branching semantics is then expressed by the following theorem.

Theorem 5. *For all game structure $\mathcal{G}$, all $i \in \{1,2\}$, and all closed μ-calculus formulas $\phi \in \mathcal{PMC}_S$ in strongly deterministic form and containing only the function symbol pre_i, we have that $[\![\phi]\!]_{\mathcal{G}}^{\text{p}} = \langle i \rangle_{\mathcal{G}}^{\text{q}} [\phi]^{\text{plins}_{\mathcal{G}}}$.*

For player 1, the above theorem states that for all $s \in S_{\mathcal{G}}$,

$$[\![\phi]\!]_{\mathcal{G}}^{\text{p}}(s) = \sup_{\pi_1 \in \Pi_1} \inf_{\pi_2 \in \Pi_2} \text{E}_s^{\pi_1,\pi_2}([\phi]^{\text{plins}_{\mathcal{G}}}). \tag{21}$$

This equation can be read as follows: the value of a control μ-calculus formula $[\![\phi]\!]_{\mathcal{G}}^{\text{p}}$ is equal to the maximal expectation that player 1 can guarantee for the same formula, evaluated on linear traces. The theorem relates not only the branching and the linear probabilistic semantics, but also a global optimization problem to a local one. In (21) the right-hand side represents a global optimization problem: player 1 is trying to maximize the value of the function $[\phi]^{\text{plins}}$ over traces, and player 2 is trying to oppose this. On the left-hand side, on the other hand, the optimization is local, being performed through the evaluation of the operator $[\![pre_1]\!]_{\mathcal{G}}^{\text{p}}$ at all states of $\mathcal{G}$.

5 Discounted Verification and Control

In the boolean and probabilistic settings, properties are specified as ω-regular languages, and algorithms are encoded as fixpoint expressions in the μ-calculus. The main theorems, such as Theorem 1 and Theorem 3, express the relationship between the properties and the μ-calculus fixpoints that solve the verification and control problems. The correspondence between the branching and linear semantics serves mainly to clarify the relationship between the local optimization that takes place in the branching semantics, and the global optimization that takes place in the linear semantics. In the discounted setting, on the other hand, we choose not have an independent notion of discounted property: rather, discounted properties are specified by the linear semantics of formulas of the discounted μ-calculus. The main results for the discounted setting concern thus the relationship between the linear semantics (used to express properties) and the branching semantics (which represents algorithms) of discounted μ-calculus, as well as the relationship between the discounted setting and the undiscounted one. As both properties and algorithms are defined in terms of the μ-calculus, we begin by introducing discounted μ-calculus.

5.1 Discounted μ-Calculus

Given a set S of states and a set Υ of *discount factors*, the set $\mathcal{DMC}_{S,\Upsilon}$ of discounted μ-calculus formulas consists of all the formulas defined over the set of predicates $\mathcal{P}_S = 2^S$ and the set of functions

$$\mathcal{F}_\Upsilon = \{\ \alpha pre_i,\ (1-\alpha) + \alpha pre_i \ \mid i \in \{1,2\}, \alpha \in \Upsilon\},$$

where $\text{Dual}(\alpha pre_1) = (1-\alpha) + \alpha pre_2$ and $\text{Dual}(\alpha pre_2) = (1-\alpha) + \alpha pre_1$. As in the probabilistic case, given a game structure $\mathcal{G} = \langle S, \mathcal{M}, \Gamma_1, \Gamma_2, \delta\rangle$, we interpret these formulas over the lattice $\mathbb{L}(S \mapsto [0,1], \leq)$ of functions $S \mapsto [0,1]$, ordered pointwise. Again, we define negation by $\sim f = \lambda s \in S.1 - f(s)$. The interpretation of predicates and functions is parameterized by a *discount factor interpretation* $\eta : \Upsilon \mapsto [0,1]$, that assigns to each discount factor $\alpha \in \Upsilon$ its value $\eta(\alpha) \in [0,1]$. As in the probabilistic semantics, we interpret the predicates $p \in \mathcal{P}$ as their characteristic function, i.e., $[\![p]\!]^{\text{d}}_{\mathcal{G},\eta} = \chi(p)$. For all $\eta \in (\Upsilon \mapsto [0,1])$ and all $i \in \{1,2\}$, we let:

$$[\![\alpha pre_i]\!]^{\text{d}}_{\mathcal{G},\eta}(X) = \lambda s \in S.\eta(\alpha)[\![pre_i]\!]^{\text{P}}_{\mathcal{G}}(X)(s) \tag{22}$$

$$[\![(1-\alpha) + \alpha pre_i]\!]^{\text{d}}_{\mathcal{G},\eta}(X) = \lambda s \in S.(1-\eta(\alpha)) + \eta(\alpha)[\![pre_i]\!]^{\text{P}}_{\mathcal{G}}(X)(s). \tag{23}$$

Thus, the discounted interpretation of αpre_i is equal to the probabilistic interpretation of pre_i, discounted by a factor α; the discounted interpretation of $(1-\alpha) + \alpha pre_i$ is equal to the probabilistic interpretation of pre_i, discounted by a factor of α, and with $1-\alpha$ added to it. We denote by $disc(\mathcal{G},\eta) = (\mathbb{L}(S \mapsto [0,1], \leq), [\![\cdot]\!]^{\text{d}}_{\mathcal{G},\eta})$ the resulting semantics for the μ-calculus, and we write $[\![\cdot]\!]^{\text{d}}_{\mathcal{G},\eta}$ for $[\![\cdot]\!]^{disc(\mathcal{G},\eta)}$, omitting $\mathcal{G}$ when clear from the context.

While (22) is the expected definition, (23) requires some justification. Consider a game structure $\mathcal{G} = \langle S, \mathcal{M}, \Gamma_1, \Gamma_2, \delta\rangle$. First, notice that this definition ensures that pre_1 and $(1-\alpha) + \alpha pre_2$ are dual: in fact, for $s \in S$ we have

$$\begin{aligned}
1 - [\![\alpha pre_1]\!]^{\text{d}}_{\eta}(X)(s) &= 1 - \eta(\alpha)[\![pre_1]\!]^{\text{P}}(X)(s) \\
&= 1 - \eta(\alpha) + \eta(\alpha) - \eta(\alpha)[\![pre_1]\!]^{\text{P}}(X)(s) \\
&= (1-\eta(\alpha)) + \eta(\alpha)\big[1 - [\![pre_1]\!]^{\text{P}}(X)(s)\big] \\
&= (1-\eta(\alpha)) + \eta(\alpha)[\![pre_2]\!]^{\text{P}}(\sim X)(s) \\
&= [\![(1-\alpha) + \alpha pre_2]\!]^{\text{d}}_{\eta}(\sim X)(s).
\end{aligned}$$

The definitions (22) and (23) can also be justified by showing how the resulting predecessor operators can be used to solve discounted reachability and safety games in a way analogous to (18) and (19). Let $T \subseteq S$ be a set of target states, and fix a player $i \in \{1, 2\}$. Consider a *discounted reachability game,* in which player i gets the payoff $\eta(\alpha)^k$ when the target T is reached after k steps, and the payoff 0 if T is not reached. The maximum payoff that player i can guarantee is given by

$$[\![\mu x.(T \vee \alpha pre_i(x))]\!]^{\mathrm{d}}_{\eta}. \tag{24}$$

As an example, consider again the game MATCHBIT, along with the formula $[\![\mu x.(\{s_{goal}\} \vee \alpha pre_i(x))]\!]^{\mathrm{d}}_{\eta}$, and let $r = \eta(\alpha)$. Let $X_0 = \lambda s.0$ and, for $k \in \mathbb{N}$, let $X_{k+1} = \chi(\{s_{goal}\}) \sqcup [\![\alpha pre_1]\!]^{\mathrm{d}}_{\eta}(X_k)$. We can verify that $X_0(s_{try}) = 0$, $X_1(s_{try}) = r(\frac{1}{2} \cdot 0 + \frac{1}{2} \cdot 1) = \frac{r}{2}$, $X_2(s_{try}) = r(\frac{1}{2} \cdot \frac{r}{2} + \frac{1}{2} \cdot 1) = \frac{r}{2} + \frac{r^2}{4}$, and $\lim_{k \to \infty} X_k(s_{try}) = r/(2-r) = [\![\mu x.(\{s_{goal}\} \vee \alpha pre_i(x))]\!]^{\mathrm{d}}_{\eta}(s_{try})$.

Consider now a *discounted safety game,* in which player i gets the payoff $1 - \eta(\alpha)^k$ if the game stays in T for k consecutive steps, and the payoff 1 if T is never left. The maximum payoff that player i can guarantee is given by

$$[\![\nu x.(T \wedge (1-\alpha) + \alpha pre_i(x))]\!]^{\mathrm{d}}_{\eta}. \tag{25}$$

Indeed, one can verify that for all $s \in S$, we have

$$1 - [\![\mu x.(T \vee \alpha pre_1(x))]\!]^{\mathrm{d}}_{\eta}(s) = [\![\nu x.(\neg T \wedge (1-\alpha) + \alpha pre_2(x))]\!]^{\mathrm{d}}_{\eta}(s) \tag{26}$$

indicating that the payoff player 1 can guarantee in a discounted T-reachability game is equal to 1 minus the payoff that player 2 can guarantee for the discounted $\neg T$-safety game.

Above, we have informally introduced discounted reachability and safety games in terms of payoffs associated with the traces. How are these payoffs defined, for more general goals? And what is the precise definition of the games that (24) and (25) solve? To answer these questions, we introduce the linear semantics of discounted μ-calculus, and we once more relate the linear semantics to the branching one.

5.2 The Linear Semantics of Discounted μ-Calculus

A *discounted property* is the interpretation of a discounted μ-calculus formula over linear traces. Similarly to the probabilistic case, the linear semantics of discounted μ-calculus associates with each trace a number in the interval $[0, 1]$ obtained by evaluating the μ-calculus formula over the trace, and taking the value at the initial state of the trace.

Consider a set Υ of discount factors, along with a discount factor interpretation $\eta : \Upsilon \mapsto [0, 1]$. A trace $\sigma \in S^\omega$ gives rise to an interpretation $\mathcal{I}^{\eta}_{\sigma} = (\mathbb{L}(N \mapsto [0, 1], \leq), [\![\cdot]\!]^{\mathrm{d}}_{\sigma,\eta})$ for the discounted μ-calculus formulas in $\mathcal{DMC}_{S,\Upsilon}$. As in the probabilistic case, all predicates $p \in 2^S$ are interpreted as their characteristic function, i.e., for all $k \geq 0$ we have $[\![p]\!]^{\mathrm{p}}_{\sigma}(k) = 1$ if $\sigma_k \in p$, and $[\![p]\!]^{\mathrm{p}}_{\sigma}(k) = 0$ if

$\sigma_k \notin p$. The definitions (22) and (23) can be simplified, since in a trace, every location has a single successor. For all $X : \mathbb{N} \mapsto [0,1]$ and $i \in \{1,2\}$ we let

$$\llbracket \alpha pre_i \rrbracket^{\mathrm{d}}_{\sigma,\eta}(X) = \lambda k.\big[\eta(\alpha)\, X(k+1)\big]$$
$$\llbracket (1-\alpha) + \alpha pre_i \rrbracket^{\mathrm{d}}_{\sigma,\eta}(X) = \lambda k.\big[(1-\eta(\alpha)) + \eta(\alpha)\, X(k+1)\big].$$

Given a closed μ-calculus formula $\phi \in \mathcal{DMC}_{S,\Upsilon}$, we then define its *discounted linear semantics* $[\phi]^{\mathrm{dlin}_{S,\eta}} : S^\omega \mapsto [0,1]$ by $[\phi]^{\mathrm{dlin}_{S,\eta}}(\sigma) = \llbracket \phi \rrbracket^{\mathcal{I}^\eta_\sigma}(0)$. A *discounted property* is the mapping $S^\omega \mapsto [0,1]$ defined by the linear semantics $[\phi]^{\mathrm{dlin}_{S,\eta}}$ of a closed discounted μ-calculus formula $\phi \in \mathcal{DMC}_{S,\Upsilon}$.

As an example, consider again a subset $T \subseteq S$ of target states, and a player $i \in \{1,2\}$. The payoff of the discounted reachability game considered informally in Sect. 5.1 can be defined by $[\mu x.(T \vee \alpha pre_i(x))]^{\mathrm{dlin}_{S,\eta}}$: indeed,

$$[\mu x.(T \vee \alpha pre_i(x))]^{\mathrm{dlin}_{S,\eta}}(\sigma) = \eta(\alpha)^k,$$

where $k = \min\{j \in \mathbb{N} \mid \sigma_j \in T\}$. The fact that (24) represents the maximum payoff that player 1 can achieve in a game structure $\mathcal{G}$ can be formalized as

$$\langle 1 \rangle^{\mathrm{q}}_{\mathcal{G}} [\mu x.(T \vee \alpha pre_i(x))]^{\mathrm{dlin}_{S,\eta}} = \llbracket \mu x.(T \vee \alpha pre_i(x)) \rrbracket^{\mathrm{d}}_{\mathcal{G},\eta}. \tag{27}$$

Similarly, the payoff of the discounted safety game considered informally in Sect. 5.1 can be defined by $[\nu x.(T \wedge (1-\alpha) + \alpha pre_i(x))]^{\mathrm{dlin}_{S,\eta}}$: indeed,

$$[\nu x.(T \wedge (1-\alpha) + \alpha pre_i(x))]^{\mathrm{dlin}_{S,\eta}}(\sigma) = 1 - \eta(\alpha)^k,$$

where $k = \min\{j \in \mathbb{N} \mid \sigma_j \notin T\}$. Also in this case, for all game structures $\mathcal{G}$ we have:

$$\langle 1 \rangle^{\mathrm{q}}_{\mathcal{G}} [\nu x.(T \wedge (1-\alpha) + \alpha pre_i(x))]^{\mathrm{dlin}_{S,\eta}} = \llbracket \nu x.(T \wedge (1-\alpha) + \alpha pre_i(x)) \rrbracket^{\mathrm{d}}_{\mathcal{G},\eta}. \tag{28}$$

The relations (27) and (28) are just two special cases of the general relation between the linear and branching semantics of discounted μ-calculus, expressed by the following theorem.

Theorem 6. *For all game structures $\mathcal{G}$, all players $i \in \{1,2\}$, all sets Υ of discount factors, all discount factor evaluations $\eta \in (\Upsilon \mapsto [0,1])$, and all closed μ-calculus formulas $\phi \in \mathcal{DMC}_{S,\Upsilon}$ in strongly deterministic form that contain only the function symbols αpre_i and $(1-\alpha) + \alpha pre_i$ for $\alpha \in \Upsilon$, we have that $\llbracket \phi \rrbracket^{\mathrm{d}}_{\mathcal{G},\eta} = \langle i \rangle^{\mathrm{q}}_{\mathcal{G}} [\phi]^{\mathrm{dlin}_{S,\eta}}$.*

This theorem is the main result about the verification of discounted properties, as it relates a discounted property $[\phi]^{\mathrm{dlin}_{S,\eta}}$ to the valuation $\llbracket \phi \rrbracket^{\mathrm{d}}_{\mathcal{G},\eta}$ computed by the verification algorithm ϕ over the game structure $\mathcal{G}$.

Determinacy. Since discounted properties are defines as the linear semantics of discounted μ-calculus formulas, the duality of discounted control problems can be stated as follows.

Theorem 7. *For all game structures $\mathcal{G}$, all sets Υ of discount factors, all discount factor evaluations $\eta \in (\Upsilon \mapsto [0,1])$, and all closed μ-calculus formulas $\phi \in \mathcal{DMC}_{S_\mathcal{G},\Upsilon}$ in strongly deterministic form that contain only the function symbols αpre_1 and $(1-\alpha)+\alpha pre_1$ for $\alpha \in \Upsilon$, we have that*

$$1 - \langle 1 \rangle^{\mathrm{q}}_{\mathcal{G}} [\phi]^{\mathrm{dlins}_{\mathcal{G},\eta}} = \langle 2 \rangle^{\mathcal{G}}_{q} [\neg\phi]^{\mathrm{dlins}_{\mathcal{G},\eta}}.$$

5.3 Relation between Discounted and Probabilistic μ-Calculus

Given $r \in [0,1]$, denote by $E_r(\Upsilon) : \Upsilon \mapsto [0,r]$ the set of all discount factor interpretations bound by r. If $\eta \in E_r(\Upsilon)$ for $r < 1$, we say that η is *contractive.* A fixpoint quantifier μx or νx occurs *guarded* in a formula ϕ if a function symbol *pre* occurs on every syntactic path from the quantifier to a quantified occurrence of the variable x. For example, in the formula $\mu x.(T \vee \alpha pre_i(x))$ the fixpoint quantifier occurs guarded; in the formula $(1-\alpha)+\alpha pre_i(\mu x.(T \vee x))$ it does not. Under a contractive discount factor interpretation, every guarded occurrence of a fixpoint quantifier defines a contractive operator on the values of the free variables that are in the scope of the quantifier. Hence, by the Banach fixpoint theorem, the fixpoint is unique. In such cases, we need not distinguish between μ and ν quantifiers, and we denote both by κ.

If $\eta(\alpha) = 1$, then both $[\![\alpha pre_i]\!]^{\mathrm{d}}_{\mathcal{G},\eta}$ and $[\![(1-\alpha)+\alpha pre_i]\!]^{\mathrm{d}}_{\mathcal{G},\eta}$ reduce to the undiscounted function $[\![pre_i]\!]^{\mathrm{p}}_{\mathcal{G}}$, for $i \in \{1,2\}$. The following theorem extends this observation to the complete μ-calculus, showing how the semantics of the discounted μ-calculus converges to the semantics of the undiscounted μ-calculus as the discount factors approach 1. To state the result, we extend the semantics of discounted μ-calculus to interpret also the functions pre_1, pre_2, letting $[\![pre_i]\!]^{\mathrm{d}}_{\mathcal{G},\eta} = [\![pre_i]\!]^{\mathrm{p}}_{\mathcal{G}}$ for all $i \in \{1,2\}$, game structures $\mathcal{G}$, and all discount interpretations η. We also let $\eta[\alpha := a]$ be the discount factor interpretation defined by $\eta[\alpha := a](\alpha) = a$ and $\eta[\alpha := a](\alpha') = \eta(\alpha')$ for $\alpha \neq \alpha'$.

Theorem 8. [dAHM03] *For all game structures $\mathcal{G}$, Let $\phi(x) \in \mathcal{DMC}_{S_\mathcal{G},\Upsilon}$ be a μ-calculus formula with free variable x, and discount factor α. The following assertions hold:*

1. *If x and α always and only occur in the context $\alpha pre_i(x)$, for $i \in \{1,2\}$, then*

$$\lim_{a \to 1} [\![\lambda x.\phi(\alpha pre_i(x))]\!]^{\mathrm{d}}_{\mathcal{G},\eta[\alpha:=a],e} = [\![\mu x.\phi(pre_i(x))]\!]^{\mathrm{d}}_{\mathcal{G},\eta,e}.$$

2. *If x and α always and only occur in the context $(1-\alpha)+\alpha pre_i(x)$, then*

$$\lim_{a \to 1} [\![\lambda x.\phi((1-\alpha)+\alpha pre_i(x))]\!]^{\mathrm{d}}_{\mathcal{G},\eta[\alpha:=a],e} = [\![\nu x.\phi(pre_i(x))]\!]^{\mathrm{d}}_{\mathcal{G},\eta,e}.$$

The remarkable fact is that the order of quantifiers in probabilistic μ-calculus corresponds to the order in which the limits are taken in discounted μ-calculus. For instance, for a game structure $\mathcal{G}$ and $T \subseteq S_{\mathcal{G}}$, let $\phi = \lambda y.\lambda x.((\neg T \wedge \alpha pre_i(x)) \vee (T \wedge (1-\beta) + \beta pre_i(y)))$. We have that

$$\lim_{a \to 1} \lim_{b \to 1} [\![\phi]\!]^{\mathrm{d}}_{\mathcal{G},\eta[\alpha:=a,\beta:=b]} = [\![\mu x.\nu y.((\neg T \wedge pre_i(x)) \vee (T \wedge pre_i(y)))]\!]^{\mathrm{P}}_{\mathcal{G}} \quad (29)$$

$$\lim_{b \to 1} \lim_{a \to 1} [\![\phi]\!]^{\mathrm{d}}_{\mathcal{G},\eta[\alpha:=a,\beta:=b]} = [\![\nu y.\mu x.((\neg T \wedge pre_i(x)) \vee (T \wedge pre_i(y)))]\!]^{\mathrm{P}}_{\mathcal{G}}. \quad (30)$$

Formula (29) is the solution of probabilistic co-Büchi games with goal $\Diamond\Box T$, while (30) is the solution of probabilistic Büchi games with goal $\Box\Diamond T$.

6 Equivalence Metrics

To complete the extension of the classical boolean framework for specification, verification, and control to the quantitative case, we show how the classical notion of bisimulation can be extended to the quantitative setting, and how our quantitative μ-calculi characterize quantitative bisimulation, just as the boolean μ-calculus, like CTL, characterizes bisimulation.

6.1 Alternating Bisimulation

In the boolean setting, and for deterministic game structures, the notion of bisimulation for games is called *alternating simulation* [AHKV98]. Fix a deterministic game structure $\mathcal{G} = \langle S, \mathcal{M}, \Gamma_1, \Gamma_2, \delta \rangle$, along with a set $\mathcal{P} \subseteq 2^S$ of predicates. A relation $R \subseteq S \times S$ is a *player-1 alternating bisimulation* if, for all $s, t \in S$, $(s,t) \in R$ implies that $s \in p \leftrightarrow t \in p$ for all $p \in \mathcal{P}$, and if $(s,t) \in R$, then

$$\forall a_1 \in \Gamma_1(s).\exists b_1 \in \Gamma_1(t).\forall b_2 \in \Gamma_2(t).\exists a_2 \in \Gamma_2(s).\hat{R}(\tau(s,a_1,a_2), \tau(s,b_1,b_2)),$$

$$\forall b_1 \in \Gamma_1(t).\exists a_1 \in \Gamma_1(s).\forall a_2 \in \Gamma_2(s).\exists b_2 \in \Gamma_2(t).\hat{R}(\tau(s,a_1,a_2), \tau(s,b_1,b_2)),$$

where $\hat{R}(\{t_1\}, \{t_2\})$ iff $(t_1, t_2) \in R$, for all $t_1, t_2 \in S$. The definition of a *player-2 alternating bisimulation* is obtained by exchanging in the above definition the roles of players 1 and 2. A relation R is an *alternating bisimulation* if it is both a player-1 and a player-2 alternating bisimulation.

To obtain the coarsest player 1 alternating bisimulation, i.e., the largest relation that is a player-1 alternating bisimulation, we can use a symbolic fixpoint approach [Mil90], which in view of our extension to the quantitative case, we state as follows. A *binary distance function* is a function $d : S \times S \mapsto \{0,1\}$ that maps each pair of states $s, t \in S$ to their distance $d(s,t) \in \{0,1\}$, and such that for all $s, t, u \in S$, we have $d(s,t) = d(t,s)$ and $d(s,t) \leq d(s,u) + d(u,t)$. For distance functions d, d' we let $d \leq d'$ iff $d(s,t) \leq d'(s,t)$ for all $s, t \in S$. We define the functor F_1 mapping binary distance functions to binary distance functions: for all binary distance functions d and all $s, t \in S$, we let $F_1(d)(s,t) = 1$ if there is $p \in \mathcal{P}$ such that $s \in p \not\leftrightarrow t \in p$, and we let

$$F_1(d)(s,t) = \max \left\{ \begin{array}{l} \max\limits_{a_1 \in \Gamma_1(s)} \min\limits_{b_1 \in \Gamma_1(t)} \max\limits_{b_2 \in \Gamma_2(t)} \min\limits_{a_2 \in \Gamma_2(s)} \hat{d}(\tau(s,a_1,a_2), \tau(s,b_1,b_2)), \\ \max\limits_{b_1 \in \Gamma_1(t)} \min\limits_{a_1 \in \Gamma_1(s)} \max\limits_{a_2 \in \Gamma_2(s)} \min\limits_{b_2 \in \Gamma_2(t)} \hat{d}(\tau(s,a_1,a_2), \tau(s,b_1,b_2)) \end{array} \right\}$$

otherwise, where $\hat{d}(\{t_1\},\{t_2\}) = d(t_1,t_2)$ for all $t_1, t_2 \in S$. A player-1 alternating bisimulation R is simply a relation whose characteristic function is a fixpoint of F_1, i.e., it is a subset $R \subseteq S \times S$ such that $\chi(R) = F_1(\chi(R))$, where $\chi(R)(s,t)$ is 0 if $(s,t) \in R$, and is 1 otherwise. In particular, the coarsest player-1 bisimulation B_1^{bin} is given by $d_1^* = \chi(B_1^{\mathrm{bin}})$, where d_1^* is least fixpoint of the functor F_1, i.e., the least distance function that satisfies $d_1^* = F_1(d_1^*)$. We define the coarsest player-2 alternating bisimulation B_2^{bin} in an analogous fashion, with respect to a functor F_2 obtained by swapping the roles of players 1 and 2 in the definition of F_1. Finally, the coarsest alternating bisimulation B^{bin} is given by $\chi(B^{\mathrm{bin}}) = d^*$, where d^* is the least distance function that satisfies both $d^* = F_1(d^*)$ and $d^* = F_2(d^*)$. When we wish to make explicit the dependence of the bisimulation relations on the game and on $\mathcal{P}$, we write $B_{\mathcal{G},\mathcal{P}}^{\mathrm{bin}}$, $B_{1,\mathcal{G},\mathcal{P}}^{\mathrm{bin}}$, and $B_{2,\mathcal{G},\mathcal{P}}^{\mathrm{bin}}$ for B^{bin}, B_1^{bin}, and B_2^{bin}. The following theorem, derived from [AHKV98], relates alternating bisimulation and boolean μ-calculus.

Theorem 9. *For a deterministic game structure $\mathcal{G}$, the following assertions hold:*

1. *For all $i \in \{1,2\}$, we have that $(s,t) \notin B_{i,\mathcal{G},\mathcal{P}}^{\mathrm{bin}}$ iff there is a closed μ-calculus formula $\phi \in \mathcal{BMC}_{S_\mathcal{G}}$ containing only predicates in $\mathcal{P}$ and functions in $\{pre_i, dpre_i\}$ such that $s \in [\![\phi]\!]_\mathcal{G}^{\mathrm{b}}$ and $t \notin [\![\phi]\!]_\mathcal{G}^{\mathrm{b}}$.*
2. *$(s,t) \notin B_{\mathcal{G},\mathcal{P}}^{\mathrm{bin}}$ iff there is a closed μ-calculus formula $\phi \in \mathcal{BMC}_{S_\mathcal{G}}$ containing only predicates in $\mathcal{P}$ such that $s \in [\![\phi]\!]_\mathcal{G}^{\mathrm{b}}$ and $t \notin [\![\phi]\!]_\mathcal{G}^{\mathrm{b}}$.*

6.2 Game Bisimulation Distance

To obtain a quantitative version of alternating bisimulation, we adapt the definition of F_i to the case of probabilistic game structures and quantitative distance functions [dAHM03]. Fix a game structure $\mathcal{G} = \langle S, \mathcal{M}, \Gamma_1, \Gamma_2, \delta \rangle$, along with a set $\mathcal{P} \subseteq 2^S$ of predicates. A *distance function* is a mapping $d : S \times S \mapsto [0,1]$ such that for all $s,t,u \in S$ we have $d(s,t) = d(t,s)$ and $d(s,t) \leq d(s,u) + d(u,t)$. We define *discounted game bisimulation* [dAHM03] with respect to a discount factor $r \in [0,1]$; the undiscounted case corresponds to $r = 1$. Given $r \in [0,1]$, we define the functor G_r mapping distance functions to distance functions: for every distance function d and all states $s,t \in S$, we define $G_r(d)(s,t) = 1$ if there is $p \in \mathcal{P}$ such that $s \in p \not\leftrightarrow t \in p$, and

$$G_r(d)(s,t) =$$

$$r \cdot \max \left\{ \begin{array}{l} \sup\limits_{\zeta_1 \in \mathcal{D}_1(s)} \inf\limits_{\xi_1 \in \mathcal{D}_1(t)} \sup\limits_{\xi_2 \in \mathcal{D}_2(t)} \inf\limits_{\zeta_2 \in \mathcal{D}_2(s)} D(d)(\hat{\delta}(s,\zeta_1,\zeta_2), \hat{\delta}(t,\xi_1,\xi_2)), \\ \sup\limits_{\xi_1 \in \mathcal{D}_1(t)} \inf\limits_{\zeta_1 \in \mathcal{D}_1(s)} \sup\limits_{\zeta_2 \in \mathcal{D}_2(s)} \inf\limits_{\xi_2 \in \mathcal{D}_2(t)} D(d)(\hat{\delta}(s,\zeta_1,\zeta_2), \hat{\delta}(t,\xi_1,\xi_2)) \end{array} \right\}$$

otherwise. For a distance function d and distributions ζ_1 and ζ_2, we let $D(d)(\zeta_1, \zeta_2)$ be the extension of the function d from states to distributions [vBW01a] given by the solution to the linear program $\max \sum_{s \in Q} (\zeta_1(s) - \zeta_2(s)) k_s$ where the variables $\{k_s\}_{s \in Q}$ are subject to $k_s - k_t \leq d(s,t)$ for all $s, t \in Q$. The least distance function that is a fixpoint of G_r is called *r-discounted game bisimilarity,* and denoted B_r^{disc}. On MDPs (one-player game structures), for $r < 1$, discounted game bisimulation coincides with the discounted distance metrics of [vBW01a]. Again, we write $B_{r,\mathcal{G},\mathcal{P}}^{\mathrm{disc}}$ when we wish to make explicit the dependency of B_r^{disc} from the game $\mathcal{G}$ and from the subset of predicates $\mathcal{P}$.

By the minimax theorem [vN28], we can exchange the two middle sup and inf operators in the definition of G_r; as a consequence, it is easy to see that the definition is symmetrical with respect to players 1 and 2. Thus, there is only one version of (un)discounted game bisimulation, in contrast to the two distinct player-1 and player-2 alternating bisimulations. Indeed, comparing the definition of F_i and G_r, we see that alternating bisimulation is defined with respect to *deterministic* move distributions, and the minimax theorem does not hold if the players are forced to use deterministic distributions. The following theorem relates game bisimilarity with quantitative and discounted μ-calculus.

Theorem 10. [dAHM03] *The following assertions hold for all game structures $\mathcal{G}$.*

1. *Let $\mathcal{PMC}_{S_\mathcal{G},\mathcal{P}}$ be the set of closed μ-calculus formulas in $\mathcal{PMC}_{S_\mathcal{G}}$ that contain only predicates in $\mathcal{P}$. For all $s, t \in S_\mathcal{G}$ we have*

$$B_{1,\mathcal{G},\mathcal{P}}^{\mathrm{disc}}(s,t) = \sup_{\phi \in \mathcal{PMC}_{S_\mathcal{G},\mathcal{P}}} \left| [\![\phi]\!]_\mathcal{G}^{\mathrm{p}}(s) - [\![\phi]\!]_\mathcal{G}^{\mathrm{p}}(t) \right|.$$

2. *Let $\mathcal{DMC}_{S_\mathcal{G},\Upsilon,\mathcal{P}}$ be the set of closed μ-calculus formulas in $\mathcal{DMC}_{S_\mathcal{G},\Upsilon}$ that contain only predicates in $\mathcal{P}$. For all $s, t \in S_\mathcal{G}$ and all $r \in [0,1]$, we have*

$$B_{r,\mathcal{G},\mathcal{P}}^{\mathrm{disc}}(s,t) = \sup_{\phi \in \mathcal{DMC}_{S_\mathcal{G},\mathcal{P}}} \sup_{\eta \in E_r(\Upsilon)} \left| [\![\phi]\!]_{\mathcal{G},\eta}^{\mathrm{d}}(s) - [\![\phi]\!]_{\mathcal{G},\eta}^{\mathrm{d}}(t) \right|.$$

It is possible to extend the connection between discounted μ-calculus and equivalence relations further, including results about the stability of bisimulation and discounted μ-calculus with respect to perturbations in the game structure [DGJP02,dAHM03].

References

[AHK02] R. Alur, T.A. Henzinger, and O. Kupferman. Alternating time temporal logic. *J. ACM*, 49:672–713, 2002.

[AHKV98] R. Alur, T.A. Henzinger, O. Kupferman, and M.Y. Vardi. Alternating refinement relations. In *CONCUR 97: Concurrency Theory. 8th Int. Conf.*, volume 1466 of *Lect. Notes in Comp. Sci.*, pages 163–178. Springer-Verlag, 1998.

[BC96] G. Bhat and R. Cleaveland. Efficient model checking via the equational μ-calculus. In *Proc. 11th IEEE Symp. Logic in Comp. Sci.*, pages 304–312, 1996.

[Ber95] D.P. Bertsekas. *Dynamic Programming and Optimal Control.* Athena Scientific, 1995. Volumes I and II.

[BL69] J.R. Büchi and L.H. Landweber. Solving sequential conditions by finite-state strategies. *Trans. Amer. Math. Soc.*, 138:295–311, 1969.

[Bry86] R.E. Bryant. Graph-based algorithms for boolean function manipulation. *IEEE Transactions on Computers*, C-35(8):677–691, 1986.

[Con92] A. Condon. The complexity of stochastic games. *Information and Computation*, 96:203–224, 1992.

[dAH00] L. de Alfaro and T.A. Henzinger. Concurrent omega-regular games. In *Proc. 15th IEEE Symp. Logic in Comp. Sci.*, pages 141–154, 2000.

[dAHK98] L. de Alfaro, T.A. Henzinger, and O. Kupferman. Concurrent reachability games. In *Proc. 39th IEEE Symp. Found. of Comp. Sci.*, pages 564–575. IEEE Computer Society Press, 1998.

[dAHM01a] L. de Alfaro, T.A. Henzinger, and R. Majumdar. From verification to control: Dynamic programs for omega-regular objectives. In *Proc. 16th IEEE Symp. Logic in Comp. Sci.*, pages 279–290. IEEE Press, 2001.

[dAHM01b] L. de Alfaro, T.A. Henzinger, and R. Majumdar. Symbolic algorithms for infinite-state games. In *CONCUR 01: Concurrency Theory. 12th Int. Conf.*, volume 2154 of *Lect. Notes in Comp. Sci.*, pages 536–550. Springer-Verlag, 2001.

[dAHM03] L. de Alfaro, T.A. Henzinger, and R. Majumdar. Discounting the future in systems theory. In *Proc. 30th Int. Colloq. Aut. Lang. Prog.*, Lect. Notes in Comp. Sci. Springer-Verlag, 2003.

[dAM01] L. de Alfaro and R. Majumdar. Quantitative solution of omega-regular games. In *Proc. 33rd ACM Symp. Theory of Comp.*, pages 675–683. ACM Press, 2001.

[DEP02] J. Desharnais, A. Edalat, and P. Panangaden. Bisimulation for labelled markov processes. *Information and Computation*, 179(2):163–193, 2002.

[Der70] C. Derman. *Finite State Markovian Decision Processes.* Academic Press, 1970.

[DGJP99] J. Desharnais, V. Gupta, R. Jagadeesan, and P. Panangaden. Metrics for labelled markov systems. In *CONCUR'99: Concurrency Theory. 10th Int. Conf.*, volume 1664 of *Lect. Notes in Comp. Sci.*, pages 258–273. Springer, 1999.

[DGJP02] J. Desharnais, V. Gupta, R. Jagadeesan, and P. Panangaden. The metric analogue of weak bisimulation for probabilistic processes. In *Proc. 17th IEEE Symp. Logic in Comp. Sci.*, pages 413–422, 2002.

[EJ91] E.A. Emerson and C.S. Jutla. Tree automata, mu-calculus and determinacy (extended abstract). In *Proc. 32nd IEEE Symp. Found. of Comp. Sci.*, pages 368–377. IEEE Computer Society Press, 1991.

[EL86] E.A. Emerson and C.L. Lei. Efficient model checking in fragments of the propositional μ-calculus. In *Proc. First IEEE Symp. Logic in Comp. Sci.*, pages 267–278, 1986.

[Eve57] H. Everett. Recursive games. In *Contributions to the Theory of Games III*, volume 39 of *Annals of Mathematical Studies*, pages 47–78, 1957.

[Fel83] Y.A. Feldman. A decidable propositional probabilistic dynamic logic. In *Proc. 15th ACM Symp. Theory of Comp.*, pages 298–309, 1983.

[FH82] Y.A. Feldman and D. Harel. A probabilistic dynamic logic. In *Proc. 14th ACM Symp. Theory of Comp.*, pages 181–195, 1982.

[FV97] J. Filar and K. Vrieze. *Competitive Markov Decision Processes.* Springer-Verlag, 1997.

[HK97] M. Huth and M. Kwiatkowska. Quantitative analysis and model checking. In *Proc. 12th IEEE Symp. Logic in Comp. Sci.*, pages 111–122, 1997.

[HM00] T.A. Henzinger and R. Majumdar. A classification of symbolic transition systems. In *Proc. of 17th Annual Symp. on Theor. Asp. of Comp. Sci.*, volume 1770 of *Lect. Notes in Comp. Sci.*, pages 13–34. Springer-Verlag, 2000.

[Koz83a] D. Kozen. A probabilistic PDL. In *Proc. 15th ACM Symp. Theory of Comp.*, pages 291–297, 1983.

[Koz83b] D. Kozen. Results on the propositional μ-calculus. *Theoretical Computer Science*, 27(3):333–354, 1983.

[KSK66] J.G. Kemeny, J.L. Snell, and A.W. Knapp. *Denumerable Markov Chains.* D. Van Nostrand Company, 1966.

[Mar98] D.A. Martin. The determinacy of Blackwell games. *The Journal of Symbolic Logic*, 63(4):1565–1581, 1998.

[Mil90] R. Milner. Operational and algebraic semantics of concurrent processes. In J. van Leeuwen, editor, *Handbook of Theoretical Computer Science*, volume B, pages 1202–1242. Elsevier Science Publishers (North-Holland), Amsterdam, 1990.

[MP91] Z. Manna and A. Pnueli. *The Temporal Logic of Reactive and Concurrent Systems: Specification.* Springer-Verlag, New York, 1991.

[Rei80] J.H. Reif. Logic for probabilistic programming. In *Proc. 12th ACM Symp. Theory of Comp.*, pages 8–13, 1980.

[Ros90] K.I. Rosenthal. *Quantales and Their Applications*, volume 234 of *Pitman Research Notes in Mathematics Series.* Longman Scientific & Technical, Harlow, 1990.

[Sha53] L.S. Shapley. Stochastic games. *Proc. Nat. Acad. Sci. USA*, 39:1095–1100, 1953.

[Tho90] W. Thomas. Automata on infinite objects. In J. van Leeuwen, editor, *Handbook of Theoretical Computer Science*, volume B, chapter 4, pages 135–191. Elsevier Science Publishers (North-Holland), Amsterdam, 1990.

[Tho95] W. Thomas. On the synthesis of strategies in infinite games. In *Proc. of 12th Annual Symp. on Theor. Asp. of Comp. Sci.*, volume 900 of *Lect. Notes in Comp. Sci.*, pages 1–13. Springer-Verlag, 1995.

[vBW01a] F. van Breugel and J. Worrel. An algorithm for quantitative verification of probabilistic transition systems. In *CONCUR 01: Concurrency Theory. 12th Int. Conf.*, volume 2154 of *Lect. Notes in Comp. Sci.*, pages 336–350, 2001.

[vBW01b] F. van Breugel and J. Worrel. Towards quantitative verification of probabilistic systems. In *Proc. 28th Int. Colloq. Aut. Lang. Prog.*, volume 2076 of *Lect. Notes in Comp. Sci.*, pages 421–432. Springer-Verlag, 2001.

[vN28] J. von Neumann. Zur Theorie der Gesellschaftsspiele. *Math. Annal*, 100:295–320, 1928.

Playing Games with Boxes and Diamonds* **

Rajeev Alur[1], Salvatore La Torre[2], and P. Madhusudan[1]

[1] University of Pennsylvania
[2] Università degli Studi di Salerno

Abstract. Deciding infinite two-player games on finite graphs with the winning condition specified by a linear temporal logic (LTL) formula, is known to be 2EXPTIME-complete. The previously known hardness proofs encode Turing machine computations using the *next* and/or *until* operators. Furthermore, in the case of model checking, disallowing next and until, and retaining only the *always* and *eventually* operators, lowers the complexity from PSPACE to NP. Whether such a reduction in complexity is possible for deciding games has been an open problem. In this paper, we provide a negative answer to this question. We introduce new techniques for encoding Turing machine computations using games, and show that deciding games for the LTL fragment with only the *always* and *eventually* operators is 2EXPTIME-hard. We also prove- that if in this fragment we do not allow the *eventually* operator in the scope of the *always* operator and vice-versa, deciding games is EXPSPACE-hard, matching the previously known upper bound. On the positive side, we show that if the winning condition is a Boolean combination of formulas of the form "eventually p" and "infinitely often p," for a state-formula p, then the game can be decided in PSPACE, and also establish a matching lower bound. Such conditions include safety and reachability specifications on game graphs augmented with fairness conditions for the two players.

1 Introduction

Linear temporal logic (LTL) is a specification language for writing correctness requirements of reactive systems [13,11], and is used by verification tools such as SPIN [8]. The most studied decision problem concerning LTL is *model checking*: given a finite-state abstraction G of a reactive system and an LTL formula φ, do all infinite computations of G satisfy φ? The corresponding *synthesis* question is: given a game graph G whose states are partitioned into system states and environment states, and an LTL formula φ, consider the infinite game in which the protagonist chooses the successor in all system states and the adversary chooses the successor in all environment states; then, does the the protagonist

* Detailed proofs are available at http://www.cis.upenn.edu/ madhusud/

** Supported in part by ARO URI award DAAD19-01-1-0473, NSF awards CCR9970925 and ITR/SY 0121431. The second author was also supported by the MIUR grant project "Metodi Formali per la Sicurezza e il Tempo" (MEFISTO) and MIUR grant ex-60% 2002.

R. Amadio, D. Lugiez (Eds.): CONCUR 2003, LNCS 2761, pp. 128–143, 2003.

have a strategy to ensure that all the resulting computations satisfy φ? Such a game-based interpretation for LTL is useful in many contexts: for synthesizing controllers from specifications [14], for formalizing compositionality requirements such as *realizability* [1] and *receptiveness* [6], for specification and verification of open systems [3], for modular verification [10], and for construction of the most-general environments for automating assume-guarantee reasoning [2]. In the contexts of open systems and modular verification, this game is played in the setting where a module is considered as the protagonist player, and its environment, which may consist of other concurrent modules in the system that interact with this module, is taken as the adversary.

An LTL formula is built from state predicates (Π), Boolean connectives, and temporal operators such as *next*, *eventually*, *always*, and *until*. While the model checking problem for the full LTL is known to be PSPACE-complete, the fragment $L_{\Box,\Diamond,\wedge,\vee}(\Pi)$ that allows only *eventually* and *always* operators (but no *next* or *until*), has a small model property with NP-complete model checking problem [15]. Deciding games for the full LTL is known to be 2EXPTIME-complete [14]. The hardness proof, like many lower bound proofs for LTL, employs the until/next operators in a critical way to relate successive configurations. This raises the hope that deciding games for $L_{\Box,\Diamond,\wedge,\vee}(\Pi)$ has a lower complexity than the full LTL. In this paper, we provide a negative answer to this question by proving a 2EXPTIME lower bound.

The proof of 2EXPTIME-hardness is by reduction of the halting problem for alternating exponential-space Turing machines to deciding games with winning condition specified by formulas that use only the always and eventually operators. The reduction introduces some new techniques for counting and encoding configurations in game graphs and formulas. We believe that these techniques are of independent interest. Using these techniques we show another hardness result: deciding games for the fragment $\mathcal{B}(L_{\Diamond,\wedge,\vee}(\Pi))$ is EXPSPACE-hard. This fragment contains top-level Boolean combinations of formulas from $L_{\Diamond,\wedge,\vee}(\Pi)$, the logic of formulas built from state predicates, conjunctions, disjunctions, and eventually operators. $\mathcal{B}(L_{\Diamond,\wedge,\vee}(\Pi))$ is known to be in EXPSPACE [4], while $L_{\Diamond,\wedge,\vee}(\Pi)$ is known to be in PSPACE [12], so our result closes this complexity gap.

Finally, we consider the fragment $\mathcal{B}(L_{\Box\Diamond}(\Pi))$ that contains Boolean combinations of formulas of the form $\Box\Diamond\, p$, where p is a state predicate. Complexity for formulas of specific form in this class is well-known: generalized Büchi games (formulas of the form $\wedge_i \Box\Diamond p_i$) are solvable in polynomial time, and Streett games ($\wedge_i(\Box\Diamond p_i \rightarrow \Box\Diamond q_i)$) are CO-NP-complete (the dual, Rabin games are NP-complete) [7]. We show that the Zielonka-tree representation of the winning sets of vertices [17] can be exploited to get a PSPACE-procedure to solve the games for the fragment $\mathcal{B}(L_{\Box\Diamond}(\Pi))$. This logic is of relevance in modeling *fairness* assumptions about components of a reactive system. A typical fairness requirement such as "if a choice is enabled infinitely often then it must be taken infinitely often," corresponds to a Streett condition [11]. Such conditions are common in the context of concurrent systems where it is used to capture the fairness in the scheduling of processes. In games with fairness constraints, the winning condition is modified to "if the adversary satisfies all the fairness constraints then the protagonist satisfies its fairness constraints and meets the specification" [3]. Thus, adding fairness changes the winning conditions for specifications from a logic $\mathcal{L}$ to Boolean combinations of $L_{\Box\Diamond}(\Pi)$ and $\mathcal{L}$ formulas. We show that the PSPACE upper bound holds for fair games for specifications in $\mathcal{B}(L_{\Diamond,\wedge}(\Pi))$ containing Boolean combinations of formulas that are built from state predicates, conjunctions, and eventually operators, and can specify combinations of invariant and termination properties. This result has been used to show that the model checking of the game-based temporal logic Alternating Temporal Logic is in PSPACE under the strong fairness requirements [3]. We conclude by showing that deciding games for formulas of the form "Streett implies Streett" (that is, $\bigwedge_i(\Box\Diamond p_i \rightarrow \Box\Diamond q_i) \rightarrow \bigwedge_j(\Box\Diamond r_j \rightarrow \Box\Diamond s_j)$) is PSPACE-hard.

2 LTL Fragments and Game Graphs

2.1 Linear Temporal Logic

We first recall the syntax and the semantics of linear temporal logic. Let $\mathcal{P}$ be a set of propositions. Then the set of *state predicates* Π is the set of boolean formulas over $\mathcal{P}$. We define temporal logics by assuming that the atomic formulas are state predicates. A *linear temporal logic* (LTL) formula is composed of state predicates (Π), the Boolean connectives *negation* ($\neg$), *conjunction* ($\wedge$) and *disjunction* ($\vee$), the temporal operators *next* ($\bigcirc$), *eventually* ($\Diamond$), *always* ($\Box$), and *until* ($\mathcal{U}$). Formulas are built up in the usual way from these operators and connectives, according to the grammar

$$\varphi := p \,|\, \neg\varphi \,|\, \varphi \wedge \varphi \,|\, \varphi \vee \varphi \,|\, \bigcirc\varphi \,|\, \Diamond\varphi \,|\, \Box\varphi \,|\, \varphi\,\mathcal{U}\,\varphi,$$

where $p \in \Pi$. LTL formulas are interpreted on ω-words over $2^{\mathcal{P}}$ in the standard way [13].

In the rest of the paper we consider some fragments of LTL. For a set of LTL formulas Γ, we denote by $L_{op_1,\ldots,op_k}(\Gamma)$ the logic built from the formulas in Γ by using only the operators in the list $op_1, \ldots, op_k$. When the list of operators contains only the Boolean connectives we use the notation $\mathcal{B}(\Gamma)$, i.e., $\mathcal{B}(\Gamma) = L_{\neg,\wedge}(\Gamma)$. In particular, $\Pi = \mathcal{B}(\mathcal{P})$. As an example, the logic $L_{\Diamond,\wedge}(\Pi)$ is the one defined by the grammar $\varphi := p \,|\, \varphi \wedge \varphi \,|\, \Diamond\varphi$, where $p \in \Pi$, and the logic $\mathcal{B}(L_{\Diamond,\wedge}(\Pi))$ contains Boolean combinations of the formulas from $L_{\Diamond,\wedge}(\Pi)$.

2.2 LTL Games

A *game graph* is a tuple $G = (\Sigma, V, V_0, V_1, \gamma, \mu)$ where Σ is a finite set of labels, V is a finite set of vertices, V_0 and V_1 define a partition of V, $\gamma : V \rightarrow 2^V$ is a function giving for each vertex $u \in V$ the set of its *successors* in G, and $\mu : V \rightarrow \Sigma$ is a labeling function. For $i = 0, 1$, the vertices in V_i are those from which only player i can move and the allowed moves are given by γ. A *play* starting at x_0 is a sequence $x_0x_1\ldots$ in V^* or V^ω such that $x_j \in \gamma(x_{j-1})$, for every j. A *strategy* for player i is a total function $f : V^*V_i \rightarrow V$ mapping each finite play ending in V_i into V (it gives the moves of player i in any play ending in V_i). A play $x_0 x_1 \ldots$ is consistent with f if for all $x_j \in V_i$ with $j \geq 0$, $f(x_0 \ldots x_j) = x_{j+1}$.

In this paper, we focus on determining the existence of strategies for player 0. For this reason, player 0 is called the *protagonist*, while player 1 is the *adversary*. Unless specified otherwise, by 'strategy' we mean a strategy for the protagonist. Moreover, we consider game graphs along with winning conditions expressed by LTL formulas (LTL *games*). Formally, an LTL game is a triple (G, φ, u), where G is a game graph with vertices labeled with subsets of atomic propositions, φ is an LTL formula and u is a vertex of G. A strategy f in (G, φ, u) is winning if all infinite plays consistent with f and starting at u satisfy φ. The decision problem for an LTL game (G, φ, u) is to determine if there exists a winning strategy for the protagonist.

3 Lower Bound Results

When proving lower bounds for LTL games, the usual technique is to code the acceptance problem for alternating Turing machines [16,12]. The crux in such a proof is to detect, using the LTL specification, that the content of the ith cell in a configuration is in accordance with the $(i-1)^{th}$, i^{th} and $(i+1)^{th}$ cells of the previous configuration. In a reduction from 2EXPTIME (i.e. alternating EXPSPACE), typically, the cell numbers are explicitly encoded using a sub-word of bits in the configuration sequence; these numbers can be read by zooming into the ith cell in some configuration using the $\Diamond$ operator, using the $\bigcirc$ operator to read the cell numbers and using $\mathcal{U}$ operator to access the next configuration. Here the $\mathcal{U}$ operator can be used instead of $\bigcirc$, but the $\mathcal{U}$ cannot be replaced by $\Diamond$. In an EXPSPACE reduction (i.e. alternating EXPTIME), the configuration numbers are also encoded explicitly, and one can just use the $\Diamond$ and $\bigcirc$ operators to access three cells of a configuration and the appropriate position in the next configuration. Hence both proofs use the $\bigcirc$ operator and the 2EXPTIME reduction uses the $\mathcal{U}$ operator crucially.

The primary difficulty in the lower bounds we present is in dealing with reductions in the absence of the $\bigcirc$ and $\mathcal{U}$ operators. The $\Diamond$ operator can basically check only for subsequences and can hence "jump" arbitrarily far making it difficult to read the cell and configuration numbers. The main idea to solve this, which we call the *matching trick* below, is to introduce a "sandwiched" encoding of addresses which forces the reading of cell numbers to be contiguous. This then yields an EXPSPACE lower bound for $\mathcal{B}(L_{\Diamond,\wedge,\vee}(\Pi))$.

Then we consider $L_{\Box,\Diamond,\wedge,\vee}(\Pi)$ where one can nest $\Diamond$ and $\Box$ operators. Though this does not allow us to check the required property on an entire sequence of configurations, it allows us to check it for the *last two* configurations in a sequence. By giving the adversary the ability to stop the sequence of configurations at any point, we can ensure that the entire sequence satisfies the property, which leads to a 2EXPTIME lower bound for this fragment.

3.1 Alternating Turing Machines

An *alternating Turing machine* on words over an alphabet Σ is a Turing machine $M = (Q, Q_\exists, Q_\forall, q_{in}, q_f, \delta)$, where $Q_\exists$ and $Q_\forall$ are disjoint sets of respectively existential and universal states that form a partition of Q, q_{in} is the initial state, q_f is the final state and $\delta : Q \times \Sigma \times \{D_1, D_2\} \longrightarrow Q \times \Sigma \times \{L, R\}$. For each pair $(q, \sigma) \in Q \times \Sigma$, there are exactly two transitions that we denote respectively as the D_1-transition and the D_2-transition. Suppose q is the current state and the tape head is reading the symbol σ on cell i, if the d-transition $\delta(q, \sigma, d) = (q', \sigma', L)$ is taken, M writes σ' on cell i, enters state q' and moves the read head to the left (L) to cell $(i-1)$. A configuration of M is a word $\sigma_1 \ldots \sigma_{i-1}(q, \sigma_i) \ldots \sigma_n$ where $\sigma_1 \ldots \sigma_n$ is the content of the tape and where M is at state q with the tape head at cell i. The *initial configuration* contains the word w and the initial state. An *outcome* of M is a sequence of configurations, starting from the initial configuration, constructed as a play in the game where the $\exists$-player picks the next transition (i.e. D_1 or D_2) when the play is in a state of $Q_\exists$, and the $\forall$-player picks the next transition when the play is in a state of $Q_\forall$. A *computation* of M is a strategy of the $\exists$-player, and an input word w is accepted iff there exists a computation such that all plays according to it reach a configuration with state q_f. We recall that an alternating Turing machine is $g(n)$ time-bounded if it halts on all input words of length n within $g(n)$ steps, and $g(n)$ space-bounded if it halts on all input words of length n using at most $g(n)$ tape cells (see [9]). We also recall that the acceptance problem is EXPSPACE-complete for exponentially time-bounded alternating Turing machines, and is 2EXPTIME-complete for exponentially space-bounded alternating Turing machines [5].

3.2 The Matching Trick

Fix n (which we assume is a power of 2) and let $m = \log_2 n$. Let us fix a set of propositions $\{p_1^\top, \ldots, p_m^\top, p_1^\perp, \ldots p_m^\perp\}$ and another set $\{q_1^\top, \ldots, q_m^\top, q_1^\perp, \ldots q_m^\perp\}$. Let us also fix a finite alphabet Σ disjoint from these. The gadget we describe allows us to access the ith element of a sequence of n letters over Σ, using a formula which is polynomial in log n and which uses only the $\Diamond$ modality. Let $[i, j]$ denote the set of numbers from i to j, both inclusive. Let $[i]$ denote $[1, i]$.

Let $\boldsymbol{u} = u_m \ldots u_1$ denote a sequence of length m such that u_b is either $\{p_b^\top\}$ or $\{p_b^\perp\}$, for $b \in [m]$. Similarly, let $\boldsymbol{v} = v_1 \ldots v_m$ (note the reversal in indices) denote a sequence where each v_b is either $\{q_b^\top\}$ or $\{q_b^\perp\}$. We call these sequences u-addresses and v-addresses, respectively. A u-address $\boldsymbol{u} = u_m \ldots u_1$ is to be seen as the binary representation of a number from 0 to $n-1$: the proposition $p_b^\top$ belongs to u_b if the bth bit of the number is 1, otherwise $p_b^\perp$ belongs to u_b (u_m encodes the most-significant bit). Similarly, a v-address $\boldsymbol{v}$ also represents a number, but note that the representation is reversed with the most-significant bit v_m at the end of the sequence. For $i \in \{0, \ldots, n-1\}$, let $u[i]$ and $v[i]$ denote the u-address and v-address representing the number i. For example, if $m = 4$, $u[5] = \{p_4^\perp\} \cdot \{p_3^\top\} \cdot \{p_2^\perp\} \cdot \{p_1^\top\}$ and $v[5] = \{q_1^\top\} \cdot \{q_2^\perp\} \cdot \{q_3^\top\} \cdot \{q_4^\perp\}$.

We encode a letter $a \in \Sigma$ at a position $i \in [0, n-1]$ as the string $\langle a \rangle_i = u[i] \cdot a \cdot v[n-1-i]$, i.e., $\langle a \rangle_i$ has 'a' sandwiched between a u-address representing

i and a v-address representing $n-1-i$. Note that in $\langle a\rangle_i$, $v_b = \{q_b^\top\}$ iff $u_b = \{p_b^\perp\}$, for every $b \in [m]$. Now consider a sequence of such encodings in ascending order $\langle a_0\rangle_0 \cdot \langle a_1\rangle_1 \cdot \ldots \cdot \langle a_{n-1}\rangle_{n-1}$. We call such a sequence a *proper* sequence. Note that while the u-addresses run from 0 to $n-1$, the v-addresses run from $n-1$ to 0. Let w be a proper sequence, $a \in \Sigma$ and $i \in [0, n-1]$. We say that a *matches* i *in* w if $\langle a\rangle_i$ is a (not necessarily contiguous) subsequence of w. The main property of this encoding is that it allows us to check whether a letter a is encoded at a position i by simply checking whether $\langle a\rangle_i$ is a subsequence of the proper sequence w, i.e. by checking if a matches i in w. For example, when $m = 4$, consider $w = u[0]a_0v[15]\ldots u[15]a_{15}v[0]$. Let us consider for which letters $a \in \Sigma$, the string $u[5]av[10]$ is a subsequence of w. $\{p_3^\top\}$ is in $u[5]$ and the first place where it occurs in w is in the address $u[4]$. After this, the first place where $\{p_1^\top\}$ occurs in w is in $u[5]$. Hence the shortest prefix w' of w such that $u[5]$ is a subsequence of w is $u[0]a_0v[15]\ldots u[5]$. Similarly the shortest suffix w'' of w such that $v[10]$ is a subsequence of w'' is $v[10]\ldots u[15]a_{15}v[0]$. Hence if $u[5]av[10]$ is a subsequence of w, then $a = a_5$. The following lemma captures this:

Lemma 1. *Let* $w = \langle a_0\rangle_0 \cdot \langle a_1\rangle_1 \cdot \ldots \cdot \langle a_{n-1}\rangle_{n-1}$ *be a proper sequence,* $a \in \Sigma$ *and* $i \in [0, n-1]$. *Then,* a *matches* i *in* w *iff* $a = a_i$.

Finally, we show how to check whether a matches i in w using a formula in $\mathcal{B}(L_{\Diamond,\wedge,\vee}(\Pi))$. If $\beta_1, \ldots, \beta_k$ are state predicates, let $Seq(\beta_1, \ldots, \beta_k)$ stand for the formula $\Diamond(\beta_1 \wedge \Diamond(\beta_2 \wedge \ldots \Diamond(\beta_k)\ldots))$. Intuitively, this checks if there is a subsequence along which β_1 through β_k hold, in that order. Let $a \in \Sigma$ and $i \in [0, n-1]$. Let $x_m, \ldots, x_1$ be such that $x_b = true$ iff the bth bit in the binary representation of i is 1. Let $Same(p_b, x_b)$ stand for the formula $(p_b^\top \wedge x_b) \vee (p_b^\perp \wedge \neg x_b)$, which asserts that the value of p_b is the same as that of x_b, where $b \in [m]$. Similarly, let $Diff(q_b, x_b)$ be the formula $(q_b^\top \wedge \neg x_b) \vee (q_b^\perp \wedge x_b)$, which asserts that the value of q_b is the negation of x_b. With $Match(a, i)$ we denote the formula $Seq(Same(p_m, x_m), \ldots, Same(p_1, x_1), a, Diff(q_1, x_1), \ldots, Diff(q_m, x_m))$. It is then easy to see that for any proper sequence w, w satisfies $Match(a, i)$ iff $\langle a\rangle_i$ is a subsequence of w, i.e iff a matches i in w.

3.3 Lower Bound for $\mathcal{B}(L_{\Diamond,\wedge,\vee}(\Pi))$

We show in this section that deciding $\mathcal{B}(L_{\Diamond,\wedge,\vee}(\Pi))$ games is EXPSPACE-hard. The reduction is from the membership problem for alternating exponential-time Turing machines. We show that for such a Turing machine M, given an input word w, we can construct an instance of a game and a $\mathcal{B}(L_{\Diamond,\wedge,\vee}(\Pi))$ specification in polynomial time such that M accepts w if and only if the protagonist has a winning strategy in the game.

Let us fix an alternating exponential-time Turing machine $M = (\Sigma, Q, Q_\exists, Q_\forall, q_{in}, q_f, \delta)$ and an input word $w \in \Sigma^*$. Let us assume that M takes at most m units of time on w (m is exponential in $|w|$). The game we construct will be such that the protagonist generates sequences of configurations beginning with the initial configuration (with w on its tape). During the game, after an existential configuration the protagonist gets to choose the transition (i.e. D_1 or D_2) while at a universal configuration the adversary gets to pick the transition. The specification will demand that successive configurations do indeed correspond to moves of M. Hence strategies of the protagonist correspond to runs on w.

Let π be any play according to a computation. Then π is a sequence of configurations of length m and each configuration can be represented using a sequence of at most m symbols (since M cannot use space more than m). We record π by encoding each cell of each configuration by explicitly encoding the number of the configuration and the number of the cell within the configuration. Configurations are represented as strings over the alphabet $\Sigma' = \Sigma \cup (Q \times \Sigma)$, namely from the set $\Sigma^* \cdot (Q \times \Sigma) \cdot \Sigma^*$.

In order to describe the configuration number and the cell number, each of which ranges from 0 to $m-1$, we need k bits where $m^2 = 2^k$. Let us fix a set of p-bits $\{p_l^z \mid l \in [1,k], z \in \{\top, \bot\}\}$ and a set of q-bits $\{q_l^z \mid l \in [1,k], z \in \{\top, \bot\}\}$. We employ the matching trick using these p-bits, q-bits and Σ' (see Sect. 3.2 and recall the definitions).

For any $i \in [0, m^2-1]$, the $k/2$ less significant bits of i will represent the cell number and the $k/2$ more significant bits will represent the configuration number. Let $u'[conf, cell]$, where $conf, cell \in [0, m-1]$ denote the u-address $u[2^{k/2}.conf + cell]$. Hence $u'[conf, cell]$ encodes that the current configuration number is $conf$ and the current cell number is $cell$. Similarly, define $v'[conf, cell]$ as the v-address $v[2^{k/2}.conf + cell]$. A proper sequence hence is of the form:

$$u'[0,0]a_{(0,0)}v'[m-1,m-1] \ldots u'[0,m-1]a_{(0,m-1)}v'[m-1,0]\ u'[1,0]a_{(1,0)}v'[m-2,m-1] \\ \ldots\ldots\ldots\ldots u'[m-1,m-1]a_{(m-1,m-1)}v'[0,0].$$

The game graph we construct is composed of three parts: a main part and two sub-graphs G_1 and G_2. In the main part, the protagonists aim is to generate sequences of letters from Σ' sandwiched between a u-address and a v-address that form a proper sequence. The adversary's aim is to check that the current sequence is proper and conforms to the behavior of M. If the adversary claims that one of these is not true, the game moves to the subgraphs G_1 and G_2 where the adversary will have to provide witnesses to prove these claims.

The set of propositions we use includes the p-bits, q-bits and those in Σ', as well as a set of r-bits, s-bits, t-bits and e-bits $\{r_b^z, s_b^z, t_b^z, e_b^z \mid b \in [1,k], z \in \{\top, \bot\}\}$, and other new propositions $\{D_1, D_2, ok, obj_1, obj_2\}$. The game graph typically allows plays that look like:

$$\begin{array}{llll}
 & \nearrow obj_1 & \nearrow obj_1 & \nearrow obj_1 \\
u_0a_0v_0 & ok \ldots\ldots u_ya_yv_y & ok\ d\ u'_0a'_0v'_0 \ldots\ldots\ldots u_fa_fv_f & ok\ ok\ ok \ldots \\
 & & & \searrow obj_2
\end{array}$$

The central line above shows how a play normally proceeds. The protagonist generates sequences of triples consisting of a u-address, a letter in Σ' and a v-address. Note that each triple has $2k+1$ letters. After every such triple, the adversary gets a chance where it can continue the normal course by choosing ok or can generate an objection by choosing obj_1. Objection obj_1 leads the play to the subgraph G_1. Whenever the protagonist generates an address where the last $k/2$ bits of the u-address is 1 (denoted $u_y a_y v_y$ above), this denotes the end of a configuration, and the play reaches a protagonist state if the configuration generated was existential and an adversary state if it was universal. Accordingly, the protagonist or the adversary choose the next letter $d \in \{D_1, D_2\}$ which denotes the transition they wish to take from the current configuration.

At the end of the whole sequence, when a u-address with all its k bits set to 1 is generated (denoted $u_f a_f v_f$ above), the adversary, apart from being able to raise the first kind of objection, can also raise a second kind of objection by choosing the action obj_2 that leads the play to the sub-graph G_2. If the adversary instead chooses not to raise an objection, the game enters a state where the action *ok* occurs infinitely often and no other actions are permitted.

Note that the game graph does not ensure that the sequence generated is a proper sequence; in fact, it even allows triples of the form $u[i] \cdot a \cdot v[j]$ where $j \neq 2^k{-}1{-}i$. The objections obj_1 and obj_2 will take care of this and make sure that a proper sequence needs to be generated for the protagonist to win.

On the objection obj_1, the play moves to G_1 where the adversary claims that in the sequence generated thus far, there was a u-address for i which was not followed by a u-address for $i+1$, or there was a u-address for i which was not followed by a v-address for $2^k{-}1{-}i$. The adversary chooses as a witness a sequence of k r-bits $x_k \ldots x_1$, where $x_b = \{r_b^\top\}$ or $x_b = \{r_b^\perp\}$. This denotes the binary representation of a number $\bar{r}$, with $x_b = \{r_b^\top\}$ iff the bth bit in the binary representation of $\bar{r}$ is 1 (with x_k encoding the most significant bit). Next, the adversary chooses a sequence of k s-bits, encoding a number $\bar{s}$. The adversary should choose these sequences such that $\bar{s} = \bar{r} + 1$.

The fact that $\bar{s} = \bar{r} + 1$ can be checked using the formula $succ(\bar{r}, \bar{s})$:

$$\bigvee_{j=1}^{k} \left(\bigwedge_{h=1}^{j-1} (\Diamond r_h^\top \wedge \Diamond s_h^\perp) \wedge (\Diamond r_j^\perp \wedge \Diamond s_j^\top) \wedge \bigwedge_{h=j+1}^{k} ((\Diamond r_h^\top \wedge \Diamond s_h^\top) \vee (\Diamond r_h^\perp \wedge \Diamond s_h^\perp)) \right)$$

Let $same(p_i, r_i)$ stand for the formula $((p_i^\top \wedge \Diamond r_i^\top) \vee (p_i^\perp \wedge \Diamond r_i^\perp))$. Similarly define $same(p_i, s_i)$. Also, let $diff(q_i, r_i)$ stand for the formula $((q_i^\top \wedge \Diamond r_i^\perp) \vee (q_i^\perp \wedge \Diamond r_i^\top))$, which checks if the ith q-bit is the complement of the ith r-bit. The specification formula then has the following conjunct ψ_1:

$$\begin{aligned}
\psi_1 &= \Diamond obj_1 \rightarrow ([(succ(\bar{r}, \bar{s}) \wedge \varphi_1) \rightarrow \varphi_2] \wedge (\varphi_1' \rightarrow \varphi_2')) \;\; \textit{where} \\
\varphi_1 &= Seq(same(p_k, r_k), \ldots, same(p_1, r_1), (p_1^\top \vee p_1^\perp)) \\
\varphi_2 &= Seq(same(p_k, r_k), \ldots, same(p_1, r_1), same(p_k, s_k), \ldots, same(p_1, s_1)) \\
\varphi_1' &= Seq(same(p_k, r_k), \ldots, same(p_1, r_1)) \\
\varphi_2' &= Seq(same(p_k, r_k), \ldots, same(p_1, r_1), diff(q_1, r_1), \ldots, diff(q_k, r_k))
\end{aligned}$$

φ_1 says that there is a subsequence of p-bits matching $\bar{r}$ and after this matching there is a future point where some p-bit (and hence a u-address) is defined, i.e. $u[\bar{r}].p_1^\top$ or $u[\bar{r}].p_1^\perp$ is a subsequence of the play. φ_2 demands that there is a subsequence of p-bits that match $\bar{r}$ followed by a sequence of p-bits matching $\bar{s}$, i.e. $u[\bar{r}] \cdot u[\bar{s}]$ is a subsequence.

The formula φ_1' checks whether there is a subsequence of p-bits matching $\bar{r}$ and φ_2' checks if there is a subsequence of p-bits matching r followed by a subsequence of q-bits matching $2^k-1-\bar{r}$, i.e. whether $u[\bar{r}]\cdot v[2^k-1-\bar{r}]$ is a subsequence of the play.

Consider a strategy for the protagonist such that all plays according to the strategy satisfy ψ_1. If the sequence $u[i_0]a_0v[j_0]$ *ok* $u[i_1]a_1v[j_1]$ *ok* ... is a play according to this strategy, we can prove that this must be a proper sequence. One can also show that if the protagonist plays only proper sequences, then she cannot lose because of the conjunct ψ_1.

In summary, we have so far managed to construct a game graph that lets the protagonist generate cell contents at various addresses ranging from 0 to m^2-1. The specification ψ_1 forces the protagonist to generate this in increasing order. The game graph forces each contiguous block of declared cells to be a configuration and when a configuration is finished, the game graph ensures that the correct player gets to choose the direction of how the computation will proceed.

Let us now turn to the second objection obj_2 that ensures that the configuration sequences generated do respect the transitions of the Turing machine. After raising objection obj_2, the play reaches the subgraph G_2, where the adversary picks four numbers $\bar{r}$, $\bar{s}$, $\bar{t}$ and $\bar{e}$ (using the r-, s-, t- and e-bits). The adversary should generate these such that $\bar{s}=\bar{r}+1$, $\bar{t}=\bar{s}+1$ and $\bar{e}=\bar{s}+2^{k/2}$. Also, $\bar{r}$, $\bar{s}$ and $\bar{t}$ will point to three consecutive cells of a particular configuration and hence $\bar{e}$ will point to a cell in the *successor configuration* where the cell number is the same as that pointed to by $\bar{s}$. The adversary claims that the cell content defined at $\bar{e}$ is not correct (note that the cell content at $\bar{e}$ solely depends on the cell contents at $\bar{r}$, $\bar{s}$ and $\bar{t}$).

First, the correctness of the values of $\bar{r}$, $\bar{s}$, $\bar{t}$ and $\bar{e}$ can be ensured using a formula φ_3, similar to the way we check whether a number is the successor of another. Also, let Δ_{D_1} be the set of all elements of the form $\langle a_1,a_2,a_3,a_2'\rangle$, where $a_1,a_2,a_3,a_2'\in\Sigma'$ and a_2' is the expected value of the cell number *cell*, if a_1, a_2 and a_3 are the cell contents of the cells (*cell*-1), *cell* and (*cell*$+1$) of the previous configuration and the machine took the D_1 transition. Similarly, define Δ_{D_2}. Let $match(a,r,\varphi)$, where φ is a temporal formula, denote the formula which checks whether a matches $\bar{r}$ in w (where $\bar{r}$ is the number encoded by the r-bits that occur somewhere in the future of where the string is matched) and after matching, the suffix of w from that point satisfies φ. More precisely, let $same(p_b,r_b)=((p_b^\top\wedge\Diamond\, r_b^\top)\vee(p_b^\perp\wedge\Diamond\, r_b^\perp))$, as defined before, and let $diff(q_b,r_b)=((q_b^\top\wedge\Diamond\, r_b^\perp)\vee(q_b^\perp\wedge\Diamond\, r_b^\top))$, for every $b\in[1,k]$. Then, $match(a,r,\varphi)=Seq(same(p_k,r_k),\ldots,same(p_1,r_1),a,diff(q_1,r_1),\ldots,diff(q_k,r_k),\varphi)$. Note that if φ is in $\mathcal{B}(L_{\Diamond,\wedge,\vee}(\Pi))$, then so is $match(a,r,\varphi)$. Define similarly formulas for s, t and e.

Now, we have in the specification the conjunct $\psi_2=(\Diamond obj_2\wedge\varphi_3)\rightarrow\varphi_4$, where φ_3 is as explained earlier and, denoting by $\xi(a_1,a_2,a_3,a_2',d)$ the formula $match(a_1,r,match(a_2,s,match(a_3,t,\Diamond(d\wedge match(a_2',e,\ true)))))$,

$$\varphi_4=\bigvee_{d\in\{D_1,D_2\},\langle a_1,a_2,a_3,a_2'\rangle\in\Delta_d}\xi(a_1,a_2,a_3,a_2',d).$$

φ_4 checks whether there is tuple $\langle a_1, a_2, a_3, a_2' \rangle$ in Δ_{D_1} or Δ_{D_2} such that a_1 matches $\bar{r}$ followed by a_2 matching $\bar{s}$ followed by a_3 matching $\bar{t}$ followed by the corresponding direction D_1 or D_2 followed by a_2' matching $\bar{e}$. It is easy to see that a proper sequence that encodes a list of valid configurations interspersed with direction labels satisfies ψ_2 (for all possible values of $\bar{r}$, $\bar{s}$, $\bar{t}$ and $\bar{e}$) iff it corresponds to a correct evolution of M according to the direction labels.

The complete specification is then $\psi_1 \wedge \psi_2 \wedge \psi_3$, where $\psi_3 = \Diamond\, obj_1 \vee \Diamond\, obj_2 \vee \bigvee_{a \in \Sigma} \Diamond(q_f, a)$ which demands that if no objection is raised, then the play must meet the final state.

We can show thus that there is a winning strategy for the protagonist iff M accepts w. Since the main part of G is $O(|w| + k + |M|)$, size of both G_1 and G_2 is $O(k)$, and size of $\psi_1 \wedge \psi_2 \wedge \psi_3$ is $O(k\,(k + |M|))$, we have:

Theorem 1. *Deciding $\mathcal{B}(L_{\Diamond,\wedge,\vee}(\Pi))$ games is* EXPSPACE*-hard.*

3.4 Lower Bound for $L_{\Box,\Diamond,\wedge,\vee}(\Pi)$

In this section, we show that deciding games for specifications given by formulas in $L_{\Box,\Diamond,\wedge,\vee}(\Pi)$ is 2EXPTIME-hard. The reduction is from the membership problem for alternating exponential-space Turing machines. We show that for such a Turing machine M, given an input word w, we can construct in polynomial time a game graph G' and an $L_{\Box,\Diamond,\wedge,\vee}(\Pi)$ formula φ such that M accepts w if and only if the protagonist has a winning strategy in the game.

Let G be the game graph constructed in Sect. 3.3. We give a reduction based on the construction of a game graph G' which is slightly different from G. First, the configuration numbers are not encoded explicitly (since they can be doubly exponential) but the cell numbers are encoded. However, for a configuration sequence $c_0 c_1 \ldots$, we want to count the configurations using a counter modulo 3. For this, we introduce new propositions $\mathcal{P}' = \{0, 1, 2\}$, and in the entire sequence encoding c_i, the proposition $i \bmod 3$ is true. This counter's behaviour is ensured by the design of the game graph.

The role of obj_1 is similar as in G; using this the adversary ensures that the sequence generated is proper and hence that the system does generate a sequence of configurations with proper cell numbers. However, if it wants to claim that the sequence of configurations is not according to the Turing machine, note that it cannot provide an exact witness as configurations are not numbered. We hence allow the adversary to raise the objection obj_2 after the end of *every* configuration. When the adversary chooses obj_2 it gives a cell number k and claims that the contents of cell k in the *last* configuration thus far is incorrect with respect to the corresponding.cells in the previous configuration.

The crucial point is that one can check whether the cell in the last configuration is correct with respect to the penultimate configuration by using an $L_{\Box,\Diamond,\wedge,\vee}(\Pi)$ formula that uses the modulo-3 counter. Intuitively, if we want to check a formula φ_1 on the suffix starting from the penultimate configuration, we can do so by the formula

$$\bigvee_{j \in \{0,1,2\}} (\Diamond\,(j \wedge \varphi_1 \wedge \Diamond(j+1) \wedge \neg\Diamond(j+2)))$$

Note that the formula is in $L_{\Box,\Diamond,\wedge,\vee}(\Pi)$, but not in $\mathcal{B}(L_{\Diamond,\wedge,\vee}(\Pi))$, because of the subformula $\neg\Diamond(j+2)$ (which plays a vital role). Using such a formula, we ensure that the protagonist must generate correct configuration sequences to win the game. The rest of the proof is in details and we omit them; we then have:

Theorem 2. *Deciding $L_{\Box,\Diamond,\wedge,\vee}(\Pi)$ games is 2*EXPTIME*-hard.*

4 Fairness Games

In modeling reactive systems, fairness assumptions are added to rule out infinite computations in which some choice is repeatedly ignored [11]. A typical fairness constraint is of the form "if an action is enabled infinitely often, then it is taken infinitely often," and is captured by a formula of the form $\Box\Diamond p \rightarrow \Box\Diamond p'$. In the game setting, fairness constraints can refer to both the players. Let ψ_0 be the formula expressing the fairness constraint for the protagonist and ψ_1 be the formula for the fairness constraints of the adversary. Then, the winning condition φ of a game is changed either to $\psi_1 \rightarrow (\psi_0 \wedge \varphi)$ ("if the adversary is fair then the protagonist plays fair and satisfies the specification") or $\psi_0 \wedge (\psi_1 \rightarrow \varphi)$ ("the protagonist plays fair and if the adversary plays fair then the specification is satisfied") [3]. Thus, adding fairness changes the winning conditions for specifications from a logic $\mathcal{L}$ to Boolean combinations of $L_{\Box\Diamond}(\Pi)$ and $\mathcal{L}$ formulas.

We consider adding fairness to $\mathcal{B}(L_{\Diamond,\wedge}(\Pi))$ games. The fragment $\mathcal{B}(L_{\Diamond,\wedge}(\Pi))$ contains Boolean combinations of formulas built from state predicates using eventualities and conjunctions, and includes combinations of typical invariants and termination properties. A sample formula of this fragment is $\Box p \vee \Diamond(q \wedge \Diamond r)$. In this section, we prove that games for this logic augmented with fairness constraints are still decidable in polynomial space. More precisely, we prove that deciding $\mathcal{B}(L_{\Box\Diamond}(\Pi) \cup L_{\Diamond,\wedge}(\Pi))$ games is PSPACE-complete. We begin by considering $\mathcal{B}(L_{\Box\Diamond}(\Pi))$ games.

4.1 Boolean Combinations of Büchi Conditions

In this section we give a polynomial space algorithm to solve $\mathcal{B}(L_{\Box\Diamond}(\Pi))$ games. We adapt the technique proposed by Zielonka [17] for Muller games. Muller games are game graphs with winning conditions given as a collection of sets of vertices $\mathcal{F}$ with the meaning that a play π is winning if the set of the infinitely repeating vertices in π belongs to $\mathcal{F}$.

Let V be a finite set, $\mathcal{F}$ be a subset of 2^V, and $\bar{\mathcal{F}}$ be $2^V \setminus \mathcal{F}$. A set $U \in \mathcal{F}$ is maximal for $\mathcal{F}$ if for all $U' \in \mathcal{F}$, $U \not\subset U'$. A *Zielonka tree* for the pair $(\mathcal{F}, \bar{\mathcal{F}})$ is a finite tree T with vertices labeled by pairs of the form $(0, U)$ with $U \in \mathcal{F}$ or $(1, U)$ with $U \in \bar{\mathcal{F}}$. It is inductively defined as follows. The root of T is labeled with $(0, V)$, if $V \in \mathcal{F}$, and by $(1, V)$ otherwise. Suppose x is a node labeled with $(0, U)$. If $U_1, \ldots, U_m$, $m > 0$, are the maximal subsets of U belonging to $\bar{\mathcal{F}}$, then x has m children respectively labeled with $(1, U_1), \ldots, (1, U_m)$. If all subsets of U belong to $\mathcal{F}$, then x is a leaf. The case $(1, U)$ is analogous. Notice that while

```
Z-solve(G_x, x)
    Let V_x be the set of G_x vertices and let (i, U_x) be the label of x
    W ← W′ ← ∅
    if x is not a leaf then
      repeat
        W ← W ∪ W′;  W′ ← ∅
        W ← Attractor-set(W, 1 − i)
        for each child y of x do
          Let (1 − i, U_y) be the label of y
          G_y ← Sub-game(G_x, W, U_y)
          W′ ← W′∪ Z-solve(G_y, y)
      until (W = W ∪ W′)
    return (V_x \ W)
```

Fig. 1. Algorithm for $\mathcal{B}(L_{\Box\Diamond}(\Pi))$ games.

the number of children can be exponential in $|V|$, the depth of the tree is linear in $|V|$.

Let V be the set of vertices of a game graph G and $\mathcal{F}$ denote a Muller winning condition. The algorithm in Fig. 1 implements the solution given by Zielonka [17]. Let G_x be a sub-game of G (i.e. G_x is a game graph which is a subgraph of G) and let x be a node of the Zielonka tree for $(\mathcal{F}, \bar{\mathcal{F}})$, with x labeled with (i, U_x). On the call *Z-solve*(G_x, x), the procedure computes the set of positions in G_x from which player i has a winning strategy with respect to the Muller condition restricted to G_x. The procedure is hence initially invoked with $(G, \hat{x})$ where $\hat{x}$ is the root of the Zielonka tree.

The procedure works by growing the set W of the vertices from which player $1{-}i$ has a winning strategy on G_x. The main parts of *Z-solve*(G_x, x) are the enumeration of the children of x, and the calls to procedures *Attractor-set* and *Sub-game.* A call *Attractor-set*$(G_x, W, 1{-}i)$ constructs the largest set of vertices in G_x from which player $1{-}i$ has a strategy to reach W. For a set $U \subseteq V_x$, let Z be the set of vertices constructed in the call *Attractor-set*$(V \setminus U, i)$. Then, *Sub-game*(G_x, W, U) constructs a game graph contained in G_x, that is induced by the vertices $V_x \setminus (W \cup Z)$. Each call to either *Attractor-set* or *Sub-game* takes at most polynomial time. Note that the recursive depth of calls is bounded by the depth of the tree.

It is worth noting that for an implicitly defined Muller condition, such as the one defined by a formula in $\mathcal{B}(L_{\Box\Diamond}(\Pi))$, one can use the same procedure above but without explicitly constructing the Zielonka tree. The algorithm just needs to compute the children of a node of the tree, and, as we show below, this can be done in polynomial space as long as the membership test $U \in \mathcal{F}$ can be implemented in polynomial space. Note that the recursive call depth is bounded by V and hence the algorithm will run in polynomial space.

To explain the computation of children of a node of the Zielonka tree more formally, consider a $\mathcal{B}(L_{\Box\Diamond}(\Pi))$ formula φ. For a set $U \subseteq V$, let ν_U be the mapping that assigns a sub-formula $\Box\Diamond p$ of φ to true if p holds in some vertex in U, and assigns it to false otherwise. We say that U *meets* φ if under the assignment ν_U, φ evaluates to true. Intuitively, if U is the set of the vertices that repeat infinitely often on a play π of G, then π satisfies φ if and only if U meets φ. Let $\mathcal{F}_\varphi$ be the set of $U \subseteq V$ such that U meets φ, and $\bar{\mathcal{F}}_\varphi$ be its complement with respect to 2^V. The Zielonka tree for φ is the Zielonka tree T for $(\mathcal{F}_\varphi, \bar{\mathcal{F}}_\varphi)$. We observe that each child of a node x of T can be generated in polynomial space simply from φ and the label (i, U) of x. For example, if $i = 0$, then for each $U' \subseteq U$ we can check if $(1, U')$ is a child of x by checking whether it falsifies φ and is maximal within the subsets of U that do not meet φ (which can be done in polynomial space). Thus we can enumerate the children of a node in the Zielonka tree using only polynomial space and we have the following:

Theorem 3. *Deciding $\mathcal{B}(L_{\Box\Diamond}(\Pi))$ games is in* PSPACE.

4.2 Solving Fairness Games

The result from Sect. 4.1 can be extended to prove that when the winning condition is a formula from $\mathcal{B}(L_{\Box\Diamond}(\Pi) \cup L_{\Diamond,\wedge}(\Pi))$, that is a Boolean combination of formulas from $L_{\Box\Diamond}(\Pi)$ and $L_{\Diamond,\wedge}(\Pi)$, then games can still be decided in polynomial space.

We first describe a polynomial space algorithm to solve games for the simpler fragment of the Boolean combinations of $L_{\Box\Diamond}(\Pi)$ formulas and formulas of the form $\Diamond p$ using the polynomial-space algorithm for $\mathcal{B}(L_{\Box\Diamond}(\Pi))$ above.

Let us explain the intuition behind the solution with an example. Consider the formula $\varphi = \Box\Diamond p_1 \vee (\Box\Diamond p_2 \wedge \Diamond p_3)$. The protagonist can win a play by visiting a state satisfying p_1 infinitely. However, if it meets a state satisfying p_3 then it wins using the formula $\varphi' = \Box\Diamond p_1 \vee \Box\Diamond p_2$. Now, assume that we know the exact set Z of positions from which the protagonist can win the game G with φ' as the winning condition. We construct a game graph G' from G by adding two vertices *win* and *lose*, that have self loops and let a new proposition p^{win} be true only at *win*. Now, for a vertex u in G where p_3 holds, we remove all the edges from u and instead add an edge to either *win* or *lose* — we add an edge to *win* if u is in Z, and an edge to *lose* otherwise. Then clearly the protagonist wins the overall game if and only if it wins the game $(G', \Box\Diamond p_1 \vee \Box\Diamond p^{win})$. In general, we need to define and solve many such games. Each such game corresponds to some subset X of the subformulas of the kind $\Diamond p_i$ mentioned in φ. In such a game, when we meet a state that meets a new predicate p, where $\Diamond p$ is a subformula of φ but is not in X, we jump to *win* or *lose* depending on whether the game corresponding to $X \cup \{\Diamond p\}$ is winning or losing.

Consider a $\mathcal{B}(L_{\Box\Diamond}(\Pi) \cup L_{\Diamond}(\Pi))$ game (G, φ) and let $\Diamond p_1, \ldots, \Diamond p_k$ be all the sub-formulas of φ of the form $\Diamond p$ that are not in the scope of the $\Box$ operator. For each assignment ν of truth values to $\Diamond p_1, \ldots, \Diamond p_k$, define φ_ν as the formula obtained from φ by assigning $\Diamond p_1, \ldots, \Diamond p_k$ according to ν. Clearly, φ_ν is a $\mathcal{B}(L_{\Box\Diamond}(\Pi))$ formula. For an assignment ν and a vertex u, we denote by $\nu + u$ the assignment that maps $\Diamond p_i$ to true iff either ν assigns $\Diamond p_i$ to true or p_i holds true at u. We say that a vertex u *meets* an assignment ν if whenever p_i holds true at u, then ν also assigns $\Diamond p_i$ to true.

We denote by G_ν the game graph obtained from G removing all the edges (u, v) such that u does not meet ν, and adding two new vertices *win* and *lose* along with new edges as follows. We use a new atomic proposition p^{win} that is true only at *win*. We add a self loop on both *win* and *lose*, and there are no other edges leaving from these vertices (i.e., they are both sinks). Denoting by X_ν the set of vertices that meet ν, we add an edge from each $u \notin X_\nu$ to *win* if there is a winning strategy of the protagonist in $(G_{\nu+u}, \varphi_{\nu+u} \vee \Box\Diamond p^{win}, u)$, and to *lose* otherwise.

In order to construct G_ν, we may need to make recursive calls to construct and solve $(G_{\nu+u}, \varphi_{\nu+u} \vee \Box\Diamond p^{win}, u)$, for some u. However, note that the number of elements set to true in $\nu + u$ is more than those set in ν to be true. Also, if ν assigns all elements to true, there will be no recursive call to construct games. Hence the depth of recursion is bounded by the number of subformulas of the kind $\Diamond p$ in φ. Note however that there can be an exponential number of calls required when constructing G_ν. Since any of the games (G_ν, φ'), once constructed, can be solved in polynomial space, it follows that $G_\perp$ can be constructed in polynomial space, where $\perp$ is the empty assignment where every element is set to false. Therefore, we have the following lemma.

Lemma 2. *Given a $\mathcal{B}(L_{\Box\Diamond}(\Pi) \cup L_\Diamond(\Pi))$ game (G, φ, u), there exists a winning strategy of the protagonist in (G, φ, u) if and only if there exists a winning strategy of the protagonist in $(G_\perp, \varphi_\perp \vee \Box\Diamond p^{win}, u)$.*

To decide $\mathcal{B}(L_{\Box\Diamond}(\Pi) \cup L_{\Diamond,\wedge}(\Pi))$ games we need to modify the above procedure. We know from [4] that for each formula ψ in $\mathcal{B}(L_{\Diamond,\wedge}(\Pi))$ there exists a deterministic Büchi automaton A accepting all the models of ψ such that: (1) size of A is exponential in $|\psi|$, (2) the automaton can be constructed "on-the-fly" — for any state of the automaton, the transitions from it can be found in polynomial time, (3) the length of simple paths in A is linear in $|\psi|$, and (4) the only cycles in the transition graph of A are the self-loops. For a given $\mathcal{B}(L_{\Box\Diamond}(\Pi) \cup L_{\Diamond,\wedge}(\Pi))$ formula φ, let $\psi_1, \ldots, \psi_k$ be all the sub-formulas of φ from $\mathcal{B}(L_{\Diamond,\wedge}(\Pi))$ that are not in the scope of the $\Box$ operator. Let A_i be a deterministic automaton accepting models of ψ_i and satisfying the above properties. Let Q be the product of the sets of states of $A_1, \ldots, A_k$. For a tuple $\bar{q} = \langle q_1, \ldots, q_k \rangle \in Q$, we associate a truth assignment $\nu[\bar{q}]$ such that $\nu[\bar{q}](\psi_i)$ is true if and only if q_i is an accepting state. Moreover, for $i = 1, \ldots, k$, let q'_i be the state entered by A_i starting from q_i reading the label of a vertex u of G. We denote the tuple $\langle q'_1, \ldots, q'_k \rangle$ by $\bar{q}(u)$. For a tuple of states $\bar{q} = \langle q_1, \ldots, q_k \rangle \in Q$, the graph $G_{\bar{q}}$ is constructed similar to G_ν. The main differences are that we solve games of the form $(G_{\bar{q}(v)}, \varphi_{\nu[\bar{q}(v)]} \vee \Box\Diamond p^{win}, v)$, we use the set $X_{\bar{q}} = \{v \mid \bar{q} \neq \bar{q}(v)\}$ instead of X_ν and the recursion depth is bounded by $O(k \cdot |\varphi|)$. Clearly, each such graph can be constructed in polynomial space and thus we have the following result.

Theorem 4. *Deciding $\mathcal{B}(L_{\Box\Diamond}(\Pi) \cup L_{\Diamond,\wedge}(\Pi))$ games is in* PSPACE.

We can also show that deciding $\mathcal{B}(L_{\Box\Diamond}(\Pi))$ games is PSPACE-hard and hence, from the results above, deciding $\mathcal{B}(L_{\Box\Diamond}(\Pi) \cup L_{\Diamond,\wedge}(\Pi))$ games and $\mathcal{B}(L_{\Box\Diamond}(\Pi))$ games is PSPACE-complete. We in fact show a stronger result that a fragment of $\mathcal{B}(L_{\Box\Diamond}(\Pi))$ is already PSPACE-hard.

Let $\mathcal{L}_R$ denote the set of formulas of the form $\bigvee_{i=1}^{k} \varphi_i$, where each φ_i is of the form $(\Box\Diamond p_i \wedge \Diamond\Box p'_i)$ (where each p_i and p'_i are state predicates). $\mathcal{L}_R$ is then a fragment of $\mathcal{B}(L_{\Box\Diamond}(\Pi))$ and represents Rabin conditions. Let $\mathcal{L}_S$ denote the set of formulas of the form $\bigwedge_{i=1}^{k} \varphi_i$, where each φ_i is of the form $(\Box\Diamond p_i \rightarrow \Box\Diamond p'_i)$. $\mathcal{L}_S$ is also a fragment of $\mathcal{B}(L_{\Box\Diamond}(\Pi))$ and represents Streett conditions, which are the dual of Rabin conditions. Let $\mathcal{L}_{RS}$ represent a disjunction of a Rabin and a Streett condition, i.e. $\mathcal{L}_{RS}$ contains formulas of the kind $\varphi_R \vee \varphi_S$ where $\varphi_R \in \mathcal{L}_R$ and $\varphi_S \in \mathcal{L}_S$. We can then show the that deciding games for formulas in $\mathcal{L}_{RS}$ is PSPACE-hard. Note that it is known that deciding $\mathcal{L}_R$ games is NP-complete and, hence, deciding $\mathcal{L}_S$ games is CO-NP-complete.

Theorem 5. *Deciding $\mathcal{L}_{RS}$ games is* PSPACE-*hard.*

5 Conclusions

We have shown that games for the fragment $L_{\Box,\Diamond,\wedge,\vee}(\Pi)$ are 2EXPTIME-hard, games for the fragment $\mathcal{B}(L_{\Diamond,\wedge,\vee}(\Pi))$ are EXPSPACE-hard, games for $\mathcal{B}(L_{\Box\Diamond}(\Pi))$ are PSPACE-hard, and games for $\mathcal{B}(L_{\Box\Diamond}(\Pi) \cup L_{\Diamond,\wedge}(\Pi))$ are in PSPACE. Our lower bound proofs introduce new techniques for counting and encoding using game graphs and LTL formulas without using next or until operators. Our upper bound techniques are useful for combinations of safety/reachability games on game graphs with strong fairness requirements on the choices of the two players. The results in this paper complete the picture for the complexity bounds for various fragments of LTL and is summarized in Fig. 2. Recall that m-odel-checking of $L_{\Box,\Diamond,\wedge,\vee}(\Pi)$ formulas is NP-complete, and becomes PSPACE-complete (as for the full LTL) by allowing the until and/or the next operators [15]. As shown in Fig. 2, this does not hold for games. Note that allowing nested always and eventually operators the complexity of games increases to the complexity of the whole LTL (i.e., 2EXPTIME-complete), while the use of the next and eventually

	Lower bound	Upper bound
$\mathcal{B}(L_{\Diamond,\wedge}(\Pi))$	PSPACE-complete [4]	
$\mathcal{B}(L_{\Box\Diamond}(\Pi) \cup L_{\Diamond,\wedge}(\Pi))$	**Pspace-complete**	
$L_{\Diamond,\wedge,\vee}(\Pi)$	PSPACE-hard [4]	PSPACE [12]
$\mathcal{B}(L_{\Diamond,\bigcirc,\wedge}(\Pi))$	EXPTIME-complete [4]	
$\mathcal{B}(L_{\Diamond,\wedge,\vee}(\Pi))$	**Expspace-hard**	EXPSPACE [4]
$\mathcal{B}(L_{\Diamond,\bigcirc,\wedge,\vee}(\Pi))$	EXPSPACE-hard [12]	EXPSPACE [4]
$L_{\Box,\Diamond,\wedge,\vee}(\Pi)$	**2Exptime-hard**	2EXPTIME [14]
LTL	2EXPTIME-complete [14]	

Fig. 2. Complexity of LTL games

operators (with the negation only at the top level) that makes model checking PSPACE-hard, increases the complexity of games only to EXPSPACE.

References

1. M. Abadi, L. Lamport, and P. Wolper. Realizable and unrealizable specifications of reactive systems. In *Proc. of ICALP'89*, LNCS 372, pages 1–17, 1989.
2. R. Alur, L. de Alfaro, T. Henzinger, and F. Mang. Automating modular verification. In *Proc. of CONCUR'99*, LNCS 1664, pages 82–97, 1999.
3. R. Alur, T. Henzinger, and O. Kupferman. Alternating-time temporal logic. *Journal of the ACM*, 49(5):1–42, 2002.
4. R. Alur and S. La Torre. Deterministic generators and games for LTL fragments. *Proc. of LICS'01*, pages 291–302, 2001.
5. A. Chandra, D. Kozen, and L. Stockmeyer. Alternation. *Journal of the ACM*, 28(1):114–133, 1981.
6. D. Dill. *Trace Theory for Automatic Hierarchical Verification of Speed-independent Circuits.* ACM Distinguished Dissertation Series. MIT Press, 1989.
7. E. Emerson and C. Jutla. The complexity of tree automata and logics of programs. In *Proc. of FOCS'88*, pages 328–337, 1988.
8. G. Holzmann. The model checker SPIN. *IEEE Transactions on Software Engineering*, 23(5):279–295, 1997.
9. J. Hopcroft and J. Ullman. *Introduction to Automata Theory, Languages, and Computation.* Addison-Wesley, 1979.
10. O. Kupferman and M. Vardi. Module checking. In *Proc. of CAV'96*, LNCS 1102, pages 75–86. Springer, 1996.
11. Z. Manna and A. Pnueli. *The temporal logic of reactive and concurrent systems: Specification.* Springer, 1991.
12. J. Marcinkowski and T. Truderung. Optimal complexity bounds for positive ltl games. In *Proc. of CSL 2002*, LNCS 2471, pages 262–275. Springer, 2002.
13. A. Pnueli. The temporal logic of programs. In *Proc. of FOCS'77*, pages 46–77, 1977.
14. A. Pnueli and R. Rosner. On the synthesis of a reactive module. In *Proc. of POPL'89*, pages 179–190, 1989.
15. A. Sistla and E. Clarke. The complexity of propositional linear temporal logics. *The Journal of the ACM*, 32:733–749, 1985.
16. M.Y. Vardi and L. Stockmeyer. Improved upper and lower bounds for modal logics of programs. In *Proc. 17th Symp. on Theory of Computing*, pages 240–251, 1985.
17. W. Zielonka. Infinite games on finitely coloured graphs with applications to automata on infinite trees. *Theoretical Computer Science*, 200:135–183, 1998.

The Element of Surprise in Timed Games*

Luca de Alfaro[1], Marco Faella[1,2], Thomas A. Henzinger[3], Rupak Majumdar[3], and Mariëlle Stoelinga[1]

[1] Department of Computer Engineering, UC Santa Cruz, USA
[2] Dipartimento di Informatica ed Applicazioni, Università di Salerno, Italy
[3] Department of Electrical Engineering and Computer Sciences, UC Berkeley, USA

Abstract. We consider concurrent two-person games played in real time, in which the players decide both which action to play, and when to play it. Such timed games differ from untimed games in two essential ways. First, players can take each other by surprise, because actions are played with delays that cannot be anticipated by the opponent. Second, a player should not be able to win the game by preventing time from diverging. We present a model of timed games that preserves the element of surprise and accounts for time divergence in a way that treats both players symmetrically and applies to all ω-regular winning conditions. We prove that the ability to take each other by surprise adds extra power to the players. For the case that the games are specified in the style of timed automata, we provide symbolic algorithms for their solution with respect to all ω-regular winning conditions. We also show that for these timed games, memory strategies are more powerful than memoryless strategies already in the case of reachability objectives.

1 Introduction

Games have become a central modeling paradigm in computer science. In synthesis and control, it is natural to view a system and its environment as players of a game that pursue different objectives [Chu63,RW89,PR89]. Similarly, in modular specification and verification it is often appropriate to model the components of a system as individual players that may or may not cooperate, depending on the application [AHK02,AdAHM99]. Such games are played on a state space and proceed in an infinite sequence of rounds. In each round, the players choose actions to play, and the chosen actions determine the successor state. For the synthesis and modular analysis of *real-time* systems, we need to use games where time elapses between actions [MPS95]. In such *timed* games, each player chooses both which action to play, and *when* to play it. Timed games differ from their untimed counterparts in two essential ways. First, players can take each other by surprise, because actions are played with delays that cannot be anticipated by

* Supported in part by the AFOSR MURI grant F49620-00-1-0327, the DARPA grant F33615-C-98-3614, the MARCO grant 98-DT-660, -the ONR grant N00014-02-1-0671, the NSF grants CCR-9988172, CCR-0225610, and CCR-0234690, the NSF CAREER award CCR-0132780, and the MIUR grant MEFISTO.

R. Amadio, D. Lugiez (Eds.): CONCUR 2003, LNCS 2761, pp. 144–158, 2003.

the opponent. Second, a player should not be able to win the game by preventing time from diverging [SGSAL98,AH97]. We present a model of timed games that preserves the element of surprise and accounts for the need of time divergence. We study both the properties of the winning strategies and the algorithms for their construction.

We consider two-player timed games that are played over a possibly infinite state space. In each state, each player chooses, simultaneously and independently of the other player, a move $\langle \Delta, a \rangle$, indicating that the player wants to play the action a after a delay of $\Delta \in \mathbb{R}_{\geq 0}$ time units. A special action, $\perp$, signifies the player's intention to remain idle for the specified time delay. Of the moves chosen by the two players, the one with the smaller delay is carried out and determines the successor state; if the delays are equal, then one of the chosen moves occurs nondeterministically (this models the fact that, in real-time interaction, true contemporaneity cannot be achieved). This process, repeated for infinitely many rounds, gives rise to a *run* of the game. Our definition of moves preserves the element of surprise: a player cannot anticipate when the opponent's action will occur in the current round. This contrasts with many previous definitions of timed games (e.g., [AH97,HHM99,dAHM01b,MPS95,AMPS98]), where players can only either play immediately an action a, or wait for a delay Δ. Such formulations may be simpler and more elegant for timed transition systems (i.e., one-player games), but in the case of two-player formulations, the element of surprise is lost, because after each delay both players have the opportunity to propose a new move. This allows a player to intercept the opponent's move $\langle \Delta, a \rangle$ just before the action a is carried out. We show that the element of surprise gives a distinct advantage to a player. In particular, we prove that there are simple reachability games that can be won under our formulation of moves, but not under the previous "no-surprise" versions.

The objective for a player is given by a set Φ of desired game outcomes. A player achieves this goal if all game outcomes belong to Φ. For a timed game to be physically meaningful, a player should not be able to achieve a goal by stopping the progress of time. For instance, if Φ consists of the set of runs that stay forever in a certain set U of states, and if player 2 has an action to leave U only after a delay of 4, then player 1 should not be able to win by always playing $\langle 0, \perp \rangle$. Therefore, several conditions $WC_i(\Phi)$ have been proposed in the literature to express when player $i \in \{1, 2\}$ wins a timed game with goal Φ.

In [SGSAL98,AH97] the winning condition $WC_1(\Phi)$ is defined to be $\Phi \cap (td \cup Blameless_1)$, where td is the set of runs along which time diverges, and $Blameless_1$ is the set of runs along which player 1 proposes the shorter delay only finitely often. Clearly, player 1 is not responsible if time converges along a run in $Blameless_1$. Informally, the condition states that player 1 must achieve the goal Φ, and moreover, either time diverges or player 1 is blameless for its convergence. This definition works if the goal Φ is a safety property, but not if it is a reachability or, more general, a ω-regular property. To see this, observe that player 1 must achieve the goal even if player 2 stops the progress of time. Consider a game where the goal consists of reaching a set U of states, and where player 1 has an action leading to U which is always available once time advances beyond 1. Then, player 1 cannot win: player 2 can stop time, preventing the action from ever becoming enabled, and ensuring that no run is in Φ.

In [MPS95], the winning condition $\Phi \cap td$ is proposed. This condition requires player 1 to guarantee time divergence, which is not possible in models where player 2 can block the progress of time. In [dAHS02], this condition is modified to $WC_i^*(\Phi) = (\Phi \cap td) \cup Blameless_i$ for player $i \in \{1, 2\}$. While this is appropriate in the asymmetric setting considered there, the problem in our setting, where both players are treated completely symmetrically, is that the two conditions $WC_1^*(\Phi)$ and $WC_2^*(\neg\Phi)$ are not disjoint (here $\neg\Phi$ is the complementary language of Φ). This means that there are games in which both players can win: for instance, player 1 can ensure $\Phi \cap td$, and player 2 can ensure $Blameless_2$. Other works on timed games (e.g., [AMPS98,FLM02]) have avoided the issue of time divergence altogether by putting syntactic constraints on the game structures.

We define timed games and their winning conditions in a completely symmetric fashion, and in a way that works for all goals (in particular for all ω-regular goals) and ensures that players can win only by playing in a physically meaningful way. The winning conditions we propose are $WC_i(\Phi) = (\Phi \cap td) \cup (Blameless_i \setminus td)$, for $i \in \{1, 2\}$. These winning conditions imply that $WC_1(\Phi) \cap WC_2(\neg\Phi)$ is empty, ensuring that at most one player can win. Note that there are runs that belong neither to $WC_1(\Phi)$ nor to $WC_2(\neg\Phi)$: this contrasts with the traditional formulation of untimed games, where runs are either winning for a player with respect to a goal, or winning for the opponent with respect to the complementary goal. We argue that the lack of run-level determinacy is unavoidable in timed games. To see this, consider a run $\bar{r}$ along which both players take turns in proposing moves with delay 0, thus stopping the progress of time. If we somehow assign this run to be winning for a player, say player 1, then it would be possible to construct games in which the moves with delay 0 are the *only* moves available, and in which player 1 could nevertheless win. This would go against our intention that a player can win only in a physically meaningful way. The lack of run-level determinacy also implies that there are states from which neither player can win.

The form of the winning conditions for timed games have other important implications. We show that to win with respect to a reachability goal, in contrast to the untimed case, strategies with memory may be required. For safety goals, however, memoryless strategies suffice also in the timed case. We prove several additional structural properties of the winning strategies for timed games. For instance, we define a class of *persistent* strategies, in which players do not change their mind about the time of future moves when interrupted by a $\langle \Delta, \bot \rangle$ move of the opponent. We show that persistent strategies always suffice to win games, for all possible goals.

While we define timed games at first semantically, we also offer a timed-automaton-style [AD94] syntax for a specific class of timed games. We show that for these *timed automaton games* the winning states with respect to any ω-regular goal can be computed by a symbolic algorithm that iterates a controllable predecessor operator on clock regions. In particular, we prove that timed automaton games can be won using *region strategies*, where the players need only remember the history of the game as a sequence of regions, rather than more precisely, as a sequence of states. Furthermore, the problem of solving these games is shown to be, as expected [AH97], complete for EXPTIME.

2 Timed Games

2.1 Timed Game Structures

A *timed game structure* is a tuple $\mathcal{G} = (S, Acts_1, Acts_2, \Gamma_1, \Gamma_2, \delta)$, where

- S is a set of states.
- $Acts_1$ and $Acts_2$ are two disjoint sets of actions for player 1 and player 2, respectively. We assume that $\perp \notin Acts_i$ and write $Acts_i^\perp = Acts_i \cup \{\perp\}$. The set of moves of player i is given by $M_i = \mathbb{R}_{\geq 0} \times Acts_i^\perp$.
- For $i = 1, 2$, the function $\Gamma_i : S \mapsto 2^{M_i} \setminus \emptyset$ is an enabling condition, which assigns to each state s a set $\Gamma_i(s)$ of moves available to player i in that state.
- $\delta : S \times (M_1 \cup M_2) \mapsto S$ is a destination function that, given a state and a move of either player, determines the next state in the game.

We require that the move $\langle 0, \perp \rangle$ is always enabled and does not leave the state: $\langle 0, \perp \rangle \in \Gamma_i(s)$ and $\delta(s, \langle 0, \perp \rangle) = s$ for all $s \in S$. Similarly to [Yi90], we require for all $0 \leq \Delta' \leq \Delta$ and $a \in Acts_i^\perp$, that (1) $\langle \Delta, a \rangle \in \Gamma_i(s)$ if and only if $\langle \Delta', \perp \rangle \in \Gamma_i(s)$ and $\langle \Delta - \Delta', a \rangle \in \Gamma_i(\delta(s, \langle \Delta', \perp \rangle))$, and (2) if $\delta(s, \langle \Delta', \perp \rangle) = s'$, and $\delta(s', \langle \Delta - \Delta', a \rangle) = s''$, then $\delta(s, \langle \Delta, a \rangle) = s''$.

Intuitively, at each state $s \in S$, player 1 chooses a move $\langle \Delta_1, a_1 \rangle \in \Gamma_1(s)$, and simultaneously and independently, player 2 chooses a move $\langle \Delta_2, a_2 \rangle \in \Gamma_2(s)$. If $\Delta_1 < \Delta_2$, then the move $\langle \Delta_1, a_1 \rangle$ is taken; if $\Delta_2 < \Delta_1$, then the move $\langle \Delta_2, a_2 \rangle$ is taken. If $\Delta_1 = \Delta_2$, then the game takes nondeterministically one of the two moves $\langle \Delta_1, a_1 \rangle$ or $\langle \Delta_2, a_2 \rangle$. Formally, we define the *joint destination function* $\widetilde{\delta} : S \times M_1 \times M_2 \mapsto 2^S$ by

$$\widetilde{\delta}(s, \langle \Delta_1, a_1 \rangle, \langle \Delta_2, a_2 \rangle) = \begin{cases} \{\delta(s, \langle \Delta_1, a_1 \rangle)\} & \text{if } \Delta_1 < \Delta_2, \\ \{\delta(s, \langle \Delta_2, a_2 \rangle)\} & \text{if } \Delta_1 > \Delta_2, \\ \{\delta(s, \langle \Delta_1, a_1 \rangle), \delta(s, (\Delta_2, a_2))\} & \text{if } \Delta_1 = \Delta_2. \end{cases}$$

The time elapsed when moves $m_1 = \langle \Delta_1, a_1 \rangle$ and $m_2 = \langle \Delta_2, a_2 \rangle$ are played is given by $delay(m_1, m_2) = \min(\Delta_1, \Delta_2)$. For $i \in \{1, 2\}$, the boolean predicate $bl_i(s, m_1, m_2, s')$ holds if player i is responsible for the state change from s to s'. Formally, denoting with $\sim i = 3 - i$ the opponent of player i, we define $bl_i(s, m_1, m_2, s')$ iff both $\Delta_i \leq \Delta_{\sim i}$ and $s' = \delta(s, m_i)$. Note that both $bl_1(s, m_1, m_2, s')$ and $bl_2(s, m_1, m_2, s')$ may hold at the same time.

An *infinite run* (or simply a *run*) of the timed game structure $\mathcal{G}$ is a sequence $s_0, \langle m_1^1, m_1^2 \rangle, s_1, \langle m_2^1, m_2^2 \rangle, s_2, \ldots$ such that $s_k \in S$, $m_{k+1}^1 \in \Gamma_1(s_k)$, $m_{k+1}^2 \in \Gamma_2(s_k)$, and $s_{k+1} \in \widetilde{\delta}(s_k, m_{k+1}^1, m_{k+1}^2)$ for all $k \geq 0$. A *finite run* $\bar{r}$ is a finite prefix of a run that terminates at a state s; we then set $last(\bar{r}) = s$. We denote by *FRuns* the set of all finite runs of the game structure, and by *Runs* the set of its infinite runs. A finite or infinite run $\bar{r} = s_0, \langle m_1^1, m_1^2 \rangle, s_1, \ldots$ induces a *trace* $states(\bar{r}) = s_0, s_1, \ldots$ of states occurring in $\bar{r}$. A state s' is *reachable* from another state s if there exist a finite run $s_0, \langle m_1^1, m_1^2 \rangle, s_1, \ldots, s_n$ such that $s_0 = s$ and $s_n = s'$.

A *strategy* π_i for player $i \in \{1,2\}$ is a mapping $\pi_i : FRuns \mapsto M_i$ that associates with each finite run $s_0, \langle m_1^1, m_1^2 \rangle, s_1, \ldots, s_k$ the move $\pi_i(s_0, \langle m_1^1, m_1^2 \rangle, s_1, \ldots, s_k)$ to be played at s_k. We require that the strategy only selects enabled moves, that is, $\pi_i(\bar{r}) \in \Gamma_i(last(\bar{r}))$ for all $\bar{r} \in FRuns$. For $i \in \{1,2\}$, let Π_i denote the set of all player i strategies, and $\Pi = \Pi_1 \cup \Pi_2$ the set of all strategies. For all states $s \in S$ and strategies $\pi_1 \in \Pi_1$ and $\pi_2 \in \Pi_2$, we define the set of *outcomes* $Outcomes(s, \pi_1, \pi_2)$ as the set of all runs $s_0, \langle m_1^1, m_1^2 \rangle, s_1, \ldots$ such that $s_0 = s$, and for all $k \geq 0$ and $i = 1, 2$, we have $\pi_i(s_0, \langle m_1^1, m_1^2 \rangle, s_1, \ldots, s_k) = m_{k+1}^i$. Note that in our timed games, two strategies and a start state yield a *set* of outcomes, because if the players propose moves with the same delay, a nondeterministic choice between the two moves is made. According to this definition, strategies can base their choices on the entire history of the game, consisting of both past states and moves. In Proposition 1 we show that, to win the game, strategies need only consider past states.

2.2 Timed Goals and Timed Winning Conditions

We consider winning conditions given by sets of infinite traces. A *goal* Φ is a subset of S^ω; we write $[\Phi]_r = \{\bar{r} \in Runs \mid states(\bar{r}) \in \Phi\}$. We write $\neg\Phi$ for the set $S^\omega \setminus \Phi$. We often use linear-time temporal logic formulas to specify goals; the propositional symbols of the formula consist of sets of states of the timed game [MP91]. We distinguish between the goal of a player and the corresponding *winning condition*. The goal represents the control objective that the player must attain; for instance, staying forever in a region of "safe" states. To win the game, however, a player must not only attain this goal, but also make sure that this is done in a physically meaningful way: this is encoded by the winning condition. To this end, we define the set of *time divergent runs td* as the set of all runs $s_0, \langle m_1^1, m_1^2 \rangle, s_1, \langle m_2^1, m_2^2 \rangle, s_2, \ldots$ such that $\sum_{k=1}^{\infty} delay(m_k^1, m_k^2) = \infty$. For $i \in \{1,2\}$, we define the set of *player i blameless runs* $Blameless_i$ as the set of all runs in which player i plays first (proposes a shorter delay) only finitely many times. Formally, $Blameless_i$ consists of all runs $s_0, \langle m_1^1, m_1^2 \rangle, s_1, \ldots$ such that there exists an $n \in \mathbb{N}$ with $\neg bl_i(s_k, m_{k+1}^1, m_{k+1}^2, s_{k+1})$ for all $k \geq n$. Corresponding to the goal Φ, we define the following winning condition:

$$WC_i(\Phi) : \; (td \cap [\Phi]_r) \cup (Blameless_i \setminus td).$$

Informally, this condition states that if time diverges, the goal must be met, and if time does not diverge, the player must be blameless.

Given a goal Φ and a state $s \in S$, we say that player i *wins* from s the game with goal Φ, or equivalently, wins from s the game with winning condition $WC_i(\Phi)$, if there exists a player i strategy $\pi_i \in \Pi_i$ such that for all opposing strategies $\pi_{\sim i} \in \Pi_{\sim i}$, we have $Outcomes(s, \pi_1, \pi_2) \subseteq WC_i(\Phi)$. In that case,

$\pi_i \in \Pi_i$ is called a *winning strategy*. Given a goal Φ, we let $\langle i \rangle \Phi$ be the states from which player i can win the game with goal Φ. A state s is *well-formed* if for every state s' reachable from s, and each player $i \in \{1, 2\}$, we have $s' \in \langle i \rangle S^\omega$. States that are not well-formed are "pathological": if a player cannot win the goal S^ω, then he cannot ensure that the game outcomes are physically meaningful.

3 Timed Automaton Games

In this section, we introduce *timed automaton games*, a syntax derived from timed automata [AD94] for representing timed games. As in timed automata, a finitely specified timed automaton game usually represents a timed game with infinitely many states. A *clock condition* over a set C of clocks is a boolean combination of formulas of the form $x \preceq c$ or $x - y \preceq c$, where c is an integer, $x, y \in C$, and $\preceq$ is either $<$ or $\leq$. We denote the set of all clock conditions over C by $\mathit{ClkConds}(C)$. A *clock valuation* is a function $\kappa : C \mapsto \mathbb{R}_{\geq 0}$, and we denote by $K(C)$ the set of all clock valuations for C.

A *timed automaton game* is a tuple $\mathcal{A} = (Q, C, \mathit{Acts}_1, \mathit{Acts}_2, E, \theta, \rho, \mathit{Inv}_1, \mathit{Inv}_2)$, where:

- Q is a finite set of locations.
- C is a finite set of clocks which includes the unresettable clock z, which measures the time since the start of the game.
- Acts_1 and Acts_2 are two disjoint, finite sets of actions for player 1 and player 2, respectively.
- $E \subseteq Q \times (\mathit{Acts}_1 \cup \mathit{Acts}_2) \times Q$ is an edge relation.
- $\theta : E \mapsto \mathit{ClkConds}(C)$ is a mapping that associates with each edge a clock condition that specifies when the edge can be traversed. We require that for all $(q, a, q_1), (q, a, q_2) \in E$ with $q_1 \neq q_2$, the conjunction $\theta(q, a, q_1) \wedge \theta(q, a, q_2)$ is unsatisfiable. In other words, the game move and clock values determine uniquely the successor location.
- $\rho : E \mapsto 2^{C \setminus \{z\}}$ is a mapping that associates with each edge the set of clocks to be reset when the edge is traversed.
- $\mathit{Inv}_1, \mathit{Inv}_2 : Q \to \mathit{ClkConds}(C)$ are two functions that associate with each location an invariant for player 1 and 2, respectively.

Given a clock valuation $\kappa : C \mapsto \mathbb{R}_{\geq 0}$ and $\Delta \in \mathbb{R}_{\geq 0}$, we denote by $\kappa + \Delta$ the valuation defined by $(\kappa + \Delta)(x) = \kappa(x) + \Delta$ for all clocks $x \in C$. The clock valuation $\kappa : C \mapsto \mathbb{R}_{\geq 0}$ *satisfies* the clock constraint $\alpha \in \mathit{ClkConds}(C)$, written $\kappa \models \alpha$, if the condition α holds when the clocks have the values specified by κ. For a subset $D \subseteq C$ of clocks, $\kappa[D := 0]$ denotes the valuation defined by $\kappa[D := 0](x) = 0$ if $x \in D$, and by $\kappa[D := 0](x) = \kappa(x)$ otherwise.

The timed automaton game $\mathcal{A}$ induces a timed game structure $[\![\mathcal{A}]\!]$, whose states consist of a location of $\mathcal{A}$ and a clock valuation over C. The idea is the following. A player i move $\langle \Delta, \perp \rangle$ is enabled in state $\langle q, \kappa \rangle$ if either $\Delta = 0$ or the invariant $\mathit{Inv}_i(q)$ holds continuously when we let Δ time units pass, that is, $\kappa + \Delta' \models \mathit{Inv}_i(q)$ for all $\Delta' \leq \Delta$. Taking the move $\langle \Delta, \perp \rangle$ leads to the state $\langle q, \kappa + \Delta \rangle$. For $a \in Acts_i$, the move $\langle \Delta, a \rangle$ is enabled in $\langle q, \kappa \rangle$ if (1) the invariant $Inv_i(q)$ holds continuously when we let Δ time units pass, (2) there is a transition (q, a, q') in E which is enabled in the state $\langle q, \kappa + \Delta \rangle$, and (3) the invariant $Inv_i(q')$ holds when the game enters location q'. The move $\langle \Delta, a \rangle$ leads to the state $\langle q', \kappa' \rangle$, where κ' is obtained from $\kappa + \Delta$ by resetting all clocks in $\rho(q, a, q')$.

Formally, the timed automaton game $\mathcal{A} = (Q, C, Acts_1, Acts_2, E, \theta, \rho, Inv_1, Inv_2)$ induces the timed game structure $[\![\mathcal{A}]\!] = (S, Acts_1, Acts_2, \Gamma_1, \Gamma_2, \delta)$. Here, $S = Q \times K(C)$ and for each state $\langle q, \kappa \rangle \in S$, the set $\Gamma_i(\langle q, \kappa \rangle)$ is given by:

$$\begin{aligned}\Gamma_i(\langle q, \kappa \rangle) = \{\langle \Delta, a \rangle \in M_i \mid \forall \Delta' \in [0, \Delta] \,.\, \kappa + \Delta' \models Inv_i(q) \;\wedge \\ (a \neq \bot \Rightarrow \exists q' \in Q \,.\, ((q, a, q') \in E \wedge (\kappa + \Delta) \models \theta(q, a, q') \;\wedge \\ (\kappa + \Delta)[\rho(q, a, q') := 0] \models Inv_i(q')))\} \quad \cup \quad \{\langle 0, \bot \rangle\}.\end{aligned}$$

The destination function δ is defined by $\delta(\langle q, \kappa \rangle, \langle \Delta, \bot \rangle) = \langle q, \kappa + \Delta \rangle$, and for $a \in Acts_1 \cup Acts_2$, by $\delta(\langle q, \kappa \rangle, \langle \Delta, a \rangle) = \langle q', \kappa' \rangle$, where q' is the unique location such that $(q, a, q') \in E$ and $(\kappa + \Delta) \models \theta(q, a, q')$, and $\kappa' = (\kappa + \Delta)[\rho(q, a, q') := 0]$. A state, a run, and a player i strategy of $\mathcal{A}$ are, respectively, a state, a run, and a player i strategy of $[\![\mathcal{A}]\!]$. We say that player i wins the goal $\Phi \subseteq S^\omega$ from state $s \in S$ in $\mathcal{A}$ if he wins Φ from s in $[\![\mathcal{A}]\!]$. We say that s is *well-formed* in $\mathcal{A}$ if it is so in $[\![\mathcal{A}]\!]$.

Regions. Timed automaton games, similarly to timed automata, can be analyzed with the help of an equivalence relation of finite index on the set of states. Given a timed automaton game $\mathcal{A}$, for each clock $x \in C$, let c_x be the largest constant in the guards and invariants of $\mathcal{A}$ that involve x, where $c_x = 0$ if x does not occur in any guard or invariant of $\mathcal{A}$. Two clock valuations κ_1, κ_2 are *clock equivalent* if (1) for all $x \in C$, either $\lfloor \kappa_1(x) \rfloor = \lfloor \kappa_2(x) \rfloor$ or both $\lfloor \kappa_1(x) \rfloor > c_x$ and $\lfloor \kappa_2(x) \rfloor > c_x$, (2) the ordering of the fractional parts of the clock variables in the set $\{z\} \cup \{x \in C \mid \kappa_1(x) < c_x\}$ is the same in κ_1 and κ_2, and (3) for all $x \in (\{z\} \cup \{y \in C \mid \kappa_1(y) < c_y\})$, the clock value $\kappa_1(x)$ is an integer if and only if $\kappa_2(x)$ is an integer. A *clock region* is a clock equivalence class, and we write $[\kappa]$ for the clock equivalence class of the clock valuation κ. Two states $\langle q_1, \kappa_1 \rangle$ and $\langle q_2, \kappa_2 \rangle$ are *region equivalent*, written $\langle q_1, \kappa_1 \rangle \equiv \langle q_2, \kappa_2 \rangle$, if (1) $q_1 = q_2$ and (2) κ_1 and κ_2 are clock equivalent. A *region* is an equivalence class with respect to $\equiv$; we write $[s]$ for the region containing state s.

4 Structural Properties of Winning Strategies

We now consider structure theorems for strategies in timed automaton games. Throughout this section, a_1 is an action for player 1, and a_2 one for player 2. For a location p in a timed automaton game $\mathcal{A}$ with clock set C, we let $\Diamond p =$

$\Diamond\{\langle p,\kappa\rangle \mid \kappa \in K(C)\}$ and $\Box p = \Box\{\langle p,\kappa\rangle \mid \kappa \in K(C)\}$.[1] Moreover, $\mathbf{0}$ denotes the valuation that assigns 0 to all clocks in C.

Determinacy. A class $\mathcal{C}$ of timed game structures is *strongly determined* (respectively, *weakly determined*) for a class $\mathcal{F}$ of goals if the following holds for every structure $\mathcal{G} \in \mathcal{C}$, every goal $\Phi \in \mathcal{F}$, all well-formed states s, and each player $i \in \{1,2\}$: if player i cannot win $WC_i(\Phi)$ from s, then there exists a player $\sim i$ strategy $\pi_{\sim i} \in \Pi_{\sim i}$ such that for all player i strategies $\pi_i \in \Pi_i$, we have $Outcomes(s,\pi_1,\pi_2) \cap WC_{\sim i}(\neg\Phi) \neq \emptyset$ (respectively, $Outcomes(s,\pi_1,\pi_2) \not\subseteq WC_i(\Phi)$). Note that this condition is trivially false for non-well-formed states, because one player cannot win the goal S^ω, and the other player surely cannot win the goal $\emptyset$. We let the class of reachability goals be all goals of the form $\Diamond T$.

Theorem 1. *The timed automaton games (and hence, the timed game structures) are neither weakly, nor strongly, determined for the class of reachability goals.*

The following example exhibits a timed automaton game and a goal Φ such that player 1 cannot win $\langle 1\rangle\Phi$, but player 2 does not have a strategy to enforce $WC_2(\neg\Phi)$ (strong) or $\neg WC_1(\Phi)$ (weak), even if player 2 can use the nondeterministic choices to his advantage.

Example 1. Consider Fig. 1(a). It is clear that player 1 does not have a winning strategy for $WC_1(\Diamond q)$ from state $\langle p,\mathbf{0}\rangle$. To prove that this game is not strongly determined, we show that no matter which strategy π_2 is played by player 2, player 1 always has a strategy π_1 such that $Outcomes(\langle p,\mathbf{0}\rangle,\pi_1,\pi_2) \cap WC_2(\neg\Diamond q) = \emptyset$. If π_2 proposes a delay $\Delta_2 > 1$, then π_1 plays the move $\langle \Delta_1, a_1\rangle$ for $\Delta_1 = 1 + (\Delta_2 - 1)/2$; if π_2 proposes a delay $\Delta_2 \leq 1$, then π_1 proposes move $\langle 1,\perp\rangle$. Let $\bar{r} \in Outcomes(\langle p,\mathbf{0}\rangle,\pi_1,\pi_2)$. Then, either $\bar{r}$ contains a player 2 move with a positive delay, in which case q is reached, or player 2 plays $\langle 0,\perp\rangle$ moves forever and is not blameless, i.e., $\bar{r} \notin Blameless_2$. In either case, $\bar{r} \notin WC_2(\neg\Diamond q)$. In a similar way, one shows that the game is not weakly determined.

Memoryless Strategies. Memoryless strategies are strategies that only depend on the last state of a run. Formally, a strategy $\pi \in \Pi$ is *memoryless* if, for all $\bar{r},\bar{r}' \in FRuns$, we have that $last(\bar{r}) = last(\bar{r}')$ implies $\pi(\bar{r}) = \pi(\bar{r}')$. For $i \in \{1,2\}$, we often treat a memoryless strategy π_i for player i as a function in $S \mapsto M_i$ by writing $\pi_i(last(\bar{r}))$ instead of $\pi_i(\bar{r})$. In the untimed case, memoryless strategies are sufficient to win safety and reachability games. In timed games, memoryless strategies suffice to win safety games, i.e., goals of the form $WC_i(\Box T)$; however, winning strategies in reachability games (goals of the form $WC_i(\Diamond T)$) in general do require memory.

[1] We use the standard LTL operators $\Diamond T$ and $\Box T$ to denote, respectively, the set of traces that eventually reach some state in T, and the set of traces that always stay in T [MP91].

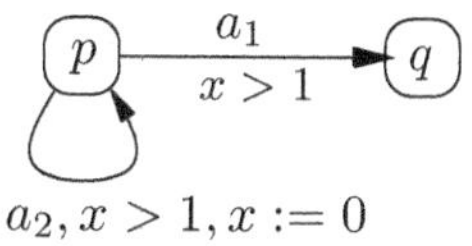

(a) Undetermined.

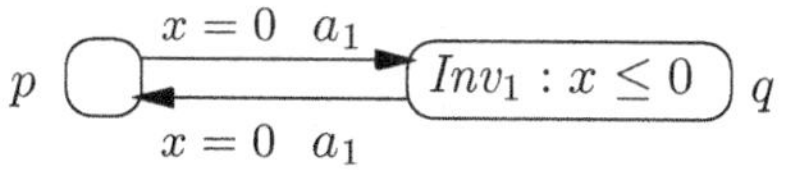

(b) Memory needed.

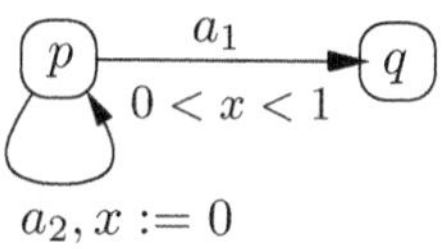

(c) Surprise needed.

Fig. 1. Games with winning condition $WC_1(\Diamond q)$, where $a_1 \in Acts_1$ and $a_2 \in Acts_2$

Theorem 2.

1. *For every well-formed state s of a timed game structure $\mathcal{G}$, and every set T of states of $\mathcal{G}$, if player i has a strategy to win $WC_i(\Box T)$ from s, then player i has a memoryless strategy for winning $WC_i(\Box T)$ from s.*
2. *There exists a timed automaton game $\mathcal{A}$, a state s of $\mathcal{A}$, and a set T of states of $\mathcal{A}$ such that player i has a strategy to win $WC_i(\Diamond T)$ from s, but no memoryless strategy for winning $WC_i(\Diamond T)$ from s.*

The following example proves part 2.

Example 2. Consider the game in Fig. 1(b). Player 1 has a winning strategy for $WC_1(\Diamond q)$ from $\langle p, \mathbf{0}\rangle$, but not a memoryless one: to win, he needs to remember whether q has been visited already. If so, then he has to let time pass, and if not, a visit to q has to be made before letting time pass. Let $\pi : S \mapsto M_1$ be a memoryless strategy for player 1. It is easy to see that, if $\pi(\langle p, \mathbf{0}\rangle) = \langle \Delta, \bot\rangle$, then q will never be reached, and otherwise, if $\pi(\langle p, \mathbf{0}\rangle) = \langle 0, a_1\rangle$, then time will not progress, while π does not ensure that player 1 is blameless. Hence, player 1 cannot win $WC_1(\Diamond q)$ with a memoryless strategy.

No-Surprise Strategies. A no-surprise strategy is a strategy that plays only two kinds of moves: either time steps (action $\bot$, with any delay), or actions with delay 0. Formally, a strategy $\pi \in \Pi$ is *no-surprise* if for all $\bar{r} \in FRuns$ either $\pi(\bar{r}) = \langle 0, a\rangle$ with $a \in Acts$, or $\pi(\bar{r}) = \langle \Delta, \bot\rangle$ with $\Delta \in \mathbb{R}_{\geq 0}$. The following theorem shows that there are cases where surprise is necessary to win, even when the goal is a reachability property, and player 2 is restricted to no-surprise strategies as well.

Theorem 3. *There is a timed automaton game $\mathcal{A}$, a state s of $\mathcal{A}$, and a goal Φ such that player 1 has a strategy to win $WC_1(\Phi)$ from s, but there is no no-surprise strategy $\pi_1 \in \Pi_1$ such that for all no-surprise strategies $\pi_2 \in \Pi_2$, we have $Outcomes(s, \pi_1, \pi_2) \subseteq WC_1(\Phi)$.*

The proof is given by the following example.

Example 3. Consider Fig. 1(c). Player 1 has a strategy to win $WC_1(\Diamond q)$ from state $\langle p, \mathbf{0}\rangle$. For instance, he can play $\pi_1(\bar{r}) = \langle \frac{1}{2^{n+1}}, a_1\rangle$ if $\bar{r}$ contains n visits to p and it ends in $\langle p, \kappa \rangle$ with $\kappa(x) + \frac{1}{2^{n+1}} < 1$; and play $\pi_1(\bar{r}) = \langle 1, \perp \rangle$ in all other cases. Let $\pi_2 \in \Pi_2$ and $\bar{r}$ be a run in $Outcomes(\langle p, \mathbf{0} \rangle, \pi_1, \pi_2)$. If one of his moves $\langle \frac{1}{2^n}, a_1 \rangle$ is taken in $\bar{r}$, then player 1 clearly wins, that is, $\bar{r} \in WC_1(\Diamond q)$. Otherwise, if none of these moves is ever carried out in $\bar{r}$, then player 1 is blameless and, as $\sum_{i=1}^{\infty} \frac{1}{2^i} = 1$, time does not diverge, so $\bar{r} \in WC_1(\Diamond q)$ as well.

However, player 1 does not have a no-surprise strategy to win $WC_1(\Diamond q)$ from $\langle p, \mathbf{0} \rangle$. All no-surprise player-1 strategies π_1 lose against player 2 playing the no-surprise strategy π_2 defined by $\pi_2(\bar{r}) = \langle 0, a_2 \rangle$ if $\bar{r} = \bar{r}' \langle m_1, m_2 \rangle s$ and $m_1 = \langle \Delta, \perp \rangle$; and $\pi_2(\bar{r}) = \langle 1, \perp \rangle$ otherwise. This is because, in order to enable a_1, player 1 has to increase x by taking some move $\langle \Delta, \perp \rangle$ first. However, immediately after he does so, player 2 plays $\langle 0, a_2 \rangle$, thus resetting x. As a result, q is never reached, and both players play infinitely often, so π_1 cannot ensure that player 1 is blameless.

Move Independence. A strategy $\pi \in \Pi$ is *move independent* if, for all $\bar{r}, \bar{r}' \in FRuns$, we have that $states(\bar{r}) = states(\bar{r}')$ implies $\pi(\bar{r}) = \pi(\bar{r}')$. We show that move independent strategies suffice to win a timed automaton game. Note that, for ω-regular goals, this result follows immediately from the strategies derived from the μ-calculus solution for these games; see Sect. 5.

Proposition 1. *Let $\mathcal{A}$ be a timed automaton game and s be a state of $\mathcal{A}$. For every goal Φ, if player i has a strategy to win $WC_i(\Phi)$ from s, then player i has a move independent strategy for winning $WC_i(\Phi)$ from s.*

Persistence. Persistent strategies are strategies that stick with their choices, even if they are interrupted by a move $\langle \Delta, \perp \rangle$ (or another move with the same effect) of the opponent. Formally, a *persistent* player 1 strategy is a strategy $\pi \in \Pi_1$ such that for all finite runs $\bar{r} = \bar{r}' s \langle m_1, m_2 \rangle s'$ with $m_1 = \langle \Delta_1, a_1 \rangle$, $m_2 = \langle \Delta_2, a_2 \rangle$, and $s' = \delta(s, \langle \Delta_2, \perp \rangle)$, we have (1) if $\Delta_2 < \Delta_1$, then $\pi(\bar{r}) = \langle \Delta_1 - \Delta_2, a_1 \rangle$, and (2) if $a_1 \neq \perp$ and $\Delta_1 = \Delta_2$, then $\pi(\bar{r}) = \langle 0, a_1 \rangle$. The persistent player 2 strategies are defined symmetrically. Consider a finite run $\bar{r} = \bar{r}' s \langle m_1, m_2 \rangle s'$. Assume that, in $\bar{r}' s$, player 1 likes to play the move $\pi(\bar{r}' s) = \langle \Delta_1, a_1 \rangle$, but is interrupted because player 2 plays a move $\langle \Delta_2, \perp \rangle$ with $\Delta_2 \leq \Delta_1$. After $\langle \Delta_2, a_2 \rangle$ has been taken, a persistent strategy requires player 1 to play the portion of his previous move $\langle \Delta_1, a_1 \rangle$ which was not carried out; that is, player 1 must play $\langle \Delta_1 - \Delta_2, a_1 \rangle$, unless $\Delta_1 = \Delta_2$ and $a_1 = \perp$. Persistent strategies suffice to win timed games.

Theorem 4. *Let $\mathcal{G}$ be a timed game structure and s be a state of $\mathcal{G}$. For every goal Φ, if player i has a strategy to win $WC_i(\Phi)$ from s, then player i has a persistent strategy for winning $WC_i(\Phi)$ from s.*

5 Solving Timed Automaton Games

In this section, we show how timed automata games can be solved with respect to ω-regular goals via the equational μ-calculus. We consider a goal that is specified by an parity automaton over the set of locations of the timed automaton game, and based on this, we construct another parity automaton that encodes the winning condition. Finally, from the automaton that encodes the winning condition we obtain a μ-calculus formula that, evaluated over the timed automaton game, defines the winning states of the game. Since the μ-calculus formula preserves the regions of the timed automaton game, it provides an algorithm for solving timed automaton games.

5.1 Representing Goals and Winning Conditions

Consider a timed automaton game $\mathcal{A}$ with locations Q and clocks C. A goal $\Phi \subseteq (Q \times K(C))^\omega$ of $\mathcal{A}$ is a *location goal* if it is independent of clock valuations; that is, if $\langle q_0, \kappa_0 \rangle \langle q_1, \kappa_1 \rangle \cdots \in \Phi$, then for all κ'_0, κ'_1, ..., we have $\langle q_0, \kappa'_0 \rangle \langle q_1, \kappa'_1 \rangle \cdots \in \Phi$. Since location goals depend only on the sequence of locations, we view, with abuse of notation, a location goal to be a subset of Q^ω. We consider in this section location goals Φ that are *ω-regular* subsets of Q^ω [Tho90]. Such location goals can be specified by means of deterministic parity automata over the alphabet Q [EJ91]. A *parity automaton* (also known as *Rabin-chain automaton*) of order k over the alphabet Σ is a tuple $H = (P, P_0, \Sigma, \tau, \ell, \Omega)$, where P is the set of locations of the automaton, $P_0 \subseteq P$ is the set of initial locations, $\tau : P \mapsto 2^P$ is the transition relation, $\ell : P \mapsto \Sigma$ assigns to each location $p \in P$ a symbol $\ell(p)$ of the alphabet Σ, and $\Omega : P \mapsto \{0, \ldots, 2k-1\}$ assigns to each location $p \in P$ an index $\Omega(p)$.

An *execution* of H from a source location $p_0 \in P$ is an infinite sequence $p_0, p_1, p_2, \ldots$ of automaton locations such that $p_{j+1} \in \tau(p_j)$ for all $j \geq 0$; if $p_0 \in P_0$, then the execution is *initialized.* The execution $\alpha = p_0, p_1, p_2, \ldots$ *generates* the trace $\ell(\alpha) = \ell(p_0), \ell(p_1), \ell(p_2), \ldots$ of symbols of Σ. Given an execution $\alpha = p_0, p_1, p_2, \ldots$, we denote by $\mathit{MaxIndex}(\Omega, \alpha)$ the largest $j \in \{0, \ldots, 2k-1\}$ such that $\Omega(p_i) = j$ for infinitely many i. The execution α is *accepting* if $\mathit{MaxIndex}(\Omega, \alpha)$ is even. The *language* $\mathcal{L}(H)$ is the set of traces $\rho \in \Sigma^\omega$ such that H has an initialized accepting execution α that generates ρ. The automaton H is *deterministic and total* if (1a) for all locations $p_1, p_2 \in P_0$, if $p_1 \neq p_2$, then $\ell(p_1) \neq \ell(p_2)$; (1b) for all symbols $\sigma \in \Sigma$, there is a location $p \in P_0$ such that $\ell(p) = \sigma$; (2a) for all locations $p_1 \in P$ and $p_2, p_3 \in \tau(p_1)$, if $p_2 \neq p_3$, then $\ell(p_2) \neq \ell(p_3)$; (2b) for all locations $p_1 \in P$ and all symbols $\sigma \in \Sigma$, there is a location $p_2 \in \tau(p_1)$ such that $\ell(p_2) = \sigma$. If H is deterministic and total, then we write $\tau(p_1, \sigma)$ for the unique location p_2 with $\ell(p_2) = \sigma$. Deterministic and total parity automata suffice for recognizing all ω-regular languages [Tho90]. We denote by $|H| = |P|$ the *size* of the automaton, measured as its number of locations, and by $|H|_*$ its order k.

Let $\mathcal{A}$ be a timed automaton game with the set Q of locations, and let Φ be a goal that is specified by means of a deterministic and total parity automaton $H_\Phi = (P, P_0, Q, \tau, \ell, \Omega)$ over the alphabet Q such that $\mathcal{L}(H_\Phi) = \Phi$. The first step towards deriving a μ-calculus formula for computing the winning states of $\mathcal{A}$ with respect to Φ represents the conditions *td* and $\mathit{Blameless}_1$ as ω-regular conditions. To this end, we consider an enlarged state space $\widehat{S} = S \times \{\mathrm{T}, \mathrm{F}\}^2$, and an augmented transition relation $\widehat{\delta} : \widehat{S} \times M_1 \times M_2 \mapsto 2^{\widehat{S}}$. Intuitively, in an augmented state $\langle s, \mathit{tick}, \mathit{bl} \rangle \in \widehat{S}$, the component $s \in S$ is a state of the original game structure $[\![\mathcal{A}]\!]$, *tick* is true if in the last transition the global clock z has crossed an integer boundary, and *bl* is true if player 1 is to blame for the last transition. Precisely, we let $\langle\langle q', \kappa'\rangle, tick', bl'\rangle \in \widehat{\delta}(\langle\langle q, \kappa\rangle, tick, bl\rangle, m_1, m_2)$ iff $\langle q', \kappa'\rangle \in \delta(\langle q, \kappa\rangle, m_1, m_2)$, $tick' = \text{T}$ iff there is $n \in \mathbb{N}$ such that $\kappa(z) \leq n < \kappa'(z)$, and $bl' = \text{T}$ iff $bl_1(\langle q, \kappa\rangle, m_1, m_2, \langle q', \kappa'\rangle)$. The set *td* corresponds to the runs along which *tick* is true infinitely often, and the set $Blameless_1$ corresponds to the runs along which *bl* is true only finitely often. Once time divergence and blame are thus encoded, the winning condition $WC_1(\Phi)$ can be specified by a parity automaton $H_{WC_1(\Phi)}$ with the alphabet $\widehat{\Sigma} = Q \times \{\text{T}, \text{F}\}^2$ and language

$$\mathcal{L}(H_{WC_1(\Phi)}) = \left\{ \begin{array}{l|c} \langle q_0, tick_0, bl_0\rangle, & (q_0, q_1, \ldots \in \mathcal{L}(H_\Phi) \wedge \forall k \in \mathbb{N} \,.\, \exists j \geq k \,.\, tick_j) \\ \langle q_1, tick_1, bl_1\rangle, & \vee \\ \ldots & \exists k \in \mathbb{N}.\forall j \geq k \,.\, (\neg bl_j \wedge \neg tick_j) \end{array} \right\} \quad (1)$$

The automaton $H_{WC_1(\Phi)} = (\widehat{P}, \widehat{P}_0, \widehat{\Sigma}, \widehat{\tau}, \widehat{\ell}, \widehat{\Omega})$ is derived from the automaton H_Φ as follows. Let k be the order of H_Φ. We have $\widehat{P} = P \times \{\text{T}, \text{F}\}^2 \times \{0, \ldots, 2k-1\}$; intuitively, a location $\langle p, tick, bl, h\rangle \in \widehat{P}$ is composed of a location $p \in P$, of two boolean symbols representing the value of *tick* and *bl* at the location, and of an integer h that keeps track of the maximum index of the locations of H_Φ that have been visited between two occurrences of $tick = \text{T}$. For $\langle p, tick, bl, h\rangle \in \widehat{P}$, we define $\widehat{\ell}(\langle p, tick, bl, h\rangle) = \langle \ell(p), tick, bl\rangle$, and we let $\langle p, tick, bl, h\rangle \in \widehat{P}_0$ iff $p \in P_0$. For all $p \in P$, $bl \in \{\text{T}, \text{F}\}$, and $h \in \{0, \ldots, 2k-1\}$, we have $\langle p', tick', bl', h'\rangle \in \widehat{\tau}(\langle p, \text{F}, bl, h\rangle)$ iff $p' \in \tau(p)$ and $h' = \max\{h, \Omega(p')\}$, and we have $\langle p', tick', bl', h'\rangle \in \widehat{\tau}(\langle p, \text{T}, bl, h\rangle)$ iff $p' \in \tau(p)$ and $h' = \Omega(p')$. The index function $\widehat{\Omega} : \widehat{P} \mapsto \{0, \ldots, 2k+1\}$ is defined, for all $p \in P$, all $bl \in \{\text{T}, \text{F}\}$, and all $h \in \{0, \ldots, 2k-1\}$, by $\widehat{\Omega}(\langle p, \text{F}, \text{F}, h\rangle) = 0$, $\widehat{\Omega}(\langle p, \text{F}, \text{T}, h\rangle) = 1$, and $\widehat{\Omega}(\langle p, \text{T}, bl, h\rangle) = h+2$. For all executions $\widehat{\alpha} = \langle p_0, tick_0, bl_0, h_0\rangle, \langle p_1, tick_1, bl_1, h_1\rangle, \langle p_2, tick_2, bl_2, h_2\rangle, \ldots$ of $H_{WC_1(\Phi)}$, let $\alpha = p_0, p_1, p_2, \ldots$ be the corresponding execution in H_Φ. We can show that (a) if there are infinitely many j such that $tick_j = \text{T}$, then $MaxIndex(\widehat{\Omega}, \alpha) = MaxIndex(\Omega, \alpha) + 2$; (b) if there is $k \in \mathbb{N}$ such that $tick_j = bl_j = \text{F}$ for all $j \geq k$, then $MaxIndex(\widehat{\Omega}, \alpha) = 0$; and (c) in all other cases (i.e., when $tick_j$ holds for only finitely many values of j, but bl_j holds for infinitely many values of j), we have $MaxIndex(\widehat{\Omega}, \alpha) = 1$. Together, these facts lead to (1).

Lemma 1. *Given H_Φ, we can construct a deterministic and total parity automaton $H_{WC_1(\Phi)}$ satisfying (1) such that $|H_{WC_1(\Phi)}| = 4 \cdot |H_\Phi| \cdot |H_\Phi|_*$ and $|H_{WC_1(\Phi)}|_* = |H_\Phi|_* + 1$.*

5.2 A μ-Calculus Formula for the Winning States

For all $\langle p, tick, bl, h\rangle \in \widehat{P}$, we let $\widehat{\ell}_Q(\langle p, tick, bl, h\rangle) = \ell(p) \in Q$, $\widehat{\ell}_t(\langle p, tick, bl, h\rangle) = tick$, and $\widehat{\ell}_b(\langle p, tick, bl, h\rangle) = bl$. The fixpoint formula ψ_Φ that solves the game with goal Φ is constructed as follows [dAHM01a]. The formula ψ_Φ is composed of blocks $\mathcal{B}_0, \dots, \mathcal{B}_{2k+1}$, where $\mathcal{B}_0$ is the innermost block and $\mathcal{B}_{2k+1}$ the outermost block. The formula uses the set of variables $\{x_j^{\widehat{p}} \mid \widehat{p} \in \widehat{P}, j \in \{0, \dots, 2k+1\}\} \cup \{y\}$, which take values in 2^S, where S is the set of states of the game structure $\mathcal{A}$. The block $\mathcal{B}_0$ is a ν-block which consists of all equations of the form

$$x_0^{\widehat{p}} = (\widehat{\ell}_Q(\widehat{p}) \times K(C)) \cap CPre_1\left(\bigvee_{\widehat{p}' \in \tau(\widehat{p})} x_{\widehat{\Omega}(\widehat{p})}^{\widehat{p}'} \times \widehat{\ell}_t(\widehat{p}') \times \widehat{\ell}_b(\widehat{p}') \right)$$

for $\widehat{p} \in \widehat{P}$, where C is the set of clocks of $\mathcal{A}$. For $0 < j < 2k+1$, the block $\mathcal{B}_j$ is a μ-block if j is odd, and a ν-block if j is even; in either case it consists of the set of equations $\{x_j^{\widehat{p}} = x_{j-1}^{\widehat{p}} \mid \widehat{p} \in \widehat{P}\}$. The block $\mathcal{B}_{2k+1}$ consists of the set of equations $\{x_{2k+1}^{\widehat{p}} = x_{2k}^{\widehat{p}} \mid \widehat{p} \in \widehat{P}\} \cup \{y = \bigvee_{\widehat{p} \in \widehat{P}_0} x_{2k+1}^{\widehat{p}}\}$. The output variable is y.

The operator $CPre_1 : \widehat{S} \mapsto S$ is the *controllable predecessor operator*, defined by $\exists m_1 \in \Gamma_1(s) . \forall m_2 \in \Gamma_2(s) . \widetilde{\delta}(s, m_1, m_2) \in X$. Intuitively, for $s \in S$ and $\widehat{X} \subseteq \widehat{S}$, we have that $s \in CPre_1(\widehat{X})$ if player 1 can force the augmented game to $\widehat{X}$ in one move. As an example, consider the set $\widehat{X} = (X_1 \times \{\text{F}\} \times \{\text{T}\}) \cup (X_2 \times \{\text{F}\} \times \{\text{F}\})$ for some $X \subseteq S$. Then, $s \in CPre_1(\widehat{X})$ if player 1 has a move such that, whatever the move played by player 2: either (a) the game proceeds to X_1, the global clock z does not advance beyond an integer boundary ($tick = \text{F}$), and player 1 is blamed ($bl = \text{T}$); or (b) the game proceeds to X_2, the global clock z does not advance beyond an integer boundary, and player 1 is not blamed. The implementation and properties of operator $CPre_1$ are discussed below. Note that the formula ψ_Φ depends only on H_Φ, but not on the timed game structure over which it is evaluated (except trivially via the product with $K(C)$, which is simply the set of all clock valuations). Denote by $[\![y]\!]_{\mathcal{A}}^{\psi_\Phi} \subseteq S$ the fixpoint valuation of y over the timed game structure $[\![\mathcal{A}]\!]$. Lemma 2 enables the computation of the winning states of the game with respect to player 1; the winning states with respect to player 2 can be computed in a symmetrical fashion.

Lemma 2. *We have* $\langle 1 \rangle \Phi = [\![y]\!]_{\mathcal{A}}^{\psi_\Phi}$.

5.3 The Controllable Predecessor Operator

The operator $CPre_1$ can be computed as follows. For $X \subseteq \widehat{S}$, write $X = (X_{\text{T}} \times \{\text{T}\}) \cup (X_{\text{F}} \times \{\text{F}\})$, for $X_{\text{T}}, X_{\text{F}} \subseteq S \times \{\text{T}, \text{F}\}$. Intuitively, X_{T} (resp. X_{F}) represents the portion of X that corresponds to the case where bl is T (resp. F). Then, $s \in CPre_1(X)$ if and only if:

$$\begin{aligned} &\exists \langle \Delta_1, a_1 \rangle \in \Gamma_1(s) . \\ &\quad \forall \langle \Delta_2, a_2 \rangle \in \Gamma_2(s) . \Big(\Delta_2 \leq \Delta_1 \implies (\delta(s, \langle \Delta_2, a_2 \rangle), tick(s, \Delta_2)) \in X_{\text{F}} \Big) \wedge \\ &\quad \Big((\delta(s, \langle \Delta_1, a_1 \rangle), tick(s, \Delta_1)) \in X_{\text{T}} \vee \forall \langle \Delta_2, a_2 \rangle \in \Gamma_2(s) . \Delta_2 < \Delta_1 \Big), \end{aligned}$$

where $tick(\langle q, \kappa \rangle, \Delta)$ is T iff $\kappa(z) \leq n < \kappa(z) + \Delta$, for some integer n. In words, the above formula states that there is a player 1 action that, played with delay Δ_1, leads to X_T; moreover, all actions of player 2, if played with delay up to Δ_1, lead to X_F. The following lemma states that the controllable predecessor operator preserves regions for timed automaton games.

Lemma 3. *For $n \geq 0$, consider $X = \bigcup_{j=1}^{n} (X_j \times \{tick_j\} \times \{bl_j\})$, where for $1 \leq j \leq n$, the set X_j is a region, and $tick_j, bl_j \in \{\text{T}, \text{F}\}$. Then, $CPre_1(X)$ is a union of regions.*

5.4 Putting It All Together

From the constructions of the previous subsections, we obtain the following decidability result for timed automaton games with ω-regular location goals.

Theorem 5. *Consider a timed automaton game $\mathcal{A}$ with the set Q of locations, and a parity automaton H_Φ that specifies a location goal $\Phi \subseteq Q^\omega$. Let C be the set of clocks of $\mathcal{A}$, let $m = |C|$, and let $c = \max\{c_x \mid x \in C\}$. Then, the set of winning states $\langle 1 \rangle \Phi$ can be computed in time $O((|Q| \cdot m! \cdot 2^m \cdot (2c+1)^m \cdot |H_\Phi| \cdot |H_\Phi|_*)^{(|H_\Phi|_*+1)})$.*

Corollary 1. *The problem of solving a timed automaton game for a location goal specified by a parity automaton is EXPTIME-complete.*

EXPTIME-hardness follows from the EXPTIME-hardness for alternating reachability on timed automata [HK99]. Membership in EXPTIME is shown by the exponential-time algorithm outlined above. The algorithm for solving timed automaton games can also be used to simultaneously construct a winning strategy for player 1, as in [dAHM01b]. The winning strategies thus constructed have the following finitary structure. Two finite runs $\bar{r} = s_0, \langle m_1^1, m_1^2 \rangle, s_1, \ldots, s_k$ and $\bar{r}' = s'_0, \langle m'^1_1, m'^2_1 \rangle, s'_1, \ldots, s'_k$ of the same length are *region equivalent*, written $\bar{r} \equiv \bar{r}'$, if for all $0 \leq j \leq k$, we have $[s_j] = [s'_j]$. A strategy is a *region strategy* if, for region equivalent finite runs, it prescribes moves to the same region. Formally, a strategy $\pi \in \Pi$ is a *region strategy* if for all $\bar{r}, \bar{r}' \in FRuns$, we have that $\bar{r} \equiv \bar{r}'$ implies $\delta(last(\bar{r}), \pi(\bar{r})) \equiv \delta(last(\bar{r}), \pi(\bar{r}'))$. Since the $CPre_1$ operator preserves regions, we can show that the strategy constructed by the above algorithm does not distinguish between region equivalent runs, and hence, the constructed strategy is a region strategy.

Theorem 6. *Let $\mathcal{A}$ be a timed automaton game and s a state of $\mathcal{A}$. For every ω-regular location goal Φ, if player i has a strategy to win $WC_i(\Phi)$ from s, then player i has a region strategy for winning $WC_i(\Phi)$ from s.*

References

[AD94] R. Alur and D.L. Dill. A theory of timed automata. *Theor. Comp. Sci.*, 126:183–235, 1994.

[AdAHM99] R. Alur, L. de Alfaro, T.A. Henzinger, and F.Y.C. Mang. Automating modular verification. In *Concurrency Theory*, Lect. Notes in Comp. Sci. 1664, pages 82–97. Springer, 1999.

[AH97] R. Alur and T.A. Henzinger. Modularity for timed and hybrid systems. In *Concurrency Theory*, Lect. Notes in Comp. Sci. 1243, pages 74–88. Springer, 1997.

[AHK02] R. Alur, T.A. Henzinger, and O. Kupferman. Alternating-time temporal logic. *J. ACM*, 49:672–713, 2002.

[AMPS98] E. Asarin, O. Maler, A. Pnueli, and J. Sifakis. Controller synthesis for timed automata. In *Proc. IFAC Symp. System Structure and Control*, pages 469–474. Elsevier, 1998.

[Chu63] A. Church. Logic, arithmetics, and automata. In *Proc. Int. Congress of Mathematicians, 1962*, pages 23–35. Institut Mittag-Leffler, 1963.

[dAHM01a] L. de Alfaro, T.A. Henzinger, and R. Majumdar. From verification to control: Dynamic programs for omega-regular objectives. In *Proc. Symp. Logic in Comp. Sci.*, pages 279–290. IEEE, 2001.

[dAHM01b] L. de Alfaro, T.A. Henzinger, and R. Majumdar. Symbolic algorithms for infinite-state games. In *Concurrency Theory*, Lect. Notes in Comp. Sci. 2154, pages 536–550. Springer, 2001.

[dAHS02] L. de Alfaro, T.A. Henzinger, and M.I.A. Stoelinga. Timed interfaces. In *Embedded Software*, Lect. Notes in Comp. Sci. 2491, pages 108–122. Springer, 2002.

[EJ91] E.A. Emerson and C.S. Jutla. Tree automata, mu-calculus, and determinacy. In *Proc. Symp. Foundations of Comp. Sci.*, pages 368–377. IEEE, 1991.

[FLM02] M. Faella, S. La Torre, and A. Murano. Dense real-time games. In *Proc. Symp. Logic in Comp. Sci.*, pages 167–176. IEEE, 2002.

[HHM99] T.A. Henzinger, B. Horowitz, and R. Majumdar. Rectangular hybrid games. In *Concurrency Theory*, Lect. Notes in Comp. Sci. 1664, pages 320–335. Springer, 1999.

[HK99] T.A. Henzinger and P.W. Kopke. Discrete-time control for rectangular hybrid automata. *Theor. Comp. Sci.*, 221:369–392, 1999.

[MP91] Z. Manna and A. Pnueli. *The Temporal Logic of Reactive and Concurrent Systems: Specification.* Springer, 1991.

[MPS95] O. Maler, A. Pnueli, and J. Sifakis. On the synthesis of discrete controllers for timed systems. In *Theor. Aspects of Comp. Sci.*, Lect. Notes in Comp. Sci. 900, pages 229–242. Springer, 1995.

[PR89] A. Pnueli and R. Rosner. On the synthesis of a reactive module. In *Proc. Symp. Principles of Programming Languages*, pages 179–190. ACM, 1989.

[RW89] P.J.G. Ramadge and W.M. Wonham. The control of discrete-event systems. *IEEE Transactions on Control Theory*, 77:81–98, 1989.

[SGSAL98] R. Segala, G. Gawlick, J. Søgaard-Andersen, and N. Lynch. Liveness in timed and untimed systems. *Info. and Comput.*, 141:119–171, 1998.

[Tho90] W. Thomas. Automata on infinite objects. In J. van Leeuwen, ed., *Handbook Theor. Comp. Sci.*, vol. B, pages 135–191. Elsevier, 1990.

[Yi90] W. Yi. Real-time behaviour of asynchronous agents. In *Concurrency Theory*, Lect. Notes in Comp. Sci. 458, pages 502–520. Springer, 1990.

Deciding Bisimilarity between BPA and BPP Processes*

Petr Jančar[1], Antonín Kučera[2], and Faron Moller[3]

[1] Dept. of Computer Science, FEI, Technical University of Ostrava
17. listopadu 15, 70833 Ostrava, Czech Republic
Petr.Jancar@vsb.cz

[2] Faculty of Informatics, Masaryk University
Botanická 68a, 60200 Brno, Czech Republic
email:tony@fi.muni.cz

[3] Dept. of Computer Science, University of Wales Swansea
Singleton Park, Swansea SA2 8PP, Wales
F.G.Moller@swansea.ac.uk

Abstract. We identify a necessary condition for when a given BPP process can be expressed as a BPA process. We provide an effective procedure for testing if this condition holds of a given BPP, and in the positive case we provide an effective construction for a particular form of one-counter automaton which is bisimilar to the given BPP. This in turn provides the mechanism to decide bisimilarity between a given BPP process and a given BPA process.

1 Introduction

During the last decade, a great deal of research effort has been devoted to the study of decidability and complexity issues for checking semantic equivalences, in particular bisimilarity, between various classes of processes. There have been several surveys presenting this work (eg, [16,13,12]), including a major Handbook chapter [3]. There is even now a project devoted to maintaining an up-to-date comprehensive overview of the state-of-the-art in this dynamic research topic [19].

Example classes of processes of particular interest in this study are pushdown automata, Petri nets, and the process algebra PA. Some milestones in the study, beginning with the decidability of bisimilarity over normed BPA [1] include the undecidability of bisimilarity over Petri nets [10]; the decidability of bisimilarity over normed PA [7]; and the decidability of bisimilarity over the class of strict deterministic grammars (a particular formulation of deterministic pushdown automata) [23]. This final result reinforces Sénizergues' solution [18] to the long-standing equivalence problem for deterministic pushdown automata. A closely related result is the decidability of bisimilarity over state-extended BPA [17,22].

The motivation for the present study is to work towards a generalization of the above decidability result for normed PA to the whole class of PA processes. The process algebra PA includes operators for composing terms both sequentially and in parallel, and as described by Hirshfeld and Jerrum in [7], there are surprising interactions between

* The first two authors are supported by the Grant Agency of the Czech Republic, grant No. 201/03/1161.

R. Amadio, D. Lugiez (Eds.): CONCUR 2003, LNCS 2761, pp. 159–173, 2003.

sequential and parallel compositions. Indeed, one can express the sequential composition $X_1 \cdot X_2$ of two terms X_1 and X_2 as a parallel composition $Y_1 || Y_2$ of two other terms Y_1 and Y_2 in infinitely-many ways, using terms of unbounded complexity. By restricting to normed process terms, Hirshfeld and Jerrum were able to develop a structural theory which allowed them, in effect, to finitely characterize the infinite set of solutions to the equivalence $X_1 \cdot X_2 = Y_1 || Y_2$, and then use this characterization to provide their decidability result. However, it remains open as to how to extend their techniques to the unnormed PA case.

The process class BPA represents the subset of PA involving only sequential composition, while BPP represents the subset involving only parallel composition. As such, in light of the above observations, it becomes natural to consider the problem of comparing an arbitrary BPA term with an arbitrary BPP term. Decidability between such a pair of terms in the normed case of course follows from the above result, though this problem was already settled in [2,5].

Bisimulation checking over normed process classes has typically proven to be far more tractable than over unnormed processes. For example, over both the BPA and BPP process classes, unique decomposability results in the spirit of [15] hold over the normed subclasses which allow for polynomial decision procedures in each case [8,9]. Though decidability of bisimilarity in the unnormed cases has been known for some time, bounds on their complexity have been elusive. Recently both problems have been shown to be PSPACE hard [20,21], and even more recently the problem for BPP has been shown to be PSPACE-complete [11]. Various novel techniques are developed in each of the above papers which contribute towards an understanding of the nature of these classes of sequential and parallel process.

In this paper we continue the exposition of these classes of processes and consider the problem of when an arbitrary BPP process can be expressed as a BPA process term. To this end, we identify a property of (arbitrary) processes which cannot be modelled "sequentially"; essentially this property entails encoding two distinct, unrelated and sufficiently-large integer values. If our given BPP process can be expressed as a BPA term, then clearly it is necessary that the process does not possess this property. Furthermore, we demonstrate how to test if a given BPP process possesses this property, and in the case that it does not, we provide an effective construction of an equivalent one-counter automaton. As one-counter automata and BPA both constitute subclasses of state-extended BPA, we arrive at the decidability of bisimilarity between BPA and BPP processes from the afore-mentioned decidability of bisimilarity over state-extended BPA.

The structure of the paper is as follows. In Sect. 2 we present various preliminary definitions, and in Sect. 3 we explore the structure of BPA processes and provide the crucial technical result of the paper. Finally, in Sect. 4 we prove our decidability result by characterizing when a BPP process abides by the structural restrictions of BPA processes exposed in Sect. 3, and then demonstrating how to decide equivalence to a given BPA process in the case where this is true.

2 Preliminary Definitions and Results

2.1 Processes and Norms

Formally, a *process* is represented by (a state in) a *labelled transition system* defined as follows.

Definition 1. *A* labelled transition system (LTS) *is a triple* $\mathcal{S} = (S, Act, \rightarrow)$ *where* S *is a set of* states, Act *is a finite set of* actions, *and* $\rightarrow \subseteq S \times Act \times S$ *is a* transition relation.

We write $s \xrightarrow{a} \bar{s}$ instead of $(s, a, \bar{s}) \in \rightarrow$ and we extend this notation to elements of Act^* in the natural way. We also use $s \rightarrow \bar{s}$ to mean $s \xrightarrow{a} \bar{s}$ for some $a \in Act$. A state $\bar{s}$ is *reachable* from a state s if $s \rightarrow^* \bar{s}$, that is, if $s \xrightarrow{w} \bar{s}$ for some $w \in Act^*$.

The notion of "behavioural sameness" between two processes can be formally captured in many different ways (see, e.g., [6] for an overview). Among those behavioural equivalences, *bisimulation equivalence* enjoys special attention. Its formal definition is as follows.

Definition 2. *Let* $\mathcal{S} = (S, Act, \rightarrow_{\mathcal{S}})$ *and* $\mathcal{T} = (T, Act, \rightarrow_{\mathcal{T}})$ *be transition systems defined over the same action set Act. A binary relation* $\mathcal{R} \subseteq S \times T$ *is a* bisimulation relation *iff whenever* $(s, t) \in R$*, we have that*

- *for each transition* $s \xrightarrow{a}_{\mathcal{S}} \bar{s}$ *there is a transition* $t \xrightarrow{a}_{\mathcal{T}} \bar{t}$ *such that* $(\bar{s}, \bar{t}) \in \mathcal{R}$*; and*
- *for each transition* $t \xrightarrow{a}_{\mathcal{T}} \bar{t}$ *there is a transition* $s \xrightarrow{a}_{\mathcal{S}} \bar{s}$ *such that* $(\bar{s}, \bar{t}) \in \mathcal{R}$.

Processes s *and* t *are* bisimulation equivalent (bisimilar), *written* $s \sim t$*, iff they are related by some bisimulation.*

An important subclass of processes are the *normed* processes, which are those for which from any state there is a sequence of transitions leading to a state having no transitions leading out of it; the *norm* of a process state is then traditionally defined to be the length of a shortest sequence of transitions leading to such a deadlocked state. We can generalize the notion of a norm as follows. (We let $\mathbb{N} = \{0, 1, 2, \ldots\}$ represent the set of natural numbers, $\mathbb{N}_\omega = \mathbb{N} \cup \{\omega\}$, and $\mathbb{N}_{\omega,-1} = \mathbb{N} \cup \{\omega, -1\}$, where $\omega + \omega = \omega + k = \omega - k = \omega - \omega = \omega$, and $k < \omega, \omega \leq \omega$ for every $k \in \mathbb{N} \cup \{-1\}$.)

Definition 3. *Let* $\mathcal{S} = (S, Act, \rightarrow)$ *be a transition system, and* $d : S \rightarrow \mathbb{N}_\omega$ *a function. We say that* d *is a* norm *iff for all* $s \in S$ *we have the following:*

- *If* $s \rightarrow s'$*, then* $d(s') \geq d(s) - 1$*; and*
- *If* $0 < d(s) < \omega$*, then there is* $s \rightarrow s'$ *such that* $d(s') = d(s) - 1$.

In the latter clause, we call such a transition a d-reducing transition.

It is possible to construct new norms out of already existing ones; for example, if d and d' are norms, then so is $\min(d, d')$. For our purposes, the following construction is particularly important. Firstly, for all $s, s' \in S$ we define the *distance* from s to s', denoted $dist(s, s')$, to be the length of a shortest sequence of transitions leading from state s to state s':

$$dist(s, s') = \min \{ \, length(w) : s \stackrel{w}{\rightarrow} s' \, \}.$$

Adhering to the convention that $\min \emptyset = \omega$, we note that $dist(s, s') = \omega$ when there is no such sequence.

Given a tuple of norms $\mathcal{F} = (d_1, \ldots, d_k)$, each transition $s \stackrel{a}{\rightarrow} s'$ determines a unique *change* of $\mathcal{F}$, denoted $\delta^{\mathcal{F}}(s \stackrel{a}{\rightarrow} s')$, which is a k-tuple of values from $\mathbb{N}_{\omega,-1}$ defined by $\delta^{\mathcal{F}}(s \stackrel{a}{\rightarrow} s') = \big(d_1(s') - d_1(s), \ldots, d_k(s') - d_k(s)\big)$. For each triple $(a, \mathcal{F}, \delta)$, where $a \in Act$, $\mathcal{F} = (d_1, \ldots, d_k)$ is a tuple of norms, and $\delta \in \mathbb{N}^k_{\omega,-1}$, we define the function $dd_{(a,\mathcal{F},\delta)} : S \rightarrow \mathbb{N}_\omega$ to be the distance to a state for which all norms of $\mathcal{F}$ are finite, and for which there is no a-transition with the change δ:

$$dd_{(a,\mathcal{F},\delta)}(s) = \min \{ \, dist(s, s') : d_i(s') \neq \omega \text{ for all } i, \text{ and } \\ \delta^{\mathcal{F}}(s' \stackrel{a}{\rightarrow} s'') \neq \delta \text{ for all } s' \stackrel{a}{\rightarrow} s'' \, \}.$$

Obviously, each such $dd_{(a,\mathcal{F},\delta)}$ is a norm.

Some norms are not semantically relevant in the sense that bisimilarity does not necessarily preserve them. In this paper we are mainly interested in *bisimulation-invariant* norms.

Definition 4. *We say that a given norm d is* bisimulation-invariant *if $s \sim s'$ implies $d(s) = d(s')$.*

A simple example of a bisimulation-invariant norm is the function dd_a (where $a \in Act$) defined as follows:

$$dd_a(s) = \min \{ \, dist(s, s') : s' \stackrel{a}{\not\rightarrow} \, \}.$$

Definition 5. *The set of* DD-functions *is defined inductively as follows:*

- *dd_a is a DD-function for every $a \in Act$;*
- *if $\mathcal{F} = (d_1, \cdots, d_k)$ is a tuple of DD-functions, $\delta \in \mathbb{N}^k_{\omega,-1}$, and $a \in Act$, then $dd_{(a,\mathcal{F},\delta)}$ is also a DD-function.*

A simple observation is that if $d_1, \cdots, d_k$ are bisimulation-invariant norms, then $dd_{(a,\mathcal{F},\delta)}$ is also a bisimulation-invariant norm for each triple $(a, \mathcal{F}, \delta)$. This in turn implies that all DD-functions are bisimulation-invariant. For each DD-function d we further define the sets $\mathcal{D}(d)$ and $\mathcal{C}(d)$ of all DD-functions and changes which are employed during the construction of d:

- $\mathcal{D}(dd_a) = \mathcal{C}(dd_a) = \emptyset$ for every action a;
- if $d = dd_{(a,\mathcal{F},\delta)}$, where $\mathcal{F} = (d_1, \cdots, d_k)$, then
  - $\mathcal{D}(d) = \mathcal{D}(d_1) \cup \cdots \cup \mathcal{D}(d_k) \cup \{d_1, \cdots, d_k\}$,
  - $\mathcal{C}(d) = \mathcal{C}(d_1) \cup \cdots \cup \mathcal{C}(d_k) \cup \{\delta\}$.

2.2 BPA, BPP, Petri Nets, and One Counter Automata

A BPA process is defined by a context-free grammar in Greibach normal form. Formally this is given by a triple $G = (V, A, \Gamma)$, where V is a finite set of *variables* (*nonterminal symbols*), A is a finite set of *labels* (*terminal symbols*), and $\Gamma \subseteq V \times A \times V^*$ is a finite set of *rewrite rules* (*productions*); it is assumed that every variable has at least one associated rewrite rule. Such a grammar gives rise to the LTS $\mathcal{S}_G = (V^*, A, \rightarrow)$ in which the states are sequences of variables, the actions are the labels, and the transition relation is given by the rewrite rules extended by the *prefix rewriting rule*: if $(X, a, \alpha) \in \Gamma$ then $X\beta \xrightarrow{a} \alpha\beta$ for all $\beta \in V^*$. In this way, concatenation of variables naturally represents sequential composition.

A BPP process is defined in exactly the same fashion from such a grammar. However, in this case elements of V^* are read modulo commutativity of concatenation, so that concatenation is interpretted as parallel composition rather than sequential composition. The states of the BPP process associated with a grammar are thus given not by sequences of variables but rather by multisets of variables.

In either case, BPA or BPP, the usual notion of the *norm* of a state $\alpha \in V^*$, denoted $|\alpha|$, is the length of a shortest path to the empty process ε: $|\alpha| = dist(\alpha, \varepsilon)$. If all variables of the underlying grammar have finite norm, then the process is said to be *normed*; otherwise it is *unnormed*.

As an example, Fig. 1 depicts the (normed) BPA and BPP processes defined by the same grammar given by the three rules $A \xrightarrow{a} AB$, $A \xrightarrow{c} \varepsilon$ and $B \xrightarrow{b} \varepsilon$. It can easily be shown [3] that the BPA process cannot be expressed by any BPP process, and equally the BPP process cannot be expressed by any BPA process. Fig. 2 on the other hand depicts an example of an (unnormed) process which is definable both as a BPA process and as a BPP process.

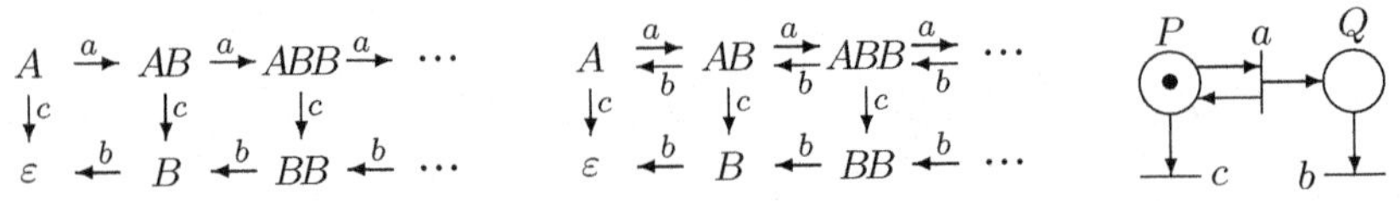

Fig. 1. BPA and BPP processes defined by the grammar $A \xrightarrow{a} AB$, $A \xrightarrow{c} \varepsilon$, $B \xrightarrow{b} \varepsilon$, along with a BPP net equivalent to the BPP

$$A \xrightarrow{a} BA \quad A \xrightarrow{a} BBA \quad A \xrightarrow{c} A \qquad P \xrightarrow{a} PQ \quad P \xrightarrow{a} PR \quad P \xrightarrow{c} P$$
$$B \xrightarrow{a} BB \quad B \xrightarrow{a} BBB \quad B \xrightarrow{c} B \qquad Q \xrightarrow{b} \varepsilon \quad R \xrightarrow{b} S \quad S \xrightarrow{b} \varepsilon$$
$$B \xrightarrow{b} \varepsilon$$

$\mathcal{R} = \left\{(PQ^iR^jS^k, B^{i+2j+k}A) : i, j, k \geq 0\right\}$ is a bisimulation showing that $P \sim A$.

Fig. 2. A BPA process, and an equivalent BPP process with its associated BPP net

An equivalent formulation of BPP, and one which we adopt to aid in distinguishing between BPA and BPP processes, is as labelled Petri nets in which each transition has a unique input place. Formally, a labelled Petri net is a tuple $\mathcal{N} = (P, T, F, A, \ell)$ where P and T are finite and disjoint sets of *places* and *transitions*, respectively, $F : (P \times T \cup T \times P) \to \mathbb{N}$ is a *flow function*, A is a set of *labels*, and $\ell : T \to A$ is a *labelling*. A *marking* is a function $M : P \to \mathbb{N}$ which associates to each place a finite number of *tokens*. A transition t is *enabled* at a marking M if $M(p) \geq F(p,t)$ for each place p. If t is enabled at M, it can be *fired* from M, producing a new marking M' defined by $M'(p) = M(p) - F(p,t) + F(t,p)$. This is written as $M \xrightarrow{t} M'$. To every labelled Petri net $\mathcal{N}$ we associate a transition system where the set of states is the set of all markings, A is the set of all labels, and $M \xrightarrow{a} M'$ iff there is transition t such that $\ell(t) = a$ and $M \xrightarrow{t} M'$.

Petri nets are often depicted as graphs with two kinds of nodes (corresponding to places and transitions) where the flow function is indicated by (multiple) arcs between places and transitions. A *BPP net* is a Petri net where for every $t \in T$ there is exactly one place $Pre(t)$ such that $F(Pre(t), t) = 1$ and $F(p,t) = 0$ for every other place p. The equivalence of these BPP nets to BPP is easily seen from the Petri net presented in Figs. 1 and 2.

We shall find the following Petri net concepts useful.

Definition 6. *For a set Q of places of a Petri net, we define* $\text{NORM}(Q)(M)$ *to be the length of a shortest sequence of transitions from M which leaves all of the places of Q empty:*

$$\text{NORM}(Q)(M) \;=\; \min\{\, dist(M, M') : M'(p) = 0 \textit{ for every } p \in Q \,\}.$$

We may readily observe that for every set Q of places, $\text{NORM}(Q)$ is a norm in the sense of Definition 3. In the case of BPP nets, $\text{NORM}(Q)(M)$ is just a weighted sum of tokens in M; that is, to each place p we can effectively associate some $c_p \in \mathbb{N}_\omega$ (which depends only on Q) so that $\text{NORM}(Q)(M) = \sum_{p \in P} c_p \cdot M(p)$ for every marking M.

Definition 7. *A set Q of places of a Petri net is called a* trap *iff for all transitions t we have that if $\sum_{p \in Q} F(p,t) > 0$ then $\sum_{p \in Q} F(t,p) > 0$.*

Thus a "marked" trap, that is, a trap containing at least one token, can never become unmarked. This then implies that for any trap Q, $\text{NORM}(Q)(M)$ is either 0 or ω.

We can note that $\emptyset$ is a trap, and that the union of traps is again a trap. This justifies the following definition.

Definition 8. $\text{MAXTRAP}(Q)$ *denotes the maximal trap contained in the set of places Q.*

Finally, we shall have cause to consider the following class of one-counter automata.

Definition 9. *A* one-counter automaton with resets *(OCR) is a tuple $\mathcal{P} = (Q, A, I, Z, \delta)$ where Q is a finite set of* control states, *A is a finite* input alphabet, *I and Z are* counter symbols *and δ is a finite set of* rules *which are of one of the following three forms:*

- $pZ \xrightarrow{a} qI^kZ$ *where $p, q \in Q$, $a \in A$, and $k \in \mathbb{N}$; these rules are called* zero rules.

- $pI \xrightarrow{a} qI^k$ *where* $p, q \in Q$, $a \in A$, *and* $k \in \mathbb{N}$; *these rules are called* positive *rules.*
- $pI \xrightarrow{a} qZ$ *where* $p, q \in Q$ *and* $a \in A$; *these rules are called* resets.

Hence, Z acts as a bottom symbol (which cannot be removed), and the number of I's which are stored in the stack above the topmost occurrence of Z represents the counter value. The reset (i.e., setting the counter back to zero) is implemented by pushing the symbol Z onto the stack.

To the OCR $\mathcal{P}$ we associate the transition system $\mathcal{S}_{\mathcal{P}}$ where $Q \times \{I, Z\}^*$ is the set of states, A is the set of actions, and the transition relation is determined by

$$pX\alpha \xrightarrow{a} q\beta\alpha \text{ iff } pX \xrightarrow{a} q\beta \in \delta.$$

3 Prefix-Encoded Norms over BPA

In this section we demonstrate that large values of DD-functions are represented by large (normed) prefixes of BPA states. This will provide a necessary condition for when a BPP process can be bisimilar to a BPA process.

Definition 10. *We define the* pseudo-norm *(or* prefix-norm*)* $pn(\alpha)$ *of a BPA process* α *as follows:*

$$pn(\alpha) = \max\{\, |\beta| : \alpha = \beta\gamma \text{ and } |\beta| < \omega \,\}.$$

We call the transition $X\beta \xrightarrow{a} \gamma\beta$ *a* pn-reducing step *iff* $|\gamma| = |X| - 1 < \omega$.

Note that pn is *not* a norm in the sense of Definition 3, and that a step $X\beta \xrightarrow{a} \gamma\beta$ such that $pn(\gamma\beta) = pn(X\beta) - 1$ is not necessarily pn-reducing.

Definition 11. *We say that a norm d is* prefix-encoded *for a given BPA process G iff there is a constant $C \in \mathbb{N}$ such that for every process α of G with $C < d(\alpha) < \omega$, the d-reducing steps from α are exactly the pn-reducing steps from α. In this case, we say that d is* prefix-encoded above C.

Given a BPA process, we define the value MAXSTEP as the maximal value $|\beta| - |X|$ where $X \xrightarrow{a} \beta$ is a rule with $|\beta| < \omega$. For any norm d we easily observe the following.

Proposition 12.

(a) $d(\alpha\beta) \le |\alpha| + d(\beta)$.
(b) If $d(\alpha\beta) < |\alpha| + d(\beta)$ *then* $\alpha \rightarrow^* \alpha'$ *with* $d(\alpha'\beta) = 0$.
(c) If d is prefix-encoded above C and $d(\alpha\beta) \ge |\alpha| + C$ *then* $d(\alpha\beta) = |\alpha| + d(\beta)$; *and if* $\alpha \rightarrow \alpha'$ *with* $|\alpha'| < \omega$ *then* $d(\alpha\beta) - 1 \le d(\alpha'\beta) \le d(\alpha\beta) + \text{MAXSTEP}$.
(d) If $d(X\alpha) < \omega$ *and there is a transition from $X\alpha$ which is d-reducing or pn-reducing but not both, then there is a maximal sequence* $X\alpha \xrightarrow{w} \beta$ *of d-reducing transitions such that* $X \xrightarrow{w} \gamma$ *with* $\gamma \ne \varepsilon$; *this implies that* $X\alpha \xrightarrow{w} \gamma\alpha$ *and* $d(\gamma\alpha) = 0$.

Lemma 13. *For any BPA process, each DD-function is prefix-encoded.*

Proof. We assume a given BPA process, and show the claim by contradiction. We first define some technical notions. For a DD-function d, we say that two states α_1 and α_2 are (d, C)-*diff-large* (for $C \in \mathbb{N}$) iff for each $d' \in \{d\} \cup \mathcal{D}(d)$ such that $d'(\alpha_1) \neq d'(\alpha_2)$ we have the following inequalities:

$$C < d'(\alpha_1) < \omega;$$
$$C < d'(\alpha_2) < \omega; \quad \text{and} \quad |d'(\alpha_2) - d'(\alpha_1)| > C.$$

We say that the states α_1 and α_2 are d-*bad* iff for some $\gamma \neq \varepsilon$:

$$0 = d(\gamma\alpha_1) < d(\gamma\alpha_2) \quad \text{or}$$
$$0 = d(\gamma\alpha_2) < d(\gamma\alpha_1).$$

Claim 1. If d is not prefix-encoded, then for any $C \in \mathbb{N}$ there are states α_1 and α_2 which are (d, C)-diff-large and d-bad.

Proof of Claim 1. If d is not prefix-encoded, then there is a sequence of states $\beta_1, \beta_2, \beta_3, \ldots$ with $d(\beta_1) < d(\beta_2) < d(\beta_3) < \cdots$ such that each β_i can make a step which is d-reducing or pn-reducing but not both.

Using the pigeonhole principle, we can assume (i.e., extract a subsequence) $\beta_1 = X\alpha_1$, $\beta_2 = X\alpha_2$, $\beta_3 = X\alpha_3$, $\ldots$ for a variable X (obviously $|X| < \omega$). We can furthermore assume (by repeated subsequence extractions) that for each $d' \in \{d\} \cup \mathcal{D}(d)$ we have either $d'(\alpha_1) = d'(\alpha_2) = d'(\alpha_3) = \cdots$, or else $d'(\alpha_1) < d'(\alpha_2) < d'(\alpha_3) < \cdots$.

Hence, for any given C there are $i < j$ such that α_i and α_j are (d, C)-diff-large. From Proposition 12(d), and the fact that $d(X\alpha_i) < d(X\alpha_j)$, we easily derive that α_i and α_j are d-bad. (□)

We now let d be a non-prefix-encoded DD-function on a minimal level. Choose $C \in \mathbb{N}$ so that:

- each $d' \in \mathcal{D}(d)$ is prefix-encoded above $(C - z_0)$, where z_0 is the maximum of the finite components of changes in $\mathcal{C}(d)$;
- $C >$ MAXSTEP; and
- $C > d'(\beta)$ whenever $d' \in \mathcal{D}(d)$ and $X \to \beta$ with $d'(\beta) < \omega = |\beta|$.

We take $d_1 \in \{d\} \cup \mathcal{D}(d)$ on a minimal level such that we can choose α_1 and α_2 which are (d_1, C)-diff-large and d_1-bad (as guaranteed by Claim 1). Assume that $0 = d_1(\gamma\alpha_1) < d_1(\gamma\alpha_2)$. If $d'(\gamma\alpha_2) = \omega$ for some $d' \in \mathcal{D}(d_1)$, then $d'(\alpha_2) = \omega = d'(\alpha_1)$. Since $d'(\gamma\alpha_1) < \omega$, there is some β such that $0 = d'(\beta\alpha_1) < d'(\beta\alpha_2)$; but this means that α_1 and α_2 are d'-bad, which contradicts the level-minimality of d_1.

Thus for some $d' \in \mathcal{D}(d_1)$ and z being a component of a change in $\mathcal{C}(d)$ there is a step $\gamma \to \gamma'$ such that

$$d'(\gamma\alpha_1) + z \neq d'(\gamma'\alpha_1) \quad \text{and} \quad d'(\gamma\alpha_2) + z = d'(\gamma'\alpha_2). \tag{1}$$

Claim 2. There is ξ such that $d'(\xi\alpha_1) \neq d'(\xi\alpha_2)$, and either $d'(\xi\alpha_1) < |\xi| + d'(\alpha_1)$ or $d'(\xi\alpha_2) < |\xi| + d'(\alpha_2)$.

Proof of Claim 2. Suppose that none of the γ and γ' from Equation (1) satisfies the claim. Since we cannot have that $|\gamma| + d'(\alpha_1) + z \neq |\gamma'| + d'(\alpha_1)$ and $|\gamma| + d'(\alpha_2) + z = |\gamma'| + d'(\alpha_2)$, it is sufficient to consider only the following cases:
(a) $d'(\gamma\alpha_1) = d'(\gamma\alpha_2)$ and $d'(\gamma'\alpha_1) \neq d'(\gamma'\alpha_2)$;
(b) $d'(\gamma\alpha_1) \neq d'(\gamma\alpha_2)$ and $d'(\gamma'\alpha_1) = d'(\gamma'\alpha_2)$.
For case (a): $d'(\gamma'\alpha_1) = |\gamma'| + d'(\alpha_1) \neq |\gamma'| + d'(\alpha_2) = d'(\gamma'\alpha_2)$ and $d'(\gamma\alpha_1) = d'(\gamma\alpha_2) = d'(\gamma'\alpha_2) - z$ (note that $z < \omega$). By Proposition 12(c) we get that $d'(\gamma'\alpha_1)$ and $d'(\gamma'\alpha_2)$ can differ by at most $\text{MAXSTEP} + 1$, which is a contradiction.
For case (b): $d'(\gamma\alpha_1) = |\gamma| + d'(\alpha_1) \neq |\gamma| + d'(\alpha_2) = d'(\gamma\alpha_2)$. Proposition 12(c) implies that we cannot have $d'(\gamma'\alpha_1) = d'(\gamma'\alpha_2)$ unless the step $\gamma \to \gamma'$ is due to a rule $X \to \beta$ with $|\beta| = \omega$. But C was chosen bigger than $d'(\beta)$. (□)

Finally we show that the existence of a (d', C)-diff-large pair α_1 and α_2, together with a ξ satisfying Claim 2 contradicts the assumption that d' is prefix-encoded above C; this will finish the proof of the Lemma.

Without loss of generality, assume $d'(\xi\alpha_1) < d'(\xi\alpha_2)$. If $d'(\xi\alpha_1) < |\xi| + d'(\alpha_1)$ then there is ξ' such that $0 = d'(\xi'\alpha_1) < d'(\xi'\alpha_2)$, meaning α_1 and α_2 are d'-bad, which is a contradiction.

It remains to consider $d'(\xi\alpha_1) = |\xi| + d'(\alpha_1) < d'(\xi\alpha_2) < |\xi| + d'(\alpha_2)$. Since necessarily $d'(\alpha_1) < d'(\alpha_2)$, we have $d'(\xi\alpha_2) > |\xi| + C$, so by Proposition 12(c), $d'(\xi\alpha_2) = |\xi| + d'(\alpha_2)$, which again is a contradiction. □

4 Bisimilarity Is Decidable on the Union of BPA and BPP

In this section we show that we can decide whether a given BPP process M_0 satisfies a necessary condition for being bisimilar to some unspecified BPA process. In the positive case we can (effectively) construct an OCR process which is bisimilar to M_0. So the decidability of the question whether $M_0 \sim \alpha_0$ (where α_0 is a BPA process) follows from the results of [17,22].

We first recall some useful results from [11] which clarify the "bisimilarity state space" for BPP processes; Firstly, by inspection of [11] we can confirm the following.

Lemma 14. *For each BPP net $\mathcal{N}$ we can effectively construct a sequence $Q_1, \ldots, Q_m$ of sets of places which are* important *in the sense that their norms capture bisimilarity:*

$$\forall M, M' : M \sim M' \;\; \textit{iff} \;\; \forall i : \text{NORM}(Q_i)(M) = \text{NORM}(Q_i)(M').$$

In fact, the collection of all $\text{NORM}(Q_i)$, $1 \leq i \leq m$, is exactly the set of all DD-functions over the state-space of $\mathcal{N}$. More precisely, for every DD-function d there is some Q_i such that $d(M) = \text{NORM}(Q_i)(M)$ for every marking M of $\mathcal{N}$. Conversely, to every Q_i one can associate a DD-function d_i so that all elements of $\mathcal{D}(d_i)$ are among the functions associated to $Q_1, \ldots, Q_{i-1}$.

We now explore further related technical notions. Let $\mathcal{N}$ be a labelled Petri net with initial marking M_0. Given $c \in \mathbb{N}_\omega$, we say that places p and q of $\mathcal{N}$ are *c-dependent* (for M_0) if for every reachable marking M we have that if $c < M(p)$ and $c < M(q)$, then $M(p) = M(q)$. Note that p and q are trivially ω-dependent (for every M_0). The *dependence level* of p, q (for M_0) is the least $c \in \mathbb{N}_\omega$ such that p and q are c-dependent for M_0.

Lemma 15. *Let $\mathcal{N}$ be a Petri net, M_0 a marking of $\mathcal{N}$, and p, q places of $\mathcal{N}$. The dependence level of p, q for M_0 is effectively computable.*

Proof. The dependence level of p, q can be computed, e.g., by employing a slightly modified version of the algorithm for constructing the coverability tree for M_0 [14]. We briefly sketch the construction, emphasizing the difference from the standard algorithm.

An *extended marking* is a function $M : P \to \mathbb{N}_\omega$. All notions introduced for "ordinary" markings also apply to extended markings by employing the standard ω-conventions introduced in Sect. 2.1. The goal is to compute a finite tree where nodes are labelled by extended markings such that the dependence level of p, q for M_0 can be "read" from the tree. It is also possible that the algorithm terminates earlier (without constructing the whole tree) and outputs ω. This happens if the part of the tree constructed so far exhibits a "pumpable" sequence of transitions witnessing the infinity of the dependence level. To simplify our notation, we introduce the following notion: Let n, n' be nodes of the tree labelled by M, M' such that n' is a descendant of n. We say that a place s is *pumped* at n' from n by k, where $0 < k < \omega$, iff $M' \geq M$ and $M'(s) - M(s) = k$. Furthermore, s is *pumpable* at n' iff s is pumped at n' from some predecessor of n' by some (positive) value.

Initially, we put M_0 to be the (label of the) root of the tree. Then, for every node n labelled by M which has not yet been processed we do the following: If the tree contains a processed node with the same label, then the node n is immediately declared as processed. Otherwise, for every transition t which is enabled at M we do the following:

- If $M(p) = M(q) = \omega$ and $F(t,p) - F(p,t) \neq F(t,q) - F(q,t)$, then the algorithm halts and outputs ω. Otherwise, a new successor n' of n with a temporary label M' (where $M \xrightarrow{t} M'$) is created.
- We check whether the following two conditions hold for every predecessor n'' of n'. If not, the algorithm halts and outputs ω.
  - If p is pumped at n' from n'' by k, then $M'(q) \neq \omega$ and q is either not pumpable at n', or it is pumped at n' from n'' by the same k.
  - If q is pumped at n' from n'' by k, then $M'(p) \neq \omega$ and p is either not pumpable at n', or it is pumped at n' from n'' by the same k.
- If the algorithm does not terminate in the previous point, we redefine $M'(r) = \omega$ for every place r pumpable at n'.

If the algorithm terminates by processing all nodes, it outputs the maximal finite value c for which there is a node n labelled by M in the constructed tree such that $M(p) \neq M(q)$, and $M(p) = c$ or $M(q) = c$. □

Now let $\mathcal{N}$ be a BPP net with initial marking M_0, and let Q and Q' be important sets of places of $\mathcal{N}$. We say that Q and Q' are *c-dependent* for a given $c \in \mathbb{N}_\omega$ if for every reachable marking M we have that if $c < \text{NORM}(Q)(M) < \omega$ and $c < \text{NORM}(Q')(M) < \omega$, then $\text{NORM}(Q)(M) = \text{NORM}(Q')(M)$. The *dependence level* of Q, Q' is the least $c \in \mathbb{N}_\omega$ such that Q and Q' are c-dependent.

Lemma 16. *Let Q and Q' be important sets of places of a BPP net $\mathcal{N}$, and let M_0 be a marking of $\mathcal{N}$. The dependence level of Q, Q' is effectively computable.*

Proof. First we extend the net $\mathcal{N}$ by two fresh places p and q. Then we remove all transitions which put a token to $\text{MAXTRAP}(Q)$ or to $\text{MAXTRAP}(Q')$, and modify the other transitions so that for every reachable marking M we have that $\text{NORM}(Q)(M) = M(p)$ and $\text{NORM}(Q')(M) = M(q)$ (i.e., we "count" $\text{NORM}(Q)$ in p and $\text{NORM}(Q')$ in q). This is easy because $\text{NORM}(Q)(M)$ and $\text{NORM}(Q')(M)$ are just weighted sums of tokens in M (cf. the remarks after Definition 6). Note that the resulting Petri net is not necessarily a BPP net. Initially, p and q contain $\text{NORM}(Q)(M_0)$ and $\text{NORM}(Q')(M_0)$ tokens, respectively. Obviously, the dependence level of Q, Q' in $\mathcal{N}$ equals to the dependence level of p, q in the modified net, and thus it is effectively computable by Lemma 15. □

The usefulness of the above explorations now becomes apparent.

Lemma 17. *Let M_0 be a marking of a BPP net $\mathcal{N}$. If Q and Q' are important sets of places with dependence level ω, then M_0 is not bisimilar to any BPA process.*

For example, the sets $\{P\}$ and $\{Q\}$ are important for the BPP net of Fig. 1; the associated DD-functions (referring to Lemma 14) are dd_a and dd_b, respectively. As these sets have a dependence level of ω, this Lemma demonstrates that there is no BPA process which is bisimilar to P.

Proof. Let $Q_1, \ldots, Q_m$ be the important sets of places associated to the BPP net $\mathcal{N}$, and let $d_1, \ldots, d_m$ be the associated DD-functions in the sense of Lemma 14. Now suppose that there are important sets Q and Q' whose dependence level for M_0 equals ω, and that there is a BPA process α_0 such that $M_0 \sim \alpha_0$. By Lemma 13, $d_1, \ldots, d_m$ are prefix-encoded on (any) BPA. Let

$$C = \max\{C_{d_i} : 1 \leq i \leq m\}$$

where C_{d_i} is the constant of Definition 11 chosen for d_i and the underlying BPA process of α_0. We also let k be the number of places of $\mathcal{N}$.

Since the dependence level of Q, Q' for M_0 equals ω, there is a reachable marking M such that

$$C + k < \text{NORM}(Q)(M) < \omega \quad \text{and} \quad C + k < \text{NORM}(Q')(M) < \omega,$$

and $\text{NORM}(Q)(M) \neq \text{NORM}(Q')(M)$. Let d and d' be the DD-functions associated to Q and Q', respectively. Since M is reachable, $M \sim \alpha$ for some α reachable from α_0. As DD-functions are bisimulation invariant, we get that

$$C + k < d(M) = d(\alpha) < \omega \quad \text{and} \quad C + k < d'(M) = d'(\alpha) < \omega\,.$$

Due to the choice of C, we know that for every sequence of (at most) k transitions, if each transition is d-reducing then each is also d'-reducing, and vice versa.

Certainly $Q \neq Q'$ as otherwise we could not have $\text{NORM}(Q)(M) \neq \text{NORM}(Q')(M)$. Hence, there is some $p \in (Q \setminus Q') \cup (Q' \setminus Q)$. Suppose, e.g., $p \in (Q \setminus Q')$. If $M(p) \geq 1$, we are done immediately, as then there is a d-reducing transition $M \stackrel{a}{\to} M'$, where $d = \text{NORM}(Q)$, which takes the token away from p, and therefore it does not decrease $\text{NORM}(Q') = d'$. The bisimilar BPA process α cannot match this transition.

If for every $p \in (Q \setminus Q') \cup (Q' \setminus Q)$ we have that $M(p) = 0$, we argue as follows: Let us assume that, e.g., $\text{NORM}(Q)(M) < \text{NORM}(Q')(M)$. Then there must be some $q \in Q \cap Q'$ such that $M(q) \geq 1$, and each sequence of d-reducing transitions, where $d = \text{NORM}(Q)$, which removes the token in q out of Q (that is, the total effect of the sequence on places in Q is that the token in q disappears) must temporarily mark some place p in Q' which is not in Q. Otherwise, we would immediately obtain that $\text{NORM}(Q')(M) \leq \text{NORM}(Q)(M)$. Now we can change the order of these transitions so that p is marked after at most $(k-1)$ d-reducing transitions. (Here we rely on a folklore result about BPP processes. Also note that any performable permutation of a sequence of d-reducing transitions also consists of d-reducing transitions). Now we can use the same argument as in the previous paragraph. □

Lemma 18. *Let M_0 be a marking of a BPP net $\mathcal{N}$. If for all important sets of places Q and Q' we have that the dependence level of Q, Q' is finite, then we can effectively construct an OCR process which is bisimilar to M_0.*

Proof. Let $Q_1, \ldots, Q_m$ be the important sets of places constructed for the BPP net $\mathcal{N}$, and let $\mathcal{F} = (d_1, \ldots, d_m)$ be the (tuple of the) associated DD-functions as in Lemma 14. Furthermore, let

$$C = \max\{dl(Q_i, Q_j) : 1 \leq i, j \leq m\}$$

where $dl(Q_i, Q_j)$ is the dependence level of Q_i, Q_j. Note that C is effectively computable due to Lemma 16. For every reachable marking M, all norms $\text{NORM}(Q_i)(M)$ which are finite and larger than C keep to coincide. More precisely, if

$$C < \text{NORM}(Q_i)(M) < \omega \quad \text{and} \quad C < \text{NORM}(Q_j)(M) < \omega$$

where $1 \leq i, j \leq m$, then $\text{NORM}(Q_i)(M) = \text{NORM}(Q_j)(M)$.

So we can construct an OCR $\mathcal{P}$, with the initial state bisimilar to M_0, which mimics the behaviour of markings of $\mathcal{N}$ in the following way: Instead of (the current marking) M, the OCR $\mathcal{P}$ records in the finite control state unit:

- for which sets Q_i the value $\text{NORM}(Q_i)(M)$ equals ω (now and forever);
- for which sets Q_i the value $\text{NORM}(Q_i)(M)$ is "small", i.e., no greater than C; the precise values of these norms are also recorded in the finite control; and
- for which sets Q_i the value $\text{NORM}(Q_i)(M)$ is larger than C.

Since the values $\text{NORM}(Q_i)(M)$ in the last collection must be the same, i.e., equal to some v, $\mathcal{P}$ can record $(v-C)$ in the counter.

Note that the configuration of $\mathcal{P}$ associated to M does not contain the "full" information about M in the sense that the exact distribution of tokens in M cannot be reconstructed from the recorded norms. It can happen that different markings M, M' have the property $\text{NORM}(Q_i)(M) = \text{NORM}(Q_j)(M')$ for every $1 \leq i \leq m$, and then (and only then) they are represented by the same configuration of $\mathcal{P}$. However, if M and M' coincide on every $\text{NORM}(Q_i)$, then (and only then) they are bisimilar—see Lemma 14.

It remains to explain how $\mathcal{P}$ performs its transitions. First, for every marking M we define the set

$$\mathit{fire}(M) = \{(a, \delta) : \exists M \xrightarrow{a} M' \text{ such that } \delta^{\mathcal{F}}(M \xrightarrow{a} M') = \delta\}.$$

Note that if $M \sim M'$, then $\mathit{fire}(M) = \mathit{fire}(M')$. Hence, for every configuration $p\alpha$ of $\mathcal{P}$ we can define $\mathit{fire}(p\alpha) = \mathit{fire}(M)$ where M is a marking to which $p\alpha$ is associated. (If there is no such M, $\mathit{fire}(p\alpha)$ is undefined.) Next we show that for every $p\alpha$ for which $\mathit{fire}(p\alpha)$ is defined, the set $\mathit{fire}(p\alpha)$ is effectively computable just from the information stored in the control state p (hence, we can denote $\mathit{fire}(p\alpha)$ just by $\mathit{fire}(p)$). It clearly suffices for our purposes, because then we can compute the sets $\mathit{fire}(q)$ for all control states q and "implement" each pair (a, δ) in a given $\mathit{fire}(q)$ in the straightforward way; in particular, if all "large" norms are set to ω, then the control state is updated and the counter is reset to zero. One can also readily verify that every marking M is bisimilar to its associated configuration of $\mathcal{P}$.

Let t be a transition of $\mathcal{N}$. In each marking M where t is enabled, the (firing of) t causes some change of $\mathcal{F}$ (i.e., of $(\textsc{norm}(Q_1), \cdots, \textsc{norm}(Q_m))$); we define

$$\delta_t(M) = \delta^{\mathcal{F}}(M \xrightarrow{t} M').$$

In general, the same transition can cause different changes in different markings $M, \bar{M}$. The (only) reason why this can happen is that some $\textsc{norm}(Q_i)$ which is finite for M can be ω for $\bar{M}$, and vice versa (note that if $\textsc{norm}(Q_i)(M) = \omega$, then $\delta_t(M)_i = \omega$). So, $\delta_t(M)$ is (completely and effectively) determined by the structure of $\mathcal{N}$ and the *mode* of M, i.e., the set of all Q_i's for which $\textsc{norm}(Q_i)(M) < \omega$. Hence, for a given a mode $\mathcal{M}$ (i.e., a subset of $\{Q_1, \cdots, Q_m\}$ whose norms are to be finite) we can effectively partition the transitions of $\mathcal{N}$ into classes $T_1, \cdots, T_k$ so that transitions in the same T_i have the same label and the same change at every marking with mode $\mathcal{M}$. Thus, to each T_i we can associate a pair (a_i, δ_i). Note that

$$\mathit{fire}(M) \subseteq \{(a_1, \delta_1), \cdots, (a_k, \delta_k)\}$$

for every marking M with mode $\mathcal{M}$. Now we show that for each T_i one can effectively construct an important set Q_{j_i} such that for every marking M with mode $\mathcal{M}$ we have that

$$(a_i, \delta_i) \in \mathit{fire}(M) \quad \text{iff} \quad \textsc{norm}(Q_{j_i})(M) > 0. \tag{2}$$

Actually, it suffices to put

$$Q_{j_i} = \bigcup_{t \in T_i} \mathit{Pre}(t) \cup \bigcup_{Q \in \mathcal{M}} \textsc{maxtrap}(Q).$$

Clearly, $\textsc{norm}(Q_{j_i})$ is a bisimulation-invariant norm; and it follows from the construction of the important sets $Q_1, \cdots, Q_m$ presented in [11] that Q_{j_i} appears in the sequence $Q_1, \cdots, Q_m$. It remains to verify that Equation (2) indeed holds. So, let M be a marking with mode $\mathcal{M}$. If $(a_i, \delta_i) \in \mathit{fire}(M)$, some of the transitions in T_i are enabled at M, and hence $\textsc{norm}(Q_{j_i})(M) > 0$. Conversely, if $\textsc{norm}(Q_{j_i})(M) > 0$, the places of $\bigcup_{Q \in \mathcal{M}} \textsc{maxtrap}(Q)$ are surely not marked (otherwise, we would have that $\textsc{norm}(Q)(M) = \omega$ for some Q in $\mathcal{M}$ which is a contradiction). Hence, some $\mathit{Pre}(t)$, where $t \in T_i$, must be marked at M which means that t is enabled at M.

Since control states of $\mathcal{P}$ carry the information about the current mode and currently positive $\textsc{norm}(Q_i)$'s, we are done. □

The previous lemmata allow us to conclude the following:

Theorem 19. *Bisimilarity between BPA and BPP processes is decidable.*

Proof. We first check if there are important sets Q and Q' whose dependence level equals ω (this is decidable by Lemmas 14 and 16). If this is the case, then M_0 cannot be bisimilar to any BPA process. Otherwise, we can effectively construct an OCR process $p\alpha$ which is bisimilar to M_0 (Lemma 18). We can then check bisimilarity between $p\alpha$ and the given BPA process (using, e.g., the algorithm of [17,22]). □

As a final remark, we note that a more detailed analysis of the properties of BPP processes which are bisimilar to BPA processes would admit a refinement of the result of Lemma 18: the OCR must be special in the sense that the first transition which increments the counter (since the last reset) "selects" the control state q to which the machine must switch by the first decrement operation. The only possibility of how to change q to some other state is to perform a reset. We conjecture that such a behaviour can be matched by an effectively constructible BPA process. This being the case, we could then use the bisimilarity-checking algorithm for BPA processes [4] instead of the one for PDA processes, which would yield an elementary upper complexity bound for bisimilarity between BPA and BPP.

References

1. J.C.M. Baeten, J.A. Bergstra and J.W. Klop. Decidability of bisimulation equivalence for processes generating context-free languages. *Journal of the ACM* **40**(3):653–682, 1993.
2. J. Blanco. Normed BPP and BPA. In *Proceedings of ACP'94*, pages 242–251, Springer 1995.
3. O. Burkart, D. Caucal, F. Moller and B. Steffen. Verification on infinite structures. In J. Bergstra, A. Ponse and S. Smolka (editors), *Handbook of Process Algebra*, chapter 9, pages 545–623. Elsevier Science, 2001.
4. O. Burkart, D. Caucal and B. Steffen. An elementary decision procedure for arbitrary context-free processes. In *Proceedings of MFCS'95, LNCS* 969, pages 423–433. Springer, 1995.
5. I. Černá, M. Křetínský and A. Kučera. Comparing expressibility of normed BPA and normed BPP processes. *Acta Informatica* **36**:233–256, 1999.
6. R. Glabbeek. The linear time – branching time spectrum I: The semantics of concrete sequential processes. In J. Bergstra, A. Ponse and S. Smolka (editors), *Handbook of Process Algebra*, chapter 1, pages 3–99. Elsevier Science, 2001.
7. Y. Hirshfeld and M. Jerrum. Bisimulation equivalence is decidable for normed process algebra. In *Proceedings of ICALP'99, LNCS* 1644, pages 412–421. Springer, 1999.
8. Y. Hirshfeld, M. Jerrum and F. Moller. A polynomial algorithm for deciding bisimilarity of normed context-free processes. *Theoretical Computer Science* **158**:143–159, 1996.
9. Y. Hirshfeld, M. Jerrum and F. Moller. A polynomial-time algorithm for deciding bisimulation equivalence of normed basic parallel processes. *Journal of Mathematical Structures in Computer Science* **6**:251–259, 1996.
10. P. Jančar. Undecidability of bisimilarity for Petri nets and some related problems. *Theoretical Computer Science* **148**(2):281–301, 1995.
11. P. Jančar. Strong bisimilarity on basic parallel processes is PSPACE-complete. In *Proceedings of LICS'03*, to appear, 2002.
12. P. Jančar and A. Kučera. Equivalence-checking with infinite-state systems: Techniques and results. In *Proceedings of SOFSEM'02, LNCS* 2540, pages 41–73. Springer, 2002.

13. P. Jančar and F. Moller. Techniques for decidability and undecidability of bisimilarity. In *Proceedings of CONCUR'99, LNCS* 1664, pages 30–45. Springer, 1999.
14. R.M. Karp and R.E. Miller. Parallel Program Schemata *Journal of Computer and Systems Sciences* **3**:147–195, 1969.
15. R. Milner and F. Moller. Unique decomposition of processes. *Theoretical Computer Science* **107**:357–363, 1993.
16. F. Moller. Infinite results. In *Proceedings of CONCUR'96, LNCS* 1119, pages 195–216. Springer, 1996.
17. G. Sénizergues. Decidability of bisimulation equivalence for equational graphs of finite out-degree. In *Proceedings of FOCS'98*, pages 120–129. IEEE Computer Society Press, 2001.
18. G. Sénizergues. L(A)=L(B)? decidability results from complete formal systems. *Theoretical Computer Science* **251**(1-2):1–166, 2001.
19. J. Srba. Roadmap of Infinite Results. http://www.brics.dk/~srba/roadmap.
20. J. Srba. Strong bisimilarity and regularity of basic process algebra is PSPACE-hard. In *Proceedings of ICALP'02, LNCS* 2380, pages 716–727. Springer, 2002.
21. J. Srba. Strong bisimilarity and regularity of basic parallel processes is PSPACE-hard. In *Proceedings of STACS'02, LNCS* 2285, pages 535–546. Springer, 2002.
22. C. Stirling. Decidability of bisimulation equivalence for pushdown processes. Research Report No. EDI-INF-RR-0005, School of Informatics, Edinburgh University. January 2000.
23. C. Stirling. Decidability of DPDA equivelance. *Theoretical Computer Science* **255**(1-2):1–31, 2001.

Verification of Parametric Concurrent Systems with Prioritized FIFO Resource Management*

Ahmed Bouajjani, Peter Habermehl, and Tomáš Vojnar

LIAFA, Paris University 7
Case 7014, 2, place Jussieu, 75251 Paris Cedex 05, France
{abou,haberm,vojnar}@liafa.jussieu.fr

Abstract. We consider the problem of parametric verification over a class of systems of processes competing for access to shared resources. We suppose the access to the resources to be controlled according to a FIFO-based policy with a possibility of distinguishing low-priority and high-priority resource requests. We propose a model of the concerned systems based on extended automata with queues. Over this model, we address verification of properties expressed in LTL\X enriched with global process quantification and interpreted on finite as well as fair behaviours of the given systems. In addition, we examine parametric verification of process deadlockability too. By reducing the parametric verification problems to finite-state model checking, we establish several decidability results for different classes of the considered properties and systems (including the special case of systems with the pure FIFO resource management). Moreover, we show that parametric verification against formulae with local process quantification is undecidable in the given context.

1 Introduction

Managing concurrent access to shared resources is a fundamental problem that appears in many contexts, e.g., operating systems, multithreaded programs, control software, etc. The critical properties to ensure are typically (1) mutual exclusion when exclusive access is required, (2) absence of starvation (a process that requires a resource will eventually get it), and (3) absence of deadlocks. Many different instances of this problem can be defined depending on the assumptions on the allowed actions for access to resources and the policies for managing the access to these resources.

In this work, we consider systems with a finite number of resources shared by a set of identical processes. These processes can require a set of resources, get access and use the requested resources, and release the used resources. The requests can be of a low-priority or a high-priority level. The access to the resources is managed by a locker according to a FIFO-based policy taking into account the priorities of the requests, i.e. a waiting high-priority request can overtake waiting low-priority ones. As a special case allowing for an optimized

* This work was supported in part by the European Commission (FET project ADVANCE, contract No. IST-1999-29082).

R. Amadio, D. Lugiez (Eds.): CONCUR 2003, LNCS 2761, pp. 174–190, 2003.

treatment, we then examine the situation when no high-priority requests are used, and the locker behaves according to the pure FIFO discipline.

As mentioned later in related work, the above framework is, in particular, inspired by a need to verify the use of shared resources in some of Ericsson's ATM switches. However, the operations for access to shared resources and the resource management policies used are quite natural in general in concurrent applications dealing with shared resources.

Verification of the described systems can, of course, be carried out using finite-state model-checking if we fix the number of processes. However, a precise number of processes present in such a system in practice is usually not known in advance, and it is thus crucial to verify that the system behaves correctly for *any* number of them. This yields a parametric verification problem that is quite nontrivial as we have to deal with an infinite number of system instances.

The aim of this paper is to study decidability of the described problem for a significant class of properties including the three most important ones given above.

For an abstract description of the concerned systems, we define a model based on extended automata with queues recording the identities of the waiting processes for each resource. Then, we address the verification problem for families of such systems with an arbitrary number of processes (called *RTR families*—RTR stands for request-take-release) against formulae of the temporal logic $LTL\setminus X$ with *global process quantification.* We consider two interpretation domains for the logic: the set of finite behaviours (which is natural for safety properties), and the set of fair behaviours (in order to cover liveness properties). In addition, we consider the parametric verification problem of process deadlockability too.

We adopt the approach of finding *cut-off bounds* to show that many interesting parametric verification problems in the given context can be reduced to finite-state model checking. This means that given a class of formulae, we prove that deciding whether all systems of a family satisfy a formula is equivalent to deciding whether some finite number of systems in the family (each of them having a fixed number of processes) satisfies this formula.

When establishing our results, we consider the question whether it is possible to find cut-off bounds that do not depend on the structure of the involved processes and the formula at hand, but only on the number of resources and the number of processes quantified in the formula. Indeed, these numbers are relatively small, especially in comparison to the size of process control automata.

We show that for RTR families where the *pure FIFO resource management* is used (i.e. no high-priority access to resources is required), parametric verification of finite as well as fair behaviour is decidable against all LTL\X formulae with global process quantification. The cut-off bound in the finite behaviour case is the number of quantified processes, whereas it is this number plus the number of resources in the fair behaviour case. These bounds lead to practical finite-state verification. Furthermore, we show that the verification of process deadlockability is decidable too (where the bound is the number of resources).

On the other hand, for the case of dealing with RTR families that *distinguish low-priority and high-priority requests*, we show that – unfortunately – general, structure-independent cut-offs do not exist neither for the interpretation of the considered logic on finite nor fair behaviours. However, we show that even for such families, parametric verification of finite behaviour is decidable, e.g., against reachability/invariance formulae, and parametric verification of fair behaviour is decidable against formulae with a single quantified process. In this way, we cover, e.g., verification of the (for the given application domain) key properties of mutual exclusion and absence of starvation. For the former case, we even obtain a structure-independent cut-off equal again to the number of quantified processes. For verification of fair behaviour against single process formulae, no general structure-independent cut-off can be found, but we provide a structure-dependent one, and in addition, we determine a significant subclass of RTR families where a structure-independent cut-off for this particular kind of properties does exist. Finally, we show that process deadlockability can be solved in the case of general RTR families via the same (structure-independent) cut-off as in the case of the families not using high-priority requests.

Lastly, we show that although the queues in RTR families are not communication queues, but just waiting queues, and the above decidability results may be established, the model is still quite powerful, and decidability may easily be lost when trying to deal with a bit more complex properties to verify. We illustrate this by proving that parametric finite-behaviour verification becomes *undecidable* (even for families not using high-priority requests) for LTL\X extended with the notion of *local process quantification* [8] allowing one to examine different processes in different encountered states.

Related Work: There exist several approaches to the parametric verification problem. We can mention, for example, the use of symbolic model checking, (automated) abstraction, or network invariants [10,1,3,14,11,12]. The idea of cut-offs has already been used in several contexts [9,6,7,5] too. However, to the best of our knowledge, there is no work covering the class of parametric systems considered here, i.e. parametric resource sharing systems with a prioritized FIFO resource management. The *two* involved obstacles (parameterization and having multiple queues over an unbounded domain of process identifiers) seem to complicate the use of any of the known methods. Using cut-offs appears to be the easiest approach here.

The work [5] targets verification of systems with shared resources (and even employs cut-offs), but the system model and the employed proof techniques differ. The involved processes need not be identical, the number of resources is not bounded, but, on the other hand, only two fixed processes may compete for a given resource, and their requests are served in random order (there are no FIFO queues in [5]). Moreover, some of the properties to be verified we consider here are different from [5] (e.g., we deal with the more realistic notion of weak/strong fairness compared to the unconditional one used there, etc.).

Finally, let us add that our work was originally motivated by [2] that concerns verification of the use of shared resources in Ericsson's AXD 301 ATM switch. In [2] finite-state model checking is used to verify some isolated instances of the given parametric system. Then, in the concluding remarks, a need for a more complete, parametric verification is mentioned, which is what our work is aiming at on the level of a general abstract model covering the given application domain.

Outline: We first formalize the notion of RTR families and define the specification logic we use. Then, we present our cut-off results for finite and fair behaviour and process deadlockability as well as the undecidability result. Due to space limitations, we provide proof ideas for some of the results only – the complete proofs can be found in [4].

2 RTR Families

2.1 The Model of RTR Families

Processes in systems of RTR families are controlled by *RTR automata*. An RTR automaton over a finite set of resources is a finite automaton with the following kinds of actions joint with transitions: skip (denoted by τ – an abstract step not changing resource utilization), request and, when it is the turn, take a set of resources at the low- or high-priority level (`rqt` /`prqt`), and, finally, release a set of resources (`rel`).

Let us, however, stress that we allow processes to block inside (`p`)`rqt` transitions[1] while waiting for the requested resources to be available for them. Therefore, a single (`p`)`rqt` transition in a model semantically corresponds to two transitions, which we denote as (`p`)`req` (request a set of resources) and (`p`)`take` (start using the requested resources when enabled to do so by the locking policy).

Definition 1. *An* RTR automaton *is a 4-tuple* $\mathcal{A} = (R, Q, q_0, T)$ *where* R *is a set of resources,* Q *is a set of control locations,* $q_0 \in Q$ *is an initial control location, and* $T \subseteq Q \times A \times Q$ *is a transition relation over the set of actions* $A = \{\tau\} \cup \{a(R') \mid a \in \{\texttt{rqt}, \texttt{prqt}, \texttt{rel}\} \wedge R' \neq \emptyset \wedge R' \subseteq R\}$. *The sets* R, Q, T, *and* A *are nonempty, finite, pairwise disjoint, and disjoint with* $\mathbb{N}$.

An *RTR family* $\mathcal{F}(\mathcal{A})$ over an RTR automaton $\mathcal{A}$ is a set of *systems* S_n consisting of $n \geq 1$ identical processes controlled by $\mathcal{A}$ and identified by elements of $P_n = \{1, ..., n\}$. (In the following, if no confusion is possible, we usually drop the reference to $\mathcal{A}$.) We denote as *RTR*$\backslash$*P families* the special cases of RTR families whose control automata contain no high-priority request actions.

2.2 Configurations

For the rest of the section, let us suppose working with an arbitrary fixed RTR family $\mathcal{F}$ over an automaton $\mathcal{A} = (R, Q, q_0, T)$ and with a system $S_n \in \mathcal{F}$.

To make the semantics of RTR families reflect the fact that processes may block in (`p`)`rqt` actions, we extend the set Q of "explicit" control locations to Q_t containing a unique internal control location q_t for each transition $t \in T$ based on a (`p`)`rqt` action. Furthermore, let T_t be the set obtained from T by preserving all

[1] We use (`p`)`rqt` when addressing both `rqt` as well as `prqt` transitions.

τ and $\texttt{rel}$ transitions and splitting each transition $t = (q_1, (\texttt{p})\texttt{rqt}(R'), q_2) \in T$ to two transitions $t_1 = (q_1, (\texttt{p})\texttt{req}(R'), q_t)$ and $t_2 = (q_t, (\texttt{p})\texttt{take}(R'), q_2)$.

We define the *resource queue alphabet* of S_n as $\Sigma_n = \{s(p) \mid s \in \{\texttt{r}, \texttt{pr}, \texttt{g}, \texttt{u}\} \wedge p \in P_n\}$. The meaning is that a process has requested a resource in the low- or high-priority way, it has been granted the resource, or it is already using the resource. A *configuration* c of S_n is then a function $c : (P_n \to Q_t) \cup (R \to \Sigma_n^*)$ that assigns the current control locations to processes and the current content of queues of requests to resources. Let C_n be the set of all such configurations.

2.3 Resource Granting and Transition Firing

We now introduce the *locker function* Λ implementing the considered FIFO resource management policy with low- and high-priority requests. This function is to be applied over configurations changed by adding/removing some requests to/from some queues in order to grant all the requests that can be granted wrt. the given strategy in the given situation. Note that in the case of RTR\P families, the resource management policy can be considered the pure FIFO policy.

A high-priority request is granted iff none of the needed resources is in use by or granted to any process, nor it is subject to any sooner raised, but not yet granted, high-priority request. A low-priority request is granted iff the needed resources are not in use nor granted and they are not subject to any sooner raised request nor any later raised high-priority request that can be granted at the given moment. (High-priority requests that currently cannot be granted do not block sooner raised low-priority requests.) Formally, for $c \in C_n$, we define $\Lambda(c)$ to be a configuration of C_n equal to c up to the following for each $r \in R$:

1. If $c(r) = w_1.\texttt{pr}(p).w_2$ for some $p \in P_n$, $w_1, w_2 \in \Sigma_n^*$ s.t. $c(p) = q_t$ for a certain $t = (q_1, \texttt{prqt}(R'), q_2) \in T$ and for all $r' \in R'$, $c(r') = w_1'.\texttt{pr}(p).w_2'$ with $w_1' \in \{\texttt{r}(p') \mid p' \in P_n\}^*$ and $w_2' \in \Sigma_n^*$, we set $\Lambda(c)(r)$ to $\texttt{g}(p).w_1.w_2$.
2. If $c(r) = \texttt{r}(p).w$ for some $p \in P_n$, $w \in \Sigma_n^*$ such that $c(p) = q_t$ for a certain $t = (q_1, \texttt{rqt}(R'), q_2) \in T$ and for all $r' \in R'$, $c(r') = \texttt{r}(p).w'$ with $w' \in \Sigma_n^*$, and the premise of case 1 is not satisfied for r', we set $\Lambda(c)(r)$ to $\texttt{g}(p).w$.

We define *enabling* and *firing of transitions* in processes of S_n via a predicate $en \subseteq C_n \times T_t \times P_n$ and a function $to : C_n \times T_t \times P_n \to C_n$.

For all transitions $t = (q_1, \tau, q_2) \in T_t$ and $t = (q_1, a(R'), q_2) \in T_t$, $a \in \{\texttt{rel}, \texttt{req}, \texttt{preq}\}$, we define $en(c, t, p) \Leftrightarrow c(p) = q_1$. For each transition $t = (q_1, (\texttt{p})\texttt{take}(R'), q_2) \in T_t$, we define $en(c, t, p) \Leftrightarrow c(p) = q_1 \wedge \forall r \in R' \; \exists w \in \Sigma_n^* : c(r) = \texttt{g}(p).w$. Intuitively, a transition is enabled in some process if the process is at the source control location of the transition and, in the case of $(\texttt{p})\texttt{take}$, if the appropriate request has been granted.

Firing of a transition $t = (q_1, \tau, q_2) \in T_t$ simply changes the control location mapping of p from q_1 to q_2, i.e. $to(c, t, p) = (c \setminus \{(p, q_1)\}) \cup \{(p, q_2)\}$.

Firing of a $(\texttt{p})\texttt{req}$ transition t corresponds to registering the request in the queues of all the involved resources and going to the internal waiting location of t. The locker is applied to (if possible) immediately grant the request. For

$t = (q_1, (\mathtt{p})\mathtt{req}(R'), q_2) \in T_t$, we define $to(c,t,p) = \Lambda((c \setminus c^-) \cup c^+)$ where $c^- = \{(p, q_1)\} \cup \{(r, c(r)) \mid r \in R'\}$ and $c^+ = \{(p, q_2)\} \cup \{(r, c(r).(\mathtt{p})\mathtt{r}(p)) \mid r \in R'\}$.

For a transition $t = (q_1, (\mathtt{p})\mathtt{take}(R'), q_2) \in T_t$, we simply change all the appropriate $\mathtt{g}$ queue items to $\mathtt{u}$ items and finish the concerned $(\mathtt{p})\mathtt{rqt}$ transition, i.e. $to(c,t,p) = (c \setminus c^-) \cup c^+$ with c^- as in the case of $(\mathtt{p})\mathtt{req}$ and $c^+ = \{(p, q_2)\} \cup \{(r, \mathtt{u}(p).w) \mid r \in R' \wedge c(r) = \mathtt{g}(p).w\}$.

Finally, a $\mathtt{rel}$ transition removes the head $\mathtt{u}$ items from the queues of the given resources provided they are owned by the given process. The locker is applied to grant all the requests that may become unblocked. Formally, for $t = (q_1, \mathtt{rel}(R'), q_2) \in T_t$, we fix $to(c,t,p) = \Lambda((c \setminus c^-) \cup c^+)$ with $c^- = \{(p, q_1)\} \cup \{(r, c(r)) \mid r \in R' \wedge \exists w \in \Sigma_n^* : c(r) = \mathtt{u}(p).w\}$ and $c^+ = \{(p, q_2)\} \cup \{(r, w) \mid r \in R' \wedge w \in \Sigma_n^* \wedge c(r) = \mathtt{u}(p).w\}$.

2.4 Behaviour of Systems of RTR Families

Let S_n be a system of an RTR family $\mathcal{F}$. We define the *initial configuration* c_0 of S_n to be such that $\forall p \in P_n : c_0(p) = q_0$ and $\forall r \in R : c_0(r) = \epsilon$. By a *finite behaviour* of S_n starting from $c_1 \in C_n$, we understand a sequence $c_1(p_1, t_1)c_2...(p_l, t_l)c_{l+1}$ such that for each $i \in \{1, ..., l\}$, $en(c_i, t_i, p_i)$ holds, and $c_{i+1} = to(c_i, t_i, p_i)$. If c_1 is the initial configuration c_0, we may drop a reference to it and speak simply about a finite behaviour of S_n. The notion of *infinite behaviours* of S_n can be defined in an analogous way. A *complete behaviour* is then either infinite or such that it cannot be extended any more.

We say a complete behaviour is *weakly (process) fair* iff each process that is eventually always enabled to fire some transitions, always eventually fires some transitions. We may call a complete behaviour *strongly (process) fair* iff each process that is always eventually enabled to fire some transitions, always eventually fires some transitions. However, we do not deal with strong fairness in the following as in our model, the notions of strong and weak fairness coincide: Due to the separation of requesting resources and starting to use them and the impossibility of cancelling issued grants of resources, a process cannot temporarily have no enabled transitions without firing anything.

For a behaviour $\beta_n = c_1(p_1, t_1)c_2(p_2, t_2)...$ of a system S_n of an RTR family $\mathcal{F}$, we call the configuration sequence $\pi_n = c_1c_2...$ a *path* of S_n corresponding to β_n and the transition firing sequence $\rho_n = (p_1, t_1)(p_2, t_2)...$ a *run* of S_n corresponding to β_n. If the behaviour is not important, we do not mention it. We denote Π_n^{fin}, $\Pi_n^{fin} \subseteq C_n^+$, the set of all finite paths of S_n and Π_n^{wf}, $\Pi_n^{wf} \subseteq C_n^+ \cup C_n^\omega$, the set of all paths of S_n corresponding to complete, weakly fair behaviours.

3 The Specification Logic

In this work, we concentrate (with the exception of process deadlockability) on verification of process-oriented, linear-time properties of systems of RTR families. For specifying the properties, we use the below described extension of LTL\X, which we denote as *MPTL* (i.e. temporal logic of many processes). We exclude the next-time operator from our framework because it sometimes allows a certain kind of counting of processes, which is undesirable when trying to limit/reduce the number of processes to be considered in verification.

We extend LTL\X by *global process quantification* in a way inspired by ICTL* (see, e.g., [8]) and allowing us to easily reason over systems composed of a parametric number of identical processes. We also allow for an explicit distinction whether a property should hold for all paths or for at least one path out of a given set. Therefore, we introduce a single top-level *path quantifier* to our formulae. We restrict quantification in the following way: (1) We implicitly require all variables to always refer to distinct processes. (2) We allow only uniformly universal (or uniformly existential) process and path quantification.

Finally, we limit *atomic formulae* to testing the current control locations of processes. We allow for referring to the internal control locations of request transitions too, which corresponds to asking whether a process has requested some resources, but has not become their user yet.

3.1 The Syntax of MPTL

Let PV, $PV \cap \mathbb{N} = \emptyset$, be a set of process variables. We first define the syntax of *MPTL path subformulae*, which we build from atomic formulae $at(p,q)$ using boolean connectives and the until operator. For $V \subseteq PV$ and $p \in V$, we have:

$$\varphi(V) ::= at(p,q) \mid \neg\varphi(V) \mid \varphi(V) \vee \varphi(V) \mid \varphi(V)\ \mathcal{U}\ \varphi(V)$$

As syntactical sugar, we can then introduce in the usual way formulae like `tt`, `ff`, $\varphi(V) \wedge \varphi(V)$, $\Box\varphi(V)$, or $\Diamond\varphi(V)$.

Subsequently, we define the syntax of *universal* and *existential MPTL formulae*, which extend MPTL path subformulae by process and path quantification used in a uniformly universal or existential way. For $V \subseteq PV$, we have $\Phi_a ::= \forall V : A\ \varphi(V)$ and $\Phi_e ::= \exists V : E\ \varphi(V)$.

In the rest of the paper, we commonly specify sets of quantified variables by listing their elements in some chosen order. Using MPTL formulae, we can then express, for example, *mutual exclusion* as $\forall p_1, p_2 : A\ \Box\ \neg(at(p_1, cs) \wedge at(p_2, cs))$ or *absence of starvation* as $\forall p : A\ \Box\ (at(p, req) \Rightarrow \Diamond\ at(p, use))$.

3.2 The Formal Semantics of MPTL

Suppose working with a set of process variables PV. As we require process quantifiers to always speak about distinct processes, we call a function $\nu_n : PV \to P_n$ a *valuation* of PV iff it is an injection.

Suppose further that we have a system S_n of an RTR family $\mathcal{F}$. Let $\pi_n \in C_n^* \cup C_n^\omega$ denote a (finite or infinite) path of S_n. For a finite (or infinite) path $\pi_n = c_1c_2...c_{|\pi_n|}$ ($\pi_n = c_1c_2...$), let π_n^l denote the suffix $c_lc_{l+1}...c_{|\pi_n|}$ ($c_lc_{l+1}...$) of π_n, respectively. (For a finite π_n with $|\pi_n| < l$, $\pi_n^l = \epsilon$.) Given a path π_n of S_n and a valuation ν_n of PV, we inductively define the semantics of MPTL path subformulae $\varphi(V)$ as follows:

- $\pi_n, \nu_n \models at(p, q)$ iff $\pi_n = c.\pi'_n$ and $c(\nu_n(p)) = q$.
- $\pi_n, \nu_n \models \neg\varphi(V)$ iff $\pi_n, \nu_n \not\models \varphi(V)$.
- $\pi_n, \nu_n \models \varphi_1(V) \vee \varphi_2(V)$ iff $\pi_n, \nu_n \models \varphi_1(V)$ or $\pi_n, \nu_n \models \varphi_2(V)$.
- $\pi_n, \nu_n \models \varphi_1(V)\ \mathcal{U}\ \varphi_2(V)$ iff there is $l \geq 1$ such that $\pi^l_n, \nu_n \models \varphi_2(V)$ and for each k, $1 \leq k < l$, $\pi^k_n, \nu_n \models \varphi_1(V)$.

As for any given behaviour β_n of S_n, there is a unique path π_n corresponding to it, we will also sometimes say in the following that β_n satisfies or unsatisfies a formula φ meaning that π_n satisfies or unsatisfies φ. We will call the processes assigned to some process variables by ν_n as processes *visible* in π_n via ν_n.

Next, let $\Pi_n \subseteq C^*_n \cup C^\omega_n$ denote any set of paths of S_n. (Later we concentrate on sets of paths corresponding to all finite or fair behaviours.) We define the semantics of MPTL universal and existential formulae as follows:

- $\Pi_n \models \forall V : A\varphi(V)$ iff for all valuations ν_n of PV and all $\pi_n \in \Pi_n$, $\pi_n, \nu_n \models \varphi(V)$.
- $\Pi_n \models \exists V : E\varphi(V)$ iff $\pi_n, \nu_n \models \varphi(V)$ for some PV valuation ν_n and some $\pi_n \in \Pi_n$.

3.3 Evaluating MPTL over Systems and Families

Let S_n be a system of an RTR family $\mathcal{F}$. Given a universal or existential MPTL formula Φ, we say the finite behaviour of S_n satisfies Φ, which we denote by $S_n \models_{fin} \Phi$, iff $\Pi^{fin}_n \models \Phi$ holds. We say the weakly fair behaviour of S_n satisfies Φ, which we denote by $S_n \models_{wf} \Phi$, iff $\Pi^{wf}_n \models \Phi$ holds.

Next, we introduce a notion of MPTL formulae satisfaction over RTR families, in which we allow for specifying the minimum size of the systems to be considered.[2] We go on with the chosen uniformity of quantification and for a universal MPTL formula Φ_a, an RTR family $\mathcal{F}$, and a lower bound l on the number of processes to be considered, we define $\mathcal{F}, l \models^a_{fin} \Phi_a$ to hold iff $S_n \models_{fin} \Phi_a$ holds for *all* systems $S_n \in \mathcal{F}$ with $l \leq n$. Dually, for an existential MPTL formula Φ_e, we define $\mathcal{F}, l \models^e_{fin} \Phi_e$ to hold iff $S_n \models_{fin} \Phi_e$ holds for *some* system $S_n \in \mathcal{F}$ with at least l processes. The same notions of MPTL formulae satisfaction over families can be introduced for weakly fair behaviour too.

4 Verification of Finite Behaviour

As we have already indicated, one of the problems we examine in this paper is verification of finite behaviour of systems of RTR families against correctness requirements expressed in MPTL. In particular, we concentrate on the *parametric finite-behaviour verification problem* of checking whether $\mathcal{F}, l \models^a_{fin} \Phi_a$ holds for a certain RTR family $\mathcal{F}$, a universal MPTL formula Φ_a, and a lower bound l

[2] Specifying the minimum size allows one to exclude possibly special behaviours of small systems. Fixing the maximum size would lead to finite-state verification. Although our results could still be used to simplify such verification, we do not discuss this case here.

on the number of processes to be considered. The problem of checking whether $\mathcal{F}, l \models^e_{fin} \Phi_e$ holds for a certain existential MPTL formula Φ_e is dual, and we will not cover it explicitly in the following.

4.1 A Cut-Off Result for RTR\P Families

We first examine the parametric finite-behaviour verification problem for the case of RTR\P families. Let $\Phi_a \equiv \forall p_1, ..., p_k : A\ \varphi(p_1, ..., p_k)$ be a universal MPTL formula with k globally quantified process variables. We show that for any RTR\P family $\mathcal{F}$, the problem of checking $\mathcal{F}, l \models^a_{fin} \Phi_a$ can be reduced to simple finite-state examination of the system $S_k \in \mathcal{F}$ with k processes. At the same time, the processes to be monitored via $p_1, ..., p_k$ may be fixed to $1, ..., k$. We denote the resulting verification problem as checking whether $S_k \models_{fin} A\ \varphi(1, ..., k)$ holds. Consequently, we can say that, e.g., to verify mutual exclusion in an RTR\P family $\mathcal{F}$, it suffices to verify it for processes 1 and 2 in the system of $\mathcal{F}$ with only these two processes.

Below, we first give a basic cut-off lemma and then we generalize it to the above.

Lemma 1. *For an RTR\P family $\mathcal{F}$ and an MPTL path formula $\varphi(p_1, ..., p_k)$, the following holds for systems of $\mathcal{F}$:*

$$\forall n \geq k : S_n \models_{fin} \forall p_1, ..., p_k : A\ \varphi(p_1, ..., p_k) \Leftrightarrow S_k \models_{fin} A\ \varphi(1, ..., k)$$

Proof. (Sketch) ($\Rightarrow$) We convert a counterexample behaviour of S_k to one of S_n by adding some processes and letting them idle at q_0. ($\Leftarrow$) To reduce a counterexample behaviour of S_n to one of S_k, we remove the invisible processes and the transitions fired by them (these processes only restrict the behaviour of others by blocking some resources) and we permute the processes to make $1, ..., k$ visible (all processes are initially equal and their names are not significant). □

Lemma 1 and properties of MPTL now easily yield the above promised result.

Theorem 1. *Let $\mathcal{F}$ be an RTR\P family and let $\Phi_a \equiv \forall p_1, ..., p_k : A\ \varphi(p_1, ..., p_k)$ be an MPTL formula. Then, checking whether $\mathcal{F}, l \models^a_{fin} \Phi_a$ holds is equal to checking whether $S_k \models_{fin} A\ \varphi(1, ..., k)$ holds.*

4.2 Inexistence of Structure-Independent Cut-Offs for RTR Families

Unfortunately, as we prove below, for families with prioritized resource management, the same reduction as above cannot be achieved even when we allow the bound to also depend on the number of available resources and fix the minimum considered number of processes to one.

Theorem 2. *For MPTL formulae Φ_a with k process variables and RTR families $\mathcal{F}$ with m resources, the parametric finite-behaviour verification problem of checking whether $\mathcal{F}, 1 \models^a_{fin} \Phi_a$ holds is not, in general, decidable by examining just the systems $S_1, ..., S_n \in \mathcal{F}$ with n being a function of k and/or m only.*

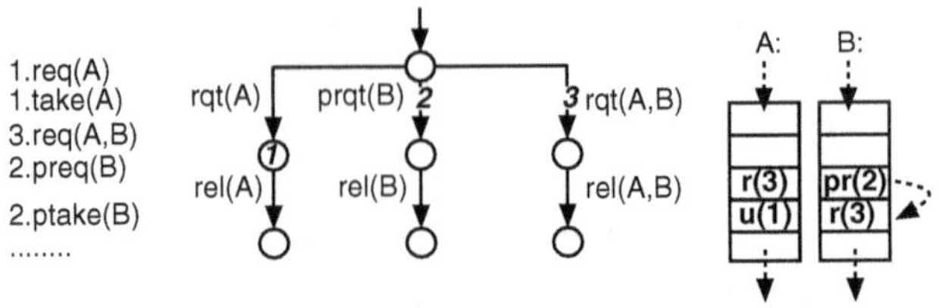

Fig. 1. A scenario problematic for the application of cut-offs (the run from the left is visualized on the RTR automaton and the appropriate resource queues)

Proof. (Idea) In the given framework, we can check whether in some system of the RTR family $\mathcal{F}$ based on the automaton from Fig. 1, some process p_1 can request A,B before some process p_2 requests B, but the wish of p_2 is granted before that of p_1. As shown in Fig. 1, the above happens in $S_n \in \mathcal{F}$ with $n \geq 3$, but not in $S_2 \in \mathcal{F}$ (the overtaking between visible processes 2 and 3 is impossible without invisible process 1). Moreover, when we start extending the B and AB branches by more and more pairs of the appropriate (p)`rqt` /`rel` actions without extending the A branch, we exclude more than one process to run in these branches via adding `rqt(C)/rel(C)` (`rqt(D)/rel(D)`) at their beginnings and ends, and we ask whether p_1 and p_2 can exhibit more and more overtaking, we will need more and more auxiliary processes in the A branch although k and m will not change. □

Despite the above result, there is still some hope that the parametric finite-behaviour verification problem for RTR and MPTL can be reduced to finite-state model checking. Then, however, the bound on the number of processes would have to also reflect the structure of the RTR automaton of the given family and/or the structure of the formula being examined. We leave the problem in its general form open for future research. Instead, we show below that for certain important subclasses of MPTL, the number of processes to be considered in parametric finite-behaviour verification can be fixed to the number of process variables in the formula at hand as in the RTR\P case (although the underlying proof construction is more complex). In this way, we cover, among others, mutual exclusion as one of the key properties of the considered class of systems.

4.3 Cut-Offs for Subclasses of MPTL

The first subclass of MPTL formulae we consider is the class of *invariance* and *reachability* formulae of the form $\Psi_a ::= \forall V : A\, \Box\psi(V)$ and $\Psi_e ::= \exists V : E\, \Diamond\psi(V)$ in which $\psi(V)$ is a boolean combination of atomic formulae $at(p,q)$. Mutual exclusion is an example of a property that falls into this class.

Let $\Psi_a \equiv \forall p_1, ..., p_k : A\, \Box\psi(p_1, ..., p_k)$ be an arbitrary invariance MPTL formula with k quantified process variables. We show that for any RTR family $\mathcal{F}$, the parametric problem of checking $\mathcal{F}, l \models^a_{fin} \Psi_a$ can be reduced to the finite-state problem of verifying $S_k \models_{fin} A\, \Box\psi(1, ..., k)$ with the number of processes fixed to k and the processes to be monitored via $p_1, ..., p_k$ fixed to $1, ..., k$. As above, we first state a basic cut-off lemma, which we subsequently generalize.

Lemma 2. *For any RTR family $\mathcal{F}$ and any nontemporal MPTL path formula $\psi(p_1, ..., p_k)$, the following holds for systems of $\mathcal{F}$:*

$$\forall n \geq k : S_n \models_{fin} \forall p_1, ..., p_k : A\ \Box\psi(p_1, ..., p_k) \Leftrightarrow S_k \models_{fin} A\ \Box\psi(1, ..., k)$$

Proof. (Sketch) We modify the proof of Lemma 1: In the ($\Leftarrow$) case, to resolve the problem with possibly disallowed overtaking among visible processes that could be enabled only due to some invisible processes blocking some low-priority visible ones (cf. Fig. 1), we postpone firing of `(p)req` transitions to be just before firing of the corresponding `(p)take` transitions (or at the very end). Then, since the preserved visible processes release resources as before and do not block them by requests till all originally overtaking requests are served, it can be shown the firability of the reduced transition sequence is guaranteed. Moreover, the behaviour is modified in a way invisible for a reachability formula (negation of $\Box\psi$). $\Box$

Theorem 3. *Let $\mathcal{F}$ be an RTR family and let $\Psi_a \equiv \forall p_1, ..., p_k : A\ \Box\psi(p_1, ..., p_k)$ be an invariance MPTL formula. Then, checking whether $\mathcal{F}, l \models^a_{fin} \Psi_a$ holds is equal to checking whether $S_k \models_{fin} A\ \Box\psi(1, ..., k)$ holds.*

Another subclass of MPTL that can be handled within parametric finite-behaviour verification of RTR in the same way as above is the class of formulae in which we allow any of the MPTL operators to be used, but we exclude distinguishing whether a process is at a location from which it can request some resources or whether it has already requested them. Using such formulae, we can, for example, check whether some overtaking among the involved processes is possible or excluded (though not on the level of particular requests). Due to space limitations, we skip a precise formulation of this result here and refer an interested reader to the full version of the paper [4].

5 Verification of Fair Behaviour

We next discuss verification of fair behaviour of systems of RTR families against correctness requirements expressed in MPTL. The results presented in this section can be applied for verification of liveness properties, such as absence of starvation, of systems of RTR families. As for finite-behaviour verification, we consider the *problem of parametric verification of weakly fair behaviour*, i.e. checking whether $\mathcal{F}, l \models^a_{wf} \Phi_a$ holds for an RTR family $\mathcal{F}$, a universal MPTL formula Φ_a, and a lower bound l on the number of processes.

We show first that under the pure FIFO resource management, considering up to $m+k$ processes – with m being the number of resources and k the number of visible processes – suffices for parametric verification of weakly fair behaviour against any MPTL formulae. By contrast, for prioritized resource management, we prove that (as for finite behaviour verification) there does not exist any general, structure-independent cut-off that would allow us to reduce parametric verification of weakly fair behaviour to finite-state verification. Moreover, we show that, unfortunately, the inexistence of a structure-independent cut-off concerns, among others, also verification of the very important property of absence of starvation. Thus, for the needs of parametric verification of fair behaviour, we subsequently examine in more detail the possibility only sketched in the previous section, i.e. trying to find a cut-off reflecting the structure of the appropriate RTR automaton and/or the structure of the formula.

5.1 A Cut-Off Result for RTR\P Families

Let $\mathcal{F}$ be an RTR\P family with m resources and $\Phi_a \equiv \forall p_1, ..., p_k : A\ \varphi(p_1, ..., p_k)$ a universal MPTL formula with k process variables. We show that the parametric verification problem of weakly fair behaviour for $\mathcal{F}$ and Φ_a can be reduced to a series of finite-state verification tasks in which we do not have to examine any systems of $\mathcal{F}$ with more than $m+k$ processes. The processes to be monitored via $p_1, ..., p_k$ may again be fixed to $1, ..., k$. We denote the thus arising finite-state verification tasks as checking whether $S_n \models_{wf} A\ \varphi(1, ..., k)$ holds.

As in Sect. 4, we now first state a basic cut-off lemma and then we generalize it. However, the way we establish the cut-off turns out to be significantly more complex, because lifting a counterexample behaviour from a small system to a big one is now much more involved than previously. To ensure weak process fairness, newly added processes must be allowed to fire some transitions, but at the same time, this cannot influence the behaviour of the visible processes.

Lemma 3. *For systems of an RTR\P family $\mathcal{F}$ with m resources and an MPTL path formula $\varphi(p_1, ..., p_k)$, the following holds:*

$$\forall n \geq m+k : S_n \models_{wf} \forall p_1, ..., p_k : A\ \varphi(p_1, ..., p_k) \Leftrightarrow S_{m+k} \models_{wf} A\ \varphi(1, ..., k)$$

Proof. (Idea) ($\Leftarrow$) Similar to Lemma 1. To eventually forever block in S_{m+k} the visible processes that forever block in S_n, we need at most one invisible process per resource. ($\Rightarrow$) To extend a counterexample behaviour β_{m+k} of S_{m+k} to one of S_n, we distinguish three cases: (1) If all original processes deadlock in β_{m+k}, the newly added ones can deadlock too. (2) If all processes run in β_{m+k}, at least one process may be shown to eventually not use any resource or always eventually release all of them. The new processes can mimic its behaviour. As they regularly do not block any resources, we may interleave them in the non-blocking phases with each other and with the original processes (without influencing the visible ones). (3) The case when some processes get blocked and some run forever in β_{m+k} may be split to subcases solvable like (1) or (2). □

Now, the theorem generalizing the lemma can be easily obtained by exploiting properties of MPTL. Note, however, that unlike in Theorems 1 and 3, it leads to a necessity of examining several systems, which is due to the difference between the cut-off bound $m+k$ and the number k of visible processes.

Theorem 4. *Let $\mathcal{F}$ be an RTR\P family with m resources and let $\Phi_a \equiv \forall p_1, ..., p_k : A\ \varphi(p_1, ..., p_k)$ be an MPTL formula. Then, checking whether $\mathcal{F}, l \models^a_{wf} \Phi_a$ holds is equal to checking whether $S_n \models_{wf} A\ \varphi(1, ..., k)$ holds for all systems $S_n \in \mathcal{F}$ such that $min(max(l, k), m+k) \leq n \leq m+k$.*

We show in [4] that examining the systems S_k (if $l \leq k$) and S_{m+k} is necessary for the above result. The question of a potential optimization of the result by not having to examine all the systems between $max(l, k)$ and $m + k$ remains open for the future, but this does not seem to be a real obstacle to practical applicability of the result.

5.2 Absence of Structure-Independent Cut-Offs for RTR Families

In verification of weakly fair behaviour of RTR families against MPTL formulae, we examine complete, usually infinite behaviours of systems of the considered families. However, to be able to examine such behaviours, we need to examine their finite prefixes as well. Then, Theorem 2 immediately shows that there does not exist any structure independent cut-off allowing us to reduce the given general problem to finite-state verification. Moreover, for the case of verifying fair behaviour of RTR families against MPTL formulae, no structure-independent cut-offs exist even for more restricted scenarios than in finite behaviour verification. Namely, the query used in the proof of Theorem 2 speaks about two processes. However, below, we give a theorem showing that for the case of parametric verification of weakly fair behaviour, no structure-independent cut-off exists even for *single-process MPTL formulae*, i.e. formulae having a single process variable and thus speaking about a single visible process. In particular, such a cut-off does not exist for a single-process formula encoding absence of starvation. The theorem is proven in [4] by giving an example family.

Theorem 5. *For RTR families $\mathcal{F}$ with m resources and the property of absence of starvation expressed as $\Phi_a \equiv \forall p : A \,\square\, (at(p, req) \Rightarrow \Diamond\, at(p, use))$, the problem of checking whether $\mathcal{F}, 1 \models^a_{wf} \Phi_a$ holds is not, in general, decidable by examining just the systems $S_1, ..., S_n \in \mathcal{F}$ with n being a function of m only.*

5.3 A Cut-Off for Single-Process MPTL Formulae

There is no simple cut-off for verification of weakly fair behaviours of RTR families against single-process MPTL formulae since a lot of invisible processes requesting resources with high priority may be needed to block a visible process. Their number depends on the structure of the control automaton. However, this number can be bounded as shown in this section.

To give the bound we need some definitions. Let $\mathcal{F}(\mathcal{A})$ be an RTR family with m resources. The set of control locations Q_t of $\mathcal{A}$ is split into two disjoint parts: Q_o (all internal control locations and those where processes own at least one resource, without loss of generality a process owns always the same resources at a given control location) and Q_n (the others). Let $F = |Q_n|$ ($F \geq 1$ as Q_n contains the initial location q_0) , $C = |2^{Q_o}| = 2^{|Q_o|}$ and $M_C = C^C$. Then, we can define the needed bound as $B_{\mathcal{F}} = CM_C(M_C{+}1)(2FC(M_C{+}1))^F{+}2C(M_C{+}1){+}2m{+}1$.

The key cut-off lemma below shows that if a formula is true in systems having between $m+1$ and $B_{\mathcal{F}}$ processes, it is also true in systems with more than $B_{\mathcal{F}}$ processes. This and Lemma 5 stating the opposite allows us to reduce the parametric verification problem to verification of systems with up to $B_{\mathcal{F}}$ processes.

Lemma 4. *Let $\mathcal{F}$ be an RTR family with m resources and $\varphi(p)$ an MPTL path formula. Then the following holds for systems of $\mathcal{F}$:*

$$\forall n \geq B_{\mathcal{F}} : (\forall n', m+1 \leq n' \leq B_{\mathcal{F}} : S_{n'} \models_{wf} \forall p : A\ \varphi(p)) \Rightarrow S_n \models_{wf} \forall p : A\ \varphi(p)$$

Proof. (Idea) The full proof is very involved. It is based on the fact that the exact identity of invisible processes is not important, only their number is. The problem of finding the bound can then be seen as finding a bounded solution of a linear equation system encoding properties of admissible counterexamples, which is possible with a lemma from Linear Integer Programming [13]. □

We now give the counterpart to Lemma 4 whose proof is similar to the proof of the appropriate direction of Lemma 3.

Lemma 5. *Let $\mathcal{F}$ be an RTR family and $\varphi(p)$ an MPTL path formula. Then, for systems of $\mathcal{F}$, we have: $\forall n' \geq m+1, n \geq n' : S_n \models_{wf} A\ \varphi(1) \Rightarrow S_{n'} \models_{wf} \forall p : A\ \varphi(p)$.*

We use Lemmas 4 and 5 to give the complete cut-off result for single-process MPTL formulae and weakly fair behaviour of systems of RTR families.

Theorem 6. *Let $\mathcal{F}$ be an RTR family with m resources and let $\Phi_a \equiv \forall p : A\ \varphi(p)$ be a single-process MPTL formula. Then, checking $\mathcal{F}, l \models_{wf}^{a} \Phi_a$ is equal to checking $S_n \models_{wf} A\ \varphi(1)$ for all $S_n \in \mathcal{F}$ with $l \leq n \leq m+1$ or $n = B_{\mathcal{F}}$.*

5.4 Simple RTR Families

Above, we have shown that parametric verification of weakly fair behaviour of RTR families against single-process MPTL formulae is decidable, but no really simple reduction to finite-state verification is possible in general. We now give a restricted subclass of RTR families for which the problem can be solved using a structure-independent cut-off bound.

An RTR family $\mathcal{F}$ is *simple* if the set of control locations Q_n contains only the initial location q_0: Processes start from it by requesting some resources (possibly in different ways) and then they may request further resources as well as release some. However, as soon as they release all of their resources, they go back to q_0. This class is not unrealistic; it corresponds to systems with a single resource-independent computational part surrounded by actions using resources. For this class we show an improved cut-off bound using $2m+2$ processes, which is better than $B_{\mathcal{F}}$ for $F = 1$. This is basically due to the fact that only m invisible processes can be simultaneously in control locations Q_o.

Theorem 7. *Let $\mathcal{F}$ be a simple RTR family with m resources and let $\Phi_a \equiv \forall p : A\ \varphi(p)$ be a single-process MPTL formula. Then, checking $\mathcal{F}, l \models_{wf}^{a} \Phi_a$ is equal to checking $S_n \models_{wf} A\ \varphi(1)$ for all $S_n \in \mathcal{F}$ with $l \leq n \leq m+1$ or $n = 2m+2$.*

Notice that an invisible process can freely move among all locations in a subcomponent Q' of $\mathcal{A}$ which is strongly connected by τ-transitions. Therefore, Theorem 7 can be generalized to families whose Q_n corresponds to such a component. Moreover, the same idea can be used to optimize the general $B_{\mathcal{F}}$ bound.

6 Process Deadlockability

Given an RTR family $\mathcal{F}$ and a system $S_n \in \mathcal{F}$, we say that a *process* p *is deadlocked* in a configuration $c \in C_n$ if there is no configuration reachable from c from which we could fire some transition in p. As is common for linear-time frameworks, process deadlockability cannot be expressed in MPTL, and so since it is an important property to check for the class of systems we consider, we now provide a specialized (structure-independent) cut-off result for dealing with it.

Theorem 8. *Let $\mathcal{F}$ be an RTR family with m resources. For any l, the systems $S_n \in \mathcal{F}$ with $l \leq n$ are free of process deadlock iff $S_{max(m,2)} \in \mathcal{F}$ is.*

Proof. (Sketch) We can encounter scenarios where a group of processes is mutually deadlocked due to some circular dependencies in queues of requests, but also situations where a process is deadlocked due to being always inevitably overtaken by processes that keep running and do not even own any resource forever. However, when we (partially) replace overtaking by postponed firing of requests (cf. Lemma 2), push blocked high-priority requests before the low-priority ones (the former block the latter, but not conversely), and preserve only the running processes that never release all resources simultaneously, we can show that we suffice with one (primary) blocked and/or blocking process per resource. □

Let us note that the possibility of inevitable overtaking examined in the proof of Theorem 8 as a possible source of process deadlocks in systems of RTR families is stronger than starvation. Starvation arises already when there is a single behaviour in which some process is eventually always being overtaken. Interestingly, as we have shown, inevitable overtaking is much easier to handle than starvation, and we obtain a cut-off bound that cannot be improved even when we restrict ourselves to RTR\P families with no overtaking.

7 RTR Families and Undecidability

Finally, we discuss an extension of MPTL by *local process quantification* [8] where processes to be monitored in a behaviour are not fixed at the beginning, but may be chosen independently in each encountered state. Local process quantification can be added to MPTL by allowing $\forall V' : \varphi(V \cup V')$ to be used in a path formula $\varphi(V)$ with the semantics $\pi_n, \nu_n \models \forall V' : \varphi(V \cup V')$ iff $\pi_n, \nu'_n \models \varphi(V \cup V')$ holds for all valuations ν'_n of PV such that $\forall p \in PV \setminus V' : \nu'_n(p) = \nu_n(p)$. Such a quantification can be used to express, e.g., the global response property $A\Box((\exists p_1 : at(p_1, req)) \Rightarrow \Diamond(\exists p_2 : at(p_2, resp)))$, which cannot be encoded with global process quantifiers if the number of processes is not known. Unfortunately, it can be shown that parametric verification of linear-time finite-behaviour properties with local process quantification is undecidable even for RTR\P.

Theorem 9. *The parametric finite-behaviour verification problem of checking $\mathcal{F}, 1 \models^a_{fin} \Phi_a$ for an RTR\P family $\mathcal{F}$ and an MPTL formula Φ_a with local process quantification is undecidable even when the only temporal operators used are $\Box$ and $\Diamond$ and no temporal operator is in the scope of any local process quantifier.*

Proof. (Idea) The proof is done via simulating two-stack push-down automata and is very complex because the queues we work with are not classical communication queues, but only waiting queues. □

8 Conclusions

In this paper, we have defined an abstract model for a significant class of parametric systems of processes competing for access to shared resources under a FIFO resource management with a possibility of distinguishing low- and high-priority requests. The primitives capturing the interaction between processes and resources and the resource management policies considered are natural and inspired by real-life applications. We have established cut-off bounds showing that many practical parametric verification problems (including verification of mutual exclusion, absence of starvation, and process deadlockability) are decidable in this context. The way the obtained results were established is sometimes technically highly involved, which is due to the fact that the considered model is quite powerful and (as we have also shown) positive decidability can easily be lost if verification of a bit more complex properties is considered.

The structure-independent cut-offs we have presented in the paper are small and – for verification of finite behaviour and process deadlockability – optimal. They provide us with practical decision procedures for the concerned parametric verification problems and, moreover, they can also be used to simplify finite-state verification for systems with a given large number of processes.

The structure-dependent cut-off for single-process formulae and verifying the fair behaviour of the general RTR families is quite big and does not yield a really practical decision procedure. One challenging problem is now to optimize this bound. Although we know that no general structure-independent cut-off exists, the bound we have provided is not optimal, and significantly improved cut-offs could be found especially for particular classes of systems as we have already shown for simple RTR families.

Another interesting issue is to improve the decidability bounds. For general RTR families and arbitrary MPTL formulae, decidability of parametric verification of finite as well as fair behaviour remains open. So far, we have only shown that these problems cannot be handled via structure-independent cut-offs. Conversely, the question of existence of practically interesting, decidable fragments of MPTL with local process quantification is worth examining too. If no (or no small) cut-off can be found, we could then try to find some adequate abstraction and/or symbolic verification techniques.

Finally, several extensions or variants of the framework can be considered. For example, the questions of nonexclusive access to resources or nonblocking requests can be examined. Moreover, several other locker policies can be considered, e.g., service in random order or a policy where any blocked process can be overtaken. We believe the results presented here and the reasoning used to establish them provide (to a certain degree) a basis for examining such questions.

Acknowledgment. We thank P.A. Abdulla and T. Arts for fruitful discussions.

References

1. P. Abdulla, A. Bouajjani, B. Jonsson, M. Nilsson. Handling Global Conditions in Parameterized System Verification. In *Proc. of CAV'99*, *LNCS* 1633. Springer, 1999.
2. T. Arts, C. Earle, J. Derrick. Verifying Erlang Code: A Resource Locker Case-Study. In *Proc. of FME'02*, *LNCS* 2391. Springer, 2002.
3. K. Baukus, S. Bensalem, Y. Lakhnech, K. Stahl. Abstracting WS1S Systems to Verify Parameterized Networks. In *Proc. of TACAS 2000*, *LNCS* 1785. Springer, 2000.
4. A. Bouajjani, P. Habermehl, T. Vojnar. Verification of Parametric Concurrent Systems with Prioritized FIFO Resource Management. Full version available at `http://verif.liafa.jussieu.fr/~vojnar/download/concur03.ps.gz`
5. E. Emerson, V. Kahlon. Model Checking Large-Scale and Parameterized Resource Allocation Systems. In *Proc. of TACAS'02*, *LNCS* 2280. Springer, 2002.
6. E. Emerson, K. Namjoshi. Reasoning about Rings. In *Proc. of POPL'95*, 1995.
7. E. Emerson, K. Namjoshi. Automatic Verification of Parameterized Synchronous Systems. In *Proc. of CAV'96*, *LNCS* 1102. Springer, 1996.
8. E. Emerson, A. Sistla. Utilizing Symmetry when Model Checking under Fairness Assumptions: An Automata-theoretic Approach. *ACM Transactions on Programming Languages and Systems*, 19(4), 1997.
9. S. German, A. Sistla. Reasoning about Systems with Many Processes. *JACM*, 39(3), 1992.
10. Y. Kesten, O. Maler, M. Marcus, A. Pnueli, E. Shahar. Symbolic Model Checking with Rich Assertional Languages. In *Proc. of CAV'97*, *LNCS* 1254, 1997.
11. R. Kurshan, K. McMillan. A Structural Induction Theorem for Processes. *Information and Computation*, 117(1), 1995.
12. A. Pnueli, S. Ruah, L. Zuck. Automatic Deductive Verification with Invisible Invariants. In *Proc. of TACAS 2001*, *LNCS* 2031. Springer, 2001.
13. J. von Zur Gathen, M. Sieveking. A Bound on Solutions of Linear Integer Equalities and Inequalities. *Proceedings of the American Mathematical Society*, 1978.
14. P. Wolper, V. Lovinfosse. Verifying Properties of Large Sets of Processes with Network Invariants. In *Autom. Verification Methods for Finite State Systems*, *LNCS* 407, 1989. Springer.

Input/Output Automata: Basic, Timed, Hybrid, Probabilistic, Dynamic,...

Nancy Lynch

MIT, Cambridge, MA 02139, USA
lynch@theory.lcs.mit.edu
http://theory.lcs.mit.edu/~lynch

Abstract. The term *Input/Output Automata* refers to a family of system modeling frameworks based on interacting infinite-state machines. The models come in several flavors, based on which features (fairness, time, continuous behavior, probability, etc.) they can express. In each of these frameworks, automata can be composed in parallel to form more complex automata, and automata can be related using levels of abstraction. Properties of automata can be proved by hand or with the assistance of theorem-proving programs.

The first framework of this kind, which appeared in 1987, was the basic fair asynchronous I/O automata modeling framework of Lynch and Tuttle. It was used originally to describe and analyze a simple network resource allocation algorithm at multiple levels of abstraction. Since then, I/O automata have been used extensively to model distributed algorithms and distributed systems, and even to prove impossibility results. For example, they have been used for algorithms that implement atomic shared memory and for systems that provide group communication services.

Next came the "timed I/O automata" framework of Lynch and Vaandrager, which augmented the "unfair" portion of the basic model with *time-passage steps*. Timed I/O automata have been used to describe a variety of timing-based algorithms, including timeout-based failure detectors and consensus algorithms, communication protocols, and clock synchronization algorithms. They have also been used to analyze performance of many algorithms and systems.

A more recent development was the "hybrid I/O automata" framework, which supports modeling and analysis of hybrid discrete/continuous systems. The main addition here is a set of *trajectories*, which can be used to describe the evolution of system state over intervals of time. Hybrid I/O automata have been used for many case studies, ranging from simple toy examples of vehicles on tracks to complex helicopter control applications.

"Probabilistic I/O automata", defined by Segala in 1995, allow probabilistic choice of the next state, in addition to nondeterministic choice. They have been used for describing and analyzing randomized distributed algorithms and security protocols. Finally, "dynamic I/O automata" were introduced recently by Attie; they add, to the basic "unfair" model, the capability for processes to create other processes and to destroy themselves.

R. Amadio, D. Lugiez (Eds.): CONCUR 2003, LNCS 2761, pp. 191–192, 2003.

In this CONCUR talk, I will define the various I/O automata modeling frameworks in some detail, will survey some of the ways in which they have been used, and will describe current research and open problems. Our current research includes re-formulating timed I/O automata as a restricted version of hybrid I/O automata, and expressing a large set of results about timed systems as theorems about the resulting framework. We are also working on developing the probabilistic I/O automata model further, emphasizing compositionality results. In the longer run, we would like to have a combined model that includes both probabilistic and hybrid continuous/discrete behavior. In fact, eventually, we would like to have a comprehensive I/O-automata-style modeling framework that can express all of the features described above – fairness, time, continuous behavior, probabilistic behavior, and dynamic behavior – but that can be specialized appropriately when only some of these features are needed.

This talk is based on work by many people, notably Mark Tuttle, Frits Vaandrager, Roberto Segala, Paul Attie, and Dilsun Kirli Kaynar.

A Process-Algebraic Language for Probabilistic I/O Automata*

Eugene W. Stark, Rance Cleaveland, and Scott A. Smolka

Department of Computer Science, State University of New York at Stony Brook
Stony Brook, NY 11794 USA
{stark,rance,sas}@cs.sunysb.edu

Abstract. We present a process-algebraic language for Probabilistic I/O Automata (PIOA). To ensure that PIOA specifications given in our language satisfy the "input-enabled" property, which requires that all input actions be enabled in every state of a PIOA, we augment the language with a set of *type inference rules*. We also equip our language with a formal operational semantics defined by a set of *transition rules*. We present a number of results whose thrust is to establish that the typing and transition rules are sensible and interact properly. The central connection between types and transition systems is that if a term is well-typed, then in fact the associated transition system is input-enabled. We also consider two notions of equivalence for our language, *weighted bisimulation equivalence* and *PIOA behavioral equivalence*. We show that both equivalences are substitutive with respect to the operators of the language, and note that weighted bisimulation equivalence is a strict refinement of behavioral equivalence.

Keywords: stochastic process algebras; typing systems and algorithms; process equivalences; continuous-time Markov chains

1 Introduction

In previous work [WSS94,WSS97] we introduced *probabilistic I/O automata* (PIOA) as a formal model for systems that exhibit concurrent and probabilistic behavior. PIOA extend the well-known I/O automaton model for nondeterministic computation [LT87] with two kinds of performance information: probability distributions representing the relative likelihood with which transitions from a given state labeled by the same input are performed; and rate information describing how long, on average, the automaton will remain in a state before taking a particular output or internal transition.

* This research was supported in part by the National Science Foundation under Grant CCR-9988155 and the Army Research Office under Grants DAAD190110003 and DAAD190110019. Any opinions, findings, and conclusions or recommendations expressed in this material are those of the author(s) and do not necessarily reflect the views of the National Science Foundation, the Army Research Office, or other sponsors.

R. Amadio, D. Lugiez (Eds.): CONCUR 2003, LNCS 2761, pp. 193–207, 2003.

PIOA are similar in many respects to *stochastic automata* [Buc99,PA91], and like stochastic automata, PIOA are associated with *continuous-time Markov chains (CTMCs)*. PIOA are also equipped with a *composition operation* by which a complex automaton can be constructed from simpler components. Both PIOA and stochastic automata can thus be seen as a formalism for describing large CTMC system models from simpler components. The composition operation for PIOA is defined in essentially the same way as for stochastic automata, however, the PIOA model draws a distinction between *input* (passive) and *output* (active) actions, and in forming the composition of automata only input/input or input/output synchronization is permitted — the output/output case is prohibited.

In [SS98] we presented algorithms for calculating certain kinds of performance parameters for systems modeled in terms of PIOA. These algorithms work in a *compositional* fashion; that is, by treating the components of a composite system in succession rather than all at once. Our implementation of these algorithms, called "PIOATool," has been integrated into the Concurrency Workbench [CPS93] (CWB), as described in [ZCS03]. The CWB provides several analysis capabilities for specifications expressed in process-algebraic language, including equivalence, preorder, and model checking. It has a retargetable front end that allows it to be applied to virtually any process-algebraic language having a formal semantics defined in the "structural operational semantics" (SOS) style. To achieve the PIOATool/CWB integration, it was necessary for us to design such a process-algebraic language for PIOA-based specifications, together with an SOS semantics for the language. This language, and associated theorems about its semantics, form the subject of the present paper.

The PIOA model exhibits certain features that differentiate it from other languages previously supported by the CWB. One such feature is the fact that each transition of a PIOA, besides being labeled by an action, is also labeled by a numeric *weight*, which can be either a probability (in the case of an input transition) or a rate (in the case of an output or internal transition). Another such feature is the so-called "input-enabled" property, which requires that all input actions be enabled in every state of a PIOA. It is the second of these features that has the most impact on the design of a process-algebraic language for PIOA, since it is necessary to ensure that the input-enabled property holds for every well-formed specification in the language. The problem is that process-algebraic specifications of desired input-enabled transition systems usually have to be built up from component specifications that are not input-enabled.

To solve the problem of guaranteeing that PIOA specifications given in our language satisfy the input-enabled property, we augment the language with a set of *type inference rules*. These rules define a set of inferable *typing judgements* of the form $t : I/J \Rightarrow O$. Such a judgement asserts that for term t, I is a set of actions for which input transitions are guaranteed to be enabled at the first step of t, J is a set of actions for which input transitions are guaranteed to be enabled at all steps of t after the first, and O is a set of actions that includes at least all the outputs that may be produced by t (but which may be larger). A closed term t is called *well-typed* if there is some typing judgement that is inferable for it. Besides enforcing input-enabledness, types are used to enforce compatibility conditions for parallel composition and they also appear as hypotheses in some of the transition rules of the language's operational semantics.

We present a number of results whose thrust is to establish that the typing and transition rules are sensible and interact properly, including a principal type theorem, a connection between the types that can be inferred for a term and the transitions that can be inferred for it, and a subject reduction theorem which establishes that well-typedness is preserved across transitions. The central connection between types and transition systems is that if a term is well-typed, then in fact the associated transition system is input-enabled.

We also define two notions of equivalence for our language, *weighted bisimulation equivalence* and *PIOA behavioral equivalence*, and investigate their properties. In particular, we observe that weighted bisimulation equivalence strictly refines behavioral equivalence (a detailed proof can be found in [Sta03]) and that both equivalences are substitutive with respect to the operators of the PIOA language.

The rest of the paper develops along the following lines. Section 2 surveys some related work by other researchers. Section 3 defines the syntax of our PIOA language. Section 4 presents the language's type-inference rules and transition rules. Section 5 gives metatheoretic results that connect the typing and transition rules. Section 6 defines the two notions of equivalence and establishes that they are substitutive. Section 7 contains our concluding remarks. Due to space limitations, all proofs are omitted.

2 Related Work

Formal languages for specifying (non-probabilistic) I/O automata have previously been proposed by several researchers. The process-algebraic languages presented in [Vaa91,DNS95] ensure input-enabledness by filling in "default transitions" for missing input transitions. In the case of [Vaa91], the default transitions are "self-loop" input transitions taken from a term to itself. In [DNS95], the default transitions lead to the "unspecified I/O automaton" Ω_S. In contrast, we have found in writing actual specifications that sometimes one wants default transitions that are self-loops and sometimes one wants default transitions that go to an error state. An automatic mechanism for filling in defaults is likely to get it wrong a significant fraction of the time, resulting in a specification language that is less transparent to the user. Thus, our language does not make any attempt to fill in default transitions, but rather it employs a notion of well-typedness of terms which guarantees that all well-typed terms are input-enabled.

Another language for describing I/O automata is the IOA language of [GL00]. IOA uses guarded-command-like "transition definitions" consisting of preconditions and effects to encode I/O automata. It also provides constructs for nondeterministic choice, composition, and action hiding. Automatic code generation from IOA specifications is also supported.

A number of process algebras capturing performance-related aspects of system behavior have been proposed in the literature; see [HH02] for a comprehensive survey. Among these, EMPA [BDG98] is perhaps most closely related to our PIOA process algebra as it makes an I/O-like master-slave distinction between "active" and "passive" actions. Active and passive actions can synchronize, with the rate of the synchronization determined by the rate of the passive action, while synchronization between active actions is disallowed. Hillston [Hil94] gives a thoughtful discussion of the issues surrounding synchronization in stochastic processes, including the role of passive and active actions. The issue has also been treated more recently by Brinksma and Hermanns [BH01].

3 Syntax

Let *Act* be a set of *actions*, and let *Var* be a set of *variables*. We use $a, b, c \ldots$ to range over *Act* and we use $X, Y, Z \ldots$ to range over *Var*. Our language has the following syntax:

$$P ::= X \mid \text{nil} \mid a_{(w)}\,?t \mid b_{(r)}\,!t \mid \tau_{(r)} \cdot t \mid \\ t_1 + t_2 \mid t_1 \;{}_{O_1}\|_{O_2}\; t_2 \mid t\,[O] \mid t\{a \leftarrow a'\} \mid \mu X.t$$

The informal meaning of the various constructs is as follows:

- X is a process variable, used in forming recursive processes.
- nil denotes a process with no actions and no transitions.
- $a_{(w)}\,?t$ denotes an *input-prefixed* process that can perform input action $a \in$ *Act* with *weight* w and then become the process t. The weight w must be a positive real number, and it is typically a probability.
- $b_{(r)}\,!t$ denotes an *output-prefixed* process that can perform output action $b \in$ *Act* with *rate* r and then become process denoted by t. The rate r must be a positive real number, which (as usual for CTMC-based models) we regard as the parameter of an exponential probability distribution that describes the length of time before term $b_{(r)}\,!t$ will perform a transition.
- $\tau_{(r)} \cdot t$ denotes an *internal-prefixed* process that can perform an internal transition with rate r and then become the process denoted by t. Here τ is a special symbol, not in *Act*, used to indicate an internal transition. The rate r must be a positive real number, as for output prefixing.
- $t_1 + t_2$ denotes a *choice* between alternatives offered by t_1 and t_2. Choices between summands prefixed by distinct input actions are determined by the environment, and amount to a form of external nondeterminism. Choices between summands prefixed by the same input action are probabilistic choices governed by the relative weights appearing in the prefixes. Choices between summands prefixed by output or internal actions are probabilistic choices governed by the usual *race condition* involving the rates appearing in the prefixes. Choices between input-prefixed summands and summands prefixed by output or internal actions are ultimately resolved by a race between the process and its environment.
- $t_1 \;{}_{O_1}\|_{O_2}\; t_2$ denotes a process that is the parallel composition of the processes denoted by t_1 and t_2. The sets O_1 and O_2 are the sets of output actions controlled by t_1 and t_2, respectively. These sets are required to be disjoint.

- $t\,[O]$ denotes a term t in which all output transitions labeled by actions not in the set O have been *hidden* by transforming them into internal transitions.
- $t\{e \leftarrow e'\}$ denotes the *renaming* of action e to e' in t. The typing rules presented below will ensure that action e' is a fresh action that is not already an input or output action for t.
- $\mu X.t$ denotes a recursively defined process in the usual way. The recursion variable X is required to be be *guarded* by input, output, or internal prefixing in the expression t.

4 Semantics

4.1 Types

Our language is equipped with a set of inference rules for inferring *typing judgements*, which take the form $t : I/J \Rightarrow O$ where I, J, and O are sets of actions. The intuitive meaning of such judgements was described in Sect. 1. We use the abbreviation $I \Rightarrow O$ for the special case $I/I \Rightarrow O$ in which the sets I and J are equal. A closed term t is *well-typed* if some typing judgement can be inferred for it.

The type-inference rules, given in Fig. 1, are expressed in a natural-deduction style. Each rule is to be applied in the context of a set $\mathcal{A}$ of assumptions about the types of the free variables appearing in the terms, where each assumption in $\mathcal{A}$ has the form $X : I/J \Rightarrow O$. Rules other than the recursion rule are applicable if under assumptions $\mathcal{A}$ the judgements in the premises can be inferred, and in that case the judgement in the conclusion can also be inferred under the same assumptions $\mathcal{A}$. The rule for recursion is slightly different, in that in order to establish the premise one is permitted to add to the set $\mathcal{A}$ an additional assumption about the recursive variable X. This additional assumption is discharged by the rule, so that the conclusion is inferable under assumptions $\mathcal{A}$ without the additional assumption on X. Since the set $\mathcal{A}$ is the same in the premises and conclusion of each rule except the rule for recursion, to avoid clutter, we have not explicitly indicated the set $\mathcal{A}$ in each case.

In the sequel, we will use the notation $\mathcal{A} \vdash t : \phi$ to assert that there is an inference of the judgement $t : \phi$ from the set of hypotheses $\mathcal{A}$. We will use $\vdash t : \phi$ to assert that a typing judgement $t : \phi$ is inferable from the empty set of assumptions. Note that this is only possible if t is closed.

It is worth pointing out that the type-inference rules do not uniquely associate a type with each well-typed term. The simplest case of this is the rule for nil, which permits any judgment of the form nil : $\emptyset \Rightarrow O$ to be inferred. However, as we will show later, if a closed term t is well-typed, then in fact there is a uniquely determined set I and a *smallest* set O such that a judgment $t : I \Rightarrow O$ is inferable.

4.2 Transitions

The transition rules for the PIOA language are used to infer transitions of one of the following three types: $t \xrightarrow[w]{a?} u$, $t \xrightarrow[r]{b!} u$, or $t \xrightarrow[r]{\tau} u$. The first of these

$$\frac{t : I \Rightarrow O \quad a \in I}{a_{(w)}\,?t : \{a\}/I \Rightarrow O} \qquad \frac{t : I \Rightarrow O \quad b \notin I}{b_{(r)}\,!t : \emptyset/I \Rightarrow O \cup \{b\}} \qquad \frac{t : I \Rightarrow O}{\tau_{(r)} \cdot t : \emptyset/I \Rightarrow O}$$

$$\frac{t_1 : I_1/J \Rightarrow O_1 \quad t_2 : I_2/J \Rightarrow O_2}{t_1 + t_2 : I_1 \cup I_2/J \Rightarrow O_1 \cup O_2} \qquad \frac{t_1 : I_1 \Rightarrow O_1' \quad t_2 : I_2 \Rightarrow O_2' \quad O_1' \subseteq O_1 \quad O_2' \subseteq O_2}{t_1 \,{}_{O_1}\|_{O_2}\, t_2 : (I_1 \cup I_2)\backslash(O_1 \cup O_2) \Rightarrow O_1 \cup O_2}$$

$$\frac{t : I \Rightarrow O \quad a \in I \quad a' \notin I \cup O}{t\{a \leftarrow a'\} : (I\backslash\{a\}) \cup \{a'\} \Rightarrow O} \qquad \frac{t : I \Rightarrow O \quad b' \notin I \cup O}{t\{b \leftarrow b'\} : I \Rightarrow (O\backslash\{b\}) \cup \{b'\}}$$

$$\frac{t : I \Rightarrow O' \quad O \subseteq O'}{t\,[O] : I \Rightarrow O} \qquad \text{nil} : \emptyset \Rightarrow O \qquad \frac{X : I \Rightarrow O \;\vdash\; t : I \Rightarrow O}{\mu X.t : I \Rightarrow O}$$

Fig. 1. Type-inference rules

denotes an *input transition* having associated action a and *weight* w. The second denotes an *output transition* having associated action b and *rate* r. The third denotes an *internal transition* having associated rate r. Both weights w and rates r are required to be positive real numbers, however we regard weights w as dimensionless quantities (such as probabilities) and we regard rates as dimensional quantities with units of 1/time. The full set of transition rules is given in Fig. 2.

There are several points to be noted about the transition rules. In the rules for a parallel composition $t_1 \,{}_{O_1}\|_{O_2}\, t_2$, an input transition for component t_1 can occur either independently, if the associated action a is in neither the input set I_2 of t_2 nor the set O_2 of outputs declared to be controlled by t_2, or as a synchronized input transition, if a is in both I_1 and I_2, or else as a synchronized output transition, if a is in I_1 and O_2. Synchronization in a parallel composition results in multiplication of the values that label the transitions. However, note that the rules only call for the multiplication of two weights, or the multiplication of a weight and a rate, but never the multiplication of two rates. This is consistent with our view of weights as dimensionless quantities (*e.g.* probabilities) and with rates as quantities with dimensions of 1/time.

In a parallel composition $t_1 \,{}_{O_1}\|_{O_2}\, t_2$ the syntax declares explicitly the sets O_1 and O_2 of outputs that are to be controlled by t_1 and t_2, respectively. The sets of outputs O_1' and O_2' that t_1 and t_2 can actually produce may be smaller. The reason for this is because as t_1 and t_2 evolve, the sets of outputs that they are capable of actually producing may diminish, though in a parallel composition they still exert control over "lost" output actions by inhibiting their occurrence as inputs in other components.

$$a_{(w)}\,?\,t \xrightarrow[w]{a?} t \qquad b_{(r)}\,!\,t \xrightarrow[r]{b!} t \qquad \tau_{(r)} \cdot t \xrightarrow[r]{\tau} t$$

$$\frac{t_1 \xrightarrow[w]{a?} t'}{t_1 + t_2 \xrightarrow[w]{a?} t'} \qquad \frac{t_2 \xrightarrow[w]{a?} t'}{t_1 + t_2 \xrightarrow[w]{a?} t'} \qquad \frac{t_1 \xrightarrow[r]{b!} t'}{t_1 + t_2 \xrightarrow[r]{b!} t'} \qquad \frac{t_2 \xrightarrow[r]{b!} t'}{t_1 + t_2 \xrightarrow[r]{b!} t'}$$

$$\frac{t_1 \xrightarrow[r]{\tau} t'}{t_1 + t_2 \xrightarrow[r]{\tau} t'} \qquad \frac{t_2 \xrightarrow[r]{\tau} t'}{t_1 + t_2 \xrightarrow[r]{\tau} t'}$$

$$\frac{t_1 \xrightarrow[w]{a?} t_1' \quad t_2 : I_2 \Rightarrow O_2' \quad a \notin I_2 \cup O_2}{t_1 \;{}_{O_1}\|_{O_2}\; t_2 \xrightarrow[w]{a?} t_1' \;{}_{O_1}\|_{O_2}\; t_2} \qquad \frac{t_1 : I_1 \Rightarrow O_1' \quad a \notin I_1 \cup O_1 \quad t_2 \xrightarrow[w]{a?} t_2'}{t_1 \;{}_{O_1}\|_{O_2}\; t_2 \xrightarrow[w]{a?} t_1 \;{}_{O_1}\|_{O_2}\; t_2'}$$

$$\frac{t_1 \xrightarrow[w_1]{a?} t_1' \quad t_2 \xrightarrow[w_2]{a?} t_2'}{t_1 \;{}_{O_1}\|_{O_2}\; t_2 \xrightarrow[w_1 w_2]{a?} t_1' \;{}_{O_1}\|_{O_2}\; t_2'}$$

$$\frac{t_1 \xrightarrow[r]{b!} t_1' \quad t_2 : I_2 \Rightarrow O_2' \quad b \notin I_2}{t_1 \;{}_{O_1}\|_{O_2}\; t_2 \xrightarrow[r]{b!} t_1' \;{}_{O_1}\|_{O_2}\; t_2} \qquad \frac{t_1 : I_1 \Rightarrow O_1' \quad b \notin I_1 \quad t_2 \xrightarrow[r]{b!} t_2'}{t_1 \;{}_{O_1}\|_{O_2}\; t_2 \xrightarrow[r]{b!} t_1 \;{}_{O_1}\|_{O_2}\; t_2'}$$

$$\frac{t_1 \xrightarrow[r]{a!} t_1' \quad t_2 \xrightarrow[w]{a?} t_2'}{t_1 \;{}_{O_1}\|_{O_2}\; t_2 \xrightarrow[wr]{a!} t_1' \;{}_{O_1}\|_{O_2}\; t_2'} \qquad \frac{t_1 \xrightarrow[w]{a?} t_1' \quad t_2 \xrightarrow[r]{a!} t_2'}{t_1 \;{}_{O_1}\|_{O_2}\; t_2 \xrightarrow[wr]{a!} t_1' \;{}_{O_1}\|_{O_2}\; t_2'}$$

$$\frac{t_1 \xrightarrow[r]{\tau} t_1'}{t_1 \;{}_{O_1}\|_{O_2}\; t_2 \xrightarrow[r]{\tau} t_1' \;{}_{O_1}\|_{O_2}\; t_2} \qquad \frac{t_2 \xrightarrow[r]{\tau} t_2'}{t_1 \;{}_{O_1}\|_{O_2}\; t_2 \xrightarrow[r]{\tau} t_1 \;{}_{O_1}\|_{O_2}\; t_2'}$$

$$\frac{t \xrightarrow[w]{a?} t'}{t\,[O] \xrightarrow[w]{a?} t'\,[O]} \qquad \frac{t \xrightarrow[r]{b!} t' \quad b \in O}{t\,[O] \xrightarrow[r]{b!} t'\,[O]} \qquad \frac{t \xrightarrow[r]{b!} t' \quad b \notin O}{t\,[O] \xrightarrow[r]{\tau} t'\,[O]} \qquad \frac{t \xrightarrow[r]{\tau} t'}{t\,[O] \xrightarrow[r]{\tau} t'\,[O]}$$

$$\frac{t \xrightarrow[w]{a?} t'}{t\{a \leftarrow a'\} \xrightarrow[w]{a'?} t'\{a \leftarrow a'\}} \qquad \frac{t \xrightarrow[w]{a?} t' \quad a \neq e}{t\{e \leftarrow e'\} \xrightarrow[w]{a?} t'\{e \leftarrow e'\}}$$

$$\frac{t \xrightarrow[r]{b!} t'}{t\{b \leftarrow b'\} \xrightarrow[r]{b'!} t'\{b \leftarrow b'\}} \qquad \frac{t \xrightarrow[r]{b!} t' \quad b \neq e}{t\{e \leftarrow e'\} \xrightarrow[r]{b!} t'\{e \leftarrow e'\}} \qquad \frac{t \xrightarrow[r]{\tau} t'}{t\{e \leftarrow e'\} \xrightarrow[r]{\tau} t'\{e \leftarrow e'\}}$$

$$\frac{t[\mu X.t/X] \xrightarrow[w]{a?} t'}{\mu X.t \xrightarrow[w]{a?} t'} \qquad \frac{t[\mu X.t/X] \xrightarrow[r]{b!} t'}{\mu X.t \xrightarrow[r]{b!} t'} \qquad \frac{t[\mu X.t/X] \xrightarrow[r]{\tau} t'}{\mu X.t \xrightarrow[r]{\tau} t'}$$

Fig. 2. Transition rules

5 Metatheory

In this section, we present a number of results targeted at showing that the typing and transition rules presented in the previous section are sensible and interact properly. In particular, we have the following:

- A principal type theorem (Theorem 1).
- A connection between the types that can be inferred for a term and the transitions that can be inferred for it. (Theorems 2 and 3).
- A subject reduction theorem (Theorem 4): well-typedness is preserved across inferable transitions.

5.1 Principal Types

Our first result in this section states that inferable types have disjoint sets of inputs and outputs, and that the set of inputs available on the first transition is contained in the set of inputs available on subsequent transitions.

Lemma 1. *Suppose* $\vdash t : I/J \Rightarrow O$. *Then* $I \subseteq J$ *and* $J \cap O = \emptyset$.

It is tempting to think that if $\vdash t : I/J \Rightarrow O$, then $\vdash t : I/J \Rightarrow O'$ for all $O' \supseteq O$ such that $J \cap O' = \emptyset$. However this result does *not* hold for our type system. As a trivial example, if t is the term "nil $[\emptyset]$", then although $\vdash t : \emptyset \Rightarrow \emptyset$, we do not have $\vdash t : \emptyset \Rightarrow O$ for any nonempty O.

Theorem 1 (Principal Type Theorem). *If* $\vdash t : I/J \Rightarrow O$ *for some* I, J, *and* O, *then there exists* $\hat{O}$ *such that* $\vdash t : I/J \Rightarrow \hat{O}$, *and such that whenever* $\vdash t : I'/J' \Rightarrow O'$ *then* $I' = I$, $J' = J$, *and* $O' \supseteq \hat{O}$.

For a given closed, well-typed term t, define the *principal type* of t to be the type $I/J \Rightarrow \hat{O}$ given by Theorem 1. Let $\mathrm{Proc}_{I,O}$ denote the set of all well-typed closed terms t having principal type $I \Rightarrow O'$ for some $O' \subseteq O$.

5.2 Types and Transitions

We next establish connections between the types inferable for a term and the transitions inferable for that term. In particular, if a judgement $t : I/J \Rightarrow O$ is inferable, then I is precisely the set of actions a for which a transition of the form $t \xrightarrow[w]{a?} t'$ is inferable, and O contains all actions b for which a transition of the form $t \xrightarrow[r]{b!} t'$ is inferable. Moreover, well-typedness is preserved across transitions, although inferable types are not preserved exactly due to the possibility that the capacity of producing a particular output action can be lost as a result of taking a transition.

A term t such that $t : I/J \Rightarrow O$ is called *input-enabled* if for all actions $e \in I$ some transition of the form $t \xrightarrow[w]{e?} t'$ is inferable.

Theorem 2 (Input Enabledness Theorem). *Suppose $t : I/J \Rightarrow O$. Then for all actions e, $e \in I$ if and only if a transition of the form $t \xrightarrow[w]{e?} t'$ is inferable.*

Theorem 3. *Suppose $t : I/J \Rightarrow O$. Then for all actions e, if a transition of the form $t \xrightarrow[r]{e!} t'$ is inferable, then $e \in O$.*

Theorem 4 (Subject Reduction Theorem). *Suppose $\vdash t : I/J \Rightarrow O$. If for some term t' a transition of the form $t \xrightarrow[w]{e?} t'$, $t \xrightarrow[r]{e!} t'$, or $t \xrightarrow[r]{\tau} t'$ is inferable, then $\vdash t' : J/J \Rightarrow O'$ for some $O' \subseteq O$. In particular,* $\mathrm{Proc}_{I,O}$ *is closed under transitions.*

5.3 Total Transition Weight/Rate

For given terms t and t' and action e, the transition inference rules may yield zero or more distinct inferences of transitions of one of the forms: $t \xrightarrow[w]{e?} t'$, $t \xrightarrow[r]{e!} t'$, or $t \xrightarrow[r]{\tau} t'$, where w and r vary depending on the specific inference. However, it is a consequence of the requirement that all recursive variables be guarded by a prefixing operation that there can be only finitely many such inferences. We write $t \overset{e?}{\underset{w}{\longmapsto}} t'$ to assert that w is the sum of all the weights w_i appearing in distinct inferences of transitions of the form $t \xrightarrow[w_i]{e?} t'$. We call such an expression a *total transition.* Since there are only finitely many such inferences, the sum w is finite. In case there are no inferable transitions $t \xrightarrow[w_i]{e?} t'$ we write $t \overset{e?}{\underset{0}{\longmapsto}} t'$. For output and internal transitions, the notations $t \overset{e!}{\underset{r}{\longmapsto}} t'$ and $t \overset{\tau}{\underset{r}{\longmapsto}} t'$ are defined similarly.

A related notation will also be useful. Suppose t and t' are closed terms in $\mathrm{Proc}_{I,O}$. Then for all $e \in \mathit{Act} \cup \{\tau\}$ define $\Delta^O_e(t,t')$ as follows:

1. If $e \in I$, then $\Delta^O_e(t,t')$ is the unique weight w for which $t \overset{e?}{\underset{w}{\longmapsto}} t'$.
2. If $e \in O$, then $\Delta^O_e(t,t')$ is the unique rate r for which $t \overset{e!}{\underset{r}{\longmapsto}} t'$.
3. If $e = \tau$, then $\Delta^O_e(t,t')$ is the unique rate r for which $t \overset{\tau}{\underset{r}{\longmapsto}} t'$.
4. If $e \notin I \cup O \cup \{\tau\}$ then $\Delta^O_e(t,t) = 1$ and $\Delta^O_e(t,t') = 0$ if $t' \neq t$.

The *derivative* of term t by action e is the mapping $\Delta^O_e t : \mathrm{Proc}_{I,O} \to [0,\infty)$ defined so that the relation $(\Delta^O_e t)(t') = \Delta^O_e(t,t')$ holds identically for all terms t'. If S is a set of terms, then we use $\Delta^O_e(t,S)$ or $(\Delta^O_e t)(S)$ to denote the sum $\sum_{t' \in S} \Delta^O_e(t,t')$, which is finite.

Note that the reason why we retain the superscripted O in the Δ^O_e notation is because the terms t and t' do not uniquely determine the set O, therefore whether clause (2) or (4) in the definition applies for a given action e depends on the set O.

Define the class of *input-stochastic* terms to be the largest subset of $\mathrm{Proc}_{I,O}$ such that if t is input-stochastic then the following conditions hold:

1. For all $e \in I$ we have $\sum_{t'} \Delta_e^O(t,t') = 1$.
2. Whenever $\Delta_e^O(t,t') > 0$ then t' is also input-stochastic.

Input-stochastic terms are those for which the weights associated with input transitions can be interpreted as probabilities. These are the terms that are naturally associated with PIOA, in the sense that the set of all stochastic terms in $\mathrm{Proc}_{I,O}$ is the set of states of a PIOA with input actions I, output actions O, and the single internal action τ, and with Δ_e^O as the "transition matrix" for action e.

In a later section, we will require the notion of the *total rate* $\mathrm{rt}(t)$ of a closed, well-typed term t such that $\vdash t : I \Rightarrow O$. This quantity is defined as follows:

$$\mathrm{rt}(t) = \sum_{e \in O \cup \{\tau\}} \sum_{t'} \Delta_e^O(t,t').$$

It is a consequence of the fact that only finitely many actions e can appear in term t that $\mathrm{rt}(t)$ is finite. Note also that $\mathrm{rt}(t)$ does not depend on O.

6 Equivalence of Terms

In this section, we define two notions of equivalence for our language, and investigate their properties. The first equivalence, which we call *weighted bisimulation equivalence*, is a variant of bisimulation that is based on the same ideas as *probabilistic bisimulation* [LS91], Hillston's "strong equivalence" [Hil96], and "strong Markovian bisimulation" [HH02]. The second equivalence, called *PIOA behavior equivalence*, is based on the notion of the "behavior map" associated with a PIOA, which has appeared in various forms in our previous work [WSS94, WSS97,SS98], along with motivation for the concept. Additional motivation and a detailed comparison of probabilistic bisimulation equivalence and PIOA behavior equivalence can be found in [Sta03]. In the present paper we focus primarily on congruence properties of these equivalences with respect to the operators of our language.

6.1 Weighted Bisimulation Equivalence

A *weighted bisimulation* is an equivalence relation R on $\mathrm{Proc}_{I,O}$ such that whenever $t \; R \; t'$ then for all actions e and all equivalence classes $\mathcal{C}$ of R we have $\Delta_e^O(t,\mathcal{C}) = \Delta_e^O(t',\mathcal{C})$. Clearly, the identity relation is a weighted bisimulation. It is a standard argument to prove that the transitive closure of the union of an arbitrary collection of weighted bisimulations is again a weighted bisimulation. Thus, there exists a largest weighted bisimulation $\underset{I,O}{\sim}$ on $\mathrm{Proc}_{I,O}$. We call $\underset{I,O}{\sim}$ the *weighted bisimulation equivalence* relation.

Define a *weighting* on terms to be a function μ from $\mathrm{Proc}_{I,O}$ to the non-negative real numbers, such that that $\mu(t) = 0$ for all but finitely many terms $t \in \mathrm{Proc}_{I,O}$. Suppose R is an equivalence relation on $\mathrm{Proc}_{I,O}$. Define the *lifting* of R to weightings to be the relation $\overline{R}$ on weightings defined by the following condition: $\mu \overline{R} \mu'$ if and only if $\mu(\mathcal{C}) = \mu'(\mathcal{C})$ for all equivalence classes $\mathcal{C}$ of R.

The following result (*cf.* [JLY01]) simply restates the definition of weighted bisimulation in terms of weightings.

Lemma 2. *An equivalence relation R on* $\mathrm{Proc}_{I,O}$ *is a weighted bisimulation if and only if $t \; R \; u$ implies $\Delta_e^O t \; \overline{R} \; \Delta_e^O u$ for all terms t, u and all actions e.*

Lemma 3. *Let R be a symmetric relation on terms. If for all terms t, u and all actions e we have*

$$t \; R \; u \text{ implies } \Delta_e^O t \; \overline{(R \cup \underset{I,O}{\sim})^*} \; \Delta_e^O u,$$

then $R \subseteq \underset{I,O}{\sim}$.

Lemma 3 can be used to establish that weighted bisimilarity is substitutive with respect to the operators of our language.

Theorem 5. *The following hold, whenever the sets of inputs and outputs are such that the terms are well-typed and the indicated relations make sense:*

1. *If $t \underset{I',O'}{\sim} t'$, then*
   a) $a_{(w)} ? t \underset{I,O}{\sim} a_{(w)} ? t'$
   b) $b_{(r)} ! t \underset{I,O}{\sim} b_{(r)} ! t'$
   c) $\tau_{(r)} \cdot t \underset{I,O}{\sim} \tau_{(r)} \cdot t'$
2. *If $t_1 \underset{I_1,O_1}{\sim} t_1'$ and $t_2 \underset{I_2,O_2}{\sim} t_2'$, then $t_1 + t_2 \underset{I,O}{\sim} t_1' + t_2'$*
3. *If $t_1 \underset{I_1,O_1}{\sim} t_1'$ and $t_2 \underset{I_2,O_2}{\sim} t_2'$, then $t_1 \,{}_{O_1}\|_{O_2}\, t_2 \underset{I,O}{\sim} t_1' \,{}_{O_1}\|_{O_2}\, t_2'$.*
4. *If $t \underset{I',O'}{\sim} t'$, then $t\,[O] \underset{I,O}{\sim} t'\,[O]$.*
5. *If $t \underset{I',O'}{\sim} t'$, then $t\{e \leftarrow e'\} \underset{I,O}{\sim} t'\{e \leftarrow e'\}$.*

6.2 Behavior Equivalence

In this section, we restrict our attention to the fragment of the language obtained by omitting internal actions and hiding. Let $\mathrm{Proc}_{I,O}^-$ denote the portion of $\mathrm{Proc}_{I,O}$ contained in this fragment. The full language can be treated, but the definition of behavior equivalence becomes more complicated and requires the use of fixed-point techniques, rather than the simple inductive definition given below.

Behavior equivalence is defined by associating with each closed term t with $\vdash t : I \Rightarrow O$ a certain function $\mathcal{B}_t^O$ which we call the *behavior* of t. Terms t and t' will be called *behavior equivalent* if their associated behaviors are identical.

To define $\mathcal{B}_t^O$, some preliminary definitions are required. A *rated action* is a pair $(r, e) \in [0, \infty) \times Act$. Rather than the somewhat heavy notation (r, e), we usually denote a rated action by an expression ${}_re$ in which the rate appears as a subscript preceding the action. A finite sequence

$${}_{r_1}e_1{}_{r_2}e_2 \cdots {}_{r_n}e_n$$

of rated actions is called a *rated trace*. We use ϵ to denote the empty rated trace.

An *observable* is a mapping from rated traces to real numbers. We use Obs to denote the set of all observables. The *derivative* of an observable Φ by a rated action ${}_r e$ is the observable Ψ defined by $\Psi(\alpha) = \Phi({}_r e\ \alpha)$ for all rated traces α. Borrowing notation from the literature on formal power series (of which observables are an example), we write ${}_r e^{-1}\Phi$ to denote the derivative of Φ by the rated action ${}_r e$.

To each term t in $\mathrm{Proc}^{-}_{I,O}$ we associate a *transformation of observables*:

$$\mathcal{B}^O_t : \mathrm{Obs} \to \mathrm{Obs}$$

according to the following inductive definition:

$$\mathcal{B}^O_t[\Phi](\epsilon) = \Phi(\epsilon)$$

$$\mathcal{B}^O_t[\Phi]({}_r e\ \alpha) = \sum_{t'} \Delta^O_e(t,t')\ \mathcal{B}^O_{t'}[{}_{r+\mathrm{rt}(t)}e^{-1}\Phi](\alpha).$$

Terms t and t' in $\mathrm{Proc}^{-}_{I,O}$ are called *behavior equivalent*, and we write $t \underset{I,O}{\equiv} t'$, if $\mathcal{B}^O_t = \mathcal{B}^O_{t'}$.

Intuitively, in the definition of $\mathcal{B}^O_t[\Phi](\alpha)$, one should think of the rated trace α as giving certain partial information about a particular set of execution trajectories that might be traversed by a process t in combination with its environment. In particular, if $\alpha = {}_{r_1}e_1{}_{r_2}e_2 \ldots {}_{r_n}e_n$, then $e_1e_2\ldots e_n$ is the sequence of actions performed in such a trajectory (including both input and output actions) and $r_1r_2\ldots r_n$ is the sequence of output rates associated with the successive states visited by the environment in such a trajectory. The observable Φ should be thought of as a way of associating some numeric measure, or reward, with trajectories. By "unwinding" the definition of $\mathcal{B}^O_t[\Phi](\alpha)$, one can see that it amounts to a weighted summation of the rewards $\Phi(\alpha')$ associated with trajectories α' that start from t and that "match" α, in the sense that α and α' have the same sequence of actions, but the rates of actions in α' are obtained by adding to the rate of the corresponding action in α the total rate $\mathrm{rt}(u)$ of a state u reachable by process t. Further explanation and examples of what can be done with behavior maps can be found in [SS98,Sta03].

The next result states that behavior equivalent terms have the same total rate, and the same total transition weight for each individual action.

Lemma 4. *Suppose* $t \underset{I,O}{\equiv} t'$. *Then*

1. $\sum_u \Delta^O_e(t,u) = \sum_u \Delta^O_e(t',u)$ *for all* $e \in Act$.
2. $\mathrm{rt}(t) = \mathrm{rt}(t')$.

A mistake that we made repeatedly while developing the language and these results was to suppose that the choice operator in the language ought to correspond to sum of behavior maps. This is wrong. The following result shows the correct relationship.

Lemma 5. *Suppose t_1 and t_2 are terms, such that $\vdash t_1 + t_2 : I \Rightarrow O$. Then for all observables Φ, rated actions ${}_r e$, and rated traces α':*

$$\mathcal{B}^O_{t_1+t_2}[\Phi](\epsilon) = \Phi(\epsilon)$$
$$\mathcal{B}^O_{t_1+t_2}[\Phi]({}_r e\ \alpha') = \mathcal{B}^O_{t_1}[\Phi]({}_{r+\mathrm{rt}(t_2)} e\ \alpha') + \mathcal{B}^O_{t_2}[\Phi]({}_{r+\mathrm{rt}(t_1)} e\ \alpha').$$

The following result states that behavior maps are compositional with respect to the parallel operator. We have proved this result in various forms in our previous papers [WSS94,WSS97,SS98]. A proof of the result based on the specific definition of behavior map given here appears in [Sta03].

Lemma 6. *Suppose t_1 and t_2 are terms, such that $\vdash t_1\ {}_{O_1}\|_{O_2}\ t_2 : I \Rightarrow O$. Then*

$$\mathcal{B}^O_{t_1\ {}_{O_1}\|_{O_2}\ t_2} = \mathcal{B}^{O_1}_{t_1} \circ \mathcal{B}^{O_2}_{t_2}.$$

Lemma 7. *Suppose $\vdash t\{e \leftarrow e'\} : I \Rightarrow O$. Let mapping h on rated traces be the string homomorphism that interchanges ${}_r e'$ and ${}_r e$ and is the identity mapping on all other rated actions. Then*

$$\mathcal{B}^O_{t\{e\leftarrow e'\}}[\Phi] = \mathcal{B}^O_t[\Phi \circ h] \circ h,$$

where $O' = O$ if $e' \in I$, and $O' = (O \setminus \{e'\}) \cup \{e\}$ if $e' \in O$.

The preceding lemmas can be used to show that behavior equivalence is substitutive with respect to the operations of our language (exclusive of internal prefixing and hiding). The proofs are all ultimately by induction on the length of the rated trace α supplied as argument, though in the cases of parallel composition and renaming we have been able to hide this "operational" induction inside the more "denotational" Lemmas 6 and 7.

Theorem 6. *The following hold, whenever the sets of inputs and outputs are such that the terms are well-typed and the indicated relations make sense:*

1. *If $t \underset{I',O'}{\equiv} t'$, then*
   a) *$a_{(w)}\,?t \underset{I,O}{\equiv} a_{(w)}\,?t'$*
   b) *$b_{(r)}\,!t \underset{I,O}{\equiv} b_{(r)}\,!t'$*
2. *If $t_1 \underset{I_1,O_1}{\equiv} t_1'$ and $t_2 \underset{I_2,O_2}{\equiv} t_2'$, then $t_1 + t_2 \underset{I,O}{\equiv} t_1' + t_2'$*
3. *If $t_1 \underset{I_1,O_1}{\equiv} t_1'$ and $t_2 \underset{I_2,O_2}{\equiv} t_2'$, then $t_1\ {}_{O_1}\|_{O_2}\ t_2 \underset{I,O}{\equiv} t_1'\ {}_{O_1}\|_{O_2}\ t_2'$.*
4. *If $t \underset{I',O'}{\equiv} t'$, then $t\{e\leftarrow e'\} \underset{I,O}{\equiv} t'\{e\leftarrow e'\}$.*

6.3 Comparison of the Equivalences

The following result is a consequence of characterizations, obtained in [Sta03], of weighted bisimulation equivalence and behavior equivalence.

Theorem 7. *Suppose t and t' are in $\mathrm{Proc}^-_{I,O}$. If $t \underset{I,O}{\sim} t'$, then $t \underset{I,O}{\equiv} t'$.*

In addition, if $I = \emptyset$ and $O = \{a, b, c\}$ then we have

$$a_{(1)} \,!\, b_{(2)} \,!\, \mathrm{nil} + a_{(1)} \,!\, c_{(2)} \,!\, \mathrm{nil} \underset{I,O}{\equiv} a_{(2)} \,!\, (b_{(1)} \,!\, \mathrm{nil} + c_{(1)} \,!\, \mathrm{nil}),$$

but the same two terms are not related by $\underset{I,O}{\sim}$. Thus, weighted bisimulation equivalence is a strict refinement of behavior equivalence.

7 Conclusion

We have presented a process-algebraic language having input, output, and internal transitions, where input actions are labeled by weights and output and internal actions are labeled by rates. A set of typing rules is employed to define the sets $\mathrm{Proc}_{I,O}$ of well-typed terms, which are guaranteed to have transitions enabled for all actions $a \in I$. A readily identifiable subset of the well-typed terms are the input-stochastic terms, in which input weights can be interpreted as probabilities. The input-stochastic terms are therefore the states of a PIOA, so that the language is suitable for writing PIOA-based specifications. We have defined two equivalences on the language, a weighted bisimulation equivalence defined in the same pattern as the classical probabilistic bisimulation equivalence, and a so-called "behavior equivalence" whose definition is motivated by our previous work on PIOA. Both equivalences were shown to be congruences, and we noted that weighted bisimulation equivalence is a strict refinement of behavior equivalence.

A natural direction for future work is to axiomatize the equational theories of the two congruences. For weighted bisimulation equivalence, a standard equational axiomatization should be possible, and is not likely to yield any surprises. The situation for behavior equivalence is a bit different, however. Weighted bisimulation equivalence is the largest equivalence on terms that respects transition weights in the sense of Lemma 2. Since behavior equivalence relates terms that are not weighted bisimulation equivalent, it will not be possible to obtain an equational axiomatization of behavior equivalence, at least in the context of a theory of equations between terms. However, it appears that it is possible to obtain an axiomatization of behavior equivalence in the context of a theory of equations between *weightings*, rather than terms. We are currently working out the details of this idea.

References

[BDG98] M. Bernardo, L. Donatiello, and R. Gorrieri. A formal approach to the integration of performance aspects in the modeling and analysis of concurrent systems. *Information and Computation*, 144(2):83–154, 1998.

[BH01] E. Brinksma and H. Hermanns. Process algebra and Markov chains. In E. Brinksma, H. Hermanns, and J.-P. Katoen, editors, *FMPA 2000: Euro-Summerschool on Formal Methods and Performance Analysis*, volume 2090 of *Lecture Notes in Computer Science*, pages 183–231. Springer-Verlag, 2001.

[Buc99] P. Buchholz. Exact performance equivalence: An equivalence relation for stochastic automata. *Theoretical Computer Science*, 215:263–287, 1999.

[CPS93] R. Cleaveland, J. Parrow, and B. U. Steffen. The Concurrency Workbench: A semantics-based tool for the verification of concurrent systems. *ACM TOPLAS*, 15(1), 1993.

[DNS95] R. De Nicola and R. Segala. A process algebraic view of Input/Output Automata. *Theoretical Computer Science*, 138(2), 1995.

[GL00] S.J. Garland and N.A. Lynch. Using I/O automata for developing distributed systems. In Gary T. Leavens and Murali Sitaraman, editors, *Foundations of Component-Based Systems*, pages 285–312. Cambridge University Press, 2000.

[HH02] J.-P. Katoen H. Hermanns, U. Herzog. Process algebra for performance evaluation. *Theoretical Computer Science*, 274:43–97, 2002.

[Hil94] J. Hillston. The nature of synchronization. In U. Herzog and M. Rettelbach, editors, *Proceedings of the 2nd Workshop on Process Algebra and Performance Modeling*, pages 51–70, University of Erlangen, July 1994.

[Hil96] J. Hillston. *A Compositional Approach to Performance Modelling.* Cambridge University Press, 1996.

[JLY01] B. Jonsson, K. G. Larsen, and W. Yi. Probabilistic extensions of process algebras. In J.A. Bergstra, A. Ponse, and S.A. Smolka, editors, *Handbook of Process Algebra.* Elsevier, 2001.

[LS91] K.G. Larsen and A. Skou. Bisimulation through probabilistic testing. *Information and Computation*, 94(1):1–28, September 1991.

[LT87] N.A. Lynch and M. Tuttle. Hierarchical correctness proofs for distributed algorithms. In *Proceedings of the 6th Annual ACM Symposium on Principles of Distributed Computing*, pages 137–151, 1987.

[PA91] B. Plateau and K. Atif. Stochastic automata networks for modeling parallel systems. *IEEE Transactions on Software Engineering*, 17:1093–1108, 1991.

[SS98] E.W. Stark and S. Smolka. Compositional analysis of expected delays in networks of probabilistic I/O automata. In *Proc. 13th Annual Symposium on Logic in Computer Science*, pages 466–477, Indianapolis, IN, June 1998. IEEE Computer Society Press.

[Sta03] E. Stark. On behavior equivalence for probabilistic I/O automata and its relationship to probabilistic bisimulation. *Journal of Automata, Languages, and Combinatorics*, 8(2), 2003. to appear.

[Vaa91] F.W. Vaandrager. On the relationship between process algebra and input/output automata. In *Sixth Annual Symposium on Logic in Computer Science (LICS '91)*, pages 387–398, Amsterdam, July 1991. Computer Society Press.

[WSS94] S.-H. Wu, S.A. Smolka, and E.W. Stark. Compositionality and full abstraction for probabilistic I/O automata. In *Proceedings of CONCUR '94 — Fifth International Conference on Concurrency Theory*, Uppsala, Sweden, August 1994.

[WSS97] S.-H. Wu, S.A. Smolka, and E.W. Stark. Composition and behaviors of probabilistic I/O automata. *Theoretical Computer Science*, 176(1-2):1–38, 1997.

[ZCS03] D. Zhang, R. Cleaveland, and E.W. Stark. The integrated CWB-NC/PIOATool for functional and performance analysis of concurrent systems. In *Proceedings of the Ninth International Conference on Tools and Algorithms for the Construction and Analysis of Systems (TACAS 2003).* Lecture Notes in Computer Science, Springer-Verlag, April 2003.

Compositionality for Probabilistic Automata

Nancy Lynch[1,*], Roberto Segala[2,**], and Frits Vaandrager[3,***]

[1] MIT Laboratory for Computer Science
Cambridge, MA 02139, USA
lynch@theory.lcs.mit.edu

[2] Dipartimento di Informatica, Università di Verona
Strada Le Grazie 15, 37134 Verona, Italy
roberto.segala@univr.it

[3] Nijmegen Institute for Computing and Information Sciences
University of Nijmegen
P.O. Box 9010, 6500 GL Nijmegen, The Netherlands
fvaan@cs.kun.nl

Abstract. We establish that on the domain of probabilistic automata, the trace distribution preorder coincides with the simulation preorder.

1 Introduction

Probabilistic automata [9,10,12] constitute a mathematical framework for modeling and analyzing probabilistic systems, specifically, systems consisting of asynchronously interacting components that may make nondeterministic and probabilistic choices. They have been applied successfully to distributed algorithms [3,7,1] and practical communication protocols [13].

An important part of a system modeling framework is a notion of *external behavior* of system components. Such a notion can be used to define implementation and equivalence relationships between components. For example, the external behavior of a nondeterministic automaton can be defined as its set of *traces*—the sequences of external actions that arise during its executions [5]. Implementation and equivalence of nondeterministic automata can be defined in terms of inclusion and equality of sets of traces. By analogy, Segala [9] has proposed defining the external behavior of a probabilistic automaton as its set of *trace distributions*, and defining implementation and equivalence in terms of inclusion and equality of sets of trace distributions. Stoelinga and Vaandrager have proposed a simple testing scenario for probabilistic automata, and have proved that the equivalence notion induced by their scenario coincides with Segala's trace distribution equivalence [14].

* Supported by AFOSR contract #F49620-00-1-0097, NSF grant #CCR-0121277, and DARPA/AFOSR MURI #F49620-02-1-0325.

** Supported by MURST projects MEFISTO and CoVer.

*** Supported by PROGRESS project TES4999: Verification of Hard and Softly Timed Systems (HaaST) and DFG/NWO bilateral cooperation project 600.050.011.01 Validation of Stochastic Systems (VOSS).

R. Amadio, D. Lugiez (Eds.): CONCUR 2003, LNCS 2761, pp. 208–221, 2003.

However, a problem with these notions is that trace distribution inclusion and equivalence are not compositional. To address this problem, Segala [9] defined more refined notions of implementation and equivalence. In particular, he defined the *trace distribution precongruence*, $\leq_{DC}$, as the coarsest precongruence included in the trace distribution inclusion relation. This yields compositionality by construction, but does not provide insight into the nature of the $\leq_{DC}$ relation. Segala also provided a characterization of $\leq_{DC}$ in terms of the set of trace distributions observable in a certain *principal context*—a rudimentary probabilistic automaton that makes very limited nondeterministic and probabilistic choices. However, this indirect characterization still does not provide much insight into the structure of $\leq_{DC}$, for example, it does not explain its branching structure.

In this paper, we provide an explicit characterization of the trace distribution precongruence, $\leq_{DC}$, for probabilistic automata, that completely explains its branching structure. Namely, we show that $\mathcal{P}_1 \leq_{DC} \mathcal{P}_2$ if and only if there exists a *weak probabilistic (forward) simulation relation* from $\mathcal{P}_1$ to $\mathcal{P}_2$. Moreover, we provide a similar characterization of $\leq_{DC}$ for nondeterministic automata in terms of the existence of a weak (non-probabilistic) simulation relation. It was previously known that simulation relations are sound for $\leq_{DC}$ [9], for both nondeterministic and probabilistic automata; we show the surprising fact that they are also *complete*. That is, we show that, for both nondeterministic and probabilistic automata, probabilistic contexts can observe all the distinctions that can be expressed using simulation relations.

Sections 2 and 3 contain basic definitions and results for nondeterministic and probabilistic automata, respectively, and for the preorders we consider. These sections contain no new material, but recall definitions and theorems from the literature. For a more leisurely introduction see [5,12]. Sections 4 and 5 contain our characterization results for nondeterministic and probabilistic automata. Section 6 contains our conclusions.

A full version of this paper, including all proofs, appears in [4].

2 Definitions for Nondeterministic Automata

A *(nondeterministic) automaton* is a tuple $\mathcal{A} = (Q, \bar{q}, E, H, D)$, where Q is a set of *states*, $\bar{q} \in Q$ is a *start state*, E is a set of *external actions*, H is a set of *internal (hidden) actions* with $E \cap H = \emptyset$, and $D \subseteq Q \times (E \cup H) \times Q$ is a *transition relation*. We denote $E \cup H$ by A and we refer to it as the set of *actions*. We denote a transition (q, a, q') of D by $q \xrightarrow{a} q'$. We write $q \rightarrow q'$ if $q \xrightarrow{a} q'$ for some a, and we write $q \rightarrow$ if $q \rightarrow q'$ for some q'. We assume finite branching: for each state q the number of pairs (a, q') such that $q \xrightarrow{a} q'$ is finite. We denote the elements of an automaton $\mathcal{A}$ by $Q_{\mathcal{A}}, \bar{q}_{\mathcal{A}}, E_{\mathcal{A}}, H_{\mathcal{A}}, D_{\mathcal{A}}, A_{\mathcal{A}}, \xrightarrow{a}_{\mathcal{A}}$. Often we use the name $\mathcal{A}$ for a generic automaton; then we usually omit the subscripts, writing simply Q, $\bar{q}$, E, H, D, A, and $\xrightarrow{a}$. We extend this convention to allow indices and primes as well; thus, the set of states of automaton $\mathcal{A}'_i$ is denoted by Q'_i.

An *execution fragment* of an automaton $\mathcal{A}$ is a finite or infinite sequence $\alpha = q_0 a_1 q_1 a_2 q_2 \cdots$ of alternating states and actions, starting with a state and, if the sequence is finite, ending in a state, where each $(q_i, a_{i+1}, q_{i+1}) \in D$. State q_0, the first state of α, is denoted by *fstate*(α). If α is a finite sequence, then the last state of α is denoted by *lstate*(α). An *execution* is an execution fragment whose first state is the start state $\bar{q}$. We let *frags*$(\mathcal{A})$ denote the set of execution fragments of $\mathcal{A}$ and *frags*$^*(\mathcal{A})$ the set of finite execution fragments. Similarly, we let *execs*$(\mathcal{A})$ denote the set of executions of $\mathcal{A}$ and *execs*$^*(\mathcal{A})$ the set of finite executions.

Execution fragment α is a *prefix* of execution fragment α', denoted by $\alpha \leq \alpha'$, if sequence α is a prefix of sequence α'. Finite execution fragment $\alpha_1 = q_0 a_1 q_1 \cdots a_k q_k$ and execution fragment α_2 can be concatenated if $fstate(\alpha_2) = q_k$. In this case the *concatenation* of α_1 and α_2, $\alpha_1 \frown \alpha_2$, is the execution fragment $q_0 a_1 q_1 \cdots a_k \alpha_2$. Given an execution fragment α and a finite prefix α', $\alpha \rhd \alpha'$ (read α after α') is defined to be the unique execution fragment α'' such that $\alpha = \alpha' \frown \alpha''$.

The *trace* of an execution fragment α of an automaton $\mathcal{A}$, written $trace_{\mathcal{A}}(\alpha)$, or just *trace*$(\alpha)$ when $\mathcal{A}$ is clear from context, is the sequence obtained by restricting α to the set of external actions of $\mathcal{A}$. For a set S of executions of $\mathcal{A}$, $traces_{\mathcal{A}}(S)$, or just *traces*(S) when $\mathcal{A}$ is clear from context, is the set of traces of the executions in S. We say that β is a trace of $\mathcal{A}$ if there is an execution α of $\mathcal{A}$ with $trace(\alpha) = \beta$. Let *traces*$(\mathcal{A})$ denote the set of traces of $\mathcal{A}$. We define the *trace preorder* relation on automata as follows: $\mathcal{A}_1 \leq_T \mathcal{A}_2$ iff $E_1 = E_2$ and $traces(\mathcal{A}_1) \subseteq traces(\mathcal{A}_2)$. We use $\equiv_T$ to denote the kernel of $\leq_T$.

If $a \in A$, then $q \overset{a}{\Longrightarrow} q'$ iff there exists an execution fragment α such that $fstate(\alpha) = q$, $lstate(\alpha) = q'$, and $trace(\alpha) = trace(a)$. (Here and elsewhere, we abuse notation slightly by extending the *trace* function to arbitrary sequences.) We call $q \overset{a}{\Longrightarrow} q'$ a *weak transition.* We let tr range over either transitions or weak transitions. For a transition $tr = (q, a, q')$, we denote q by *source*(tr) and q' by *target*(tr).

Composition: Automata $\mathcal{A}_1$ and $\mathcal{A}_2$ are *compatible* if $H_1 \cap A_2 = A_1 \cap H_2 = \emptyset$. The *composition* of compatible automata $\mathcal{A}_1$ and $\mathcal{A}_2$, denoted by $\mathcal{A}_1 \| \mathcal{A}_2$, is the automaton $\mathcal{A} \triangleq (Q_1 \times Q_2, (\bar{q}_1, \bar{q}_2), E_1 \cup E_2, H_1 \cup H_2, D)$ where D is the set of triples (q, a, q') such that, for $i \in \{1, 2\}$:

$$a \in A_i \Rightarrow (\pi_i(q), a, \pi_i(q')) \in D_i \text{ and } a \notin A_i \Rightarrow \pi_i(q) = \pi_i(q').$$

Let α be an execution fragment of $\mathcal{A}_1 \| \mathcal{A}_2$, $i \in \{1, 2\}$. Then $\pi_i(\alpha)$, the ith projection of α, is the sequence obtained from α by projecting each state onto its ith component, and removing each action not in A_i together with its following state. Sometimes we denote this projection by $\alpha \lceil \mathcal{A}_i$.

Proposition 1. *Let $\mathcal{A}_1$ and $\mathcal{A}_2$ be automata, with $\mathcal{A}_1 \leq_T \mathcal{A}_2$. Then, for each automaton $\mathcal{C}$ compatible with both $\mathcal{A}_1$ and $\mathcal{A}_2$, $\mathcal{A}_1 \| \mathcal{C} \leq_T \mathcal{A}_2 \| \mathcal{C}$.*

Simulation Relations: Below we define two kinds of simulation relations: a forward simulation, which provides a step-by-step correspondence, and a weak forward simulation, which is insensitive to the occurrence of internal steps.

Namely, relation $R \subseteq Q_1 \times Q_2$ is a *forward simulation* (resp., *weak forward simulation*) from $\mathcal{A}_1$ to $\mathcal{A}_2$ iff $E_1 = E_2$ and both of the following hold:

1. $\bar{q}_1 \; R \; \bar{q}_2$.
2. If $q_1 \; R \; q_2$ and $q_1 \xrightarrow{a} q_1'$, then there exists q_2' such that $q_2 \xrightarrow{a} q_2'$ (resp., $q_2 \overset{a}{\Longrightarrow} q_2'$) and $q_1' \; R \; q_2'$.

We write $\mathcal{A}_1 \leq_F \mathcal{A}_2$ (resp., $\mathcal{A}_1 \leq_{wF} \mathcal{A}_2$) when there is a forward simulation (resp., a weak forward simulation) from $\mathcal{A}_1$ to $\mathcal{A}_2$.

Proposition 2. *Let $\mathcal{A}_1$ and $\mathcal{A}_2$ be automata. Then:*

1. *If $\mathcal{A}_1 \leq_F \mathcal{A}_2$ then $\mathcal{A}_1 \leq_{wF} \mathcal{A}_2$.*
2. *If $H_1 = H_2 = \emptyset$, then $\mathcal{A}_1 \leq_F \mathcal{A}_2$ iff $\mathcal{A}_1 \leq_{wF} \mathcal{A}_2$.*
3. *If $\mathcal{A}_1 \leq_{wF} \mathcal{A}_2$ then $\mathcal{A}_1 \leq_T \mathcal{A}_2$.*

Proof. Standard; for instance, see [6].

Tree-Structured Automata: An automaton is *tree-structured* if each state can be reached via a unique execution. The *unfolding* of automaton $\mathcal{A}$, denoted by *Unfold*$(\mathcal{A})$, is the tree-structured automaton $\mathcal{B}$ obtained from $\mathcal{A}$ by unfolding its transition graph into a tree. Formally, $Q_\mathcal{B} = execs^*(\mathcal{A})$, $\bar{q}_\mathcal{B} = \bar{q}_\mathcal{A}$, $E_\mathcal{B} = E_\mathcal{A}$, $H_\mathcal{B} = H_\mathcal{A}$, and $D_\mathcal{B} = \{(\alpha, a, \alpha a q) \mid (lstate(\alpha), a, q) \in D_\mathcal{A}\}$.

Proposition 3. $\mathcal{A} \equiv_F$ *Unfold*$(\mathcal{A})$.

Proof. See [6]. It is easy to check that the relation R, where $\alpha \; R \; q$ iff $lstate(\alpha) = q$, is a forward simulation from *Unfold*$(\mathcal{A})$ to $\mathcal{A}$ and that the inverse relation of R is a forward simulation from $\mathcal{A}$ to *Unfold*$(\mathcal{A})$.

Proposition 4. $\mathcal{A} \equiv_T$ *Unfold*$(\mathcal{A})$.

Proof. By Proposition 3 and Proposition 2, Parts 1 and 3.

3 Definitions for Probabilistic Automata

A *discrete probability measure* over a set X is a measure μ on $(X, 2^X)$ such that $\mu(X) = 1$. A *discrete sub-probability measure* over X is a measure μ on $(X, 2^X)$ such that $\mu(X) \leq 1$. We denote the set of discrete probability measures and discrete sub-probability measures over X by *Disc*(X) and *SubDisc*(X), respectively. We denote the support of a discrete measure μ, i.e., the set of elements that have non-zero measure, by $supp(\mu)$. We let $\delta(q)$ denote the *Dirac measure* for q, the discrete probability measure that assigns probability 1 to $\{q\}$. Finally, if X is finite, then $\mathcal{U}(X)$ denotes the *uniform distribution* over X, the measure that assigns probability $1/|X|$ to each element of X.

A *probabilistic automaton (PA)* is a tuple $\mathcal{P} = (Q, \bar{q}, E, H, D)$, where all components are exactly as for nondeterministic automata, except that D, the *transition relation*, is a subset of $Q \times (E \cup H) \times Disc(Q)$. We define A as before.

We denote transition (q, a, μ) by $q \xrightarrow{a} \mu$. We assume finite branching: for each state q the number of pairs (a, μ) such that $q \xrightarrow{a} \mu$ is finite. Given a transition $tr = (q, a, \mu)$ we denote q by *source*(tr) and μ by *target*(tr).

Thus, a probabilistic automaton differs from a nondeterministic automaton in that a transition leads to a probability measure over states rather than to a single state. A nondeterministic automaton is a special case of a probabilistic automaton, where the last component of each transition is a Dirac measure. Conversely, we can associate a nondeterministic automaton with each probabilistic automaton by replacing transition relation D by the relation D' given by

$$(q, a, q') \in D' \Leftrightarrow \exists \mu : (q, a, \mu) \in D \wedge \mu(q') > 0.$$

Using this correspondence, notions such as execution fragments and traces carry over from nondeterministic automata to probabilistic automata.

A *scheduler* for a PA $\mathcal{P}$ is a function $\sigma : \mathit{frags}^*(\mathcal{P}) \rightarrow \mathit{SubDisc}(D)$ such that $tr \in \mathit{supp}(\sigma(\alpha))$ implies $\mathit{source}(tr) = \mathit{lstate}(\alpha)$. A scheduler σ is said to be *deterministic* if for each finite execution fragment α, either $\sigma(\alpha)(D) = 0$ or else $\sigma(\alpha) = \delta(tr)$ for some $tr \in D$.

A scheduler σ and a state q_0 induce a measure μ on the σ-field generated by cones of execution fragments as follows. If $\alpha = q_0 a_1 q_1 \cdots a_k q_k$ is a finite execution fragment, then the *cone* of α is defined by $C_\alpha = \{\alpha' \in \mathit{frags}(\mathcal{P}) \mid \alpha \leq \alpha'\}$, and the measure of C_α is defined by

$$\mu(C_\alpha) = \prod_{i \in \{0,k-1\}} \left(\sum_{(q_i, a_{i+1}, \mu') \in D} \sigma(q_0 a_1 \cdots a_i q_i)((q_i, a_{i+1}, \mu'))\mu'(q_{i+1}) \right).$$

Standard measure theoretical arguments ensure that μ is well defined. We call the measure μ a *probabilistic execution fragment* of $\mathcal{P}$ and we say that μ is *generated* by σ and q_0. We call state q_0 the *first state* of μ and denote it by *fstate*(μ). If *fstate*(μ) is the start state $\bar{q}$, then μ is called a *probabilistic execution.*

The trace function is a measurable function from the σ-field generated by cones of execution fragments to the σ-field generated by cones of traces. Given a probabilistic execution fragment μ, we define the *trace distribution* of μ, *tdist*(μ), to be the image measure of μ under *trace*. We denote the set of trace distributions of probabilistic executions of a PA $\mathcal{P}$ by *tdists*$(\mathcal{P})$. We define the *trace distribution preorder* relation on probabilistic automata by: $\mathcal{P}_1 \leq_D \mathcal{P}_2$ iff $E_1 = E_2$ and $\mathit{tdists}(\mathcal{P}_1) \subseteq \mathit{tdists}(\mathcal{P}_2)$.

Combined Transitions: Let $\{q \xrightarrow{a} \mu_i\}_{i \in I}$ be a collection of transitions of $\mathcal{P}$, and let $\{p_i\}_{i \in I}$ be a collection of probabilities such that $\sum_{i \in I} p_i = 1$. Then the triple $(q, a, \sum_{i \in I} p_i \mu_i)$ is called a *combined transition* of $\mathcal{P}$.

Consider a probabilistic execution fragment μ that assigns probability 1 to the set of all finite execution fragments with trace a. Let μ' be the measure defined by $\mu'(q) = \mu(\{\alpha \mid \mathit{lstate}(\alpha) = q\})$. Then $\mathit{fstate}(\mu) \stackrel{a}{\Longrightarrow} \mu'$ is a *weak combined transition* of $\mathcal{P}$. If μ can be generated by a deterministic scheduler, then $\mathit{fstate}(\mu) \stackrel{a}{\Longrightarrow} \mu'$ is a *weak transition.*

Proposition 5. *Let $\{tr_i\}_{i\in I}$ be a collection of weak combined transitions of a PA $\mathcal{P}$, all starting in the same state q, and all labeled by the same action a, and let $\{p_i\}_{i\in I}$ be probabilities such that $\sum_{i\in I} p_i = 1$. Then $\sum_{i\in I} p_i tr_i$ is a weak combined transition of $\mathcal{P}$ labeled by a.*

Proof. See [9] or [11].

Composition: Two PAs, $\mathcal{P}_1$ and $\mathcal{P}_2$, are *compatible* if $H_1 \cap A_2 = A_1 \cap H_2 = \emptyset$. The *composition* of two compatible PAs $\mathcal{P}_1, \mathcal{P}_2$, denoted by $\mathcal{P}_1 \| \mathcal{P}_2$, is the PA $\mathcal{P} = (Q_1 \times Q_2, (\bar{q}_1, \bar{q}_2), E_1 \cup E_2, H_1 \cup H_2, D)$ where D is the set of triples $(q, a, \mu_1 \times \mu_2)$ such that, for $i \in \{1, 2\}$:

$$a \in A_i \Rightarrow (\pi_i(q), a, \mu_i) \in D_i \text{ and } a \notin A_i \Rightarrow \mu_i = \delta(\pi_i(q)).$$

The trace distribution preorder is not preserved by composition [10,11]. Thus, we define the *trace distribution precongruence*, $\leq_{DC}$, to be the coarsest precongruence included in the trace distribution preorder $\leq_D$. This relation has a simple characterization:

Proposition 6. *Let $\mathcal{P}_1$ and $\mathcal{P}_2$ be PAs. Then $\mathcal{P}_1 \leq_{DC} \mathcal{P}_2$ iff for every PA $\mathcal{C}$ that is compatible with both $\mathcal{P}_1$ and $\mathcal{P}_2$, $\mathcal{P}_1 \| \mathcal{C} \leq_D \mathcal{P}_2 \| \mathcal{C}$.*

Simulation Relations: The definitions of forward simulation and weak forward simulation in Sect. 2 can be extended naturally to PAs [10]. However, Segala has shown [8] that the resulting simulations are not complete for $\leq_{DC}$, and has defined new candidate simulations. These new simulations relate states to probability distributions on states.

In order to define formally the new simulations we need three new concepts. First we show how to lift a relation between sets to a relation between distributions over sets [2]. Let $R \subseteq X \times Y$. The *lifting* of R is a relation $R' \subseteq Disc(X) \times Disc(Y)$ such that $\mu_X \; R' \; \mu_Y$ iff there is a function $w : X \times Y \to [0, 1]$ that satisfies:

1. If $w(x, y) > 0$ then $x \; R \; y$.
2. For each $x \in X$, $\sum_{y\in Y} w(x, y) = \mu_X(x)$.
3. For each $y \in Y$, $\sum_{x\in X} w(x, y) = \mu_Y(y)$.

We abuse notation and denote the lifting of a relation R by R as well.

Next we define a flattening operation that converts a measure μ contained in $Disc(Disc(X))$ into a measure $\mathit{flatten}(\mu)$ in $Disc(X)$. Namely, we define

$$\mathit{flatten}(\mu) = \sum_{\rho\in supp(\mu)} \mu(\rho)\rho \; .$$

Finally, we lift the notion of a transition to a *hyper-transition* [11] that begins and ends with a probability distributions over states. Thus, let $\mathcal{P}$ be a PA and let $\mu \in Disc(Q)$. For each $q \in supp(\mu)$, let $q \xrightarrow{a} \mu_q$ be a combined transition of $\mathcal{P}$. Let μ' be $\sum_{q\in supp(\mu)} \mu(q)\mu_q$. Then $\mu \xrightarrow{a} \mu'$ is called a *hyper-transition* of $\mathcal{P}$.

Also, for each $q \in supp(\mu)$, let $q \stackrel{a}{\Longrightarrow} \mu_q$ be a weak combined transition of $\mathcal{P}$. Let μ' be $\sum_{q \in supp(\mu)} \mu(q)\mu_q$. Then $\mu \stackrel{a}{\Longrightarrow} \mu'$ is called a *weak hyper-transition* of $\mathcal{P}$.

We now define simulations for probabilistic automata. A relation $R \subseteq Q_1 \times Disc(Q_2)$ is a *probabilistic forward simulation* (resp., *weak probabilistic forward simulation*) from PA $\mathcal{P}_1$ to PA $\mathcal{P}_2$ iff $E_1 = E_2$ and both of the following hold:

1. $\bar{q}_1 \; R \; \delta(\bar{q}_2)$.
2. For each pair q_1, μ_2 such that $q_1 \; R \; \mu_2$ and each transition $q_1 \stackrel{a}{\rightarrow} \mu_1'$ there exists a distribution $\mu_2' \in Disc(Disc(Q_2))$ such that $\mu_1' \; R \; \mu_2'$ and such that $\mu_2 \stackrel{a}{\rightarrow} \mathit{flatten}(\mu_2')$ (resp., $\mu_2 \stackrel{a}{\Longrightarrow} \mathit{flatten}(\mu_2')$) is a hyper-transition (resp., a weak hyper-transition) of D_2.

We write $\mathcal{P}_1 \leq_{PF} \mathcal{P}_2$ (resp., $\mathcal{P}_1 \leq_{wPF} \mathcal{P}_2$) whenever there is a probabilistic forward simulation (resp., a weak probabilistic forward simulation) from $\mathcal{P}_1$ to $\mathcal{P}_2$. Note that a forward simulation between nondeterministic automata is a probabilistic forward simulation between the two automata viewed as PAs:

Proposition 7. *Let $\mathcal{A}_1$ and $\mathcal{A}_2$ be nondeterministic automata. Then:*

1. *$\mathcal{A}_1 \leq_F \mathcal{A}_2$ implies $\mathcal{A}_1 \leq_{PF} \mathcal{A}_2$, and*
2. *$\mathcal{A}_1 \leq_{wF} \mathcal{A}_2$ implies $\mathcal{A}_1 \leq_{wPF} \mathcal{A}_2$.*

Proposition 8. *Let $\mathcal{P}_1$ and $\mathcal{P}_2$ be PAs. Then:*

1. *If $\mathcal{P}_1 \leq_{PF} \mathcal{P}_2$ then $\mathcal{P}_1 \leq_{wPF} \mathcal{P}_2$.*
2. *If $H_1 = H_2 = \emptyset$ then $\mathcal{P}_1 \leq_{PF} \mathcal{P}_2$ iff $\mathcal{P}_1 \leq_{wPF} \mathcal{P}_2$.*
3. *If $\mathcal{P}_1 \leq_{wPF} \mathcal{P}_2$ then $\mathcal{P}_1 \leq_{DC} \mathcal{P}_2$.*

Proof. See [9].

Tree-Structured Probabilistic Automata: The *unfolding* of a probabilistic automaton $\mathcal{P}$, denoted by $Unfold(\mathcal{P})$, is the tree-structured probabilistic automaton $\mathcal{Q}$ obtained from $\mathcal{P}$ by unfolding its transition graph into a tree. Formally, $Q_{\mathcal{Q}} = execs^*(\mathcal{P})$, $\bar{q}_{\mathcal{Q}} = \bar{q}_{\mathcal{P}}$, $E_{\mathcal{Q}} = E_{\mathcal{P}}$, $H_{\mathcal{Q}} = H_{\mathcal{P}}$, and $D_{\mathcal{Q}} = \{(\alpha, a, \mu) \mid \exists_{\mu'} (lstate(\alpha), a, \mu') \in D_{\mathcal{P}}, \forall_q \mu'(q) = \mu(\alpha a q)\}$.

Proposition 9. $\mathcal{P} \equiv_{PF} Unfold(\mathcal{P})$.

Proof. It is easy to check that the relation R where $\alpha \; R \; \delta(q)$ iff $lstate(\alpha) = q$ is a probabilistic forward simulation from $Unfold(\mathcal{P})$ to $\mathcal{P}$ and that the "inverse" of R is a probabilistic forward simulation from $\mathcal{P}$ to $Unfold(\mathcal{P})$.

Proposition 10. $\mathcal{P} \equiv_{DC} Unfold(\mathcal{P})$.

Proof. By Proposition 9, and Proposition 8, Parts 1 and 3.

4 Characterizations of $\leq_{DC}$: Nondeterministic Automata

In this section, we prove our characterization theorems for $\leq_{DC}$ for nondeterministic automata: Theorem 1 characterizes $\leq_{DC}$ in terms of $\leq_F$, for automata without internal actions, and Theorem 2 characterizes $\leq_{DC}$ in terms of $\leq_{wF}$, for arbitrary nondeterministic automata. In each case, we prove the result first for tree-structured automata and then extend it to the non-tree-structured case via unfolding. The interesting direction for these results is the completeness direction, showing that $\mathcal{A}_1 \leq_{DC} \mathcal{A}_2$ implies the existence of a simulation relation from $\mathcal{A}_1$ to $\mathcal{A}_2$.

Our proofs of completeness for nondeterministic automata use the simple characterization in Proposition 6, applied to a special context for $\mathcal{A}_1$ that we call the dual probabilistic automaton of $\mathcal{A}_1$. Informally speaking, the *dual probabilistic automaton* of a nondeterministic automaton $\mathcal{A}$ is a probabilistic automaton $\mathcal{C}$ whose traces contain information about states and transitions of $\mathcal{A}$. $\mathcal{C}$'s states and start state are the same as those of $\mathcal{A}$. For every state q of $\mathcal{A}$, $\mathcal{C}$ has a self-loop transition labeled by q. Also, if Tr is the (nonempty) set of transitions from q in $\mathcal{A}$, then from state q, $\mathcal{C}$ has a uniform transition labeled by ch to $\{target(tr) \mid tr \in Tr\}$.

Definition 1. *The* dual probabilistic automaton *of an automaton $\mathcal{A}$ is a PA $\mathcal{C}$ such that*

- $Q_{\mathcal{C}} = Q_{\mathcal{A}}$, $\bar{q}_{\mathcal{C}} = \bar{q}_{\mathcal{A}}$,
- $E_{\mathcal{C}} = Q_{\mathcal{A}} \cup \{ch\}$, $H_{\mathcal{C}} = \emptyset$,
- $D_{\mathcal{C}} = \{(q, ch, \mathcal{U}(\{q' \mid q \rightarrow_{\mathcal{A}} q'\})) \mid q \rightarrow_{\mathcal{A}}\} \cup \{(q, q, q) \mid q \in Q_{\mathcal{A}}\}$.

Since $\mathcal{C}$ and $\mathcal{A}$ share no actions, $\mathcal{C}$ cannot ensure that its traces faithfully emulate the behavior of $\mathcal{A}$. However, an appropriate scheduler can synchronize the two automata and ensure such an emulation.

4.1 Automata without Internal Actions

We first consider tree-structured automata.

Proposition 11. *Let $\mathcal{A}_1$ and $\mathcal{A}_2$ be tree-structured nondeterministic automata without internal actions, such that $\mathcal{A}_1 \leq_{DC} \mathcal{A}_2$. Then $\mathcal{A}_1 \leq_F \mathcal{A}_2$.*

Proof. Assume that $\mathcal{A}_1 \leq_{DC} \mathcal{A}_2$. Let $\mathcal{C}$ be the dual probabilistic automaton of $\mathcal{A}_1$. Without loss of generality, we assume that the set of actions of $\mathcal{C}$ is disjoint from those of $\mathcal{A}_1$ and $\mathcal{A}_2$. This implies that $\mathcal{C}$ is compatible with both $\mathcal{A}_1$ and $\mathcal{A}_2$.

Consider the scheduler σ_1 for $\mathcal{A}_1 \| \mathcal{C}$ that starts by scheduling the self-loop transition labelled by the start state of $\mathcal{C}$, leading to state $(\bar{q}_1, \bar{q}_1)$, which is of the form (q, q). Then σ_1 repeats the following as long as $q \rightarrow_1$:

1. Schedule the ch transition of $\mathcal{C}$, thus choosing a new state q' of $\mathcal{A}_1$.

2. Schedule (q, a, q') in $\mathcal{A}_1$, where a is uniquely determined by the selected state q' (recall that $\mathcal{A}_1$ is a tree).
3. Schedule the self-loop transition of $\mathcal{C}$ labeled by q', resulting in the state (q', q'), which is again of the form (q, q).

Scheduler σ_1 induces a trace distribution μ_T. Observe that μ_T satisfies the following three properties, for all finite traces β and for all states q:

$$\mu_T(C_{\bar{q}_1}) = 1 \tag{1}$$

$$q \rightarrow_1 \quad \Rightarrow \quad \mu_T(C_{\beta q ch}) = \mu_T(C_{\beta q}) \tag{2}$$

$$\mu_T(C_{\beta q ch}) > 0 \quad \Rightarrow \quad \sum_{a, q' | q \xrightarrow{a}_1 q'} \mu_T(C_{\beta q ch a q'}) = \mu_T(C_{\beta q ch}) \tag{3}$$

Since $\mathcal{A}_1 \leq_{DC} \mathcal{A}_2$, Proposition 6 implies that μ_T is also a trace distribution of $\mathcal{A}_2 \| \mathcal{C}$. That is, there exists a probabilistic execution μ of $\mathcal{A}_2 \| \mathcal{C}$, induced by some scheduler σ_2, whose trace distribution is μ_T. Now we define a relation R: $q_1 \; R \; q_2$ if and only if there exists an execution α of $\mathcal{A}_2 \| \mathcal{C}$ such that:

1. $\mathit{lstate}(\alpha) = (q_2, q_1)$,
2. $\mu(C_\alpha) > 0$, and
3. $\sigma_2(\alpha)$ assigns a non-zero probability to a transition labeled by q_1.

We claim that R is a forward simulation from $\mathcal{A}_1$ to $\mathcal{A}_2$. For the start condition, we must show that $\bar{q}_1 \; R \; \bar{q}_2$. Define execution α to be the trivial execution consisting of the start state $(\bar{q}_2, \bar{q}_1)$. Conditions 1 and 2 are clearly satisfied. For Condition 3, observe that, by Equation (1), $\mu_T(C_{\bar{q}_1}) = 1$. Therefore, since there are no internal actions in $\mathcal{A}_2$ or $\mathcal{C}$, the only action that can be scheduled initially by σ_2 is $\bar{q}_1$. Therefore, $\sigma_2(\alpha)$ assigns probability 1 to the unique transition whose label is $\bar{q}_1$, as needed.

For the step condition, assume $q_1 \; R \; q_2$, and let $q_1 \xrightarrow{a}_1 q_1'$. By definition of R, there exists a finite execution α of $\mathcal{A}_2 \| \mathcal{C}$, with last state (q_2, q_1), such that $\mu(C_\alpha) > 0$ and $\sigma_2(\alpha)$ assigns a non-zero probability to a transition labeled by q_1. Therefore, the sequence $\alpha' = \alpha q_1 (q_2, q_1)$ is an execution of $\mathcal{A}_2 \| \mathcal{C}$ such that $\mu(C_{\alpha'}) > 0$. Therefore, $\mu_T[C_{\beta q_1}] > 0$, where $\beta = \mathit{trace}(\alpha)$. Since q_1 enables at least one transition in $\mathcal{A}_1$, Equation (2) implies that $\mu_T(C_{\beta q_1 ch}) = \mu_T(C_{\beta q_1})$. Then since $\mathcal{A}_2$ and $\mathcal{C}$ have no internal actions, σ_2 must schedule action ch from α' with probability 1.

Since action ch leads to state q_1' of $\mathcal{C}$ with non-zero probability, which enables only actions q_1' and ch, by Equation (3), σ_2 schedules at least one transition labeled by a, followed by a transition labeled by q_1'. Observe that the transition labeled by a is a transition of $\mathcal{A}_2$. Let (q_2, a, q_2') be such a transition. Then, the sequence $\alpha'' = \alpha' ch (q_2, q_1') a (q_2', q_1')$ is an execution of $\mathcal{A}_2 \| \mathcal{C}$ such that $\mu(C_{\alpha''}) > 0$ and such that $\sigma_2(\alpha'')$ assigns a non-zero probability to a transition labeled by q_1'. This shows that $q_1' \; R \; q_2'$ and completes the proof since we have found a state q_2' such that $q_2 \xrightarrow{a}_2 q_2'$ and $q_1' \; R \; q_2'$.

Now we present our main result, for general (non-tree-structured) nondeterministic automata without internal actions.

Theorem 1. *Let $\mathcal{A}_1$, $\mathcal{A}_2$ be nondeterministic automata without internal actions. Then $\mathcal{A}_1 \leq_{DC} \mathcal{A}_2$ if and only if $\mathcal{A}_1 \leq_F \mathcal{A}_2$.*

Proof. First we prove soundness of forward simulations:

$$\begin{array}{ll} \mathcal{A}_1 \leq_F \mathcal{A}_2 & \Rightarrow \text{(Proposition 7, Part 1)} \\ \mathcal{A}_1 \leq_{PF} \mathcal{A}_2 & \Rightarrow \text{(Proposition 8, Part 1)} \\ \mathcal{A}_1 \leq_{wPF} \mathcal{A}_2 & \Rightarrow \text{(Proposition 8, Part 3)} \\ \mathcal{A}_1 \leq_{DC} \mathcal{A}_2 \ . & \end{array}$$

Completeness is established by:

$$\begin{array}{ll} \mathcal{A}_1 \leq_{DC} \mathcal{A}_2 & \Rightarrow \text{(Proposition 10)} \\ \mathit{Unfold}(\mathcal{A}_1) \leq_{DC} \mathcal{A}_1 \leq_{DC} \mathcal{A}_2 \leq_{DC} \mathit{Unfold}(\mathcal{A}_2) & \Rightarrow (\leq_{DC} \text{ is transitive}) \\ \mathit{Unfold}(\mathcal{A}_1) \leq_{DC} \mathit{Unfold}(\mathcal{A}_2) & \Rightarrow \text{(Proposition 11)} \\ \mathit{Unfold}(\mathcal{A}_1) \leq_F \mathit{Unfold}(\mathcal{A}_2) & \Rightarrow \text{(Proposition 3)} \\ \mathcal{A}_1 \leq_F \mathit{Unfold}(\mathcal{A}_1) \leq_F \mathit{Unfold}(\mathcal{A}_2) \leq_F \mathcal{A}_2 & \Rightarrow (\leq_F \text{ is transitive}) \\ \mathcal{A}_1 \leq_F \mathcal{A}_2 \ . & \end{array}$$

4.2 Automata with Internal Actions

Next we extend the results of Sect. 4.1 to automata that include internal actions. The proofs are analogous to those in Sect. 4.1, and use the same dual probabilistic automaton. The difference is that, in several places in the proof of Proposition 12, we need to reason about multi-step extensions of executions instead of single-step extensions. Again, we begin with tree-structured automata.

Proposition 12. *Let $\mathcal{A}_1$, $\mathcal{A}_2$ be tree-structured nondeterministic automata such that $\mathcal{A}_1 \leq_{DC} \mathcal{A}_2$. Then $\mathcal{A}_1 \leq_{wF} \mathcal{A}_2$.*

Proof. Assume that $\mathcal{A}_1 \leq_{DC} \mathcal{A}_2$. Let $\mathcal{C}$ be the dual probabilistic automaton of $\mathcal{A}_1$, and define scheduler σ_1 exactly as in the proof of Proposition 11. Equations (1), (2) and (3) hold in this case as well. We redefine relation R: $q_1 \ R \ q_2$ iff there exists an execution α of $\mathcal{A}_2 \| \mathcal{C}$ such that:

1. $\mathit{lstate}(\alpha) = (q_2, q_1)$,
2. $\mu(C_\alpha) > 0$, and
3. there exists an execution fragment, α', of $\mathcal{A}_2 \| \mathcal{C}$, such that $\mathit{trace}(\alpha') = q_1$ and $\mu(C_{\alpha \frown \alpha'}) > 0$.

We claim that R is a weak forward simulation from $\mathcal{A}_1$ to $\mathcal{A}_2$. For the start condition, we show that $\bar{q}_1 \ R \ \bar{q}_2$. Define α to be the trivial execution consisting of the start state $(\bar{q}_2, \bar{q}_1)$; this clearly satisfies Conditions 1 and 2. For Condition 3, observe that, by Equation (1), $\mu_T(C_{\bar{q}_1}) = 1$. The inverse image under the trace mapping for $\mathcal{A}_2 \| \mathcal{C}$, of $C_{\bar{q}_1}$, is a union of cones of the form $C_{\alpha'}$, where α' is an execution of $\mathcal{A}_2 \| \mathcal{C}$ with trace $\bar{q}_1$; therefore, there exists such an α' with $\mu(C_{\alpha'}) > 0$. Since the first state of α' is $(\bar{q}_2, \bar{q}_1)$, $\alpha \frown \alpha' = \alpha'$. Thus, $\mu(C_{\alpha \frown \alpha'}) > 0$, as needed.

For the step condition, assume $q_1 \; R \; q_2$, and let $q_1 \xrightarrow{a}_1 q_1'$. By definition of R, there exists a finite execution α of $\mathcal{A}_2 \| \mathcal{C}$, with last state (q_2, q_1), such that $\mu(C_\alpha) > 0$ and there exists an execution fragment, α', of $\mathcal{A}_2 \| \mathcal{C}$, such that $\mathit{trace}(\alpha') = q_1$ and $\mu(C_{\alpha \frown \alpha'}) > 0$. Let $\beta = \mathit{trace}(\alpha)$; then $\mathit{trace}(\alpha \frown \alpha') = \beta q_1$, and so $\mu_T(C_{\beta q_1}) > 0$. Since q_1 enables at least one transition in $\mathcal{A}_1$, Equation (2) implies that $\mu_T(C_{\beta q_1 \mathit{ch}}) = \mu_T(C_{\beta q_1})$. Thus there exists an execution fragment α'' of $\mathcal{A}_2 \| \mathcal{C}$ with trace ch such that $\mu(C_{\alpha \frown \alpha' \frown \alpha''}) > 0$. Furthermore, since the transition of $\mathcal{C}$ labeled by ch leads to state q_1' with non-zero probability, we can assume that the last state of α'' is of the form (q'', q_1') for some state q''.

Since $\mu_T(C_{\beta q_1 \mathit{ch}}) > 0$, Equation (3) applies. Furthermore, since from the last state of α'' the only external actions of $\mathcal{C}$ that are enabled are ch and q_1', there exists an execution fragment α''' with trace $a q_1'$ (a is uniquely determined by q_1' since $\mathcal{A}_1$ is tree-structured), such that $\mu(C_{\alpha \frown \alpha' \frown \alpha'' \frown \alpha'''}) > 0$.

Now we split α''' into $\alpha_1''' \frown \alpha_2'''$, where $\mathit{trace}(\alpha_1''') = a$. Then the last state of α_1''' is of the form (q''', q_1'). We claim that $q_1' \; R \; q'''$. Indeed, the execution $\alpha \frown \alpha' \frown \alpha'' \frown \alpha_1'''$ ends with state (q''', q_1') (Condition 1) and satisfies $\mu(C_{\alpha \frown \alpha' \frown \alpha'' \frown \alpha_1'''}) > 0$ (Condition 2). Furthermore, α_2''' is an execution fragment that satisfies Condition 3.

It remains to show that $q_2 \stackrel{a}{\Rightarrow} q'''$. For this, it suffices to observe that the execution fragment $(\alpha' \frown \alpha'' \frown \alpha_1''') \lceil \mathcal{A}_2$ has trace a, first state q_2, and last state q'''.

Theorem 2. *Let $\mathcal{A}_1$, $\mathcal{A}_2$ be nondeterministic automata. Then $\mathcal{A}_1 \leq_{DC} \mathcal{A}_2$ if and only if $\mathcal{A}_1 \leq_{wF} \mathcal{A}_2$.*

Proof. Analogous to the proof of Theorem 1.

5 Characterizations of $\leq_{DC}$: Probabilistic Automata

Finally, we present our characterization theorems for $\leq_{DC}$ for probabilistic automata: Theorem 3 characterizes $\leq_{DC}$ in terms of $\leq_{PF}$, for PAs without internal actions, and Theorem 4 characterizes $\leq_{DC}$ in terms of $\leq_{wPF}$, for arbitrary PAs. Again, we give the results first for tree-structured automata and extend them by unfolding.

Our proofs of completeness for PAs are analogous to those for nondeterministic automata. We define a new kind of *dual probabilistic automaton* $\mathcal{C}$ for a PA $\mathcal{P}$, which is slightly different from the one for nondeterministic automata. The main differences are that the new $\mathcal{C}$ keeps track, in its state, of transitions as well as states of the given PA $\mathcal{P}$, and that the new $\mathcal{C}$ has separate transitions representing nondeterministic and probabilistic choices within $\mathcal{P}$. Specifically, the states of $\mathcal{C}$ include a distinguished start state, all the states of $\mathcal{P}$, and all the transitions of $\mathcal{P}$. $\mathcal{C}$ has a special transition from its own start state $\bar{q}_\mathcal{C}$ to the start state of $\mathcal{P}$, $\bar{q}_{\mathcal{P}}$, labeled by $\bar{q}_{\mathcal{P}}$. Also, from every state q of $\mathcal{P}$, $\mathcal{C}$ has a uniform transition labeled by ch to the set of transitions of $\mathcal{P}$ that start in state q. Finally, for every transition tr of $\mathcal{P}$, and every state q in the support of $target(tr)$, $\mathcal{C}$ has a transition labeled by q from tr to q.

Definition 2. *The* dual probabilistic automaton *of a PA $\mathcal{P}$ is a PA $\mathcal{C}$ such that*

- $Q_{\mathcal{C}} = \{\bar{q}_{\mathcal{C}}\} \cup Q_{\mathcal{P}} \cup D_{\mathcal{P}}$,
- $E_{\mathcal{C}} = Q_{\mathcal{P}} \cup \{ch\}$, $H_{\mathcal{C}} = \emptyset$,
- $D_{\mathcal{C}} = \{(\bar{q}_{\mathcal{C}}, \bar{q}_{\mathcal{P}}, \bar{q}_{\mathcal{P}})\} \cup$
  $\{(q, ch, \mathcal{U}(\{tr \in D_{\mathcal{P}} \mid source(tr) = q\})) \mid q \in Q_{\mathcal{P}}\} \cup$
  $\{(tr, q, q) \mid tr \in D_{\mathcal{P}}, q \in supp(target(tr))\}$.

Proposition 13. *Let $\mathcal{P}_1$, $\mathcal{P}_2$ be tree-structured probabilistic automata without internal actions, such that $\mathcal{P}_1 \leq_{DC} \mathcal{P}_2$. Then $\mathcal{P}_1 \leq_{PF} \mathcal{P}_2$.*

Proof. (Sketch:) Assume that $\mathcal{P}_1 \leq_{DC} \mathcal{P}_2$. Let $\mathcal{C}$ be the dual probabilistic automaton of $\mathcal{P}_1$. Consider the scheduler σ_1 for $\mathcal{P}_1 \| \mathcal{C}$ that starts by scheduling the transition of $\mathcal{C}$ from the start state of $\mathcal{C}$ to the start state of $\mathcal{P}_1$, leading to state $(\bar{q}_1, \bar{q}_1)$, which is of the form (q, q). Then σ_1 repeats the following as long as $q \rightarrow_1$:

1. Schedule the ch transition of $\mathcal{C}$, thus choosing a transition tr of $\mathcal{P}_1$.
2. Schedule transition tr of $\mathcal{P}_1$, leading $\mathcal{P}_1$ to a new state q'.
3. Schedule the transition of $\mathcal{C}$ labeled by the state q', resulting in the state (q', q'), which is again of the form (q, q).

Scheduler σ_1 induces a trace distribution μ_T. Since $\mathcal{P}_1 \leq_{DC} \mathcal{P}_2$, Proposition 6 implies that μ_T is also a trace distribution of $\mathcal{P}_2 \| \mathcal{C}$. That is, there exists a probabilistic execution μ of $\mathcal{P}_2 \| \mathcal{C}$, induced by some scheduler σ_2, whose trace distribution is μ_T.

For each state q_1 in Q_1, let Θ_{q_1} be the set of finite executions of $\mathcal{A}_2 \| \mathcal{C}$ whose last transition is labeled by q_1. For each state q_2 of $\mathcal{P}_2$, let Θ_{q_1,q_2} be the set of finite executions in Θ_{q_1} whose last state is the pair (q_2, q_1). Now define relation R: $q_1 \; R \; \mu_2$ iff for each state q_2 of Q_2,

$$\mu_2(q_2) = \frac{\sum_{\alpha \in \Theta_{q_1,q_2}} \mu(C_\alpha)}{\sum_{\alpha \in \Theta_{q_1}} \mu(C_\alpha)} . \tag{4}$$

We claim that R is a probabilistic forward simulation from $\mathcal{P}_1$ to $\mathcal{P}_2$. The proof of this claim appears in [4].

Theorem 3. *Let $\mathcal{P}_1$, $\mathcal{P}_2$ be probabilistic automata without internal actions. Then $\mathcal{P}_1 \leq_{DC} \mathcal{P}_2$ if and only if $\mathcal{P}_1 \leq_{PF} \mathcal{P}_2$.*

Proposition 14. *Let $\mathcal{P}_1$, $\mathcal{P}_2$ be tree-structured probabilistic automata such that $\mathcal{P}_1 \leq_{DC} \mathcal{P}_2$. Then $\mathcal{P}_1 \leq_{wPF} \mathcal{P}_2$.*

Proof. (Sketch:) We use the same dual automaton $\mathcal{C}$. Define scheduler σ_1 and relation R exactly as in the proof of Proposition 13. Now R is a weak probabilistic forward simulation, as shown in [4].

Theorem 4. *Let $\mathcal{P}_1$, $\mathcal{P}_2$ be probabilistic automata. Then $\mathcal{P}_1 \leq_{DC} \mathcal{P}_2$ if and only if $\mathcal{P}_1 \leq_{wPF} \mathcal{P}_2$.*

6 Concluding Remarks

We have characterized the trace distribution precongruence for nondeterministic and probabilistic automata, with and without internal actions, in terms of four kinds of simulation relations, $\leq_F$, $\leq_{wF}$, $\leq_{PF}$, and $\leq_{wPF}$. In particular, this shows that probabilistic contexts are capable of observing all the distinctions that can be expressed using these simulation relations. Some technical improvements are possible. For example, our finite branching restriction can be relaxed to countable branching, simply by replacing uniform distributions in the dual automata by other distributions such as exponential distributions.

For future work, it would be interesting to try to restrict the class of schedulers used for defining the trace distribution precongruence, so that fewer distinctions are observable by probabilistic contexts. It remains to define such restrictions and to provide explicit chacterizations of the resulting new notions of $\leq_{DC}$, for instance in terms of button pushing scenarios.

References

1. S. Aggarwal. Time optimal self-stabilizing spanning tree algorithms. Master's thesis, Department of Electrical Engineering and Computer Science, Massachusetts Institute of Technology, May 1994. Available as Technical Report MIT/LCS/TR-632.
2. K.G. Larsen B. Jonsson. Specification and refinement of probabilistic processes. In *Proceedings* 6th *Annual Symposium on Logic in Computer Science,* Amsterdam, pages 266–277. IEEE Press, 1991.
3. N.A. Lynch, I. Saias, and R. Segala. Proving time bounds for randomized distributed algorithms. In *Proceedings of the 13th Annual ACM Symposium on the Principles of Distributed Computing,* pages 314–323, Los Angeles, CA, August 1994.
4. N.A. Lynch, R. Segala, and F.W. Vaandrager. Compositionality for probabilistic automata. Technical Report MIT-LCS-TR-907, MIT, Laboratory for Computer Science, 2003.
5. N.A. Lynch and M.R. Tuttle. An introduction to input/output automata. *CWI Quarterly,* 2(3):219–246, September 1989.
6. N.A. Lynch and F.W. Vaandrager. Forward and backward simulations, I: Untimed systems. *Information and Computation,* 121(2):214–233, September 1995.
7. A. Pogosyants, R. Segala, and N.A. Lynch. Verification of the randomized consensus algorithm of Aspnes and Herlihy: a case study. *Distributed Computing,* 13(3):155–186, 2000.

8. R. Segala. Compositional trace–based semantics for probabilistic automata. In *Proc. CONCUR'95*, volume 962 of *Lecture Notes in Computer Science*, pages 234–248, 1995.
9. R. Segala. *Modeling and Verification of Randomized Distributed Real-Time Systems.* PhD thesis, Department of Electrical Engineering and Computer Science, Massachusetts Institute of Technology, June 1995. Available as Technical Report MIT/LCS/TR-676.
10. R. Segala and N.A. Lynch. Probabilistic simulations for probabilistic processes. *Nordic Journal of Computing*, 2(2):250–273, 1995.
11. M.I.A. Stoelinga. *Alea jacta est: Verification of Probabilistic, Real-Time and Parametric Systems.* PhD thesis, University of Nijmegen, April 2002.
12. M.I.A. Stoelinga. An introduction to probabilistic automata. *Bulletin of the European Association for Theoretical Computer Science*, 78:176–198, October 2002.
13. M.I.A. Stoelinga and F.W. Vaandrager. Root contention in IEEE 1394. In J.-P. Katoen, editor, *Proceedings 5th International AMAST Workshop on Formal Methods for Real-Time and Probabilistic Systems,* Bamberg, Germany, volume 1601 of *Lecture Notes in Computer Science*, pages 53–74. Springer-Verlag, 1999.
14. M.I.A. Stoelinga and F.W. Vaandrager. A testing scenario for probabilistic automata. In J.C.M. Baeten, J.K. Lenstra, J. Parrow, and G.J. Woeginger, editors, *Proceedings* 30th *ICALP*, volume 2719 of *Lecture Notes in Computer Science*, pages 407–418. Springer-Verlag, 2003.

Satisfiability and Model Checking for MSO-Definable Temporal Logics Are in PSPACE

Paul Gastin[1] and Dietrich Kuske[2]

[1] LIAFA, Université Paris 7, 2, place Jussieu, F-75251 Paris Cedex 05
Paul.Gastin@liafa.jussieu.fr
[2] Institut für Algebra, TU Dresden, D-01062 Dresden
kuske@math.tu-dresden.de

Abstract. Temporal logics over Mazurkiewicz traces have been extensively studied over the past fifteen years. In order to be usable for the verification of concurrent systems they need to have reasonable complexity for the satisfiability and the model checking problems. Whenever a new temporal logic was introduced, a new proof (usually non trivial) was needed to establish the complexity of these problems. In this paper, we introduce a unified framework to define local temporal logics over traces. We prove that the satisfiability problem and the model checking problem for asynchronous Kripke structures for local temporal logics over traces are decidable in PSPACE. This subsumes and sometimes improves all complexity results previously obtained on local temporal logics for traces.

1 Introduction

Over the past fifteen years, a lot of papers have been devoted to the study of temporal logics over partial orders and in particular over Mazurkiewicz traces. This is motivated by the need for specification languages that are suited for concurrent systems where a property should not depend on the ordering between independent events. Hence logics over linearizations of behaviors are not adequate and logics over partial orders were developed. In order to be useful for the verification of concurrent systems, these specification languages should enjoy reasonable complexity for the satisfiability and the model checking problems.

Temporal logics over traces can be classified in *global* ones and *local* ones. Here we are interested in the latter. They are evaluated at single events corresponding to local views of processes. Process based logics [13,14,11] were introduced by Thiagarajan and shown to be decidable in EXPTIME using difficult results on *gossip* automata. A specific feature of process based logics is the until modality that can only walk along a single process. Another approach was taken in [1] were the until is existential and walks along some path in the Hasse diagram of the partial order. The decidability in PSPACE of this logic was shown using a tableau construction. Due to this existential until, this logic is not contained

R. Amadio, D. Lugiez (Eds.): CONCUR 2003, LNCS 2761, pp. 222–236, 2003.

in first order logic of traces [4]. In the quest for an expressively complete local temporal logic over traces, a universal until was introduced in [4] and filtered variants together with past modalities were needed in [7]. Again these logics were proved to be decidable in PSPACE using alternating automata. For each local logic, a specific proof has to be developed for the complexity of the satisfiability or the model checking problem. Such proofs are usually difficult and span over several pages.

In this paper, we introduce a unified framework to define local temporal logics over traces (Sect. 5). This approach is inspired from [12]. Basically, a local temporal logic is given by a finite set of modalities whose semantics is given by a monadic second order (MSO) formula having a single individual free variable. We call these logics MSO-definable. We show that all local temporal logics considered so far (and much more) are MSO-definable. Then we show that the satisfiability problem and the model checking problem for asynchronous Kripke structures for MSO-definable temporal logics over traces are decidable in PSPACE (Sect. 6). This subsumes and sometimes improves all the complexity results over local logics discussed above. We would like to stress that the proofs for our main results are actually simpler than some proofs specific to some local logics and even from a practical point of view, our decision procedures are as efficient as specific ones could be. Also, our results may be surprising at first since the satisfiability problem for MSO is non elementary, but because we use a finite set of MSO-definable modalities our decision problems stay in PSPACE.

Actually, we start by introducing our MSO-definable temporal logics for words (Sect. 3) and we prove that the satisfiability and the model checking problems are decidable in PSPACE (Sect. 4). Though words are special cases of traces, we believe that the paper is easier to follow in this way and that results for words are interesting by themselves. A reader that is not familiar with traces can easily understand the results for words. Other general frameworks for temporal logics over words have been studied [17,16,9]. In [17] the modalities are defined by right linear grammars extended to infinite words while in [16,9] the modalities are defined by various kinds of automata (either non-deterministic Büchi, or alternating or two-way alternating). Note that in these approaches, the automata that define the modalities are part of the formulas. In all cases, the satisfiability problem is proved to be decidable in PSPACE. Our approach is indeed similar but differs by the way modalities are defined. We have chosen MSO modalities because this is how the semantics of local temporal logics over traces is usually defined. In this way, we trivially obtain as corollaries of our main theorems the complexity results for local temporal logics over traces. It is also possible to give automata for the local modalities over traces and apply the results of [16,9]. This is basically what is done in [5] but such a reduction is difficult and long.

2 Monadic Second Order Logic

Let Σ be an alphabet. *Monadic second order logic* (MSO) is a formalism to speak about the properties of words over Σ. It is based on individual variables

$x, y, z, \ldots$ that range over positions in the word (*i.e.*, over elements of $\mathbb{N}$) and on set variables $X, Y, Z, \ldots$ that range over sets of positions (*i.e.*, over subsets of $\mathbb{N}$). Its atomic formulas are $x \leq y$, $P_a(x)$ for $a \in \Sigma$ and $X(x)$ where x, y are individual variables and X is a set variable. The use of Boolean connectives $\wedge, \vee, \neg, \rightarrow$ *etc* and quantification $\exists x$ and $\exists X$ over individual and set variables allows to build more complex formulas. We denote by $\mathrm{MSO}_\Sigma(<)$ the set of MSO formulas over the alphabet Σ.

To define the semantics of a formula, let $w = w_0 w_1 \cdots \in \Sigma^\infty = \Sigma^+ \cup \Sigma^\omega$. We denote by $|w|$ the length of w which may be finite or infinite. A *position* in w is an integer p with $0 \leq p < |w|$. A *valuation* in w for the formula φ is a mapping ν that assigns positions in w to the free individual variables of φ and sets of positions in w to the free set variables of φ.

$w, \nu \models_{\mathrm{MSO}} x \leq y$ if $\nu(x) \leq \nu(y)$
$w, \nu \models_{\mathrm{MSO}} P_a(x)$ if $w_{\nu(x)} = a$
$w, \nu \models_{\mathrm{MSO}} X(x)$ if $\nu(x) \in \nu(X)$
$w, \nu \models_{\mathrm{MSO}} \exists x \varphi$ if $w, \nu[x \mapsto p] \models_{\mathrm{MSO}} \varphi$ for some position p in w
$w, \nu \models_{\mathrm{MSO}} \exists X \varphi$ if $w, \nu[X \mapsto P] \models_{\mathrm{MSO}} \varphi$ for some set P of positions in w

Here, $\nu[x \mapsto p]$ is the mapping that coincides with ν except for the value of x which is p; $\nu[X \mapsto P]$ is defined similarly. If φ is an MSO formula with free variables $X_1, \ldots, X_\ell, x_1, \ldots, x_k$ and ν is a valuation in a word w then we also write $w \models_{\mathrm{MSO}} \varphi(\nu(X_1), \ldots, \nu(X_\ell), \nu(x_1), \ldots, \nu(x_k))$ for $w, \nu \models_{\mathrm{MSO}} \varphi$.

3 A Uniform Framework for Temporal Logics over Words

We introduce our approach on an example. We use PLTL (linear temporal logic with past) because it is well-known and allows us to introduce easily the main definitions. We start with a finite alphabet Σ and recall that the syntax of PLTL is given by

$$\varphi ::= a \mid \neg\varphi \mid \varphi \vee \varphi \mid \mathsf{X}\,\varphi \mid \mathsf{Y}\,\varphi \mid \varphi\,\mathsf{U}\,\varphi \mid \varphi\,\mathsf{S}\,\varphi$$

where a ranges over Σ. We assume the reader is familiar with the semantics of PLTL over words: $w, p \models_{\mathrm{PLTL}} \varphi$ means that the formula φ holds in the word w at position p. Here $w = w_0 w_1 \cdots \in \Sigma^\infty$ and $0 \leq p < |w|$. For instance,

$w, p \models_{\mathrm{PLTL}} a$ if $w_p = a$
$w, p \models_{\mathrm{PLTL}} \mathsf{Y}\,\varphi$ if $p > 0$ and $w, p-1 \models_{\mathrm{PLTL}} \varphi$
$w, p \models_{\mathrm{PLTL}} \varphi\,\mathsf{U}\,\psi$ if $\exists k (p \leq k$ and $w, k \models_{\mathrm{PLTL}} \psi$ and
$w, j \models_{\mathrm{PLTL}} \varphi$ for all $p \leq j < k$

In order to define PLTL in our framework, we start with a vocabulary B of modality names and a mapping arity : $B \rightarrow \mathbb{N}$ giving the arity of each modality. The modality names of arity 0 are the atomic formulas of $\mathrm{TL}(B)$. Other formulas are obtained from atomic formulas by the application of modalities. The syntax of the temporal logic $\mathrm{TL}(B)$ based on the vocabulary B is then

$$\varphi ::= \sum_{M \in B} M(\underbrace{\varphi, \ldots, \varphi}_{\mathrm{arity}(M)}).$$

For PLTL we consider $B_{PLTL} = \Sigma \cup \{\neg, \mathsf{X}, \mathsf{Y}, \vee, \mathsf{U}, \mathsf{S}\}$ and the arity is 0 for elements in Σ, 1 for $\neg, \mathsf{X}, \mathsf{Y}$ and 2 for $\vee, \mathsf{U}, \mathsf{S}$. The syntax of $\mathrm{TL}(B_{PLTL})$ is then precisely that of $PLTL$.

In order to define the semantics of $\mathrm{TL}(B)$ we consider a mapping $[\![-]\!] : B \to \mathrm{MSO}_\Sigma(<)$ in such a way that if $M \in B$ is of arity ℓ then $[\![M]\!]$ is an ℓ-ary MSO modality, that is, an MSO formula with ℓ free set variables $X_1, \ldots, X_\ell$ and one free individual variable x. The intuition is that a word w at position p satisfies $M(\varphi_1, \ldots, \varphi_\ell)$ if $w, \nu \models_{\mathrm{MSO}} [\![M]\!](X_1, \ldots, X_\ell, x)$ when $\nu(x) = p$ and for each i, $\nu(X_i)$ is the set of positions in w where φ_i holds. For PLTL, the mapping $[\![-]\!]$ is given by

$$
\begin{array}{lll}
[\![a]\!](x) & = P_a(x) \quad \text{for } a \in \Sigma & \\
[\![\neg]\!](X_1, x) & = \neg X_1(x) & \\
[\![\mathsf{X}]\!](X_1, x) & = X_1(x+1) & = \exists z(x < z \wedge X_1(z) \wedge \forall y(x < y \to z \leq y)) \\
[\![\mathsf{Y}]\!](X_1, x) & = X_1(x-1) & = \exists z(z < x \wedge X_1(z) \wedge \forall y(y < x \to y \leq z)) \\
[\![\vee]\!](X_1, X_2, x) & = X_1(x) \vee X_2(x) & \\
[\![\mathsf{U}]\!](X_1, X_2, x) & = \exists z(x \leq z \wedge X_2(z) \wedge \forall y(x \leq y < z \to X_1(y))) & \\
[\![\mathsf{S}]\!](X_1, X_2, x) & = \exists z(z \leq x \wedge X_2(z) \wedge \forall y(z < y \leq x \to X_1(y))) &
\end{array}
$$

Finally, given a word $w \in \Sigma^\infty$ and a formula $\varphi \in \mathrm{TL}(B)$, we define inductively the set φ^w of position in w where φ holds. If $\varphi = M(\varphi_1, \ldots, \varphi_\ell)$ where $M \in B$ is of arity $\ell \geq 0$, then

$$\varphi^w = \{p < |w| \mid w \models_{\mathrm{MSO}} [\![M]\!](\varphi_1^w, \ldots, \varphi_\ell^w, p)\}.$$

Proposition 1. *Let $\varphi \in \mathrm{TL}(B_{PLTL}) = PLTL$ and $w \in \Sigma^\infty$. Then,*

$$\varphi^w = \{p < |w| \mid w, p \models_{\mathrm{PLTL}} \varphi\}.$$

The proof of this proposition is easy and omitted. What is interesting is that it exhibits an alternative definition of PLTL using a vocabulary B (with arity) and a semantic map $[\![-]\!]$. By varying the vocabulary and the semantic map we have a very general way to define temporal logics for words and therefore a formal framework to state complexity results for a large class of temporal logics. This is exactly what we were looking for.

For convenience, we summarize below the definition of an MSO temporal logics over words.

Definition 2. *We start with a set B consisting of modality names together with a mapping* $\mathrm{arity} : B \to \mathbb{N}$ *giving the arity of each modality. Then the syntax of the temporal logic* $\mathrm{TL}(B)$ *is defined by the grammar*

$$\varphi ::= \sum_{M \in B} M(\underbrace{\varphi, \ldots, \varphi}_{\mathrm{arity}(M)}).$$

Consider a mapping $[\![-]\!] : B \to \mathrm{MSO}_\Sigma(<)$ such that $[\![M]\!]$ is an ℓ-ary MSO modality, that is, an MSO formula with ℓ free set variables $X_1, \ldots, X_\ell$ and one *free individual variable x. Given a word $w \in \Sigma^\infty$ and a formula $\varphi \in \mathrm{TL}(B)$, the semantics is given by the set φ^w of position in w where φ holds. The inductive definition is as follows. If $\varphi = M(\varphi_1, \ldots, \varphi_\ell)$ where $M \in B$ is of arity $\ell \geq 0$, then*

$$\varphi^w = \{p < |w| \mid w \models_{\mathrm{MSO}} [\![M]\!](\varphi_1^w, \ldots, \varphi_\ell^w, p)\}.$$

We also write $w, p \models \varphi$ for $p \in \varphi^w$.

If we fix the triple (B,arity, $[\![-]\!]$) once for ever, the expressive power of $\mathrm{TL}(B)$ is limited. For instance, the expressive power of PLTL is known to be strictly weaker than that of monadic second order logic [8]. We can extend its expressive power introducing a new modality name **even** of arity 1 with associated MSO-modality

$$[\![\mathbf{even}]\!] = (\exists Y(|Y| \text{ is even } \wedge \forall y(Y(y) \leftrightarrow (X_1(y) \wedge y \geq x)))).$$

The formula $\mathbf{even}(a) \in \mathrm{TL}(\{\mathbf{even}, a\})$ is satisfied by a word w in position p if and only if the word w contains an even number of occurrences of the letter a to the right of p. Recall that this property is not expressible in PLTL [8].

4 Complexity of Temporal Logics for Words

In this section, we show that, whatever the finite set B of modality names and associated MSO-modalities is, the satisfiability and the model checking problems for $\mathrm{TL}(B)$ are decidable in PSPACE.

Satisfiability Problem for $\mathrm{TL}(B)$ *over Words:* Given a formula $\xi \in \mathrm{TL}(B)$, does there exist a word $w \in \Sigma^\infty$ and a position p in w such that $w, p \models \xi$?

Remark 3. One may also consider *initial* satisfiability of a given formula $\xi \in \mathrm{TL}(B)$, i.e., does there exists a word $w \in \Sigma^\infty$ such that $w, 0 \models \xi$. This problem can be easily reduced to the general satisfiability. Add a modality name init of arity 1 to B with associated MSO-modality $[\![\mathrm{init}]\!](X_1, x) = \exists y(y \leq x \wedge X_1(y) \wedge y \text{ minimal})$. Now, a formula $\xi \in \mathrm{TL}(B)$ is *initially* satisfiable if and only if the formula $\mathrm{init}(\xi)$ is satisfiable.

For a word $w = a_0 a_1 \cdots \in \{0,1\}^\infty$, let $\mathrm{supp}(w) = \{p < |w| \mid a_p = 1\}$ denote the *support* of w. For $\ell \in \mathbb{N}$, we consider the alphabet $\Sigma_\ell = \Sigma \times \{0,1\}^\ell$. A letter $a \in \Sigma_\ell$ will be written $a = (a_0, a_1, \ldots, a_\ell)$ and a word $w \in \Sigma_\ell^\infty$ will be identified with a tuple of words of same length in the obvious way: $w = (w_0, w_1, \ldots, w_\ell) \in \Sigma^\infty \times (\{0,1\}^\infty)^\ell$ with $|w| = |w_i|$ for $0 \leq i \leq \ell$.

Recall the following result that can easily be extracted from the proof of Büchi's theorem.

Theorem 4 ([2]). *Let M be an ℓ-ary modality name and $[\![M]\!]$ its associated MSO-modality. Then there exists a Büchi-automaton $\mathcal{B}_M$ over the alphabet $\Sigma_{\ell+1}$ such that $w = (w_0, w_1, \ldots, w_{\ell+1}) \in \mathcal{L}(\mathcal{B}_M)$ if and only if* $\mathrm{supp}(w_{\ell+1}) = \{p < |w| \mid w_0 \models_{\mathrm{MSO}} [\![M]\!](\mathrm{supp}(w_1), \ldots, \mathrm{supp}(w_\ell), p)\}$.

Proof. Consider the MSO formula

$$\overline{[\![M]\!]}(X_1,\dots,X_{\ell+1}) = \forall x(X_{\ell+1}(x) \leftrightarrow [\![M]\!](X_1,\dots,X_\ell,x)).$$

From the proof of Büchi's theorem (see e.g. [15]), we find an automaton $\mathcal{B}_M$ over $\Sigma_{\ell+1}$ such that a word $w = (w_0, w_1, \dots, w_{\ell+1}) \in \mathcal{L}(\mathcal{B}_M)$ if and only if $w_0 \models_{\mathrm{MSO}} \overline{[\![M]\!]}(\mathrm{supp}(w_1),\dots,\mathrm{supp}(w_{\ell+1}))$. This is equivalent with $\mathrm{supp}(w_{\ell+1}) = \{p < |w| \mid w_0 \models_{\mathrm{MSO}} [\![M]\!](\mathrm{supp}(w_1),\dots,\mathrm{supp}(w_\ell),p)\}$ by definition of $\overline{[\![M]\!]}$. □

As examples, we give the automata $\mathcal{B}_\vee$ and $\mathcal{B}_\mathsf{U}$:

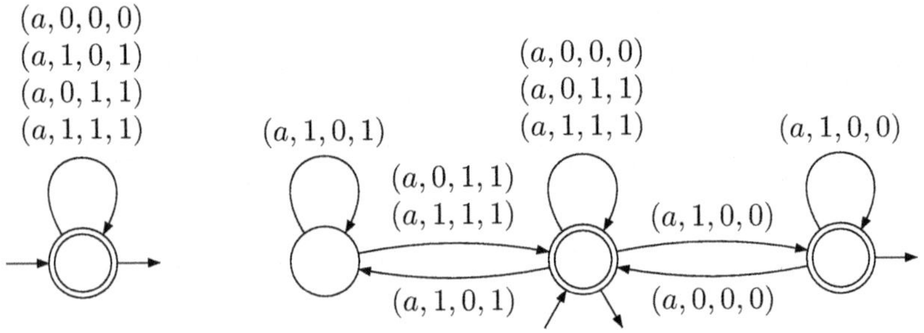

For formulas φ and ψ, we write $\varphi \leq \psi$ if φ is a subformula of ψ (this includes the case $\varphi = \psi$). Let ξ be a formula from $\mathrm{TL}(B)$ and let $\mathrm{Sub}(\xi) = \{\varphi \in \mathrm{TL}(B) \mid \varphi \leq \xi\}$. In the sequel, we will consider words over the alphabet $\overline{\Sigma} = \Sigma \times \{0,1\}^{\mathrm{Sub}(\xi)}$. Typically, the elements of $\overline{\Sigma}$ are of the form $\overline{a} = (a, (a_\varphi)_{\varphi\leq\xi})$ with $a \in \Sigma$ and $a_\varphi \in \{0,1\}$ for $\varphi \leq \xi$. As above, we identify a word $\overline{w} \in \overline{\Sigma}^\infty$ with a tuple of words of same length: $\overline{w} = (w, (w_\varphi)_{\varphi\leq\xi})$ with $w \in \Sigma^\infty$, $w_\varphi \in \{0,1\}^\infty$ for $\varphi \leq \xi$ and $|\overline{w}| = |w| = |w_\varphi|$.

Now let $\psi = M(\varphi_1,\dots,\varphi_\ell) \leq \xi$. Then $\overline{a}{\restriction}\psi := (a, a_{\varphi_1}, \dots, a_{\varphi_\ell}, a_\psi) \in \Sigma_{\ell+1}$. Accordingly, for $\overline{w} \in \overline{\Sigma}^\infty$ we let $\overline{w}{\restriction}\psi = (w, w_{\varphi_1}, \dots, w_{\varphi_\ell}, w_\psi) \in \Sigma_{\ell+1}^\infty$.

The Construction. For a formula $\varphi \in \mathrm{TL}(B)$, let $\mathrm{top}(\varphi)$ be the outermost modality name of φ. Let $Q = \prod_{\varphi\leq\xi} Q_{\mathrm{top}(\varphi)}$ be the set of states of the automaton $\mathcal{A}_\xi$ where $Q_{\mathrm{top}(\varphi)}$ is the set of states of the Büchi-automaton $\mathcal{B}_{\mathrm{top}(\varphi)}$. The alphabet of $\mathcal{A}_\xi$ is $\overline{\Sigma}$. For a letter $\overline{a} \in \overline{\Sigma}$ and states $p = (p_\varphi)_{\varphi\leq\xi}$ and $q = (q_\varphi)_{\varphi\leq\xi}$, we have a transition $p \xrightarrow{\overline{a}} q$ in $\mathcal{A}_\xi$ if and only if, for all $\varphi \leq \xi$, we have $p_\varphi \xrightarrow{\overline{a}\restriction\varphi} q_\varphi$ in the automaton $\mathcal{B}_{\mathrm{top}(\varphi)}$. Note that a sequence of states $p^0, p^1, \dots$ defines a run of $\mathcal{A}_\xi$ for a word $\overline{w} \in \overline{\Sigma}^\infty$ if and only if for each $\varphi \leq \xi$, its projection $p^0_\varphi, p^1_\varphi, \dots$ on φ is a run of $\mathcal{B}_{\mathrm{top}(\varphi)}$ for the word $\overline{w}{\restriction}\varphi$. A run of $\mathcal{A}_\xi$ is accepting if and only if for each $\varphi \leq \xi$, its projection on $\mathcal{B}_{\mathrm{top}(\varphi)}$ is accepting.

Lemma 5. *Let $\overline{w} = (w, (w_\varphi)_{\varphi\leq\xi}) \in \overline{\Sigma}^\infty$. Then, $\overline{w} \in \mathcal{L}(\mathcal{A}_\xi)$ if and only if for each $\varphi \leq \xi$ we have* $\mathrm{supp}(w_\varphi) = \varphi^w = \{p < |w| \mid w, p \models \varphi\}$.

Proof. Assume $\overline{w} \in \mathcal{L}(\mathcal{A}_\xi)$. We show that $\varphi^w = \mathrm{supp}(w_\varphi)$ by structural induction on $\varphi \leq \xi$. So let $\varphi = M(\varphi_1,\dots,\varphi_\ell) \leq \xi$ such that $\varphi_i^w = \mathrm{supp}(w_{\varphi_i})$

holds for $1 \leq i \leq \ell$. Since $\overline{w}$ is accepted by the automaton $\mathcal{A}_\xi$, the word $\overline{w}\restriction\varphi = (w, w_{\varphi_1}, \ldots, w_{\varphi_\ell}, w_\varphi)$ is accepted by $\mathcal{B}_M$. Hence, using Theorem 4 and the hypothesis we get

$$\begin{aligned}\mathrm{supp}(w_\varphi) &= \{p < |w| \mid w \models_{\mathrm{MSO}} [\![M]\!](\mathrm{supp}(w_{\varphi_1}), \ldots, \mathrm{supp}(w_{\varphi_\ell}), p)\} \\ &= \{p < |w| \mid w \models_{\mathrm{MSO}} [\![M]\!](\varphi_1^w, \ldots, \varphi_\ell^w, p)\} \\ &= \varphi^w.\end{aligned}$$

For the other direction, assume that $\varphi^w = \mathrm{supp}(w_\varphi)$ for all $\varphi \leq \xi$. Let $\varphi = M(\varphi_1, \ldots, \varphi_\ell) \leq \xi$. We have $\varphi^w = \{p < |w| \mid w \models_{\mathrm{MSO}} [\![M]\!](\varphi_1^w, \ldots, \varphi_\ell^w, p)\}$ and we get $\mathrm{supp}(w_\varphi) = \{p < |w| \mid w \models_{\mathrm{MSO}} [\![M]\!](\mathrm{supp}(w_{\varphi_1}), \ldots, \mathrm{supp}(w_{\varphi_\ell}), p)\}$ using our hypothesis. Since $\overline{w}\restriction\varphi = (w, w_{\varphi_1}, \ldots, w_{\varphi_\ell}, w_\varphi)$ we deduce from Theorem 4 that $\overline{w}\restriction\varphi$ is accepted by $\mathcal{B}_M$. Since this holds for each $\varphi \leq \xi$ we obtain $\overline{w} \in \mathcal{L}(\mathcal{A}_\xi)$. □

Proposition 6. *The formula ξ is satisfiable if and only if there exists $\overline{w} \in \mathcal{L}(\mathcal{A}_\xi)$ with* $\mathrm{supp}(w_\xi) \neq \emptyset$.

Proof. Assume that ξ is satisfiable. There exist a word $w \in \Sigma^\infty$ and a position p in w with $w, p \models \xi$. For each $\varphi \in \mathrm{TL}(B)$, there is a unique word $w_\varphi \in \{0,1\}^\infty$ with $|w| = |w_\varphi|$ and $\mathrm{supp}(w_\varphi) = \varphi^w$. Let $\overline{w} = (w, (w_\varphi)_{\varphi \leq \xi}) \in \overline{\Sigma}^\infty$. By Lemma 5 we get $\overline{w} \in \mathcal{L}(\mathcal{A}_\xi)$. Moreover, we have $p \in \xi^w = \mathrm{supp}(w_\xi) \neq \emptyset$.

Conversely let $\overline{w} = (w, (w_\varphi)_{\varphi \leq \xi}) \in \mathcal{L}(\mathcal{A}_\xi)$ with $\mathrm{supp}(w_\xi) \neq \emptyset$. By Lemma 5 we get $\emptyset \neq \mathrm{supp}(w_\xi) = \xi^w = \{p < |w| \mid w, p \models \xi\}$. Therefore, ξ is satisfiable. □

Theorem 7. *Let B be a finite set of modality names with associated MSO-modalities. Then the satisfiability problem for* $\mathrm{TL}(B)$ *is in* PSPACE.

Proof. Let ξ be some formula from $\mathrm{TL}(B)$ whose satisfiability we want to check. By Proposition 6, we have to decide whether $\mathcal{A}_\xi$ accepts some word $\overline{w}$ with $\mathrm{supp}(w_\xi) \neq \emptyset$. Recall that a state of $\mathcal{A}_\xi$ is a tuple of states from the automata $\mathcal{B}_M$ whose length is bounded by the size of the formula ξ. Hence a state of $\mathcal{A}_\xi$ requires space polynomial in the size of ξ and the same holds for any letter from $\overline{\Sigma}$. Given two states q and q' of $\mathcal{A}_\xi$ and a letter $a \in \overline{\Sigma}$, one can check in polynomial space whether $q \xrightarrow{a} q'$ in $\mathcal{A}_\xi$. Note that the automata $\mathcal{B}_M$ are fixed and need not be computed. Hence the search for an accepting run can be performed by a nondeterministic Turing machine using space polynomial in the size of ξ. □

A *Kripke structure* is transition system $K = (S, \rightarrow, s, \sigma)$ with S a finite set of states, $\rightarrow \subseteq S^2$ the transition function, $s \in S$ the initial state and $\sigma : S \rightarrow \Sigma$ the labeling function. A formula $\xi \in \mathrm{TL}(B)$ holds in K (written $K \models \xi$) if for all maximal paths $s_0, s_1, \ldots$ in K with $s_0 = s$ we have $\sigma(s_0)\sigma(s_1)\ldots, 0 \models \xi$.

Model Checking Problem for $\mathrm{TL}(B)$ *over Words:* Given a Kripke structure K and a formula $\xi \in \mathrm{TL}(B)$, do we have $K \models \xi$?

Theorem 8. *Let B be a finite set of modality names with associated MSO-modalities. Then the model checking problem for* TL(B) *is in* PSPACE.

Proof. Let $\xi \in \mathrm{TL}(B)$. The formula $\neg\xi$ is in $\mathrm{TL}(B \cup \{\neg\})$ and we consider the automaton $\mathcal{A}$ obtained from $\mathcal{A}_{\neg\xi}$ by projecting the transition labels to Σ, *i.e.*, $p \xrightarrow{a} q$ in $\mathcal{A}$ if there exists $\overline{a} = (a, (a_\varphi)_{\varphi \le \neg\xi}) \in \overline{\Sigma}$ with $p \xrightarrow{\overline{a}} q$ in $\mathcal{A}_{\neg\xi}$. Again, a state of $\mathcal{A}$ can be stored in polynomial space and one can check whether $p \xrightarrow{a} q$ in $\mathcal{A}$ in polynomial space. Therefore, applying the usual technique we get a PSPACE algorithm for the model checking problem. □

The actual performance of the algorithms for satisfiability and model checking depend on the basic automata $\mathcal{B}_M$ for $M \in B$. For PLTL, these basic automata have very few states: $\mathcal{B}_a$ for $a \in \Sigma$, $\mathcal{B}_\neg$ and $\mathcal{B}_\vee$ have just one state, $\mathcal{B}_\mathsf{U}$ has three states, and all the other automata have two states. Thus, the automaton $\mathcal{A}_\xi$ has at most $2^m \cdot 3^n$ states where m is the number of occurrences of temporal operators different from U and n is the number of occurrences of U in ξ.

5 Local Temporal Logic over Traces

We briefly recall some notions about Mazurkiewicz traces (see [6] for background). A *dependence alphabet* is a pair (Σ, D) where the alphabet Σ is a finite set of actions and the *dependence relation* $D \subseteq \Sigma \times \Sigma$ is reflexive and symmetric.

For a partial order $(V, \le)$, let $\lessdot$ denote the successor relation $\lessdot \; = \; < \setminus <^2$. Further, $\|$ denotes incomparability, *i.e.*, $\| = V^2 \setminus (\le \cup \ge)$. A *(Mazurkiewicz) trace* is a finite or infinite labeled partial order $t = (V, \le, \lambda)$ where V is a set of vertices labeled by $\lambda : V \to \Sigma$ and $\le$ is a partial order relation on V satisfying the following conditions:

1. for all $y \in V$, the set $\downarrow y = \{x \in V \mid x \le y\}$ is finite,
2. $x \parallel y$ implies $(\lambda(x), \lambda(y)) \notin D$ for all $x, y \in V$, and
3. $x \lessdot y$ implies $(\lambda(x), \lambda(y)) \in D$ for all $x, y \in V$.

The set of all traces is denoted $\mathbb{R}(\Sigma, D)$.

We now interpret monadic second order formulas over traces. The semantics for traces is defined as for words in Sect. 2. Let $t = (V, \le, \lambda)$ be a trace. A *valuation in t* for the formula φ is now a mapping ν that assigns elements of V to free individual variables of φ and subsets of V to free set variables of φ. The definition of satisfaction $t, \nu \models_{\mathrm{MSO}} \varphi$ can be taken verbatim from Sect. 2 with the only exception that $t, \nu \models_{\mathrm{MSO}} P_a(x)$ if and only if $\lambda(\nu(x)) = a$. It should be noted that $\nu(x) \le \nu(y)$ refers now to the partial order of the trace.

Similarly, the temporal logic TL(B) is defined as in Definition 2. The only difference is that the semantics φ^t is now defined for a trace t:

$$\varphi^t = \{p \in V \mid t \models_{\mathrm{MSO}} [\![M]\!](\varphi_1^t, \ldots, \varphi_\ell^t, p)\}$$

and as before, we write $t, p \models \varphi$ for $p \in \varphi^t$.

In the next section we show that the satisfiability problem and the model checking problem are decidable in PSPACE for TL(B) when B is finite. But first, we show that all modalities that were considered so far in local logics for traces can be defined in our setting. As a corollary, we get that all local temporal logics for traces considered so far are decidable in PSPACE.

We start with event based temporal logics and will consider later process based ones. In addition to the constants Σ and the boolean connectives $\neg$ and $\vee$, these logics are build using various temporal modalities described below.

Universal Until. The simplest logic $\mathrm{LocTL}_\Sigma(\mathsf{EX}, \mathsf{U})$ studied in [4] uses only two modalities EX of arity 1 and U of arity 2 (there are some technical subtleties about initial modalities or initial satisfiability of a formula that will be discussed later). Intuitively, $\mathsf{EX}\,\varphi$ means that there is an immediate successor of the current vertex where φ holds. The universal until $\varphi \,\mathsf{U}\, \psi$ claims the existence of a vertex z in the future of the current one x such that ψ holds at z and φ holds for all vertices between x and z. Formally, we have $\mathrm{LocTL}_\Sigma(\mathsf{EX}, \mathsf{U}) = \mathrm{TL}(\Sigma \cup \{\neg, \vee, \mathsf{EX}, \mathsf{U}\})$ if EX and U are defined by the following MSO-modalities.

$$\begin{aligned}
[\![\mathsf{EX}]\!](X_1, x) &= \exists z(x < z \wedge X_1(z) \wedge \forall y(x < y \leq z \rightarrow y = z)) \\
[\![\mathsf{U}]\!](X_1, X_2, x) &= \exists z(x \leq z \wedge X_2(z) \wedge \forall y(x \leq y < z \rightarrow X_1(y)))
\end{aligned}$$

The logic $\mathrm{LocTL}_\Sigma(\mathsf{EX}, \mathsf{U})$ is expressively complete with respect to $\mathrm{FO}_\Sigma(<)$, the first order theory of traces if and only if the dependence alphabet is a cograph [4]. The satisfiability problem was shown to be PSPACE-complete. The hardness follows from the corresponding result on words. The PSPACE algorithm is obtained using alternating automata. Though not all details were given, the proof of this upper bound was more than 4 pages long in [5]. Since $\mathrm{LocTL}_\Sigma(\mathsf{EX}, \mathsf{U}) = \mathrm{TL}(\Sigma \cup \{\neg, \vee, \mathsf{EX}, \mathsf{U}\})$, it is a trivial corollary of Theorem 9.

Filtered Until. In order to obtain expressive completeness for arbitrary dependence alphabets, [7] considered $\mathrm{LocTL}_\Sigma(\mathsf{EX}, \mathsf{EY}, \mathsf{U}_C, \mathsf{S}_C)$ where $C \subseteq \Sigma$. Compared to the universal until U, the filtered universal until U_C adds an alphabetic requirement on the vertices that are below z but not below x. The modalities EY and S_C are the past versions of EX and U_C. We can express this logic in our framework, $\mathrm{LocTL}_\Sigma(\mathsf{EX}, \mathsf{EY}, \mathsf{U}_C, \mathsf{S}_C) = \mathrm{TL}(\Sigma \cup \{\neg, \vee, \mathsf{EX}, \mathsf{EY}, \mathsf{U}_C, \mathsf{S}_C\})$ if we associate with EY, U_C and S_C the following MSO-modalities.

$$\begin{aligned}
[\![\mathsf{EY}]\!](X_1, x) &= \exists z(z < x \wedge X_1(z) \wedge \forall y(z \leq y < x \rightarrow y = z)) \\
[\![\mathsf{U}_C]\!](X_1, X_2, x) &= \exists z(x \leq z \wedge X_2(z) \wedge \forall y(x \leq y < z \rightarrow X_1(y)) \\
&\qquad \wedge \forall y(y \leq z \wedge \textstyle\bigvee_{c \in C} P_c(y) \rightarrow y \leq x)) \\
[\![\mathsf{S}_C]\!](X_1, X_2, x) &= \exists z(z \leq x \wedge X_2(z) \wedge \forall y(z < y \leq x \rightarrow X_1(y)) \\
&\qquad \wedge \forall y(y \leq x \wedge \textstyle\bigvee_{c \in C} P_c(y) \rightarrow y \leq z))
\end{aligned}$$

In [7], the logic $\mathrm{LocTL}_\Sigma(\mathsf{EX}, \mathsf{EY}, \mathsf{U}_C, \mathsf{S}_C)$ was shown to be expressively complete with respect to $\mathrm{FO}_\Sigma(<)$ for arbitrary dependence alphabets. The satisfiability problem was also shown to be decidable in PSPACE using two-way alternating automata, the proof being long and non trivial. Again this complexity upper bound becomes a trivial corollary of Theorem 9.

We say that EX, EY, U_C and S_C are *first order modalities* because $[\![\mathsf{EX}]\!]$, $[\![\mathsf{EY}]\!]$, $[\![\mathsf{U}_C]\!]$ and $[\![\mathsf{S}_C]\!]$ use quantification over individual variables only. The temporal logics defined with FO-modalities are thus trivially contained in $\mathrm{FO}_\Sigma(<)$. We will see now a temporal logic using some modalities that are not FO-definable.

Existential Until. The temporal logic for causality (TLC) was introduced in [1]. In our framework, it can be defined by $\mathrm{TL}(\Sigma \cup \{\neg, \vee, \mathsf{EX}, \mathsf{EY}, \mathsf{Eco}, \mathsf{EG}, \mathsf{EU}, \mathsf{ES}\})$. Intuitively, $\mathsf{Eco}\,\varphi$ claims that φ holds for some vertex concurrent to the current one. The formula $\varphi\,\mathsf{EU}\,\psi$ holds if there is a path starting at the current vertex in the Hasse diagram of the trace such that φ holds along the path until ψ holds. Similarly, $\mathsf{EG}\,\varphi$ claims the existence of a maximal path in the Hasse diagram of the trace, starting from the current vertex, where φ always holds. Finally, ES is the past version of EU. Formally, the semantics of TLC is obtained with the following MSO-modalities.

$$
\begin{aligned}
[\![\mathsf{Eco}]\!](X_1, x) &= \exists z(\neg(x \le z) \wedge \neg(z \le x) \wedge X_1(z))\\
[\![\mathsf{EU}]\!](X_1, X_2, x) &= \exists z(x \le z \wedge X_2(z) \wedge \exists Y(\forall y(Y(y) \wedge y < z \rightarrow X_1(y)) \wedge\\
&\qquad Y \text{ is a maximal totally ordered set contained in } {\uparrow}x \cap {\downarrow}z))\\
[\![\mathsf{ES}]\!](X_1, X_2, x) &= \exists z(z \le x \wedge X_2(z) \wedge \exists Y(\forall y(Y(y) \wedge z < y \rightarrow X_1(y)) \wedge\\
&\qquad Y \text{ is a maximal totally ordered set contained in } {\downarrow}x \cap {\uparrow}z))\\
[\![\mathsf{EG}]\!](X_1, x) &= \exists Y(\forall y(Y(y) \rightarrow X_1(y)) \wedge\\
&\qquad Y \text{ is a maximal totally ordered set contained in } {\uparrow}x)
\end{aligned}
$$

TLC was proved to be decidable in PSPACE in [1] using a tableau construction. Again, this upper bound becomes a corollary of Theorem 9. The expressiveness results for TLC were established in [4]. For cograph dependence alphabets TLC has the same expressive power as $\mathrm{FO}_\Sigma(<)$, but due to the claim of the existence of a path in the modality EU it is not contained in FO for arbitrary dependence alphabets.

Initial Satisfiability. A given formula $\xi \in \mathrm{TL}(B)$ is satisfiable over traces if there exists a trace $t \in \mathbb{R}(\Sigma, D)$ and some position p in t such that $t, p \models \xi$. Since a trace does not necessarily have a unique minimal position, there is no canonical way to define *initial* satisfiability over traces. Two approaches have been considered.

In [4], an initial modality $\mathsf{EM}\,\varphi$ was introduced with the meaning $t \models \mathsf{EM}\,\varphi$ if there is a minimal position p in t with $t, p \models \varphi$. Then, an initial formula α is a boolean combination of initial modalities and the initial satisfiability problem is to know whether there exists a trace $t \in \mathbb{R}(\Sigma, D)$ with $t \models \alpha$. To cope with this approach, we associate with EM the MSO modality

$$[\![\mathsf{EM}]\!](X_1, x) = \exists y(X_1(y) \wedge \neg\exists z(z < y)).$$

Then, the formula $\alpha \in \mathrm{LocTL}_\Sigma(\cdots)$ is initially satisfiable over traces if and only if the formula $\alpha \in \mathrm{TL}(B)$ is satisfiable (with $[\![-]\!]$) over traces.

In [1] a dual approach is taken which can be dealt with in the same way. Here, it is said that a a local formula φ is initially satisfiable if there exists a trace t such that φ holds at all minimal vertices of t, i.e., $t \models \neg\,\mathsf{EM}\,\neg\varphi$.

The other approach used in [3] is to consider rooted traces. Let $\# \notin \Sigma$ and $t = (V, \leq, \lambda) \in \mathbb{R}(\Sigma, D)$. The rooted trace associated with t is $\# \cdot t = (V \cup \{\#\}, \leq \cup (\{\#\} \times (V \cup \{\#\})), \lambda \cup (\# \mapsto \#)$. It is a trace over the alphabet $\Sigma' = \Sigma \cup \{\#\}$ and the dependence relation $D' = D \cup (\{\#\} \times \Sigma) \cup (\Sigma \times \{\#\})$. Then, we say that a local formula $\varphi \in \mathrm{LocTL}_\Sigma(\cdots)$ is initially satisfiable if there exists a trace $t \in \mathbb{R}(\Sigma, D)$ such that $\# \cdot t, \# \models \varphi$. To cope with this approach, we add a modality name init of arity 1 to B with associated MSO-modality

$$[\![\mathrm{init}]\!](X_1, x) = \exists y(X_1(y) \wedge P_\#(y) \wedge \forall z(y \leq z) \wedge \forall z(P_\#(z) \to z = y)).$$

Then, the formula $\varphi \in \mathrm{LocTL}_\Sigma(\cdots)$ is initially satisfiable over $\mathbb{R}(\Sigma, D)$ if and only if the formula $\mathrm{init}(\varphi) \in \mathrm{TL}(B)$ is satisfiable (with $[\![-]\!]$) over $\mathbb{R}(\Sigma', D')$.

Process-Based Modalities. We conclude the section by showing that the temporal logic over traces TrPTL introduced by Thiagarajan [13] can also be dealt with in our framework. The underlying idea is that the actions of the dependence alphabet are executed by independent processes. Communication between these processes is possible by the execution of joint actions. Hence, with any action $a \in \Sigma$, we associate a nonempty set of processes $p(a) \subseteq \{1, 2, \ldots, n\}$ in such a way that $(a, b) \in D$ iff $p(a) \cap p(b) \neq \emptyset$. This ensures that events performed by process i are linearly ordered in any trace t. With this additional information, one can define modalities that speak about the location of an action. The logic TrPTL is based on modalities p_i, $\mathcal{O}_i$ and U_i ($i \in \{1, \ldots, n\}$) of arity 0, 1 and 2 respectively.

The semantics given in [13] is that of a global temporal logic. Hence it may come as a surprise that we can deal with it in our framework. But actually, apart initially, formulas are evaluated at *prime* configurations, i.e., configurations having exactly one maximal element. By identifying a prime configuration with its maximal vertex we see that the logic is actually local. Intuitively, p_i holds if the current vertex is located on process i and $\mathcal{O}_i\varphi$ means that φ holds at the first vertex of process i which is not below the current one. Finally, $\varphi \mathsf{U}_i \psi$ means that we have φ until ψ on the sequence of vertices located on process i and starting from the last vertex of process i which is below the current one. Formally, the semantics is defined as follows using the macro $P_i(x) = \bigvee_{\{c \mid i \in p(c)\}} P_c(x)$:

$$\begin{aligned}
[\![p_i]\!](x) &= P_i(x) \\
[\![\mathcal{O}_i]\!](X_1, x) &= \exists y(X_1(y) \wedge P_i(y) \wedge \neg(y \leq x) \wedge \forall z(P_i(z) \to (z \leq x \vee y \leq z))) \\
[\![\mathsf{U}_i]\!](X_1, X_2, x) &= \exists y(P_i(y) \wedge y \leq x \wedge \forall z(P_i(z) \wedge z \leq x \to z \leq y) \\
&\qquad \wedge \exists z(P_i(z) \wedge y \leq z \wedge X_2(z) \\
&\qquad\qquad \wedge \forall u((P_i(u) \wedge y \leq u < z) \to X_1(u))))
\end{aligned}$$

TrPTL was proved to be decidable in EXPTIME in [13] using a difficult result on *gossip* automata over traces [10]. As a corollary of Theorem 9, we can improve this upper bound to PSPACE. Since the logic TrPTL is defined by FO-modalities, it is contained in $\mathrm{FO}_\Sigma(<)$ but the precise expressive power of TrPTL is still unknown.

6 Complexity of Local Temporal Logics for Traces

We want to show that the following problem is decidable in PSPACE.

Satisfiability Problem for TL(B) *over Traces:* Given a formula $\xi \in \mathrm{TL}(B)$, does there exist a trace $t \in \mathbb{R}(\Sigma, D)$ and some position p in t such that $t, p \models \xi$?

This will be done by a reduction to Theorem 7. For this reason, we first recall the relation between words and traces, more details can be found in [6].

Let $t = (V, \leq, \lambda)$ be a trace and let $\sqsubseteq$ be any linear extension of $\leq$ of order type at most ω. Then we can view $(V, \sqsubseteq, \lambda)$ as a word $w \in \Sigma^\infty$. The set of *linearizations* $\mathrm{Lin}(t) \subseteq \Sigma^\infty$ of t is the set of all words $w \in \Sigma^\infty$ that arise in this way. Conversely, each word $w \in \Sigma^\infty$ is the linearization of a unique trace $t \in \mathbb{R}(\Sigma, D)$.

In the following, we will evaluate MSO formulas over words and over traces. To make this clear, we use $\models^t_{\mathrm{MSO}}$ for traces and $\models^w_{\mathrm{MSO}}$ for words (though the context is sufficient to distinguish between the two). There exists a FO formula $\eta(x, y)$ with two free individual variables such that for all traces $t \in \mathbb{R}(\Sigma, D)$, words $w \in \mathrm{Lin}(t)$ and vertices $p, q \in V$, we have $t \models^t_{\mathrm{MSO}} p \leq q$ if and only if $w \models^w_{\mathrm{MSO}} \eta(p, q)$. Let φ be an MSO formula. We denote by $\overline{\varphi}$ the MSO formula obtained by replacing in φ any subformula of the form $x \leq y$ by $\eta(x, y)$. Then, we have for all traces $t \in \mathbb{R}(\Sigma, D)$, words $w \in \mathrm{Lin}(t)$ and valuations ν in V, $t, \nu \models^t_{\mathrm{MSO}} \varphi$ if and only if $w, \nu \models^w_{\mathrm{MSO}} \overline{\varphi}$.

After these preliminary remarks, fix some set B of modality names together with their arity function and associated MSO-modality defined by the mapping $[\![-]\!] : B \to \mathrm{MSO}_\Sigma(<)$. This defines a temporal logic TL(B) whose interpretation over traces with $[\![-]\!]$ is denoted $\models^t_{[\![-]\!]}$. We also consider the mapping $\overline{[\![-]\!]} : B \to \mathrm{MSO}_\Sigma(<)$ so that for $M \in B$, $\overline{[\![M]\!]}$ is obtained by replacing in $[\![M]\!]$ any subformula of the form $x \leq y$ by $\eta(x, y)$. The interpretation of TL(B) over words with $\overline{[\![-]\!]}$ is denoted $\models^w_{\overline{[\![-]\!]}}$. We obtain the following essential link between the two semantics: for all $\xi \in \mathrm{TL}(B)$, for all traces $t \in \mathbb{R}(\Sigma, D)$, all words $w \in \mathrm{Lin}(t)$ and all positions p in t, we have $t, p \models^t_{[\![-]\!]} \xi$ if and only if $w, p \models^w_{\overline{[\![-]\!]}} \xi$.

Therefore, the formula ξ is satisfiable over traces with the MSO-modalities $[\![-]\!]$ if and only if it is satisfiable over words with the MSO-modalities $\overline{[\![-]\!]}$. Since, by Theorem 7, this latter question is decidable in space polynomial in the size of ξ, we obtain the following

Theorem 9. *Let* (Σ, D) *be a dependence alphabet,* B *a finite set of modality names with associated MSO-modalities. Then the satisfiability problem for* TL(B) *over traces is decidable in PSPACE.*

We turn now to the model checking problem. In order to give its definition, we first introduce *asynchronous Kripke structures.* We need to fix some notation. Let Loc be a finite set of locations and let Q_i be a finite set for each $i \in \mathrm{Loc}$. We let $Q_I = \prod_{i \in I} Q_i$ for $I \subseteq \mathrm{Loc}$ and if $q = (q_i)_{i \in \mathrm{Loc}} \in Q_{\mathrm{Loc}}$ then we let $q_I = (q_i)_{i \in I}$ for $I \subseteq \mathrm{Loc}$. An *asynchronous Kripke structure* (AKS for short) is a tuple $\mathrm{AK} = ((Q_i)_{i\in\mathrm{Loc}}, (\delta_I)_{I\subseteq\mathrm{Loc}}, q^0, (\sigma_i)_{i\in\mathrm{Loc}})$ where Q_i is a finite set of local states for process i, $\delta_I \subseteq Q_I \times Q_I$ is a local transition relation, $q^0 \in Q_{\mathrm{Loc}}$ is the global initial state, and $\sigma_i : Q_i \to 2^{\mathrm{AP}_i}$ assigns to each local state the set of atomic propositions from the finite set AP_i that holds in this states.

A run of AK is (an isomorphism class of) a labelled partial order $\rho = (V, \leq, \ell, W)$ where a vertex $v \in V$ represents the occurrence of a transition, $\leq$ is the ordering between transitions, $\ell : V \to 2^{\mathrm{Loc}} \setminus \{\emptyset\}$ gives for each transition v the nonempty set $\ell(v)$ of processes taking part in it and W assigns to each transition $v \in V$ the tuple $W(v) \in Q_{\ell(v)}$ of updated states for the processes in $\ell(v)$. We require that

1. for all $v \in V$, the set $\downarrow v = \{u \in V \mid u \leq v\}$ is finite,
2. $u \parallel v$ implies $\ell(u) \cap \ell(v) = \emptyset$ for all $u, v \in V$, and
3. $u \lessdot v$ implies $\ell(u) \cap \ell(v) \neq \emptyset$ for all $u, v \in V$.

This implies in particular that two transitions cannot read or write simultaneously the same process. Finally, the transition relations of AK must be satisfied: for $v \in V$, let $R(v) = (R_i(v))_{i\in\ell(v)}$ be defined by $R_i(v) = q_i^0$ if $\{u < v \mid i \in \ell(u)\} = \emptyset$ and $R_i(v) = W_i(\max(\{u < v \mid i \in \ell(u)\}))$ otherwise. Then, we must have $(R(v), W(v)) \in \delta_{\ell(v)}$ for all $v \in V$.

If $\rho = (V, \leq, \ell, W)$ is a run of AK and $U \subseteq V$ is such that $U = \downarrow U = \{v \in V \mid v \leq u$ for some $u \in U\}$ then the restriction $(U, \leq, \ell, W)$ of ρ to U is also a run of AK which is called a prefix of ρ. A run of AK is maximal if it is not a strict prefix of some other run of AK.

Without loss of generality, we may assume that $\sigma_i(q_i) \neq \emptyset$ for all $q_i \in Q_i$ and that the sets AP_i are pairwise disjoint. Let $\mathrm{AP} = \biguplus_{i\in\mathrm{Loc}} \mathrm{AP}_i$ and $\Sigma = 2^{\mathrm{AP}} \setminus \{\emptyset\}$. For $a \in \Sigma$ we let $\mathrm{loc}(a) = \{i \in \mathrm{Loc} \mid \mathrm{AP}_i \cap a \neq \emptyset\}$. The dependence relation over Σ is defined by $(a, b) \in D$ if $\mathrm{loc}(a) \cap \mathrm{loc}(b) \neq \emptyset$. With each run $\rho = (V, \leq, \ell, W)$ of AK we associate $\tau(\rho) = (V, \leq, \lambda)$ where $\lambda(v) = \bigcup_{i\in\ell(v)} \sigma_i(W_i(v))$. It is not hard to see that $\tau(\rho)$ is a trace over (Σ, D).

An asynchronous Kripke structure AK satisfies a temporal formula $\xi \in \mathrm{TL}(B)$ ($\mathrm{AK} \models \xi$) if, for any maximal run ρ of AK, we have $\# \cdot \tau(\rho), \# \models \xi$.

Model Checking Problem for TL(B) *and AKS:* Given an asynchronous Kripke structure AK and a formula $\xi \in \mathrm{TL}(B)$, do we have $\mathrm{AK} \models \xi$?

Theorem 10. *Let* $(\mathrm{AP}_i)_{i\in\mathrm{Loc}}$ *and* (Σ, D) *be as above. Let* B *be a finite set of modality names with associated MSO-modalities over the alphabet* Σ. *Then the model checking problem for* TL(B) *and AKS is decidable in PSPACE.*

Proof. Let $\mathrm{AK} = ((Q_i)_{i\in\mathrm{Loc}}, (\delta_I)_{I\subseteq\mathrm{Loc}}, q^0, (\sigma_i)_{i\in\mathrm{Loc}})$ be an AKS. We define an associated sequential (global) Kripke structure $K = (S, \delta, s_0, \sigma)$. The set of global states is $S = Q_{\mathrm{Loc}} \times 2^{\mathrm{Loc}}$ and $s_0 = (q^0, \mathrm{Loc})$ is the initial global state. The transition relation $\delta \subseteq S \times S$ is defined by $((p, I), (q, J)) \in \delta$ if $J \neq \emptyset$, $(p_J, q_J) \in \delta_J$ and $p_{\overline{J}} = q_{\overline{J}}$ where $\overline{J} = \mathrm{Loc} \setminus J$. Finally, the labelling $\sigma : S \to \Sigma$ is given by $\sigma(q, I) = \bigcup_{i\in I} \sigma_i(q_i)$.

Runs of K correspond to linearizations of runs of AK. More precisely, let $\rho = (V, \leq, \ell, W)$ be a run of AK and let $\sqsubseteq$ be any linear extension of $\leq$ of order type at most ω. We can write $V = \{v_1, v_2, \dots\}$ with $v_{n-1} \sqsubseteq v_n$. We define a sequence of global states $s_n = (q^n, I_n)$ by $I_0 = \text{Loc}$ and for $n > 0$, $I_n = \ell(v_n)$, $q^n_{I_n} = W(v_n)$ and $q^n_{\overline{I_n}} = q^{n-1}_{\overline{I_n}}$. Then, $s_0 s_1 \cdots$ is a run of K which is a linearization of ρ. Moreover, the word $\sigma(s_0)\sigma(s_1)\ldots \in \Sigma^\infty$ is a linearization of the trace $\tau(\rho)$. Conversely, any run of K is a linearization of some run of AK.

For the model checking problem, we are interested in maximal runs. Clearly, a linearization of a maximal run of AK is a maximal run of K. Conversely, a maximal *finite* run of K is a linearization of a maximal finite run of AK. Now, an infinite run $(q^0, \text{Loc})(q^1, I_1)(q^2, I_2)\ldots$ of K is a linearization of a maximal run of AK if and only if eventually, there is no enabled transition involving a set of processes that participate in finitely many transitions of the run: there exists $N \geq 0$ such that for all $\emptyset \neq J \subseteq \text{Loc}$ with $J \cap I_n = \emptyset$ for all $n > N$, we have $(\{q^N_J\} \times Q_J) \cap \delta_J = \emptyset$. We call a run of K accepting if it is either finite and maximal or infinite and satisfies the above condition (which by the way can be described with a Muller table). Hence, accepting runs of K correspond to maximal runs of AK.

Now, let $\xi \in \text{TL}(B)$. We use the notation introduced for the satisfiability. Then AK $\models^t_{[\![-]\!]} \xi$ if and only if for all accepting runs $s_0 s_1 \ldots$ of K we have $\sigma(s_0)\sigma(s_1)\ldots, 0 \models^w_{\overline{[\![-]\!]}} \xi$. Therefore, we are reduced to a model checking problem of a Kripke structure K with some acceptance condition on infinite runs.

Note that a state of K can be stored in space polynomial in the size of AK. Also, the same space bound suffices to decide whether a pair of states (s, s') forms a transition of K and to compute $\sigma(s)$. Finally searching for a loop that satisfies the acceptance condition can also be done in space polynomial in the size of AK. One just has to guess at the beginning of the loop the set J of processes that will not participate in the transitions of the loop. This guess is easy to check within the polynomial space bound as well as the fact that no transition involving a set of processes contained in J is enabled at the beginning of the loop. Therefore, using for ξ (interpreted with $\overline{[\![-]\!]}$) the technique described in the proof of Theorem 8, a slight modification of the usual model checking procedure allows to solve our problem in PSPACE. □

The theorems above show that for any of the local temporal logics introduced in Sect. 5, the satisfiability and the model checking problems become decidable in PSPACE. For some of these logics, this result was known, for TrPTL [13], it seems to be new.

7 Generalizations

The framework of MSO-definable local temporal logics extends verbatim to more general partial orders than Mazurkiewicz traces. The difficulty is to find reasonable classes of partial orders such that complexity results can be obtained for the satisfiability and the model checking problems. For instance, we can show that for the class of all Message sequence charts (MSCs), the satisfiability for a very restricted local temporal logic (namely, a small fragment of TLC^-) is undecidable. On the other hand, there are natural subclasses of MSCs for which the satisfiability problem is decidable in PSPACE. These results will appear in a forthcoming paper.

References

1. R. Alur, R. Peled, and W. Penczek. Model checking of causality properties. In *LICS 95*, pages 90–100. IEEE Computer Society Press, 1995.
2. J.R. Büchi. On a decision method in restricted second order arithmetics. In E. Nagel et al., editors, *Proc. Intern. Congress on Logic, Methodology and Philosophy of Science*, pages 1–11. Stanford University Press, Stanford, 1960.
3. V. Diekert. A pure future local temporal logic beyond cograph-monoids. In M. Ito, editor, *Proc. of the RIMS Symposium on Algebraic Systems, Formal Languages and Conventional and Unconventional Computation Theory, Kyoto, Japan 2002*, 2002.
4. V. Diekert and P. Gastin. Local temporal logic is expressively complete for cograph dependence alphabets. In *LPAR 01*, Lecture Notes in Artificial Intelligence vol. 2250, pages 55–69. Springer, 2001.
5. V. Diekert and P. Gastin. Local temporal logic is expressively complete for cograph dependence alphabets. Tech. Rep. LIAFA, Université Paris 7 (France), 2003. http://www.liafa.jussieu.fr/~gastin/Articles/diegas03.html.
6. V. Diekert and G. Rozenberg. *The Book of Traces.* World Scientific Publ. Co., 1995.
7. P. Gastin and M. Mukund. An elementary expressively complete temporal logic for Mazurkiewicz traces. In *Proc. of ICALP'02*, number 2380 in LNCS, pages 938–949. Springer Verlag, 2002.
8. H.W. Kamp. *Tense logic and the theory of linear order.* PhD thesis, University of California, Los Angeles, USA, 1968.
9. O. Kupferman, N. Piterman, and M.Y. Vardi. Extended temporal logic revisited. In *Proc. of CONCUR'01*, number 2154 in LNCS, pages 519–535. Springer Verlag, 2001.
10. M. Mukund and M. Sohoni. Keeping trace of the latest gossip: bounded time-stamps suffice. In *Proc. of FST&TCS'93*, number 761 in LNCS, pages 388–399. Springer Verlag, 1993.
11. M. Mukund and P.S. Thiagarajan. Linear time temporal logics over Mazurkiewicz traces. In *Proc. of MFCS'96*, number 1113 in LNCS, pages 62–92. Springer Verlag, 1996.
12. A. Rabinovich and S. Maoz. An infinite hierarchy of temporal logics over branching time. *Information and Computation*, 171(2):306–332, 2001.
13. P.S. Thiagarajan. A trace based extension of linear time temporal logic. In *Proc. of LICS'94*, pages 438–447. IEEE Computer Society Press, 1994.
14. P.S. Thiagarajan. A trace consistent subset of PTL. In *Proc. of CONCUR'95*, number 962 in LNCS, pages 438–452, 1995.
15. W. Thomas. Automata on infinite objects. In J. van Leeuwen, editor, *Handbook of Theoretical Computer Science*, pages 133–191. Elsevier Science Publ. B.V., 1990.
16. M.Y. Vardi and P. Wolper. Reasonning about infinite computations. *Information and Computation*, 115:1–37, 1994.
17. P. Wolper. Temporal logic can be more expressive. *Inf. and Control*, 56:72–99, 1983.

Equivalence Checking of Non-flat Systems Is EXPTIME-Hard*

Zdeněk Sawa

Dept. of Computer Science, FEI
Technical University of Ostrava
17. listopadu 15
CZ-708 33 Ostrava, Czech Republic
Zdenek.Sawa@vsb.cz

Abstract. The equivalence checking of systems that are given as a composition of interacting finite-state systems is considered. It is shown that the problem is *EXPTIME*-hard for any notion of equivalence that lies between bisimulation equivalence and trace equivalence, as conjectured by Rabinovich (1997). The result is proved for parallel composition of finite-state systems where hiding of actions is allowed, and for 1-safe Petri nets. The technique of the proof allows to extend this result easily to other types of 'non-flat' systems.

Keywords: equivalence checking, finite-state systems, complexity

1 Introduction

One problem that naturally arises in the area of automatic verification of systems is the problem of equivalence checking. This problem can be stated as follows: given two descriptions of labelled transition systems, decide if the systems behave equivalently.

Many different types of equivalences were proposed in the literature, and some of the most prominent were organized by van Glabbeek [10] into linear time – branching time spectrum. All these equivalences lie between bisimulation equivalence (which is the finest of these equivalences) and trace equivalence (which is the coarsest).

We call a finite transition system that is given explicitly a *flat* transition system. A *non-flat* system is a system given as a composition of interacting flat systems. The set of global states of a non-flat system can be exponentially larger than the sum of sizes of its parts. This phenomenon is known as a state explosion and presents the main challenge in the design of efficient algorithms for verification of non-flat systems.

Overview of Existing Results. Rabinovich [6] considered a composition of finite-state systems that synchronize on identical actions and where some actions may

* This work was sponsored by the Grant Agency of the Czech Republic, Grant No. 201/03/1161

R. Amadio, D. Lugiez (Eds.): CONCUR 2003, LNCS 2761, pp. 237–250, 2003.

be 'hidden' in the sense that they are replaced with invisible τ actions. He proved that equivalence checking is *PSPACE*-hard for such systems for any relation between bisimilarity and trace equivalence, and that the problem is *EXPSPACE*-complete for trace equivalence. He also mentioned that the problem is *EXPTIME*-complete for bisimilarity and conjectured that the problem is in fact *EXPTIME*-hard for any relation between bisimilarity and trace equivalence.

Laroussinie and Schnoebelen [5] approved the Rabinovich's conjecture for all relations that lie between bisimilarity and simulation preorder. The non-flat systems, used in their proof, synchronize on identical actions and do not use hiding. It is not possible to extend their result to all equivalences between bisimilarity and trace equivalence, because for example trace equivalence can be decided in *PSPACE* for this model, as was proved in [8]. See also [9] for results for other types of 'trace-like' equivalences and non-flat systems. Other type of non-flat systems are 1-safe Petri nets. See [4] for some results concerning them, in particular, deciding of bisimilarity is *EXPTIME*-complete for 1-safe Petri nets.

Our Contribution. The Rabinovich's conjecture is approved in this paper for *all* relations between bisimilarity and trace preorder, not only for relations between bisimilarity and simulation preorder. We show that equivalence checking is *EXPTIME*-hard for any such relation if the considered model is a parallel composition of finite-state systems with hiding, the model for which Rabinovich formulated his conjecture in [6].

To simplify the proof, a new auxiliary model called reactive linear bounded automaton (RLBA) is introduced in this paper. Reactive linear bounded automata can be easily modeled by different types of non-flat systems, for example by parallel compositions of finite-state systems with hiding, or by 1-safe Petri nets. The *EXPTIME*-hardness result is shown for RLBA first, and then it is extended to other types of non-flat systems that are able to model RLBA.

From the construction in the proof we also obtain a simpler proof of the result, shown in [7], that equivalence checking is *PTIME*-hard for flat systems for every relation between bisimilarity and trace preorder.

Overview of the Paper. Section 2 contains some necessary definitions. The outline of the proof is presented in Sect. 3. Reactive linear bounded automata are introduced in Sect. 4, together with the description how they can be transformed into other non-flat systems. The proof of *PTIME*-hardness of equivalence checking in case of flat systems is presented in Sect. 5. The construction in this proof forms a base of the more complicated construction described in Sect. 6 where we show that equivalence checking is *EXPTIME*-hard for reactive linear bounded automata.

2 Basic Definitions

2.1 Labelled Transition Systems

A *labelled transition system* (LTS) is a tuple $\mathcal{T} = (S, \mathcal{A}, \longrightarrow)$ where S is a set of *states*, $\mathcal{A}$ is the finite set of *actions*, and $\longrightarrow \subseteq S \times \{\mathcal{A} \cup \{\tau\}\} \times S$ is a *transition relation* where $\tau \notin \mathcal{A}$ is a special *invisible* action. We write $s \xrightarrow{a} s'$ instead of $(s, a, s') \in \longrightarrow$. Only *finite-state* LTSs where S is finite are considered in this paper.

An LTS $\mathcal{T} = (S, \mathcal{A}, \longrightarrow)$ where S, $\mathcal{A}$ and $\longrightarrow$ are given explicitly is called *flat system* (FS) and the size $|\mathcal{T}|$ of FS $\mathcal{T}$ is $|S| + |\mathcal{A}| + |\longrightarrow|$.

More complicated LTSs can be created from FSs by parallel composition and hiding. In the parallel composition a visible action a is executed iff every LTS that has a in its alphabet executes it. Invisible actions are not synchronized, that is, when an LTS executes the invisible action τ, other LTSs do nothing. Formally, the *parallel composition* $\mathcal{T}_1 \parallel \cdots \parallel \mathcal{T}_n$ of LTSs $\mathcal{T}_1, \ldots, \mathcal{T}_n$ where $\mathcal{T}_i = (S_i, \mathcal{A}_i, \longrightarrow_i)$ for each $i \in I$ where $I = \{1, \ldots, n\}$, is the LTS $(S, \mathcal{A}, \longrightarrow)$ where $S = S_1 \times \cdots \times S_n$, $\mathcal{A} = \mathcal{A}_1 \cup \cdots \cup \mathcal{A}_n$, and $(s_1, \ldots, s_n) \xrightarrow{a} (s'_1, \ldots, s'_n)$ iff either

- $a \in \mathcal{A}$ and for every $i \in I$: if $a \in \mathcal{A}_i$, then $s_i \xrightarrow{a} s'_i$, and if $a \notin \mathcal{A}_i$, then $s_i = s'_i$,
- $a = \tau$ and for some $i \in I$ is $s_i \xrightarrow{\tau} s'_i$ and $s_j = s'_j$ for each $j \in I$ such that $j \neq i$.

Tuples from $S_1 \times \cdots \times S_n$ are called *global states*. In this paper only *binary* synchronizations are needed, where any $a \in \mathcal{A}$ belongs to at most two different $\mathcal{A}_i$.

Hiding of actions removes a set of visible actions from the alphabet of an LTS and relabels corresponding transitions with the invisible action τ. Formally, *hide* $\mathcal{B}$ *in* $\mathcal{T}_1$, where $\mathcal{T}_1$ is an LTS $(S_1, \mathcal{A}_1, \longrightarrow_1)$ and $\mathcal{B} \subseteq \mathcal{A}_1$, denotes the LTS $(S, \mathcal{A}, \longrightarrow)$ where $S = S_1$, $\mathcal{A} = \mathcal{A}_1 - \mathcal{B}$, and $s \xrightarrow{a} s'$ iff there is some $a' \in (\mathcal{A}_1 \cup \{\tau\})$ such that $s \xrightarrow{a'} s'$ and either $a \notin \mathcal{B}$ and $a = a'$ or $a' \in \mathcal{B}$ and $a = \tau$.

A *parallel composition with hiding* (PCH) is an LTS $\mathcal{T}$ given in the form *hide* $\mathcal{B}$ *in* $(\mathcal{T}_1 \parallel \cdots \parallel \mathcal{T}_n)$ where $\mathcal{T}_1, \ldots, \mathcal{T}_n$ are FSs. The size $|\mathcal{T}|$ of PCH $\mathcal{T}$ is $|\mathcal{T}_1| + \cdots + |\mathcal{T}_n| + |\mathcal{B}|$.

Other type of non-flat systems are 1-safe Petri nets. A *labelled net* is a tuple $\mathcal{N} = (P, T, F, \lambda)$, where P and T are finite sets of *places* and *transitions*, $F \subseteq (S \times T) \cup (T \times S)$ is the *flow relation*, and $\lambda : T \to \mathcal{A}$ is a labelling function that associates to each transition t a label $\lambda(t)$ taken from some given set of actions $\mathcal{A}$. Pairs from F are called *arcs*. We identify F with its characteristic function $(P \times T) \cup (T \times P) \to \{0, 1\}$. A *marking* is a mapping $M : P \to \mathbb{N}$. A *labelled Petri net* is a pair $N = (\mathcal{N}, M_0)$ where $\mathcal{N}$ is a labelled net and M_0 is the initial marking. A transition t is *enabled* at a marking M if $M(p) > 0$ for every p such that $(p, t) \in F$. If t is enabled in M, then it can *fire* and its firing leads to the successor marking M' which is defined for every place p by $M'(p) = M(p) + F(t, p) - F(p, t)$. Given $a \in \mathcal{A}$, we denote by $M \xrightarrow{a} M'$ that there is some transition t such that M enables transition t, the marking reached by the firing of t is M', and $\lambda(t) = a$. A Petri net is *1-safe* if $M(p) \leq 1$ for every place p and every reachable marking M.

Let $\mathcal{T} = (S, \mathcal{A}, \longrightarrow)$ be an LTS. A *trace* from $s \in S$ is any $w = a_1 \ldots a_n \in \mathcal{A}^*$ such that there is a sequence $s_0, s_1, \ldots, s_n \in S$ where $s_0 = s$ and $s_{i-1} \xrightarrow{a_i} s_i$ for every $1 \leq i \leq n$. The set of all traces from s is denoted $\mathit{Tr}(s)$. States s, s'

are in *trace preorder*, written $s \sqsubseteq_{tr} s'$, iff $Tr(s) \subseteq Tr(s')$. States s, s' are *trace equivalent* iff $s \sqsubseteq_{tr} s'$ and $s' \sqsubseteq_{tr} s$.

A bisimulation over $\mathcal{T}$ is any relation $\mathcal{R} \subseteq S \times S$ satisfying the following two conditions for each $s, t \in S$ such that $s\mathcal{R}t$:

- if $s \xrightarrow{a} s'$ for some s', then $t \xrightarrow{a} t'$ for some t' such that $s'\mathcal{R}t'$, and
- if $t \xrightarrow{a} t'$ for some t', then $s \xrightarrow{a} s'$ for some s' such that $s'\mathcal{R}t'$.

(It is said that $s \xrightarrow{a} s'$ is *matched* by $t \xrightarrow{a} t'$, resp. $t \xrightarrow{a} t'$ by $s \xrightarrow{a} s'$.) States s, s' are *bisimilar*, written $s \sim s'$, iff there exists some bisimulation $\mathcal{R}$ such that $s\mathcal{R}s'$.

Let $\mathcal{R}$ be some binary relation defined over states of LTSs, such that $s \sim s'$ implies $s\mathcal{R}s'$, and $s\mathcal{R}s'$ implies $s \sqsubseteq_{tr} s'$, i.e., $\sim \subseteq \mathcal{R} \subseteq \sqsubseteq_{tr}$. The problem FS-EQ$_{\mathcal{R}}$ is defined as:

INSTANCE: An FS $\mathcal{T}$ and its two states s and s'.
QUESTION: Is $s\,\mathcal{R}\,s'$?

the problem PCH-EQ$_{\mathcal{R}}$ as:

INSTANCE: A PCH $\mathcal{T}$ and its two global states $(s_1, \ldots, s_n)$ and $(s'_1, \ldots, s'_n)$.
QUESTION: Is $(s_1, \ldots, s_n)\,\mathcal{R}\,(s'_1, \ldots, s'_n)$?

and the problem PN-EQ$_{\mathcal{R}}$ as:

INSTANCE: A labelled net $\mathcal{N}$ with two markings M, M', such that $(\mathcal{N}, M)$ and $(\mathcal{N}, M')$ are 1-safe Petri nets.
QUESTION: Is $M\,\mathcal{R}\,M'$?

The main results of the paper show that FS-EQ$_{\mathcal{R}}$ is *PTIME*-hard, and PCH-EQ$_{\mathcal{R}}$ and PN-EQ$_{\mathcal{R}}$ are *EXPTIME*-hard for any $\mathcal{R}$ satisfying $\sim \subseteq \mathcal{R} \subseteq \sqsubseteq_{tr}$.

2.2 Alternating Graphs

In the proof of *PTIME*-hardness of FS-EQ$_{\mathcal{R}}$ we show a logspace reduction from the *Alternating Graph Problem* (AGP) that is known to be *PTIME*-complete, see for example [2]. The definition of this problem follows.

An *alternating graph* is a directed graph $G = (V, E, t)$ where V is a finite set of nodes, $E \subseteq V \times V$ is a set of edges, and $t : V \to \{\wedge, \vee\}$ is a labelling function that partitions V into sets $V_\wedge$ and $V_\vee$ of disjunctive and conjunctive nodes. The set of successors of a node v, i.e., the set $\{v' \in V \mid (v, v') \in E\}$, is denoted by $\sigma(v)$.

The set of *successful* nodes W is the least subset of V with the property that if a node v is conjunctive and all nodes from $\sigma(v)$ are in W, or disjunctive and at least one node from $\sigma(v)$ is in W, then v also belongs to W. AGP is the problem whether a given node is successful:

INSTANCE: An alternating graph $G = (V, E, t)$ and a node $v \in V$.
QUESTION: Is v successful?

Let $\mathcal{P}(V)$ be the power set of V. Notice that W is the least fixed point of a function $f : \mathcal{P}(V) \to \mathcal{P}(V)$ where for $U \subseteq V$ is $v \in f(U)$ iff either $v \in V_\wedge$ and $(\forall v' \in \sigma(v) : v' \in U)$, or $v \in V_\vee$ and $(\exists v' \in \sigma(v) : v' \in U)$.

Let us have a node that has no successors. If this node is conjunctive, it is called an *accepting* node, and otherwise it is called a *rejecting* node. Notice that accepting nodes are always successful, rejecting nodes are never successful, and that W is nonempty iff G contains at least one accepting node.

2.3 Alternating Linear Bounded Automata

In the proof we use a logspace reduction from a well known *EXPTIME*-complete problem that is called ALBA-ACCEPT in this paper. It is a problem of deciding if a given alternating linear bounded automaton accepts a given word. Its definition follows, see [1] for further details.

A *linear bounded automaton* (LBA) is a tuple $A = (Q, \Sigma, \Gamma, \delta, q_0, q_{acc}, q_{rej})$, where Q is a set of control states, Σ is an input alphabet, Γ is a tape alphabet, $\delta \subseteq (Q - \{q_{acc}, q_{rej}\}) \times \Gamma \times Q \times \Gamma \times \{-1, 0, +1\}$ is a set of transitions, $q_0, q_{acc}, q_{rej} \in Q$ are an initial, accepting and rejecting state. The alphabet Γ contains left and right endmarkers $\vdash$ and $\dashv$.

A *configuration* of A is a triple $\alpha = (q, w, i)$ where q is the current state, $w = a_1 a_2 \dots a_n$ is the tape content, and $1 \leq i \leq |w|$ is the head position. Only configurations where $w = \vdash w' \dashv$ and endmarkers do not occur in w' are allowed. The size $|\alpha|$ of α is $|w|$. A configuration $\alpha' = (q', w', i')$ is a *successor* of $\alpha = (q, w, i)$, written $\alpha \vdash_A \alpha'$ (or just $\alpha \vdash \alpha'$ when A is obvious), iff $(q, a, q', a', d) \in \delta$, w contains a on position i, $i' = i + d$, and w' is obtained from w by writing a' on position i. Endmarkers may not be overwritten, and the machine is constrained never to move left of the $\vdash$ nor right of the $\dashv$. Notice that when $\alpha \vdash \alpha'$, then $|\alpha| = |\alpha'|$. The initial configuration for an input $w \in \Sigma^*$ is $\alpha_{ini}(w) = (q_0, \vdash w \dashv, 1)$. A configuration is *accepting* iff $q = q_{acc}$, and *rejecting* iff $q = q_{rej}$.

An *alternating LBA* (ALBA) is an LBA extended with a function $l : Q \to \{\wedge, \vee\}$ that labels each state as either conjunctive or disjunctive. We extend l to configurations in an obvious manner and so also configurations are labeled as conjunctive and disjunctive. A configuration is *successful* iff it is either accepting, or disjunctive with at least one successful successor, or conjunctive with all successors successful. An ALBA A accepts an input $w \in \Sigma^*$ iff $\alpha_{ini}(w)$ is successful.

The problem ALBA-ACCEPT is defined as:

INSTANCE: An ALBA A and a word $w \in \Sigma^*$.
QUESTION: Does A accept w?

Notice that there is a close relationship between AGP and ALBA-ACCEPT. A computation of an ALBA can be viewed as an alternating graph where successful nodes correspond to successful configurations. The size of this graph can be exponentially larger than the size of the corresponding instance of ALBA-ACCEPT.

3 Outline of the Proof

Reactive linear bounded automata (RLBA) are introduced in Sect. 4 and it is shown that they can be modeled by other types of non-flat systems, in particular by PCH and by 1-safe Petri nets. An RLBA is similar to a usual LBA, but is intended to generate an LTS, not to accept or reject an input. The equivalence checking problem where the instance is an RLBA and two of its configurations is denoted RLBA-EQ$_{\mathcal{R}}$ in this paper.

The main technical result of the paper shows that RLBA-EQ$_{\mathcal{R}}$ is *EXPTIME*-hard for any relation $\mathcal{R}$ satisfying $\sim\subseteq \mathcal{R} \subseteq\sqsubseteq_{tr}$. From this follows *EXPTIME*-hardness of equivalence checking for every model that is able to model an RLBA.

To show *EXPTIME*-hardness of RLBA-EQ$_{\mathcal{R}}$, we present a logspace reduction from ALBA-ACCEPT. The construction in the proof is based on a simpler construction that can be used to show *PTIME*-hardness of FS-EQ$_{\mathcal{R}}$. This simpler construction is presented in Sect. 5, where we show a logspace reduction from AGP to the complement of FS-EQ$_{\mathcal{R}}$ that works for any $\mathcal{R}$ such that $\sim\subseteq \mathcal{R} \subseteq\sqsubseteq_{tr}$. The basic idea is to construct an LTS with two distinguished states s, s', such that $s \not\sqsubseteq_{tr} s'$ if the answer to the original problem is YES, and $s \sim s'$ otherwise. The same construction can be used for every $\mathcal{R}$, because $s \not\sqsubseteq_{tr} s'$ implies that not $s\mathcal{R}s'$, and $s \sim s'$ implies $s\mathcal{R}s'$. The same idea was also used for example in [3] and [6]. We can conclude that the complement of FS-EQ$_{\mathcal{R}}$ is *PTIME*-hard for any $\mathcal{R}$, and so also FS-EQ$_{\mathcal{R}}$ is *PTIME*-hard because *PTIME* is closed under complement. *PTIME*-hardness of FS-EQ$_{\mathcal{R}}$ was already proved in [7], but the reduction presented here is simpler.

Now consider ALBA-ACCEPT, a well known *EXPTIME*-complete problem. A computation of an ALBA can be viewed as an alternating graph, where successful nodes correspond to successful configurations, and this allows us to 'shift' the previous result 'higher' in the complexity hierarchy. We will construct an RLBA that will model the LTS which we would obtain when we would apply the above mentioned reduction to the alternating graph corresponding to the computation of the ALBA. Moreover, logarithmic space will be sufficient for the construction of this RLBA from the instance of ALBA-ACCEPT.

4 Reactive Linear Bounded Automata

Reactive linearly bounded automata are introduced in this section. A *reactive linear bounded automaton* (RLBA) is like a usual LBA, but it has special control states, called *reactive* states, where it can perform actions from some given set of actions $\mathcal{A}$. Only the control state is changed after performing such actions, neither the tape content nor the head position is modified. The other control states are called *computational* and RLBA performs steps as a usual LBA in them. Each such step is represented as the invisible action τ.

Formally, an RLBA is a tuple $B = \{Q, \Gamma, \delta, \mathcal{A}, l, \longrightarrow\}$, where the meaning of Q, Γ and δ is the same as in a usual LBA, $\mathcal{A}$ is the finite set of actions, the function $l : Q \rightarrow \{r, c\}$ partitions Q into sets Q_r and Q_c of reactive and computational states, and $\longrightarrow \subseteq Q_r \times (\mathcal{A} \cup \{\tau\}) \times Q$ is the transition relation (we write $q \stackrel{a}{\longrightarrow} q'$ instead of $(q, a, q') \in \longrightarrow$). It is also required that if $(q, b, q', b', d) \in \delta$ then $q \in Q_c$. The definition of a configuration and a successor relation is the same as for a usual LBA.

An RLBA B generates an LTS $\mathcal{T}(B) = (S, \mathcal{A}, \longrightarrow)$, where S is the set of configurations of B, and $(q, w, i) \stackrel{a}{\longrightarrow} (q', w', i')$ iff either $q \in Q_c$, $(q, w, i) \vdash (q', w', i')$ and $a = \tau$, or $q \in Q_r$, $q \stackrel{a}{\longrightarrow} q'$, $w = w'$ and $i = i'$.

For each $\mathcal{R}$, such that $\sim \subseteq \mathcal{R} \subseteq \sqsubseteq_{tr}$, we can define the problem RLBA-EQ$_{\mathcal{R}}$:

INSTANCE: An RLBA B and its two configurations α, α' of size n.
QUESTION: Is $\alpha \mathcal{R} \alpha'$?

An RLBA with a configuration of size n can be easily modeled by various non-flat systems, as two following lemmas show.

Lemma 1. *There is a logspace reduction from* RLBA-EQ$_{\mathcal{R}}$ *to* PCH-EQ$_{\mathcal{R}}$.

Proof. Let us have an RLBA B and two its configurations of size n. We construct a PCH $\mathcal{T}$ of the form *hide* $\mathcal{B}$ *in* $(\mathcal{T}_c \parallel \mathcal{T}_1 \parallel \cdots \parallel \mathcal{T}_n)$ which models the LTS generated by B. In particular, $\mathcal{T}_c$ models the control unit, and $\mathcal{T}_1, \ldots, \mathcal{T}_n$ model the tape cells of B. A state of $\mathcal{T}_c$ represents the current control state and head position, and a state of $\mathcal{T}_i$ represents the symbol on the i-th position of the tape.

Let $I = \{1, \ldots, n\}$ be the set of all possible positions of the head. For each $i \in I$ is $\mathcal{T}_i = (S_i, \mathcal{A}_i, \longrightarrow_i)$ where $S_i = \Gamma$, $\mathcal{A}_i = \{\langle b, b', i\rangle \mid b, b' \in \Gamma\}$, and $\longrightarrow_i$ contains transitions $b \stackrel{\langle b,b',i\rangle}{\longrightarrow} b'$ for each $b, b' \in \Gamma$.

In $\mathcal{T}_c = (S_c, \mathcal{A}_c, \longrightarrow_c)$ is $S_c = \{\langle q, i\rangle \mid q \in Q, i \in I\}$ and $\mathcal{A}_c = \mathcal{A} \cup \mathcal{A}_0 \cup \cdots \cup \mathcal{A}_n$ (w.l.o.g. we can assume that $\mathcal{A} \cap \mathcal{A}_i = \emptyset$ for each $i \in I$). To $\longrightarrow_c$ we add for each $(q, b, q', b', d) \in \delta$ and $i \in I$, such that $i + d \in I$, a transition $\langle q, i\rangle \stackrel{\langle b,b',i\rangle}{\longrightarrow} \langle q', i+d\rangle$, and for each $q \in Q_r$, $q' \in Q$, $a \in (Act \cup \{\tau\})$ and $i \in I$ where $q \stackrel{a}{\longrightarrow} q'$ we add a transition $\langle q, i\rangle \stackrel{a}{\longrightarrow} \langle q', i\rangle$.

The set $\mathcal{B}$ in $\mathcal{T}$ is defined as $\mathcal{A}_1 \cup \cdots \cup \mathcal{A}_n$. Each configuration $\alpha = (q, a_1 a_2 \cdots a_n, i)$ has a corresponding global state $g(\alpha) = (\langle q, i\rangle, a_1, a_2, \ldots, a_n)$. As can be easily checked, $\alpha \stackrel{a}{\longrightarrow} \alpha'$ in B iff $g(\alpha) \stackrel{a}{\longrightarrow} g(\alpha')$ in $\mathcal{T}$, and so $\alpha \mathcal{R} \alpha'$ in B iff $g(\alpha) \mathcal{R} g(\alpha')$ for any $\mathcal{R}$ such that $\sim \subseteq \mathcal{R} \subseteq \sqsubseteq_{tr}$. It is obvious that $\mathcal{T}$ can be constructed from B in a logarithmic space. □

Lemma 2. *There is a logspace reduction from* RLBA-EQ$_{\mathcal{R}}$ *to* PN-EQ$_{\mathcal{R}}$.

Proof. Let us have an instance of RLBA-EQ$_{\mathcal{R}}$, i.e. an RLBA B and two of its configurations of size n. We construct a labelled net as follows. Let $I = \{1, \ldots, n\}$. The set of places will be $Q \cup \{\langle a, i\rangle \mid a \in \Gamma, i \in I\} \cup I$.

For each $(q, b, q', b', d) \in \delta$ and $i \in I$ where $q \in Q_c$ and $i + d \in I$ we add a transition $t = \langle q, b, q', b', i, i + d\rangle$ labelled with τ together with incoming arcs (q, t), $(\langle b, i\rangle, t)$, and (i, t), and outgoing arcs (t, q'), $(t, \langle s', i\rangle)$, and $(t, i + d)$.

For each $q, q' \in Q$ and $a \in (\mathcal{A} \cup \{\tau\})$ where $q \in Q_r$ and $q \stackrel{a}{\longrightarrow} q'$ we add a new transition $t = \langle q, a, q'\rangle$ labelled with a together with an incoming arc $\langle q, t\rangle$ and an outgoing arc $\langle t, q'\rangle$.

For a configuration $\alpha = \{q, a_1 a_2 \cdots a_n, i\}$ we define a corresponding marking M_α where $M_\alpha(p) = 1$ if p is q, i, or $\langle a_j, j\rangle$ where $j \in I$, and $M_\alpha(p) = 0$ otherwise. It is easy to check that $\alpha \xrightarrow{a} \alpha'$ iff $M_\alpha \xrightarrow{a} M_{\alpha'}$. □

5 Construction for Flat Systems

A logspace reduction from AGP to the complement of FS-EQ$_\mathcal{R}$ is presented in this section. For a given alternating graph $G = (V, E, t)$ with a distinguished node x we construct a corresponding LTS $\mathcal{T}_G = (S, \mathcal{A}, \longrightarrow)$ with two distinguished states $s, s' \in S$ such that $s \not\sqsubseteq_{tr} s'$ if x is successful, and $s \sim s'$ otherwise.

The set of states S is V. For each $v \in V$ we define a set of corresponding actions $Act(v)$. If $v \in V_\wedge$, then $Act(v) = \{\langle v\rangle\}$, and if $v \in V_\vee$, then $Act(v) = \{\langle v, i\rangle \mid 1 \leq i \leq |\sigma(v)|\}$. The set of actions $\mathcal{A}$ is $\bigcup_{v\in V} Act(v)$. We assume w.l.o.g. that successors of each node are ordered in some fixed order. The i-th successor of v where $i \in \{1, \ldots, |\sigma(v)|\}$ is denoted by $\sigma_i(v)$.

The transition relation contains transitions of three types:

1. $v \xrightarrow{\langle a\rangle} v$ for each $v \in V$ and $a \in \mathcal{A}$ such that $a \notin Act(v)$.
2. $v \xrightarrow{\langle v,i\rangle} v'$ for each $v \in V_\vee$ and $i \in \{1, \ldots, |\sigma(v)|\}$ where $v' = \sigma_i(v)$.
3. $v \xrightarrow{\langle u\rangle} u'$ for each $v \in V$, $u \in V_\wedge$ and $u' \in V$ such that $u' \in \sigma(u)$.

We may assume w.l.o.g. that G contains at least one rejecting node z. The instance of FS-EQ$_\mathcal{R}$ then consists of $\mathcal{T}_G$ and states z and x, where x is the distinguished node from the instance of AGP.

Proposition 3. *If $v \in V$ is not successful then $z \sim v$.*

Proof. It is sufficient to show that $\{(z, v) \mid v \in (V - W)\} \cup Id$ is a bisimulation (Id denotes the identity relation $\{(v, v) \mid v \in V\}$). Let us consider some pair (z, v) where $v \in (V - W)$, and a transition $v \xrightarrow{a} v'$. This transitions is either of:

- type 1 and then it is matched by $z \xrightarrow{a} z$ of type 1, because $Act(z) = \emptyset$ and so $z \xrightarrow{a} z$ for every $a \in \mathcal{A}$,
- type 2 and then $v \in V_\vee$ and because v is unsuccessful, each $v' \in \sigma(v)$ is also unsuccessful, and so $v \xrightarrow{a} v'$ is matched by $z \xrightarrow{a} z$,
- type 3 and then it can be matched by $z \xrightarrow{a} v'$ of type 3.

Now consider a transition of the form $z \xrightarrow{a} z'$. It is of:

- type 1 and then $z' = z$ and either $a \notin Act(v)$, and $z \xrightarrow{a} z$ is matched by $v \xrightarrow{a} v$ of type 1, or $a \in Act(v)$ and there are two possibilities:
  - if $v \in V_\vee$ then each $v' \in \sigma(v)$ is unsuccessful since v is unsuccessful, and so $z \xrightarrow{a} z$ can be matched by $v \xrightarrow{a} v'$ of type 2,
  - if $v \in V_\wedge$ then there is at least one unsuccessful $v' \in \sigma(v)$, and so $z \xrightarrow{a} z$ can be matched by $v \xrightarrow{a} v'$ of type 3,
- type 2, but this is not possible as $Act(z) = \emptyset$,

– type 3 and it can be matched by $v \xrightarrow{a} z'$ of type 3. □

Proposition 4. *There is $w \in \mathcal{A}^*$ such that if $v \in V$ is successful, then $w \notin Tr(v)$.*

Proof. As W can be computed as the least fixed point of f, we can define a sequence $W_0 \subseteq W_1 \subseteq W_2 \subseteq \cdots$ of subsets of W where $W_0 = \emptyset$ and $W_{i+1} = f(W_i)$ for $i > 0$. For each $v \in W$ there is some least i such that $v \in W_i$. This i is denoted $rank(v)$. Let $m = |W|$, and let $v_1, v_2, \ldots, v_m$ be the nodes in W ordered by their rank, i.e., if $i < j$ then $rank(v_i) \leq rank(v_j)$.

Let us consider a word $w_m = a_m a_{m-1} \cdots a_1$ where $a_i = \langle v_i \rangle$ if $v_i \in V_\wedge$, and if $v_i \in V_\vee$ then $a_i = \langle v_i, k \rangle$ where we choose k such that $v' = \sigma_k(v_i)$ is successful and $rank(v') < rank(v_i)$ (obviously there is at least one such v'). We show that $w_m \notin Tr(v)$ for any successful node v. In particular, for each $i \leq m$ we show that $w_i = a_i a_{i-1} \cdots a_1 \notin Tr(v_j)$ if $j \leq i$. We proceed by induction on i and in the proof we use the following simple observation: $w_i \notin Tr(v)$ iff for each v' such that $v \xrightarrow{a_i} v'$ is $w_{i-1} \notin Tr(v')$.

The base case ($i = 0$) is trivial. In the induction step we consider $i > 0$ and show that the proposition holds for every v_j where $1 \leq j \leq i$.

If $v_i \in V_\vee$ then $a_i = \langle v_i, k \rangle$. Any transition of the form $v_j \xrightarrow{a_i} v'$ is either of type 1, and then $v' = v_j$ and $j < i$, and by induction hypothesis $w_{i-1} \notin Tr(v')$, or of type 2, and then $v' = \sigma_k(v_i)$, so v' is successful and $rank(v') < rank(v)$, and by induction hypothesis $w_{i-1} \notin Tr(v')$.

If $v_i \in V_\wedge$ then $a_i = \langle v_i \rangle$. Any transition of the form $v_j \xrightarrow{a_i} v'$ is either of type 1, and then $v' = v_j$ and $j < i$, and by induction hypothesis $w_{i-1} \notin Tr(v')$, or of type 3, and then $v' \in \sigma(v_i)$ and so v' is successful and $rank(v') < rank(v_i)$, so by induction hypothesis $w_{i-1} \notin Tr(v')$. □

Notice that $z \xrightarrow{a} z$ for each $a \in \mathcal{A}$, because $Act(z) = \emptyset$, and so $Tr(z) = \mathcal{A}^*$. From this and Proposition 4 we have that $z \not\sqsubseteq_{tr} x$ if x is successful. On the other hand, from Proposition 3 we have that $z \sim x$ if x is not successful, and so the described construction is correct.

The described reduction can be obviously performed in a logarithmic space. Since the problem AGP is *PTIME*-complete and *PTIME* is closed under complement, we obtain the following result:

Lemma 5. *FS-EQ$_\mathcal{R}$ is PTIME-hard for any $\mathcal{R}$ such that $\sim \subseteq \mathcal{R} \subseteq \sqsubseteq_{tr}$.*

6 Construction for Non-flat Systems

The description of the reduction from ALBA-ACCEPT to the complement of RLBA-EQ$_\mathcal{R}$ consists of several steps that are summarized in the following figure where $\overline{\text{FS-EQ}_\mathcal{R}}$ and $\overline{\text{RLBA-EQ}_\mathcal{R}}$ denote the complements of FS-EQ$_\mathcal{R}$ and RLBA-EQ$_\mathcal{R}$:

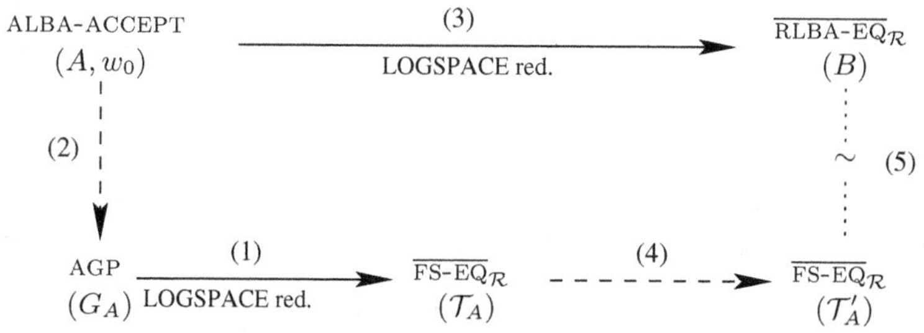

The reduction (1) from AGP to $\overline{\text{FS-EQ}_{\mathcal{R}}}$ can be applied to the alternating graph G_A that corresponds (2) to the ALBA A in the instance of ALBA-ACCEPT. We obtain an LTS $\mathcal{T}_A$. From the instance of ALBA-ACCEPT we construct (3) a RLBA B that models $\mathcal{T}_A$ in the sense, that after we apply a certain kind of transformation (4) to $\mathcal{T}_A$, we obtain an LTS $\mathcal{T}'_A$ bisimilar (5) with B. It will be proved that the transformation (4) preserves some important properties, in particular, states that were bisimilar are bisimilar after the transformation, and states that were not in trace preorder are not in trace preorder after the transformation. Bisimilarity (5) implies that the same is true for corresponding configurations of B, from which the correctness of the construction (3) follows. The *EXPTIME*-hardness of $\overline{\text{RLBA-EQ}_{\mathcal{R}}}$ implies the *EXPTIME*-hardness of RLBA-EQ$_{\mathcal{R}}$ since *EXPTIME* is closed under complement.

The rest of the paper is devoted to the description of a logspace reduction from ALBA-ACCEPT to $\overline{\text{RLBA-EQ}_{\mathcal{R}}}$.

Let an ALBA $A = (Q, \Sigma, \Gamma, \delta, q_0, q_{acc}, q_{rej})$ with a word $w_0 \in \Sigma^*$ be an instance of ALBA-ACCEPT. We can assume that transitions in δ are ordered and that this ordering determines the order of successors of a configuration. For simplicity we can assume w.l.o.g. that each configuration of A, which is not accepting nor rejecting, has exactly two successors, and that $l(q_{acc}) = \wedge$ and $l(q_{rej}) = \vee$. Let *Conf* be the set of all configurations of A of size $n = |w_0| + 2$, and let $Conf_\wedge$, $Conf_\vee$, and $Conf_{rej}$ be the sets of conjunctive, disjunctive, and rejecting configurations of size n, respectively. Notice that any configuration reachable from $\alpha_0 = \alpha_{ini}(w_0)$ is of size n.

The ALBA A has a corresponding alternating graph $G_A = (V, E, t)$, where $V = Conf$, $(\alpha, \alpha') \in E$ iff $\alpha \vdash \alpha'$, and $t(\alpha) = l(\alpha)$ for each $\alpha \in Conf$. Notice that a configuration α is successful in A iff the node α is successful in G_A, and that A accepts w iff the node α_0 is successful.

When we apply the logspace reduction described in Sect. 5 to G_A with a node α_0, we obtain the LTS $\mathcal{T}_A = (S_A, \mathcal{A}_A, \longrightarrow_A)$, where $S_A = Conf$, $Act(\alpha) = \{\langle\alpha\rangle\}$ for each $\alpha \in Conf_\wedge$, $Act(\alpha) = \{\langle\alpha, i\rangle \mid i \in \{1, 2\}\}$ for each $\alpha \in Conf_\vee - Conf_{rej}$, $Act(\alpha) = \emptyset$ for each $\alpha \in Conf_{rej}$, $\mathcal{A}_A = \bigcup_{\alpha \in Conf} Act(\alpha)$, and $\longrightarrow_A$ contains the following transitions for each $\alpha \in Conf$:

1. $\alpha \xrightarrow{x}_A \alpha$ for each $x \in (\mathcal{A}_A - Act(\alpha))$,
2. $\alpha \xrightarrow{\langle\alpha,i\rangle}_A \alpha'$ if $\alpha \in Conf_\vee$, and α' is the i-th successor of α,
3. $\alpha \xrightarrow{\langle\beta\rangle}_A \beta'$ for each $\beta \in Conf_\wedge$ and $\beta' \in Conf$ such that $\beta \vdash \beta'$.

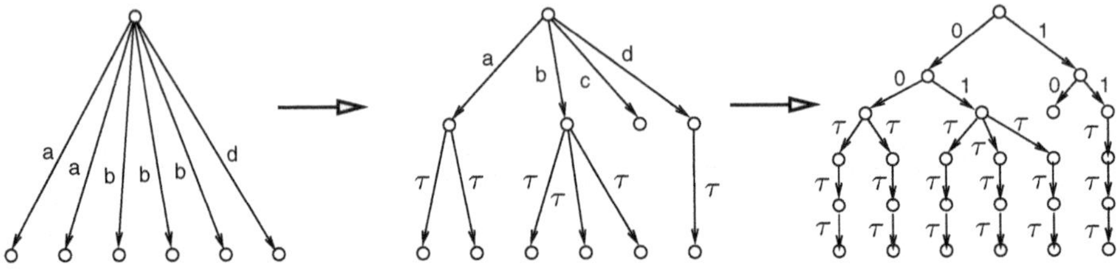

Fig. 1. The transformation performed on $\mathcal{T}_A$

Let $\alpha_{rej} \in Conf_{rej}$ be some rejecting configuration. The states α_{rej} and α_0 are the two distinguished states with the property, that if A accepts w_0, then $\alpha_{rej} \not\sqsubseteq_{tr} \alpha_0$, and $\alpha_{rej} \sim \alpha_0$ otherwise.

An RLBA $B = (Q_B, \Gamma_B, \delta_B, \mathcal{A}_B, l_B, \longrightarrow_B)$ that in some sense 'models' $\mathcal{T}_A$ will be constructed. The RLBA B will be described only informally, but it should be clear from this description how to construct it. In fact B models an LTS that we obtain from $\mathcal{T}_A$ by a transformation illustrated in Fig. 1.

Figure 1 shows only transitions going from one state, but the same transformation is performed for all states and transitions. In this simplified example is $\mathcal{A}_A = \{a, b, c, d\}$. At first, the non-deterministic choice is postponed. Notice that that a new state is added for each action in $\mathcal{A}_A$. Next, each action from $\mathcal{A}_A$ is replaced by sequence of actions from some 'small' alphabet $\mathcal{A}_B$. In our example is $\mathcal{A}_B = \{0, 1\}$ and a, b, c, d are replaced with 00, 01, 10 and 11. Invisible actions representing non-deterministic choice are replaced with sequences of τ actions of some fixed length m (in our example $m = 3$). This kind of transformation is described more formally in the next subsection.

Configurations of A can be written as words in an alphabet $\Delta = (Q \times \Gamma) \cup \Gamma$, where occurrence of the symbol from $Q \times \Gamma$ denotes the position of the head (there must be exactly one such symbol in the word). A word from Δ^* corresponding to a configuration α is denoted by $desc(\alpha)$. Actions from $Act(\alpha)$ are replaced with sequences of actions corresponding to $desc(\alpha)$ in B. In particular, $\mathcal{A}_B = \Delta_{\mathcal{A}} \cup \{1, 2\}$ where $\Delta_{\mathcal{A}} = \Delta - \{(q_{rej}, a) \mid a \in \Gamma\}$. Actions from $\{1, 2\}$ are used to identify a successor of a disjunctive configuration.

B has a tape with two tracks, denoted track 1 and 2, respectively. A current state α of $\mathcal{T}_A$ is is stored as a word $desc(\alpha)$ on track 1. B also needs to store information about the label of a transition that $\mathcal{T}_A$ performs. The configuration from the label of the transition is stored on track 2. Formally this means that $\Gamma_B = (\Delta \times \Delta) \cup \{\vdash, \dashv\}$. See Fig. 2 for an example:

As mentioned above, a transition of $\mathcal{T}_A$ labelled with an action from $Act(\beta)$ is represented in B as a sequence of transitions. Each such sequence start and ends in a configuration where track 1 contains the current state α of $\mathcal{T}_A$ and where the head of B points to the first symbol of $desc(\alpha)$, i.e., it is on position 2. The contents of track 2 is not important, since it will be overwritten. The sequence of transitions of B corresponding to one transition of $\mathcal{T}_A$ has two phases (denoted as phase 1 and phase 2):

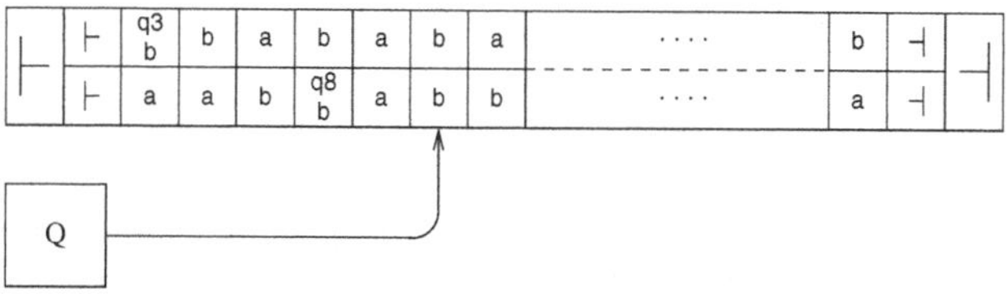

Fig. 2. An example of a configuration of the RLBA B

1. An actions representing symbols of $desc(\beta)$ are performed one by one and the corresponding symbols are stored on track 2. The head of B goes from left to right.
2. One of the possibilities is chosen non-deterministically (possibilities depend on some properties of α and β that are described below, information about these properties can be kept in the control unit of B):
   (a) The head of B moves back to the left endmarker without changing anything.
   (b) A chosen successor of β is stored on track 1 while the head returns back to the left endmarker. This involves copying of track 2 to track 1 with the necessary modifications on positions where β and its successor differ.

The three following steps are performed for each symbol a of $desc(\beta)$ during phase 1:

- the symbol from track 1 is read into the control unit,
- an action a is performed, and remembered in the control unit,
- a is written on track 2 and the head moves to the next cell.

This means that actions $\tau a \tau$ are performed for each symbol a. Phase 1 ends when the right endmarker $\dashv$ is reached. If $\beta \in Conf_{\vee}$, then phase 1 includes also an action $a \in \{1, 2\}$ identifying a successor of β. This number is stored in the control unit of B.

The possible choices at the start of phase 2 depend only on whether $\alpha = \beta$, and on the type of β (if it is accepting, conjunctive or disjunctive). This information can be stored in the control unit of B. To find out if $\alpha = \beta$, notice that we can compare symbols on tracks 1 and 2 during phase 1. The possible non-deterministic choices are the following: if β is disjunctive, the successor of β that was chosen at the end of phase 1 can be stored on track 1, and if β is conjunctive, the non-deterministically chosen successor of β can be stored on track 1. The choice (a), i.e., to keep track 1 intact, is possible only when $\alpha \neq \beta$. Notice that if β is accepting, there are no successors of β and so there are no transitions possible when $\alpha = \beta$.

B can be constructed in such a way that only valid configurations can be written on track 2 during phase 1, and some fixed number m of steps is always performed during phase 2, where $m \in \mathcal{O}(n)$. In particular, we can put $m = 2n+4$, because two steps are needed to copy one symbol from track 2 to track 1, and we need two additional steps to modify track 1 to reflect one step of A. We also need additional steps at the beginning and at the end of phase 2.

6.1 Decomposition of Transitions

In this subsection we describe the transformation performed on $\mathcal{T}_A$ more formally and we show that it preserves some important properties of the original LTS.

Let us have an LTS $\mathcal{T} = (S, \mathcal{A}, \longrightarrow)$, a set of actions $\mathcal{A}'$, some positive integer m and a mapping $h : \mathcal{A} \to \mathcal{A}'^*$ such that $h(a)$ is not prefix of $h(a')$ if $a \neq a'$. Let $H = \{h(a) \mid a \in \mathcal{A}\}$ and let $\mathit{Pref}(H)$ be the set of all prefixes of words from H.

We can construct a new LTS $\mathcal{T}' = (S', \mathcal{A}', \longrightarrow')$ where $S' = \{\langle s, w\rangle \mid s \in S, w \in \mathit{Pref}(H)\} \cup \{\langle s, i\rangle \mid s \in S, 0 \leq i < m\}$, where we identify the the states $\langle s, 0\rangle$ and $\langle s, \varepsilon\rangle$ (i.e., $\langle s, 0\rangle$ and $\langle s, \varepsilon\rangle$ are the same state), and where $\longrightarrow'$ contains transitions:

- $\langle s, w\rangle \stackrel{a}{\longrightarrow} \langle s, wa\rangle$ for each $s \in S$, $w \in \mathcal{A}'^*$ and $a \in \mathcal{A}'$ such that $wa \in \mathit{Pref}(H)$,
- $\langle s, w\rangle \stackrel{\tau}{\longrightarrow} \langle s', m-1\rangle$ for each $s, s' \in S$ and $a \in \mathcal{A}$ such that $s \stackrel{a}{\longrightarrow} s'$ and $h(a) = w$,
- $\langle s, i\rangle \stackrel{\tau}{\longrightarrow} \langle s, i-1\rangle$ for each $s \in S$ and $0 < i < m$.

For each state $s \in S$ in $\mathcal{T}$ there is a corresponding state $\langle s, \varepsilon\rangle \in S'$ in $\mathcal{T}'$.

Lemma 6. *For each $s, s' \in S$: if $s \sim s'$, then $\langle s, \varepsilon\rangle \sim \langle s', \varepsilon\rangle$, and if $s \not\sqsubseteq_{tr} s'$, then $\langle s, \varepsilon\rangle \not\sqsubseteq_{tr} \langle s', \varepsilon\rangle$.*

Proof (sketch). To prove the first part of the lemma, it is sufficient to show that $R = \{(\langle s, w\rangle, \langle t, w\rangle) \mid s \sim t, w \in \mathit{Pref}(H)\} \cup \{(\langle s, i\rangle, \langle t, i\rangle) \mid s \sim t, 0 < i < m\}$ is a bisimulation.

To prove the second part, let us define a mapping $\hat{h} : \mathcal{A}^* \to \mathcal{A}'^*$ such that $\hat{h}(\varepsilon) = \varepsilon$, and $\hat{h}(aw) = h(a)\tau^m\hat{h}(w)$. By induction on $|w|$ we can show that for every $s \in S$ and $w \in \mathcal{A}^*$ is $w \in \mathit{Tr}(s)$ iff $\hat{h}(w) \in \mathit{Tr}(\langle s, \varepsilon\rangle)$.

If $s \not\sqsubseteq_{tr} s'$ then there is some $w \in \mathcal{A}^*$ such that $w \in \mathit{Tr}(s)$ and $w \notin \mathit{Tr}(s')$. This implies that $\hat{h}(w) \in \mathit{Tr}(\langle s, \varepsilon\rangle)$ and $\hat{h}(w) \notin \mathit{Tr}(\langle s', \varepsilon\rangle)$. □

6.2 Correctness of the Construction of the RLBA

Theorem 7. *The problem* RLBA-EQ$_{\mathcal{R}}$ *is EXPTIME-hard for any $\mathcal{R}$ such that $\sim \subseteq \mathcal{R} \subseteq \sqsubseteq_{tr}$.*

Proof. Let us return to the construction of B and consider the corresponding $\mathcal{T}_A$. We define a mapping $h : \mathcal{A}_A \to \mathcal{A}_B^*$ such that $h(\langle\alpha\rangle) = \tau a_1 \tau\tau a_2 \tau \cdots \tau a_n \tau$ for $\alpha \in \mathit{Conf}_\wedge$, and $h(\langle\alpha, i\rangle) = \tau a_1 \tau\tau a_2 \tau \ldots \tau a_n \tau i$ for $\alpha \in \mathit{Conf}_\vee - \mathit{Conf}_{rej}$, where $\mathit{desc}(\alpha) = a_1 a_2 \cdots a_n$. We apply the transformation described in the previous subsection with h and $m = 2n + 4$ to $\mathcal{T}_A$, and we obtain $\mathcal{T}_A'$. It is straightforward to create a bisimulation that relates configurations of B and states of $\mathcal{T}_A'$.

States α_{rej} and α_0 from $\mathcal{T}_A$ correspond to a rejecting, resp. initial, configuration of A. If A accepts w, then $\alpha_{rej} \not\sqsubseteq_{tr} \alpha_0$ in $\mathcal{T}_A$, and so $\langle\alpha_{rej}, \varepsilon\rangle \not\sqsubseteq_{tr} \langle\alpha_0, \varepsilon\rangle$ in $\mathcal{T}_A'$, and if A does not accept w, then $\alpha_{rej} \sim \alpha_0$ in $\mathcal{T}_A$ and $\langle\alpha_{rej}, \varepsilon\rangle \sim \langle\alpha_0, \varepsilon\rangle$ in $\mathcal{T}_A'$ by Lemma 6. The same holds for the corresponding configurations of B. This shows that the described construction is correct.

RLBA B with two configurations can be constructed from an instance of ALBA-ACCEPT in a logarithmic space, since it is obvious that some fixed number of pointers pointing to symbols in the instance would be sufficient for the construction. The problem ALBA-ACCEPT is *EXPTIME*-complete and *EXPTIME* is closed under complement. □

So from Theorem 7 and Lemmas 1 and 2 we obtain the main result of the paper:

Theorem 8. *The problems* PCH-EQ$_{\mathcal{R}}$ *and* PN-EQ$_{\mathcal{R}}$ *are EXPTIME-hard for any* $\mathcal{R}$ *such that* $\sim \subseteq \mathcal{R} \subseteq \sqsubseteq_{tr}$.

Acknowledgements. I would like to thank P. Schnoebelen for drawing my attention to equivalence checking of non-flat systems, and P. Jančar for fruitful discussions and comments.

References

1. A.K. Chandra, D.C. Kozen, and L.J. Stockmeyer. Alternation. *Journal of the Association for Computing Machinery*, 28(1):114–133, January 1981.
2. N. Immerman. *Descriptive Complexity*, pages 53–54. Springer-Verlag, 1998.
3. P. Jančar. Nonprimitive recursive complexity and undecidability for petri net equivalences. *Theoretical Computer Science*, 256:23–30, 2001.
4. L. Jategaonkar and A. R. Meyer. Deciding true concurrency equivalences on safe, finite nets. *Theoretical Computer Science*, 154(1):107–143, January 1996.
5. F. Laroussinie and P. Schnoebelen. The state explosion problem from trace to bisimulation equivalence. In *Proc. FOSSACS'2000 (Berlin, Germany, Mar.-Apr. 2000)*, volume 1784, pages 192–207. Springer, 2000.
6. A. Rabinovich. Complexity of equivalence problems for concurrent systems of finite agents. *Information and Computation*, 139(2):111–129, 15 December 1997.
7. Z. Sawa and P. Jančar. *P*-hardness of equivalence testing on finite-state processes. In *Proc. SOFSEM 2001 (Piestany, Slovak Rep., November 2001)*, volume 2234 of *Lecture Notes in Computer Science*, page 326. Springer, 2001.
8. S. K. Shukla, H. B. Hunt, D. J. Rosenkrantz, and R. E. Stearns. On the complexity of relational problems for finite state processes. In *Proc. ICALP'96 (Paderborn, Germany)*, volume 1099 of *Lecture Notes in Computer Science*, pages 466–477. Springer-Verlag, 1996.
9. A. Valmari and A. Kervinen. Alphabet-based synchronisation is exponentially cheaper. In *Proc. CONCUR 2002*, volume 2421 of *Lecture Notes in Computer Science*, page 161. Springer, 2002.
10. R.J. van Glabbeek. The Linear Time - Branching Time Spectrum. In J.C.M. Baeten and J.W. Klop, editors, *Proceedings of CONCUR '90, Theories of Concurrency: Unification and Extension*, volume 458 of *Lecture Notes in Computer Science*, pages 278–297. Springer-Verlag, Berlin, 1990.

Model Checking a Path
(Preliminary Report)

N. Markey[1,2] and P. Schnoebelen[2]

[1] Lab. d'Informatique Fondamentale d'Orléans
Univ. Orléans & CNRS FRE 2490
markey@lifo.univ-orleans.fr
[2] Lab. Spécification & Vérification
ENS de Cachan & CNRS UMR 8643
markey|phs@lsv.ens-cachan.fr

Abstract. We consider the problem of checking whether a finite (or ultimately periodic) run satisfies a temporal logic formula. This problem is at the heart of "runtime verification" but it also appears in many other situations. By considering several extended temporal logics, we show that the problem of model checking a path can usually be solved efficiently, and profit from specialized algorithms. We further show it is possible to efficiently check paths given in compressed form.

1 Introduction

Model checking, introduced in the early 80's, has now become a widely used approach to the verification of all kinds of systems [CGP99,BBF$^+$01]. The name "model checking" covers a variety of techniques dealing with various subproblems: how to model systems by some kind of Kripke structures?, how to express properties in temporal logics or some other formalisms?, how to use symbolic techniques for dealing with large state spaces?, and, most importantly, how to algorithmically check that a model satisfies a property?

These techniques rest upon a solid body of foundational knowledge regarding the expressive power of temporal logics and the computational complexity of their model checking problems [Sch03].

In this paper, we consider the problem of *model checking a single path.* This problem appears in several situations, most notably in *runtime verification* [Dru00,Hav00,FS01]. There are situations where thousands of paths are checked one by one, e.g. the Monte-Carlo approach for assessing the probability that a random run satisfies some property [YS02,LP02]. Less standard situations exist: [RG01] advocates using temporal logic for describing patterns of intrusive behaviors recorded in log files. Such a log file, where a series of system events are recorded, is just a long path on which the temporal formula will be evaluated.

We do not restrict to finite paths and also consider checking ultimately periodic paths (given as finite "lasso-shaped" loop). Checking a path is much simpler

R. Amadio, D. Lugiez (Eds.): CONCUR 2003, LNCS 2761, pp. 251–265, 2003.

than checking a Kripke structure, so much so that the problem may appear trivial: using standard dynamic programming methods "*à la CTL* model checking", a path can obviously be checked in bilinear, i.e. $O(|\text{model}| \times |\text{formula}|)$, time.

This may explain why the problem, while ubiquitous, has not been isolated and studied from a theoretical viewpoint. For example, it is not known whether checking a simple temporal formula over a finite path can be done more efficiently than with the "bilinear time" method , e.g. with memory-efficient algorithms in SC, or with fast parallel algorithms in NC. Indeed this open problem was only identified recently [DS02].

With this paper, we aim to show that the problem is worthy of more fundamental investigations. Of course, the problem is a generic one, with many variants (which temporal logic? what kind of paths?) and here we only start scratching its surface.

More specifically, we present results (some of them folklore) showing that

Checking a Path is Easier: As we show in this paper, model checking a path is often much easier than checking a Kripke structure. We exhibit examples of richly expressive temporal logics that allow polynomial-time algorithms for checking a single path, while checking all paths of a Kripke structure is highly untractable. It is even possible to achieve polynomial-time when checking compressed paths (i.e. exponentially long paths that are given and stored in compressed form).

Checking a Path Relies on Specific Techniques: These efficient algorithms rely on specific aspects of the problem. Checking a path definitely comes with its own set of notions, technical tricks, and conceptual tools. For example, all our algorithms for checking ultimately periodic paths rely on a specific reduction technique to checking some kind of short finite prefix of the infinite path.

Outline of the Paper. We define the basic problem of model checking *LTL* formulae over finite or ultimately periodic paths (Sect. 2). This problem is still not satisfactorily solved, but we argue that its intrinsic difficulty is already present in the case of finite paths (Sect. 3). We then show that model checking a path is much easier than model checking a Kripke structure by looking at various rich temporal logics: the monadic first-order logic of order, the extension of *LTL* with *Chop*, or the extension of *LTL* with forgettable past (Sect. 4). We provide polynomial-time algorithms for the last two instances. Finally we look at the problem of checking paths given in compressed form (Sect. 5).

Related Works. Model checking a path is a central problem in runtime verification. In this area, the problem is seen through some specific practical applications, sometimes with an emphasis on online algorithms, with the result that the fundamental complexity analysis has not received enough attention.

Dynamic programming algorithms for checking finite and ultimately periodic paths are also used in *bounded model checking* [BCC+03]. In this area, the relevant measures for efficiency are not the classical notions of running time and memory space, but have more to do with, say, the number of Boolean variables introduced by the algorithm, or the pattern of dependencies between them.

Model checking a path has a lot in common with algorithmics on words (witness Section 5). However, our concern with temporal logics and ultimately periodic paths is not standard in that other area.

2 Linear-Time Temporal Logic and Paths

We assume familiarity with temporal logics (mainly *LTL*) and model checking: see [Eme90,CGP99,BBF+01].

Syntax of LTL + Past. Let $AP = \{p_0, p_1, p_2, \dots\}$ be a countably infinite set of *atomic propositions*. The formulae of *LTL + Past*, are given by the following grammar:

$$\varphi, \psi ::= \neg\varphi \mid \varphi \wedge \psi \mid \mathsf{X}\,\varphi \mid \mathsf{X}^{-1}\,\varphi \mid \varphi\,\mathsf{U}\,\psi \mid \varphi\,\mathsf{S}\,\psi \mid p_0 \mid p_1 \mid p_2 \mid \cdots$$

S (*Since*) and X^{-1} (*Previously*) are the past-time mirrors of the well-known U (*Until*) and X (*Next*). We shall freely use the standard abbreviations $\top$, $\varphi \Rightarrow \psi$, $\varphi \vee \psi$, $\mathsf{F}\,\varphi$ ($\overset{\text{def}}{\Leftrightarrow} \top\,\mathsf{U}\,\varphi$), $\mathsf{G}\,\varphi$ ($\overset{\text{def}}{\Leftrightarrow} \neg\mathsf{F}\neg\varphi$), $\mathsf{F}^{-1}\varphi$ ($\overset{\text{def}}{\Leftrightarrow} \top\,\mathsf{S}\,\varphi$) and $\mathsf{G}^{-1}\varphi$ ($\overset{\text{def}}{\Leftrightarrow} \neg\mathsf{F}^{-1}\neg\varphi$).

LTL, the well-known *propositional linear-time temporal logic*, is the fragment where S and X^{-1} are not used, also called the *pure-future* fragment. While *LTL+Past* is not more expressive than *LTL* [GPSS80,Rab02], and not harder to verify [SC85], it can be (at least) exponentially more succinct than *LTL* [LMS02].

Semantics. Linear-time formulae are evaluated along paths. Formally, a *path* is a sequence $\pi = s_0, s_1, \dots$, finite or infinite, of states, where a *state* is a valuation $s \in 2^{AP}$ of the atomic propositions. $|\pi| \in \mathbb{N} \cup \{\omega\}$ denotes the length of π and, for a position $l < |\pi|$, one defines when a formula holds at position i of $\pi = (s_l)_{l<|\pi|}$ by induction on the structure of formulae:

$$\begin{array}{lll}
\pi, i \models p & \text{iff } p \in s_i & \text{for } p \in AP \\
\pi, i \models \mathsf{X}\,\varphi & \text{iff } \pi, i+1 \models \varphi & (\text{hence } i+1 < |\pi|) \\
\pi, i \models \mathsf{X}^{-1}\varphi & \text{iff } \pi, i-1 \models \varphi & (\text{hence } i > 0) \\
\pi, i \models \varphi\,\mathsf{U}\,\psi & \text{iff } \exists j \geq i : \begin{pmatrix} \pi, j \models \psi, \text{ and} \\ \forall i \leq k < j : \pi, k \models \varphi \end{pmatrix} & (\text{hence } j < |\pi|) \\
\pi, i \models \varphi\,\mathsf{S}\,\psi & \text{iff } \exists j \leq i : \begin{pmatrix} \pi, j \models \psi, \text{ and} \\ \forall j < k \leq i : \pi, k \models \varphi \end{pmatrix} & (\text{hence } j \geq 0)
\end{array}$$

omitting the usual clauses for negation and conjunction. We say a non-empty path π satisfies φ, written $\pi \models \varphi$, when $\pi, 0 \models \varphi$, i.e. when φ holds at the beginning of π.

Since only propositions that appear in φ are relevant for deciding whether $\pi \models \varphi$, we usually assume that paths only carry valuations for the finite number of propositions that will be used later on them.

Model Checking. We are interested in the computational problem of model checking a path against an *LTL* formula. This requires that the path argument be given in some finite way. In classical model checking, where we evaluate a temporal formula along all the paths of a finite Kripke structure (KS), the KS is the finite input that describe an infinite set of infinite paths. Here we assume that the given path is *finite*, or is *ultimately periodic.*

Ultimately periodic, or *u.p.*, paths, are given via a pair (u, v) of two finite paths, called a *loop* for short. A loop (u, v) denotes the infinite path $\pi = u.v^\omega$, called its *unfolding*, where an initial u prefix is followed by repeated copies of v. For uniformity, we shall assume finite paths are given via loops too, only they have empty v. We say loop (u, v) *has type* (m, p) when m is the length $|u|$ of u and p is the length of v.

Model Checking a Path. The generic computational problem we are considering is:

> **PMC**(L) **(Path Model Checking for** L**).**
> **Input:** two finite paths u, v and a temporal formula φ of L.
> **Output:** yes iff $u.v^\omega \models \varphi$, no otherwise.

Here L can be any temporal logic (but it is not meaningful to consider branching-time logics). We shall consider several problems: **PMC**(*LTL*), **PMC**(*LTL* + *Past*), etc. We denote by $\mathbf{PMC_f}(L)$ the restricted problem when only finite paths are considered (*i.e.* when $v = \varepsilon$). A recurring pattern in our results is that **PMC**(L) reduces to $\mathbf{PMC_f}(L)$ (by default, we consider logspace reductions).

3 How Efficient Can Path Model Checking Be?

We mentioned in the introduction that the following holds:

Theorem 3.1. **PMC**(*LTL*) *can be solved in time* $O(|uv| \times |\varphi|)$.

Proof. Obvious since, over paths, *CTL* and *LTL* coincide. So that the well-known bilinear algorithm for *CTL* model checking can be used. □

That polynomial-time algorithms also exist for *LTL* + *Past* is less obvious:

Theorem 3.2. **PMC**(*LTL* + *Past*) *can be solved in time* $O(|uv| \times |\varphi|^2)$.

This can be obtained as a corollary of Theorem 4.5 but it is instructive to look at a direct proof, since it illustrates a recurring pattern.

We start with the simpler case where the path is finite:

Proposition 3.3. $\mathbf{PMC_f}$(*LTL* + *Past*) *can be solved in time* $O(|u| \times |\varphi|)$.

Proof (Sketch). The obvious dynamic programming algorithm works: starting from the innermost subformulae, we recursively fill a Boolean table $T[i, \psi]$, where i is a position in the finite path, and ψ is a subformula of φ, in such a way that $T[i, \psi] = \top$ iff $\pi, i \models \psi$. □

This algorithm is too naive for u.p. paths: one cannot label uniformly states inside the loop. A state in the loop corresponds to different positions in the unfolded path, *and these positions have different pasts.*

However it is only necessary to unfold the loop a small number of times, something we present as a reduction from **PMC**(*LTL*+ *Past*) to $\mathbf{PMC}_\mathrm{f}$(*LTL*+ *Past*).

Let φ be an *LTL*+*Past* formula, and (u, v) a loop of type (m, p). In the sequel, we write $h_F(\varphi)$ for the *future temporal height* of φ, i.e. its maximum number of nested future-time modalities. Similarly, $h_P(\varphi)$ denotes the *past temporal height* of φ. We write $H(\varphi)$ for $h_F(\varphi) + h_P(\varphi)$. E.g., for $\varphi = \mathsf{F}\mathsf{F}^{-1}\mathsf{X}\mathsf{F}p_1 \vee \mathsf{F}^{-1}\mathsf{G}^{-1}p_2$, we have $h_F(\varphi) = 3$, $h_P(\varphi) = 2$ and $H(\varphi) = 5$.

Lemma 3.4 ([Mar02]). *For all subformulas ψ of φ, and $k \geq m + h_P(\varphi)p$*

$$u.v^\omega, k \models \psi \text{ iff } u.v^\omega, k + p \models \psi. \tag{1}$$

This may be proved by structural induction on formula ψ.

We now reduce model checking of φ on the loop (u, v) to a finite path model checking problem. We assume that $p \neq 0$ (otherwise the result is obvious), and build the finite path $\pi' = uv'^{H(\varphi)+1}$ where v' is like v except that v'_0 carries a new proposition $q \notin AP$ (and we replace AP by $AP' = AP \cup \{q\}$). Formally, v' is given by:

$$|v'| = p \qquad v'_0 \stackrel{\text{def}}{=} v_0 \cup \{q\} \qquad v'_i \stackrel{\text{def}}{=} v_i \quad \text{for } i > 0.$$

We also recursively build a set of formulae χ_k as follows:

$$\chi_0 \stackrel{\text{def}}{=} \top \qquad \chi_k \stackrel{\text{def}}{=} \mathsf{F}(q \wedge \mathsf{X}\chi_{k-1}).$$

Obviously, $\pi', i \models \chi_k$ iff $i \leq m + (H(\varphi)+1-k)p$. Now $\overline{\varphi}$ is inductively given by:

$$\begin{aligned} \overline{p} &= p & & \\ \overline{\neg\psi} &= \neg\overline{\psi} & \overline{\psi_1 \vee \psi_2} &= \overline{\psi_1} \vee \overline{\psi_2} \\ \overline{\mathsf{X}\psi} &= \mathsf{X}\overline{\psi} & \overline{\psi_1 \mathsf{U} \psi_2} &= \overline{\psi_1}\mathsf{U}(\overline{\psi_2} \wedge \chi_{h_F(\psi_1 \mathsf{U} \psi_2)}) \\ \overline{\mathsf{X}^{-1}\psi} &= \mathsf{X}^{-1}\overline{\psi} & \overline{\psi_1 \mathsf{S} \psi_2} &= \overline{\psi_1}\mathsf{S}\overline{\psi_2} \end{aligned}$$

Lemma 3.5. *For all subformulae ψ of φ, for all $i < m + (H(\varphi) - h_F(\psi))p$, we have:*

$$u.v^\omega, i \models \psi \text{ iff } \pi', i \models \overline{\psi}.$$

A direct corollary is the reduction we announced:

Theorem 3.6. *For any LTL + Past formula φ, and loop (u, v), one can build in logspace a formula $\overline{\varphi}$ and a finite path π' s.t.*

$$u.v^\omega \models \varphi \quad \text{iff} \quad \pi' \models \overline{\varphi}.$$

Since $|\pi'|$ is in $O(|uv||\varphi|)$, we obtain Theorem 3.2 by combining with Proposition 3.3.

An Open Question and a Conjecture. We do not know whether the upper bounds given in Theorems 3.1 and 3.2 are tight. There is an obvious NC_1 lower bound that has nothing to do with model checking: evaluating a Boolean expression is NC_1-hard. But the gap between NC_1 and PTIME is (assumed to be) quite large.

We have been unable to prove even LOGSPACE-hardness for $\mathbf{PMC}(\mathit{LTL} + \mathit{Past})$, or to find an algorithm for $\mathbf{PMC}(\mathit{LTL})$ (even for $\mathbf{PMC}_\mathrm{f}(L(\mathsf{F}))$ that would be memory efficient (e.g. requiring only polylog-space) or that would be considered as an efficient parallel algorithm (e.g. in NC).

We consider that the open question of assessing the precise complexity of $\mathbf{PMC}(\mathit{LTL})$ and $\mathbf{PMC}(\mathit{LTL} + \mathit{Past})$ is one of the important open problems in model checking [DS02, Sect. 4]. In view of how Theorem 3.6 reduces $\mathbf{PMC}(\dots)$ to $\mathbf{PMC}_\mathrm{f}(\dots)$, one thing we can tell about the open problem is that the difficulty does not come from allowing u.p. paths.

Our conjecture is that $\mathbf{PMC}(\mathit{LTL})$ is not PTIME-hard. This conviction is grounded in our experience with all the PTIME-hardness proofs we can obtain for richer logics (see next sections) and the way they always exploit some powerful trick or other that *LTL* and *LTL* + *Past* do not allow.

4 Richly Expressive Temporal Logics

Many temporal logics are more expressive, or more succinctly expressive, than *LTL* [Eme90,Rab02]. However this increased expressivity usually comes with an increased cost for verification, which explains why they are not so popular in the model checking community.

In this section we consider a few such temporal logics. Since we focus on logics with first-order definable modalities, the "rich expressiveness" should be understood as "succinct expressiveness".

Our first example is *FOMLO*, the first-order logic of order with monadic predicates. This formalism is not really a *modal* logic, like *LTL* is, but it is fundamental since it encompasses all natural temporal logics. We show that model checking *FOMLO* on paths is PSPACE-complete, hence is much easier on paths than on Kripke structures (where it is nonelementary [Sto74]).

We then look at two more specific extensions of *LTL*. *LTL* + *Chop* has a non-elementary model checking problem on Kripke structures [RP86], but we show it leads to a PTIME-complete path model checking problem. Another PTIME-complete problem on paths appears with *LTL* + *Past* + *Now*, an extension of *LTL* + *Past* that has an EXPSPACE-complete model checking problem on Kripke structures [LMS02].

Figure 1 summarizes our results. The obvious conclusion is that, when model checking paths, there is no reason to restrict oneself to *LTL*: much more expressive formalisms can be handled at (more or less) no additional price.

4.1 *FOMLO*, the First-Order Monadic Logic of Order

We will not recall here all the basic definitions and notations for *FOMLO*. Let us simply say that we use $qh(\varphi)$ to denote the *quantifier-height* of φ, that we write

logic	checking Kripke structures	checking paths
FOMLO	nonelementary	PSPACE-complete
LTL	PSPACE-complete	PTIME-easy
LTL + Past	PSPACE-complete	PTIME-easy
LTL + Past + Now	EXPSPACE-complete	PTIME-complete
LTL + Chop	nonelementary	PTIME-complete

Fig. 1. Checking richly expressive logics on paths

$\varphi(x_1, \ldots, x_n)$ to stress that the free variables in φ are among $\{x_1, \ldots, x_n\}$, and that $\pi, a_1, \ldots, a_n \models \varphi(x_1, \ldots, x_n)$ denotes that path π with selected positions $a_1, \ldots, a_n \in \mathbb{N}$ is a model of $\varphi(x_1, \ldots, x_n)$ when the x_i's are interpreted by the a_i's.

Theorem 4.1. PMC(*FOMLO*) *is* PSPACE-*complete.*

PSPACE-hardness already occurs with finite paths of length two where there is an obvious reduction from Quantified Boolean Formula (QBF).

Proving membership in PSPACE is more involved. But if we restrict to finite paths, there is no difficulty since the naive evaluation of first-order formulae over finite first-order structures only needs polynomial-space [CM77]. Therefore, the difficult part in Theorem 4.1 is the proof that model checking *FOMLO* formulae over *ultimately periodic paths* $u.v^\omega$ can be done in polynomial-space.

We now prove

Proposition 4.2. *Telling whether* $u.v^\omega \models \varphi$ *for* φ *a closed FOMLO formula can be done in space* $O(|uv| \times |\varphi|^2)$.

Assume $u.v^\omega$ is an u.p. path of type (m, p) with $p > 0$. We say two positions $a, b \in \mathbb{N}$ are *congruent*, written $a \equiv b$, if $a = b$, or $a \geq m \leq b$ and $a \mod p = b \mod p$ (i.e. they point to equal valuations on $u.v^\omega$). Two tuples $\langle a_1, \ldots, a_n \rangle$ and $\langle b_1, \ldots, b_n \rangle$ of natural numbers are *k-equivalent*, written $\langle a_1, \ldots, a_n \rangle \sim_k \langle b_1, \ldots, b_n \rangle$, when $a_i \equiv b_i$ for all $1 \leq i \leq n$ and $(a_i - a_j \neq b_i - b_j) \Rightarrow |a_i - a_j| \geq 2^k p$ for all $1 \leq i \leq j, \leq n$.

Lemma 4.3. *If* $\langle m, a_1, \ldots, a_n \rangle \sim_k \langle m, b_1, \ldots, b_n \rangle$ *and* $qh(\varphi) \leq k$, *then*

$$u.v^\omega, a_1, \ldots, a_n \models \varphi(x_1, \ldots, x_n) \textit{ iff } u.v^\omega, b_1, \ldots, b_n \models \varphi(x_1, \ldots, x_n).$$

Proof (Idea). A standard use of Ehrenfeucht-Fraïssé games on linear orderings [Ros82]. □

For a closed *FOMLO* formula φ we let $\widetilde{\varphi}$ be the *relativized variant* obtained from φ by replacing every quantification "$\exists x$" in φ by "$\exists x < m + p(2^k - 2^{h-1})$ where k is $qh(\varphi)$ and h is the height of the "$\exists x$" occurrence we replace.

For example, assuming $m = 10$ and $p = 3$, the formula

$$\exists x\, \forall y\, (x > y \wedge p_0(y) \Rightarrow \exists z\, (y < z < x \wedge p_1(z))) \qquad (\varphi)$$

is relativized as

$$\exists x < 22\, \forall y < 28\, (x > y \wedge p_0(y) \Rightarrow \exists z < 31\, (y < z < x \wedge p_1(z))). \qquad (\widetilde{\varphi})$$

A corollary of Lemma 4.3 is the following

Lemma 4.4. *$u.v^\omega \models \varphi$ iff $u.v^\omega \models \widetilde{\varphi}$.*

We can now evaluate whether $u.v^\omega \models \varphi$ in polynomial-space, proving Proposition 4.2. Lemma 4.4 reduces this question to a bounded problem, where only a finite prefix of $u.v^\omega$ has to be examined. That prefix still has exponential size $O(m + p2^{qh(\varphi)})$ but we do not have to build it. Rather, we only go over all values for the position variables in $\widetilde{\varphi}$, storing them in binary notation (say) to ensure polynomial-space. Then it is easy to evaluate the predicates on these binary notations: the only dyadic predicate is $<$, and the monadic predicates reduces to simple arithmetical computations to find a congruent position in $u.v$.

4.2 *LTL* with Forgettable Past

"*LTL* with forgettable past" is *LTL* + *Past* + *Now*, i.e. *LTL* + *Past* where we add a new unary modality N (for "*from Now on*"). The semantics of N is given by

$$\pi, i \models \mathsf{N}\, \varphi \text{ iff } \pi_{\geq i}, 0 \models \varphi.$$

We refer to [LMS02] for motivations on *LTL* + *Past* + *Now*: that logic can be exponentially more succinct than *LTL* + *Past*, and its model checking problem is EXPSPACE-complete. *LTL* + *Past* + *Now* is included in Fig. 1 because Theorem 4.5 was the first hint that **PMC** allows dealing efficiently with rich logics.

Theorem 4.5 ([LMS02]). **PMC**(*LTL* + *Past* + *Now*) *is* PTIME-*complete.*

4.3 The *Chop* Modality

"*Chop*", introduced in [HKP80] and studied in [RP86], is a two-place modality whose semantics is defined as follows:

$$\pi, 0 \models \varphi\, \mathsf{C}\, \psi \ \text{ iff } \ \exists k \geq 0 \text{ s.t. } \pi_{\geq k+1} \models \psi \text{ and } \pi^{\leq k} \models \varphi$$

where $\pi_{\geq k+1}$ is the suffix of π starting at (and including) position $k+1$, and $\pi^{\leq k}$ is the prefix of π up to (and including) position k. It is useful in cases we want to see subruns inside a run (e.g. sessions, or specific segments) and state their temporal specifications [RP86].

Satisfiability for *LTL* + *Chop* is non elementary [RP86]. Hence model checking *LTL* + *Chop* properties on Kripke structures is non elementary too (there exists a polynomial-space reduction from satisfiability to model checking, see [DS02, Prop. 3.1]).

However, model checking a path is easier:

Proposition 4.6. **PMC**$(LTL + Chop)$ *can be solved in time* $O(|uv|^2 \times |\varphi|^3)$.

The outline of the proof is similar to the case of $LTL+Past$. First, we observe that, for a finite path π, the following holds:

Proposition 4.7. $\mathbf{PMC}_\mathrm{f}(LTL + Chop)$ *can be solved in time* $O(|\pi|^2 |\varphi|)$.

Proof (Sketch). Again dynamic programming techniques suffice. We fill a Boolean table $T[i, j, \psi]$, for each positions $i \leq j$ in π, and subformula ψ of φ, in such a way that

$$T[i, j, \psi] = \top \text{ iff } \pi_{[i,j]} \models \psi$$

where $\pi_{[i,j]} = (\pi_{\leq j})_{\geq i}$. This can be done in quadratic time in the size of the path, and linear time in the size of the formula. □

We now consider u.p. paths. The next lemma states that some transformations on paths do not affect the truth value of $LTL+ Chop$ formulae:

Lemma 4.8. *Let* $\varphi \in LTL+ Chop$, *and* $m, n \geq |\varphi|$. *Let* u, v *be two finite paths, and* w *be a (finite or infinite) path. Then*

$$u.v^m.w \models \varphi \textit{ iff } u.v^n.w \models \varphi.$$

We now perform the reduction from **PMC**$(LTL + Chop)$ to $\mathbf{PMC}_\mathrm{f}(LTL + Chop)$. We first exclude the trivial case when v is empty. We keep the notations of the proof of Theorem 3.6, and define new path and formulae: $\pi' = uv'^{|\varphi|}$ and

$$\overline{\psi_1 \mathsf{U} \psi_2} = \overline{\psi_1} \mathsf{U}(\overline{\psi_2} \wedge \chi_{|\psi_1 \mathsf{U} \psi_2|}) \qquad \overline{\psi_1 \mathsf{C} \psi_2} = \psi_1 \mathsf{C}(\overline{\psi_2} \wedge \chi_{|\psi_2|})$$

For this path π', we now have $\pi', i \models \chi_k$ iff $i \leq m + (|\varphi| - k)p$.

With Lemma 4.8 we can prove the following by induction on the structure of ψ:

Lemma 4.9. *For all subformula* ψ *of* φ, *for all* $i < m + (|\varphi| - |\psi|)p$, *we have:*

$$u.v^\omega, i \models \psi \textit{ iff } \pi', i \models \overline{\psi}.$$

It now suffices to observe that $|\pi'|$ is in $O(|uv| \times |\varphi|)$, and we obtain Proposition 4.6 from Proposition 4.7.

It turns out that, as with $LTL + Past + Now$, we have a case where model checking a path is PTIME-complete:

Theorem 4.10. **PMC**$(LTL + Chop)$ *is* PTIME-*complete.*

In fact, PTIME-hardness already occurs with finite paths, i.e. for $\mathbf{PMC}_\mathrm{f}(LTL+ Chop)$. We prove this by a reduction from the (Synchronous Alternating Monotone) CIRCUIT-VALUE Problem [GHR95, problem A.1.6]. We illustrate the reduction on a example and consider the circuit C of figure 2.

Let E_C denote the set of links (pairs of gates) in C. With C we associate the finite path π_C given in figure 3.

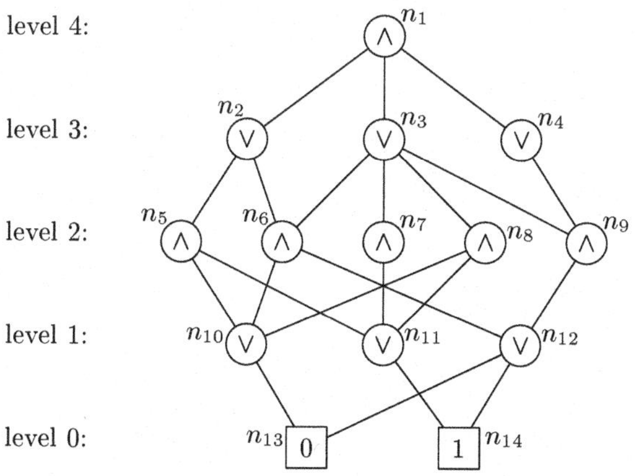

Fig. 2. An instance of CIRCUIT-VALUE

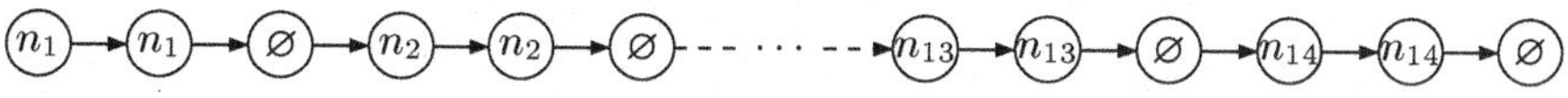

Fig. 3. The path π_C associated to our instance of CIRCUIT-VALUE

Finally, we define the following sequence of formulae:

$$\varphi_0 \stackrel{\text{def}}{\equiv} n_{14} \qquad \begin{aligned} \varphi_{2k+1} &\stackrel{\text{def}}{\equiv} \Big[\bigvee\nolimits_{\substack{i \text{ at level } 2k+1 \\ (i,j)\in E_C}} i \wedge \mathsf{F}\mathsf{G} j\Big] \,\mathsf{C}\, (\mathsf{X}\varnothing \wedge \varphi_{2k}) \\ \varphi_{2k+2} &\stackrel{\text{def}}{\equiv} \neg\Big(\Big[\bigvee\nolimits_{\substack{i \text{ at level } 2k+2 \\ (i,j)\in E_C}} i \wedge \mathsf{F}\mathsf{G} j\Big] \,\mathsf{C}\, (\mathsf{X}\varnothing \wedge \neg\varphi_{2k+1})\Big) \end{aligned}$$

Lemma 4.11. *For any gate n_i at level p in $\mathcal{C}$, n_i evaluates to* `true` *in $\mathcal{C}$ iff $\pi_{\mathcal{C}}, (3i-2) \models \varphi_p$.*

Proof (Idea). By induction on p. The base case where $p = 0$ is obvious. For the induction step we first consider the case where $p = 2k+1$ is odd. Hence n_i is a disjunctive gate. The right-hand side of φ_{2k+1} requires that we "chop" the path between two nodes labeled by a same n_j, and that this node satisfies the formula corresponding to the level below (hence gate n_j evaluates to `true` by ind. hyp.). The left-hand side of φ_{2k+1} ensures that gate n_j is a child of the current n_i (a finite path satisfies some $\mathsf{FG}\psi$ iff its last state satisfies ψ). Thus $\pi_{\mathcal{C}}, 3i-2 \models \varphi_p$ iff a child of n_i evaluates to `true` iff n_i evaluates to `true`.

When p is even, a dual reasoning applies. □

Thus $\mathcal{C}$ evaluates to `true` iff $\pi_{\mathcal{C}}, 1 \models \varphi_l$, where l is the height of $\mathcal{C}$. Hence PTIME-hardness.

5 Model Checking Compressed Paths

In this section we show that it is possible to efficiently model check paths that are given in compressed form, i.e. via some succinct encoding. One such encoding is the *exponent notation*, e.g. writing paths as

$$u = ([(s_1.s_2.s_3)^3(s_4)^{10}]^{12}(s_5.s_6)^7)^{1000}.$$

5.1 Compressed Words

Working directly on compressed words is a standard technique in some fields, e.g. when handling long DNA strings in gene-mapping applications. Several encoding schemes are possible, and the more interesting ones are those where a compressed word can be exponentially more succinct than the described word.

Here we follow [PR99] and adopt the standard framework of *straight-line programs*, or SLP's: these are context-free grammars where the non-terminals $N_1, \ldots, N_k$ are ordered (N_1 being the axiom), and where every non-terminal has a single production of the form $N_i \longrightarrow a$ for a terminal a, or $N_i \longrightarrow N_j N_k$ for some $j, k > i$ [PR99]. For an SLP P, we write $w(P)$ for the unique word described by P.

SLP's are equivalent (polynomial-time inter-reducible) to Lempel-Ziv compression schemes but are mathematically nicer. They are more general than the exponent notation. The algorithms we give for SLP's easily adapt to these other compression schemes.

5.2 Model Checking Compressed Paths

A *compressed path* is a pair (P_1, P_2) of two SLP's, encoding the u.p. path $w(P_1).w(P_2)^\omega$. Since compressed paths are succinct descriptions, we should expect that model-checking paths given in compressed form is hard. This is indeed the case with *LTL* model checking:

Theorem 5.1. *Model checking LTL formulae on compressed paths is* PSPACE-*complete.*

However, the difficulty has more to do with the *LTL* formulae than with the compressed paths, as our next result shows:

Theorem 5.2. *Checking whether a compressed path is recognized by a Büchi automaton is* PTIME-*complete.*

Checking whether a compressed path (P_1, P_2) satisfies an *LTL* formula φ can be done in time $(|P_1| + |P_2|)2^{O(|\varphi|)}$, hence for long paths and simple fixed formulae, model-checking compressed paths is essentially "linear-time".

The rest of the section proves the above two theorems. We observe the usual pattern: u.p. paths are not harder than finite paths.

	checking paths	checking compressed paths
LTL formulae	PTIME-easy	PSPACE-complete
Finite state automata	NL-complete	PTIME-complete

Fig. 4. Checking compressed paths

5.3 Compressed Paths Accepted by Büchi Automata

Proposition 5.3. [PR99] *Saying whether $w(P)$ is recognized by $\mathcal{A}$ (for P an SLP, and $\mathcal{A}$ a finite-state automaton) can be done in time $O(|P| \times |\mathcal{A}|^3)$.*

Proof. [PR99] describes a simple dynamical programming solution. For two states r, s of $\mathcal{A}$ and non-terminal X_i of P, set $T[r,s,i] = \texttt{true}$ iff $w(X_i)$ labels a path going from r to s in $\mathcal{A}$. Obviously, if $X_i \longrightarrow X_j\, X_k$ is a rule in P, then $T[r,s,i] = \bigvee_u T[r,u,j] \wedge T[u,s,k]$. Hence the table $T[\dots]$ is easy to fill. Then we can use $T[\dots]$ to see whether $w(P)$, i.e. $w(X_1)$, labels an accepting path. □

Corollary 5.4. *Saying whether a compressed path (P_1, P_2) is recognized by $\mathcal{A}$ (a Büchi automaton) can be done in time $O((|P_1| + |P_2|) \times |\mathcal{A}|^3)$.*

Proof (Idea). This is a simple extension of the previous algorithm. For example, one can build a second table $T'[\dots]$ s.t. $T[r,s,i] = \texttt{true}$ iff a power of $w(X_i)$ labels a path going from r to s *and visiting an accepting state of $\mathcal{A}$*. □

Proposition 5.5 ([MS03]). *Saying whether $w(P)$ is recognized by a deterministic finite-state automaton $\mathcal{A}$, is* PTIME*-hard (hence* PTIME*-complete).*

This shows a situation where compressed words are harder than uncompressed words since recognizability by a FSA is NL-complete for uncompressed words.

5.4 Compressed Paths Satisfying *LTL* Formulae

The easy part of Theorem 5.1 is the upper bound:

Proposition 5.6. *Deciding whether a compressed path satisfies an LTL formula can be done in polynomial-space.*

Proof (Idea). Model checking *LTL* formulae on products of concurrent Kripke structures is PSPACE-complete [HKV02], and compressed paths can easily be encoded in such products. □

PSPACE-hardness is more involved. Note it already occurs with finite paths:

Proposition 5.7. *Deciding whether a finite compressed path satisfies an LTL formula is* PSPACE-*hard.*

We now sketch the proof. It is by reduction from Quantified Boolean Formula (QBF). Assume $\mathcal{I}$ is a QBF instance of the form $\exists v_1 \forall v_2 \exists v_3 \ldots \forall x_n (\bigwedge_i \bigvee_j l_{i,j})$ where every $l_{i,j}$ is $\pm_{i,j} v_{n_{i,j}}$, i.e. a Boolean variable from $V = \{v_1, \ldots, v_n\}$ or its negation.

With $\mathcal{I}$ we associate a compressed word that lists all valuations for the V-variables in lexicographical order. This is given by the following SLP:

$$P_{\mathcal{I}} = \begin{cases} N_1 = \mathtt{b}_1.N_2.\mathtt{t}_1.\mathtt{e}_1.\mathtt{b}_1.N_2.\mathtt{f}_1.\mathtt{e}_1 \\ N_2 = \mathtt{b}_2.N_3.\mathtt{t}_2.\mathtt{e}_2.\mathtt{b}_2.N_3.\mathtt{f}_2.\mathtt{e}_2 \\ \ldots \\ N_n = \mathtt{b}_n.\mathtt{t}_n.\mathtt{e}_n.\mathtt{b}_n\mathtt{f}_n.\mathtt{e}_n \end{cases}$$

Here letters $\mathtt{t}_i$ and $\mathtt{f}_i$ state that we v_i is <u>t</u>rue and, resp., <u>f</u>alse. Letters $\mathtt{b}_i$ and $\mathtt{e}_i$ are <u>b</u>egin and <u>e</u>nd markers.

We now encode $\mathcal{I}$ via $\varphi_{\mathcal{I}}$, the following *LTL* formula:

$$[\mathsf{b}_1 \wedge ([\mathsf{b}_2 \Rightarrow ([\mathsf{b}_3 \wedge \ldots [\mathsf{b}_n \Rightarrow \bigwedge_i \bigvee_j \pm_{i,j}(\mathtt{t}_{n_{i,j}} \mathsf{B}\, \mathsf{e}_n)] \mathsf{U}\, \mathsf{e}_{n-1} \ldots] \mathsf{B}\, \mathsf{e}_2)] \mathsf{U}\, \mathsf{e}_1)] \mathsf{B} \perp$$

where $\varphi \mathsf{B} \psi$, defined as $(\neg\psi) \mathsf{U} \varphi$, is short for "*$\varphi$ at least once before a ψ*". In the above formula, $\mathsf{b}_1 \mathsf{B} \perp$ encodes "there is a position where a value for v_1 is picked", $\mathsf{b}_2 \mathsf{U}\, \mathsf{e}_1$ encodes "for all positions where a value for v_2 is picked *before we change the value for* v_1", etc. When we look at a position where v_n receives a value, the current valuation for all of V can be recovered by writing "$\mathtt{t}_k \mathsf{B}\, \mathsf{e}_n$" for of v_k.

Finally, the QBF instance $\mathcal{I}$ is true iff $w(P_{\mathcal{I}}) \models \varphi_{\mathcal{I}}$. Hence we have provided a logspace reduction from QBF to model-checking *LTL* formulae on finite compressed paths.

6 Conclusions

We considered the problem of model checking a finite (or ultimately periodic) path. This is a fundamental problem in runtime verification, and it occurs in many other verification situations. This problem has not yet been the subject of serious fundamental investigation,probably because it looks like it is trivial.

We argue that "model checking a path" should be recognized as an interesting problem, and identified as such whenever it occurs. The main benefits one can expect are specialized algorithms that are more efficient than the usual algorithms we use (algorithms that were designed for the general case of model checking Kripke structures). We illustrate this with two kinds of specialized algorithms: model checking a path can be done efficiently (sometimes in polynomial-time) even when using richly expressive temporal logics that would usually be considered as highly intractable, and model checking a path can be done efficiently (sometimes in polynomial-time) even when the path is given in compressed form.

We feel this opens the door to a whole line of investigations, aiming at finding efficient algorithms for the whole variety of path model checking problems that naturally occur in practice.

From a more theoretical viewpoint, the basic problem of model checking *LTL* formulae over finite or ultimately periodic paths should be considered as an important open problem. It is not known whether the problem is PTIME-hard, or whether it admits efficient parallel algorithms (e.g. in NC), or memory-efficient algorithms (e.g. in SC). The gap between the known upper and lower bounds is quite large, but we have been unable to narrow it.

Acknowledgement. The anonymous referees made several suggestions that greatly helped improve this paper.

References

[BBF+01] B. Bérard, M. Bidoit, A. Finkel, F. Laroussinie, A. Petit, L. Petrucci, and Ph. Schnoebelen. *Systems and Software Verification. Model-Checking Techniques and Tools.* Springer, 2001.

[BCC+03] A. Biere, A. Cimatti, E. M. Clarke, O. Strichman, and Yunshan Zhu. Bounded model checking. In *Highly Dependable Software*, volume 58 of *Advances in Computers.* Academic Press, 2003. To appear.

[CGP99] E.M. Clarke, O. Grumberg, and D.A. Peled. *Model Checking.* MIT Press, 1999.

[CM77] A.K. Chandra and P.M. Merlin. Optimal implementation of conjunctive queries in relational databases. In *Proc. 9th ACM Symp. Theory of Computing (STOC'77), Boulder, CO, USA, May 1977*, pages 77–90, 1977.

[Dru00] D. Drusinsky. The Temporal Rover and the ATG Rover. In *SPIN Model Checking and Software Verification, Proc. 7th Int. SPIN Workshop, Stanford, CA, USA, Aug. 2000*, volume 1885 of *Lecture Notes in Computer Science*, pages 323–330. Springer, 2000.

[DS02] S. Demri and Ph. Schnoebelen. The complexity of propositional linear temporal logics in simple cases. *Information and Computation*, 174(1):84–103, 2002.

[Eme90] E.A. Emerson. Temporal and modal logic. In J. van Leeuwen, editor, *Handbook of Theoretical Computer Science*, volume B, chapter 16, pages 995–1072. Elsevier Science, 1990.

[FS01] B. Finkbeiner and H. Sipma. Checking finite traces using alternating automata. In *Proc. 1st Int. Workshop on Runtime Verification (RV'01), Paris, France, DK, July 2001*, volume 55(2) of *Electronic Notes in Theor. Comp. Sci.* Elsevier Science, 2001.

[GHR95] R. Greenlaw, H.J. Hoover, and W.L. Ruzzo. *Limits to Parallel Computation: P-Completeness Theory.* Oxford Univ. Press, 1995.

[GPSS80] D.M. Gabbay, A. Pnueli, S. Shelah, and J. Stavi. On the temporal analysis of fairness. In *Proc. 7th ACM Symp. Principles of Programming Languages (POPL'80), Las Vegas, NV, USA, Jan. 1980*, pages 163–173, 1980.

[Hav00] K. Havelund. Using runtime analysis to guide model checking of Java programs. In *SPIN Model Checking and Software Verification, Proc. 7th Int. SPIN Workshop, Stanford, CA, USA, Aug. 2000*, volume 1885 of *Lecture Notes in Computer Science*, pages 245–264. Springer, 2000.

[HKP80] D. Harel, D.C. Kozen, and R. Parikh. Process logic: Expressiveness, decidability, completeness. In *Proc. 21st IEEE Symp. Foundations of Computer Science (FOCS'80), Syracuse, NY, USA, Oct. 1980*, pages 129–142, 1980.

[HKV02] D. Harel, O. Kupferman, and M.Y. Vardi. On the complexity of verifying concurrent transition systems. *Information and Computation*, 173(2):143–161, 2002.

[LMS02] F. Laroussinie, N. Markey, and Ph. Schnoebelen. Temporal logic with forgettable past. In *Proc. 17th IEEE Symp. Logic in Computer Science (LICS'2002), Copenhagen, Denmark, July 2002*, pages 383–392. IEEE Comp. Soc. Press, 2002.

[LP02] R. Lassaigne and S. Peyronnet. Approximate verification of probabilistic systems. In *Proc. 2nd Joint Int. Workshop Process Algebra and Probabilistic Methods, Performance Modeling and Verification (PAPM-PROBMIV'2002), Copenhagen, Denmark, July 2002*, volume 2399 of *Lecture Notes in Computer Science*, pages 213–214. Springer, 2002.

[Mar02] N. Markey. Past is for free: on the complexity of verifying linear temporal properties with past. In *Proc. 9th Int. Workshop on Expressiveness in Concurrency (EXPRESS'2002), Brno, Czech Republic, Aug. 2002*, volume 68.2 of *Electronic Notes in Theor. Comp. Sci.* Elsevier Science, 2002.

[MS03] N. Markey and Ph. Schnoebelen. A PTIME-complete problem for SLP-compressed words. Submitted for publication, 2003.

[PR99] W. Plandowski and W. Rytter. Complexity of language recognition problems for compressed words. In J. Karhumaki, H. Maurer, G. Păun, and G. Rozenberg, editors, *Jewels are Forever*, pages 262–272. Springer, 1999.

[Rab02] A. Rabinovich. Expressive power of temporal logics. In *Proc. 13th Int. Conf. Concurrency Theory (CONCUR'2002), Brno, Czech Republic, Aug. 2002*, volume 2421 of *Lecture Notes in Computer Science*, pages 57–75. Springer, 2002.

[RG01] Muriel Roger and Jean Goubault-Larrecq. Log auditing through model checking. In *Proc. 14th IEEE Computer Security Foundations Workshop (CSFW'01), Cape Breton, Nova Scotia, Canada, June 2001*, pages 220–236. IEEE Comp. Soc. Press, 2001.

[Ros82] J.G. Rosenstein. *Linear Orderings*. Academic Press, 1982.

[RP86] R. Rosner and A. Pnueli. A choppy logic. In *Proc. 1st IEEE Symp. Logic in Computer Science (LICS'86), Cambridge, MA, USA, June 1986*, pages 306–313. IEEE Comp. Soc. Press, 1986.

[SC85] A.P. Sistla and E.M. Clarke. The complexity of propositional linear temporal logics. *Journal of the ACM*, 32(3):733–749, 1985.

[Sch03] Ph. Schnoebelen. The complexity of temporal logic model checking. In *Advances in Modal Logic, papers from 4th Int. Workshop on Advances in Modal Logic (AiML'2002), Sep.-Oct. 2002, Toulouse, France.* World Scientific, 2003. To appear.

[Sto74] L.J. Stockmeyer. *The complexity of decision problems in automata and logic.* PhD thesis, MIT, 1974.

[YS02] H.L.S. Younes and R.G. Simmons. Probabilistic verification of discrete event systems using acceptance sampling. In *Proc. 14th Int. Conf. Computer Aided Verification (CAV'2002), Copenhagen, Denmark, July 2002*, volume 2404 of *Lecture Notes in Computer Science*, pages 223–235. Springer, 2002.

Multi-valued Model Checking via Classical Model Checking

Arie Gurfinkel and Marsha Chechik

Department of Computer Science, University of Toronto
Toronto, ON M5S 3G4, Canada
{arie,chechik}@cs.toronto.edu

Abstract. Multi-valued model-checking is an extension of classical model-checking to reasoning about systems with uncertain information, which are common during early design stages. The additional values of the logic are used to capture the degree of uncertainty. In this paper, we show that the multi-valued μ-calculus model-checking problem is reducible to several classical model-checking problems. The reduction allows one to reuse existing model-checking tools and algorithms to solve multi-valued model-checking problems. This paper generalizes, extends and corrects previous work in this area, done in the context of 3-valued models, symbolic model-checking, and De Morgan algebras.

1 Introduction

Temporal logic model-checking [9] is one of the most widely used automated verification techniques. Its strength lies in its "push-button" approach to reasoning. Once a user has specified a model K, usually as a finite-state transition system, and a property in some temporal logic L, a model-checker returns true if the model satisfies the property and false otherwise.

In this paper, we assume that the temporal logic used to specify properties is the temporal μ-calculus [25]. Classical model-checking is defined over concrete models that explicitly allow some behaviors and prohibit others. This makes it well suited for analyzing systems at the end of the design cycle when all of the information is known. However, it is inconvenient for models that contain uncertain information, which is common during early design stages.

Sources of uncertainty come from partial information about the system, or internal inconsistencies. The former include partial systems where some behaviors are neither explicitly allowed nor prohibited, and abstracted systems where an abstraction results in the loss of information. Another source of uncertain information can be the property itself. For example, in temporal logic query-checking [4], temporal logic is extended with unknowns (called *placeholders*) that indicate a user's uncertainty about the correct formulation of the property. Inconsistent models come from representing a system as a composition of several (usually consistent) modules. Such modules can be features, with the goal of discovering feature interaction, or partial descriptions of the system, contributed by different stakeholders. In both cases, inconsistencies are inevitable.

Multi-valued logics provide a unifying framework for reasoning about systems with uncertain information [14,13,5]. Additional logic values are used to capture the degree

R. Amadio, D. Lugiez (Eds.): CONCUR 2003, LNCS 2761, pp. 266–280, 2003.

of uncertainty, and are used to construct the model of the system. For example, partial information can be represented using a 3-valued logic [22] with values T, M and F, where T and F represent definite information, and M ("maybe") represents partial knowledge [1].

Multi-valued logics are typically defined using finite De Morgan algebras [29], also known as quasi-boolean algebras. This ensures that many laws of classical logic, such as idempotance, associativity, distributivity, De Morgan laws, involution of negation ($\neg\neg a = a$), are preserved. Laws that are not necessarily preserved include non-contradiction ($a \wedge \neg a = \bot$) and excluded middle ($a \vee \neg a = \top$). *Multi-valued model-checking* [5] is defined as a procedure that receives a multi-valued transition system K (where either propositions are multi-valued, the transition relation is multi-valued, or both) and a formula φ from a temporal logic defined over some De Morgan algebra $\mathcal{L}$, and returns the degree to which φ holds on K.

The multi-valued model-checking problem can be decided directly, using a specialized tool [7]. Yet, it is appealing to reduce it to several classical model-checking problems. Such a reduction allows one to check correctness of the direct approach and opens venues to use the mature classical model-checking technology. It also provides a connection between multi-valued and classical model-checking and allows to lift theoretical results from classical model-checking to multi-valued. For example, it is used in [20,19] to show that the refinement relation over 3-valued models is an extended version of the bisimulation relation [27], and in [18] to show that query-checking is an instance of multi-valued model-checking.

Reduction algorithms for 3-valued logic [2,1,16,20,21] have been well understood. Such reductions typically involve two independent checks to classical models, and the negation is handled on the level of atomic propositions. Konikowska and Penczek [23, 24] provided reductions for several other logics for the negation-free fragment of mv-CTL$^*(\mathcal{L})$. The contribution of this paper is in generalizing these reductions to μ-calculus over arbitrary finite De Morgan algebras. The solution is effectively to reduce multi-valued model-checking to $|\mathcal{J}(\mathcal{L})|$ 2-valued models, where $\mathcal{J}(\mathcal{L})$ is the set of join-irreducible elements of a given De Morgan algebra $\mathcal{L}$. Each 2-valued model is encoded to be able to decide both universal and existential temporal logic properties. A similar approach, although in the context of *classical* temporal logic questions, was proposed by Huth and Pradhan [21].

The rest of the paper is organized as follows. After giving the necessary background in Sect. 2, we systematically develop and analyze the reduction algorithm in Sect. 3 and compare it with related work in Sect. 4. Section 5 concludes the paper.

2 Background

In this section, we give a brief introduction to lattice theory, De Morgan algebras, and multi-valued model-checking.

2.1 Lattices and De Morgan Algebras

A *lattice* is a partial order $(\mathcal{L}, \sqsubseteq)$, where every finite subset $B \subseteq \mathcal{L}$ has a least upper bound (called "join" and written $\sqcup B$) and a greatest lower bound (called "meet" and written $\sqcap B$). $\top$ and $\bot$ are the maximal and the minimal elements of a lattice, respectively. For

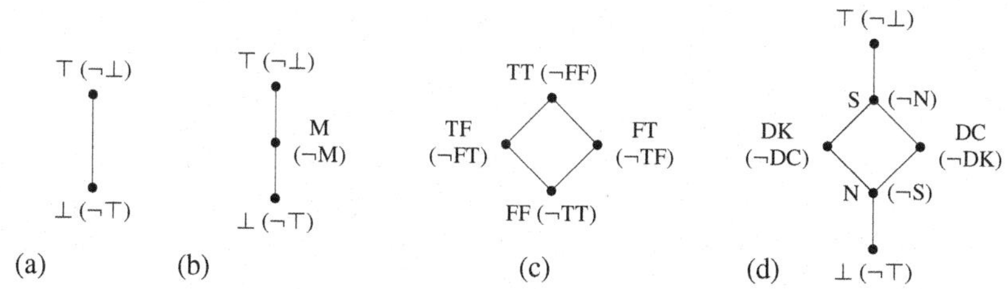

Fig. 1. Examples of a few distributed lattices and the corresponding De Morgan algebras (bracketed values describe the negation): (a) lattice **2**; (b) lattice **3**; (c) cross-product lattice **2x2**; and (d) lattice $\mathbf{2x2} + \mathbf{2}$

notational convenience, we often refer to a lattice $(\mathcal{L}, \sqsubseteq)$ by its carrier set $\mathcal{L}$. A lattice is called *distributive* if meet and join distribute over each other, i.e., $a \sqcap (b \sqcup c) = (a \sqcap b) \sqcup (a \sqcap c)$. A few examples of distributive lattices are given in Fig. 1.

Definition 1. *[11] An element j in a lattice $\mathcal{L}$ is* join-irreducible *iff $j \neq \bot$ and for any x and y in $\mathcal{L}$, $j = x \sqcup y$ implies $j = x$ or $j = y$.*

In other words, j is a join-irreducible if it cannot be further decomposed into a join of other elements in the lattice. For example, the join-irreducibles of the lattices in Fig. 1a,b,c,d are $\{\top\}$, $\{\top, \mathrm{M}\}$, {TF,FT}, $\{\top$, DK, DC, N$\}$, respectively. The set of all join-irreducibles of $\mathcal{L}$ is denoted by $\mathcal{J}(\mathcal{L})$.

Every element of a finite lattice can be uniquely decomposed as a join of all join-irreducible elements below it:

Theorem 1. *[11] For any $\ell \in \mathcal{L}$, $\ell = \bigsqcup\{j \in \mathcal{J}(\mathcal{L}, \sqsubseteq) \mid j \sqsubseteq \ell\}$.*

For any join-irreducible element $j \in \mathcal{J}(\mathcal{L})$, the function $\cdot \sqsupseteq j$ distributes over meets and joins:

$$(a \sqcap b) \sqsupseteq j = (a \sqsupseteq j) \sqcap (b \sqsupseteq j) \qquad (a \sqcup b) \sqsupseteq j = (a \sqsupseteq j) \sqcup (b \sqsupseteq j)$$

For any lattice $\mathcal{L}$ and a collection of $\mathcal{L}$ elements B, the *downward closure* of B, written $\downarrow B$, is the set of all elements of $\mathcal{L}$ that are below some elements of B:

$$\downarrow B \triangleq \{\ell \in \mathcal{L} \mid \exists b \in B \cdot \ell \sqsubseteq b\}$$

Definition 2. *A De Morgan algebra is a tuple $(\mathcal{L}, \sqsubseteq, \neg)$, where $(\mathcal{L}, \sqsubseteq)$ is a finite distributive lattice and $\neg$ is any operation that preserves involution ($\neg\neg\ell = \ell$) and De Morgan laws.*

De Morgan algebras provide a natural model for De Morgan logics where the logical conjunction ($\wedge$) and disjunction ($\vee$) are interpreted as meet and join of the algebra, respectively. In De Morgan algebras, we get $\neg\top = \bot$ and $\neg\bot = \top$, but not necessarily the law of non-contradiction ($\ell \sqcap \neg\ell = \bot$) or excluded middle ($\ell \sqcup \neg\ell = \top$). For notational convenience, we write $\Rightarrow$ for material implication: $a \Rightarrow b \triangleq \neg a \sqcup b$.

We can define several De Morgan algebras using the lattices given in Fig. 1. The domain of logical values of the classical logic, referred to as **2**, is the lattice in Fig. 1(a). The three-valued algebra **3** (Kleene logic [22]) is defined on the lattice in Fig. 1(b), where $\neg\top = \bot$, $\neg\bot = \top$, $\neg$M = M. The four-valued algebra **2x2** is defined on the lattice in Fig. 1(c). A logic based on this algebra can be used for reasoning about inconsistency. Note that $\top$ and $\bot$ elements of an algebra are interpreted as values true and false of the logic, respectively. When the negation and the ordering operators of an algebra $(\mathcal{L}, \sqsubseteq, \neg)$ are clear from the context, we refer to it by its carrier set $\mathcal{L}$.

Given a set S, and a De Morgan algebra $\mathcal{L}$, we denote the set of functions from S to $\mathcal{L}$ by $\mathcal{L}^S$. If $\mathcal{L}$ is a De Morgan algebra, then so is $(\mathcal{L}^S, \sqsubseteq, \neg)$, where $\sqsubseteq$ and $\neg$ are pointwise extensions of the corresponding operators of $\mathcal{L}$. That is, for $\mathbb{F}, \mathbb{G} \in \mathcal{L}^S$,

$$\mathbb{F} \sqsubseteq \mathbb{G} = \forall s \in S \cdot \mathbb{F}(s) \sqsubseteq \mathbb{G}(s) \qquad \mathbb{G} = \neg\mathbb{F} \textbf{ iff } \forall s \in S \cdot \mathbb{G}(s) = \neg\mathbb{F}(s)$$

Theorem 2. *[6] For any De Morgan algebra $(\mathcal{L}, \sqsubseteq, \neg)$ there exists a function* neg : $\mathcal{J}(\mathcal{L}) \to \mathcal{J}(\mathcal{L})$ *defined as* $\mathsf{neg}(j) \triangleq \sqcap(\mathcal{L} \setminus \downarrow \neg j)$, *such that*

$$\forall \ell \in \mathcal{L} \cdot \forall j \in \mathcal{J}(\mathcal{L}) \cdot \neg\ell \sqsupseteq j \;=\; \neg(\ell \sqsupseteq \mathsf{neg}(j))$$

Note that neg maps join-irreducible elements to join-irreducible elements and can be easily [6]. For example, for the algebra **3**, $\mathsf{neg}(\top) = $ M and $\mathsf{neg}(\text{M}) = \top$. For the algebras **2** and **2x2**, neg is the identity function.

2.2 Multi-valued Model-Checking

Multi-valued model-checking [8] is a generalization of the temporal logic model-checking problem to arbitrary De Morgan logics. A multi-valued model-checker receives a De Morgan algebra, a multi-valued model, and a temporal property, and determines the value with which this property holds in the model. Multi-valued models are defined over χKripke structures – generalizations of Kripke structures, where each atomic proposition and each transition between a pair of states are labeled with values from the algebra. Formally, a χKripke structure is a tuple $K = (S, s_0, \mathbb{R}, I, A, \mathcal{L})$, where S is a finite set of states; $\mathcal{L}$ is a De Morgan algebra; A is a set of atomic propositions; $s_0 \in S$ is the initial state; $\mathbb{R} : S \times S \to \mathcal{L}$ is a multi-valued transition relation; $I : S \times A \to \mathcal{L}$ is a (total) labeling function, such that for each atomic proposition $a \in A$, $I(s, a) = \ell$ means that variable a has value ℓ in state s. Thus, any Kripke structure is also a χKripke structure over the algebra **2**. An example χKripke structure for the algebra **3** is given in Fig. 2(a). To avoid clutter when presenting finite-state machines graphically, we follow the convention of not showing $\bot$ transitions and not labeling $\top$ transitions.

Temporal logic properties are specified in $L_\mu^+(A, \mathcal{L})$ – a generalization of μ-calculus [25] to arbitrary De Morgan algebras.

Definition 3. *Let* Var *be a set of fixpoint variable names, A be a set of propositions, and $\mathcal{L}$ be a De Morgan algebra. The logic $L_\mu^+(A, \mathcal{L})$ is the set of formulas defined as:*

$$\varphi \triangleq Z \mid \ell \mid \neg\ell \mid p \mid \neg p \mid \varphi \wedge \varphi \mid \varphi \vee \varphi \mid \Box\varphi \mid \Diamond\varphi \mid \nu Z \cdot \varphi \mid \mu Z \cdot \varphi$$

where $\ell \in \mathcal{L}$, $p \in A$, and $Z \in$ Var.

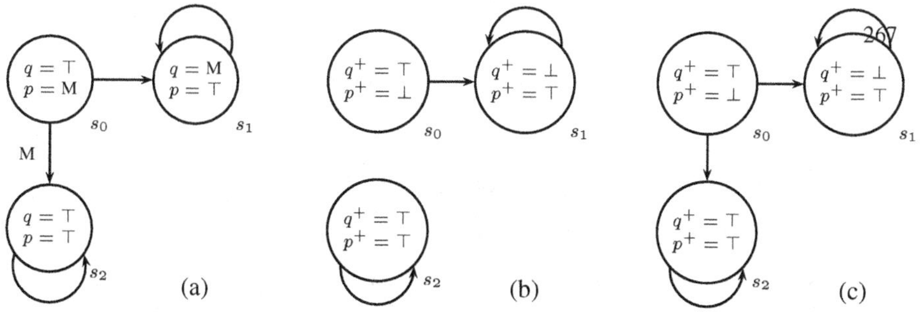

Fig. 2. (a) A χKripke structure K over the algebra **3**; (b) a Kripke structure $K_\top$ used to check truth of existential properties over K; (c) a Kripke structure $K'_\top$ for universal properties

$\Diamond$ and $\Box$ are the next-state operators, with the intuitive meaning "there exists a next state" and "for all next states", respectively. This gives rise to $\Diamond L_\mu^+(A, \mathcal{L})$ (*existential*) and $\Box L_\mu^+(A, \mathcal{L})$ (*universal*) fragments in which the only allowed next-state operators are $\Diamond$ and $\Box$, respectively.

We write $\varphi(Z)$ for a $L_\mu^+(A, \mathcal{L})$ formula φ that *may* contain a free occurrence of Z, and $\varphi(\psi)$ for a formula obtained from φ by replacing *all* free occurrences of Z by ψ. μ and ν denote the least and the greatest fixpoint operators, respectively.

Note that in $L_\mu^+(A, \mathcal{L})$, the negation operator $\neg$ is restricted to elements of $\mathcal{L}$ and propositions. Alternatively, we can define a logic $L_\mu(A, \mathcal{L})$ by relaxing this restriction. In this case, for a formula $\varphi \in L_\mu(A, \mathcal{L})$, $\nu Z \cdot \varphi(Z)$ and $\mu Z \cdot \varphi(Z)$ are in $L_\mu(A, \mathcal{L})$ if and only if Z occurs under an even scope of negations in φ. For example, $\mu Z \cdot \neg\Diamond\neg Z$ is in $L_\mu(A, \mathcal{L})$, but $\mu Z \cdot \neg\Diamond Z$ is not. The traditional definition of μ-calculus is equivalent to $L_\mu(A, \mathbf{2})$. To simplify the notation, we often write L_μ when parameters A and $\mathcal{L}$ are clear from the context, or $L_\mu(A)$ and $L_\mu(\mathcal{L})$ when we want to emphasize only one of the parameters.

The semantics of L_μ is given by the function $||\cdot||$ that, for each formula φ and a state s of a χKripke structure, returns the value of φ in s. Note that $||\cdot||$ takes an additional parameter called an *environment* that is used to interpret the fixpoint variables.

Definition 4. *Let ℓ be an element of $\mathcal{L}$, $p \in A$, $s \in S$, $\varphi, \psi \in L_\mu$, and $\rho : \mathsf{Var} \to \mathcal{L}^S$. Then the function $||\cdot|| : L_\mu \times (\mathsf{Var} \to \mathcal{L}^S) \to \mathcal{L}^S$ is defined as follows:*

$$
\begin{aligned}
||Z||_\rho(s) &\triangleq \rho(Z)(s) & ||\ell||_\rho(s) &\triangleq \ell \\
||p||_\rho(s) &\triangleq I(s,p) & ||\neg\varphi||_\rho(s) &\triangleq \neg||\varphi||_\rho(s) \\
||\varphi \wedge \psi||_\rho(s) &\triangleq ||\varphi||_\rho(s) \sqcap ||\psi||_\rho(s) & ||\varphi \vee \psi||_\rho(s) &\triangleq ||\varphi||_\rho(s) \sqcup ||\psi||_\rho(s) \\
||\Diamond\varphi||_\rho(s) &\triangleq \sqcup_{t \in S}(\mathbb{R}(s,t) \sqcap ||\varphi||_\rho(t)) & ||\Box\varphi||_\rho(s) &\triangleq \sqcap_{t \in S}(\mathbb{R}(s,t) \Rightarrow ||\varphi||_\rho(t))
\end{aligned}
$$
$$
\begin{aligned}
||\mu Z \cdot \varphi(Z)||_\rho(s) &\triangleq \sqcap\{\mathbb{C} \in \mathcal{L}^S \mid ||\varphi||_{\rho[Z \to \mathbb{C}]} \sqsubseteq \mathbb{C}\} \\
||\nu Z \cdot \varphi(Z)||_\rho(s) &\triangleq \sqcup\{\mathbb{C} \in \mathcal{L}^S \mid \mathbb{C} \sqsubseteq ||\varphi||_{\rho[Z \to \mathbb{C}]}\}
\end{aligned}
$$

where $\rho[Z \to \mathbb{C}]$ is an environment like ρ except that it maps Z to $\mathbb{C}$.

An environment that maps every $Z \in \mathsf{Var}$ to $\bot$ is denoted by $\bot$. For a closed L_μ formula φ, we write $||\varphi||$ to stand for $||\varphi||_\bot$. The value of a closed L_μ formula φ on a

χKripke structure K is given by the value of φ in the initial state of K, i.e., $||\varphi||(s_0)$, and is often written as $||\varphi||^K$.

Note that under our definition, the next-time operators $\Diamond$ and $\Box$ are duals of each other, i.e., $\neg\Diamond\neg\varphi = \Box\varphi$. This also ensures the duality of the least and the greatest fixpoint operators, i.e. $\neg\mu Z \cdot \varphi(\neg Z) = \nu Z \cdot \varphi(Z)$. Combining the above results with the involution property of the negation operator, we obtain the following theorem.

Theorem 3. *The negation-free fragment L_μ^+ of L_μ is as expressive as L_μ.*

Both CTL [10] and its multi-valued extension χCTL($\mathcal{L}$) [5] can be expressed in $L_\mu(A, \mathcal{L})$ as follows:

$$
\begin{array}{rlrl}
EXp &= \Diamond p & AXp &= \Box p \\
E[\varphi U \psi] &= \mu Z \cdot \psi \vee \varphi \wedge \Diamond Z & A[\varphi U \psi] &= \mu Z \cdot \psi \vee \varphi \wedge \Box Z \\
EG\varphi &= \nu Z \cdot \varphi \wedge \Diamond\varphi & AG\varphi &= \nu Z \cdot \varphi \wedge \Box\varphi
\end{array}
$$

So, the reduction technique developed later in this paper for $L_\mu(\mathcal{L})$ is directly applicable to χCTL($\mathcal{L}$) as well.

3 Reduction

In this section, we systematically decompose a multi-valued μ-calculus (L_μ) model-checking problem into several classical μ-calculus ($L_\mu(\mathbf{2})$) model-checking problems. One approach to the reduction, particularly prevalent for model-checking over the algebra **3** (e.g. [1]), is to reduce the model only, while keeping the formula the same. In the 3-valued case, one constructs two models, corresponding to "the best" and "the worst" possible behaviors, with respect to the given temporal property. Of course, the definition of "best" and "worst" depends on quantifiers used in the property; further, for a property containing both universal and existential quantifiers, one must reduce the formula as well!

We start by showing that a model-checking problem for $L_\mu(A, \mathcal{L})$ is reducible to several model-checking problems over a different logic, which we call $L_\mu^{\sqsupseteq}(A, \mathcal{L})$ (Sect. 3.1). We then show how to change a χKripke structure so that each of the resulting problems can be solved using a single call to a classical model-checker on a Kripke structure (Sect. 3.2). We put the two reductions together and illustrate them on an example in Sect. 3.3. Section 3.4 summarizes consequences of the reduction and analyzes its complexity.

3.1 Model-Checking: From Multi-valued to Boolean

This section introduces the first step of the reduction, by showing that a multi-valued model-checking problem can be reduced to several boolean model-checking problems. Note that this step only changes the property, while leaving the model unchanged.

For any De Morgan algebra $\mathcal{L}$, any element $\ell \in \mathcal{L}$ is uniquely represented by the join of join-irreducible elements below it (Theorem 1). For any $L_\mu(\mathcal{L})$ formula φ and state s, $||\varphi||(s)$ is simply an element of $\mathcal{L}$, so we can extend Theorem 1 to $L_\mu(\mathcal{L})$ as follows:

Theorem 4. *Let φ be a $L_\mu(\mathcal{L})$ formula, and s be a state of a χKripke structure over $\mathcal{L}$. Then $||\varphi||_\rho(s) = \bigsqcup_{j \in \mathcal{J}(\mathcal{L})}(j \sqcap (||\varphi||_\rho(s) \sqsupseteq j))$.*

This theorem provides the basis for our reduction technique. The expression $||\varphi||_\rho(s) \sqsupseteq j$ is interpreted over a χKripke structure, but it always evaluates to either $\top$ or $\bot$. Thus, it allows us to reduce a multi-valued model-checking problem for $L_\mu(\mathcal{L})$ to $|\mathcal{J}(\mathcal{L})|$ boolean model-checking problems.

In order to express the statement $||\varphi||_\rho(s) \sqsupseteq j$ by a single temporal logic formula, we introduce the logic $L_\mu^{\sqsupseteq}$.

Definition 5. *Let* Var *be a set of fixpoint variable names, A be a set of propositions, and $\mathcal{L}$ be a De Morgan algebra. The logic $L_\mu^{\sqsupseteq}(A, \mathcal{L})$ is the set of formulas defined as:*

$$\varphi \triangleq Z \sqsupseteq j \mid \top \mid \bot \mid p \sqsupseteq j \mid \neg p \sqsupseteq j \mid \varphi \wedge \varphi \mid \varphi \vee \varphi \mid [\sqsupseteq j]\varphi \mid \langle \sqsupseteq j \rangle \varphi \mid \nu Z \cdot \varphi \mid \mu Z \cdot \varphi$$

where $j \in \mathcal{L}$, $p \in A$, and $Z \in$ Var.

Furthermore, in expressions $Z \sqsupseteq \ell_1, \ldots Z \sqsupseteq \ell_n$, we require that all algebra values be the same, i.e. $\forall i, j \cdot \ell_i = \ell_j$. The semantics of $L_\mu^{\sqsupseteq}(A, \mathcal{L})$ is given with respect to χKripke structures and is defined as follows:

Definition 6. *Let j be elements of $\mathcal{L}$, $p \in A$, $s \in S$, $\varphi, \psi \in L_\mu^{\sqsupseteq}$, and $\rho : \mathsf{Var} \to \mathcal{L}^S$. Then the function $||\cdot|| : L_\mu^{\sqsupseteq} \times (\mathsf{Var} \to \mathcal{L}^S) \to \mathcal{L}^S$ is defined as:*

$$||\top||_\rho(s) \triangleq \top \qquad ||\bot||_\rho(s) \triangleq \bot$$
$$||p \sqsupseteq j||_\rho(s) \triangleq I(s,p) \sqsupseteq j \qquad ||\neg p \sqsupseteq j||_\rho(s) \triangleq \neg I(s,p) \sqsupseteq j$$
$$||Z \sqsupseteq j||_\rho(s) \triangleq \rho(Z)(s) \sqsupseteq j$$
$$||\langle \sqsupseteq j \rangle \varphi||_\rho(s) \triangleq \sqcup_{t \in S}((\mathbb{R}(s,t) \sqsupseteq j) \sqcap ||\varphi||_\rho(t))$$
$$||[\sqsupseteq j]\varphi||_\rho(s) \triangleq \sqcap_{t \in S}((\mathbb{R}(s,t) \sqsupseteq j) \Rightarrow ||\varphi||_\rho(t))$$

with the semantics of $\wedge$, $\vee$, μ and ν operators being the same as in Definition 4.

Finally, we show that for any $L_\mu(\mathcal{L})$ formula φ, and any join-irreducible element j, the statement $||\varphi||_\rho \sqsupseteq j$ is expressible in $L_\mu^{\sqsupseteq}$.

Theorem 5. *Let φ be a $L_\mu(\mathcal{L})$ formula, $j \in \mathcal{J}(\mathcal{L})$, and s be a state of a χKripke structure. Then there exists a $L_\mu^{\sqsupseteq}$ formula $\varphi \Uparrow j$, called the* cut *of φ with respect to j, such that $||\varphi||_\rho(s) \sqsupseteq j \;=\; ||\varphi \Uparrow j||_\rho(s)$.*

The proof of this theorem is available in the full version of this paper. As a direct consequence of the proof, we obtain the following procedure for constructing $\varphi \Uparrow j$. Given a formula $\varphi \in L_\mu(\mathcal{L})$ and a join-irreducible element j of $\mathcal{L}$, $\varphi \Uparrow j$ is constructed by recursively applying the transformation $\cdot \Uparrow j$ that distributes over $\wedge$, $\vee$, and greatest and least fixpoints μ and ν:

$$p \Uparrow j = p \sqsupseteq j \qquad (\neg p) \Uparrow j = \neg p \sqsupseteq j$$
$$(\Diamond\varphi) \Uparrow j = \langle \sqsupseteq j \rangle(\varphi \Uparrow j) \qquad (\Box\varphi) \Uparrow j = [\sqsupseteq \mathsf{neg}(j)](\varphi \Uparrow j)$$

For example, if $\varphi = \mu Z \cdot p \vee (\Box Z \wedge \Diamond\top)$, then its cut with respect to a join-irreducible element $\top$ of the algebra **3** is given by

$$\varphi \Uparrow \top = \mu Z \cdot (p \sqsupseteq \top) \vee ([\sqsupseteq \mathrm{M}](Z \sqsupseteq \top) \wedge \langle \sqsupseteq \top \rangle(\top \sqsupseteq \top))$$

Combining Theorem 4 and Theorem 5, we obtain the following theorem.

Theorem 6. *A multi-valued model-checking problem for the logic $L_\mu(\mathcal{L})$ is reducible to $\mathcal{J}(|\mathcal{L}|)$ boolean model-checking problems for $L_\mu^{\sqsupseteq}(\mathcal{L})$.*

3.2 Reducing $L_\mu^{\sqsupseteq}$

We now show that the problem of model-checking a formula $\varphi \in L_\mu^{\sqsupseteq}$ on a χKripke structure is reducible to a *single* model-checking problem for classical μ-calculus $L_\mu(\mathbf{2})$ on a Kripke structure.

Theorem 7. *Let φ be a $L_\mu^{\sqsupseteq}(A, \mathcal{L})$ formula, $K = (S, s_0, \mathbb{R}, I, A, \mathcal{L})$ be a χKripke structure, and ρ be an environment. Then there exists a $L_\mu(A', \mathbf{2})$ formula $T(\varphi)$, a Kripke structure $K' = (S', s_0', \mathbb{R}', I', A', \mathbf{2})$, and an environment ρ' such that $||\varphi||_\rho^K = ||T(\varphi)||_{\rho'}^{K'}$. Moreover, $|S'|$ is in $O(|\mathcal{J}(\mathcal{L})| \times |S|)$.*

Thus, given an algebra $\mathcal{L}$, the model-checking problem for $L_\mu^{\sqsupseteq}(\mathcal{L})$ is reducible to model-checking a $L_\mu(\mathbf{2})$ formula at the expense of a linear increase in the size of the statespace. The Kripke structure K' is obtained from K by first constructing a Kripke Transition system (KTS) [28], treating algebra values on transitions as actions, and then converting the resulting KTS into a Kripke structure.

Instead of proving Theorem 7 in the general case, we prove it for two fragments of $L_\mu^{\sqsupseteq}$ that are used in the reduction in Sect. 3.1.

Temporal Logic $L_\mu^{\sqsupseteq j}$. Let j be an element of $\mathcal{L}$. Then the fragment $L_\mu^{\sqsupseteq j}$ is defined to be the set of all formulas of $L_\mu^{\sqsupseteq}$ where only j can appear on the right-hand side of $\sqsupseteq$, and only $\langle \sqsupseteq j \rangle$ and $[\sqsupseteq j]$ are allowed. For example, $\nu Z \cdot (p \sqsupseteq \mathrm{M}) \vee \langle \sqsupseteq \mathrm{M} \rangle (Z \sqsupseteq \mathrm{M})$ is in $L_\mu^{\sqsupseteq \mathrm{M}}(\{p\}, \mathbf{3})$, but $(p \sqsupseteq \top)$ is not. This fragment is used to reduce existential properties.

Given a formula $\varphi \in L_\mu^{\sqsupseteq j}(A, \mathcal{L})$ and a χKripke structure $K = (S, s_0, \mathbb{R}, I, A, \mathcal{L})$, we construct a Kripke structure $K' = (S, s_0, \mathbb{R}', I', A', \mathbf{2})$ as follows:

$$A' \triangleq \{p^+ \mid p \in A\} \cup \{p^- \mid p \in A\} \qquad I'(s, p^+) \triangleq I(s, p) \sqsupseteq j$$
$$\mathbb{R}'(s, t) \triangleq \mathbb{R}(s, t) \sqsupseteq j \qquad I'(s, p^-) \triangleq \neg I(s, p) \sqsupseteq j$$

Note that K' has the same statespace as K, but twice as many propositions. For every proposition $p \in A$, it has a pair of propositions p^+ and p^- corresponding to p and $\neg p$, respectively. The transition relation of K' consists of all transitions of K whose value is above j. The reduced formula $T(\varphi)$ is obtained from φ by recursively removing $\sqsupseteq j$ and replacing every occurrence of $\langle \sqsupseteq j \rangle$ and $[\sqsupseteq j]$ by $\Diamond$ and $\Box$, respectively. For example, a formula $\varphi = \nu Z \cdot (p \sqsupseteq j) \wedge [\sqsupseteq j](Z \sqsupseteq j)$ is reduced to $T(\varphi) = \nu Z \cdot p^+ \wedge \Box Z$. Finally, an environment ρ is replaced by $\rho' \triangleq \rho \sqsupseteq j$. The fact that $||\varphi||_\rho^K = ||T(\varphi)||_{\rho'}^{K'}$ follows trivially from the construction.

Temporal Logic $L_\mu^{j,i}$. Let j and i be elements of $\mathcal{L}$. Then the fragment $L_\mu^{j,i}$ is defined to be the set of all formulas obtained from $L_\mu^{\sqsupseteq j}$ by replacing the universal next-time operator $[\sqsupseteq j]$ with $[\sqsupseteq i]$. For example, $[\sqsupseteq M](p \sqsupseteq \top)$ is in $L_\mu^{\top, M}(\{p\}, \mathbf{3})$, but $[\sqsupseteq \top](p \sqsupseteq \top)$ is not. This fragment is used to reduce properties that contain both universal and existential next-time operators.

Given a formula $\varphi \in L_{\mu}^{j,i}(A, \mathcal{L})$ and a χKripke structure $K = (S, s_0, \mathbb{R}, I, A, \mathcal{L})$, we construct a Kripke structure $K' = (S \times \{\top, \bot\}, (s_0, \top), \mathbb{R}', I', A', \mathbf{2})$ as follows:

$$
\begin{array}{ll}
I'((s,a),p^+) \triangleq I(s,p) \sqsupseteq j & \mathbb{R}'((s,a),(t,\top)) \triangleq \mathbb{R}(s,t) \sqsupseteq j \\
I'((s,a),p^-) \triangleq \neg I(s,p) \sqsupseteq j & \mathbb{R}'((s,a),(t,\bot)) \triangleq \mathbb{R}(s,t) \sqsupseteq i \\
I'((s,a),tval^j) \triangleq a & I'((s,a),tval^i) \triangleq \neg a
\end{array}
$$
$$A' \triangleq \{p^+ \mid p \in A\} \cup \{p^- \mid p \in A\} \cup \{tval^j, tval^i\}$$

The Kripke structure K' can be seen as having two distinct transition relations $\mathbb{R} \sqsupseteq j$ and $R \sqsupseteq i$. To encode this, its statespace is extended to *double* the size of the statespace of K such that there exists a transition between (s,a) and $(t,\top)$ if and only if there is a transition between s and t in K with value above j; and there exists a transition between (s,a) and $(t,\bot)$ if and only if there exists a transition between s and t in K with value above i. Similarly, its set of propositions A' is extended with additional propositions $tval^j$ and $tval^i$, where $tval^j$ is true in a state (s,a) if and only if s is reachable by a transition in K with value above j, and $tval^i$ is true if and only if s is reachable by a transition whose value is above i.

The reduced formula $T(\varphi)$ is obtained by eliminating all occurrences of $\sqsupseteq j$ from φ and replacing the temporal next-time operators $\langle \sqsupseteq j \rangle$ and $[\sqsupseteq i]$ as follows:

$$T(\langle \sqsupseteq j \rangle \varphi) = \Diamond(tval^j \wedge T(\varphi)) \qquad T([\sqsupseteq i]\varphi) = \Box(tval^i \Rightarrow T(\varphi))$$

Finally, an environment ρ is replaced by $\rho' = \rho \sqsupseteq j$. The proof that $||\varphi||_{\rho}^{K} = ||T(\varphi)||_{\rho'}^{K'}$ follows trivially from the construction.

Note that the logic $L_{\mu}^{j,i}$ is as expressive as $L_{\mu}^{\sqsupseteq j}$ for χKripke structures whose transition relation is boolean, i.e., $\forall s, t \in S \cdot \mathbb{R}(s,t) \in \{\top, \bot\}$. In this case, the operators $[\sqsupseteq j]$ and $[\sqsupseteq i]$ are identical for all i and j.

3.3 Model-Checking: From Multi-valued to Classical

Here, we combine Theorems 6 and 7 to yield the overall reduction and illustrate it on an example.

Theorem 8. *A multi-valued model-checking problem for the logic $L_\mu(\mathcal{L})$ is reducible to $\mathcal{J}(|\mathcal{L}|)$ classical model-checking problems for $L_\mu(\mathbf{2})$.*

Theorem 8 leads to the following reduction algorithm. Given an $L_\mu(\mathcal{L})$ formula φ and a χKripke structure K, for every join-irreducible j we (a) construct the j-cut $\varphi \Uparrow j$, and (b) use one of the reductions of Sect. 3.2 to reduce checking $||\varphi \Uparrow j||^K$ to a classical model-checking problem. The choice of the reduction to use depends on the structure of φ: if φ is existential, its cut is expressible in $L_{\mu}^{\sqsupseteq j}$; otherwise, it is expressible in $L_{\mu}^{j,i}$.

Example. To illustrate the reduction algorithm, we apply it to an existential property $\varphi = \mu Z \cdot p \vee \Diamond Z$ and state s_0 of the χKripke structure K shown in Fig. 2a. We start by constructing cuts of φ with respect to the two join-irreducible elements of algebra $\mathbf{3}$:

$$\varphi \Uparrow \top = \mu Z \cdot p \sqsupseteq \top \vee \langle \sqsupseteq \top \rangle (Z \sqsupseteq \top)$$
$$\varphi \Uparrow \mathrm{M} = \mu Z \cdot p \sqsupseteq \mathrm{M} \vee \langle \sqsupseteq \mathrm{M} \rangle (Z \sqsupseteq \mathrm{M})$$

Next, for each cut $\varphi \Uparrow j$, we construct a Kripke structure K_j and a μ-calculus formula corresponding to the cut. Following the construction outlined in Sect. 3.2, both cuts are reduced to $\mu Z \cdot p^+ \vee \Diamond Z$. The Kripke structure $K_\top$ shown in Fig. 2b is obtained from K by eliminating all non-$\top$ transitions, and replacing atomic propositions by their positive and negative versions. For conciseness we only show positive propositions in the figure. Finally, we model-check the μ-calculus formula in state s_0 of $K_\top$. In our example, the property $\mu Z \cdot p^+ \vee \Diamond Z$ is true on $K_\top$ which implies that $\varphi \Uparrow \top$ holds on the χKripke structure K, and therefore the value of the original property φ on K is $\top$:

$$\begin{aligned} & ||\mu Z \cdot p^+ \vee \Diamond Z||^{K_\top} = \textsf{true} \\ \Leftrightarrow\ & ||\varphi \Uparrow \top||^K = \top \\ \Leftrightarrow\ & ||\varphi||^K \sqsupseteq \top \\ \Leftrightarrow\ & ||\varphi||^K = \top \end{aligned}$$

The case of $\varphi \Uparrow$ M is similar, except that K_M is constructed from K by treating all non-$\bot$ transitions as $\top$.

For another example, let ψ be a universal property $\psi = \nu Z \cdot p \ \wedge\ \Box Z$. After computing the cuts and performing the reduction, we obtain $\nu Z \cdot p^+ \ \wedge \Box Z$. The Kripke structure $K'_\top$ corresponding to the $\top$-cut (see Fig. 2c) is obtained from K by treating all non-$\bot$ transitions as $\top$. Model-checking the property on $K'_\top$ yields false, which implies that the cut $\psi \Uparrow \top$ evaluates to $\bot$ on the χKripke structure K, and therefore the value of ψ on K is less then $\top$. Since the algebra **3** has only three elements, this means that ψ evaluates to either M or $\bot$ on it.

$$||\nu Z \cdot p^+ \ \wedge\ \Box Z||^{K'_\top} = \textsf{false} \Leftrightarrow ||\psi \Uparrow \top||^K = \bot \Leftrightarrow ||\psi||^K \not\sqsupseteq \top \Leftrightarrow ||\psi||^K \in \{\bot, \mathrm{M}\}$$

In this particular example, the value of ψ on K is M, and is obtained by checking the second cut, $\psi \Uparrow$ M. □

Note that in the example above, the cut properties $\varphi \Uparrow j$ for both join-irreducible elements were syntactically equivalent. Thus, we only had to reduce the χKripke structure, once for each join-irreducible element. However, the Kripke structures $K_\top$ and $K'_\top$, corresponding to the join-irreducible $\top$, were different: although they had the same statespace and the labeling function, the transition relation of $K_\top$ was that of K'_M. The reason is that $K_\top$ and $K'_\top$ were used to decide existential and universal formulas, respectively. In general, an existential part of a mixed formula should be checked over $K_\top$, and its universal part – over $K'_\top$. Then the Kripke structure corresponding to the join-irreducible $\top$ contains transition relations of *both* $K_\top$ and $K'_\top$, as in the second reduction in Sect. 3.2. This construction gets reflected in cut formulas, which are different for each join-irreducible.

3.4 Discussion and Complexity

We summarize the consequence of Theorem 8 for several fragments of $L_\mu(\mathcal{L})$ in Table 1. The first column of the table indicates the fragment of $L_\mu(\mathcal{L})$ used to specify the property; the second describes the restrictions placed on χKripke structures; the third specifies the number of $L_\mu(\mathbf{2})$ model-checking problems required; and the last indicates the ratio between the size of the statespace S' of the Kripke structures used by the reduction,

Table 1. Reducing multi-valued model-checking to classical

Property	χKripke restrictions	# of $L_\mu(\mathbf{2})$ problems	$\frac{\lvert S'\rvert}{\lvert S\rvert}$
$L_\mu(\mathcal{L})$	none	$\lvert\mathcal{J}(\mathcal{L})\rvert$	2
	$\{\top,\bot\}$ transition relation		1
$\Diamond L_\mu(\mathcal{L})$	none		1
$\Box L_\mu(\mathcal{L})$	none		1

and the statespace S of the original χKripke structure. For example, model-checking an arbitrary $L_\mu(\mathcal{L})$ property on a χKripke structure K is reducible to $|\mathcal{J}(\mathcal{L})|$ classical model-checking problems, each over a Kripke structure whose statespace is twice that of K. On the other hand, if the property is expressed in either existential or universal fragments of $L_\mu(\mathcal{L})$, then the multi-valued model-checking problem is reducible to $|\mathcal{J}(\mathcal{L})|$ classical model-checking problems, each over a Kripke structure with the statespace identical to the statespace of the original χKripke structure.

Note that we have only considered reductions for the negation-free fragment $L_\mu^+(\mathcal{L})$ of $L_\mu(\mathcal{L})$. This is not a limitation of our approach since the negation-free fragment is as expressive as $L_\mu(\mathcal{L})$ (Theorem 3). Alternatively, it is easy to show that $||\neg\varphi \Uparrow j||(s) = \neg||\varphi \Uparrow \mathsf{neg}(j)||(s)$ directly:

$$\begin{array}{lr} ||\neg\varphi \Uparrow j||(s) & \text{(Definition of } \cdot \Uparrow j) \\ = ||\neg\varphi||(s) \sqsupseteq j & \text{(Definition of } ||\cdot||) \\ = \neg||\varphi||(s) \sqsupseteq j & \text{(Theorem 2)} \\ = \neg(||\varphi||(s) \sqsupseteq \mathsf{neg}(j)) & \text{(Definition of } \cdot \Uparrow j) \\ = \neg(||\varphi \Uparrow \mathsf{neg}(j)||(s)) & \end{array}$$

This, however, does not yield an elegant reduction algorithm.

4 Related Work

In this section, we compare the reduction presented in Sect. 3 to the work of others.

Multi-valued Models (Fitting). Fitting [13,14] introduced a concept of multi-valued models and extended the propositional modal logic (i.e. $L_\mu(\mathcal{L})$ without the fixpoint operators) to them. In his models, the values of propositions and transitions come from a Heyting instead of a De Morgan algebra.

Definition 7. *[14] A* Heyting algebra *is a tuple* $(\mathcal{L}, \sqsubseteq, \rightarrow, -)$, *where* $(\mathcal{L}, \sqsubseteq)$ *is a distributive lattice;* $\rightarrow$ *is a* relative pseudo-complement *defined as* $a \rightarrow b \triangleq \bigsqcup\{c \mid (c \sqcap a) \sqsubseteq b\}$; *and "$-$" is the negation operator defined as* $-a \triangleq a \rightarrow \bot$.

Heyting algebras are traditionally used as models for intuitionistic logic [15], with the relative pseudo-complement operator $\rightarrow$ used to model intuitionistic implication. The negation operator "$-$" satisfies the law of non-contradiction ($a \wedge -a = \bot$), but not necessarily the law of excluded middle, the involution of negation, or any of the De Morgan laws.

Since the definition of $\Diamond L_\mu(\mathcal{L})$ only depends on the fact that $\mathcal{L}$ is distributive, it extends the propositional modal logic of [14] with the fixpoint operators. For this logic, the reduction from multi-valued to two-valued semantics suggested by Fitting is identical to ours, with the only exception that he uses an equivalent concept of *proper prime filters* [11] instead of join-irreducible elements.

The definition of the universal next-time operator $\Box$ in [14] is the same as ours syntactically ($||\Box\varphi||(s) \triangleq \sqcap_{t\in S}(\mathbb{R}(s,t) \Rightarrow ||\varphi||(t))$). However, Fitting interprets the implication $\Rightarrow$ operator as the relative pseudo-complement, whereas we interpret it as material implication. The two definitions coincide for Boolean algebras – algebras that are both Heyting and De Morgan.

Definition 8. A Boolean algebra *is a Heyting algebra* $(\mathcal{L}, \sqsubseteq, \rightarrow, -)$ *such that* $(\mathcal{L}, \sqsubseteq, -)$ *is a De Morgan algebra.*

The relative pseudo-complement $\rightarrow$ operator of a Boolean algebra is equivalent to the material implication $a \rightarrow b \;=\; a \Rightarrow b \;=\; -a \vee b$.

Note that in the special case of Boolean algebras, Theorem 2 can be strengthened as follows:

Theorem 9. *Let* $(\mathcal{L}, \sqsubseteq, \rightarrow, \neg)$ *be a Boolean algebra. Then for any join-irreducible element* $j \in \mathcal{J}(\mathcal{L})$*, and* $\ell \in \mathcal{L}$*:* $\neg\ell \sqsupseteq j \;=\; \neg(\ell \sqsupseteq j)$

A stronger version of Theorem 8 for Boolean algebras is given below.

Theorem 10. *For a Boolean algebra* $\mathcal{L}$*, multi-valued model-checking for the temporal logic* $L_\mu(\mathcal{L})$ *is reducible to* $|\mathcal{J}(\mathcal{L})|$ *model-checking problems for* $L_\mu^{\sqsupseteq j}$*.*

That is, for a Boolean algebra $\mathcal{L}$, we get $|\mathcal{J}(\mathcal{L})|$ classical model-checking problems, each over a Kripke structure with the statespace *identical* to the statespace of K.

Reducing Multi-valued Model-Checking to Classical Model-Checking. Konikowska and Penczek [23,24] introduced a multi-valued temporal logic mv-CTL$^*(\mathcal{L})$ – an extension of CTL* [12] to De Morgan algebras – and defined its semantics on χKripke structures. An interesting consequence of their definition is that it does not preserve the duality of a universal path quantifier A and an existential path quantifier E. For example, for an mv-CTL$^*(\mathcal{L})$ state formula φ

$$
\begin{aligned}
& ||A\varphi||(s) && \text{(Definition of } ||\cdot|| \text{ from [24])}\\
&= \sqcap_{t\in S}(\mathbb{R}(s,t) \sqcap ||\varphi||(t)) \\
&\neq \neg \sqcup_{t\in S}(\mathbb{R}(s,t) \sqcap ||\neg\varphi||(t)) && \text{(Definition of } ||\cdot|| \text{ from [24])}\\
&= ||\neg E \neg\varphi||(s)
\end{aligned}
$$

The consequences of this observation are: (a) when restricted to the classical logic, mv-CTL$^*(\mathcal{L})$ is equivalent to the classical CTL* *only* on *total* Kripke structures; (b) contrary to the claim made in [24], the negation-free fragment of mv-CTL$^*(\mathcal{L})$ is *less* expressive than the full mv-CTL$^*(\mathcal{L})$; and (c) mv-CTL$^*(\mathcal{L})$ does not subsume commonly used multi-valued temporal logics such as χCTL($\mathcal{L}$) and the 3-valued μ-calculus of [1].

In [24], Konikowska and Penczek also identified sufficient conditions for reducing a model-checking problem for the negation-free fragment of mv-CTL$^*(\mathcal{L})$ into several mv-CTL* model-checking problems over sub-algebras of $\mathcal{L}$. However, neither a proof of the existence of the reduction, nor a constructive algorithm for it are provided. Instead, the reduction is developed for three classes of De Morgan algebras: a finite total order, a product of finite total orders, and a De Morgan algebra depicted in Fig. 1d. Theorem 8 in the current paper provides the missing existence proof and the reduction algorithm.

Chechik et al. [6] developed a symbolic model-checking algorithm for χCTL($\mathcal{L}$) based on Binary Decision Diagrams (BDD) [3]. They showed that for a given set S and a De Morgan algebra $\mathcal{L}$, $S \to \mathcal{L}$ functions (an mv-set in the terminology of [6]) can be represented and manipulated using a collection of $|\mathcal{J}(\mathcal{L})|$ boolean functions (or sets). For a given mv-set $\mathbb{Z} : S \to \mathcal{L}$, a set $\mathbb{Z} \Uparrow j$ of the collection corresponding to a join-irreducible element $j \in \mathcal{J}(\mathcal{L})$ is defined as $\mathbb{Z} \Uparrow j \triangleq \{x \in S \mid \mathbb{Z}(x) \sqsupseteq j\}$. In the current paper, we have extended the reduction technique of [6] to a richer logic $L_\mu(\mathcal{L})$ and decoupled the reduction from any particular implementation of the model-checking algorithm. Thus, our reduction technique is applicable to any current or future algorithm for classical model-checking together with any optimizations. The particular implementation of Chechik et al. [6] can be seen as an application of the reduction technique presented in the current paper in the context of symbolic model-checking.

Reductions for 3-Valued Reasoning. Bruns and Godefroid [2,1], Godefroid et. al [16], and Huth et al. [20,19] studied the problem of model-checking over the algebra **3** on a variety of 3-valued finite-state transition systems. Bruns and Godefroid [1,2] investigated 3-valued model-checking on Partial Kripke structures, where propositions are 3-valued but the transition relation is in $\{\top, \bot\}$. Godefroid et al. [16] provided an extension of the algorithm to Modal Transition Systems – a generalization of Labeled Transition Systems of [26], in which a transition relation is allowed to become 3-valued. The idea is further extended by Huth et al. [20] to Kripke Modal Transition Systems which are equivalent to our χKripke structures when the algebra is **3**. In all of these cases, it is shown that 3-valued model-checking is reducible to two classical model-checking problems. This is not surprising since all of the modeling formalisms have been shown to be equivalent [17]. Our logic $L_\mu(\mathbf{3})$ is equivalent to the 3-valued μ-calculus of [2,16,20] (and thus subsumes 3-valued CTL [1]). So the multi-valued model-checking reduction technique presented here can be seen as an extension of the reduction for the 3-valued model-checking beyond the algebra **3**.

Partial Model Checking from Multiple Viewpoints. The work of Huth and Pradhan [21] is the closest to ours. In this work, each of C stakeholders, arranged in a partial order, submits a partial model, consisting of valid ("must") and consistent ("may") statements about states and transitions. Given a first-order property, the model-checking problem is to determine sets of stakeholders for which the property is valid or consistent, respectively. The stakeholders correspond to join-irreducibles in our framework. The reduction described in [21] results in $|C|$ single-view partial models. Verification on each model is performed by switching between "valid" and "consistent" interpretations of satisfiability of properties.

5 Conclusion

In this paper, we studied the problem of multi-valued μ-calculus model-checking. Instead of solving the problem directly, we reduced it to several classical problems that can be solved using existing model-checking tools. The number of such problems depends on the number of join-irreducible elements of the logic, and each problem is linear in the size of the property and the model. We have also put numerous existing work in the area of non-classical verification into the context of our work.

Our results enable construction of clever algorithms that use results obtained from classical problems and the order of join-irreducibles to minimize the number of redundant checks. Yet, optimality can be achieved only by solving these problems in parallel, which is done by "true" multi-valued model-checkers such as χChek [7].

Acknowledgment. We thank Michael Huth, Benet Devereux and the anonymous referees for helping improve the presentation of this paper. Financial support for this research has been provided by NSERC and CITO.

References

1. G. Bruns and P. Godefroid. "Model Checking Partial State Spaces with 3-Valued Temporal Logics". In *Proceedings of 11th International Conference on Computer-Aided Verification (CAV'99)*, volume 1633 of *Lecture Notes in Computer Science*, pages 274–287, Trento, Italy, 1999. Springer.
2. G. Bruns and P. Godefroid. "Generalized Model Checking: Reasoning about Partial State Spaces". In C. Palamidessi, editor, *Proceedings of 11th International Conference on Concurrency Theory (CONCUR'00)*, volume 1877 of *Lecture Notes in Computer Science*, pages 168–182, University Park, PA, USA, August 2000. Springer.
3. R. E. Bryant. "Graph-based algorithms for boolean function manipulation.". *Transactions on Computers*, 8(C-35):677–691, 1986.
4. W. Chan. "Temporal-Logic Queries". In E. Emerson and A. Sistla, editors, *Proceedings of the 12th Conference on Computer Aided Verification (CAV'00)*, volume 1855 of *Lecture Notes in Computer Science*, pages 450–463, Chicago, IL, USA, July 2000.
5. M. Chechik, B. Devereux, S. Easterbrook, and A. Gurfinkel. "Multi-Valued Symbolic Model-Checking". *ACM Transactions on Software Engineering and Methodology*, 2003. (Accepted for publication.).
6. M. Chechik, B. Devereux, S. Easterbrook, A. Lai, and V. Petrovykh. "Efficient Multiple-Valued Model-Checking Using Lattice Representations". In K.G. Larsen and M. Nielsen, editors, *Proceedings of 12th International Conference on Concurrency Theory (CONCUR'01)*, volume 2154 of *Lecture Notes in Computer Science*, pages 451–465, Aalborg, Denmark, August 2001. Springer.
7. M. Chechik, B. Devereux, and A. Gurfinkel. "χChek: A Multi-Valued Model-Checker". In *Proceedings of 14th International Conference on Computer-Aided Verification (CAV'02)*, volume 2404 of *Lecture Notes in Computer Science*, pages 505–509, Copenhagen, Denmark, July 2002. Springer.
8. M. Chechik, S. Easterbrook, and V. Petrovykh. "Model-Checking Over Multi-Valued Logics". In *Proceedings of Formal Methods Europe (FME'01)*, volume 2021 of *Lecture Notes in Computer Science*, pages 72–98. Springer, March 2001.
9. E. Clarke, O. Grumberg, and D. Peled. *Model Checking*. MIT Press, 1999.

10. E.M. Clarke, E.A. Emerson, and A.P. Sistla. "Automatic Verification of Finite-State Concurrent Systems Using Temporal Logic Specifications". *ACM Transactions on Programming Languages and Systems*, 8(2):244–263, April 1986.
11. B.A. Davey and H.A. Priestley. *Introduction to Lattices and Order*. Cambridge University Press, 1990.
12. E.A. Emerson. "Temporal and Modal Logic". In J. van Leeuwen, editor, *Handbook of Theoretical Computer Science*, volume B, pages 995–1072. Elsevier Science Publishers, Amsterdam, 1990.
13. M. Fitting. "Many-Valued Modal Logics". *Fundamenta Informaticae*, 15(3-4):335–350, 1991.
14. M. Fitting. "Many-Valued Modal Logics II". *Fundamenta Informaticae*, 17:55–73, 1992.
15. M.C. Fitting. *Intuitionistic Logic Model Theory and Forcing*. North-Holand Publishing Co., Amsterdam, 1969.
16. P. Godefroid, M. Huth, and R. Jagadeesan. "Abstraction-Based Model Checking Using Modal Transition Systems". In K.G. Larsen and M. Nielsen, editors, *Proceedings of 12th International Conference on Concurrency Theory (CONCUR'01)*, volume 2154 of *Lecture Notes in Computer Science*, pages 426–440, Aalborg, Denmark, 2001. Springer.
17. P. Godefroid and R. Jagadeesan. "On the Expressiveness of 3-Valued Models". In *Proceedings of 4th International Conference on Verification, Model Checking, and Abstract Interpretation (VMCAI'03)*, volume 2575 of *Lecture Notes in Computer Science*, pages 206–222, New York, USA, January 2003. Springer.
18. A. Gurfinkel, M. Chechik, and B. Devereux. "Temporal Logic Query Checking: A Tool for Model Exploration". *IEEE Transactions on Software Engineering*, 2003. (Accepted for publication.).
19. M. Huth, R. Jagadeesan, and D. Schmidt. "A Domain Equation for Refinement of Partial Systems". *Mathematical Structures in Computer Science*, 2003. (Accepted for publication).
20. M. Huth, R. Jagadeesan, and D. A. Schmidt. "Modal Transition Systems: A Foundation for Three-Valued Program Analysis". In *Proceedings of 10th European Symposium on Programming (ESOP'01)*, volume 2028 of *Lecture Notes in Computer Science*, pages 155–169. Springer, 2001.
21. M. Huth and S. Pradhan. "An Ontology for Consistent Partial Model Checking". *Electronic Notes in Theoretical Computer Science*, 23, 2003.
22. S. C. Kleene. *Introduction to Metamathematics*. New York: Van Nostrand, 1952.
23. B. Konikowska and W. Penczek. "Model Checking for Multi-Valued CTL*". In M. Fitting and E. Orlowska, editors, *Multi-Valued Logics*. 2002. (To appear.).
24. B. Konikowska and W. Penczek. "Reducing Model Checking from Multi-Valued CTL* to CTL*". In *Proceedings of 13 International Conference on Concurrency Theory (CONCUR'02)*, volume 2421 of *Lecture Notes in Computer Science*, Brno, Czech Republic, August 2002. Springer.
25. D Kozen. "Results on the Propositional μ-calculus". *Theoretical Computer Science*, 27:334–354, 1983.
26. K.G. Larsen and B. Thomsen. "A Modal Process Logic". In *Proceedings of 3rd Annual Symposium on Logic in Computer Science (LICS'88)*, pages 203–210. IEEE Computer Society Press, 1988.
27. R. Milner. *Communication and Concurrency*. Prentice-Hall, New York, 1989.
28. M. Müller-Olm, D. Schmidt, and B. Steffen. "Model-Checking: A Tutorial Introduction". In *Proceedings of 6th International Static Analysis Symposium (SAS'99*, volume 1694 of *Lecture Notes in Computer Science*, pages 330–354, Berlin, September 1999.
29. H. Rasiowa. *An Algebraic Approach to Non-Classical Logics. Studies in Logic and the Foundations of Mathematics*. Amsterdam: North-Holland, 1978.

An Extension of Pushdown System and Its Model Checking Method

Naoya Nitta[1] and Hiroyuki Seki[1]

Nara Institute of Science and Technology
8916-5 Takayama, Ikoma, Nara 630-0192, Japan
{naoya-n,seki}@is.aist-nara.ac.jp

Abstract. In this paper, we present a class of infinite transition systems which is an extension of pushdown systems (PDS), and show that LTL (linear temporal logic) model checking for the class is decidable. Since the class is defined as a subclass of term rewriting systems, pushdown stack of PDS is naturally extended to tree structure. By this extension, we can model recursive programs with exception handling.

1 Introduction

Model checking [2] is a well-known technique which automatically verifies whether a system satisfies a given specification. Most of existing model checking methods and tools assume that a system to be verified has finite state space. This is a serious restriction when we apply model checking to software verification since a program is usually modeled as a system with infinite state space. There are two approaches to resolving the problem. One is that if a system to be verified has infinite state space, then the system is transformed into an abstract system with finite state space [4,11]. However, the abstract system does not always retain the desirable property which the original system has, in which case the verification fails.

Another approach is to introduce a new subclass of transition systems which is wider than finite state systems. Pushdown system (abbreviated as PDS) is such a subclass that is wider than finite state systems and yet has decidable properties on model checking. A PDS can model a system which has well-nested structure such as a program involving recursive procedure calls. Recently, efficient algorithms of LTL and CTL* model checking for PDS have been proposed in [5,6] (also see related works). The transition relation of a PDS is defined by transition rules which rewrite the finite control and a prefix of the string in the pushdown stack. Thus, if we model a program as a PDS, we are forced to define the behavior of the program by transition rules on strings.

In this paper, we focus on term rewriting system (abbreviated as TRS), which is one of the well-known general computation models, and define the model checking problem for TRS. For simplicity, we consider the rewrite relation induced by the rewriting only at the root position of a term (root rewriting). Since a transition in a PDS changes the finite control and a prefix of the strings

R. Amadio, D. Lugiez (Eds.): CONCUR 2003, LNCS 2761, pp. 281–295, 2003.

in the stack, PDS can be regarded as a TRS with root rewriting. Next, a new subclass of TRS, called generalized-growing TRS (GG-TRS) is defined. GG-TRS properly includes growing TRS [17] of Nagaya and Toyama. We present a necessary and sufficient condition for a left-linear(LL-)GG-TRS $\mathcal{R}$ to have an infinite rewrite sequence which visits terms in a given set infinitely often. Based on this condition, we then present a condition for $\mathcal{R}$ to satisfy a given LTL formula ϕ. The latter condition is decidable if $\mathcal{R}$ has a property called pre-(or post-)recognizability preserving property. Lastly, we introduce a subclass of TRS called LL-SPO-TRS and show that every TRS in this subclass has pre-recognizability preserving property. Every PDS belongs to both of GG-TRS and LL-SPO-TRS. Furthermore, we show that a program with recursive procedure and exception handling can be naturally modeled as a TRS in both GG-TRS and LL-SPO-TRS, which is not strongly bisimilar to any PDS. In this sense, the decidability results on LTL model checking in this paper is an extension of the results in [5,6]. Detailed proofs are omitted due to space limitation (see [18]).

Related Works. The model checking problem for PDS and the modal μ-calculus is studied in [24]. For LTL and CTL*, efficient model checking algorithms for PDS are proposed in [5,6]. Major applications of model checking for PDS are static analysis of programs and security verification. For the former, Esparza et al. [6] discuss an application of model checking for PDS to dataflow analysis of recursive programs. Some results obtained by using their verification tool are also reported in [7]. The first work which applies model checking of a pushdown-type system to security verification is Jensen et al.'s study [13], which introduces a safety verification problem for a program with access control which generalizes JDK1.2 stack inspection. Nitta et al. [19,20] improve the result of [13] and show that a safety verification problem is decidable for an arbitrary program with stack inspection. In [20], a subclass of programs which exactly represents programs with JDK1.2 stack inspection is proposed, for which the safety verification problem is decidable in polynomial time of the program size. In [6], it is shown that LTL model checking is decidable for an arbitrary programs with stack inspection. Jha and Reps show that name reduction in SPKI [22] can be represented as a PDS, and prove the decidability of a number of security problems by reductions to decidability properties of model checking for PDS [14]. Among other infinite state systems for which model checking has been studied are process rewrite system (PRS) [16] and ground TRS [15]. PRS includes PDS and Petri Net as its subclasses. However, LTL model checking is undecidable for both of PRS and ground TRS.

2 Preliminaries

2.1 Term Rewriting System

We use the usual notions for terms, substitutions, etc (see [1] for details). Let N denote the set of natural numbers. Let $\mathcal{F}$ be a *signature* and $\mathcal{V}$ be an enumerable set of *variables*. An element in $\mathcal{F}$ is called a *function symbol* and the *arity* of

$f \in \mathcal{F}$ is denoted by $arity(f)$. A function symbol c with $arity(c) = 0$ is called a *constant*. The set of *terms* generated by $\mathcal{F}$ and $\mathcal{V}$ is denoted by $\mathcal{T}(\mathcal{F}, \mathcal{V})$. The set of variables occurring in t is denoted by $\mathcal{V}ar(t)$. A term t is *ground* if $\mathcal{V}ar(t) = \emptyset$. The set of all ground terms is denoted by $\mathcal{T}(\mathcal{F})$. A term is *linear* if no variable occurs more than once in the term. A *substitution* θ is a mapping from $\mathcal{V}$ to $\mathcal{T}(\mathcal{F}, \mathcal{V})$ and written as $\theta = \{x_1 \mapsto t_1, \ldots, x_n \mapsto t_n\}$ where t_i with $1 \leq i \leq n$ is a term which substitutes for the variable x_i. The term obtained by applying a substitution θ to a term t is written as $t\theta$. We call $t\theta$ an *instance* of t. A *position* in a term t is defined as a sequence of positive integers, and the root position is the empty sequence denoted by λ. The depth of a position $p \in (N - \{0\})^*$, written as $|p|$, is the length of p (e.g. $|132| = 3$). Let $\preceq_{pref}$ denote the prefix relation on positions. The set of all positions in a term t is denoted by $Pos(t)$. Also let us define $Pos_{=n}(t) = \{p \in Pos(t) \mid |p| = n\}$ and $Pos_{\geq n}(t) = \{p \in Pos(t) \mid |p| \geq n\}$. A subterm of t at a position $p \in Pos(t)$ is denoted by $t_{|p}$. $Pos(t, s)$ is the set $\{p \mid t_{|p} = s\}$. If $t_{|p} = f(\cdots)$, then we write $lab(t,p) = f$. If a term t is obtained from a term t' by replacing the subterms of t' at positions $p_1, \ldots, p_m$ ($p_i \in Pos(t'), p_i$ and p_j are disjoint if $i \neq j$) with terms $t_1, \ldots, t_m$, respectively, then we write $t = t'[p_i \leftarrow t_i \mid 1 \leq i \leq m]$. The depth of a term t is $max\{|p| \mid p \in Pos(t)\}$. For terms s, t, let $mgu(s,t)$ denote the most general unifier of s and t if it is defined. Otherwise, let $mgu(s,t) = \perp$.

A *rewrite rule* over a signature $\mathcal{F}$ is an ordered pair of terms in $\mathcal{T}(\mathcal{F}, \mathcal{V})$, written as $l \to r$. A *term rewriting system* (*TRS*) over $\mathcal{F}$ is a finite set of rewrite rules over $\mathcal{F}$. For terms t, t' and a TRS $\mathcal{R}$, we write $t \to_{\mathcal{R}} t'$ if there exists a position $p \in \mathcal{P}os(t)$, a substitution θ and a rewrite rule $l \to r \in \mathcal{R}$ such that $t/p = l\theta$ and $t' = t[p \leftarrow r\theta]$. Define $\to^*_{\mathcal{R}}$ to be the reflexive and transitive closure of $\to_{\mathcal{R}}$. Sometimes $t \to^*_{\mathcal{R}} t'$ is called a rewrite sequence. Also the transitive closure of $\to_{\mathcal{R}}$ is denoted by $\to^+_{\mathcal{R}}$. The subscript $\mathcal{R}$ of $\to_{\mathcal{R}}$ is omitted if $\mathcal{R}$ is clear from the context. A *redex* (*in* $\mathcal{R}$) is an instance of l for some $l \to r \in \mathcal{R}$. A *normal form* (in $\mathcal{R}$) is a term which has no redex as its subterm. Let $\mathrm{NF}_{\mathcal{R}}$ denote the set of all ground normal forms in $\mathcal{R}$. A rewrite rule $l \to r$ is *left-linear*(resp. *right-linear*) if l is linear (resp. r is linear). A rewrite rule is *linear* if it is left-linear and right-linear. A TRS $\mathcal{R}$ is *left-linear* (resp. *right-linear, linear*) if every rule in $\mathcal{R}$ is left-linear (resp. right-linear, linear).

2.2 Tree Automata and Recognizability

A *tree automaton*(*TA*) [8] is defined by a 4-tuple $\mathcal{A} = (\mathcal{F}, \mathcal{Q}, \Delta, \mathcal{Q}^{final})$ where $\mathcal{F}$ is a signature, $\mathcal{Q}$ is a finite set of states, $\mathcal{Q}^{final} \subseteq \mathcal{Q}$ is a set of final states, and Δ is a finite set of transition rules of the form $f(q_1, \ldots, q_n) \to q$ where $f \in \mathcal{F}$, $arity(f) = n$, and $q_1, \ldots, q_n, q \in \mathcal{Q}$ or of the form $q' \to q$ where $q, q' \in \mathcal{Q}$. Consider the set of ground terms $\mathcal{T}(\mathcal{F} \cup \mathcal{Q})$ where we define $arity(q) = 0$ for $q \in \mathcal{Q}$. A *transition* of a TA can be regarded as a rewrite relation on $\mathcal{T}(\mathcal{F} \cup \mathcal{Q})$ by regarding transition rules in Δ as rewrite rules over $\mathcal{F} \cup \mathcal{Q}$. For terms t and t' in $\mathcal{T}(\mathcal{F} \cup \mathcal{Q})$, we write $t \vdash_{\mathcal{A}} t'$ if and only if $t \to_{\Delta} t'$. The reflexive and transitive closure and the transitive closure of $\vdash_{\mathcal{A}}$ is denoted by $\vdash^*_{\mathcal{A}}$ and $\vdash^+_{\mathcal{A}}$, respectively. For a TA $\mathcal{A}$ and $t \in \mathcal{T}(\mathcal{F})$, if $t \vdash^*_{\mathcal{A}} q_f$ for a final state $q_f \in \mathcal{Q}^{final}$, then we say t is *accepted* by $\mathcal{A}$. The set of all ground terms in $\mathcal{T}(\mathcal{F})$ accepted by $\mathcal{A}$ is denoted by $\mathcal{L}(\mathcal{A})$ and we say that $\mathcal{A}$ recognizes $\mathcal{L}(\mathcal{A})$. A subset $L \subseteq \mathcal{T}(\mathcal{F})$ of ground terms is called a tree language. A tree language L is *recognizable* if there is a TA $\mathcal{A}$ such that $L = \mathcal{L}(\mathcal{A})$.

For a TRS $\mathcal{R}$ and a tree language L, let $post^*_{\mathcal{R}}(L) = \{t \mid \exists s \in L \text{ s.t. } s \to^*_{\mathcal{R}} t\}$ and $pre^*_{\mathcal{R}}(L) = \{t \mid \exists s \in L \text{ s.t. } t \to^*_{\mathcal{R}} s\}$. A TRS $\mathcal{R}$ is said to *effectively preserve post-recognizability* (abbreviated as post-PR) if, for any TA $\mathcal{A}$, $post^*_{\mathcal{R}}(\mathcal{L}(\mathcal{A}))$ is also recognizable and we can effectively construct a TA which accepts $post^*_{\mathcal{R}}(\mathcal{L}(\mathcal{A}))$. We define pre-PR in a similar way. For a TRS $\mathcal{R}$, let $\mathcal{R}^{-1} = \{r \to l \mid l \to r \in \mathcal{R}\}$. By definition, $post^*_{\mathcal{R}^{-1}}(L) = pre^*_{\mathcal{R}}(L)$. Thus, a TRS $\mathcal{R}$ is pre-PR if and only if $\mathcal{R}^{-1}$ is post-PR. The class of recognizable tree languages is closed under boolean operations and the inclusion problem is decidable for the class [8]. Due to these properties, some important problems, e.g., reachability, joinability and local confluence are decidable for post-PR TRS [10, 12]. However, whether a given TRS is pre-PR (post-PR) is undecidable [9], and decidable subclasses of pre-PR or post-PR TRS have been proposed, some of which are listed with inclusion relation:

RL-SM(semi-monadic)-TRS[3] $\subset$ RL-GSM(generalized semi-monadic)-TRS [12] $\subset$ RL-FPO(finitely path overlapping)-TRS[23]

where RL stands for 'right-linear.' As a decidable subclass of pre-PR TRS, left-linear growing TRS (LL-G-TRS) [17] is known. A TRS $\mathcal{R}$ is a G-TRS if for every rule $l \to r$ in $\mathcal{R}$, every variable in $\mathcal{V}ar(l) \cap \mathcal{V}ar(r)$ appears at depth 0 or 1 in l. Hence, a shallow TRS is always a G-TRS. Note that $\mathcal{R}$ is an SM-TRS if and only if $\mathcal{R}^{-1}$ is a G-TRS and the left-hand side of any rule in $\mathcal{R}$ is not a constant.

2.3 Transition Systems and Linear Temporal Logic

A *transition system* is a 3-tuple $\mathcal{S} = (S, \to, s_0)$, where S is a (possibly infinite) set of states, $\to \subseteq S \times S$ is a *transition relation* and $s_0 \in S$ is an *initial state.* The transitive closure of $\to$ and the reflexive and transitive closure of $\to$ are written by $\to^+$ and $\to^*$, respectively. A *run* of $\mathcal{S}$ is an infinite sequence of states $\sigma = s_1 s_2 \ldots$ such that $s_i \to s_{i+1}$ for each $i \geq 1$. Let $At = \{\alpha_1, \alpha_2, \ldots, \alpha_k\}$ be a set of atomic propositions. The syntax of linear temporal logic (LTL) formula ϕ is defined by

$$\phi ::= tt \mid \alpha_i \mid \neg\phi \mid \phi_1 \wedge \phi_1 \mid \mathcal{X}\phi \mid \phi_1 \mathcal{U} \phi_2$$

($1 \leq i \leq k$ and ϕ_1, ϕ_2 are LTL formulas). For a transition system $\mathcal{S} = (S, \to, s_0)$, a *valuation* of $\mathcal{S}$ is a function $\nu : At \to 2^S$. The *validity* of an LTL formula ϕ for a run $\sigma = s_1 s_2 \ldots$ w.r.t. a valuation ν is denoted by $\sigma \models^\nu \phi$, and defined in the standard way [2]. We say ϕ is *valid* at s w.r.t. ν, denoted as $s \models^\nu \phi$, if and only if $\sigma \models^\nu \phi$ for each run σ starting in s.

2.4 Model Checking for TRS

Given a TRS $\mathcal{R}$ over a signature $\mathcal{F}$ and a term $t_0 \in \mathcal{T}(\mathcal{F})$, we can define a transition system $\mathcal{S}_{\mathcal{R}} = (\mathcal{T}(\mathcal{F}), \to_{\mathcal{S}_{\mathcal{R}}}, t_0)$ where $\to_{\mathcal{S}_{\mathcal{R}}} = \to_{\mathcal{R}} \cup \{(t,t) \mid t \in \mathrm{NF}_{\mathcal{R}}\}$.

Note that the reflexive relation $\{(t,t) \mid t \in \mathrm{NF}_{\mathcal{R}}\}$ is needed to make the transition relation $\to_{\mathcal{S}_{\mathcal{R}}}$ total. The validity of LTL formula ϕ at t_0 in $\mathcal{S}_{\mathcal{R}}$ w.r.t. $\nu : At \to 2^{\mathcal{T}(\mathcal{F})}$ is denoted as $\mathcal{R}, t_0 \models^{\nu} \phi$. From an LTL formula ϕ, we can construct a Büchi automaton which recognizes the set of models of $\neg\phi$. Therefore, we often assume that we are given a Büchi automaton instead of an LTL formula. In a similar way to the model checking method in [6], we define a *Büchi TRS* which synchronizes a transition system $\mathcal{S}_{\mathcal{R}}$ given by a TRS $\mathcal{R}$ with a Büchi automaton $\mathcal{B}$. First, to make the definition constructive, we make a few observations. To synchronize $\mathcal{S}_{\mathcal{R}}$ with $\mathcal{B}$, we must construct a Büchi TRS so that the redex can keep track of the information on the current state of $\mathcal{B}$ and the valuation of the current term of $\mathcal{S}_{\mathcal{R}}$. However, if we allow an arbitrary redex to be rewritten, transmitting the above information to the next redex in the Büchi TRS becomes difficult. For this reason, we consider *root rewriting*, which restricts rewriting positions to the root position.

Definition 1. (Root Rewriting) *For terms t, t' and a TRS $\mathcal{R}$, we say $t \to_{\mathcal{R}} t'$ is root rewriting, if there exist a substitution θ and a rewrite rule $l \to r \in \mathcal{R}$ such that $t = l\theta$ and $t' = r\theta$.* ∎

If we consider root rewriting, it is not difficult to see that there effectively exists a TRS of which the rewrite relation exactly corresponds to $\to_{\mathcal{S}_{\mathcal{R}}}$. Let $\{\Lambda_1, \ldots, \Lambda_m\}$ be a set of terms in $\mathcal{T}(\mathcal{F}, \mathcal{V})$ such that $\mathrm{NF}_{\mathcal{R}} = \bigcup_{1 \le i \le m}\{\Lambda_i\theta \mid \theta : \mathcal{V} \to \mathcal{T}(\mathcal{F})$ is a substitution$\}$, and $mgu(\Lambda_i, \Lambda_j) = \bot (1 \le i < j \le m)$. Also, let $\widetilde{\mathcal{R}} = \mathcal{R} \cup \{\Lambda_i \to \Lambda_i \mid 1 \le i \le m\}$. Then, $t \in \mathrm{NF}_{\mathcal{R}}$ if and only if there exists a unique Λ_i such that $t \to_{\widetilde{\mathcal{R}}} t \to_{\widetilde{\mathcal{R}}} \cdots$ where $\Lambda_i \to \Lambda_i$ is applied in each rewrite step. Hence, we know $\to_{\mathcal{S}_{\mathcal{R}}} = \to_{\widetilde{\mathcal{R}}}$, i.e., the transition relation $\to_{\mathcal{S}_{\mathcal{R}}}$ of $\mathcal{S}_{\mathcal{R}}$ can be induced by TRS $\widetilde{\mathcal{R}}$. Next, we extend the definitions of valuations of PDS [5,6].

Definition 2. (Simple Valuation)
Let $\mu : At \to \mathcal{T}(\mathcal{F}, \mathcal{V})$ be a function such that for each $\alpha \in At$ and $l \to r \in \widetilde{\mathcal{R}}$, $mgu(l, \mu(\alpha)) = l$ or $= \bot$. The simple valuation $\nu : At \to 2^{\mathcal{T}(\mathcal{F})}$ given by μ is defined as $\nu(\alpha) = \{\mu(\alpha)\theta \mid \theta$ is a substitution$\}$. ∎

In the definition, $\mu(\alpha)$ specifies a pattern of terms for which proposition α is true. For example, if $\mu(\alpha_1) = f(x, g(y))$ then $\mathcal{R}, t \models^{\nu} \alpha_1$ if and only if t is an instance of $f(x, g(y))$. The restriction that $mgu(l, \mu(\alpha)) = l$ or $= \bot$ guarantees that for a rewrite rule $l \to r$, whether $\mathcal{R}, l\theta \models^{\nu} \alpha$ is determined independent of a substitution θ.

Definition 3. (Regular Valuation)
For each atomic proposition $\alpha \in At$, a TA $\mathcal{A}_{\alpha}$ is given. The regular valuation $\nu : At \to 2^{\mathcal{T}(\mathcal{F})}$ given by $\langle \mathcal{A}_{\alpha} \rangle_{\alpha \in At}$ is defined as $\nu(\alpha) = \mathcal{L}(\mathcal{A}_{\alpha})$. ∎

Definition 3 says that $\mathcal{R}, t \models^{\nu} \alpha$ if and only if t is accepted by $\mathcal{A}_{\alpha}$. This is a natural extension of regular valuation ν of PDS, where a configuration $\langle q, w \rangle$ is a pair of a control location q and a sequence w of stack symbols and $\langle q, w \rangle \models^{\nu} \alpha$ if and only if the sequence qw is accepted by a finite state automaton $\mathcal{A}_{\alpha}$ given for α.

Definition 4. (Büchi TRS) *Let At be a set of atomic propositions, $\mathcal{R}$ be a TRS, $\mathcal{B} = (\mathcal{Q}_\mathcal{B}, \Sigma_\mathcal{B}, \Delta_\mathcal{B}, q_{0\mathcal{B}}, \mathcal{Q}_\mathcal{B}^{acc})$ ($\Sigma_\mathcal{B} = 2^{At}$, $\mathcal{Q}_\mathcal{B} \cap \mathcal{F} = \emptyset$) be a Büchi automaton, and ν be the simple valuation given by $\mu : At \to \mathcal{T}(\mathcal{F}, \mathcal{V})$. For $\mathcal{R}$, $\mathcal{B}$ and ν, we define Büchi TRS $\mathcal{BR}^\nu$ as follows: The signature of $\mathcal{BR}^\nu$ is $\mathcal{F}_{\mathcal{BR}^\nu} = \mathcal{Q}_\mathcal{B} \cup \mathcal{F}$ (for any $q \in \mathcal{Q}_\mathcal{B}$, $arity(q) = 1$), and $\mathcal{BR}^\nu$ is the minimum set of rules satisfying:*

$$q \xrightarrow{a} q' \in \Delta_\mathcal{B}, l \to r \in \widetilde{\mathcal{R}}, \text{ and } a \subseteq \{\alpha \in At \mid mgu(l, \mu(\alpha)) = l\}$$
$$\Rightarrow q(l) \to q'(r) \in \mathcal{BR}^\nu.$$

If ν is regular, then we can reduce a model checking problem w.r.t. ν to a model checking problem w.r.t. a simple valuation in a similar way to [6].

Lemma 1. *Let $\mathcal{R}$ be a TRS, $t_0 \in \mathcal{T}(\mathcal{F})$ be an initial state, ϕ be an LTL formula, $\mathcal{B} = (\mathcal{Q}_\mathcal{B}, \Sigma_\mathcal{B}, \Delta_\mathcal{B}, q_{0\mathcal{B}}, \mathcal{Q}_\mathcal{B}^{acc})$ ($\Sigma_\mathcal{B} = 2^{At}$, $\mathcal{Q}_\mathcal{B} \cap \mathcal{F} = \emptyset$) be a Büchi automaton which represents $\neg\phi$, and ν be the simple valuation given by $\mu : At \to \mathcal{T}(\mathcal{F}, \mathcal{V})$. Also, let $\mathcal{T}_{\mathrm{acc}} = \{q_\mathrm{a}(t) \mid q_\mathrm{a} \in \mathcal{Q}_\mathcal{B}^{acc}, t \in \mathcal{T}(\mathcal{F})\}$. $\mathcal{R}, t_0 \not\models^\nu \phi$ if and only if there exists an infinite root rewrite sequence of Büchi TRS $\mathcal{BR}^\nu$ starting in $q_{0\mathcal{B}}(t_0)$ and visiting $\mathcal{T}_{\mathrm{acc}}$ infinitely often.*

3 Generalized-Growing TRS and Its Model Checking

The restriction of root rewriting (Definition 1) on TRS $\mathcal{R}$ is insufficient to make the model checking problem for $\mathcal{R}$ decidable, because root rewriting TRSs are still Turing powerful. In fact, we can define an automaton with two pushdown stacks (which is Turing powerful) as a left-linear root rewriting TRS by encoding a state of the finite control as a root symbol q with arity 2 and each of the two stacks as each argument of q. The reason why root rewriting TRSs are Turing powerful is unrestricted information flow between different arguments of a function symbol such as q above. We introduce a subclass of TRS, called LL-GG-TRS, in which the information of (function symbol in) an argument is never shifted to another argument, and show that if an LL-GG-TRS $\mathcal{R}$ is post-PR (or pre-PR), then LTL model checking for $\mathcal{R}$ is decidable. For positions p_1, p_2, we define the *least common ancestor* $p_1 \sqcup p_2$ as the longest common prefix of p_1 and p_2.

Definition 5. (Left-Linear Generalized-Growing TRS (LL-GG-TRS)) *A left-linear rule $l \to r$ is generalized-growing, if every two different variables $x, y \in \mathcal{V}ar(l) \cap \mathcal{V}ar(r)$ satisfy the following condition: For the positions o_l^x, o_l^y of x, y in l and for each positions $o_r^x \in Pos(r, x), o_r^y \in Pos(r, y)$ of x, y in r,*

$$|o_l^x| - |o_l^x \sqcup o_l^y| \le |o_r^x| - |o_r^x \sqcup o_r^y|, \text{ and } |o_l^y| - |o_l^x \sqcup o_l^y| \le |o_r^y| - |o_r^x \sqcup o_r^y|.$$

$\mathcal{R}$ is left-linear generalized-growing (LL-GG), if every rule in TRS $\mathcal{R}$ is left-linear and generalized-growing.

Obviously, an LL-G-TRS (see Sect. 2.2) is always an LL-GG-TRS.

Example 1. Consider $\mathcal{R}_1 = \{\ f(g(x, y)) \to f(h(y), x)\ \}$. The position of x is 11 in l and 2 in r, and the position of y is 12 in l and 11 in r. Since $11 \sqcup 12 = 1$ and

```
    void main() {
L0:     switch (random_integer) {
        case 0:
L1:        main();
           break;
        case 1:
L2:        try {
L3:           main();
           } catch (e) {
L4:           nop;
           }
           break;
        case 2:
L5:        throw e;
        }
L6:     return;
    }
```

$\Rightarrow$

$$
\begin{array}{l}
\mathcal{R}_2 = \{ \\
\quad \left.\begin{array}{l}
L_0(x,y) \to L_1(x,y), \\
L_0(x,y) \to L_2(x,y), \\
L_0(x,y) \to L_5(x,y), \\
L_0(x,y) \to L_6(x,y), \\
L_4(x,y) \to L_6(x,y),
\end{array}\right\} \quad : \textbf{seq} \\
\quad L_6(x,y) \to x, \qquad : \textbf{ret} \\
\quad L_5(x,y) \to y, \qquad : \textbf{throw} \\
\quad \left.\begin{array}{l}
L_1(x,y) \to L_0(L_6(x,y),y), \\
L_3(x,y) \to L_0(L_6(x,y),y),
\end{array}\right\} \quad : \textbf{call} \\
\quad L_2(x,y) \to L_3(x, L_4(x,y)) \qquad : \textbf{try-catch} \\
\}
\end{array}
$$

Fig. 1. A sample program with exception handling

$2 \sqcup 11 = \lambda$, $\mathcal{R}_1$ is an LL-GG-TRS, but $\mathcal{R}_1$ is not an LL-G-TRS because variables x and y occur at depth 2 in l. On the other hand, $\mathcal{R}_1^{-1} = \{ f(h(y),x) \to f(g(x,y)) \}$ is not an LL-GG-TRS, since the difference of the depth of positions in l between y and the least common ancestor of x and y is larger than that in r. ∎

Example 2. **(Recursive Program with Exception Handling)**
It is well-known that a program with recursive procedure can be naturally modeled as a PDS, and further in [21], a PDS model of Java-like programs including exception handling was proposed. In this model, the exception handling mechanism is implemented by adding extra control states and rules which represent low-level operations of the execution environment. On the other hand, in this example, we present an LL-GG-TRS model of recursive programs, which is closer to the behavioral semantics incorporated with exception handling in the source code level. For example, a Java-like program in the left half of Fig. 1 can be directly modeled as an LL-GG-TRS $\mathcal{R}_2$ shown in the right half. Note that the class LL-GG-TRS is properly wider than the class of PDSs w.r.t. strong bisimulation equivalence, and $\mathcal{R}_2$ is an example of LL-GG-TRS which has no strongly bisimilar PDS [18]. In a Java program, try-catch-throw statements are used for specifying exception handling. By the execution of a *throw* statement, an exception is propagated in the program. If an exception occurs within a *try* block, then the control immediately moves to the *catch* statement coupled with the try statement (with unwinding the control stack). From a program *Prog* including try-catch-throw statements, we can construct an LL-GG-TRS $\mathcal{R}$ as follows. In $\mathcal{R}$, every term t has the form of $f(t_1, t_2)$ where f denotes the current program location of *Prog*, t_1 denotes the next state of t if a return statement is executed at t, and t_2 denotes the next state of t if an exception occurs at t. A constant symbol $\square$ denotes the stack bottom. Every unit execution of *Prog* belongs to one

of the five types, **seq**, **call**, **ret**, **try-catch** and **throw**, and is translated into one of the following rules according to its type:

$$\begin{aligned}
\textbf{seq: } & current(x,y) \to succ(x,y), \\
\textbf{call: } & caller(x,y) \to callee(succ(x,y),y), \\
\textbf{ret: } & ret(x,y) \to x, \\
\textbf{try-catch: } & try(x,y) \to succ(x, catch(x,y)), \\
\textbf{throw: } & throw(x,y) \to y.
\end{aligned}$$

A **seq** rule and a **call** rule represent a sequential execution in a method and a method invocation, respectively. A **try-catch** rule represents the behavior of a try-catch block, where *succ* is the entry point of the try block and *catch* is the entry point of the catch block. A **ret** rule and a **throw** rule represent a return statement and a throw statement, respectively. It is interesting to recognize a symmetry between (**call, ret**) rules and (**try-catch, throw**) rules. Recall the program in Fig. 1. Since the statement at L2 is try, the entry point of the try block is L3, and the entry point of the catch block is L4, $L_2(x,y) \to L_3(x, L_4(x,y)) \in \mathcal{R}_2$. ∎

In the following, we only consider root rewrite sequences consisting of ground terms. The first lemma for LL-GG-TRS states that for any root rewrite sequence σ if there exists a position o_0 in the first term t_0 of σ such that the depth of (a residual of) o_0 is never shortened in σ, then for every 'sufficiently deep' position p_0 in t_0, every residual of p_0 never be contained in any redex. For a TRS $\mathcal{R}$, let $max_v(\mathcal{R})$ be the maximum depth of positions of variables in the left-hand sides of rules in $\mathcal{R}$, and $max_f(\mathcal{R})$ be the maximum depth of positions of function symbols in both sides of rules in $\mathcal{R}$. For a rewrite sequence $\sigma : t \to^*_{\mathcal{R}} t'$ and $p \in Pos(t)$, the set of *residuals* of p in σ, denoted as $Res(p,\sigma)$, is defined as follows. $Res(p, t \to^0_{\mathcal{R}} t) = \{p\}$. Assume $t = l\theta \to_{\mathcal{R}} r\theta = t'$ for a rule $l \to r$ and a substitution θ.

$$Res(p, t \to_{\mathcal{R}} t') = \begin{cases} \{p'_1 p_2 \mid r_{|p'_1} = x\} & \text{if } p = p_1 p_2 \text{ and } l_{|p_1} = x \in Var(l), \\ \emptyset & \text{otherwise.} \end{cases}$$

For a rewrite sequence $t \to^*_{\mathcal{R}} t' \to_{\mathcal{R}} t''$, $Res(p, t \to^*_{\mathcal{R}} t' \to_{\mathcal{R}} t'') = \{p'' \mid p' \in Res(p, t \to^*_{\mathcal{R}} t') \text{ and } p'' \in Res(p', t' \to_{\mathcal{R}} t'')\}$. We abbreviate $Res(p, t \to^*_{\mathcal{R}} t')$ as $Res(p,t')$ if the sequence $t \to^*_{\mathcal{R}} t'$ is clear from the context.

Lemma 2. *Let $\mathcal{R}$ be an LL-GG-TRS and $c = max_v(\mathcal{R}) + max_f(\mathcal{R}) + 1$. Also let $\sigma = t_0 \to_{\mathcal{R}} t_1 \to_{\mathcal{R}} \cdots \to_{\mathcal{R}} t_{k-1} \to_{\mathcal{R}} t_k \to_{\mathcal{R}} t_{k+1}$ of $\mathcal{R}$ be a root rewrite sequence and $o_0 \in Pos(t_0)$ be a position. If there exists a position $o_i \in Res(o_0, t_i)$ such that $|o_0| \le |o_i|$ for each $i (1 \le i \le k)$, then every position $p_0 \in Pos_{\ge c}(t_0)$ satisfies the following (a) and (b):*

(a) For an arbitrary $p_k \in Res(p_0, t_k)$,

$$|p_k| - |o_k \sqcup p_k| \ge |p_0| - |o_0 \sqcup p_0| \quad \textit{and} \quad |p_k| > max_f(\mathcal{R}).$$

(b) For an arbitrary $s \in \mathcal{T}(\mathcal{F})$,

$$t_0[p_0 \leftarrow s] \rightarrow^*_{\mathcal{R}} t_{k+1}[Res(p_0, t_{k+1}) \leftarrow s].$$

Proof Sketch. (a) By induction on the length k of σ (see [18] for the detail). (b) By (a), each $p_i \in Res(p_0, t_i)(0 \le i \le k)$ satisfies $|p_i| > max_f(\mathcal{R})$. Hence, we can construct a rewrite sequence starting in $t_0[p_0 \leftarrow s]$, applying the rules in the same order as σ. ∎

The next lemma states that for any infinite root rewrite sequence σ of an LL-GG-TRS and any term t_n in σ, one can find a term t_m after t_n such that every 'sufficiently deep' position in t_m does not affect the rewrite sequence after t_m.

Definition 6. (Longest-Living Position) *Let* $t_0 \rightarrow_{\mathcal{R}} t_1 \rightarrow_{\mathcal{R}} \cdots$ *be a rewrite sequence and* $o_0 \in Pos(t_0)$ *be a position. The lifetime of* o_0 *(in* t_0*) is defined as* k*, if there exists* k *such that* $Res(o_0, t_i) \neq \emptyset$ $(0 \le i \le k)$ *and* $Res(o_0, t_i) = \emptyset$ $(i > k)$*. Otherwise (*$Res(o_0, t_i) \neq \emptyset$ *for any* $i \ge 0$*), the lifetime of* o_0 *is undefined. A position which has the maximum lifetime in* t_0 *is called the longest-living position, if the lifetime of every position in* t_0 *is defined.*

Lemma 3. *Let* $\mathcal{R}$ *be an LL-GG-TRS and* $c = max_v(\mathcal{R}) + max_f(\mathcal{R}) + 1$*. If there exists an infinite root rewrite sequence* $\sigma = t_0 \rightarrow_{\mathcal{R}} t_1 \rightarrow_{\mathcal{R}} \cdots$ *of* $\mathcal{R}$*, then for any* $n \ge 0$*, there exists* $m > n$ *such that for every* $p_m \in Pos_{\ge c}(t_m)$*,* $k > m$ *and* $s \in \mathcal{T}(\mathcal{F})$*,* $t_m[p_m \leftarrow s] \rightarrow^*_{\mathcal{R}} t_k[Res(p_m, t_k) \leftarrow s]$ *holds.*

Proof Sketch. Assume that there exists a position p_n in t_n of which the lifetime is undefined (the proof for the other case is given in [18]). Let p_i be the deepest residual of p_n in $t_i (i > n)$, and m be the minimum $j (> n)$ such that $|p_j| \le |p_i|$ for each $i (> j)$. Note that m is always defined since $t_n \rightarrow_{\mathcal{R}} t_{n+1} \rightarrow_{\mathcal{R}} \ldots$ is an infinite sequence. Also, $t_m \rightarrow_{\mathcal{R}} t_{m+1} \rightarrow_{\mathcal{R}} \ldots$ and p_m satisfy the hypothesis of Lemma 2. Hence, by Lemma 2(b), the lemma holds. ∎

Definition 7. (Inclusion Order $\sqsupseteq_a$) *The inclusion order* $\sqsupseteq_a$ *w.r.t. constant* a *is the least relation satisfying the following condition:*

- *For any term* t, $t \sqsupseteq_a a$.
- *If* $t_1 \sqsupseteq_a t'_1, t_2 \sqsupseteq_a t'_2, \ldots t_n \sqsupseteq_a t'_n$*, then* $f(t_1, t_2, \ldots, t_n) \sqsupseteq_a f(t'_1, t'_2, \ldots, t'_n)$. ∎

In the rest of this section, we assume a is a new constant which is not a member of $\mathcal{F}$. For a term $t \in \mathcal{T}(\mathcal{F} \cup \{a\})$, let $|t|_a$ denote $|Pos(t, a)|$. When a tuple of terms $\boldsymbol{\theta} = \langle \theta_1, \ldots, \theta_n \rangle \in \mathcal{T}^n(\mathcal{F} \cup \{a\})$ is given where $n = |t|_a$, let $t\boldsymbol{\theta}$ denote $t[p_i \leftarrow \theta_i \mid 1 \le i \le n]$ for $Pos(t, a) = \{p_1, \ldots, p_n\}$, by slightly abusing the notation. The following lemma states that every infinite root rewrite sequence of an LL-GG-TRS has a kind of cyclic property.

Lemma 4. *Let* $\mathcal{R}$ *be an LL-GG-TRS and* $c = max_v(\mathcal{R}) + max_f(\mathcal{R}) + 1$*. For an infinite root rewrite sequence* $\sigma = t_0 \rightarrow_{\mathcal{R}} t_1 \rightarrow_{\mathcal{R}} \cdots$ *of* $\mathcal{R}$*, there exist a term*

$t_R \in \mathcal{T}(\mathcal{F} \cup \{a\})(n = |t_R|_a)$ of which the depth is c or less and tuples of terms $\boldsymbol{\theta} \in \mathcal{T}^n(\mathcal{F} \cup \{a\}), \boldsymbol{\theta}' \in \mathcal{T}^n(\mathcal{F})$ such that $t_R \in pre^*(\{t_R\boldsymbol{\theta}\})$ and $t_0 \in pre^*(\{t_R\boldsymbol{\theta}'\})$ hold.

Let $\mathcal{T}_G \subseteq \mathcal{T}(\mathcal{F} \cup \{a\})$ be a set of terms, which is downward-closed w.r.t $\sqsubseteq_a$. If terms in $\mathcal{T}_G$ appear infinitely often in σ, then $t_R \in pre^*(\mathcal{T}_G \cap pre^+(\{t_R\boldsymbol{\theta}\}))$ and $t_0 \in pre^*(\{t_R\boldsymbol{\theta}'\})$ hold.

Proof. We define an infinite sequence $\sigma_0, \sigma_1, \sigma_2, \ldots$ of infinite sequences and a function $f : N \to N$ as follows (Fig. 2). The kth element of σ_i is denoted as $\sigma_i(k)$.

- $i = 0$: $\sigma_0 = \sigma$.
- $i > 0$: $f^i(0)$ is defined as m in Lemma 3 when infinite root rewrite sequence σ_{i-1} and $n = f^{i-1}(0)$ are given. $\sigma_i(k)$ is defined according to k as follows:
  - $k < f^i(0)$: $\sigma_i(k)$ is undefined.
  - $k = f^i(0)$:
  $$\sigma_i(k) = \sigma_{i-1}(k)[Pos_{=c}(\sigma_{i-1}(k)) \leftarrow a]. \tag{3.1}$$
  - $k > f^i(0)$: By the definition of $f^i(0)$, we can use Lemma 3 and obtain:
  $$\sigma_{i-1}(f^i(0))[Pos_{=c}(\sigma_{i-1}(f^i(0))) \leftarrow a] \quad \to^*_{\mathcal{R}} \sigma_{i-1}(k)[\mathcal{P}^{i,k} \leftarrow a], \tag{3.2}$$
  where:
  $\mathcal{P}^{i,k} = Res(Pos_{=c}(\sigma_{i-1}(f^i(0))), \sigma_{i-1}(k)) \subseteq Pos_{\geq(max_f(\mathcal{R})+1)}(\sigma_{i-1}(k))$.
  Now, let
  $$\sigma_i(k) = \sigma_{i-1}(k)[\mathcal{P}^{i,k} \leftarrow a], \tag{3.3}$$
  then (3.2) can be written as $\sigma_i(f^i(0)) \to^*_{\mathcal{R}} \sigma_i(k)$ by (3.1).

For the infinite sequence $\sigma_0, \sigma_1, \sigma_2, \ldots$,

$$\sigma_0(k) \sqsupseteq_a \sigma_1(k) \sqsupseteq_a \sigma_2(k) \sqsupseteq_a \cdots \sqsupseteq_a \sigma_j(k) \; (f^j(0) \leq k < f^{j+1}(0)) \tag{3.4}$$

holds by (3.3). Now, we consider the infinite sequence $\sigma_1(f(0)), \sigma_2(f^2(0)), \ldots$ by picking up the 'diagonal' terms. Then, the depths of these terms are always c or less. By this fact, we can see that there exist an integer i and an infinite sequence $i < j_0 < j_1 < j_2 < \cdots$ of numbers such that for every $j_h (h \geq 0)$,

$$\sigma_i(f^i(0)) = \sigma_{j_h}(f^{j_h}(0)). \tag{3.5}$$

By (3.4) and (3.5), $\sigma_i(f^i(0)) \sqsubseteq_a \sigma_i(f^{j_0}(0))$. Hence, for $t_R = \sigma_i(f^i(0))$, there exists $\boldsymbol{\theta} \in \mathcal{T}^n(\mathcal{F} \cup \{a\})$ such that $\sigma_i(f^{j_0}(0)) = t_R\boldsymbol{\theta}$. Since $\sigma_i(f^i(0)) \to^*_{\mathcal{R}} \sigma_i(f^{j_0}(0))$, $t_R \in pre^*(\{t_R\boldsymbol{\theta}\})$ holds. Similarly, by (3.4), we can obtain $\sigma_i(f^i(0))(= t_R) \sqsubseteq_a \sigma_0(f^i(0))(= t_R\boldsymbol{\theta}')$ for some $\boldsymbol{\theta}' \in \mathcal{T}^n(\mathcal{F})$, and thus $\sigma_0(0)(= t_0) \in pre^*(\{t_R\boldsymbol{\theta}'\})$ holds. By (3.1), the depth of t_R is c or less. Next, we consider the case that terms in $\mathcal{T}_G$ appear infinitely often in σ. We can easily see that there exist integers l, m and $\sigma_0(l) \in \mathcal{T}_G$ such that $f^i(0) \leq l < f^{j_m}(0)$ holds. By (3.4), $\sigma_0(l) \sqsupseteq_a \sigma_i(l)$, and thus $\sigma_i(l) \in \mathcal{T}_G$ because $\mathcal{T}_G$ is a downward-closed set. On the other hand, since $t_R = \sigma_i(f^i(0)) \to^*_{\mathcal{R}} \sigma_i(l) \to^+_{\mathcal{R}} \sigma_i(f^{j_m}(0))$, we can obtain $\sigma_i(l) \in \mathcal{T}_G \cap pre^+(\{t_R\boldsymbol{\theta}\})$ and $t_R \in pre^*(\{\sigma_i(l)\})$ in a similar way to the above case. Hence, $t_R \in pre^*(\mathcal{T}_G \cap pre^+(\{t_R\boldsymbol{\theta}\}))$. ∎

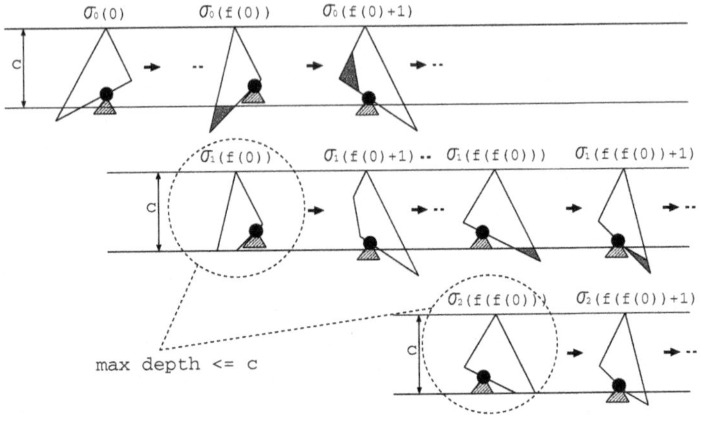

Fig. 2. Proof of Lemma 5: infinite sequence $\sigma_0, \sigma_1, \cdots$

Theorem 1. *Let $\mathcal{R}$ be an LL-GG-TRS and $c = max_v(\mathcal{R}) + max_f(\mathcal{R}) + 1$. Let $\mathcal{T}_G \subseteq \mathcal{T}(\mathcal{F} \cup \{a\})$ be a set of terms, which is upward-closed and downward-closed w.r.t $\sqsubseteq_a$. There exists an infinite root rewrite sequence of $\mathcal{R}$ starting in t_0 in which terms in $\mathcal{T}_G$ appear infinitely often if and only if there exist $t_\mathrm{R} \in \mathcal{T}(\mathcal{F} \cup \{a\})(|t_\mathrm{R}|_a = n)$ of which the depth is c or less and tuples of terms $\boldsymbol{\theta} \in \mathcal{T}^n(\mathcal{F} \cup \{a\}), \boldsymbol{\theta}' \in \mathcal{T}^n(\mathcal{F})$ such that $t_\mathrm{R} \in pre^*(\mathcal{T}_G \cap pre^+(\{t_\mathrm{R}\boldsymbol{\theta}\}))$ and $t_0 \in pre^*(\{t_\mathrm{R}\boldsymbol{\theta}'\})$ (or equivalently, $t_\mathrm{R}\boldsymbol{\theta} \in post^+(post^*(\{t_\mathrm{R}\}) \cap \mathcal{T}_G)$ and $t_\mathrm{R}\boldsymbol{\theta}' \in post^*(\{t_0\})$) hold.*

Proof. The *only if* part of this theorem follows from Lemma 4.

The *if* part is proved as follows. If $t_0 \in pre^*(\{t_\mathrm{R}\boldsymbol{\theta}'\})$ and $t_\mathrm{R} \in pre^*(\mathcal{T}_G \cap pre^+(\{t_\mathrm{R}\boldsymbol{\theta}\}))$, then there exists a term $t_\mathrm{G} \in \mathcal{T}_G$ such that $t_\mathrm{R} \in pre^*(\{t_\mathrm{G}\})$ and $t_\mathrm{G} \in pre^+(\{t_\mathrm{R}\boldsymbol{\theta}\})$ hold. By these facts, we can construct infinite root rewrite sequence $t_0 \to_{\mathcal{R}}^* t_\mathrm{R}\boldsymbol{\theta}' \to_{\mathcal{R}}^* t_\mathrm{G}\boldsymbol{\theta}' \to_{\mathcal{R}}^+ t_\mathrm{R}\boldsymbol{\theta}\boldsymbol{\theta}' \to_{\mathcal{R}}^* t_\mathrm{G}\boldsymbol{\theta}\boldsymbol{\theta}' \to_{\mathcal{R}}^+ t_\mathrm{R}\boldsymbol{\theta}^2\boldsymbol{\theta}' \to_{\mathcal{R}}^* \cdots$, where $\boldsymbol{\theta}' = \langle \theta'_1, \ldots, \theta'_n \rangle$ and $\boldsymbol{\theta}\boldsymbol{\theta}'$ is a term obtained by replacing a in $\boldsymbol{\theta}$ by one of $\theta'_1, \ldots, \theta'_n$. Since $t_\mathrm{G}\boldsymbol{\theta}^n\boldsymbol{\theta}' \sqsupseteq_a t_\mathrm{G}$ and $\mathcal{T}_G$ is upward-closed, $t_\mathrm{G}\boldsymbol{\theta}^n\boldsymbol{\theta}' \in \mathcal{T}_G$. Therefore, terms in $\mathcal{T}_G$ appear infinitely often in the above sequence. ∎

Theorem 2. *Let $\mathcal{R}$ be an LL-GG-TRS, $t_0 \in \mathcal{T}(\mathcal{F})$, ϕ be an LTL formula, ν be a simple valuation. There exists a term $t_\mathrm{R} \in \{q(t') \mid q \in \mathcal{Q}_\mathcal{B}, t' \in \mathcal{T}(\mathcal{F} \cup \{a\})\}$ of which the depth is c or less, and*

$$\begin{aligned}\mathcal{R}, t_0 \not\models^\nu \phi &\Leftrightarrow t_\mathrm{R} \in pre^*_{[\mathcal{BR}^\nu]}(\mathcal{T}_\mathrm{acc} \cap pre^+_{[\mathcal{BR}^\nu]}(\mathcal{T}'_\mathrm{R})) \text{ and } q_{0\mathcal{B}}(t_0) \in pre^*_{[\mathcal{BR}^\nu]}(\mathcal{T}_\mathrm{R}) \\ &\Leftrightarrow post^+_{[\mathcal{BR}^\nu]}(post^*_{[\mathcal{BR}^\nu]}(\{t_\mathrm{R}\}) \cap \mathcal{T}_\mathrm{acc}) \cap \mathcal{T}'_\mathrm{R} \neq \emptyset \\ &\quad \text{and } \mathcal{T}_\mathrm{R} \cap post^*_{[\mathcal{BR}^\nu]}(\{q_{0\mathcal{B}}(t_0)\}) \neq \emptyset,\end{aligned}$$

where $\mathcal{B}$ is a Büchi automaton representing $\neg\phi$, $q_{0\mathcal{B}}$ and $\mathcal{Q}_\mathcal{B}^{acc}$ are the initial state and accepting states of $\mathcal{B}$, $\mathcal{T}_\mathrm{acc} = \{q_\mathrm{a}(t) \mid q_\mathrm{a} \in \mathcal{Q}_\mathcal{B}^{acc}, t \in \mathcal{T}(\mathcal{F} \cup \{a\})\}$, $\mathcal{T}_\mathrm{R} = \{t_\mathrm{R}\boldsymbol{\theta} \mid \boldsymbol{\theta} \in \mathcal{T}^n(\mathcal{F})\}$, $\mathcal{T}'_\mathrm{R} = \{t_\mathrm{R}\boldsymbol{\theta} \mid \boldsymbol{\theta} \in \mathcal{T}^n(\mathcal{F} \cup \{a\})\}$.

Proof. If $\mathcal{R}$ is an LL-GG-TRS, then $\mathcal{BR}^\nu$ is also an LL-GG-TRS. $\mathcal{T}_{\text{acc}}$ is upward-closed and downward-closed w.r.t $\sqsubseteq_a$. Therefore, by Lemma 1 and Theorem 1, the theorem holds. ∎

Corollary 1. *Let $\mathcal{R}$ be an LL-GG-TRS, $t_0 \in \mathcal{T}(\mathcal{F})$, ϕ be an LTL formula, ν be a simple valuation. If $\mathcal{BR}^\nu$ is pre-PR or post-PR, then $\mathcal{R}, t_0 \models^\nu \phi$ is decidable.*

Proof. The corollary follows from the facts that the number of candidates for t_{R} in Theorem 2 is finite, that we can construct TAs which recognize $\mathcal{T}_{\text{acc}}$, $\{q_{0\mathcal{B}}(t_0)\}$, $\mathcal{T}_{\text{R}}$ and $\mathcal{T}'_{\text{R}}$. ∎

4 Computing *pre**

By Corollary 1, if an LL-GG-TRS $\mathcal{R}$ is post-PR or pre-PR, then LTL model checking for $\mathcal{R}$ is decidable. Unfortunately, an LL-GG-TRS is not always post-PR. For example, $\mathcal{R} = \{f(x,y) \to f(g(x), g(y))\}$ is an LL-GG-TRS. However, $post^+_{\mathcal{R}}(\{f(a,a)\})$ is not recognizable and thus $\mathcal{R}$ is not post-PR. It is unknown whether every LL-GG-TRS is pre-PR. In this section, we propose a decidable subclass of pre-PR TRS. Let $\mathcal{R}$ be a TRS. By the definition of pre-PR, for a given TA $\mathcal{A}$, if we can extend $\mathcal{A}$ so that $t \to_{\mathcal{R}} s \in pre^*_{\mathcal{R}}(\mathcal{L}(\mathcal{A}))$ implies $t \in pre^*_{\mathcal{R}}(\mathcal{L}(\mathcal{A}))$ (backward closedness w.r.t.$\to_{\mathcal{R}}$) then $\mathcal{R}$ is pre-PR. This requires us to add to $\mathcal{A}$ new states and transition rules to satisfy the condition that $t \to_{\mathcal{R}} s \vdash^*_{\mathcal{A}} q$ implies $t \vdash^*_{\mathcal{A}} q$. For example, let $f(g(x,y)) \to g(h(y),x) \in \mathcal{R}$, $t = f(g(a,b))$, $s = f(h(b),a)$, and $s \vdash^*_{\mathcal{A}} f(h(q_2),q_1) \vdash^*_{\mathcal{A}} q$ for states q_1, q_2 and q of a TA $\mathcal{A}$. Note that $t \to_{\mathcal{R}} s$ with substitution $\theta = \{x \mapsto a, y \mapsto b\}$. Then, we add the following states and transition rules to $\mathcal{A}$ so that $t \vdash^*_{\mathcal{A}} q$.

states: $\langle g(q_1,q_2)\rangle$, $\langle f(g(q_1,q_2))\rangle$.
rules: $g(q_1,q_2) \to \langle g(q_1,q_2)\rangle$, $f(\langle g(q_1,q_2)\rangle) \to \langle f(g(q_1,q_2))\rangle$, $\langle f(g(q_1,q_2))\rangle \to q$.

That is, we use a subterm of the left-hand side of the rewrite rule as a state to keep track of the position where the head of $\mathcal{A}$ is located. However, states substituted into variables such as q_1, q_2, q above may recursively be subterms, and hence the above construction does not always halt. The condition for a TRS $\mathcal{R}$ to be an LL-SPO-TRS stated below is a condition for $\mathcal{R}$ not to have a kind of overlapping between subterms of rewrite rules, which guarantees that the above construction always halts.

4.1 LL-SPO-TRS

For an ordinary rewrite relation not limited to root rewriting, LL-FPO^{-1}-TRS is known as a decidable subclass of pre-PR TRS (see Sect. 2.2). Based on the definition of LL-FPO^{-1}-TRS, we define a new subclass called LL-SPO-TRS and show that every LL-SPO-TRS is pre-PR with respect to root rewriting.

Definition 8. (Sticking Out Relation)
Let s and t be terms in $\mathcal{T}(\mathcal{F},\mathcal{V})$. We say s sticks out of t if $t \notin \mathcal{V}$ and there exists a position $o_{var} \in Pos(t)$ ($lab(t, o_{var}) \in \mathcal{V}$) such that

– *for any positions* o $(\lambda \preceq_{pref} o \prec_{pref} o_{var})$*,* $o \in Pos(s)$ *and* $lab(s,o) = lab(t,o)$*, and*
– $o_{var} \in Pos(s)$ *and* $s_{|o_{var}}$ *is not a ground term.*

When the position o_{var} *is of interest, we say that* s *sticks out of* t *at* o_{var}*. If* s *sticks out of* t *at* o_{var} *and* $lab(s, o_{var})$ *is not a variable, then we say that* s *properly sticks out of* t *(at* o_{var}*).* ∎

For example, $f(g(x), a)$ sticks out of $f(g(y), b)$ at 11 and $f(g(g(x)), a)$ properly sticks out of $f(g(y), b)$ at 11. Remember that a configuration of a PDS is a pair $\langle q, w\rangle$ of a control location (finite control) q and a sequence w of symbols stored in the pushdown stack. In the rest of this section, we assume that a signature $\mathcal{F}$ is decomposed into Π and Σ, that is, $\mathcal{F} = \Pi \cup \Sigma$ and $\Pi \cap \Sigma = \emptyset$. For each $\pi \in \Pi$, we assume $arity(\pi) = 1$. Each $\pi \in \Pi$ is called a control symbol and each $f \in \Sigma$ is called a data symbol.

Definition 9. (Simply Path Overlapping TRS (SPO-TRS))
A TRS $\mathcal{R}$ *is SPO if every rule in* $\mathcal{R}$ *has the form either* $\pi_1(l) \to \pi_2(r)$*,* $\pi_1(l) \to r$ *or* $l \to r$ *where* $\pi_1, \pi_2 \in \Pi$ *and* $l, r \in \mathcal{T}(\Sigma, \mathcal{V})$*, and the sticking-out graph* $G_{\mathcal{R}}$ *of* $\mathcal{R}$ *has no cycle with weight one or more. The sticking-out graph of a TRS* $\mathcal{R}$ *is a weighted directed graph* $G_{\mathcal{R}} = (\mathcal{R}, E)$*. Let* $v_1 \overset{i}{\to} v_2$ *denote a directed edge from a node* v_1 *to a node* v_2 *with weight* i*.* E *is defined as follows. Let* $v_1 : l_1 \to r_1$ *(or* $\pi_{11}(l_1) \to r_1$ *or* $\pi_{11}(l_1) \to \pi_{12}(r_1)$*) and* $v_2 : l_2 \to r_2$ *(or* $\pi_{21}(l_2) \to r_2$ *or* $\pi_{21}(l_2) \to \pi_{22}(r_2)$*) be rules in* $\mathcal{R}$*. Replace each variable in* $\mathcal{V}ar(l_1)\backslash\mathcal{V}ar(r_1)$ *or* $\mathcal{V}ar(l_2)\backslash\mathcal{V}ar(r_2)$ *with a constant not in* $\mathcal{F}$*, say* $\diamond$*.*

(1) If l_1 *properly sticks out of* r_2*, then* $v_1 \overset{1}{\to} v_2 \in E$*.*
(2) If r_2 *sticks out of* l_1*, then* $v_1 \overset{0}{\to} v_2 \in E$*.* ∎

If $\mathcal{R}$ is an LL-SPO-TRS, then for any TA $\mathcal{A}$, we can construct a TA $\mathcal{A}_*$ such that $\mathcal{L}(\mathcal{A}_*) = pre^*_{\mathcal{R}}(\mathcal{L}(\mathcal{A}))$ (see [18]). That is, every LL-SPO-TRS is pre-PR.

Theorem 3. *For every recognizable tree language* L *and LL-SPO-TRS* $\mathcal{R}$*,* $pre^*_{\mathcal{R}}(L)$ *is also recognizable.* ∎

Corollary 2. *Assume* $t_0 \in \mathcal{T}(\mathcal{F})$*,* ϕ *is an LTL formula and* ν *is a simple valuation. If* $\mathcal{R} \in$ *LL-GG-TRS* $\cap$ *SPO-TRS, then* $\mathcal{R}, t_0 \models^{\nu} \phi$ *is decidable.*

Proof. Let $\mathcal{B} = (\mathcal{Q}_{\mathcal{B}}, \Sigma_{\mathcal{B}}, \Delta_{\mathcal{B}}, q_{0\mathcal{B}}, \mathcal{Q}^{acc}_{\mathcal{B}})$ be a Büchi automaton representing $\neg\phi$ and Π be the set of control symbols of $\mathcal{R}$. Consider the construction of Büchi TRS $\mathcal{BR}^{\nu}$ from $\mathcal{R}$, $\mathcal{B}$ and ν. If $\mathcal{R}$ is an SPO-TRS, then by constructing $\langle q, p\rangle(l) \to \langle q', p'\rangle(r) \in \mathcal{BR}^{\nu}$ instead of $q(p(l)) \to q'(p'(r)) \in \mathcal{BR}^{\nu}$ for each rule $p(l) \to p'(r) \in \mathcal{R}$ $(p, p' \in P)$, $\mathcal{BR}^{\nu}$ becomes an SPO-TRS. By Corollary 1 and Theorem 3, $\mathcal{R}, t_0 \models^{\nu} \phi$ is decidable. ∎

4.2 Application

As mentioned in Sect. 3, we can model a recursive program with exception handling by an LL-GG-TRS. If the LL-GG-TRS is always an SPO-TRS, then

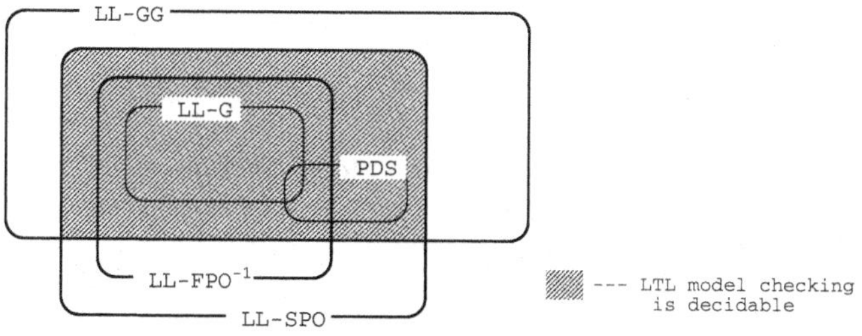

Fig. 3. The relation between TRS subclasses

LTL model checking for the TRS is decidable. Recall $\mathcal{R}_2$ in Example 2. Since for any two rules $l_1 \rightarrow r_1$ and $l_2 \rightarrow r_2$ in $\mathcal{R}_2$, l_1 never properly sticks out of r_2, $\mathcal{R}_2$ is an SPO-TRS. Similarly, we can easily see that every LL-GG-TRS constructed by the method in Example 2 is always an SPO-TRS. Thus, LTL model checking problem is decidable for recursive programs with exception handling.

5 Conclusion

In this paper, we introduced two classes of TRS, LL-GG-TRS and SPO-TRS, and showed that for a TRS in LL-GG-TRS $\cap$ SPO-TRS, LTL model checking is decidable. Since every PDS is a member of LL-GG-TRS $\cap$ SPO-TRS, this model checking is considered as an extension of LTL model checking for PDS. In fact, a recursive program with exception handling can be modeled as a TRS to which this model checking method can be applied and to which no PDS is strongly bisimilar.

We can reduce some decision problems of TRS to LTL model checking problems. For example, let ν be a regular valuation and α_{NF} be an atomic proposition such that $\nu(\alpha_{NF}) = \mathrm{NF}_{\mathcal{R}}$. Whether there exists no infinite rewrite sequence starting in t_0 (strongly normalizing) is checked by $\mathcal{R}, t_0 \models^{\nu} \Diamond(\alpha_{NF})$, and whether there exists a finite rewrite sequence starting in t_0 (weakly normalizing) is checked by $\mathcal{R}, t_0 \not\models^{\nu} \Box(\neg\alpha_{NF})$.

The following problems remain as future study:

- finding a wider subclass of TRS in which LTL model checking is solvable,
- developing an efficient LTL model checking method w.r.t. regular valuation,
- and finding other applications of this model checking method.

Acknowledgments. The authors would like to thank Professor Igor Walukiewicz of Université Bordeaux I for his valuable comments and discussions.

References

1. F. Baader and T. Nipkow: *Term Rewriting and All That*, Cambridge University Press, 1998.
2. E.M. Clarke, O. Grumberg and D. Peled: *Model Checking,* MIT Press, 2000.
3. J.L. Coquidé, M. Dauchet, R. Gilleron and S. Vágvölgyi: Bottom-up tree pushdown automata: classification and connection with rewrite systems, *Theoretical Computer Science*, **127**, 69–98, 1994.
4. J.C. Corbett, M.B. Dwyer, J. Hatcliff, S. Laubach, C.S. Păsăreanu, Robby and H. Zheng: Bandera: Extracting finite-state models from Java source code, Int'l Conf. on Software Engineering, 439–448, 2000.
5. J. Esparza, D. Hansel, P. Rossmanith and S. Schwoon: Efficient algorithms for model-checking pushdown systems, CAV2000, LNCS **1855**, 232–247, 2000.
6. J. Esparza, A. Kučera and S. Schwoon: Model-checking LTL with regular variations for pushdown systems, TACS01, LNCS **2215**, 316–339, 2001.
7. J. Esparza and S. Schwoon: A BDD-based model checker for recursive programs, CAV2001, LNCS **2102**, 324–336, 2001.
8. F. Gécseq and M. Steinby: *Tree Automata*, Académiai Kiadó, 1984.
9. R. Gilleron: Decision problems for term rewriting systems and recognizable tree languages, STACS'91, LNCS **480**, 148–159, 1991.
10. R. Gilleron and S. Tison: Regular tree languages and rewrite systems, *Fundamenta Informaticae*, **24**, 157–175, 1995.
11. S. Graf and H. Saïdi: Construction of abstract state graphs with PVS, CAV97, LNCS **1254**, 72–83, 1997.
12. P. Gyenizse and S. Vágvölgyi: Linear generalized semi-monadic rewrite systems effectively preserve recognizability, *Theoretical Computer Science*, **194**, 87–122, 1998.
13. T. Jensen, D. Le Métayer and T. Thorn: Verification of control flow based security properties, IEEE Symp. on Security and Privacy, 89–103, 1999.
14. S. Jha and T. Reps: Analysis of SPKI/SDSI certificates using model checking, IEEE Computer Security Foundations Workshop, 129–144, 2002.
15. C. Löding: Model-checking infinite systems generated by ground tree rewriting, FOSSACS, LNCS **2303**, 280–294, 2002.
16. R. Mayr: Process rewrite systems, *Inform. & Comput.*, **156**, 264–286, 1999.
17. T. Nagaya and Y. Toyama: Decidability for left-linear growing term rewriting systems, RTA99, LNCS **1631**, 256–270, 1999.
18. N. Nitta and H. Seki: An extension of pushdown system and its model checking method, Technical Report, Nara Institute of Science and Technology, 2003.
19. N. Nitta, Y. Takata and H. Seki: Security verification of programs with stack inspection, 6th ACM Symp. on Access Control Models and Technologies, 31–40, 2001.
20. N. Nitta, Y. Takata and H. Seki: An efficient security verification method for programs with stack inspection, 8th ACM Conf. on Computer and Communication Security, 68–77, 2001.
21. J. Obdržálek: Model checking Java using pushdown systems, ECOOP Workshop on Formal Techniques for Java-like Programs, 2002.
22. `http://www.pobox.com/~cme/spki.txt`
23. T. Takai, Y. Kaji and H. Seki: Right-linear finite path overlapping term rewriting systems effectively preserve recognizability, RTA2000, LNCS **1833**, 246–260, 2000.
24. I. Walukiewicz: Pushdown processes: Games and model-checking, CAV96, LNCS **1102**, 62–74, 1996.

Netcharts: Bridging the Gap between HMSCs and Executable Specifications

Madhavan Mukund[1], K. Narayan Kumar[1], and P.S. Thiagarajan[2]

[1] Chennai Mathematical Institute, 92 G N Chetty Road, Chennai 600 017, India
{madhavan,kumar}@cmi.ac.in
[2] School of Computing, National University of Singapore, Singapore
thiagu@comp.nus.edu.sg

Abstract. We define a new notation called *netcharts* for describing sets of message sequence chart scenarios (MSCs). Netcharts correspond to a distributed version of High-level Message Sequence Charts (HMSCs). Netcharts improve on HMSCs in two respects.

(i) Netcharts admit a natural and direct translation into communicating finite-state machines, unlike HMSCs, for which the *realization* problem is nontrivial.

(ii) Netcharts can describe all regular MSC languages (sets of MSCs in which channel capacities have a finite upper bound), unlike HMSCs, which can only describe finitely-generated regular MSC languages.

1 Introduction

Message sequence charts (MSCs) are an appealing visual formalism used to capture system requirements in the early stages of design. They are particularly suited for describing scenarios for distributed telecommunication software [10, 15]. They also appear in the literature as sequence diagrams, message flow diagrams and object interaction diagrams and are used in a number of software engineering methodologies including UML [3,7,15]. In its basic form, an MSC depicts the exchange of messages between the processes of a distributed system along a single partially-ordered execution. A collection of MSCs is used to capture the scenarios that a designer might want the system to exhibit (or avoid).

Given the requirements in the form of MSCs, one can hope to discover errors at the early stages of design. One question that naturally arises in this context is: What constitutes a reasonable collection of MSCs on which one can hope to do formal analysis? In [9], the notion of a regular collection of MSCs is introduced and shown to be robust in terms of logical and automata-theoretic characterizations. In particular, regular collections of MSCs can be implemented using a natural model of message-passing automata with bounded queues [12].

A standard way to specify a set of MSCs is to use a High-level Message Sequence Chart (HMSC) [10,11]. An HMSC is a finite directed graph in which each node is itself labelled by an HMSC, with a finite level of nesting.

HMSCs are an attractive visual formalism for describing collections of MSCs. An HMSC is not, however, an executable model. To obtain one, one must extract

R. Amadio, D. Lugiez (Eds.): CONCUR 2003, LNCS 2761, pp. 296–310, 2003.

from the inter-object specification provided by an HMSC, an *intra-object* specification (say in the form of a finite state automaton) for each process together with a communication mechanism. This is non-trivial because the structure of an HMSC provides implicit global control over the behaviour of the processes in the system. This allows for specifications that are not realizable in practice because of un-implementable global choices. Even for the class of so-called bounded HMSCs [2], the problem of realizing them as, say, a network of finite state automata communicating via bounded queues is non-trivial. Admittedly, the sub-class of bounded HMSCs that are *weakly realizable* can implemented easily, but this is an undecidable property of bounded HMSCs! [1]. A detailed study of the realization problems associated with HMSCs can be found in [4].

HMSCs also have a limitation with respect to expressiveness—they can only describe finitely generated MSC languages. But there are natural regular MSC languages, such as the set of scenarios corresponding to the alternating bit protocol, that are not finitely generated [6].

In this paper, we introduce a new visual formalism for specifying collections of MSCs, called *netcharts*. Netcharts are distributed versions of HMSCs in much the same way that Petri nets are distributed versions of finite-state automata. Netcharts improve on HMSCs in two respects. First, due to their natural — control flow — semantics, netcharts can be directly translated into communicating finite-state machines. In this sense the "realization" problem for netcharts is easy. Second, netcharts can describe all regular MSC languages, including those that are not finitely-generated.

An important aspect of the netchart model is that the compound MSCs that are defined by a netchart are built up from "complete" MSCs. As a result, this formalism can be used to capture system requirements in an intuitive fashion. The distributed structure allows these scenarios to be intertwined together to form complicated new scenarios.

The netchart model is related to Communication Transaction Processes (CTP) [14]. The main difference is that in a CTP, each transition in the high-level control flow model is labelled as a guarded choice of MSCs, where the guards are formed using atomic propositions associated with the processes. The focus in the study of the CTP model is the modelling of complex non-atomic interactions between communicating processes and not on the collections of MSCs that are definable. A second piece of related work is [16] in which it is suggested that the transitions of a 1-safe Petri net be labelled by arbitrary MSCs. The realization problem for this model is also difficult and the authors propose a number of restrictions for their model and study the realization problem in terms of composing a fixed set of basic Petri net templates to obtain a Petri net model.

The paper is organized as follows. After some background on MSCs and HMSCs, we introduce the netchart model in Sect. 3. In Sect. 4, we show how to transform netcharts into message-passing automata. Next, we discuss when a netchart definable MSC language is regular. Sect. 6 compares the expressive power of HMSCs and netcharts. In Sect. 7, we show that netcharts can describe all regular MSC languages.

2 Background

2.1 Message Sequence Charts

Throughout this paper, we let $\mathcal{P}$ denote a finite set of processes. A *message sequence chart (MSC)* over $\mathcal{P}$ is a graphical representation of a pattern of interactions (often called a *scenario*) consisting of an exchange of messages between the processes in $\mathcal{P}$. In an MSC, each process is represented by a vertical line with time flowing from top to bottom. Messages are drawn as directed edges from the source process to the target process, with an annotation denoting the message type, if any. In addition, internal events are marked along the process lines.

Figure 1 is an MSC involving processes $\{p_1, p_2, p_3\}$. In this scenario, clients p_1 and p_3 both request a resource from the server, p_2. The server p_2 first responds to p_3. After p_3 notifies p_2 that it has returned the resource, p_3 hands over the resource to p_1, but this message crosses a "reminder" from p_1 to p_3.

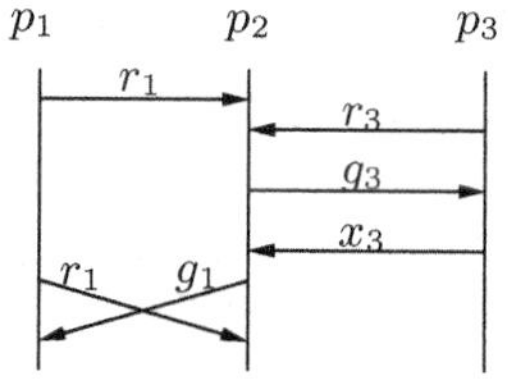

Fig. 1. An MSC

The semantics of an MSC is given in terms of labelled partial orders. Formally, an MSC over $\mathcal{P}$ is a structure $M = (E, \leq, \tau, \varphi, \pi)$, where:

(i) E is a finite set of *events*.
(ii) $\pi : E \to \mathcal{P}$ associates a unique process with each event.
(iii) $\tau : E \to \{\mathsf{send}, \mathsf{receive}, \mathsf{internal}\}$ specifies the nature of each event.
(iv) φ is a bijection from the set of send events $\{e \mid \tau(e) = \mathsf{send}\}$ to the set of receive events $\{e \mid \tau(e) = \mathsf{receive}\}$. Each pair in the set $\{(e, \varphi(e)) \mid \tau(e) = \mathsf{send}\}$, constitutes a *message*. Messages are often labelled using finite set Δ of message types, with typical elements m, m' etc.
(v) E is partially ordered by $\leq$, such that for each $p \in \mathcal{P}$, the set $E_p = \{e \mid \pi(e) = p\}$ is linearly ordered by $\leq$. Let us denote this linear order $\leq_p$. Further, $\leq$ is the reflexive, transitive closure of $\bigcup_{p\in\mathcal{P}} \leq_p \cup \{(e, \varphi(e)) \mid \tau(e) = \mathsf{send}\}$.
(vi) Messages between each pair of processes are delivered in a FIFO fashion.[1] More formally, let e, e' be a pair of events such that $\pi(e) = \pi(e')$, $\tau(e) = \tau(e') = \mathsf{send}$ and $\pi(\varphi(e)) = \pi(\varphi(e'))$. Then, $e \leq e'$ iff $\varphi(e) \leq \varphi(e')$.

Let M be a message sequence chart. The *type* of M is the set $\pi(M) = \{p \in \mathcal{P} \mid \exists e \in M.\ \pi(e) = p\}$ of processes that are *active* in M.

[1] We will see how to relax this restriction later.

For $X \subseteq \mathcal{P}$, $\mathcal{M}_X$ denotes the set of MSCs of type X. The set of all possible MSCs over the set of processes $\mathcal{P}$ is given by $\mathcal{M} = \bigcup_{\emptyset \neq X \subseteq \mathcal{P}} \mathcal{M}_X$.

2.2 Regular MSC Languages

A communicating system is normally specified using a set of scenarios or, equivalently, a set of MSCs. An *MSC language* is a (finite or infinite) collection of MSCs. We can also represent MSC languages in terms of word languages.

Each send or receive event e in an MSC is characterized by the values $\pi(e)$, $\tau(e)$ and the message type, if any. Let $p!q(m)$ denote the action associated with sending message m from p to q and let $q?p(m)$ denote the corresponding receive action. We then have an alphabet of *communication actions* $\Gamma = \{p!q(m) \mid p, q \in \mathcal{P}, m \in \Delta\} \cup \{p?q(m) \mid p, q \in \mathcal{P}, m \in \Delta\}$. For an MSC M, viewed as Γ- labelled poset, the set $lin(M)$ of valid linearizations of M describes a word language over Γ. An MSC language $L \subseteq \mathcal{M}$ can thus be represented by the word language $\bigcup\{lin(M) \mid M \in L\}$. Notice that internal events do not play any role in defining MSC languages.

We say that an MSC language L is *regular* if the corresponding word language over Γ is regular.

Let $M = (E, \leq, \tau, \varphi, \pi)$ be an MSC. A configuration is a set of events $c \subseteq E$ such that c is closed with respect to $\leq$. The channel capacity of c, $\chi(c)$, specifies the number of unmatched send events in c—that is, $\chi(c) = |\{e \in c \mid \pi(e) = \mathsf{send}, \varphi(e) \notin c\}|$. We define the channel capacity of the MSC M to be maximum value of $\chi(c)$ over all configurations c of M. The following is easy to show.

Proposition 1. *Let L be a regular MSC language. Then, there is a uniform upper bound $B \in \mathbb{N}_0$ such that for every MSC $M \in L$, $\chi(M) \leq B$.*

The converse statement is not true, in general. We shall see a counterexample.

2.3 High-Level Message Sequence Charts

The standard mechanism to generate an MSC language is a High-level Message Sequence Chart (HMSC). A simple type of HMSC is a Message Sequence Graph (MSG). An MSG is a finite, directed graph with designated initial and terminal vertices. Each vertex in an MSG is labelled by an MSC. The collection of MSCs represented by an MSG consists of all those MSCs obtained by tracing a path in the MSG from an initial vertex to a terminal vertex and concatenating the MSCs that are encountered along the path.

An HMSC is just an MSG in which a vertex can, in turn, be labelled by another HMSC, with the restriction that the overall nesting depth be finite. Thus, an HMSC can always be "flattened out" into an MSG. Henceforth, when we use the term HMSC we always assume the flat structure of an MSG.

The edges in an HMSC represent the natural operation of MSC concatenation which can be defined as follows. Let $M_i = (E_i, \leq_i, \tau_i, \varphi_i, \pi_i)$, $i \in \{1, 2\}$, be a pair for MSCs such that $E_1 \cap E_2 = \emptyset$. The *(asynchronous) concatenation* of M_1 and M_2 is the MSC $M_1 \circ M_2 = (E, \leq, \tau, \varphi, \pi)$ where $E = E_1 \cup E_2$, $\pi(e) = \pi_i(e)$ and $\tau(e) = \tau_i(e)$ for $e \in E_i$, $i \in \{1, 2\}$, $\varphi = \varphi_1 \cup \varphi_2$ and $\leq$ is the reflexive, transitive closure of $\leq_1 \cup \leq_2 \cup \{(e, e') \mid e \in E_1, e' \in E_2, \pi_1(e) = \pi_2(e')\}$.

Formally, an HMSC is a structure $\mathcal{H} = (S, \longrightarrow, S_{in}, F, \Phi)$, where:

(i) S is a finite and nonempty set of states.
(ii) $\longrightarrow \subseteq S \times S$.
(iii) $S_{in} \subseteq S$ is a set of initial states.
(iv) $F \subseteq S$ is a set of final states.
(v) $\Phi : S \to \mathcal{M}$ is a (state-)labelling function.

A *path* $\sigma = s_0 \longrightarrow s_1 \longrightarrow \cdots \longrightarrow s_n$ through an HMSC $\mathcal{H}$ generates the MSC $M(\sigma) = \Phi(s_0) \circ \Phi(s_1) \circ \cdots \circ \Phi(s_n)$. A path $\sigma = s_0 \longrightarrow s_1 \longrightarrow \cdots \longrightarrow s_n$ is a *run* if $s_0 \in S_{in}$ and $s_n \in F$. The language of MSCs accepted by $\mathcal{H}$ is $L(\mathcal{H}) = \{M(\sigma) \in \mathcal{M} \mid \sigma \text{ is a run through } \mathcal{H}\}$.

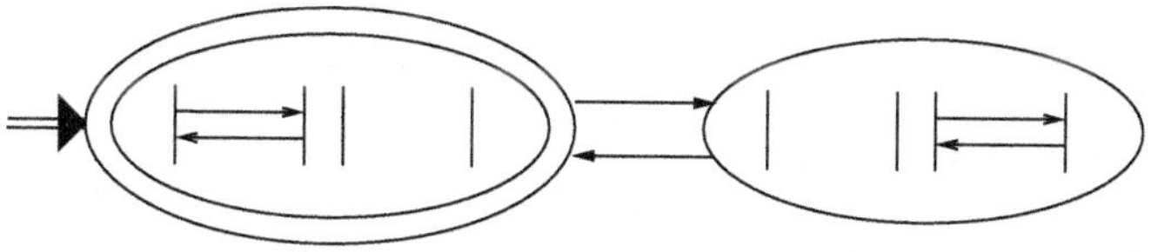

Fig. 2. An HMSC

Figure 2 is an example of an HMSC. The initial state is marked with an incoming arrow and the final state is marked with a double outline. It is easy to see that the language $\mathcal{L}$ defined by this HMSC is *not* regular. Though all the channels are uniformly bounded, there is a "context-free" correlation between the number of iterations of two disjoint sets of communications.

A sufficient criterion for an HMSC to generate a regular MSC language is for it to be *locally synchronized* [13].[2]

For an MSC $M = (E, \leq, \tau, \varphi, \pi)$, *the communication graph of* M is the directed graph $CG_M = (\mathcal{P}, \mapsto)$ where $\mathcal{P}$ is the set of processes of the system and $(p, q) \in \mapsto$ if p sends a message to q in M. More formally, $(p, q) \in \mapsto$ if there exists an $e \in E$ with $\pi(e) = p$, $\tau(e) = \mathsf{send}$ and $\pi(\varphi(e)) = q$.

The MSC M is said to be *connected* if CG_M consists of one nontrivial strongly connected component and isolated vertices. An MSC language $L \subseteq \mathcal{M}$ is connected in case each MSC $M \in L$ is connected.

The HMSC $\mathcal{H}$ is *locally synchronized* if for every loop $\sigma = q \to q_1 \to \cdots \to q_n \to q$, the MSC $M(\sigma)$ is connected. Every locally synchronized HMSC defines a regular MSC language [2] (though the converse does not hold). The HMSC in Fig. 2 is *not* locally synchronized.

[2] This notion is called "bounded" in [2].

2.4 Finitely Generated MSC Languages

Every MSC generated by an HMSC is, by definition, a concatenation of the MSCs that label the vertices of the HMSC. We say an MSC M is *atomic* if M cannot be written as the concatenation $M_1 \circ M_2$ of two nonempty MSCs. An MSC language L is said to be *finitely generated* if there is a finite set of atomic MSCs $Atoms = \{M_1, M_2, \ldots, M_k\}$ such that every MSC $M \in L$ can be decomposed as $M_{i_1} \circ M_{i_2} \circ \cdots \circ M_{i_\ell}$, where each $M_{i_j} \in Atoms$.

Clearly, every HMSC language is finitely generated. However, there exist regular MSC languages that are *not* finitely generated [6,9]. Every finitely generated regular MSC language can be represented as a locally synchronized HMSC [8].

3 Netcharts

Netcharts are distributed versions of HMSCs in much the same way that Petri nets are distributed versions of finite-state automata. (We refer the reader to [5] for basic concepts and results on Petri nets.)

3.1 Nets

- A net is a triple (S, T, F) where S is a set of *places* (or local states), T is a set of *transitions* and $F \subseteq (S \times T) \cup (T \times S)$ is the *flow relation.* For an element $x \in S \cup T$, we use ${}^\bullet x$ and $x^\bullet$, as usual, to denote the immediate predecessors and successors, respectively, of x with respect to the flow relation F.
- An S-net is a net (S, T, F) in which $|{}^\bullet t| = 1 = |t^\bullet|$ for every $t \in T$.

At the top level, a netchart has the structure of a safe Petri net that can be decomposed in a natural way into sequential components. Transitions synchronize distinct components and are annotated by MSCs involving those components.

3.2 The Netchart Model

A *netchart* is a structure $((S, T, F), M_{in}, loc, \lambda)$ where:

(i) (S, T, F) is a net.
(ii) $M_{in} \subseteq S$ is the *initial marking.*
(iii) The function *loc* maps each state $s \in S$ to a process in $\mathcal{P}$. For $s \in S$, we refer to $loc(s)$ as the *location* of s. For $p \in \mathcal{P}$, let $S_p = \{s \mid loc(s) = p\}$
(iv) For $t \in T$, let $loc(t) = \{loc(s) \mid s \in {}^\bullet t \cup t^\bullet\}$. The labelling function λ associates an MSC M with each transition t such that $loc(t) = \pi(M)$.
(v) We impose the following restriction on the structure of a netchart:
   - For each $t \in T$, for each $p \in loc(t)$, $|{}^\bullet t \cap S_p| = |t^\bullet \cap S_p| = 1$. In other words, each transition t has exactly one input place and one output place for each process that participates in t.
   - For each $p \in \mathcal{P}$, $|M_{in} \cap S_p| = 1$. In other words, M_{in} places exactly one token in each component S_p.

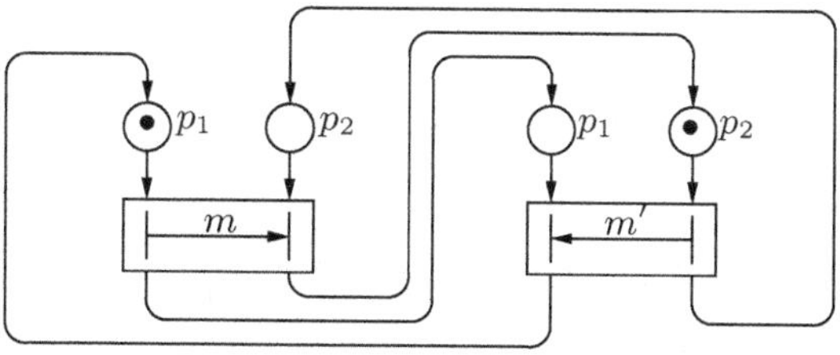

Fig. 3. A netchart

– For $p \in \mathcal{P}$, let $T_p = \{t \mid p \in loc(t)\}$ denote the set of transitions that p participates in. For each $p \in \mathcal{P}$, the net (S_p, T_p, F_p), where $F_p = F \cap ((S_p \times T_p) \cup (T_p \times S_p))$ is an S-net.

Figure 3 shows an example of a netchart. A more elaborate example modelling the alternating bit protocol is shown in Fig. 6.

3.3 Semantics

The operational semantics of a netchart is obtained by converting each MSC that labels a transition into a Petri net and "plugging in" these nets into the top-level safe net of the netchart.

The crucial feature of our semantics is that each MSC that is used to label a high-level transition has a private set of channels. Thus, if a message m is sent from p to q in the MSC labelling transition t, it can only be read when q participates in the same high level transition t.

We convert an MSC M into a net (S_M, T_M, F_M) in an obvious way. The set of transitions T_M corresponds to the events of the MSC. Since each process is sequential, we insert a place between every adjacent pair of events along each process line. In addition, for each pair of processes (p, q), we introduce a buffer place $b_{(p,q)}$. For each send event of a message from p to q, the corresponding transition in T_M feeds into the place $b_{(p,q)}$. For each receive event of a message from p to q, the corresponding transition in T_M has $b_{(p,q)}$ as an input place.

In the netchart, for a high-level transition t labelled by the MSC M , for each process $p \in loc(t)$, we connect the transition corresponding to the $\leq_p$-minimum p-event in the MSC M to the (unique) place in S_p that feeds into t. Similarly, we connect the transition corresponding to the $\leq_p$-maximum p-event in the MSC M to the (unique) output place in S_p of t. Observe that we do not need to model the channels between processes as *queues*. It suffices to maintain a count of the messages in transit between each pair of processes. The structure of the MSC and the labelling of the events ensures that messages are consumed in the order in which they are generated.

The behaviour of the netchart is now given by the normal token game on the "low level" net that we have generated by plugging in a net for each MSC in the netchart. In the low level net, the control places that are inherited from the

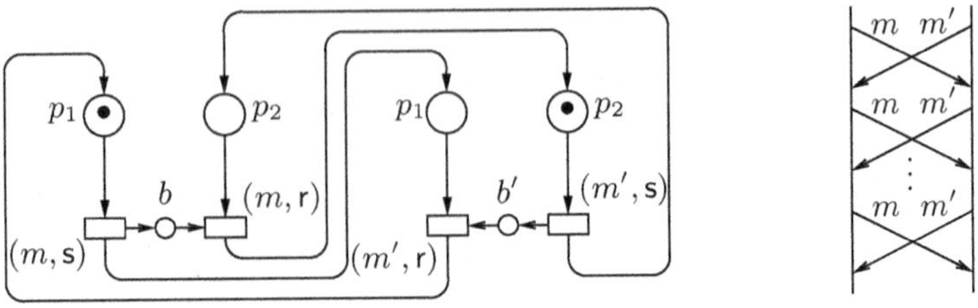

Fig. 4. The low level net for Fig. 3 and a typical MSC that it generates

original high level net are safe. However, the buffer places can have an arbitrary number of tokens. Thus, the low level net is not, in general, safe.

All processes need not traverse the high level transitions in the same order. Each process proceeds asynchronously and only gets blocked if, within an MSC, it requires a token from a buffer place and the buffer place is empty.

As we fire the "low level" transitions corresponding to the events of the MSCs labelling the high level transitions, we build up in the obvious way a partial order that can be regarded as an MSC. A complete MSC is generated when all send events have been matched up with corresponding receive events. This is captured by declaring a marking to be an accepting one if all the buffer places are empty at the marking. Thus, the language of the netchart is defined to be the set of MSCs generated by all firing sequences leading to accepting markings. (Note that there are only a finite number of such markings.) We can further control the language of a netchart by defining a set $\mathcal{F}$ of final control markings and insist that a marking is accepting only if all buffer places are empty *and* the control marking belongs to $\mathcal{F}$. Figure 4 describes the low-level net associated with the netchart in Fig. 3 and also exhibits a typical MSC in the language defined by this netchart.

4 From Netcharts to Message-Passing Automata

The low level net associated with a netchart makes it an executable specification. We now show that a netchart can also easily be transformed into the executable model often used in connection with HMSCs [1,12].

4.1 Message Passing Automata

A natural implementation model for MSC languages is a message-passing automaton. Recall that the set of processes $\mathcal{P}$ determines the communication alphabet Γ. For $p \in \mathcal{P}$, let $\Gamma_p = \{p!q(m), p?q'(m') \mid q, q' \in \mathcal{P}, m, m' \in \Delta\}$ denote the actions that process p participates in.

A *message-passing automaton* over Γ is a structure $\mathcal{A} = (\{\mathcal{A}_p\}_{p\in\mathcal{P}}, Aux, s_{in}, \mathcal{F})$ where:

- Aux is a finite alphabet of (auxiliary) messages.

- Each component $\mathcal{A}_p$ is of the form $(S_p, \longrightarrow_p)$ where
  - S_p is a finite set of p-local states.
  - $\longrightarrow_p \subseteq S_p \times \Gamma_p \times Aux \times S_p$ is the p-local transition relation.
- $s_{in} \in \prod_{p \in \mathcal{P}} S_p$ is the global initial state.
- $\mathcal{F} \subseteq \prod_{p \in \mathcal{P}} S_p$ is the set of global final states.

The local transition relation $\longrightarrow_p$ specifies how the process p sends and receives messages. The transition $(s, p!q(m), \mu, s')$ specifies that in state s, p can send the message m augmented with auxiliary information μ to q by executing the communication action $p!q(m)$ and go to the state s'. Similarly, the transition $(s, p?q(m), \mu, s')$ signifies that at the state s, p can receive the message (m, μ) from q by executing the action $p?q(m)$ and go to the state s'.

The behaviour of a message-passing automaton is described in terms of *configurations*. Channels are modelled as queues. A configuration specifies the local state of each process together with the contents of each queue. A transition of the form $(s, p!q(m), \mu, s')$ appends the message (m, μ) to the queue (p, q). Similarly, a transition of the form $(s, p?q(m), \mu, s')$ removes the message (m, μ) from the head of the queue (q, p)—this transition is blocked if such a message is not available. A final configuration is one in which the local states of the processes constitute a final state of the message-passing automaton and all queues are empty. We can define a run of such an automaton, as usual. Runs that lead to final configurations are called *accepting*. The language accepted by the automaton is the set of words that admit accepting runs. The structure of the automaton ensures that this word language corresponds to the set of linearizations of an MSC language, so we can associate an MSC language in a natural way with a message-passing automaton. For more details, the reader is referred to [9,12].

4.2 From Netcharts to Automata

Each non-buffer place in the low-level net corresponding to a netchart can be uniquely assigned a location in $\mathcal{P}$. The places that belong to process p define, in a natural way, a local sequential component of a message-passing automaton.

The resulting message-passing automaton differs in one important aspect from the original netchart—each MSC that labels a transition in a netchart writes into a private set of channels, while there is only one common set of channels in a message-passing automaton. So long as the MSCs generated by the netchart obey the FIFO restriction on each channel, this collapsing of channels does not affect the MSC language and the language accepted by the message-passing automaton is the same as the one defined by the netchart.

However, netcharts can define MSCs that violate the FIFO restriction on channels, as seen in Fig. 5. We cannot directly translate such a netchart into a message-passing automaton. The solution is to weaken the definition of an MSC to permit multiple channels between processes. Different channels may carry messages of the same type. Each channel is FIFO, but messages on different channels can overtake each other (even between the same pair of processes).

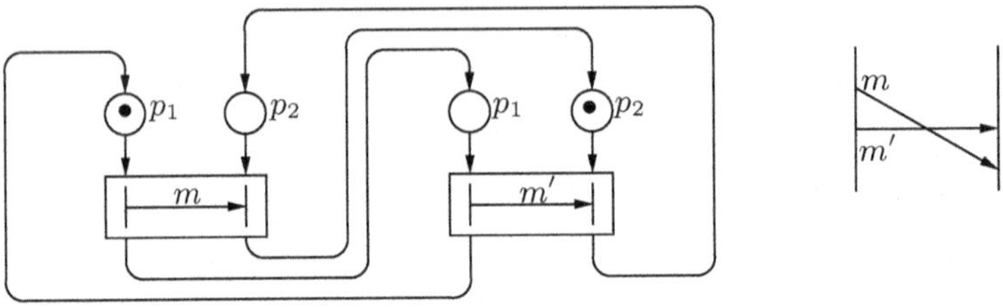

Fig. 5. A netchart that generates a non FIFO MSC

We can then augment message-passing automata so that each send and receive specifies the channel name in addition to the message type. With this extension, netcharts can always be converted into message-passing automata. However, for simplicity, we shall assume that we only deal with netcharts that generate MSCs that respect the FIFO restriction and work with the simpler definition of message-passing automata that we presented initially.

5 Regular Netchart Languages

We begin with a simple observation about Petri nets. Let $N = ((S, T, F), M_{in})$ be a Petri net and $L(N)$ denote the set of firing sequences of N. Recall that a Petri net is bounded if there exists a uniform upper bound B such that at any reachable marking, every place contains at most B tokens.

Proposition 2. *If N is a bounded Petri net then, $L(N)$ is regular.*

This easily yields a sufficient criterion for a netchart language to be regular.

Lemma 3. *If the low level net associated with a netchart is bounded, the MSC language of the netchart is regular.*

Since boundedness is a decidable property for nets, this condition can be effectively checked. Since the converse of the preceding lemma does not hold, deciding the regularity of a netchart language is not straightforward. For the netchart language to be non-regular, we have to find an unbounded family of markings, each of which can be reduced to a legal final marking where all buffer places are empty.

However, for the special case where each component of a netchart is a cyclic process, regularity is decidable.

Theorem 4. *Consider a netchart in which each component is a cyclic process. It is decidable if the language of such a netchart is regular.*

Proof. We construct the *communication graph* for the entire netchart. This graph has the set of processes as vertices and an edge from p to q if there is a transition in the netchart labelled by an MSC with a message from p to q.

Let C, C' be maximal strongly connected components (scc's) of this communication graph. We draw an edge from C to C' if there exists $p \in C$ and $q \in C'$ such that there is an edge in the communication graph from p to q. This induced graph of scc's is a dag. This dag may itself break up into a number of connected components. We can analyze each of these components separately, so, without loss of generality, assume that the entire dag is connected.

The buffer places that connect processes within each scc are automatically bounded (for the same reason that a locally synchronized HMSC generates a regular MSC language). Thus, the only buffer places that can be unbounded are those that connect processes in different scc's.

Suppose $p \in C$, $q \in C'$ and there is an edge from p to q in the communication graph. Then, by the structure of a netchart, each time p enters a high-level transition where it sends a message to q, there must be a matching instance where q enters so that the buffer places are cleared out. This has two implications.

(i) If p is not live—that is, it can execute only a finite number of actions—then q is not live. Indeed, all of C and C' are not live.
(ii) If q is not live, then any messages generated by p after q quits will never be consumed, so the resulting behaviour will not lead to an MSC in the language. Thus, effectively the buffer place between p and q is bounded.

Thus, if a process p is not live, all processes in the scc C containing p as well as all scc's that are descendants of C in the dag are not live. Moreover, after a bounded initial period, any message produced by an ancestor scc will never be consumed. So, as far as the MSC language is concerned, the ancestors of C also have only a bounded life. In other words, if even a single process is not live, the entire dag has a bounded behaviour and the resulting MSC language is bounded. Conversely, the language is unbounded iff all the scc's in the dag are live.

Thus the problem reduces to checking whether all the scc's in the dag are live. Observe that the minimal scc's in the dag have no input buffer places, so their liveness can be analyzed in isolation. This is not difficult to check given the simple cyclic structure of the components. If all the immediate predecessors of an scc are live, then the liveness of the scc again depends only on its internal structure, so we perform the same check as for the minimal scc's. Thus, we can systematically check the liveness of all the scc's in the dag by sorting them topologically and analyzing them in this sorted sequence. □

6 HMSCs vs. Netcharts

Every HMSC language is finitely generated (and all finitely generated regular MSC languages are HMSC representable). It turns out that netcharts can also generate *regular* MSC languages that are not finitely generated. An example of this is the alternating bit protocol in which two processes communicate over a lossy channel. The sender alternately tags the data it sends with bits 0 and 1. The receiver acknowledges the bit corresponding to each data item it receives.

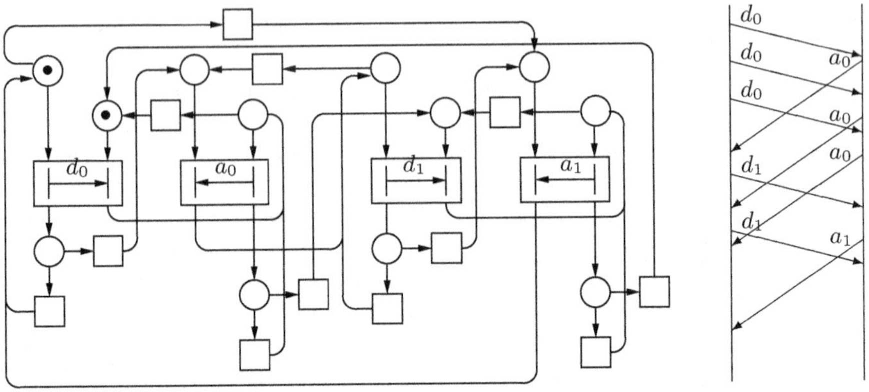

Fig. 6. A netchart for the alternating bit protocol, with a typical MSC

The sender flips its bit each time it gets the acknowledgement it is waiting for (and ignores any incorrect acknowledgement that it may receive).

Figure 6 shows a netchart corresponding to the alternating bit protocol. In the figures, messages d_0 and d_1 represent data sent with bit 0 and 1, respectively, and messages a_0 and a_1 represent the corresponding acknowledgements.

The first HMSC we encountered, Fig. 2, generates a non-regular language in which all channels are bounded. It is not difficult to show that any netchart implementation of this HMSC (i.e., its associated low level Petri net) would have bounded buffers. For netcharts, bounded buffers imply regularity. Hence, this HMSC is not representable via netcharts. However, from the result we will prove in the next section it follows that every HMSC that defines a regular MSC language has an equivalent netchart.

7 From Regular MSC Languages to Netcharts

We now show that every regular MSC language can be represented as a netchart. We begin by recalling the characterization of regular MSC languages in terms of message-passing automata. A B-bounded message-passing automaton is one in which in every reachable configuration, there are at most B messages in transit in any queue. We then have the following result from [12].

Theorem 5. *An MSC language L is regular iff there is a B-bounded message-passing automaton that accepts L, for some bound $B \in \mathbb{N}_0$.*

We now show that netcharts with final control markings can generate all regular MSC languages.

Theorem 6. *Every regular MSC language can be represented as a netchart.*

Proof Sketch: We can simulate a B-bounded message-passing automaton using a netchart.

In a B-bounded message-passing automaton, each queue is uniformly bounded by B. We can naïvely regard each channel as a cyclic buffer of size B with slots labelled 0 to $B-1$. Each process begins by reading from or writing into slot 0 along each channel that it is connected to. After a process reads from (or writes to) slot i of a channel c, when it next access the same channel it will read from (or write to) slot $i+1$ mod B, which we denote $i \oplus 1$.

In this framework, the complete configuration of a process is given by its local state s and the slot k_i that it will next read from or write to for each channel c_i that it is connected to. Initially, a process is in the configuration $(s^p_{in}, 0, \ldots, 0)$.

Each local transition $s \stackrel{c_i!(m,\mu)}{\longrightarrow} s'$ (or $s \stackrel{c_i?(m,\mu)}{\longrightarrow} s'$) generates a family of moves of the form $(s, k_1, k_2, \ldots, k_i, \ldots, k_n) \stackrel{c_i!(m,\mu)}{\longrightarrow} (s, k_1, k_2, \ldots, k_i \oplus 1, \ldots, k_n)$ (or $(s, k_1, k_2, \ldots, k_i, \ldots, k_n) \stackrel{c_i?(m,\mu)}{\longrightarrow} (s, k_1, k_2, \ldots, k_i \oplus 1, \ldots, k_n)$, for each choice of $k_1, k_2, \ldots, k_n$, corresponding to the n channels that the process reads or writes.

We construct a netchart in which, for each process p, there is separate place corresponding to each configuration $(s, k_1, k_2, \ldots, k_i, \ldots, k_n)$ of p in the B-bounded message-passing automaton. Consider a message (m, μ) that is transmitted on channel c_i during a run of the message-passing automaton. When this message is sent, the sending process is some configuration of the form $(s, k_1, k_2, \ldots, k_i, \ldots, k_n)$ and when this message is received, the receiving process is in some configuration of the form $(t, \ell_1, \ell_2, \ldots, k_i, \ldots, \ell_n)$ (notice that the value of k_i in the two configurations is necessarily the same). For each such pair of sending and receiving configurations, we create a separate transition in the netchart representing c_i, labelled with an MSC consisting of a single message (m, μ) (see Fig. 7). We use internal transitions to let the sending process guess the context in which the message will be received and fire the corresponding transition in the netchart. Symmetrically, the receiving process guesses the context in which the transition was sent and enters the corresponding transition in the netchart.

We can then associate a set of global final control states with this netchart corresponding to the accepting states of the original message-passing automaton. It is not difficult to see that this netchart accepts the same MSC language as the original message-passing automaton.

The crucial observation is that any inconsistent choice among nondeterministic transitions in the netchart must lead to deadlock. Suppose that for channel c_i, the sending and receiving processes make incorrect guesses about each other's contexts for message (m, μ) sent in slot k_i along c_i. This would result in the receiving process getting stuck in the wrong netchart transition. The only way for it to make progress is for the sending process to eventually cycle through all the other slots along channel c_i, return to slot k_i and send a fresh copy of the message (m, μ) within the netchart transition where the receiving process is stuck. This can be mapped back into a run of the message-passing automaton in which the channel c_i is not B-bounded, which is a contradiction. □

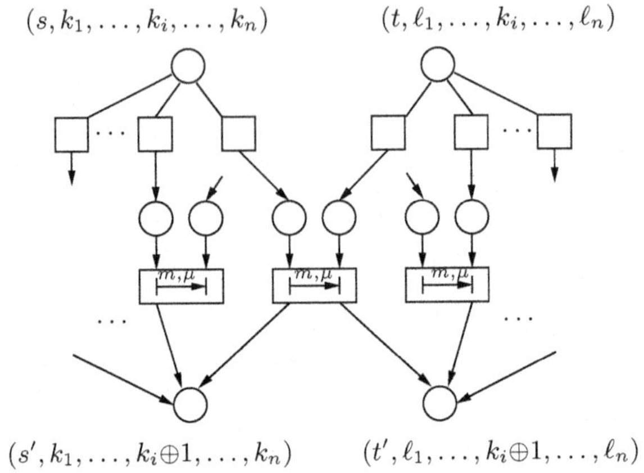

Fig. 7. The translation from B-bounded message-passing automata to netcharts

8 Discussion

We have shown that by distributing the control in an HMSC in appropriate way, we can derive a model that is much more easily implementable than HMSCs while retaining the appeal of a visual formalism. In terms of the regular MSC languages they can represent, netcharts are more expressive than HMSCs. In general, the expressive power of the two formalisms is incomparable.

As mentioned in the introduction, an important aspect of the netchart model is that the compound MSCs that are defined by a netchart are built up from "complete" MSCs. This is important from the point of view of using such a formalism to capture system requirements at a reasonably intuitive level. An alternative approach, suggested in [6], is to label HMSC nodes with *compositional* MSCs that may have unmatched sends and receives. A single send or receive action is a trivial example of a compositional MSC. Thus, a specification in terms of compositional MSCs very quickly becomes indistinguishable from a concrete implementation in terms of distributed automata.

One natural direction in which to extend this work is to use netcharts for capturing requirements in realistic settings.

At a technical level, the following interesting questions remain open:

- What is the exact relationship between the class of Netchart languages and the class of HMSC representable languages?
- Is the problem of checking whether the language of a netchart is regular decidable?

References

1. Alur, R., Etessami, K. and Yannakakis, M.: Realizability and verification of MSC graphs. *Proc. ICALP'01*, LNCS **2076**, Springer-Verlag (2001) 797–808.

2. Alur, R. and Yannakakis, M.: Model checking of message sequence charts. *Proc. CONCUR'99*, LNCS **1664**, Springer-Verlag (1999) 114–129.
3. Booch, G., Jacobson, I. and Rumbaugh, J.: *Unified Modeling Language User Guide*. Addison-Wesley (1997).
4. Caillaud, B., Darondeau, P., Helouet, L., and Lesventes, G.: HMSCs as partial specifications ... with PNs as completions. *Modeling and Verification of Parallel Processes 4th Summer School, MOVEP 2000*, LNCS **2067**, Springer-Verlag (2001) 125–152.
5. J. Desel and J. Esparza: Free Choice Petri Nets. *Cambridge Tracts in Theoretical Computer Science 40*, Cambridge University Press (1995).
6. Gunter, E., Muscholl, A. and Peled, D.: Compositional message sequence charts. *Proc. TACAS'01*, LNCS **2031**, Springer-Verlag (2001) 496–511.
7. Harel, D. and Gery, E.: Executable object modelling with statecharts. *IEEE Computer*, July 1997 (1997) 31–42.
8. Henriksen, J.G., Mukund, M., Narayan Kumar K., and Thiagarajan, P.S.: On Message Sequence Graphs and Finitely Generated Regular MSC Languages. *Proc. ICALP'00*, LNCS **1854**, Springer-Verlag (2000) 675–686.
9. Henriksen, J.G., Mukund, M., Narayan Kumar K., and Thiagarajan, P.S.: Regular Collections of Message Sequence Charts. *Proc. MFCS'00*, LNCS **1893**, Springer-Verlag (2000) 405–414.
10. ITU-TS Recommendation Z.120: *Message Sequence Chart (MSC)*. ITU-TS, Geneva (1997).
11. Mauw, S. and Reniers, M. A.: High-level message sequence charts, *Proc. 8th SDL Forum, SDL'97: Time for Testing — SDL, MSC and Trends*, Elsevier (1997) 291–306.
12. Mukund, M., Narayan Kumar, K. and Sohoni, M.: Synthesizing distributed finite-state systems from MSCs. *Proc CONCUR 2000*, LNCS **1877**, Springer-Verlag (2000) 521–535.
13. Muscholl, A. and Peled, D.: Message sequence graphs and decision problems on Mazurkiewicz traces. *Proc. MFCS'99*, LNCS **1672**, Springer-Verlag (1999) 81–91.
14. Roychoudhury, A. and Thiagarajan, P. S.: Communicating transaction processes. *Proc. ACSD'03*, IEEE Press, 2003 (to appear).
15. Rudolph, E., Graubmann, P. and Grabowski, J.: Tutorial on message sequence charts. *Computer Networks and ISDN Systems — SDL and MSC* **28** (1996).
16. Sgroi, M., Kondratyev, A., Watanabe, Y. and Sangiovanni-Vincentelli, A. : Synthesis of Petri nets from MSC-based specifications. Unpublished manuscript.

High-Level Message Sequence Charts and Projections

Blaise Genest[1], Loïc Hélouët[2,*], and Anca Muscholl[1]

[1] LIAFA, Université Paris VII, 2 place Jussieu, 75251 Paris, France
[2] IRISA, Campus de Beaulieu, 35042 Rennes Cedex, France
{genest,anca}@liafa.jussieu.fr,loic.helouet@irisa.fr

Abstract. Abstraction is a key issue in automatic verification, and it is often performed by a projection on a subsystem that is relevant for the property to check. This paper defines projections for the scenario language of High-level Message Sequence Charts (HMSC). We first show that the projection of an HMSC is not representable as an HMSC, in general. However, we show how that projections of HMSCs can be represented by a larger class of scenario languages, namely by (realizable) compositional HMSCs (cHMSCs). Moreover, we propose an algorithm that checks whether the projection of an HMSC can be represented by an HMSC, constructing the HMSC representation, when possible. This can be used in model-checking the projection of an HMSC specification.

1 Introduction

Scenario languages such as Harel's Live Sequence Charts [7], UML sequence diagrams, interworkings, etc. have seen a growing interest this last decade. Among them, the ITU standardized notation of Message Sequence Charts (MSCs, [10]) has received a lot of attention, both in the area of formal methods and in automatic verification [1,2,8,13,11,15]. MSCs can be considered as an abstract representation of communications between asynchronous processes. They are usually used as requirements, documentations, abstract test cases, and so on. A common approach for modeling the behavior of distributed systems is to describe them by means of parallel composition of communicating instances. MSCs and high-level MSCs (HMSCs for short) propose a pictorial way of modeling behaviors, combining parallel composition (processes) with sequential composition (transition system). The main advantage of such a representation is to have a local, explicit description of the communication and the causalities appearing in the system. Even if HMSCs seem to be a very simple formalism, the explicit use of parallel composition leads to model-checking being in general undecidable ([2, 13]). By model-checking we mean validating HMSC specifications versus simple properties, that are either given in some sequential logics/automata or as HMSCs. Some weaker decidable comparison criteria have been proposed, such

* The work of this author has been partially supported by CAFE Eureka ITEA ip00004 Σ! 2023

R. Amadio, D. Lugiez (Eds.): CONCUR 2003, LNCS 2761, pp. 311–326, 2003.

as matching with gaps [12,14]. Notice also that model-checking partial order properties (i.e., specified using partial order logics) is decidable, albeit of high complexity, [11,15].

Abstraction often appears as a central issue for automatic verification of large systems [3,5,16], since it can decrease the complexity of the verification process. In general, the issue is how to obtain an abstraction that allows to transfer the result back to the initial system. Scenarios are supposed to remain rather concise, but in some cases they may have been designed with too many details, which are not relevant for the property to check. Moreover, the details may hide important information concerning the causalities. Abstraction for HMSCs can be performed by collapsing nodes of the graph, as for Kripke structures. The problem is that such an abstraction can produce more behaviors than the initial model. Thus the result of model-checking is only an approximation for the real model. Abstraction can also be defined by projecting away some events or instances, without changing the graph. This abstraction of an HMSC is still an HMSC, but has as a negative side effect the loss of certain causalities between events. For example, if one wants to know whether an event a on process p appears before b on process q, then she might want to verify it only on the subsystem obtained by projecting away all events not on $\{p, q\}$. However, in this subsystem, the events a, b might become unordered, which means that the abstraction will not preserve the property.

The first motivation of this paper is to give an exact abstraction of HMSCs, i.e. with no approximation. Our abstraction hides (projects away) specific events while preserving the causalities. This can have several benefits. First, it can provide a better comprehension of the interactions between particular instances (see Figure 3 for a concrete example). Second, when comparing two MSCs, a designer might be interested in comparing the behaviors involving common features of both scenarios. Hiding information while preserving causal dependencies becomes then a central point for this kind of comparison. Our abstraction will preserve the causal order, hence the property $(\neg b)\mathcal{U}a$ of the example above. More generally, if the projection is defined on events on a given set of processes P, then any formula defined by $\phi = a \mid \neg\phi \mid \phi\mathcal{U}_P\phi$ is satisfied by the projection if and only if it is satisfied by the concrete system. Here, the atomic propositions a concern only processes in P. The operator $\mathcal{U}_P$ is a restriction of $\mathcal{U}$ meaning that the 'until' is taken among the processes in P. That is, $\phi\mathcal{U}_P\psi = (\phi \vee \neg P)\mathcal{U}\psi$. Finally, a motivation for hiding actions in an HMSC is to be able to verify properties on a model that is hopefully smaller.

The main problem raised by projections of HMSCs that preserve the causalities is the representation of the projected HMSC. First, the projection of an MSC is not always an MSC, as hiding may produce events that represent at the same time sends and receives. A more severe problem is that, projected HMSCs (even bounded ones, [2,13]) cannot be represented by means of a *finite HMSC*, in general. The first main result of the paper is that we can always represent the projection of an HMSC by a realizable *compositional* HMSC (cHMSCs, [6]). Moreover, we give an algorithm that tests whether the projection of an HMSC can be represented by an HMSC. The second main result is an effective construction of an HMSC representing the projection of HMSC, whenever this is possible.

This paper is organized as follows. Section 2 introduces basic notions related to MSCs, and section 3 defines the projections. Section 4 shows on a concrete protocol (RMTP2) the reasons that can prevent an HMSC projection to be represented as an HMSC. Section 5 establishes the effective equivalence between projections of HMSCs and realizable cHMSCs. The main result of the paper is given in section 6. It states that we can decide in polynomial space whether a realizable cHMSCs (in particular, the projection of an HMSC) can be represented by an HMSC. In the affirmative case, we can build effectively such an HMSC. Finally, in section 7 we show that model-checking projections of HMSCs is decidable under some reasonable assumptions. Due to space limitations most proofs are omitted.

2 Preliminaries

Message Sequence Charts (MSC for short) is a scenario language standardized by the ITU ([10]). They represent simple diagrams depicting the activity and communications in a distributed system. The entities participating in the interactions are called instances (or processes) and are represented by vertical lines. Message exchanges are depicted by arrows from the sender to the receiver. In addition to messages, atomic actions and timer operations (set, reset and timeout) can also be represented. Figure 1-a gives an example of an MSC modeling interactions between a *Sender*, a *Medium* and a *Receiver*.

Formally, an *MSC* M is a partially ordered set (poset) described by a tuple $M = (E, \leq, A, \mathcal{P}, t, P, m)$, where E is a finite set of events, $\leq \subseteq E \times E$ is a partial order on E and $A = A^s \cup A^r \cup A^{at}$ is a set of actions, which can be sending, receiving or atomic actions (internal operations or events related to timers). $\mathcal{P}$ is a finite set of $\wp$ processes (instances), $t : E \longrightarrow A$ associates an action with each event and $P : E \longrightarrow \mathcal{P}$ associates a process with each event. The message relation $m \subseteq E \times E$ pairs up sending and receiving events such that each send has a unique associated receive, and vice-versa. A send action is denoted by $p!q(a)$, meaning that p sends to q the message a. A receive action is denoted by $q?p(a)$, meaning that q receives message a from p. The message relation m is consistent with the mapping t, i.e., if $(e, f) \in m$, then $t(e) = p!q(a)$ and $t(f) = q?p(a)$ for some p, q, a.

The *visual order* of M is given by a total order on each process $p \in \mathcal{P}$, i.e. on each set of events $E \cap P^{-1}(p)$ (process ordering) and by the message ordering $e \leq f$ for every message $(e, f) \in m$. The graphical representation of an MSC diagram actually corresponds to the visual order, consisting of vertical lines (process ordering) and message arrows (message ordering). The relation $\leq$ is the partial order generated by the visual order. We write $e \lessdot f$ when $e < f$ and there is no event g with $e < g < f$. It is a common assumption that inter-process communication is FIFO, i.e., there is no overtaking of messages on any channel. Figure 1-b depicts the Hasse diagram of the partial order of the MSC in Figure 1-a.

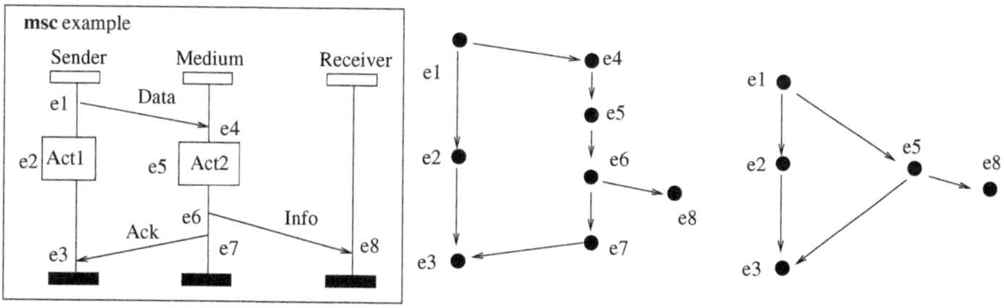

Fig. 1. a) MSC example b) partial order associated c) projection

MSCs are not expressive enough for specifying sets of scenarios and need operators such as choice or iteration to model interesting behaviors. This leads to High-level Message Sequence Charts (HMSCs) [10], that are just transition systems with nodes labeled by MSCs. Their semantics is defined by the *sequential composition* of two MSCs M_1, M_2. Let $M_i = (E_i, \leq_i, A_i, \mathcal{P}_i, t_i, P_i, m_i)$. Their sequential composition is the MSC $M_1 \circ M_2 = (E_1 \uplus E_2, \leq, A_1 \cup A_2, \mathcal{P}_1 \cup \mathcal{P}_2, t_1 \cup t_2, P_1 \cup P_2, m_1 \cup m_2)$, where $\leq$ is the transitive closure of $\leq_1 \cup \leq_2 \cup \{(e_1, e_2) \in E_1 \times E_2 \mid P_1(e_1) = P_2(e_2)\}$ defined on the disjoint union of events $E_1 \uplus E_2$. Intuitively, the sequential composition glues two MSCs along their common instance axis.

An *HMSC* is a tuple $H = (V, \to, v^0, F, \lambda)$ where $(V, \to, v^0, F)$ is a transition system with set of nodes V, transition relation $\to$, initial node v^0 and set of final nodes F. Each node v is labeled by an MSC, denoted $\lambda(v)$. An initial path of H is defined as a sequence of transitions $\rho = (v_1 \to v_2 \cdots \to v_k)$ with $v_1 = v^0$. If moreover $v_k \in F$, then ρ is an accepting path. The MSC $\lambda(\rho)$ associated with a path ρ is the sequential composition of the MSCs labeling the nodes, $\lambda(\rho) = \lambda(v_1) \circ \cdots \circ \lambda(v_k)$. An HMSC H defines a set of (finite) MSCs $\mathcal{L}(H) = \{\lambda(\rho) \mid \rho \text{ is an accepting path of } H\}$. Figure 2-a depicts an HMSC involving processes A, B, C.

Compositional MSCs (cMSC for short) is a notation that extends MSCs [6]. The difference between a cMSC and an MSC is that the message function m is a partial function, i.e., there can be sends or receives for which no matching event is defined. Such an event is called here *isolated.* The sequential composition of cMSCs is defined as above, with the additional requirement that the message function of M eventually matches isolated sends of M_1 with isolated receives of M_2 in such a way that M respects the FIFO condition. Hence, if two isolated sends from process 1 to 2 are concatenated with two isolated receives from process 1 to 2, then the resulting MSC consists of two successive messages from 1 to 2. See [6] for details.

A *compositional* HMSC H is an HMSC with nodes labeled by cMSCs. It defines a set of cMSCs $\mathcal{L}(H) = \{\lambda(\rho) \mid \rho \text{ is an accepting path of } H\}$. Moreover, a cHMSC is called *realizable* in [6], if there is no accepting path with isolated events.

We end this section by some properties needed later in our algorithms. An MSC M is called *atomic*, if it cannot be written as $M = M_1 \circ M_2$, where M_1, M_2 are non-empty MSCs. [9] describes an algorithm for testing whether an MSC M is atomic. This algorithm tests in linear time whether the following graph is strongly connected:

Definition 1. *Let $M = (E, \leq, A, \mathcal{P}, t, P, m)$ be an MSC over set of events E. The connection graph $CG(M) = (E, \rightarrow_{CG})$ of M is defined by $v_1 \rightarrow_{CG} v_2$ if either $P(v_1) = P(v_2)$ and $v_1 \lessdot v_2$, or one of (v_1, v_2), (v_2, v_1) is a message in M (edges are added from receives to associated sends).*

Proposition 1. *[9] An MSC M is atomic if and only if the connection graph $CG(M)$ is strongly connected.*

3 Projections of MSCs

Consider an MSC $M = (E, \leq, A, \mathcal{P}, t, P, m)$ and a subset of events $E' \subseteq E$. The *projection* of M on E' is noted $\pi_{E'}(M)$, and is obtained by erasing the events in $E \setminus E'$, and inheriting the causal dependencies from M. The set E' can represent for example all the events located on a subset of processes (instances). Formally, $\pi_{E'}(M)$ is the restriction of the poset M to E', defined as $\pi_{E'}(M) = (E', \leq', \mathcal{P}, P', m')$, where $\leq', P'$ are the restrictions of $\leq, P$ to E'. Events from the set E' will be called *non-erased* events. In the same way as for MSCs we depict the causal dependency between different processes by a causality (aka message) relation m'. We let $(e, f) \in m'$ for two non-erased events $e, f \in E'$, if $e \lessdot' f$ and $P(e) \neq P(f)$. That is, e and f are events located on different processes, $e < f$ in M and there is no intermediate non-erased event $g \in E'$ with $e < g < f$. The projection of an MSC will be called a pMSC for short.

Note that a pMSC is not necessarily an MSC, since an event in the projection may gather several actions of the initial MSC (these events will be called *multi-type events*). The example of Figure 1-c shows the projection of the MSC in Figure 1-a on $E' = \{e_1, e_2, e_3, e_5, e_8\}$. Since $e_1 \lessdot e_5$, we have to create a message between e_1 and e_5 to keep ordering. Similarly, as $e_5 \lessdot e_8$, we have to create a message between e_5 and e_8. In the projection, event e_5 has several types, a receive from the *Sender* process and two sends to *Receiver* and *Sender*. However, multi-type events are not a real problem for modeling, as pMSCs are still partially ordered event sets.

We can define *atomic* pMSCs similarly to MSCs: A pMSC M is atomic if the connection graph $CG(M)$ is strongly connected. For example, the pMSC $M' = \pi_{\{e_1, e_2, e_3, e_5, e_8\}}(M)$ in Figure 1-c is atomic. Thus, in addition to the edges represented in the Hasse diagram of Figure 1-c the connection graph $CG(M')$ contains the back edges $(e_5, e_1), (e_3, e_5)$ and (e_8, e_5), and is strongly connected. Note that projections do not preserve atomicity. In general, atoms of $\pi_E(M)$ can be larger than those of M.

For an HMSC $H = (V, \rightarrow, v^0, F, \lambda)$, the projection $\pi_{E'}(H)$ is defined as the projection of each MSC obtained from an accepting path of H on all occurrences of events of E'. Let v be a node of H, and let us denote by E_v the events associated to the MSC $\lambda(v)$. Let $E'' = \biguplus_{v \in V} (E_v \cap E')$ be the set of occurrences of events of E'. Then, the set of pMSCs defined by the pHMSC $H' = \pi_{E'}(H)$ is the set $\mathcal{L}(H') = \{\pi_{E''}(\lambda(v_1) \circ \cdots \circ \lambda(v_k)) \mid v_1 \to \cdots \to v_k$ is an accepting path of $H\}$. We will call the projection of an HMSC a *pHMSC*.

4 Comparing pHMSCs with HMSCs

The description of a pHMSC by an HMSC, together with a projection function, has several drawbacks. First, since causal dependencies are only implicitly given by the HMSC, a projected scenario is difficult to understand. Second, an implicit representation is not convenient for algorithmic manipulations. Third, by projecting an HMSC we usually want to obtain a smaller object, with a more compact representation. An immediate question appears when projecting an HMSC H to some pHMSC $H' = \pi_E(H)$, namely whether there exists some equivalent HMSC G, i.e., such that $\mathcal{L}(G) = \mathcal{L}(H')$? In particular, if H' is equivalent to some HMSC, then there exists a finite set X of generators for $\mathcal{L}(H')$. That is, there exists a finite set X of MSCs such that every $M \in \mathcal{L}(\pi_E(H))$ is a product of elements from X. We show below two situations that can prevent the existence of such a set X.

The first case is called an *unbounded crossing*. Intuitively, a pHMSC contains an unbounded crossing if there is a communication pattern that can be iterated an arbitrary number of times between two events situated on different processes that are causally related. For example, the HMSC of Figure 2-a generates an unbounded crossing for a projection on the instances A and B. Figure 2-b shows the partial orders generated by the HMSC of Figure 2-a, and Figure 2-c shows the partial orders after the projection on A and B. The MSCs in the projection are all atomic, hence there is no finite set generating them.

A second situation ruling out a finite representation is called a *crown*. Let us illustrate the presence of a crown on an example. Consider the HMSC of Figure 3. This HMSC describes scenarios for data transmission and acknowledgment for a multi-cast protocol called RMTP2. The RMTP2 network is organized as a tree, propagating data packets from a data source in the network, and aggregating acknowledgments in order to retransmit missed packets. Some nodes are designated to store a copy of the data sent, and retransmit each missed packet to the child subnetwork, if necessary. When a child receives a data packet, it may send an acknowledgment message *Hack* to its parent. A receiver may also decide to send an acknowledgment after a certain delay *tHackval*. The situation depicted by HMSC of Figure 3 shows communications between a node and two children. Child_1 always sends an acknowledgment to its parent upon data reception, while Child_2 always acknowledges data packets after the delay expiration. Furthermore, packets are never missed, but Child_1 can receive corrupted data, and retransmission and *Hack* packets may cross. The left part of Figure 3 shows the partial order associated with $(\textit{Hack\_incomplete} \circ \textit{Crossing})^*$. The right part of Figure 3 shows the same order after hiding the *Parent*, and all timer events.

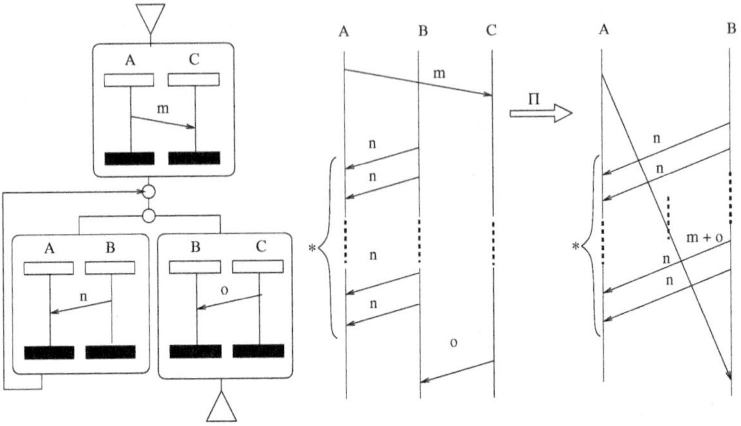

Fig. 2. a) HMSC generating an unbounded crossing b) MSC c) pMSC

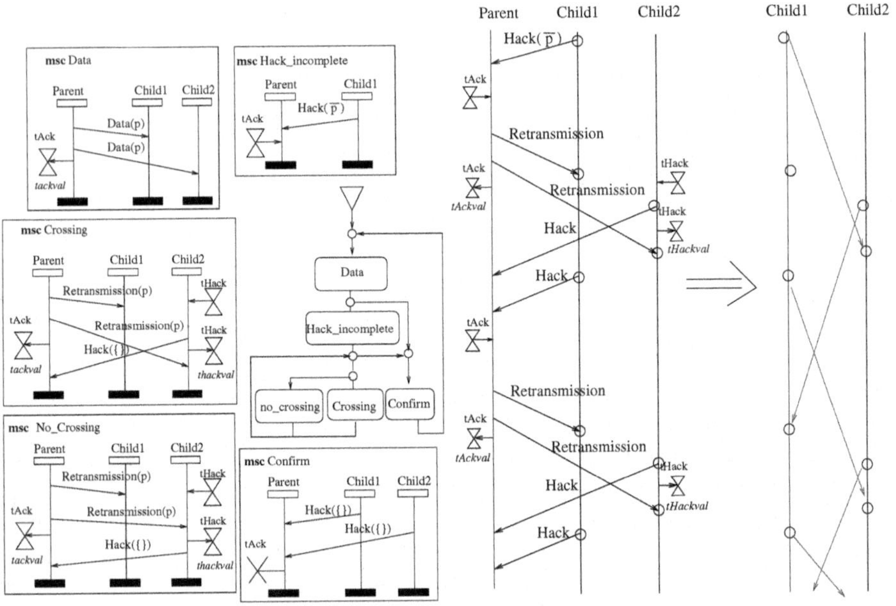

Fig. 3. A part of RMTP2 protocol and a crown generated by hiding instance Parent

It is clear that without breaking messages (or more precisely causal dependencies between distinct instances) the order obtained after projection can not be defined as a composition of finite communication patterns.

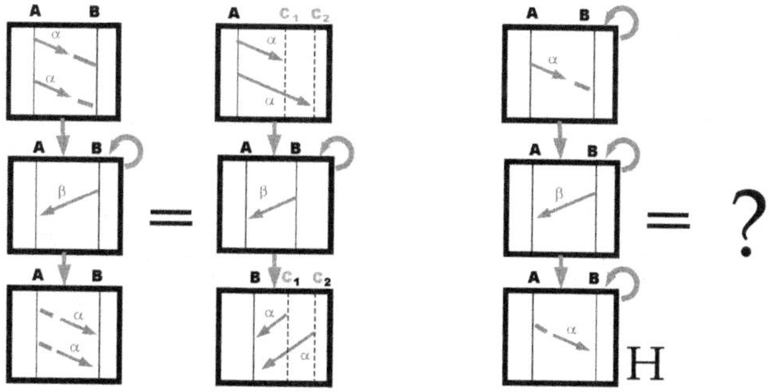

Fig. 4. The cHMSC in the right part is not a pHMSC.

5 Projections and Compositional HMSCs

In the previous section we saw that pHMSCs can describe behaviors that cannot be captured by HMSCs. Still, we might ask whether there is some graphical model for pHMSCs that involves only the non-erased events. Natural candidates for such a model are *compositional HMSCs* (cHMSC) [6], which correspond to a message relation that is defined on a given path, rather than on each node. Actually, both situations described in section 4 can be easily described by cHMSCs. The main result of this section states that pHMSCs have equal expressive power with realizable cHMSCs.

Since pHMSCs involve in general multi-type events we actually need enriched versions of MSCs and cMSCs, by allowing multi-type events. For simplicity we will assume in this section that the projections do not generate multi-type events. However, our constructions can be easily adapted to this case.

We first note that cHMSCs are at least as expressive as pHMSCs. A formal proof will be given by showing later that one can decide whether a given pHMSC is equivalent to an HMSC, whereas this question is undecidable for cHMSCs.

Figure 4 below gives a rough idea of what kind of behavior of cHMSCs cannot be expressed as the projection of an HMSC. In the cHMSC in the left part of Figure 4 two or more isolated send events α are matched after an arbitrary number of β messages. If we have to describe this behavior using a pHMSC, we would need for each event α a new process, that disappears through projection (processes C_1, C_2 in the middle part of Figure 4). More formally, for matching a sequence $e_1 < \cdots < e_n$ of sends on process A with a sequence $f_1 < \cdots < f_n$ of receives on process B, we need for each pair e_i, f_i a new process C_i. If for instance $C_1 = C_2$, then $e_1 < e_2 < f_1 < f_2$, hence we do not have $e_2 \lessdot f_2$, that is we do not create a message from e_2 to f_2. Since the number of processes is fixed we therefore cannot describe the behavior of the cHMSC in the right part of Figure 4 by a pHMSC.

5.1 Comparing pHMSCs and cHMSCs

From the previous examples, cHMSCs seem good candidates for describing pHMSCs without multi-type events. We show in this section that pHMSCs correspond precisely to a subclass of cHMSCs, that of realizable cHMSCs. We recall that a cHMSC is called *realizable*, if every accepting path is labeled by an MSC. Throughout this section, H will denote an HMSC and $H' = \pi(H)$ its projection. An event of the pHMSC H' will be denoted as *non-erased* event.

Given a pHMSC, we will construct an equivalent cHMSC by guessing the type of the non-erased events. We check the types by restricting the transition relation of the cHMSC. Let M be an MSC with event set E and let $M' = \pi_{E'}(M)$ be the projection of M onto the set of events $E' \subseteq E$. We have to guess the type of some events $e \in M'$ (send/receive/local, to which process etc). For example, in Figure 3 we have to guess that the first event on Child_1 becomes a send to Child_2 in the projection. In order to verify that the guess was correct, we need to keep track of the processes occurring in the future of e within M. Among these processes we need also to know which processes are seen by a non-erased event $e' > e$ (such processes are called *dead*, since no event f with $e \lessdot f$ can occur on such a process). Hence, let us define for each non-erased event $e \in E'$:

$$\begin{aligned}
\mathrm{F}(e) &= \{P(f) \mid e \leq f \in E\} \quad \text{(future processes)}\\
\mathrm{DeadF}(e) &= \bigcup_{e<e' \in E'} F(e') \quad \text{(dead processes)}\\
\mathrm{LiveF}(e) &= \mathrm{F}(e) \setminus \mathrm{DeadF}(e) \quad \text{(live processes)}
\end{aligned}$$

A non-erased event $e \in E'$ is called *unchecked* if $\mathrm{LiveF}(e) \neq \emptyset$. When an event e is created, it is unchecked. Intuitively, e becomes checked as soon as we have the proof that it can have no more immediate successor in the projection. It is easy to test the guess for a checked event. If the guess is correct, then we can forget the event, else the guess was wrong and the current path is not accepting. We show that $\mathrm{LiveF}(e) = \emptyset$ means that e has no more immediate successor. Let $e < e'$ be non-erased and $p \in F(e')$. Assume by contradiction that e is matched by some non-erased event f after M with $P(f) = p$. We have that $e < e' < f$, hence a contradiction.

The set of unchecked events of pMSC M' is denoted $\mathrm{ToCheck}(M')$. The next lemma bounds the number of unchecked events on any path of a pHMSC $H' = \pi(H)$ polynomially in the number of processes.

Lemma 1. *Let ρ be an initial path of an HMSC H over $\wp$ processes, and M' the pMSC defined by projecting ρ. Then $|ToCheck(M')| \leq \wp^2$.*

Theorem 1. *Let H be an HMSC with n nodes over $\wp$ processes, and consider a projection H' of H. Then we can construct a realizable cHMSC G that is equivalent to H', of size $n2^{O(\wp^3)}$.*

Our construction can be easily modified in the presence of multi-type events, for obtaining an extended realizable cHMSC.

A cHMSC is called *b-bounded* if for each initial path ρ, each prefix of ρ is labeled by a cMSC with at most b unmatched sends. Moreover, a realizable cHMSCs of size s is s-bounded. Although the cHMSC obtained in Theorem 1 has exponential size, the number of unmatched sends is polynomial in the size of H (see Lemma 1). We can restate Lemma 1 as follows:

Proposition 2. *Let $H' = \pi(H)$ be a pHMSC and let G be the realizable cHMSC constructed in Theorem 1. Then G is $\wp^2$-bounded, where $\wp$ is the number of processes of H.*

For the converse direction, we construct a pHMSC from a cHMSC by introducing new processes for isolated events:

Theorem 2. *Realizable cHMSCs and pHMSCs (with no multi-type events) have the same expressive power.*

For example, for an HMSC H defined by a unique node with a self loop, labeled by a single message (s, r) from p to q, the cHMSC equivalent to the projection of H on $\{p\}$ is a self loop on a local event of p corresponding to s. This event is not a send anymore, since it has no immediate successor on a different process. We actually need unmatched sends only when dependencies are not preserved by the trivial projection (without rebuilding the order).

6 Atoms as Generators

In this section we show how to construct a compact representation of the generators (atoms) of a given realizable cHMSC G. This problem is directly related to the construction for a given realizable cHMSC G (or a pHMSC) of an equivalent HMSC, if it exists, and it involves the computation of the smallest MSCs that are factors of some MSC $M \in \mathcal{L}(G)$. Formally, given a realizable cHMSC G we want to compute the set $\mathrm{Gen}(G)$ of *generators* of G defined as follows. An *atomic* MSC M belongs to $\mathrm{Gen}(G)$ if $N = N_1 M N_2$ for some $N \in \mathcal{L}(G)$ and some MSCs N_1, N_2.

Clearly, for a realizable cHMSC to be equivalent to an HMSC, $\mathrm{Gen}(G)$ needs to be finite. At the end of the section we show that if $\mathrm{Gen}(G)$ is finite, then we can construct effectively an equivalent HMSC.

6.1 An Automata-Based Representation of Generators

We show now how to construct for a given realizable cHMSC G a finite automaton $\mathcal{A}(G)$ that accepts only linearizations of $\mathrm{Gen}(G)$ and such that for every $M \in \mathrm{Gen}(G)$, at least one linearization of M is accepted by $\mathcal{A}(G)$. The idea is to have a non-deterministic automaton $\mathcal{A}(G)$ that works as follows. Given an execution $z \in \mathcal{L}(G)$, the automaton guesses a factor w of z containing a generator $M \in \mathrm{Gen}(G)$ and extracts the events of M from w. Note that since we cannot choose the linearization z, such a generator M won't be itself a factor of z.

The automaton $\mathcal{A}(G)$ must check two conditions in the definition of the generating set Gen(G). First, we guess for each event e of the linearization z above a type $k \in \{0,1,2\}$, telling whether e belongs to N_1 ($k=0$), to M ($k=1$) or to N_2 ($k=2$). In addition, we check whether the guessed type is consistent, that is, that every send is of the same type as its associated receive. This will guarantee that N_1, N_2 and M are all complete MSCs. The strategy is that $\mathcal{A}(G)$ will read the HMSC G and write or not an event e read according to its guess whether $e \in M$ or not.

The harder part is that $\mathcal{A}(G)$ must check that M is atomic. Proposition 1 gives an algorithm for verifying that a given MSC M is atomic, checking that the connection graph $CG(M)$ is strongly connected.

We have now to test whether an MSC M is atomic using a finite automaton, that takes as input some arbitrary linearization of M. However, we cannot use directly the algorithm of [9], since the number of connected components is not bounded. In order to handle this, we restate the result of [9] as follows:

Proposition 3. *Let M be an MSC. Then M is atomic if and only if for every pair of processes p, q, there is a path in $CG(M)$ from the last event of p to the first event of q.*

We say that an event e of M *sees* another event f if there exists a path from e to f in $CG(M)$.

The main idea behind the construction of $\mathcal{A}(G)$ is to keep some information for two kinds of events. First, $\mathcal{A}(G)$ needs to record the last event in M (type $k=1$) on each process. Let last_p denote the last event on process p seen so far by $\mathcal{A}(G)$. Moreover, $\mathcal{A}(G)$ needs to take into account the unmatched send events in M (type $k=1$). Let $\mathcal{S}$ denote the set of these sends. The set $\mathcal{S}$ contains at most b sends, where b is the size of the cHMSC G.

Let $X = \{\text{last}_p \mid 1 \le p \le \wp\} \cup \mathcal{S}$. Now, for each pair $(x,p) \in X \times \{1,\dots,\wp\}$ we record an integer $T(x,p) \in \{0,1,2\}$ telling whether x sees the first p-event in M ($T(x,p)=2$), or whether x sees last_p but not the first p-event in M ($T(x,p)=1$), or whether it sees no p-event in M at all ($T(x,p)=0$). Actually, $\mathcal{A}(G)$ uses a "vision" function $T : x,p \in X \times \mathcal{P} \mapsto T(x,p)$, and a function $S : x \in X \mapsto S(x)$ where $S(x) \subseteq \mathcal{S}$ denotes the unmatched sends seen by x.

Clearly, computing the vision function of each event last_p suffices for deciding whether $CG(M)$ is strongly connected. This idea is used in the following construction.

Theorem 3. *Let G be a realizable cHMSC. Then we can construct effectively a finite automaton $\mathcal{A}(G)$ that accepts only linearizations of Gen(G), such that for every $M \in$ Gen(G) at least one linearization of M is accepted by $\mathcal{A}(G)$.*

Using the automaton constructed in Theorem 3 we can test whether a given realizable cHMSC (or a pHMSC) is equivalent to an HMSC in polynomial space:

Theorem 4. *Checking whether Gen(G) is finite for a given realizable cHMSC G (pHMSC G, resp.) can be done in PSPACE. Moreover, the problem is co-NP-hard.*

6.2 From Realizable cHMSCs to HMSCs

We have an algorithm for checking whether the MSC executions of a realizable cHMSC G are finitely generated. We show in this section how to construct effectively an equivalent HMSC H, in case that the answer is positive. For simplicity, we will assume that states of G are labeled by single events.

Let Gen(G) be the finite set of atoms of the realizable cHMSC G. We denote by *maxsize* the size of the largest atom of Gen(G) and by b the bound on unmatched sends on paths of G.

We first describe intuitively the construction. The HMSC H will have states labeled by the atomic MSCs in Gen(G). In addition, we will label each state with some information concerning paths of G that can correspond to the sequence of atoms read so far in H. This additional information consists of a sequence of segments of a path of G (i.e. a *sub-path*), that match this sequence of atoms. That is, each time we read an atom $A \in \text{Gen}(G)$, we guess new segments of the path of G that correspond to the MSC A. We keep track of path segments by recording only the first/last node of each segment and the processes occurring in the segment. Hence, all we need is that the number of segments is bounded (see the claim below).

We first define the HMSC H. Let Path be the set of sub-paths of G consisting of at most $(b+1)\cdot$maxsize segments. The set of nodes of H is $V = \text{Gen}(G) \times \text{Path}$. A node $(A, \rho) \in V$ is labeled by A. Moreover, there is an edge in H from a node (A, ρ) to (A', ρ') iff $\rho \subsetneq \rho'$, and $\lambda(\rho') = \lambda(\rho) \circ A$. Here, we write $\rho \subseteq \rho'$ if ρ is included in ρ', i.e. each segment of ρ is included in a segment of ρ', and in the same order. The initial node is $(\emptyset, \epsilon)$, and the final nodes (A, ρ) are those where ρ is a one-segment, accepting path of G.

Theorem 5. *Let G be a cHMSC where Gen(G) is finite. Let n be the number of nodes of G, e the number of events, maxsize the maximal size of MSCs in Gen(G) and b the maximal number of unmatched sends on initial paths of G. Then we can construct an equivalent HMSC H from G of size $O((n^{2b} \cdot 2^{\wp b} \cdot e)^{maxsize})$.*

Proof. Let $M \in \mathcal{L}(H)$, then there exists an initial path of G labeled by M. Conversely, let us consider an MSC $M = A_1 \cdots A_n$, where each A_i belongs to Gen(G). This MSC labels an accepting path $\rho = v_1 \to \cdots \to v_k$ of G.

For simplicity, we extend the visual order of M to the atoms A_i by letting $A_i \leq A_j$ if there are some events e in A_i, f in A_j with $e < f$. Now, we will assume w.l.o.g. that for every $i < j$ such that $A_i \not\leq A_j$ the first event of A_i in ρ comes before the first event of A_j in ρ. That is, we choose an ordering of the atoms of M according to their first occurrence in ρ.

Claim: Let $l \in \{1, \ldots, n\}$ and let ρ_l be the sequence of segments of ρ labeled by $A_1 \cdots A_l$. Then ρ_l consists of at most $(b+1) \cdot$ maxsize segments.

proof of the claim: We denote by $\widehat{\rho_j}$ the longest prefix of ρ that contains no event of A_j. Let also m be such that $\widehat{\rho_m}$ is the longest prefix $\widehat{\rho_j}$ with $j \leq l$.

The pMSC labeling $\widehat{\rho_m}$ has at most b unmatched sends, among which b' sends belong to ρ_l. Thus, $b'' = b - b'$ is the number of unmatched sends in $\widehat{\rho_m} \setminus \rho_l$ (the difference of two paths is obtained by deleting the nodes of the second path from the first one).

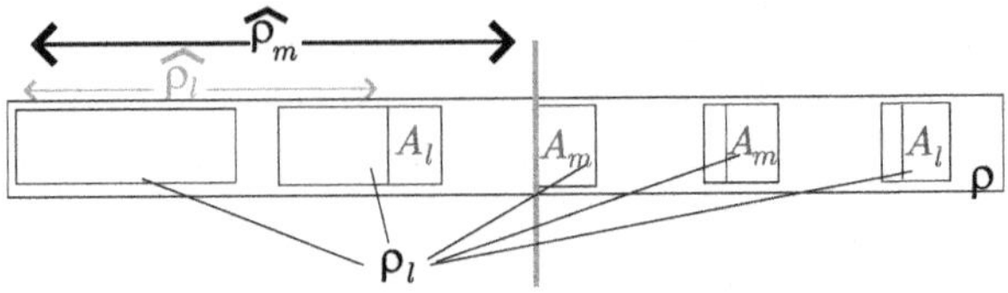

We claim first that there is no complete atomic MSC A_p in $\widehat{\rho_m} \setminus \rho_l$. To see this, note first that such a complete MSC A_p must satisfy $p > l$. However, it can satisfy neither ($A_p \not\leq A_m$ and $A_m \not\leq A_p$), by the choice of the ordering $A_1 \cdots A_n$, nor $A_m < A_p$, since $\widehat{\rho_m}$ has an empty intersection with A_m. Since each incomplete atom of $\widehat{\rho_m} \setminus \rho_l$ contributes with at least one unmatched send in $\widehat{\rho_m}$, we obtain that there are at most $b'' \cdot (\text{maxsize} - 1)$ events in $\widehat{\rho_m} \setminus \rho_l$, thus at most $b'' \cdot (\text{maxsize} - 1) + 1$ segments in $\rho_l \cap \widehat{\rho_m}$.

Moreover, by definition of m there is just one new atom starting in $\rho_l \setminus \widehat{\rho_m}$, namely A_m. Hence there are at most $b' + 1$ different atoms in $\rho_l \setminus \widehat{\rho_m}$ (b' that started already in $\widehat{\rho_m}$ plus A_m). This yields at most $(b' + 1) \cdot \text{maxsize}$ events in $\rho_l \setminus \widehat{\rho_m}$, hence at most $(b' + 1) \cdot \text{maxsize}$ segments. Therefore, we conclude that ρ_l contains at most $(b' + b'' + 1) \cdot \text{maxsize} = (b + 1) \cdot \text{maxsize}$ segments.

Concerning the size of H, for each path segment it suffices to remember the first/last node, and the processes that occurred in the segment. This gives at most $(|G|^2 \cdot 2^{\wp})^{b \cdot \text{maxsize}} = 2^{2b(\log(|G|)+\wp)\text{maxsize}}$ paths consisting of less than $b \cdot \text{maxsize}$ segments. □

Example: Let G be a cHMSC with 4 states Q, R, S, T, labeled respectively by s, s, r, r where s is a send on process 1 corresponding to a receive r on process 2. There are edges from Q to R, R to S, S to R, and S to T. Let A be the atom consisting of the message (s, r). The set of atoms of G is $\{A\}$, hence each state of H is labeled by A (or $\emptyset$). We describe some of the resulting states of H:

- s_1 is the state $[Q, \{1\}]; [S, \{2\}]$, consisting of two segments, one being a path from Q to Q and the other from S to S,
- s_2 is the state $[Q, \{1\}]; [T, \{2\}])$,
- s_3 is the state $([Q, S, \{1, 2\}]; [S, \{2\}])$ consisting of two segments, one being a path from Q to S, the other the path from S to S,
- s_4 is the (final) state $([Q, T, \{1, 2\}])$.

We have for instance edges from $(\epsilon, \emptyset)$ to s_1 and s_2. However, s_2 is a bad guess (that is, it will never reach the final state s_4). The reason is it cannot be extended by A: the segment $[T, \{2\}]$ forbids using the node v_3, since it already contains process 2. There are edges from s_1 to s_3, s_3 to s_4, s_1 to s_4 and a loop on s_3. The loop on s_3 corresponds to the guess of a scattered sub-path

$[R, \{1\}]; [S, \{2\}]$ labeled by A, and to the guess $[R, \{1\}]$ making the connection between $([Q, S, \{1, 2\}]$ and $[S, \{2\}])$, while $[S, \{2\}]$ is a disjoint segment.

Since there are at most $2^{O(\text{maxsize})}$ atoms, H is of exponential size. A priori, maxsize can be exponential in the size of the pHMSC, yielding an HMSC of doubly exponential size, but we believe this to be very unlikely. Actually, showing that maxsize is polynomial is as hard as showing that "Gen(G) finite?" is NP-complete.

7 Model-Checking HMSCs against HMSCs

Validating HMSCs specifications allows detection of inconsistencies or undesired behaviors at early design stages. However, model-checking HMSCs specifications against properties specified by HMSCs is undecidable, in general (see e.g. [2, 13]). Several papers considered restrictions of HMSCs [2,13,8], for which model-checking becomes decidable. The first positive results were obtained for *bounded* HMSCs [2,13], for which the set of MSC-linearizations is regular. A large family of HMSCs ensuring decidability of model-checking is given by the globally co-operative property [4]. An HMSC is globally cooperative if every loop is labeled by an MSC M that cannot be written as $M = M_1 \circ M_2$, where M_1, M_2 are non-empty MSCs over disjoint sets of processes.

We consider in this section model-checking for pHMSCs against HMSC properties and we show two settings for which the problem is decidable, with the same complexity as for HMSCs.

The next theorem shows that model-checking remains decidable, if the pHMSC is arbitrary, but the HMSC property is globally cooperative:

Theorem 6. *Let G be a globally cooperative HMSC, H an HMSC and $\pi_E(H)$ a projection of H. Then we can check in PSPACE whether $\mathcal{L}(\pi_E(H)) \cap \mathcal{L}(G) = \emptyset$, and in EXPSPACE whether $\mathcal{L}(\pi_E(H)) \subseteq \mathcal{L}(G)$.*

Even if the complexities we stated are rather high, note that in practice both the HMSC property and the reduced specification (the HMSC projected on a small part) can be reasonably small.

The second decidability result is based on the fact that the projection of a bounded HMSC preserves the boundedness.

Theorem 7. *Let G be a bounded HMSC, H an HMSC and $\pi_E(G), \pi_{E'}(H)$ their respective projections. Then we can check in PSPACE whether $\mathcal{L}(\pi_E(G)) \cap \mathcal{L}(\pi_{E'}(H)) = \emptyset$, and in EXPSPACE whether $\mathcal{L}(\pi_{E'}(H)) \subseteq \mathcal{L}(\pi_E(G))$.*

The last result shows that we can compare two projections of HMSCs (e.g. in order to find common parts), as long as one of them is bounded, with the same complexity as for bounded HMSCs.

8 Conclusion

In this paper we defined projections of HMSCs and showed how to decide whether an HMSC projection can be represented as an HMSC. This notion of projection can then be used to perform model-checking when at least one of the HMSCs considered is either globally cooperative or bounded. Even when model-checking is not possible, HMSC projections may still be useful for a designer for extracting causality information from scenario descriptions. As already pointed out, the projection of an HMSC may result in a larger HMSC. However, this is just a worst-case estimation that is unlikely in practice. For example, hiding a complete instance would probably have a greater impact on the shape of the projection than a random choice of the hidden events. Moreover, the presence of unbounded crossings or crowns may reveal some properties of the system under study. For example, when a projection hides the communication medium, the presence of a crown can indicate the impossibility to save a consistent global snapshot of the system in some executions.

References

1. R. Alur, G. Holzmann, and D. Peled. An Analyser for Message Sequence Charts. In *Proceedings of TACAS'96*, LNCS 1055, pages 35–48, 1996.
2. R. Alur and M. Yannakakis. Model Checking of Message Sequence Charts. In *Proceedings of CONCUR'99*, LNCS 1664, pages 114–129, 1999.
3. E. Clarke, O. Grumberg, S. Jha, Y. Lu, and H. Veith. Counterexample-guided abstraction refinement. In *Proceedings of CAV'2000*, pages 154–169. LNCS 1855, 2000.
4. B. Genest, A. Muscholl, H. Seidl, and M. Zeitoun. Infinite-state HMSCs: Model-checking and realizability. In *Proceedings of ICALP'2002*, pages 657–668. LNCS 2380, 2002.
5. S. Govindaraju and D. Dill. Counterexample-guided choice of projections in approximate symbolic model checking. In *Proceedings of ICCAD'00*, 2000.
6. E. Gunter, A. Muscholl, and D. Peled. Compositional Message Sequence Charts. In *Proceedings of TACAS'01*, pages 496–511. LNCS 2031, 2001.
7. D. Harel and W. Damm. LSCs: breathing life into Message Sequence Charts. Technical Report CS98-09, Weizmann Institute, Avril 1998.
8. J. Henriksen, M. Mukund, K. Kumar, and P. Thiagarajan. On Message Sequence Graphs and finitely generated regular MSC languages. In *Proceedings of ICALP'99*. LNCS 1644, 1999.
9. L. Hélouët and P. Le Maigat. Decomposition of Message Sequence Charts. In *Proceedings of SAM2000(2nd conference on SDL and MSCs)*, Grenoble, Juin 2000.
10. ITU-TS. *ITU-TS Recommendation Z.120: Message Sequence Chart (MSC)*. ITU-TS, Geneva, September 1993.
11. P. Madhusudan. Reasoning about sequential and branching behaviours of Message Sequence Graphs. In *Proceedings of ICALP'01*, page 809. LNCS 2076, 2001.
12. A. Muscholl. Matching specifications for Message Sequence Charts. In *Proceedings of FoSSaCS'99*, LNCS 1578, pages 273–287, 1999.
13. A. Muscholl and D. Peled. Message Sequence Graphs and decision problems on Mazurkiewicz traces. In *Proceedings of MFCS'99*, LNCS 1672, 1999.

14. A. Muscholl, D. Peled, and Z. Su. Deciding properties for Message Sequence Charts. In *Proceedings of FoSSaCS'98*, LNCS 1378, page 226, 1998.
15. D. Peled. Specification and verification of Message Sequence Charts. In *Proceedings of FORTE'00*, pages 139–154. IFIP 183, 2000.
16. A. Pnueli. Abstraction, composition, symmetry, and a little deduction: The remedies to state explosion. In *Proceedings of CAV'2000*. LNCS 1855, 2000.

Composition of Cryptographic Protocols in a Probabilistic Polynomial-Time Process Calculus

P. Mateus[1,*], J. Mitchell[2,**], and A. Scedrov[3,***]

[1] Center for Logic and Computation, IST, Lisbon, Portugal
[2] Department of Computer Science, Stanford University, Stanford, USA
[3] Department of Mathematics, University of Pennsylvania, Philadelphia, USA

Abstract. We describe a probabilistic polynomial-time process calculus for analyzing cryptographic protocols and use it to derive compositionality properties of protocols in the presence of computationally bounded adversaries. We illustrate these concepts on oblivious transfer, an example from cryptography. We also compare our approach with a framework based on interactive Turing machines.

Keywords: cryptographic protocols, probabilistic process calculus, computational security, composition theorem

1 Introduction

The design and verification of security protocols is a difficult problem. Some of the difficulties come from subtleties of cryptographic primitives. Further difficulties arise because security protocols are required to work properly when multiple instances of the protocol are carried out in parallel, where a malicious intruder may combine data from separate sessions in order to confuse honest participants. Moreover, although the protocols themselves are often very simple, the security properties they are supposed to achieve are rather subtle and should be formulated with great care.

A variety of methods are used for analyzing and reasoning about security protocols. Although such methods differ in significant ways, many of them reflect the same basic assumptions about the way an adversary may interact with the protocol or attempt to decrypt encrypted messages . In the common

* Partially supported by FCT grant SFRH/BPD/5625/2001, and by FEDER/FCT project Fiblog POCTI/2001/MAT/37239.

** Partially supported by OSD/ONR MURI "Semantic Consistency in Information Exchange" as ONR Grant N00014-97-1-0505, and by OSD/ONR CIP/SW URI "Software Quality and Infrastructure Protection for Diffuse Computing" as ONR Grant N00014-01-1-0795.

*** Partially supported by OSD/ONR MURI "Semantic Consistency in Information Exchange" as ONR Grant N00014-97-1-0505, by OSD/ONR CIP/SW URI "Software Quality, Infrastructure Protection for Diffuse Computing" as ONR Grant N00014-01-1-0795, and by NSF Grant CCR-0098096.

R. Amadio, D. Lugiez (Eds.): CONCUR 2003, LNCS 2761, pp. 327–349, 2003.

model, largely derived from [10] and suggestions found in [24], a protocol adversary is allowed to choose among possible actions nondeterministically. This is a convenient idealization, intended to give the adversary a chance to find an attack if there is one. In the presence of nondeterminism, however, the set of messages an adversary may use to interfere with a protocol must be restricted severely. Although the idealized assumptions make protocol analysis tractable, they also make it possible to "verify" protocols that are in fact susceptible to simple attacks that lie outside the adversary model. Another limitation is that a deterministic or nondeterministic setting does not allow us to analyze probabilistic protocols. In other words, actual protocols use actual cryptosystems that may have their own weaknesses, or might employ probabilistic techniques not expressed in the idealized model.

Recently there have been several research efforts to relate the idealized model to cryptographic techniques and the computational model based on probabilistic polynomial-time computation, including [7,16,23,25,26,3,2,9,4]. While these efforts develop rigorous mathematical settings carried out so far only "by hand", it is hoped that they will eventually lead to a new generation of "high fidelity" automated tools for security analysis that will be able to express the methods and concepts of modern cryptography.

Our initial contribution to this line of research was a formulation of a process calculus proposed in [16,23] as the basis for a form of protocol analysis that is formal, yet closer in foundations to the mathematical setting of modern cryptography. The framework relies on a language for defining communicating polynomial-time processes. The reason we restrict processes to probabilistic polynomial time is so that we can reason about the security of protocols by quantifying over all "adversarial" processes definable in the language. In effect, establishing a bound on the running time of an adversary allows us to relax the simplifying assumptions. Specifically, it is possible to consider adversaries that might send randomly chosen messages, or perform sophisticated (yet probabilistic polynomial-time) computation to derive an attack from messages it overhears on the network. An important aspect of our framework is that we can analyze probabilistic as well as deterministic encryption functions and protocols. Without a probabilistic framework, it would not be possible to analyze an encryption function such as [11], for which a single plaintext may have more than one ciphertext.

Some of the basic ideas of our prior work are outlined in [16,23]. Further example protocols are considered in [17]. The closest technical precursor is [1], which uses observational equivalence and channel abstraction but does not involve probability or computational complexity bounds. Prior work on CSP and security protocols, *e.g.*, [28,29], also uses process calculus and security specifications in the form of equivalence or related approximation orderings on processes.

This approach is based on the intuition that security properties of a protocol P may be expressed by means of existence of an idealized protocol Q such that for any adversary M, the interactions between M and P have the same observable behavior as the interactions between M and Q. The idea of expressing security properties in terms of some comparison to an ideal protocol goes back at least to

[15,6,5,20]. Here we emphasize a formalization of this idea by using observational equivalence, a standard notion from programming language theory. That is, two protocols P and Q are observationally equivalent if any program $C[P]$ has the same observable behavior as the program $C[Q]$, with Q instead of P. The reason observational equivalence is applicable to security analysis is that it involves quantifying over all possible additional processes represented by the contexts $C[\ ]$ that might interact with P and Q, in precisely the same way that security properties involve quantifying over all possible adversaries. Our framework is a refinement of this approach. In our asymptotic formulation [16,23], observational equivalence between probabilistic polynomial-time processes coincides with the traditional notion of indistinguishability by polynomial-time statistical tests [13, 30], a standard way of characterizing cryptographic primitives.

In this paper we derive a compositionality property from inherent structural properties of our process calculus. Basically, compositionality states that composing secure protocols remains secure. We obtain a general result of this kind in two steps. We consider a notion of a secure realization, or, *emulation* of an ideal protocol, motivated by [7] but here expressed by means of asymptotic observational equivalence. We show that the notion of emulation is congruent with the primitives of the calculus. Compositionality follows because the security requirements are expressed in the form that a real protocol securely realizes an ideal protocol.

We also illustrate some of these concepts on a traditional cryptographic example of oblivious transfer [27,12,14,9]. We show how the natural security requirements may be expressed in our calculus in the form that a real protocol emulates an ideal protocol. Finally, we establish an important relationship between the process calculus framework and the interactive Turing machine framework discussed in [7,9,25,26,3]. Indeed, the work based on [7] provides an encyclopedic treatment of a number of security requirements in a compositional setting. However, the framework of interactive Turing machines, even if optimal to deal with complexity results, is rather low-level and does not seem naturally suited for specification of and reasoning about cryptographic protocols. Moreover, the framework of interactive Turing machines comes about from the connections between cryptography and complexity, and therefore, some effort must be spent to obtain structural results, such as the composition theorem.

Basic definition and properties of the process calculus are discussed in Sect. 2. In Sect. 3 we discuss the notion of emulation, prove a general composition theorem, and analyze the example of oblivious transfer. A comparison to the interactive Turing machine model is given in Sect. 4. We conclude the paper in Sect. 5.

2 Probabilistic Polynomial-Time Process Calculus

In this section, we describe a version of the probabilistic polynomial-time process calculus [16,23], with the intention of using of the calculus to derive compositionality properties of secure protocols.

2.1 Syntax

We assume as given once and for all a countable set C of *channels*. In a discussion of security protocols it is common to consider a security parameter $n \in \mathbb{N}$. From now on, the symbol n is reserved to designate such security parameter. The role of this parameter is twofold. It bounds the length of expressions that can be sent through each channel by a polynomial in $|n|$, the length of n. This is written into the syntax: we introduce a *bandwidth map* $w : C \to \mathbf{q}$, where $\mathbf{q}$ is the set of all polynomials in one variable taking positive values on natural numbers. Given a value n for the security parameter, a channel c can send messages with at most $w(c)(|n|)$ bits. It turns out that the security parameter also bounds all the computation in the calculus by probabilistic polynomial time. This property the calculus is proved in [23].

The protocol language consists of a set of *terms,* or functional expressions that do not perform any communication, and *processes,* which can communicate with each other.

We assume a set of numerical terms T (endowed with a set of variables Var) with the following two properties. For any probabilistic polynomial (in the length of the security parameter n) time function f there is a term $t \in T$ and the associated probabilistic Turing machine M_t that computes f. Furthermore, given any term $t \in T$, the associated probabilistic Turing machine M_t is always probabilistic polynomial time, PPT, with input n and the numerical values of variables of t. (An example of such a set of terms T is described in [22], but the details of the construction are not needed here.) In order to ease notation, we shall confuse a term $t(\boldsymbol{x})$ with the a probabilistic polynomial time function $f(\boldsymbol{x}, n)$ associated to the PPT machine M_t. Hence, one can envision T simply as the set of all probabilistic polynomial time functions, neglecting any further syntax. Once again, mind that all terms denote probabilistic polynomial time functions in the length of the security parameter n. After fixing n, we denote by $P(t(\boldsymbol{a}) \to a)$ the probability of $M_t(\boldsymbol{a}, n)$ converging to a and $P(t(\boldsymbol{a}) = t'(\boldsymbol{b}))$ the probability of both associated Turing machines converging to the same value.

We now present our language of process expressions, a version of Milner's Calculus of Communicating Systems, CCS [21]. Bear in mind, though, that for us the overall computation must be probabilistic polynomial time and hence we use only polynomially bounded replication.

Definition 1. The *language of process expressions* $\mathcal{L}$ is obtained inductively as follows:

1. $0 \in \mathcal{L}$ (empty process: does nothing);
2. $\nu_c.Q \in \mathcal{L}$ where $c \in C$ and $Q \in \mathcal{L}$ (private channel: do Q with channel c considered private);
3. $\langle t \rangle_c \in \mathcal{L}$ where $t \in T$ and $c \in C$ (output: transmit the value of t on channel c);
4. $(x)_c.Q \in \mathcal{L}$ where $c \in C$, $x \in Var$ and $Q \in \mathcal{L}$ (input: read value for x on channel c and do Q);
5. $[t_1 = t_2].Q \in \mathcal{L}$ where $t_1, t_2 \in T$ and $Q \in \mathcal{L}$ (match: if $t_1 = t_2$ then do Q);

6. $(Q_1|Q_2) \in \mathcal{L}$ where $Q_1, Q_2 \in \mathcal{L}$ (parallel composition: do Q_1 in parallel with Q_2);
7. $!_q Q \in \mathcal{L}$ where $Q \in \mathcal{L}$ and $q \in \mathbf{q}$ (polynomially bounded replication: execute $q(|n|)$ copies of Q in parallel).

Every input or output on a private channel must appear within the scope of a ν-operator binding that channel, that is, the channel name in the scope of a ν-operator is considered bound. A *process* is a process expression in which the security parameter is replaced with a fixed natural number. Observe that the length of any process expression in $\mathcal{L}$ is polynomial in $|n|$.

For each fixed value k of the security parameter, we can remove replication by replacing each subexpression $!_q R$ of an expression Q by $q(|k|)$ copies of R in parallel, denoted $\overline{Q}_k$.

Let us also fix the following terminology and useful conventions. We assume that in any process expression in $\mathcal{L}$ private channels are named apart from other channels, which we call *public*. Analogously to first-order logic, a variable x is said to occur *free* in a process expression $Q \in \mathcal{L}$ if there is an occurrence of x that does not appear in the scope of a binding operator $(x)_c$. The set of all free variables of Q is called the *parameters* of Q and is denoted by $\mathbb{P}_Q$. A process expression without parameters is called *closed*.

Intuitively, messages are essentially pairs consisting of a "channel name" and a data value. The expression $\langle M \rangle_c$ places a pair $\langle c, M \rangle$ onto the network. The expression $(x)_c.P$ matches any pair $\langle c, m \rangle$ and continues process P with x bound to the least significant $w(c)(|n|)$ bits of value m, because of the bandwidth restrictions. When $(x)_c.P$ corresponds to a pair $\langle c, M \rangle$, the pair $\langle c, M \rangle$ is removed from the network and is no longer available to be read by another process. Evaluation of $(x)_c.P$ does not proceed unless or until a pair $\langle c, m \rangle$ is available.

Although we use channel names to indicate the intended source and destination of a communication, any process can potentially read any message, except when a channel is bound by the ν-operator, hiding communication. However, we only intend to use private channels for ideal specifications and for modeling various initial conditions in protocols regarding secret data, but we do not use private channels for modeling actual protocols. This communication model allows an adversary (or any process) to intercept a message between any two participants in a protocol. Once read, a pair is removed so that an adversary has the power to prevent the intended recipient from receiving a message. An adversary (or other process) may overhear a transaction without interfering simply by retransmitting every communication that it intercepts.

Observe that the output primitive of the calculus allows us to compute the image of a probabilistic polynomial time function f into some value a and send a through a channel c (or part of it if the bandwidth of c is too small). Moreover, the matching condition endows the calculus with the possibility of checking whether two terms converge to the same value. As we shall see, by combining these primitives we give a lot of power to the calculus.

In order to illustrate the flexibility of the process calculus we present the following two examples:

Example 1 (RSA encryption and decryption). Start by considering a very simple process S that knows some message M and integers a, m and just outputs $M^a \mod m$, this dummy process can be presented as $S(M) := \langle M^a \mod m\rangle_u$. Let us develop this just a bit more, and consider a remote procedure that receives x and outputs $x^a \mod m$. This procedure can be modeled by the following process: $RP := (x)_c.S(x)$.

Finally, consider that p, q are primes, and a, b are integers such that $ab \equiv 1 \mod \phi(pq)$. Consider the process $RSA(a, b, pq) := \text{Send}(M) \mid \text{Rec}$, where $\text{Send}(M) := \langle M^a \mod pq\rangle_u$ and $\text{Rec} := \nu_{u'}.(x)_u.\langle x^b \mod pq\rangle_{u'}$. Here the sender sends a message encrypted with the receiver's encryption key a and the receiver decrypts with its decryption key b and stores the plaintext privately.

Example 2 (Modular sequential composition). Suppose that a process $Q(\boldsymbol{c} \to \boldsymbol{u})$ receives inputs through public channels $\boldsymbol{c}$, works these inputs in some way, and returns relevant outputs through public channels $\boldsymbol{u}$. If another process R needs at some point to use Q, R just needs to feed Q with the required inputs, say $\boldsymbol{i}$ and wait for Q to output through channels $\boldsymbol{u}$. Indeed, process R could be defined, for instance, as $R := \langle \boldsymbol{i}\rangle_{\boldsymbol{c}} | (\boldsymbol{x})_{\boldsymbol{u}}.\ R' | Q(\boldsymbol{c} \to \boldsymbol{u})$.

2.2 Semantics

The semantics of a process Q is a Markov chain $\mathcal{S}(Q)$ over multisets of a special kind of processes, which we call eligible. Intuitively, the states of $\mathcal{S}(Q)$ represent reduction stages of the process and the transitions denote probabilistic choices between reductions. Recall that only the values on public channels are observed, and thus in the semantics these channels have special status.

The initial state S^0 of $\mathcal{S}(Q)$ is the multiset consisting of certain subprocesses of Q running in parallel, that is, if $Q = Q_1 | \dots | Q_m$ then $S^0 = \{Q_1, \dots Q_m\}$ where the head operator of each Q_i is not parallel composition. This setting captures the idea that in the initial state all such subprocesses are available for reduction. Actually, one obvious exception to this construction needs to be considered: if $Q = 0$ then there is no process to be reduced, and so, S^0 is the empty multiset. At this stage, we assume that the security parameter n is fixed, and therefore, all iterations have been replaced by parallel compositions.

Taking into account the discussion above, it is clear that we have to distinguish processes with head operator different from parallel composition. We call all these processes *eligible for reduction* and they can be defined formally as follows:

Definition 2. The *set of eligible processes* $\mathcal{E}$ is defined inductively as follows:

- $0 \in \mathcal{E}$;
- $\langle t\rangle_c \in \mathcal{E}$ where $t \in T$ and $c \in C$;
- $(x)_c.Q \in \mathcal{E}$ where $c \in C$, $x \in Var$ and $Q \in \mathcal{L}$;
- $[t_1 = t_2].Q \in \mathcal{E}$ where $t_1, t_2 \in T$ and $Q \in \mathcal{L}$;

In order to present the operational semantics, we set some notation on finite multisets. A finite multiset $\mathcal{M}$ over a set L is a map $\mathcal{M} : L \to \mathbb{N}$ such that $\mathcal{M}^{-1}(\mathbb{N}^+)$ is finite. The difference of $\mathcal{M}$ and $\mathcal{M}'$ is the multiset $\mathcal{M} \setminus \mathcal{M}'$ where $(\mathcal{M} \setminus \mathcal{M}')(l) = \max(0, \mathcal{M}(l) - \mathcal{M}'(l))$. The union of two multisets $\mathcal{M}$ and $\mathcal{M}'$ is the multiset $\mathcal{M} \cup \mathcal{M}'$ where $(\mathcal{M} \cup \mathcal{M}')(l) = \mathcal{M}(l) + \mathcal{M}'(l)$. We say that $\mathcal{M} \subset \mathcal{M}'$ iff $\mathcal{M}(l) \leq \mathcal{M}'(l)$ for all $l \in L$. Furthermore, we say that $l \in \mathcal{M}$ iff $\{l\} \subset \mathcal{M}$. Finally, we call $\wp_{fm}(L)$ the set of all finite multisets over L.

As discussed above, given a process $Q = Q_1 | \ldots | Q_m$ we need to construct the initial state $S^0 = \{Q_1, \ldots Q_m\}$ where Q_i is an eligible process. This construction is also useful during reduction, since after reducing some processes more parallel compositions may appear.

To deal with the binding channel operator ν we consider a set of fresh channels F and a fresh function

$$\text{fresh} : C \to F$$

that maps a channel c to a fresh channel c' (that is, that does not occur anywhere else) such that $w(c) = w(c')$. This fresh function insures that one can find a channel c' not occurring in any other process and therefore c' can be considered private to some process at hand. As expected, the communication through these channels is not observed.

Once again, at this step we assume a fixed security parameter n and that all iterations have been replaced by parallel compositions.

Definition 3. Given a process $Q \in \mathcal{L}$ without iteration we obtain the *multiset of sequences* of Q, which we denote by $\mathcal{M}_Q$, as follows:

- $\mathcal{M}_Q = \{\}$ whenever $Q = 0$;
- $\mathcal{M}_Q = \{Q\}$ whenever Q is eligible and different from 0;
- $\mathcal{M}_Q = \mathcal{M}_{R^c_{\text{fresh}(c)}}$ whenever $Q = \nu_c.R$ and where $R^c_{\text{fresh}(c)}$ is the process where all free occurrences of c where replaced by fresh(c);
- $\mathcal{M}_Q = \mathcal{M}_{Q'} \cup \mathcal{M}_{Q''}$ whenever $Q = Q'|Q''$.

Instead of presenting the semantics with probabilistic labeled transition systems as in [16,23], here we will use an alternative: Markov chains, a well-established concept from the stochastic processes community, following the style in [19].

Recall that a Markov chain A over a set S can be modeled as a state machine with state space S where transiting from state s to state s' has probability $0 \leq A(s, s') \leq 1$. Obviously, these probabilities are such that for any $s \in S$

$$\sum_{s' \in S} A(s, s') = 1.$$

Example 3. The following simple Markov chain models the stochastic process of independently tossing a fair coin ad nauseam:

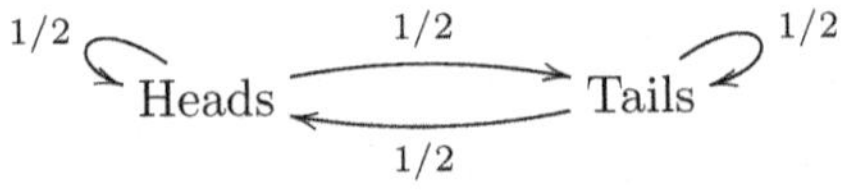

Markov chains are specially suited to model the semantics of the process algebra, since, like in [16,23], process reduction is probabilistic and depends only of the (multi)set of subprocesses that remain to be reduced. Thus, the semantics of a process Q is a Markov chain over the multisets of eligible process subprocesses of Q.

In order to establish the semantics for all processes, one can consider a huge Markov chain S, that given any multiset of eligible processes decides, accordingly to some probabilistic rules, which terms should be reduced. Such Markov chain is usually called a *scheduler*. Hence, given a multiset $\mathcal{M}$ of eligible processes there is a probability $S(\mathcal{M}, \mathcal{M}')$ of reducing $\mathcal{M}$ to $\mathcal{M}'$. The semantics of a single process Q is recovered by restricting the scheduler S to the states reachable from the multiset $\mathcal{M}_Q$ of sequences of Q.

Note that one can not accept any Markov chain as a scheduler. For instance, if the scheduler is at state $\{\}$, and therefore there is no process to reduce, the scheduler can not transit to, say, $\{Q\}$ with positive probability. In other words, $S(\{\}, \{Q\})$ must be zero. Hence, if the scheduler transits from one state to another with positive probability then at least one reduction must be enabled. For this reason, it is relevant to enumerate all possible reductions:

1. Term reduction: a term not in the scope of any input is reduced.
2. Match: a match between terms occurs.
3. Mismatch: a mismatch between terms occurs.
4. Communication: two processes communicate via an input and output.
5. Termination: none of the previous reductions is enabled (and so reduction has terminated).

Communication has lower priority than term reduction, match and mismatch. Hence, communication is only enabled when none of the previous mentioned reductions are enabled. As stated before, each reduction impose restrictions on the scheduler. For instance, *Termination* imposes that if there is no reduction enabled at state $\mathcal{M}$ then $S(\mathcal{M}, \mathcal{M}) = 1$. All other restrictions are more else obvious and are captured in the following definition.

Definition 4. A *scheduler* S for a security parameter n is a Markov chain with state space $\wp_{fm}(\mathcal{E})$ such that if $S(\mathcal{M}_1, \mathcal{M}_2) > 0$ then one of the following conditions hold:

1. *Term reduction:* $\langle t \rangle_c \in \mathcal{M}_1$ and $\mathcal{M}_2 = (\mathcal{M}_1 \setminus \{\langle t \rangle_c\}) \cup \{\langle m \rangle_c\}$; $\langle t \rangle_c$ not in the scope of any input, t does not have any free variables, t evaluates to m' with positive probability and m corresponds to the least significant $w(c)(|n|)$ bits of m'; the transition probability is given by
$$S(\{\langle t \rangle_c\}, \{\langle m \rangle_c\}) = \sum_{m' : m = m'|_{w(c)(n)}} P(t \to m').$$
2. *Match:* $[t_1 = t_2].Q \in \mathcal{M}_1$ where t_1, t_2 are closed terms and $\mathcal{M}_2 = (\mathcal{M}_1 \setminus \{[t_1 = t_2].Q\}) \cup \mathcal{M}_Q$; the transition probability is given by
$$S(\{[t_1 = t_2].Q\}, \{Q\}) = P(t_1 = t_2).$$

This transition is probable whenever there is a match expression in $\mathcal{M}_1$ and t_1 evaluates to the same value than t_2 with positive probability.

3. *Mismatch:* $[t_1 = t_2].Q \in \mathcal{M}_1$ where t_1, t_2 are closed terms and $\mathcal{M}_2 = (\mathcal{M}_1 \setminus \{[t_1 = t_2].Q\})$; the transition probability is given by

$$S(\{[t_1 = t_2].Q\}, \{\}) = P(t_1 \neq t_2).$$

This transition is probable whenever there is a match expression in $\mathcal{M}_1$ and t_1 evaluates to a different value than t_2 with positive probability.

4. *Communication*: $\{\langle m\rangle_c, (x)_c.Q\} \subset \mathcal{M}_1$; all other transitions are not probable for $\mathcal{M}_1$; $\mathcal{M}_2 = (\mathcal{M}_1 \backslash \{\langle m\rangle_c, (x)_c.Q\}) \cup \mathcal{M}_{Q^x_m}$ where Q^x_m stands for the process where we substitute all (free) occurrences of x by m in Q; This transition is probable whenever there is a pair input/output for a channel c in $\mathcal{M}_1$ and no term reduction, match or mismatch transition is probable.
5. *Termination:* $\mathcal{M}_1 = \mathcal{M}_2$; $S(\mathcal{M}_1, \mathcal{M}_2) = 1$; and all other transitions are not probable for $\mathcal{M}_1$. Whenever all reductions were made, the only enabled transition is the loop over the same state, which means that the reduction has terminated (and therefore $\mathcal{M}_1$ is an absorbing state).

Note that from a practical view point, a scheduler can be seen as a process dispatcher, that decides, based on some policy, which is the next process to reduce. Moreover, schedulers are expected to have the following good properties:

1. *Channel and variable independence*: probabilities are independent of the names of the channels and variables, that is:
   - $S(\mathcal{M}_1, \mathcal{M}_2) = S({\mathcal{M}_1}^x_y, {\mathcal{M}_2}^x_y)$ provided that x occurs free with respect to y in all processes of $\mathcal{M}_1$ and $\mathcal{M}_2$.
   - $S(\mathcal{M}_1, \mathcal{M}_2) = S({\mathcal{M}_1}^c_d, {\mathcal{M}_2}^c_d)$ where $w(d) = w(c)$ and d does not occur in all processes of $\mathcal{M}_1$ and $\mathcal{M}_2$;
2. *Environment independence*: probabilities are independent of the processes which are not involved in the transition, that is

$$S(\mathcal{M}_1, \mathcal{M}_2 | \mathcal{M} \subseteq \mathcal{M}_1 \cap \mathcal{M}_2) = S(\mathcal{M}_1 \setminus \mathcal{M}, \mathcal{M}_2 \setminus \mathcal{M}).$$

3. *Computational efficiency*: the scheduler is modeled by a probabilistic polynomial time Turing machine.

It is straightforward to check that the scheduler that gives uniform distribution to all possible transitions verifies the above properties.

As stated before, the operational semantics for a process Q is defined by restricting the scheduler.

Definition 5. Given a process Q and a scheduler S, the *operational semantics of* Q is the subMarkov chain $S(Q)$ of S consisting of all states reachable from $\mathcal{M}_Q$ such that $S(Q)^0 = \mathcal{M}_Q$ that is, the initial state of $S(Q)$ is $\mathcal{M}_Q$.

Note that the loops in $S(Q)$ are the absorbing transitions, and hence, all the states are either transient or absorbing. This fact implies that for k sufficiently large we have $P(S(Q)^k = S(Q)^{k+m}) = 1$ for all $m \in \mathbb{N}$. In other words, any random sampling of $S(Q)$ will end in an absorbing state, and therefore $S(Q)$ always terminates. This is more or less expected, since in [23] it has been shown that $S(Q)$ can be modeled by a PPT machine and so, $S(Q)$ always terminates.

2.3 Observations

In order to establish the observations of a process Q, we consider a modulated Markov process $K(Q) = (S(Q), O(Q))$ where $O(Q)$ is the stochastic process of observations of Q. The term *modulated* here means the probability distribution over the observations is computable from the distribution over the states.

This process is defined as expected: when a communication with a public channel occurs the pair (channel,output) is observed; when another type of transition occurs nothing is observed, which is modeled by the special symbol τ. Hence, the set of observations is $(C \times \mathbb{N}) \cup \{\tau\}$. Naturally, the probabilities of the observations are guided (modulated) by the probabilities on $S(Q)$. Given a scheduler S, we can obtain the global observation process $K(S,O)$ as follows:

Definition 6. Given a scheduler S and a security parameter n, we define the *observation modulated Markov process* $K = (S,O)$ where O is a stochastic process over $(U \times \mathbb{N}) \cup \{\tau\}$ such that:

- $K((\mathcal{M}_1, o_1), (\mathcal{M}_2, o_2)) = S(\mathcal{M}_1, \mathcal{M}_2)$ whenever one of the following conditions hold:
  - $o_2 = (c, m)$ and the public channel c outputs m in the transition of $\mathcal{M}_1$ to $\mathcal{M}_2$ in S (note that c can not be a fresh channel);
  - $o_2 = \tau$ and the transition from $\mathcal{M}_1$ to $\mathcal{M}_2$ in S was not a communication over a public channel.
- $K((\mathcal{M}_1, o_1), (\mathcal{M}_2, o_2)) = 0$ for all other cases.

Observe that K is indeed a Markov process modulated by S, since there exists a function f such that $K((\mathcal{M}_1, o_1), (\mathcal{M}_2, o_2)) = S(\mathcal{M}_1, \mathcal{M}_2) f(\mathcal{M}_1, \mathcal{M}_2)$. Once again, by restricting K to $S(Q)$ we obtain the required modulated stochastic process of observations $K(Q)$ for the process Q. For the sake of easing notation we denote the set of all observations by $Ob = (U \times \mathbb{N}) \cup \{\tau\}$.

In order to establish observational equivalence, we need to compute the probability of observing some output $o \in Ob$ at any point of the reduction trace. We denote this probability by $P(o \in T)$ and it can be computed as follows:

Definition 7. Given an observation process $K = (S,O)$ the probability of observing some output $o \in Ob$ at any point of the reduction trace is

$$P(o \in T) = \sum_{i=1}^{\infty} P\left(\bigwedge_{j=1}^{i} O^j \in C_j\right)$$

where $C_i = \begin{cases} \{o\} & \text{if } j = i \\ Ob \setminus \{o\} & \text{otherwise} \end{cases}$.

After fixing a scheduler S, the probability of the trace outputting o is calculated by an infinite series. Each term of the series represents the probability of o being output for the first time at the i-th reduction step. Hence, for any process Q, $T(Q)$ measures the probability of $\overline{Q}_n$ generating the output o at any point of its reduction.

Next, we present a toy example to articulate the concepts discussed above.

Example 4. Start by considering the following simple process expression Q

$$< \text{Rand} + 1 >_c \mid (x)_c.\langle x+1 \rangle_d \mid \langle 2 \rangle_d \mid (y)_d.\langle y+1 \rangle_e$$

where Rand is a uniform Bernoulli random variable taking values over $\{0,1\}$ (that is, it has $\frac{1}{2}$ probability of taking value 0 or 1). The multiset of sequences of Q is

$$\mathcal{M}_Q = \{< \text{Rand} + 1 >_c, (x)_c.\langle x+1 \rangle_d, \langle 2 \rangle_d, (y)_d.\langle y+1 \rangle_e\}.$$

We proceed by considering three different types of schedulers. For the sake of simplicity we skip over some additions.

1) Assume that the scheduler gives more priority to reducing the leftmost processes than to reducing the rightmost ones. For this particular example, we assume that $\mathcal{M}_Q$ is ordered just to express clearly which terms are reduced first by the scheduler. In that case $S(Q)$ is as follows:

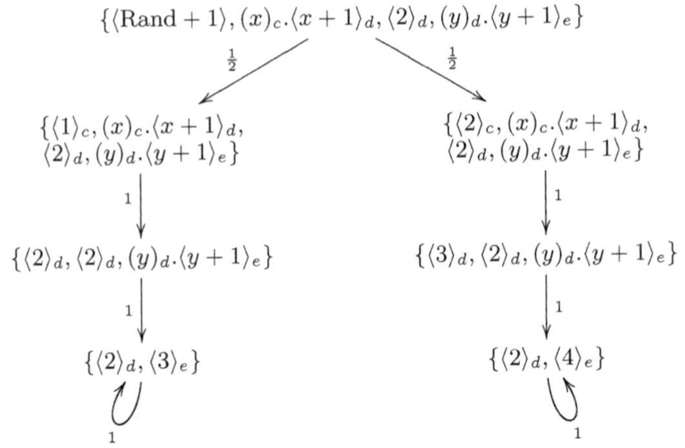

The modulated observation Markov process $K(Q) = (S(Q), O(Q))$ is:

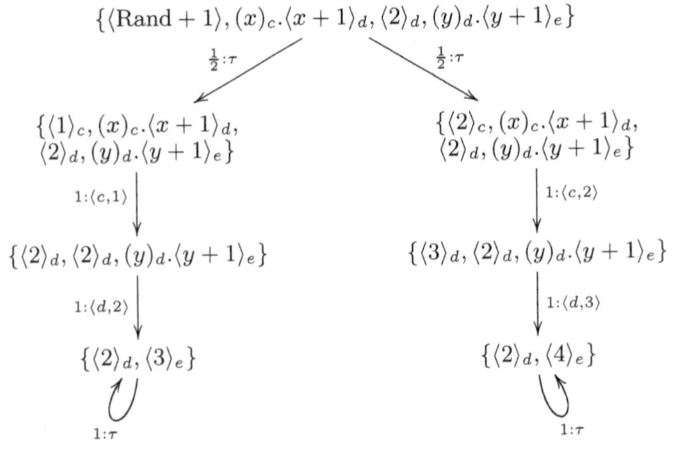

Hence, the probability of observing an output in the trace T is as presented in the following table:

$\langle c, 1\rangle$	$\langle c, 2\rangle$	$\langle d, 2\rangle$	$\langle d, 3\rangle$	τ	o
$\frac{1}{2}$	$\frac{1}{2}$	$\frac{1}{2}$	$\frac{1}{2}$	1	0

where o is any output in $Ob \setminus \{\langle c, 1\rangle, \langle c, 2\rangle, \langle d, 2\rangle, \langle d, 3\rangle, \tau\}$.

2) Now, suppose that the scheduler gives more priority to reducing the rightmost processes than to reducing the leftmost ones. Once again we assume that the multiset is ordered. In that case $S(Q)$ (together with $K(Q)$) is as follows:

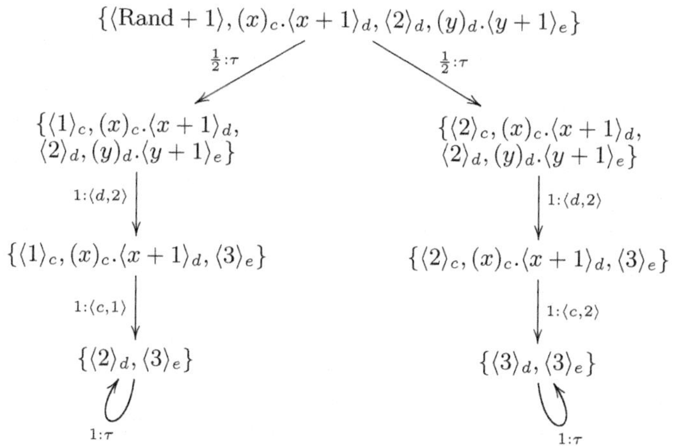

Note that even if reduction is from the right to the left, evaluating Rand has priority over any other reduction. The probability of observing an output in the trace T is:

$\langle c, 1\rangle$	$\langle c, 2\rangle$	$\langle d, 2\rangle$	τ	o
$\frac{1}{2}$	$\frac{1}{2}$	1	1	0

where o is any output in $Ob \setminus \{\langle c, 1\rangle, \langle c, 2\rangle, \langle d, 2\rangle, \tau\}$.

3) Finally, suppose that the scheduler chooses uniformly which processes to reduce. In that case $S(Q)$ (together with $K(Q)$) is of the following form:

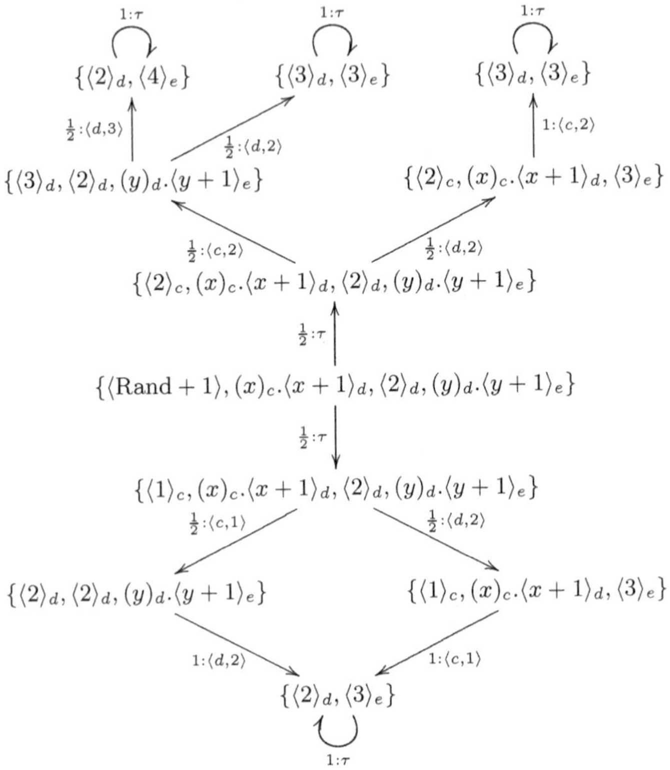

The probability of observing an output in the trace T is:

$\langle c, 1\rangle$	$\langle c, 2\rangle$	$\langle d, 2\rangle$	$\langle d, 3\rangle$	τ	o
$\frac{1}{2}$	$\frac{1}{2}$	$\frac{7}{8}$	$\frac{1}{8}$	1	0

where o is any output in $Ob \setminus \{\langle c, 1\rangle, \langle c, 2\rangle, \langle d, 2\rangle, \langle d, 3\rangle, \tau\}$.

From now on, we assume fixed once and for all a scheduler S. Thus, all random quantities mentioned in the sequel are bound to S.

2.4 Observational Equivalence

Let us now discuss the asymptotic formulation of observational equivalence for our process calculus introduced in [16,23], which draws on two sources. One source is programming language theory with its standard notion of observational equivalence of two programs P and Q, which intuitively means that in any environment, P has the same observable behavior as Q does in that same environment.

Another source of our asymptotic formulation is the notion of computational indistinguishability by polynomial-time statistical tests, standard in cryptography [13,30]. Intuitively, two probability distributions are computationally indistinguishable if it is not *feasible* to distinguish them. Stated somewhat more formally, this means that the two distributions cannot be distinguished, up to a negligible function of a security parameter, by probabilistic polynomial-time tests.

In order to present the asymptotic formulation of observational equivalence in detail in the setting of our calculus, we use the following definition of a *context*, intended to formalize an intuitive idea of a program environment:

Definition 8. The set of *contexts* **Ctx** is defined inductively as follows: $[\,] \in$ **Ctx**; $\nu_c.C[\,] \in$ **Ctx** provided that $C[\,] \in$ **Ctx**; $(x)_c.C[\,] \in$ **Ctx** provided that $C[\,] \in$ **Ctx**; $[t_1 = t_2].C[\,] \in$ **Ctx** provided that $C[\,] \in$ **Ctx**; $C[\,]|Q \in$ **Ctx** provided that $C[\,] \in$ **Ctx** and $Q \in \mathcal{L}$; $Q|C[\,] \in$ **Ctx** provided that $C[\,] \in$ **Ctx** and $Q \in \mathcal{L}$; and $!_q C[\,] \in$ **Ctx** provided that $C[\,] \in$ **Ctx** and $q \in \mathbf{q}$.

Given a context $C[\,]$ and a process expression Q, the notation $C[Q]$ means that we substitute the process Q for the $[\,]$ in $C[\,]$. We recall that for each fixed value k of the security parameter, the process $\overline{C[Q]})_k$ is obtained from $C[Q]$ by replacing each subexpression $!_q R$ by $q(|k|)$ copies of R in parallel.

Let us also recall that to establish the trace process T we rely on a probabilistic polynomial-time scheduler S and the associated observation process K. Hence, any trace process T is parameterized by a probabilistic polynomial-time scheduler S.

Definition 9. Let Q_1 and Q_2 be closed process expressions. We say that Q_1 and Q_2 are *observationally equivalent* or *computationally indistinguishable* iff for every scheduler S, every context $C[\,]$, every polynomial $q()$, every observation $(\boldsymbol{u}, \boldsymbol{m})$ and n sufficiently large

$$|P(T(\overline{C[Q_1]})_n) = (\boldsymbol{u}, \boldsymbol{m})) - P(T(\overline{C[Q_2]})_n) = (\boldsymbol{u}, \boldsymbol{m}))| \leq 1/q(n).$$

In this case we write $Q_1 \simeq Q_2$.

Therefore two closed process expressions are computationally indistinguishable iff they are indistinguishable by contexts, that is, there is no context that can distinguish, up to a negligible function of the security parameter, the observable behavior of the given process expressions in that context. Intuitively, this definition merges with the standard definition of computational indistinguishability, since process expressions can be modeled by probabilistic polynomial-time Turing machines [23] and the contexts $C[\,]$ induce the required distinguishing probabilistic polynomial-time tests. However, one benefit of our process language-based approach is the following proposition:

Proposition 1. Computational indistinguishability is a congruence relation with respect to the primitives of $\mathcal{L}$.

Proof. Both symmetry and reflexivity are trivial to check. Transitivity follows by triangular inequality, and taking into account that $\frac{1}{2}q(n)$ is a polynomial. Finally, congruence on the operators follows by noticing that for any contexts $C[\,]$ and $C'[\,]$, $C'[C[\,]]$ is also a context. □

3 Emulation and Composition Theorem

One rather flexible and expressive way of formulating the requirement that a given protocol satisfy some security property or fulfills a cryptographic objective or task is by relating the given protocol to an ideal protocol that clearly satisfies the property or fulfills the task. This idea appears in various forms already in [15,6,5,20]. In our approach [22,23,17], motivated by [28,29,1], we formulate the relationship between the given and the ideal protocol by means of observational equivalence. It is also very useful, especially for security properties that allow protocol participants to behave in an adversarial way toward each other, to structure the notion of an ideal protocol so that the generic description of the security property or the cryptographic task itself is separated from the description of the intended adversarial behavior of the participants or even external adversaries. This may be presented by means of the so-called emulation relation [7]. Let us discuss how this method may be expressed in our process calculus framework.

Let I be a generic, ideal representation some cryptographic objective or task. One can think of I as a generic process that accomplishes the objective. Such a process I is sometimes called *a functionality.* The adversarial behavior, or the threat model, may be expressed as the kind of environment B or an *ideal adversary,* in the presence of which I is intended to be executed. In our setting the description of the environment is given by means of families of contexts.

A similar distinction may be made between actual protocols, written as process expressions Q, and their intended adversaries A which are defined as certain families of contexts. We say that a protocol Q *securely realizes the functionality* I, or that Q *emulates* I, if for any real adversary, say represented by a context $A[\ ] \in \mathcal{A}$, the trace process of $A[Q]$ is observationally equivalent to the trace process of $B[I]$ for some ideal adversary, represented by a context $B[\ ] \in \mathcal{B}$, where an ideal adversary is an adversary which cannot corrupt I. This property asserts that given a real adversary $A[\ ]$ we cannot computationally distinguish the public outputs of $A[Q]$ from the public outputs of the well-behaved process $B[I]$ for some $B[\ ] \in \mathcal{B}$. Therefore, we infer that $A[Q]$ is also well-behaved. Recall that we use outputs to model what information participants possess, so if A is able to obtain some data efficiently from Q that A should have not, then A can issue an output with such information. In this case, we would not find any ideal adversary B which is able to gather from I similar information (by choosing correctly the set $\mathcal{B}$of ideal adversaries), and hence, the trace process of $A[Q]$ is not going to be observationally equivalent from $B[I]$ for any possible ideal adversary $B[\ ] \in \mathcal{B}$.

This discussion leads to the concept of emulation with respect to a set of real adversaries $\mathcal{A}$ and ideal adversaries $\mathcal{B}$.

Definition 10. Let Q and I be closed process expressions and $\mathcal{A}$ and $\mathcal{B}$ sets of contexts, then Q *emulates* I with respect to $\mathcal{A}$ and $\mathcal{B}$ iff for all contexts $A[\ ] \in \mathcal{A}$ there exists a context $B[\ ] \in \mathcal{B}$ such that $A[Q] \simeq B[I]$. In such case we write $Q \equiv_{\mathcal{A},\mathcal{B}} I$ and say that Q is an emulation of I, or that Q is a secure implementation of I with respect to $\mathcal{A}$ and $\mathcal{B}$.

A desirable property of the emulation relation is a compositionality property, informally discussed in the setting for secure computation already in [20] and more recently in [7]. Intuitively, if Q is a secure implementation of I, if R and J are two protocols that use the ideal protocol I as a component, and if R is a secure implementation of J, then R^I_Q should be a secure implementation of J. This property may be formally captured in our process calculus as follows:

Theorem 1. Let Q, I be closed process expressions, let $J[\,]$ and $R[\,]$ be contexts, and let $\mathcal{A}, \mathcal{B}, \mathcal{C}$ and $\mathcal{D}$ be sets of contexts. If $R[B[I]] \equiv_{\mathcal{C},\mathcal{D}} J[B[I]]$ for any $B[\,] \in \mathcal{B}$ and $Q \equiv_{\mathcal{A},\mathcal{B}} I$, $A[\,] \in \mathcal{A}$ there exists $B[\,] \in \mathcal{B}$ such that $R[A[Q]] \equiv_{\mathcal{C},\mathcal{D}} J[B[I]]$.

Proof. Let $A[\ ] \in \mathcal{A}$ and $B[\ ] \in \mathcal{B}$ be such that $A[Q] \simeq B[I]$. Now choose some $C[\,] \in \mathcal{C}$. Clearly $C[R[A[Q]]] \simeq C[R[B[I]]]$ since $\simeq$ is a congruence relation. Moreover, since $R[B[I]] \equiv_{\mathcal{C},\mathcal{D}} J[B[I]]$, there is a $D[\,] \in \mathcal{D}$ such that $C[R[B[I]]] \simeq D[J[B[I]]]$. Finally, by transitivity of $\simeq$, we have that $C[R[A[Q]]] \simeq D[J[B[I]]]$ and hence $R[A[Q]] \equiv_{\mathcal{C},\mathcal{D}} J[B[I]]$. □

Ideal protocols often consist of a generic, honest part I and an ideal adversary B, and are therefore of the form $B[I]$. This justifies why we consider $R[B[I]]$ in the proposition above instead of $R[I]$. Moreover, adversaries for R and J might be different from those of Q and I. Therefore, we need to consider two pairs of sets of contexts, $\mathcal{C}$, $\mathcal{D}$ and $\mathcal{A}$, $\mathcal{B}$.

3.1 Example: Oblivious Transfer

Oblivious transfer (OT) [27,12,14,9] is a two-party protocol where one agent is called the *sender* and the other the *receiver*. The sender's input is a vector of k bits $\boldsymbol{b} = b_1 \dots b_k$ and the receiver's input is a number i, $1 \le i \le k$. The purpose of the protocol is, intuitively, to transfer the i-th bit $\boldsymbol{b}_i$ of the vector $\boldsymbol{b}$ to the receiver without revealing any other bit to the receiver and without revealing the address i to the sender. We will refer to these two informal security requirements as *sender security* and *receiver security*, respectively.

Following the general paradigm just discussed at the beginning of this section, we would like to express either of these security requirements by means of observational equivalence to a certain ideal protocol, in this case, an ideal version of oblivious transfer. In an ideal setting there is a trusted and neutral third party, $T = (\boldsymbol{x})_v.(y)_{v'}.\langle \boldsymbol{x}_y \rangle_{v''}$ that expects the vector of k bits $\boldsymbol{b}$ from the sender and the value i from the receiver, and then sends $\boldsymbol{b}_i$ to the receiver, where by convention $\boldsymbol{b}_i = \boldsymbol{b}_k$ if $i \ge k$. Informally, we can think of the sender and the receiver as each communicating with T on separate private channels, or even more simply, that the sender and the receiver are subsumed into T. In any case, the only information T reveals to anyone is $\boldsymbol{x}_y$ on channel v''. T has no other outputs. T (or a copy of T with the channels renamed) is the oblivious transfer functionality.

What are the appropriate threats to consider? In the worst case the adversary may be *adaptive* [9], *i.e.*, the adversary can corrupt any of the parties at any point in response to data received during the protocol execution. We do not discuss this threat model here. Rather, we restrict ourselves to the simpler and somewhat weaker variant, the so-called *non-adaptive* or *static* adversaries [9], which can corrupt parties only once, at the beginning of the protocol execution. In this variant, it makes sense to consider several cases separately, depending on which party, the sender or the receiver, is honest and which is an adversary, and furthermore, the distinction between which is which does not change during the run of the protocol. We consider only one case, where the sender is honest and the receiver is an adversary, and in this case we are interested in sender security.

What should the static adversary receiver be able to do in the presence of the ideal oblivious transfer, the functionality T? In our setting everyone is bounded by probabilistic polynomial time, so in any case the receiver's output must be a probabilistic polynomial-time computable function, say f, of the receiver's input, i, and of any reply that the receiver can get from T, that is, one bit of the vector b. This may not be the very i-th bit because the receiver could have sent another request, j, to T in order to gain more information. But T gives only one reply so the receiver cannot learn more than one bit from T. That is, the adversary receiver's output must be of the form $f(i, b_{g(i)})$, where f and g are probabilistic polynomial-time computable functions. We will assume that an ideal adversary receiver is basically a parallel composition of processes, with one private call to a subroutine (intended to be T). The following definition describes this in a formal way:

Definition 11. An *ideal receiver adversary* is a context $B[\,]$ such that $B[T]$ is observationally equivalent to $R[T]$, where $R[\,]$ is a context of the form

$$P|(z_1)_{u_1}.\dots.(z_m)_{u_m}.(y)_{v_1}.\nu_{v'}.\nu_{v''}.(\langle t(y)\rangle_{v'}|[\,]|(z)_{v''}.Q),$$

where the other input channel of T (channel v in the description above) does not occur and where the only output not corresponding to any input is public and it occurs in Q, and this output is of the form $\langle t'(y,z)\rangle_u$, where t and t' are terms. We denote the *set of all ideal receiver adversaries* by $\mathcal{I}$.

Real adversaries that will attack a real sender have no restrictions whatsoever to the amount of information they might obtain from interacting with a real sender, other than that they must do that in probabilistic polynomial time, and that they cannot corrupt the sender. We shall assume that a real adversary is a process running in parallel with the sender, that is,

Definition 12. A *real receiver adversary* is a context of the form $[\,]|A$ We denote the set of all real receiver adversaries as $\mathcal{R}$.

We say that a protocol $Q_S|Q_R$ is a sender secure oblivious transfer protocol, if the sender Q_S running in parallel with *any* real adversary emulates the ideal setting, that is:

Definition 13. A protocol $Q_S|Q_R$ is a sender secure oblivious transfer protocol iff $Q_S \equiv_{\mathcal{R},\mathcal{I}} T$.

Note that the condition that the definition imposes only involves the real sender Q_S, not the real receiver Q_R. Furthermore, note that the correctness condition on the protocol may be expressed in a similar way, by requiring that $Q_S|Q_R$ be observationally equivalent to $R[T]$ for some *honest* ideal adversary receiver $R[\,]$ that makes an honest request to T and outputs T's reply, *i.e.*, such that $t(y) = y$ and $t'(z, y) = z_y$.

The following result is an immediate corollary of Theorem 1.

Proposition 2. The notion of sender secure is compositional. That is, let $J[\,]$ and $K[\,]$ be contexts and let $\mathcal{C}$ and $\mathcal{D}$ be sets of contexts. If $K[B[T]] \equiv_{\mathcal{C},\mathcal{D}} J[B[T]]$ for any $B[\,] \in \mathcal{I}$ and $Q_S \equiv_{\mathcal{R},\mathcal{I}} T$, then for any adversary $A \in \mathcal{R}$ there exists $B[\,] \in \mathcal{I}$ such that $K[Q_S|A] \equiv_{\mathcal{C},\mathcal{D}} J[B[T]]$.

We now present a well-known oblivious transfer protocol. We consider the version presented in [9,14], which is an adaptation of the original protocol due to Rabin [27]. In order to establish this protocol, one needs to introduce some assumptions: a *collection of trapdoor permutations* $f = \{f_\alpha : D_\alpha \to D_\alpha\}_{\alpha \in I}$, where each f_α is probabilistic polynomial-time in n but hard to invert; a *trapdoor generator* G which is is probabilistic polynomial-time in n and generates a pair (α, t) with $\alpha \in I$; and a *hard-core predicate* B on f. Mind that for a pair (α, t) there is a function $f^{-1}(t, .)$, probabilistic polynomial-time in n, such that $f^{-1}(t, .)$ is the inverse of f_α. Moreover, a hard-core predicate $B : D_\alpha \to \{0, 1\}$ is a predicate computable in polynomial time (in its input) such that knowing $f_\alpha(x)$ does not help to predict $B(x)$, that is, it is hard to predict B from an image by f_α. We ask the reader to see [13] for more details on these assumptions.

Example 5 (Rabin OT protocol). The protocol is composed of two parallel agents, the *Sender* and the *Receiver*. First, on k-bit input $b_1 \dots b_k$, the *Sender* selects a trapdoor pair (α, t) using a term G and private channel v_S and then sends α to the *Receiver*. Upon reading an input i and receiving α, the *Receiver* chooses uniformly and independently at random k elements $e_1, \dots e_k$ of D_α and then, sends the elements $y_1 \dots y_k$ to the *Sender* with $y_i = f_\alpha(e_i)$, and $y_j = e_j$ when $j \neq i$. (Thus the receiver knows $f_\alpha^{-1}(y_i) = e_i$ but cannot predict $B(f_\alpha^{-1}(y_j))$ for any $j \neq i$.) When the *Sender* receives $y_1 \dots y_k$ it sends back the tuple $(b_j \oplus B(f^{-1}(t, y_j)))_{j \in \{1,\dots,k\}}$, where $\oplus$ denotes the usual bit addition operation. (Recall that $f^{-1}(t, y_j) = f_\alpha^{-1}(y_j)$ for every $j \in \{1, \dots, k\}$.) The *Receiver* upon receiving the tuple picks the i-th element c_i and gets b_i via $c_i \oplus B(e_i) = (b_i \oplus (B(f_\alpha^{-1}(f_\alpha(e_i)))) \oplus B(e_i) = b_i$. The protocol can be written in the process calculus as follows:

– $S = (b_1, \dots, b_k)_{v_0}.\nu_{v_S}($
$\langle G \rangle_{v_S}|$
$(\alpha, t)_{v_S}.\ ($
$\langle \alpha \rangle_{v_2}|$
$(y_1, \dots, y_k)_{v_3}\langle b_1 \oplus B(f^{-1}(t, y_1)), \dots, b_k \oplus B(f^{-1}(t, y_k))\rangle_{v_4}$
$)$
$)\,;$

– $R = (i)_{v_1}.(\alpha)_{v_2}.\nu_{v_R}.\ (!_k\langle \text{Rand}(D_\alpha)\rangle_{v_R})|$
$$(e_1)_{v_R} \cdots (e_k)_{v_R}.\ (\langle e_1, \ldots, e_{i-1}, f_\alpha(e_i), e_{i+1}, \ldots, e_k\rangle_{v_3}| \ (c_1, \ldots, c_k)_{v_4}.\ \langle c_i \oplus B(e_i)\rangle_u \ ).$$

Proposition 3. The Rabin OT protocol is not sender secure.

Proof. The honest receiver in the protocol is too generous: an adversary receiver can easily do for each $j \neq i$ what the honest receiver in the Rabin protocol does only for i, and thus get from the sender all the b_j's. That is, consider the following real receiver adversary:

$$R[i] = (i)_{v_1}.(\alpha)_{v_2}.\nu_{v_R}.\ (!_k\langle \text{Rand}(D_\alpha)\rangle_{v_R})| \ (e_1)_{v_R} \cdots (e_k)_{v_R}.\ (\langle f(e_1), \ldots, f(e_k)\rangle_{v_3}| \ (c_1, \ldots, c_k)_{v_4}.\ \langle c_1 \oplus B(e_1), \ldots, c_k \oplus B(e_k)\rangle_u \ ).$$

It is easy to see that this receiver outputs *all* the b_i's. Clearly this is a successful attack by the receiver, who learns the entire input string of the sender. Formally, the output containing all the b_i's can be computationally distinguished from an output of an ideal receiver adversary, which is of the form $f(i, b_g(i))$, where f and g are probabilistic polynomial-time functions. □

In Sect. 8 of the full paper [9] and in Sect. 7.4 of [14] it is shown how to *compile* this protocol into an oblivious transfer that is sender secure as well as receiver secure. A related compilation method is discussed in [8]. The details of the compiler itself can be expressed in our process calculus, but that falls beyond the scope of this paper.

4 Related Work

We briefly compare our approach with related work based on interactive Turing machines [7,9] and secure reactive systems [25,26,3,4].

The approach in [7,9] is formulated in terms of interactive Turing machines (ITM), which are basically the familiar Turing machines with several additional tapes: read-only *random input* tape, read-and-write *switch* tape (consisting of a single cell), and a pair of *communication* tapes, one read-only and the other write-only. Several ITMs can be linked through their communication tapes. The security parameter, usually written in unary, is a shared input among a collection of linked ITMs, but each ITM may have separate, additional inputs. It is assumed that the linked ITMs are polynomial-time in the security parameter. Details may be found in [7,13].

The framework proposed in [7] involves a relationship between a *real model* representing protocol execution of actual protocols and an *ideal model* representing a generic version of a cryptographic task. A protocol in the real model *securely realizes* the task if it emulates an ideal process for the task. Either model consists of a finite set of ITMs representing the parties in the protocol, another ITM representing the protocol adversary, and yet another ITM representing the computational environment, including other protocol executions and their adversaries, other users, *etc.* The environment has external input known only to itself. The environment provides the inputs to other parties and it reads their outputs. The basic idea is that from the environment's point of view, executing the protocol in the real model should look the same as the ideal process. In somewhat more detail, a real protocol P securely realizes an ideal process I if for any real adversary A there is an ideal adversary S such that no environment can tell with non-negligible probability whether it is interacting with P and A in the real model or with I and S in the ideal model. A general composition theorem is proved in [7] and a wide variety of protocols have been studied in this framework in [7,9] and in several other papers. A related framework based on secure reactive systems, in which ITMs are seen from the perspective of input/output automata [18] is studied in [25,26,3,4].

In comparison, keeping in mind that our language of functional terms is rich enough to reflect probabilistic polynomial-time computable functions (and only those functions), functions computed by single ITMs may be represented in the framework discussed in this paper by simple process expressions, with channels representing communication tapes. Keeping in mind that our language of functional terms is rich enough to reflect probabilistic polynomial-time computable functions. Finite sets of ITMs may be represented by a parallel composition of processes, or more generally, by contexts that involve parallel composition. Adversaries and the environment are represented by contexts. In this paper we presented this in more detail in the example of oblivious transfer in the case of non-adaptive adversaries. We have investigated several other protocols in this light and we believe there is a general correspondence between our framework and the frameworks based on ITMs. In this regard it is useful to remember that any process expression in our calculus is provably executable in probabilistic polynomial-time [23]. An interesting technical point is that we consider external probabilistic polynomial-time schedulers of input/output communications on channels while in [7] the scheduling of communications is done by the adversary. It is possible, however, to force the scheduling by structuring the contexts appropriately, in particular the contexts playing the role of the adversary. This feature is already present in the specific example in the previous section.

5 Conclusions

We have expressed security requirements for cryptographic protocols in the framework of a probabilistic polynomial-time process calculus. We have also proved an abstract composition theorem for security properties. These results provide a framework for a compositional analysis of security protocols. We showed how to express an oblivious transfer protocol and its security requirements in the process calculus. Finally, we have discussed a relationship between our process calculus and the interactive Turing machine approaches in [7,25].

There are several advantages of using a process calculus instead of interactive Turing machines. Namely, the process calculus is a much more natural and clear language for specifying protocols than the low-level vocabulary of interactive Turing machines. Indeed, the precise, formal process calculus expressions remind one of high-level programming language code and are often actually shorter than even the informal descriptions of the protocol in English, let alone the low-level details of Turing machines. Another advantage lies in the fact that compositional issues are dealt with in an intrinsic, built-in way using process calculus. Indeed, in order to show the composition theorem, it is enough to prove a congruence property for the emulation relation. The candidate for the congruence relation is obvious: if A emulates B then $C[A]$ emulates $C[B]$ for any context C. Moreover, we note that probabilistic polynomial-time process calculus provides an adequate setting for the concepts related to computational security, since both the parties and the adversaries expressed in the process calculus are provably bounded by probabilistic polynomial-time algorithms. Indeed, the work presented here may be seen as a contribution to the more general effort of giving rigorous definitions of security properties independent or particular protocols.

Acknowledgments. The authors are most grateful to Ran Canetti for several helpful comments and fruitful suggestions about an early draft of this paper, especially on the relationship between the process calculus and interactive Turing machines.

References

1. M. Abadi and A. Gordon. A calculus for cryptographic protocols: the spi-calculus. *Information and Computation*, 143:1–70, 1999.
2. M. Abadi and P. Rogaway. Reconciling two views of cryptography (The computational soundness of formal encryption). *Journal of Cryptology*, 15(2):103–127, 2002.
3. M. Backes, C. Jacobi, and B. Pfitzmann. Deriving cryptographically sound implementations using composition and formally verified bisimulation. In *Formal Methods Europe*, volume 2931 of *Lecture Notes in Computer Science*, pages 310–329. Springer-Verlag, 2002.
4. M. Backes, B. Pfitzmann, and M. Waidner. Universally composable cryptographic library. Manuscript available on eprint.iacr.org as 2003/015, 2003.
5. D. Beaver. Foundations of secure interactive computing. In *Crypto'91*, volume 576 of *Lecture Notes in Computer Science*, pages 377–391. Springer-Verlag, 1991.
6. D. Beaver. Secure multiparty protocols and zero-knowledge proof systems tolerating a faulty minority. *Journal of Cryptology*, 4:75–122, 1991.
7. R. Canetti. Universally composable security: A new paradigm for cryptographic protocols. In *42-nd Annual Symposium on Foundations of Computer Science (FOCS)*, pages 136–145. IEEE Press, 2001. Full paper available at eprint.iacr.org as 2000/067.
8. R. Canetti and H. Krawczyk. Analysis of key-exchange protocols and their use for building secure channels. In *Eurocrypt'01*, volume 2045 of *Lecture Notes in Computer Science*, pages 453–474. Springer-Verlag, 2001.

9. R. Canetti, Y. Lindell, R. Ostrovsky, and A. Sahai. Universally composable two-party and multi-party secure computation. In *34-th ACM Symposium on Theory of Computing*, pages 484–503, 2002. Full paper available at eprint.iacr.org as 2002/140.
10. D. Dolev and A. Yao. On the security of public-key protocols. In *Proc. 22-nd Annual IEEE Symposium on Foundations of Computer Science (FOCS)*, pages 350–357, 1981.
11. T. ElGamal. A public-key cryptosystem and a signature scheme based on discrete logarithms. *IEEE Transactions on Information Theory*, IT-31:469–472, 1985.
12. S. Even, O. Goldreich, and A. Lempel. A randomized protocol for signing contracts. *Communications of the ACM*, 28(6):637–647, 1985.
13. O. Goldreich. *Foundations of Cryptography: Basic Tools.* Cambridge Univ. Press, 2001.
14. O. Goldreich. *Foundations of Cryptography – Volume 2.* Working Draft of Chapter 7, 2003. Available at www.wisdom.weizmann.ac.il/~oded.
15. S. Goldwasser and L. Levin. Fair computation of general functions in presence of immoral majority. In *Crypto'90*, volume 537 of *Lecture Notes in Computer Science*, pages 77–93. Springer-Verlag, 1990.
16. P. Lincoln, J. Mitchell, M. Mitchell, and A. Scedrov. Probabilistic polynomial-time framework for protocol analysis. In M. Reiter, editor, *5-th ACM Conferece on Computer and Communication Security*, pages 112–121. ACM Press, 1998.
17. P. Lincoln, J. Mitchell, M. Mitchell, and A. Scedrov. Probabilistic polynomial-time equivalence and security analysis. In *Formal Methods in the Development of Computing Systems*, volume 1708 of *Lecture Notes in Computer Science*, pages 776–793. Spriger-Verlag, 1999.
18. N. Lynch. *Distributed Algorithms.* Morgan Kaufman, 1996.
19. P. Mateus, A. Pacheco, J. Pinto, A. Sernadas, and C. Sernadas. Probabilistic situation calculus. *Annals of Mathematics and Artificial Intelligence*, 32(1):393–431, 2001.
20. S. Micali and P. Rogaway. Secure computation. In *Crypto'91*, volume 576 of *Lecture Notes in Computer Science*, pages 392–404. Springer-Verlag, 1991.
21. R. Milner. *Communication and Concurrency.* Prentice Hall, 1989.
22. J. Mitchell, M. Mitchell, and A. Scedrov. A linguistic characterization of bounded oracle computation and probabilistic polynomial time. In *39-th Annual IEEE Symposium on Foundations of Computer Science (FOCS)*, pages 725–733. IEEE Computer Society Press, 1998.
23. J. Mitchell, A. Ramanathan, A. Scedrov, and V. Teague. A probabilistic polynomial-time calculus for analysis of cryptographic protocols. *Electronic Notes in Theoretical Computer Science*, 45, 2001.
24. R. Needham and M. Schroeder. Using encryption for authentication in large networks of computers. *Communications of the ACM*, 21(12):993–9, 1978.
25. B. Pfitzmann, M. Schunter, and M. Waidner. Cryptographic security of reactive systems. *Electronic Notes in Theoretical Computer Science*, 32, 2000.
26. B. Pfitzmann and M. Waidner. Composition and integrity preservation of secure reactive systems. In *7-th ACM Conference on Computer and Communications Security*, pages 245–254. ACM Press, 2000.
27. M. Rabin. How to exchange secrets by oblivious transfer. Tech. memo TR-81, Aiken Computation Laboratory, Harvard U., 1981.
28. A. W. Roscoe. Modelling and verifying key-exchange protocols using CSP and FDR. In *8-th IEEE Computer Security Foundations Workshop (CSFW)*. IEEE Computer Society Press, 1995.

29. S. Schneider. Security properties and CSP. In *IEEE Symposium Security and Privacy*, 1996.
30. A. Yao. Theory and applications of trapdoor functions. In *23-rd IEEE Symposium on Foundations of Computer Science (FOCS)*, pages 80–91. IEEE Press, 1982.

Unifying Simulatability Definitions in Cryptographic Systems under Different Timing Assumptions*
(Extended Abstract)

Michael Backes

IBM Zurich Research Laboratory, Rüschlikon, Switzerland
mbc@zurich.ibm.com

Abstract. The cryptographic concept of simulatability has become a salient technique for faithfully analyzing and proving security properties of arbitrary cryptographic protocols. We investigate the relationship between simulatability in synchronous and asynchronous frameworks by means of the formal models of Pfitzmann et. al., which are seminal in using this concept in order to bridge the gap between the formal-methods and the cryptographic community. We show that the synchronous model can be seen as a special case of the asynchronous one with respect to simulatability, i.e., we present an embedding between both models that we show to preserve simulatability. We show that this result allows for carrying over lemmas and theorems that rely on simulatability from the asynchronous model to its synchronous counterpart without any additional work. Hence future work can concentrate on the more general asynchronous case, without having to neglect the analysis of synchronous protocols.

1 Introduction

In recent times, the analysis of cryptographic protocols has been getting more and more attention, and the demand for general frameworks for representing cryptographic protocols and the security requirements of cryptographic tasks has been rising. Existing framework are either motivated by the complexity-theoretic view on cryptography, which aims at proving cryptographic protocols with respect to the cryptographic semantics, or they are motivated by the view of the formal-methods community, which aims at capturing abstractions of cryptography in order to make such protocols accessible for formal verification. Frameworks built on abstractions will be further dealt with in the related literature along with a discussion on the cryptographic justification of these abstractions.

For living up to the probabilistic nature of cryptography, a framework for dealing with actual cryptography necessarily has to be able to deal with probabilistic behaviors. The standard understanding in well-known, non security-specific probabilistic frameworks like [31,33] is that the order of events is fixed by means of a probabilistic scheduler that has full information about the system. In contrast to that, the standard understanding in cryptology (closest to a rigorous definition in [10]) is that the adversary schedules everything, but only with realistic information. This corresponds to making a certain subclass of schedulers explicit for the model from [31]. However, if one splits a machine

* The full version is available from `http://eprint.iacr.org/2003/114`.

R. Amadio, D. Lugiez (Eds.): CONCUR 2003, LNCS 2761, pp. 350–365, 2003.

into local submachines, or defines intermediate systems for the purposes of proof only, this may introduce many schedules that do not correspond to a schedule of the original system and therefore just complicate the proofs. The typical solution is a distributed definition of scheduling which allows machines that have been scheduled to schedule certain (statically fixed) other machines themselves.

Based on these requirements, several general definitions of secure protocols were developed over the years, e.g. [15,7,28,11,30,12], which are all potential candidates for such a framework. To allow for a faithful analysis of cryptographic protocols, it is well-known that such models not only have to capture probabilistic behaviors, but also complexity-theoretically bounded adversaries as well as a reactive environment of the protocol, i.e., continuous interaction with the users and the adversary. Unfortunately, most of the above work does not live up to these requirements in spite of its generality, mainly since it concentrates on the task of secure function evaluation, which does not capture a reactive environment. Currently, the models of Pfitzmann et. al. [28,30] and Canetti [12], which have been developed concurrently but independently, stand out as the standard models for sound protocol analysis and design.

Regarding the underlying definition of time, such models can be split into synchronous and asynchronous ones. In synchronous models [28], time is assumed to be expressible in rounds, whereas asynchronous scenarios [30,12] do not impose any assumption on time. This makes asynchronous scenarios attractive since no assumption is made about network delays and the relative execution speed of the parties. Moreover, the notion of rounds is difficult to justify in practice as it seems to be very difficult to establish them for the Internet for example. This attractiveness is substantiated by a large body of literature on asynchronous cryptographic protocols, e.g., [8,14]. However, time guarantees are sometimes explicitly desired, e.g., on when a process can abort. Hence assumptions have to be made in this case, which induce a certain amount of synchrony again. This sometimes makes a synchronous assumption of time nevertheless necessary in practice, e.g., in Kerberos [23].

Hence researchers usually restrict their attention to one definition of time, or they are driving double-tracked by maintaining two separate models. However, this presupposes proving every theorem for both models. This is not nice. An alternative approach, taken in this work, is to show that the synchronous model can be regarded as a special case of an asynchronous one, and hence does not have to be considered separately, but still can be used to conveniently express synchronous protocols.

Although this idea might not be surprising, it is very difficult to achieve since it turns out that carrying over results from the asynchronous to the synchronous model presupposes the ability of (at least partially) reversing the considered embedding. Recall that suitable frameworks, especially the frameworks of Canetti and Pfitzmann et. al., have a distributed scheduling which significantly complicates this reversion.

Formally, a special case means that there is an embedding into the asynchronous model that preserves a desired property. Which property has to be preserved depends on the goals to strive for. For cryptographic protocols, the property of *simulatability* stands out. Simulatability captures the notion of a cryptographically secure implementation and serves as a link to the formal-methods community, which typically only hold a top-level view of cryptography, where cryptographic primitives are replaced by deterministic abstractions. A more comprehensive discussion of simulatability and its relationship to protocol verification work done by the formal-methods community is given in the paragraph on related literature below.

In the following, we investigate the synchronous and asynchronous models of Pfitzmann et. al. [28,30], which are seminal in using the concept of simulatability to bridge the gap between the formal-methods and the cryptographic community. We show that the synchronous model can be embedded in the asynchronous model such that simulatability is preserved by this embedding, i.e., if two systems fulfill the simulatability relation in the synchronous model, their respective images fulfill the relation in the asynchronous model and vice versa. We show that this result allows for carrying over lemmas and theorems from the asynchronous case to the synchronous case without proving them twice. We are confident that this result helps to make future protocol analysis in these models more convenient and more efficient.

Moreover, we believe that our approach for establishing the embedding and its properties can be successfully used for other models with only minor changes. Especially the asynchronous model of Canetti is surely worth to be investigated. However, his corresponding synchronous model [11] is still specific for secure function evaluation; hence adopting it to a reactive environment is a necessary prerequisite for this future work. The lack of such a reactive synchronous model was – besides the fact that the models of Pfitzmann et. al. are more rigorously defined than the one of Canetti – our main reason why we decided to base our work on the model of Pfitzmann et. al.

Related Literature. If cryptographic protocols should be verified using formal methods, some kind of abstraction is needed as the underlying reduction proofs of cryptography are still out of scope of current verification techniques. This abstraction is usually based on the so-called Dolev-Yao model [13], which considers cryptographic primitives as operators in a free algebra where only predefined cancellation rules hold. This abstraction simplifies proofs of larger protocols considerably, and it gave rise to a large body of literature on analyzing the security of protocols using techniques for formal verification of computer programs (a very partial list of work includes [22,9,20,26,1]).

Since this line of work turned out to be very successful, the interesting question arose whether these abstractions are indeed justified from the view of cryptography, i.e., whether properties proved for the abstractions are still valid for the cryptographic implementation. Abadi et. al. showed in [3,2] that the Dolev-Yao model is cryptographically faithful at least for symmetric encryption and synchronous protocols. There, however, the adversary is restricted to passive eavesdropping. Consequently, it was not necessary to choose a reactive model of a system and its honest users, and the notion of simulatability could be replaced by the weaker notion of indistinguishability [34]. Guttman et. al. showed in [17] that the probability of two executions of the same protocol – either executed in a Dolev-Yao-like framework or using real cryptographic primitives – may deviate from each other at most for a certain bound. However, their results are specific for the Wegman-Carter system so far. Moreover, as this system is information-theoretically secure, its security proof is much easier to handle than primitives with security guarantees only against computationally bounded adversaries since no reduction proofs against underlying number-theoretic assumptions have to be made. Some further approaches for special security goals or primitives are [32,19]. However, there is evidence that the original Dolev-Yao model is not justified in the presence of active attacks, even if provably secure cryptographic primitives are used, cf. [27] for an (admittedly constructed) counterexample. This exemplifies the demand for "better" abstractions which the models of Canetti and of Pfitzmann et. al. want to establish using the concept of simulatability.

Simulatability bridges this gap by serving as a cryptographically sufficient relationship between abstract specifications and cryptographic implementations, i.e., abstractions which can be shown to simulate a given implementation in a particular sense are known to be sound with respect to the security definitions of cryptography. Simulatability was first invented for multi-party function evaluation [15,7,11], i.e., systems with only one initial input set and only one output set. An extension to a reactive scenario, where participants can make new inputs many times, e.g., start new sessions like key exchanges, was first fully defined in [27], with extensions to asynchronous systems in [30,12]. Each of the three considered models has already been successfully used to built up sound abstractions of various cryptographic primitives and all of them enjoy a composition theorem, i.e., large protocols can be refined step-wise without destroying the already proven properties.

Comparing the models of Canetti and Pfitzmann et. al., we can state that Canetti's work enjoys a more general composition theorem and has moreover addressed more cryptographic primitives so far. On the other hand, the models of Pfitzmann et. al. are more rigorously defined and early examples of tool-supported proofs in their models exist [5,4], using PVS [25]. Moreover, the recently published universally composable cryptographic library [6] may pave the way to formal verification of large security protocols within their models.

2 Review of the Reactive Models in Synchronous and Asynchronous Networks

In this section we review the synchronous and the asynchronous model for probabilistic reactive systems as introduced in [28] and [30], respectively. Several definitions are only sketched, whereas those that are essential for understanding our upcoming results are given in full detail.

2.1 General System Model

In the following we consider a finite alphabet Σ and some special symbols $!, ?, ^{\leftrightarrow}, ^{\triangleleft} \notin \Sigma$ that will be used to express different ports of machines. For $s \in \Sigma^*$ and $l \in \mathbb{N}_0$, we define $s\lceil_l$ to be the l-letter prefix of s.

Our machine model is *probabilistic state-transition machines*, similar to probabilistic I/O automata as sketched by Lynch [21]. Communication between different machines is done via *ports* which are divided into *input* and *output* ports. Inspired by the CSP-Notation [18] we write input and output ports as $\mathsf{q}?$ and $\mathsf{q}!$. Let $\mathsf{q}!^C := \mathsf{q}?$ and vice versa; for a set P of input and output ports, let $P^C := \{\mathsf{q} \mid \mathsf{q}^C \in P\}$.

Ports will later be connected by naming convention, i.e., a port $\mathsf{q}!$ always sends messages to $\mathsf{q}?$. In the asynchronous model, a special machine called a *buffer* will further be inserted in each connection to ensure asynchronous behavior. A buffer stores all of its inputs in an internal list. If a machine wants to schedule the i-th message of

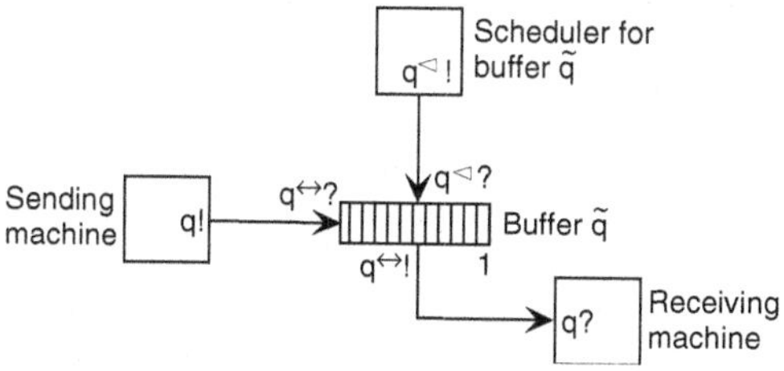

Fig. 1. Ports and buffers

buffer $\tilde{q}$ (this machine must have the unique clock-out port $q^{\triangleleft}!$), it simply sends i at $q^{\triangleleft}!$, cf. Fig. 1. The buffer then outputs the i-th message and removes it from its internal list. Neither buffers nor clock ports occur in the synchronous model; they are just included to establish a distributed scheduling in the asynchronous case.

A *machine* M has a *name* $name_{\mathsf{M}}$, a sequence $Ports_{\mathsf{M}}$ of *ports*, containing both input ports and output ports, and a set $States_{\mathsf{M}}$ of *states*, comprising sets Ini_{M} and Fin_{M} of *initial* and *final states*, respectively. If a machine is switched, it receives an input tuple at its input ports and performs its *transition function* δ_{M} yielding a new state and an output tuple in the deterministic case, or a finite distribution over the set of states and possible outputs in the probabilistic case. Furthermore, each machine has a bound l_{M} on the length of the considered inputs which allows time bounds on the computation time independent of the environment. The parts of an input that are beyond the length bound are ignored, i.e., incoming strings are only processed up to a predefined length. In particular, this is used to ensure polynomial runtime of individual machines.

For a set $\hat{M}$ of machines, let $\mathsf{ports}(\hat{M})$ denote the set of ports of all machines $\mathsf{M} \in \hat{M}$. Machines usually start with one initial input, i.e., the starting state is parameterized. Complexity is measured in terms of the length of this initial input, usually a security parameter k given in unary representation; in particular, polynomial-time is meant in this sense. We only briefly state here, that these machines have a natural realization as a probabilistic interactive Turing machine as introduced in [16]. We call a machine M a *black-box submachine* of a machine M' if the machine M' has access to the state-transition function δ_{M} of M, i.e., it can execute δ_{M} for the current state of the machine and arbitrary inputs.

A *collection* C of machines is a finite set of machines with pairwise different machine names and disjoint sets of ports. In the asynchronous model, the *completion* $[C]$ of a collection C is the union of all machines of C and the buffers needed for every channel. A port of a collection is called *free* if its connecting port is not in the collection. These port will be connected to the users and the adversary. The free ports of a collection C are denoted as $\mathsf{free}(C)$. In the asynchronous model, a collection C is called *closed* if its completion $[C]$ has no free ports except a special master clock-in port $\mathsf{clk}^{\triangleleft}?$, i.e., $\mathsf{free}([C]) = \{\mathsf{clk}^{\triangleleft}?\}$. When we define the interaction of several machines, this port will be used to resolve situations where the interaction cannot proceed. In the synchronous case, we demand $\mathsf{free}(C) = \emptyset$ instead.

For security purposes, special collections are needed, because an adversary may have taken over parts of the initially intended system, e.g., different situations have to be captured depending on which and how many users are considered as being malicious. Therefore, a system consists of several possible remaining structures.

Definition 1. *(Structures and Systems) A* structure *is a pair* $struc = (\hat{M}, S)$ *where* $\hat{M}$ *is a collection of non-buffer machines called* correct machines, *and* $S \subseteq \mathsf{free}(\hat{M})$ *is called* specified ports. *If* $\hat{M}$ *is clear from the context, let* $\bar{S} := \mathsf{free}(\hat{M}) \setminus S$. *We call* $\mathsf{forb}(\hat{M}, S) := \mathsf{ports}(\hat{M}) \cup \bar{S}^C$ *the* forbidden ports, *i.e., those ports that the honest user is forbidden to have. A* system Sys *is a set of structures. It is polynomial-time iff all machines in all its collections* $\hat{M}$ *are polynomial-time.* ◇

The separation of the free ports into specified ports and others is an important feature of the upcoming security definition. The specified ports are those where a certain service is guaranteed. Note that this definition is valid for both the synchronous and the asynchronous case. In particular, buffers do not have to be explicitly included in the specification of a system, e.g., in the specification of a cryptographic protocol that one wants to analyze. The different timing assumption stem from the different definitions of runs which we will introduce below.

A structure can be completed to a *configuration* by adding machines H and A, modeling the joint honest users and the adversary, respectively. The machine H is restricted to the specified ports S, A connects to the remaining free ports of the structure and both machines can interact, e.g., in order to model active attacks.

Definition 2. *(Configurations) A* configuration *of a system* Sys *is a tuple* $conf = (\hat{M}, S, \mathsf{H}, \mathsf{A})$ *where* $(\hat{M}, S) \in Sys$ *is a structure,* $\hat{M} \cup \{\mathsf{H}, \mathsf{A}\}$ *is a closed collection, and* $\mathsf{ports}(\mathsf{H}) \cap \mathsf{forb}(\hat{M}, S) = \emptyset$. *The set of configurations is written* $\mathsf{Conf}(Sys)$. *The set of configurations with polynomial-time user* H *and adversary* A *is called* $\mathsf{Conf}_{\mathsf{poly}}(Sys)$. *The index* $_{\mathsf{poly}}$ *is omitted if it is clear from the context. The initial states of all machines in a configuration are a common security parameter* k *in unary representation.* ◇

2.2 Capturing Asynchronous Runs

For a configuration, both models define a probability space of runs (sometimes called *traces* or *executions*). In the asynchronous model, scheduling of machines is done sequentially, so we have exactly one active machine M at any time. If this machine has clock-out ports, it is allowed to select the next message to be scheduled as explained at the beginning of Sect. 2.1. If this message exists, it is delivered by the buffer and the unique receiving machine is the next active machine. If M tries to schedule multiple messages, only one is taken, and if it schedules none or the message does not exist, the special master scheduler is scheduled. This is formally captured as follows.

Definition 3. *(Asynchronous Runs and Views) For a given configuration* $conf = (\hat{M}, S, \mathsf{H}, \mathsf{A})$ *with master scheduler* $\mathsf{X} \in \hat{M} \cup \{\mathsf{A}\}$, *set* $\hat{C} := [\hat{M} \cup \{\mathsf{H}, \mathsf{A}\}]$. *The probability space of* runs *is defined inductively by the following algorithm. It has a variable* r *for the resulting* run, *an initially empty list, a variable* $\mathsf{M_{CS}}$ *("current scheduler") over machine names, initially* $\mathsf{M_{CS}} := \mathsf{X}$, *and treats each port as a variable over* Σ^*, *initialized with* ϵ *except for* $\mathsf{clk}^{\triangleleft}? := 1$. *Probabilistic choices only occur in Phase (1).*

1. Switch current scheduler: *Switch machine* $\mathsf{M_{CS}}$, *i.e., set* $(s', O) \leftarrow \delta_{\mathsf{M_{CS}}}(s, I)$ *for its current state* s *and input port values* I. *Then assign* ϵ *to all input ports of* $\mathsf{M_{CS}}$.

2. Termination: *If* X *is in a final state, the run stops.*
3. Buffer messages: *For each simple output port* $\mathsf{q}!$ *of* $\mathsf{M_{CS}}$ *in the given order, switch buffer* $\tilde{\mathsf{q}}$ *with input* $\mathsf{q}^{\leftrightarrow}? := \mathsf{q}!$, *cf. Fig. 1. Then assign* ϵ *to all these ports* $\mathsf{q}!$ *and* $\mathsf{q}^{\leftrightarrow}?$.
4. Clean up scheduling: *If at least one clock-out port of* $\mathsf{M_{CS}}$ *has a value* $\neq \epsilon$, *let* $\mathsf{q}^{\triangleleft}!$ *denote the first such port and assign* ϵ *to the others. Otherwise let* $\mathsf{clk}^{\triangleleft}? := 1$ *and* $\mathsf{M_{CS}} := \mathsf{X}$ *and go back to Phase (1).*
5. Scheduled message: *Switch* $\tilde{\mathsf{q}}$ *with input* $\mathsf{q}^{\triangleleft}? := \mathsf{q}^{\triangleleft}!$ *(cf. Fig. 1), set* $\mathsf{q}? := \mathsf{q}^{\leftrightarrow}!$ *and then assign* ϵ *to all ports of* $\tilde{\mathsf{q}}$ *and to* $\mathsf{q}^{\triangleleft}!$. *Let* $\mathsf{M_{CS}} := \mathsf{M}'$ *for the unique machine* M' *with* $\mathsf{q}? \in \mathsf{ports}(\mathsf{M}')$. *Go back to Phase (1).*

Whenever a machine (this may be a buffer) with name $name_{\mathsf{M}}$ *is switched from* (s, I) *to* (s', O), *we add a* step $(name_{\mathsf{M}}, s, I', s', O)$ *with* $I' := I\lceil_{I_{\mathsf{M}}}$ *to the run* r, *except if* s *is final or* $I' = (\epsilon, \ldots, \epsilon)$. *For each value of the security parameter, this gives a random variable denoted as* $run_{conf,k}$, *hence we obtain a family of random variables*

$$run_{conf} = (run_{conf,k})_{k \in \mathbb{N}}.$$

The view *of a subset* $M \subset \hat{C}$ *in a run* r *is the restriction of* r *to* M, *i.e., the subsequence of all steps* $(name_{\mathsf{M}}, s, I, s', O)$, *where* $name_{\mathsf{M}}$ *is the name of a machine* $\mathsf{M} \in M$. *This gives a family of random variables*

$$view_{conf}(M) = (view_{conf,k}(M))_{k \in \mathbb{N}}.$$

For a singleton $M = \{\mathsf{H}\}$ *we write* $view_{conf}(\mathsf{H})$ *instead of* $view_{conf}(\{\mathsf{H}\})$. ◇

This rather informal definition of runs can naturally be formalized using transition probabilities, which induce probability spaces over the finite sequences of steps similar to Markov Chains. The extension to infinite sequences can then be achieved using well-established results of measure theory and probability theory, cf. Sect. 5 of [24]. It is further easy to show that views of polynomial-time machines are of polynomial size.

2.3 Capturing Synchronous Runs

In the synchronous model, ports, machines, collections, structures, and systems are defined similar to the asynchronous model. The only exception is that there are no clock ports and no buffers, which have only been included to model asynchronous timing, i.e., corresponding ports $\mathsf{p}?$ and $\mathsf{p}!$ are directly connected. The main difference is the definition of runs. Instead of our asynchronous run algorithm (cf. Definition 3), runs are defined using *rounds* which is the usual concept in synchronous scenarios. Every global round is again divided into n so-called subrounds, and there is a mapping κ, called *clocking scheme*, from the set $\{1, \ldots, n\}$ into the powerset of considered machines, i.e., the machines of the structure, the user, and the adversary. Here $\kappa(i)$ denotes the machines that switch in subround i. We call a clocking scheme *valid* if every machine is clocked at most once between two successive clockings of the adversary.

Definition 4. *(Synchronous Runs and Views) Given a configuration* $conf = (\hat{M}, S, \mathsf{H}, \mathsf{A})$ *along with a clocking scheme* κ *for* $n \in \mathbb{N}$, *runs are defined as follows: Each global* *round i has n subrounds. In subround $[i.j]$ all machines $\mathsf{M} \in \kappa(j)$ switch simultaneously, i.e., each state-transition function δ_{M} is applied to M's current input yielding a new state and output (probabilistically). The output at a port $\mathsf{p}!$ is available as input at $\mathsf{p}?$ until the machine with port $\mathsf{p}?$ is switched. If several inputs arrive until that time, they are concatenated. This gives a family of random variables*

$$run_{conf} = (run_{conf,k})_{k \in \mathbb{N}}.$$

More precisely, each run *is a function mapping each triple $(\mathsf{M}, i, j) \in \hat{M} \cup \{\mathsf{H}, \mathsf{A}\} \times \mathbb{N} \times \{1, \ldots, n\}$ to a quadruple (s, I', s', O) of the old and new state, inputs (with $I' := I\lceil_{l_\mathsf{M}}$ again), and outputs of machine M in subround $[i.j]$, with $I' \equiv \epsilon$, $O \equiv \epsilon$, and $s = s'$ if M is not switched in this subround. The* view *of $M \subset \hat{M} \cup \{\mathsf{H}, \mathsf{A}\}$ in a run r is the restriction of r to $M \times \mathbb{N} \times \{1, \ldots, n\}$. This gives a family of random variables*

$$view_{conf}(M) = (view_{conf,k}(M))_{k \in \mathbb{N}}.$$

◇

Again, the view of a polynomial-time machine can easily be shown to be of polynomial size. Alternatively, we can consider runs as a sequence of seven-tuples $(\mathsf{M}, i, j, s, I', s', O)$ for ascending values of i and j, where tuples with the same values i and j can be ordered arbitrarily since they switch simultaneously and do not influence each other. This characterization of runs is equivalent to the original one (we just expanded the function), but it is better suited for our upcoming proofs.

Instead of arbitrary clocking schemes as in the above definition of runs, the authors of [28] focus on only one special clocking scheme κ, given by $(\hat{M} \cup \{\mathsf{H}\}, \{\mathsf{A}\}, \{\mathsf{H}\}, \{\mathsf{A}\})$. Clocking the adversary between the correct machines is the well-known model of "rushing adversaries". In [28], it has been shown that this clocking scheme does not restrict the possibilities of the adversary. Since our upcoming results hold for all valid clocking schemes, they in particular hold for rushing adversaries.

2.4 Simulatability

The definition of one system securely implementing another one is based on the common concept of *simulatability*. Simulatability essentially means that whatever might happen to an honest user in a real system Sys_1 can also happen in an ideal system Sys_2: For every structure $struc_1 \in Sys_1$, every user H, and every adversary A_1, there exists an adversary A_2 on a corresponding ideal structure $struc_2$ such that the view of H is indistinguishable in the two configurations. Indistinguishability is a well-defined cryptographic notion from [34].

Definition 5. *(Computational Indistinguishability) Two families $(\mathsf{var}_k)_{k \in \mathbb{N}}$ and $(\mathsf{var}'_k)_{k \in \mathbb{N}}$ of random variables (or probability distributions) on common domains D_k are* computationally indistinguishable *("$\approx$") if for every algorithm* Dis *(the distinguisher) that is probabilistic polynomial-time in its first input,*

$$|P(\mathsf{Dis}(1^k, \mathsf{var}_k) = 1) - P(\mathsf{Dis}(1^k, \mathsf{var}'_k) = 1)| \in NEGL.^1$$

Intuitively, given the security parameter and an element chosen according to either var_k *or* var'_k, Dis *tries to guess which distribution the element came from.* ◇

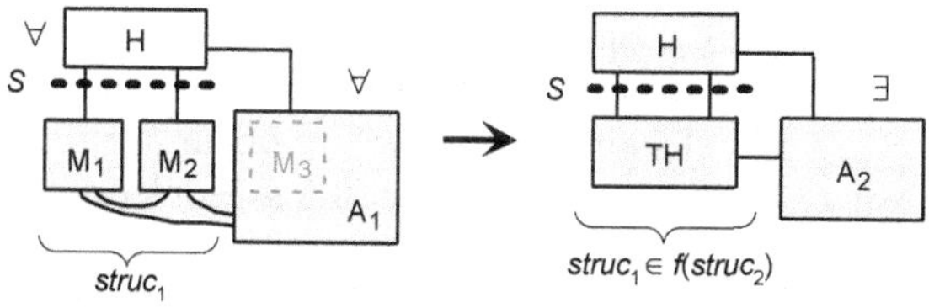

Fig. 2. Overview of the simulatability definition

Corresponding structures in the simulatability definition are designated by a function f from Sys_1 to the powerset of Sys_2. The function f is called *valid* if it maps structures with the same set of specified ports.

Definition 6. *(Simulatability) Let systems Sys_1 and Sys_2 with a valid mapping f be given. We say $Sys_1 \geq^f Sys_2$ if for every configuration $conf_1 = (\hat{M}_1, S, \mathsf{H}, \mathsf{A}_1) \in \mathsf{Conf}_{\mathsf{poly}}(Sys_1)$, there exists a configuration $conf_2 = (\hat{M}_2, S, \mathsf{H}, \mathsf{A}_2) \in \mathsf{Conf}_{\mathsf{poly}}(Sys_2)$ with $(\hat{M}_2, S) \in f(\hat{M}_1, S)$ such that $view_{conf_1}(\mathsf{H}) \approx view_{conf_2}(\mathsf{H})$.* ◇

This is shown in Fig. 2. In the following, we augment $\geq$ with a subscript $_{\mathsf{sync}}$ or $_{\mathsf{async}}$ to distinguish the definition of the synchronous and asynchronous case. In a typical ideal system, each structure contains only one machine TH called trusted host, which serves as an ideal functionality of the real system. The machine TH is usually deterministic with a very simple transition function and hence in scope of current verification techniques.

3 Idea and Definition of the Embedding

The informal idea of the embedding φ_{Sys} is to add an explicit master scheduler that should simulate the synchronous run induced by the given clocking scheme. However, due to the general distributed scheduling (cf. Definition 3), leaving the actual machines unmodified leads to non-simulatable situations, as these machines can clock themselves without ever giving control to this explicit master scheduler.

Hence, we first define a mapping φ_{M} that surrounds single synchronous machines (i.e., machines that are designed for a synchronous environment) with an "asynchronous coat". More precisely, if a synchronous machine $\mathsf{M}_{\mathsf{sync}}$ makes a transition, it obtains all inputs at once that arrived since its last scheduling, whereas in asynchronous scenarios, these inputs come one by one and have to be processed in several transitions. Thus, the surrounding asynchronous machine $\varphi_{\mathsf{M}}(\mathsf{M}_{\mathsf{sync}})$ stores all inputs internally, until it is asked to perform the transition of its synchronous submachine. This is modeled by a special port $\mathsf{p}_{\mathsf{M}_{\mathsf{sync}}}?$ of the machine $\varphi_{\mathsf{M}}(\mathsf{M}_{\mathsf{sync}})$. An input at this port causes $\varphi_{\mathsf{M}}(\mathsf{M}_{\mathsf{sync}})$ to schedule its submachine with the collected inputs and forward its outputs. As such embedded machines do not produce any clock outputs, the ports $\mathsf{p}_{\mathsf{M}_{\mathsf{sync}}}?$ can be used by the master scheduler to simulate the synchronous time by a suitable scheduling strategy.

The formal definition of φ_{M} can be found in the full version. We only briefly state that $\varphi_{\mathsf{M}}(\mathsf{M}_{\mathsf{sync}})$ is polynomial-time by construction iff $\mathsf{M}_{\mathsf{sync}}$ is polynomial-time. For a set $\hat{M}$ of synchronous machines, we further define $\varphi_{\mathsf{M}}(\hat{M}) := \bigcup_{\mathsf{M}_{\mathsf{sync}} \in \hat{M}} \varphi_{\mathsf{M}}(\mathsf{M}_{\mathsf{sync}})$.

The desired mapping φ_{Sys} on synchronous systems then simply embeds every machine and adds a specific master scheduler to each structure, i.e., for an arbitrary structure $(\mathsf{M}_{\mathsf{sync}}, S_{\mathsf{sync}})$ and clocking scheme κ, we obtain a structure $(\varphi_{\mathsf{M}}(\hat{M}_{\mathsf{sync}}) \cup \{\mathsf{X}_{\mathsf{sync},\kappa}\}, S_{\mathsf{sync}})$, where $\mathsf{X}_{\mathsf{sync},\kappa}$ is an explicit master scheduler defined as follows. Besides the master clock-in port, it has clock-out ports for clocking inputs and outputs of the given structure, for clocking the connection between A and H, and finally ports $\mathsf{p}_{\mathsf{M}_{\mathsf{sync}}}!, \mathsf{p}_{\mathsf{M}_{\mathsf{sync}}}{}^{\triangleleft}!$ for clocking, i.e., giving control to, each machine $\varphi_{\mathsf{M}}(\mathsf{M}_{\mathsf{sync}})$. Internally, it maintains a variable *local_rnd* over $\{1, \ldots, n\}$ and a variable *global_rnd* over $\mathbb{N}$ both initialized with 1. For the sake of readability, we describe the behavior of $\mathsf{X}_{\mathsf{sync},\kappa}$ using "for"-loops. This is just a notational convention that should be understood as follows: every time $\mathsf{X}_{\mathsf{sync},\kappa}$ is scheduled, it performs the next step of the loop.

1. **Schedule Current Machines:** For all machines $\mathsf{M}_{\mathsf{sync}} \in \kappa(local\_rnd)$ output $(global\_rnd, local\_rnd)$ at $\mathsf{p}_{\mathsf{M}_{\mathsf{sync}}}!$, 1 at $\mathsf{p}_{\mathsf{M}_{\mathsf{sync}}}{}^{\triangleleft}!$. The order of the switched machines can be chosen arbitrary.
2. **Schedule Outgoing Buffers:** For all $\mathsf{M}_{\mathsf{sync}} \in \kappa(local\_rnd)$ output 1 at every port $\mathsf{p}^{\triangleleft}!$ with $\mathsf{p}! \in Ports_{\mathsf{M}_{\mathsf{sync}}}$. The order of the switched machine can be chosen arbitrary except that output ports of the adversary are scheduled first if $\mathsf{A} \in \kappa(local\_rnd)$.[2]
3. **Switch to next Round:** Set $local\_rnd := local\_rnd + 1$. If $local\_rnd > n$, set $global\_rnd := global\_rnd + 1$ and $local\_rnd := 1$. Go to Phase (1).

We finally define a mapping φ_{conf} on synchronous configurations of a system Sys, i.e., configurations which consist of synchronous machines only, by

$$\varphi_{conf}(\hat{M}_{\mathsf{sync}}, S_{\mathsf{sync}}, \mathsf{H}, \mathsf{A}) := (\varphi_{\mathsf{M}}(\hat{M}_{\mathsf{sync}}) \cup \{\mathsf{X}_{\mathsf{sync},\kappa}\}, S_{\mathsf{sync}}, \varphi_{\mathsf{M}}(\mathsf{H}), \varphi_{\mathsf{M}}(\mathsf{A})),$$

with $\mathsf{X}_{\mathsf{sync},\kappa}$ given as in φ_{Sys} for the particular structure. We will in the following simply write S instead of S_{sync}; we also write φ instead of $\varphi_{Sys}, \varphi_{\mathsf{M}}$, and φ_{conf} if its meaning is clear from the context.

4 The Embedding Theorems

We now have to prove that the function φ has the desired properties with respect to simulatability, i.e., we have

$$\varphi_{Sys}(Sys_{\mathsf{sync},1}) \geq_{\mathsf{async}} \varphi_{Sys}(Sys_{\mathsf{sync},2}) \Rightarrow Sys_{\mathsf{sync},1} \geq_{\mathsf{sync}} Sys_{\mathsf{sync},2},$$

where we omitted the mappings, i.e., the superscript of both relations for the sake of readability. This captures the content of our first embedding theorem. Unfortunately, the converse direction does not hold, but our second embedding theorem will state a weaker version that is still sufficient for our purpose.

We first take a look at the runs in a synchronous system Sys_{sync} and in its asynchronous counterpart $\varphi(Sys_{\mathsf{sync}})$. We define a mapping ϕ on the runs of the asynchronous system yielding runs of the synchronous system. Intuitively, ϕ "compresses" an asynchronous run to its synchronous counterpart, which consists of much less steps.

[2] Without this restriction, the behavior of the adversary could depend on outputs of machines scheduled in the same subround, which would lead to non-simulatable situations.

4.1 Relating Asynchronous Runs and Synchronous Runs

In the following, let an arbitrary synchronous system Sys_{sync} with a clocking scheme κ and an arbitrary configuration $conf_{\mathsf{sync}} = (\hat{M}_{\mathsf{sync}}, S, \mathsf{H}_{\mathsf{sync}}, \mathsf{A}_{\mathsf{sync}}) \in \mathsf{Conf}(Sys_{\mathsf{sync}})$ be given. Moreover, let $conf_{\mathsf{async}} = (\varphi(\hat{M}_{\mathsf{sync}}) \cup \{\mathsf{X}_{\mathsf{sync},\kappa}\}, S, \varphi(\mathsf{H}_{\mathsf{sync}}), \mathsf{A}') \in \mathsf{Conf}(\varphi(Sys_{\mathsf{sync}}))$ be given (i.e., $\varphi(conf_{\mathsf{sync}})$ but with an arbitrary adversary). We call such a configuration an *embedded synchronous configuration with arbitrary adversary.*

Note that runs of $conf_{\mathsf{async}}$ always have a prescribed structure induced by the behavior of the master scheduler $\mathsf{X}_{\mathsf{sync},\kappa}$: they are built by "blocks". The steps $(\mathsf{M}_{\mathsf{sync}}, i, j, s, \mathcal{I}, s', \mathcal{O})$ of the machines $\mathsf{M}_{\mathsf{sync}} \in \hat{M}_{\mathsf{sync}} \cup \{\mathsf{H}_{\mathsf{sync}}\}$ switched in round $[i.j]$ in the synchronous run are represented by two blocks in the asynchronous run. The first block corresponds to Phase (1) in the definition of $\mathsf{X}_{\mathsf{sync},\kappa}$ and hence consists of those steps induced by clocking the machines $\varphi(\mathsf{M}_{\mathsf{sync}})$ with $\mathsf{M}_{\mathsf{sync}} \in \kappa(j)$ and A' if $\mathsf{A}_{\mathsf{sync}} \in \kappa(j)$. More precisely, it consists of $|\kappa(j)|$ sub-blocks, one for each switched machine, where a sub-block for $\mathsf{M}_{\mathsf{sync}}$ comprises the step of the master scheduler, the step of the scheduled buffer, the step of the switched machine, and steps for the receiving buffers.[3] The second block corresponds to Phase (2) in the definition of $\mathsf{X}_{\mathsf{sync},\kappa}$ and hence consists of the steps of the buffers connected to the output ports of the switched machines as well as steps of the machines receiving the scheduled message.

Informally, the mapping ϕ combines the blocks of all machines $\mathsf{M}_{\mathsf{sync}} \in \kappa(j)$ yielding the synchronous step of every machine $\mathsf{M}_{\mathsf{sync}}$ that switches in the j-th subround of the particular global round. Note that all necessary information (e.g., $\mathsf{M}_{\mathsf{sync}}, i, j, s, s'$ etc.) is already given by the blocks except for the gathered inputs I. The mapping ϕ overcomes this absence by collecting all "partial" inputs itself. It now also becomes clear why we defined the master scheduler to schedule each machine specifically with a tuple (i, j) indicating the current global and local round, since this information would otherwise not be contained in the asynchronous run. The precise definition of the mapping ϕ is omitted due to lack of space. We only state that ϕ is also well-defined on the view of arbitrary subsets of machines. The following lemma establishes a computational bound on the mapping ϕ in polynomial-time configurations:

Lemma 1. *Let $conf_{\mathsf{async}}$ be an embedded synchronous configuration with arbitrary adversary. If $conf_{\mathsf{async}}$ is polynomial-time, then ϕ applied to the view of the honest user and the adversary is computable in polynomial-time.* □

We can moreover show that the mapping ϕ compresses the view of any machine in an embedded synchronous configuration to the view of the original machine in the synchronous configuration again. Formally, this is captured in the following theorem.

Theorem 1. *Let a synchronous system Sys_{sync}, a clocking scheme κ, and a configuration $conf_{\mathsf{sync}} = (\hat{M}_{\mathsf{sync}}, S, \mathsf{H}_{\mathsf{sync}}, \mathsf{A}_{\mathsf{sync}}) \in \mathsf{Conf}(Sys_{\mathsf{sync}})$ be given, and set $conf_{\mathsf{async}} := \varphi(conf_{\mathsf{sync}})$. Then for every $\mathsf{M}_{\mathsf{sync}} \in \hat{M}_{\mathsf{sync}} \cup \{\mathsf{H}_{\mathsf{sync}}, \mathsf{A}_{\mathsf{sync}}\}$, we have*

$$view_{conf_{\mathsf{sync}}}(\mathsf{M}_{\mathsf{sync}}) = \phi(view_{conf_{\mathsf{async}}}(\varphi(\mathsf{M}_{\mathsf{sync}}))).$$

Moreover, $conf_{\mathsf{async}}$ is polynomial-time iff $conf_{\mathsf{sync}}$ is polynomial-time. □

[3] For A', we might have additionally steps because of clocked self-loops.

The following theorem finally states that the runs obtained by applying the mapping ϕ to an embedded synchronous configuration with arbitrary adversary are equal to the runs in the synchronous system for a suitable synchronous adversary.

Theorem 2. *Let a synchronous system* Sys_{sync}*, a valid clocking scheme* κ*, and a configuration* $conf_{\mathsf{async}} = (\varphi(\hat{M}_{\mathsf{sync}}) \cup \{\mathsf{X}_{\mathsf{sync},\kappa}\}, S, \varphi(\mathsf{H}_{\mathsf{sync}}), \mathsf{A}') \in \mathsf{Conf}(\varphi(Sys_{\mathsf{sync}}))$ *be given. Then there exists an adversary* $\mathsf{A}_{\mathsf{sync}}$ *using* A' *as a blackbox such that for* $conf_{\mathsf{sync}} := (\hat{M}_{\mathsf{sync}}, S, \mathsf{H}_{\mathsf{sync}}, \mathsf{A}_{\mathsf{sync}})$ *and every* $\mathsf{M}_{\mathsf{sync}} \in (\hat{M}_{\mathsf{sync}} \cup \{\mathsf{H}_{\mathsf{sync}}\})$*, we have*

$$view_{conf_{\mathsf{sync}}}(\mathsf{M}_{\mathsf{sync}}) = \phi(view_{conf_{\mathsf{async}}}(\varphi(\mathsf{M}_{\mathsf{sync}})))$$

Moreover, $conf_{\mathsf{async}}$ *is polynomial-time iff* $conf_{\mathsf{sync}}$ *is polynomial-time.* □

4.2 The Actual Embedding Theorems

This section contains our two main theorems. We start with a well-known lemma capturing some basic properties of indistinguishable random variables.

Lemma 2 (Indistinguishability). *Indistinguishability of two families of random variables implies indistinguishability of any function* δ *of them if* δ *is polynomial-time computable. Moreover, identically distributed variables are indistinguishable and indistinguishability is an equivalence relation.* □

Theorem 3. *(First Embedding Theorem) Let two arbitrary synchronous systems* $Sys_{\mathsf{sync},1}$ *and* $Sys_{\mathsf{sync},2}$ *with clocking schemes* κ_1 *and* κ_2 *be given such that* κ_2 *is valid, and let* $\varphi(Sys_{\mathsf{sync},1}) \geq^f_{\mathsf{async}} \varphi(Sys_{\mathsf{sync},2})$ *for a valid mapping* f*. Then*

$$Sys_{\mathsf{sync},1} \geq^{f'}_{\mathsf{sync}} Sys_{\mathsf{sync},2},$$

where f' *is defined as* $(\hat{M}_2, S_2) \in f'(\hat{M}_1, S_1) :\Leftrightarrow \varphi(\hat{M}_2, S_2) \in f(\varphi(\hat{M}_1, S_1))$. □

Proof. Let an arbitrary configuration $conf_{\mathsf{sync},1} = (\hat{M}_{\mathsf{sync},1}, S, \mathsf{H}_{\mathsf{sync}}, \mathsf{A}_{\mathsf{sync},1}) \in \mathsf{Conf}_{\mathsf{poly}}(Sys_{\mathsf{sync},1})$ be given. We first apply φ_{conf} on $conf_{\mathsf{sync},1}$ yielding a configuration $conf_{\mathsf{async},1} = (\varphi(\hat{M}_{\mathsf{sync},1}) \cup \{\mathsf{X}_{\mathsf{sync},1,\kappa_1}\}, S, \varphi(\mathsf{H}_{\mathsf{sync}}), \varphi(\mathsf{A}_{\mathsf{sync},1})) \in \mathsf{Conf}(\varphi(Sys_{\mathsf{sync},1}))$. According to Theorem 1, $conf_{\mathsf{async},1}$ is also polynomial-time and we have

$$view_{conf_{\mathsf{sync},1}}(\mathsf{H}_{\mathsf{sync}}) = \phi(view_{conf_{\mathsf{async},1}}(\varphi(\mathsf{H}_{\mathsf{sync}}))). \quad (1)$$

Thus, the precondition $\varphi(Sys_{\mathsf{sync},1}) \geq^f_{\mathsf{async}} \varphi(Sys_{\mathsf{sync},2})$ can be applied yielding a configuration $conf_{\mathsf{async},2} = (\varphi(\hat{M}_{\mathsf{sync},2}) \cup \{\mathsf{X}_{\mathsf{sync},2,\kappa_2}\}, S, \varphi(\mathsf{H}_{\mathsf{sync}}), \mathsf{A}_{\mathsf{async},2}) \in \mathsf{Conf}_{\mathsf{poly}}(\varphi(Sys_{\mathsf{sync},2}))$ with $view_{conf_{\mathsf{async},1}}(\varphi(\mathsf{H}_{\mathsf{sync}})) \approx view_{conf_{\mathsf{async},2}}(\varphi(\mathsf{H}_{\mathsf{sync}}))$ and $\varphi(\hat{M}_{\mathsf{sync},2}, S) \in f(\varphi(\hat{M}_{\mathsf{sync},1}, S))$. As both $conf_{\mathsf{async},1}$ and $conf_{\mathsf{async},2}$ are polynomial-time, Lemma 1 and Lemma 2 together yield

$$\phi(view_{conf_{\mathsf{async},1}}(\varphi(\mathsf{H}_{\mathsf{sync}}))) \approx \phi(view_{conf_{\mathsf{async},2}}(\varphi(\mathsf{H}_{\mathsf{sync}}))). \quad (2)$$

We now apply Theorem 2 to the configuration $conf_{\mathsf{async},2}$, which yields a configuration $conf_{\mathsf{sync},2} = (\hat{M}_{\mathsf{sync}}, S, \mathsf{H}_{\mathsf{sync}}, \mathsf{A}_{\mathsf{sync},2}) \in \mathsf{Conf}_{\mathsf{poly}}(Sys_{\mathsf{sync},2})$ with

$$\phi(view_{conf_{\mathsf{async},2}}(\varphi(\mathsf{H}_{\mathsf{sync}}))) = view_{conf_{\mathsf{sync},2}}(\mathsf{H}_{\mathsf{sync}}). \tag{3}$$

Now Equation 1, 2, and 3 together with Lemma 2 finally yield $view_{conf_{\mathsf{sync},1}}(\mathsf{H}_{\mathsf{sync}}) \approx view_{conf_{\mathsf{sync},2}}(\mathsf{H}_{\mathsf{sync}})$. Moreover, we have $\varphi(\hat{M}_{\mathsf{sync},2}, S) \in f(\varphi(\hat{M}_{\mathsf{sync},1}, S))$, i.e., $(\hat{M}_{\mathsf{sync},2}, S) \in f'(\hat{M}_{\mathsf{sync},1}, S)$ which finally yields $Sys_{\mathsf{sync},1} \geq^{f'}_{\mathsf{sync}} Sys_{\mathsf{sync},2}$. ■

So far, we have shown that asynchronous simulatability among the asynchronous representations implies synchronous simulatability, i.e.,

$$\varphi_{Sys}(Sys_{\mathsf{sync},1}) \geq_{\mathsf{async}} \varphi_{Sys}(Sys_{\mathsf{sync},2}) \Rightarrow Sys_{\mathsf{sync},1} \geq_{\mathsf{sync}} Sys_{\mathsf{sync},2}.$$

We already briefly stated that the converse implication does not hold in general. We had to show that for each configuration $conf_{\mathsf{async},1} \in \mathsf{Conf}_{\mathsf{poly}}(\varphi_{Sys}(Sys_{\mathsf{sync},1}))$ there exists an indistinguishable configuration $conf_{\mathsf{async},2} \in \mathsf{Conf}_{\mathsf{poly}}(\varphi_{Sys}(Sys_{\mathsf{sync},2}))$ provided that $Sys_{\mathsf{sync},1} \geq_{\mathsf{sync}} Sys_{\mathsf{sync},2}$. However, both the honest user and the adversary may have clock-out ports and they can alternately schedule each other (and also the system erratically), which cannot be captured by a fixed synchronous clocking scheme, so we cannot exploit our assumption $Sys_{\mathsf{sync},1} \geq_{\mathsf{sync}} Sys_{\mathsf{sync},2}$.

Anyhow, it is sufficient for our purpose to show that the claim holds for at least those configurations where the honest user $\mathsf{H}_{\mathsf{async}}$ fits the form $\varphi_{\mathsf{M}}(\mathsf{H}_{\mathsf{sync}})$ for a synchronous machine $\mathsf{H}_{\mathsf{sync}}$. We denote this version of simulatability for the restricted class of users by $\geq_{\mathsf{async},\mathsf{H}}$ in the sequel. It is immediately clear that the first embedding theorem also holds for the weaker precondition $\varphi_{Sys}(Sys_{\mathsf{sync},1}) \geq_{\mathsf{async},\mathsf{H}} \varphi_{Sys}(Sys_{\mathsf{sync},2})$, since we only have to derive an indistinguishable configuration for users of the special form $\varphi(\mathsf{H}_{\mathsf{sync}})$, and the user remains unchanged at simulatability. We can now present the second embedding theorem.

Theorem 4. *(Second Embedding Theorem) Let two arbitrary synchronous systems $Sys_{\mathsf{sync},1}$ and $Sys_{\mathsf{sync},2}$ with clocking schemes κ_1 and κ_2 be given such that κ_1 is valid, and let $Sys_{\mathsf{sync},1} \geq^{f}_{\mathsf{sync}} Sys_{\mathsf{sync},2}$ for a valid mapping f. Then*

$$\varphi(Sys_{\mathsf{sync},1}) \geq^{f'}_{\mathsf{async},\mathsf{H}} \varphi(Sys_{\mathsf{sync},2})$$

where f' is defined as $\varphi(\hat{M}_2, S_2) \in f'(\varphi(\hat{M}_1, S_1)) :\Leftrightarrow (\hat{M}_2, S_2) \in f(\hat{M}_1, S_1)$. □

5 Deriving Synchronous Theorems from Asynchronous Ones

Recall that our long-term goal is to avoid proving theorems for both models. We now briefly show how our two embedding theorems can be used to circumvent this problem. One of the most important theorems of both models is transitivity of the relation $\geq$.

Lemma 3. *(Transitivity) If $Sys_1 \geq^{f_1} Sys_2$ and $Sys_2 \geq^{f_2} Sys_3$, then $Sys_1 \geq^{f_3} Sys_3$ with $f_3(\hat{M}_1, S)$ being the union of the sets $f_2(\hat{M}_2, S)$ with $(\hat{M}_2, S) \in f_1(\hat{M}_1, S)$.* □

This has been proven in [28] for the synchronous and in [30] for the asynchronous model. We now exemplarily show how to derive the synchronous version from the asynchronous one using our previous results.

Lemma 4. *Assume that the asynchronous version of the transitivity lemma (Lemma 3) has already been proven, then the synchronous version holds as well.* □

Proof. We omit the superscripts f_i for the sake of readability. Let synchronous systems Sys_1, Sys_2, and Sys_3 be given such that $Sys_1 \geq_{\mathsf{sync}} Sys_2$ and $Sys_2 \geq_{\mathsf{sync}} Sys_3$. We have to show that $Sys_1 \geq_{\mathsf{sync}} Sys_3$ holds, provided that asynchronous transitivity has already been proven. Our second embedding theorem implies $\varphi(Sys_1) \geq_{\mathsf{async,H}} \varphi(Sys_2)$ and $\varphi(Sys_2) \geq_{\mathsf{async,H}} \varphi(Sys_3)$. Obviously, the asynchronous version of transitivity is applicable to the relation $\geq_{\mathsf{async,H}}$ instead of $\geq_{\mathsf{async}}$ as well, since it is a special case only, and the honest user remains unchanged at simulatability. Thus, we can apply our (already proven) asynchronous version of the transitivity lemma, which yields $\varphi(Sys_1) \geq_{\mathsf{async,H}} \varphi(Sys_3)$. Now, we use our first embedding theorem in conjunction with its subsequent remarks (stating that the theorem holds as well for the restricted version $\geq_{\mathsf{async,H}}$ of simulatability) yielding $Sys_1 \geq_{\mathsf{sync}} Sys_3$. ■

This proof technique is applicable to almost all theorems that rely on simulatability. As the most important example, we name the preservation theorem [29,4], which states that integrity properties expressed in linear-time logic are preserved under simulatability. The proof of this theorem is difficult and comprises several pages for both models. Using our work, the synchronous proof could as well be omitted.

References

1. M. Abadi and A.D. Gordon. A calculus for cryptographic protocols: The spi calculus. *Information and Computation*, 148(1):1–70, 1999.
2. M. Abadi and J. Jürjens. Formal eavesdropping and its computational interpretation. In *Proc. 4th International Symposium on Theoretical Aspects of Computer Software (TACS)*, pages 82–94, 2001.
3. M. Abadi and P. Rogaway. Reconciling two views of cryptography: The computational soundness of formal encryption. In *Proc. 1st IFIP International Conference on Theoretical Computer Science*, volume 1872 of *LNCS*, pages 3–22. Springer, 2000.
4. M. Backes and C. Jacobi. Cryptographically sound and machine-assisted verification of security protocols. In *Proc. 20th Annual Symposium on Theoretical Aspects of Computer Science (STACS)*, volume 2607 of *LNCS*, pages 675–686. Springer, 2003.
5. M. Backes, C. Jacobi, and B. Pfitzmann. Deriving cryptographically sound implementations using composition and formally verified bisimulation. In *Proc. 11th Symposium on Formal Methods Europe (FME 2002)*, volume 2391 of *LNCS*, pages 310–329. Springer, 2002.
6. M. Backes, B. Pfitzmann, and M. Waidner. A universally composable cryptographic library. IACR Cryptology ePrint Archive 2003/015, Jan. 2003. `http://eprint.iacr.org/`.
7. D. Beaver. Secure multiparty protocols and zero knowledge proof systems tolerating a faulty minority. *Journal of Cryptology*, 4(2):75–122, 1991.
8. M. Bellare, R. Canetti, and H. Krawczyk. A modular approach to the design and analysis of authentication and key exchange protocols. In *Proc. 30th Annual ACM Symposium on Theory of Computing (STOC)*, pages 419–428, 1998.

9. M. Burrows, M. Abadi, and R. Needham. A logic for authentication. Technical Report 39, SRC DIGITAL, 1990.
10. R. Canetti. Studies in secure multiparty computation and applications. Department of Computer Science and Applied Mathematics, The Weizmann Institute of Science, June 1995, revised March 1996, 1995.
11. R. Canetti. Security and composition of multiparty cryptographic protocols. *Journal of Cryptology*, 3(1):143–202, 2000.
12. R. Canetti. Universally composable security: A new paradigm for cryptographic protocols. In *Proc. 42nd IEEE Symposium on Foundations of Computer Science (FOCS)*, pages 136–145, 2001.
13. D. Dolev and A.C. Yao. On the security of public key protocols. *IEEE Transactions on Information Theory*, 29(2):198–208, 1983.
14. C. Dwork, M. Naor, and A. Sahai. Concurrent zero-knowledge. In *Proc. 30th Annual ACM Symposium on Theory of Computing (STOC)*, pages 409–418, 1998.
15. S. Goldwasser and L. Levin. Fair computation of general functions in presence of immoral majority. In *Advances in Cryptology: CRYPTO '90*, volume 537 of *LNCS*, pages 77–93. Springer, 1990.
16. S. Goldwasser, S. Micali, and C. Rackoff. The knowledge complexity of interactive proof systems. *SIAM Journal on Computing*, 18(1):186–207, 1989.
17. J.D. Guttman, F.J. Thayer Fabrega, and L. Zuck. The faithfulness of abstract protocol analysis: Message authentication. In *Proc. 8th ACM Conference on Computer and Communications Security*, pages 186–195, 2001.
18. C.A.R. Hoare. *Communicating Sequential Processes*. International Series in Computer Science, Prentice Hall, Hemel Hempstead, 1985.
19. P. Laud. Semantics and program analysis of computationally secure information flow. In *Proc. 10th European Symposium on Programming (ESOP)*, pages 77–91, 2001.
20. G. Lowe. Breaking and fixing the Needham-Schroeder public-key protocol using FDR. In *Proc. 2nd International Conference on Tools and Algorithms for the Construction and Analysis of Systems (TACAS)*, volume 1055 of *LNCS*, pages 147–166. Springer, 1996.
21. N. Lynch. *Distributed Algorithms*. Morgan Kaufmann Publishers, San Francisco, 1996.
22. J.K. Millen. The interrogator: A tool for cryptographic protocol security. In *Proc. 5th IEEE Symposium on Security & Privacy*, pages 134–141, 1984.
23. B. Neuman and T. Ts'o. Kerberos: An authentication service for computer networks. *IEEE Communications Magazine*, 32(9):33–38, 1994.
24. J. Neveu. *Mathematical Foundations of the Calculus of Probability*. Holden-Day, 1965.
25. S. Owre, N. Shankar, and J.M. Rushby. PVS: A prototype verification system. In *Proc. 11th International Conference on Automated Deduction (CADE)*, volume 607 of *LNCS*, pages 748–752. Springer, 1992.
26. L. Paulson. The inductive approach to verifying cryptographic protocols. *Journal of Cryptology*, 6(1):85–128, 1998.
27. B. Pfitzmann, M. Schunter, and M. Waidner. Cryptographic security of reactive systems. Presented at the DERA/RHUL Workshop on Secure Architectures and Information Flow, 1999, Electronic Notes in Theoretical Computer Science (ENTCS), March 2000. `http://www.elsevier.nl/cas/tree/store/tcs/free/noncas/pc/menu.htm`.
28. B. Pfitzmann, M. Schunter, and M. Waidner. Secure reactive systems. Research Report RZ 3206, IBM Research, 2000.
29. B. Pfitzmann and M. Waidner. Composition and integrity preservation of secure reactive systems. In *Proc. 7th ACM Conference on Computer and Communications Security*, pages 245–254, 2000.

30. B. Pfitzmann and M. Waidner. A model for asynchronous reactive systems and its application to secure message transmission. In *Proc. 22nd IEEE Symposium on Security & Privacy*, pages 184–200, 2001.
31. R. Segala and N. Lynch. Probabilistic simulation for probabilistic processes. *Nordic Journal of Computing*, 2(2):250–273, 1995.
32. D. Volpano and G. Smith. Verifying secrets and relative secrecy. In *Proc. 27th Symposium on Principles of Programming Languages (POPL)*, pages 268–276, 2000.
33. S.-H. Wu, S.A. Smolka, and E.W. Stark. Composition and behaviors of probabilistic I/O automata. *Theoretical Computer Science*, 176(1–2):1–38, 1997.
34. A.C. Yao. Theory and applications of trapdoor functions. In *Proc. 23rd IEEE Symposium on Foundations of Computer Science (FOCS)*, pages 80–91, 1982.

Contract Signing, Optimism, and Advantage*

Rohit Chadha[1,4], John C. Mitchell[2], Andre Scedrov[1], and Vitaly Shmatikov[3]

[1] University of Pennsylvania
[2] Stanford University
[3] SRI International
[4] University of Sussex

Abstract. A contract signing protocol lets two parties exchange digital signatures on a pre-agreed text. *Optimistic* contract signing protocols enable the signers to do so without invoking a trusted third party. However, an adjudicating third party remains available should one or both signers seek timely resolution. We analyze optimistic contract signing protocols using a game-theoretic approach and prove a fundamental impossibility result: in any fair, optimistic, timely protocol, an optimistic player yields an advantage to the opponent. The proof relies on a careful characterization of optimistic play that postpones communication to the third party. Since advantage cannot be completely eliminated from optimistic protocols, we argue that the strongest property attainable is the absence of *provable* advantage, *i.e.*, abuse-freeness in the sense of Garay-Jakobsson-MacKenzie.

1 Introduction

A variety of contract signing protocols have been proposed in the literature, including gradual-release two-party protocols [5,7,12] and fixed-round protocols that rely on an adjudicating "trusted third party" [2,3,18,23,26]. In this paper, we focus on fixed-round protocols that use a trusted third party optimistically, meaning that when all goes well, the third party is not needed. The reason for designing optimistic protocols is that if a protocol is widely or frequently used by many pairs of signers, the third party may become a performance bottleneck. Depending on the context, seeking resolution through the third party may delay termination, incur financial costs, or raise privacy concerns. Obviously, the value of an optimistic protocol, as opposed to one that requires a third party signature on every transaction, lies in the frequency with which "optimistic" signers can complete the protocol without using the third party.

Some useful properties of contract signing protocols are *fairness*, which means that either both parties get a signed contract, or neither does, and *timeliness*, which generally means that each party has some recourse to avoid unbounded waiting. The reason for using a trusted third party in fixed-round protocols is a basic limitation [14,24] related to the well-known impossibility of distributed consensus in the presence of faults [17]:

* The authors are partially supported by OSD/ONR CIP/SW URI "Software Quality and Infrastructure Protection for Diffuse Computing" as ONR Grant N00014-01-1-0795. Additional support for Mitchell from NSF Grant CCR-0121403, ITR/SY "Computational Logic Tools for Research and Education," for Scedrov from NSF Grant CCR-0098096, and for Shmatikov from ONR under Grants N00014-02-1-0109 and N00014-01-1-0837.

R. Amadio, D. Lugiez (Eds.): CONCUR 2003, LNCS 2761, pp. 366–382, 2003.

no fixed-length two-party protocol can be fair. Although there is a trivial protocol with a trusted third party, in which both signers always send their signatures directly to it, protocols that are fair, timely, and usefully minimize demands on the third party have proven subtle to design and verify.

This paper refines previous models, formalizes properties from the literature on fixed-round two-party contract signing protocols, and establishes relationships between them. We use the set-of-traces semantics for protocols, defining each instance of the protocol as the set of all possible execution traces arranged in a tree. Our chosen notation is multiset rewriting [10], but the results would hold for other formalisms with the same basic execution model.

Model for Optimism. One modeling innovation is an *untimed* nondeterministic setting that provides a set-of-traces semantics for optimism. Intuitively, optimistic behavior in contract signing is easily described as a temporal concept: an optimistic signer is one who waits for some period of time before contacting the trusted third party. If Alice is optimistic, and Bob chooses to continue the protocol by responding, then Alice waits for Bob's message rather than contact the third party. Since the value of an optimistic protocol lies in what it offers to an optimistic player, we evaluate protocols subject to the assumption that one of the players follows an optimistic strategy. As a direct way of mathematically characterizing optimistic play, we allow an optimistic player to give his opponent the chance to signal (out of band) whether to wait for a message. This gives us a relatively easy way to define the set of traces associated with an optimistic signer, while staying within the traditional nondeterministic, untimed setting.

Impossibility Result. In evaluating protocol performance for optimistic players, we prove that in every fair, timely protocol, an optimistic player suffers a disadvantage against a strategic adversary. The importance of this result is that optimistic protocols are only useful to the extent that signers may complete the protocol optimistically without contacting the third party. In basic terms, our theorem shows that to whatever degree a protocol allows signers to avoid the third party, the protocol proportionally gives one signer unilateral control over the outcome of the protocol.

To illustrate by example, consider an online stock trading protocol with signed contracts for each trade. Suppose the broker starts the protocol, sending her commitment to sell stock to the buyer at a specific price, and the buyer responds with his commitment. To ensure timely termination, the broker also enjoys the ability to abort the exchange by contacting the trusted third party (TTP) if the buyer has not responded. Once the buyer commits to the purchase, he cannot use the committed funds for other purposes. Even if he has the option to contact the TTP immediately, an optimistic buyer will wait for some period of time for the broker to respond, hoping to resolve the transaction amicably and avoid the extra cost or potential delay associated with contacting the TTP. This waiting period may give the broker a useful window of opportunity. Once she has the buyer's commitment, the broker can wait to see if shares are available from a selling customer at a matching or lower price. The longer the buyer is inclined to wait, the greater chance the broker has to pair trades at a profit. If the broker finds the contract unprofitable, she can abort the transaction by falsely claiming to the TTP that the buyer has not responded. This broker strategy succeeds in proportion to the time that the buyer optimistically waits for the broker to continue the protocol; this time interval, if known exactly or approximately, gives the broker a period where she can decide *unilaterally* whether to abort or complete the exchange.

Abuse-Freeness. Since advantage against an optimistic player cannot be eliminated, the most a protocol can do is prevent the opponent from *proving* that he has an advantage. For example, even though the broker in our example has control over deciding whether the sale happens, the protocol may still be able to prevent her from showing the buyer's commitment to other parties. Such protocols have been called *abuse-free* in the literature [18]. We use a formal representation of knowledge derived from epistemic logic [19, 16] to formalize the "ability to prove" and analyze abuse-freeness as the lack of *provable advantage*.

The paper is organized as follows. In Sect. 2, we briefly summarize our semantic framework and define the class of two-party contract signing protocols with trusted third party. In Sect. 3, we formalize protocol properties such as fairness, optimism, and timeliness. In Sect. 4, we formalize optimistic behavior of a participant, and show that the optimistic participant is at a disadvantage in any fair, optimistic, timely protocol. In Sect. 5, we formalize provable advantage and abuse-freeness. Related work is discussed in Sect. 6. We summarize our results in Sect. 7.

2 Model

2.1 Multiset Rewriting Formalism

Our protocol formalism is multiset rewriting with existential quantification, MSR [10], which can be seen as an extension of some standard models of computation, *e.g.*, multiset transformation [4] and chemical abstract machine [6]. This formalism faithfully expresses the underlying assumptions of the untimed, nondeterministic, asynchronous model. A protocol definition in MSR defines the *set of all possible execution traces* for any instance of the protocol. A number of other formalisms that do so, such as [1,15] and others, would have suited our purposes as well, and in this sense our main results are independent of the MSR formalism. The synchronous model with a global clock does not seem appropriate for our investigation because fixed-round contract signing protocols in the literature [2,3,18,23,26] do not rely on a global clock.

MSR syntax involves terms, facts, and rules. To specify a protocol, first choose a vocabulary, or *first-order signature*. We assume that our vocabulary contains some basic sorts such as $public\_key$ for public keys and $mssg$ for protocol messages. As usual, the *terms* over a signature are the well-formed expressions produced by applying functions to arguments of the correct sort. A *fact* is a first-order atomic formula over the chosen signature, without free variables. Therefore, a fact is the result of applying a predicate symbol to ground terms of the correct sort. A *state* is a finite multiset of facts.

A state transition is a *rule* written using two multisets of first-order atomic formulas and existential quantification, in the syntactic form $F_1, \ldots, F_k \longrightarrow \exists x_1 \ldots \exists x_j . G_1, \ldots G_n$. The meaning of this rule is that if some state S contains facts obtained by a ground substitution σ from first-order atomic formulas $F_1, \ldots, F_k$, then one possible next state is the state S^* that is similar to S, but with facts obtained by σ from $F_1, \ldots, F_k$ removed and facts obtained by σ from $G_1, \ldots, G_n$ added, where $x_1, \ldots, x_j$ are replaced by new symbols. If there are free variables in the rule $F_1, \ldots, F_k \longrightarrow \exists x_1 \ldots \exists x_j.G_1, \ldots$ G_n, these are treated as universally quantified throughout the rule. In an application of a rule, these variables may be replaced by any ground terms.

As an example, consider state $\{P(f(a)), P(b)\}$ and rule $P(x) \longrightarrow \exists z.Q(f(x), z)$. The next state is obtained by instantiating this rule to $P(f(a)) \longrightarrow \exists z.Q(f(f(a)), z)$. Applying the rule, we choose a new value c for z and replace $P(f(a))$ by $Q(f(f(a)), c)$, obtaining the new state $\{Q(f(f(a)), c), P(b)\}$.

Timer signals. In our model, timers are interpreted as *local* signals, used by participants to decide when to quit waiting for a message from the other party in the protocol. They do not refer to any global time or imply synchronicity. Timers are formalized by binary *timer predicates*, whose first argument is of the sort $public\_key$ and identifies the participant who receives its signal, while the second argument is one of the following three constants of the sort $timer\_state$: $unset$, set, and $timed\_out$.

Cryptography. Contract signing protocols usually employ cryptographic primitives. In general, the purpose of cryptography is to provide messages that are meaningful to some parties, but not subject to arbitrary (non-polynomial-time) computation by others. For example, encryption provides messages that are meaningful to any recipient with the decryption key, but not subject to decryption by any agent who does not possess the decryption key. The logic-based formalism of MSR cannot capture subtle distinctions between, for example, functions computable with high probability and functions computable with low or negligible probability. Instead, we must model functions as either feasibly computable, or not feasibly computable.

For each operation used in a protocol, we assume there is some MSR characterization of its computability properties. To give a concrete framework for presenting these rules, let us assume some set of predicates $\mathcal{P} = \{P^\alpha | \alpha \text{ is any sort}\}$. Since the sort α is determined by the sort of the arguments to P^α, we will not write the sort when it is either irrelevant, or clear from context. Intuitively, a rule of the form

$$P(s_1), \ldots, P(s_m), F_1, \ldots, F_j \longrightarrow P(t_1), \ldots, P(t_n), F_1, \ldots, F_j$$

means that if an agent possesses data $s_1, \ldots, s_m$, then under conditions specified by facts $F_1, \ldots, F_j$, it is computationally feasible for him to also learn $t_1, \ldots, t_n$. For example, here are the familiar "Dolev-Yao" [13,25] rules given in [10]:

$$P(x), P(k) \longrightarrow P(\text{encrypt}(k, x))$$
$$P(\text{encrypt}(k, x)), P(k^{-1}), \text{Keypair}(k, k^{-1}) \longrightarrow P(x)$$

In the remainder of the paper, we assume some fixed theory **Possess** of rules that characterize the computationally feasible operations on messages. As a disclaimer, we emphasize that the results in this paper are accurate statements about a protocol using cryptographic primitives only to the extent that **Possess** accurately characterizes the computationally feasible operations. In particular, protocols that distinguish between low-order polynomial computation and high-order polynomial computation, or rely on probabilistic operations in some essential way, may fall outside the scope of our analysis and may conceivably violate some of our results.

2.2 Protocol Model

We say that a protocol P is a *contract signing protocol* if it involves three parties, O (originator), R (responder), and T (trusted third party), and the goal of the protocol is to enable O (respectively, R) to obtain R's signature (respectively, O's signature) on some pre-agreed text. For brevity, we will say *signature* as a shorthand for "signature on the pre-agreed text," use terms *contract signing* and *signature exchange* interchangeably, and refer to O and R as *signers*. We specify the protocol by a set of MSR rules, which we call a *theory*. Any sequence of rules consistent with the theory corresponds to a valid execution trace of a protocol instance. If execution traces are naturally arranged in trees, then the MSR theory defines the set of all possible execution traces as a forest of trees. To obtain the impossibility result, we choose *any* contract signing protocol P and fix it. We assume that the contract text for each instance contains a unique identifier, and consider only a single instance of P.

A protocol theory $\mathbf{P}$ for the given protocol is the disjoint union of six theories: $\mathbf{O}, \mathbf{R}, \mathbf{T_0}, \mathbf{O_{timeouts}}, \mathbf{R_{timeouts}}$, and $\mathbf{T_{timeouts}}$. We will refer to $\mathbf{O}, \mathbf{R}, \mathbf{T_0}$ as *role theories*. Each role theory specifies one of the protocol roles by giving a finite list of *role state predicates* that define the internal states of the participant playing that role and the rules for advancing from state to state. Role theory also contains another, disjoint list of *timer predicates* describing the rules for the participant's timers. A participant may advance his state by "looking" at the state of his timers or the network (*i.e.*, a timer or a network predicate appears on the left side of the rule). He may also set his timer by changing the timer's state from $unset$ to set, but he may not change it to $timed\_out$.

A *timeout rule* is a rule of the form $Z(k, set) \to Z(k, timed\_out)$, where k is the public key of the participant associated with timer Z. In the protocol theory, $\mathbf{O_{timeouts}}$, $\mathbf{R_{timeouts}}$, and $\mathbf{T_{timeouts}}$ are the sets of *timeout rules* for all timers of O, R, and T, respectively. For simplicity, we will combine the role theory and the timeouts of T, and call it $\mathbf{T} = \mathbf{T_0} \cup \mathbf{T_{timeouts}}$.

Communication. Following the standard assumption that the adversary controls the network and records all messages, we model communication between O and R by a unary *network predicate* N whose argument is of the sort $mssg$. Once a fact $N(m)$ for some m is added to the state, it is never removed. As in contract signing protocols in the literature [3,18], we assume that channels between signers and T are inaccessible to the adversary and separate from the network between O and R (by contrast, [20] considers security of contract signing protocols under relaxed assumptions about channel security). Channels between signers and T are modeled by ternary $TTPchannel$ *predicates*, whose arguments are of the sort $public\_key$, $public\_key$ and $mssg$, respectively. For example, $tc_o(k_o, k_t, m)$ models the channel between O and T carrying message m.

Threat Model. We are interested in guarantees provided by contract signing protocols when one of the signers misbehaves in an arbitrary way. T is assumed to be honest. We will call the misbehaving signer the *adversary*. The adversary does not necessarily follow the protocol, and may ignore the state of the timers or stop prematurely. He may gather messages from the network, store them, decompose them into fragments and construct new messages from the fragments. These abilities are formalized by theories

$\mathbf{O_{threat}}$ and $\mathbf{R_{threat}}$ containing *dishonest rules* for O and R, respectively. Each rule models a particular dishonest operation.

A *protocol definition* consists of the protocol theory $\mathbf{P}$, theories $\mathbf{O_{threat}}$ and $\mathbf{R_{threat}}$, **Possess** theory which models computationally feasible operations on messages, and the *initial set of facts* S_0 which contains the initial states of all participants and timers. Formal definition of protocol theory can be found in appendix A. Non-probabilistic fixed-round contract signing protocols in the literature such as [3,18] can all be defined in this way.

2.3 Traces and Continuation Trees

A *state* is a finite multiset of facts. For example, the initial state S_0 may include facts $O_0(k_o, k_o^{-1}, k_r, p)$ and $R_0(k_r, k_r^{-1}, k_o, p)$ modeling the initial states of the originator and the responder in protocol p: each knows his own public and private keys, and the opponent's public key. A *trace* from state S is a chain of nodes, with the root labeled by S, each node labeled by a state, and each edge labeled by a triple $\langle t, \sigma, \mathbf{Q}\rangle$. Here Q is one of $\{\mathbf{O}, \mathbf{R}, \mathbf{T}, \mathbf{O_{timeouts}}, \mathbf{R_{timeouts}}, \mathbf{O_{threat}}, \mathbf{R_{threat}}\}$, $t \in \mathbf{Q}$ is a state transition rule, and σ is a ground substitution. If $\langle t, \sigma, \mathbf{Q}\rangle$ labels the edge from a node labeled by S_1 to a node labeled by S_2, it must be the case that the application of t to $S_1\sigma$ produces S_2. Any state labeling a node in this chain is said to be *reachable from S*. We will simply say that a state is *reachable* if it is reachable from the initial state S_0.

An edge is a *dishonest move of O* if it is labeled by some $t \in \mathbf{O_{threat}}$. O is said be *honest in the trace* if there are no dishonest moves of O in the trace. If S is reachable by a trace in which O is honest, then S is *reachable by honest O*. The definitions for R are symmetric. Assuming that dishonest participants, if any, make only a finite number of dishonest moves, let *continuation tree ctr* at state S be the finite tree of all possible traces from S. This tree can be thought of as a game tree that represents the complete set of possible plays. Let $ctr_{[O]}$ be the tree obtained from *ctr* by removing all edges in $\mathbf{O} \cup \mathbf{O_{threat}}$ along with their descendants. Intuitively, $ctr_{[O]}$ is the set of all possible plays if O stops participating in the protocol. Definition of $ctr_{[R]}$ is symmetric. We will say that any edge e in *ctr* that is labeled by a rule in $\mathbf{O}$ or $\mathbf{O_{threat}}$ (respectively, $\mathbf{R}$ or $\mathbf{R_{threat}}$), is under *O's control* (respectively, R's control). To model optimism of honest signers (see Sect. 4), we will also assume that some edges in $\mathbf{O_{timeouts}} \cup \mathbf{R_{timeouts}}$ are under control of the dishonest participant.

3 Properties of Contract Signing Protocols

MSR definition of the protocol defines the set of all possible execution traces in the form of a continuation tree. To define protocol properties such as fairness, optimism, timeliness, and advantage, we view the continuation tree as a game tree containing all possible plays, and adapt the notion of strategy from classical game theory.

For the remainder of the paper, we will assume that only one of the signers is honest. We will use A to refer to the honest signer, *i.e.*, A refers to either O, or R, depending on which of them is honest. We'll use B to refer to the other, dishonest signer.

3.1 Strategies

Following [11], we formalize strategies as truncated continuation trees. Given a set of edges E, let $ctr\backslash E$ be the tree obtained from continuation tree ctr by removing the edges in E along with their descendants. Intuitively, if E is a subset of edges of ctr under A's control, then $ctr\backslash E$ is the set of possible plays that result if A does not use transitions in E. Similarly, we can define $ctr_{[A]}\backslash E$ (recall that $ctr_{[A]}$ is the tree of all plays if A stops participating in the protocol).

Definition 1. *Let S be a reachable state and let ctr be the continuation tree from S. Let $X \subseteq \{A, B, T\}$.*
1. If E is a subset of edges of ctr such that each edge in E is under the control of some $p \in X$, then $ctr\backslash E$ is said to be a strategy for the coalition X. *If there are no dishonest moves of any $p \in X$ in $ctr\backslash E$, then $ctr\backslash E$ is said to be an* honest strategy.
2. If E is a subset of edges of $ctr_{[A]}$ such that each edge in E is under the control of some $p \in X$, then $ctr_{[A]}\backslash E$ is said to be an A-silent strategy for the coalition X.

This definition corresponds to the standard game-theoretic notion of strategy. E represents the plays that the coalition X considers unfavorable, and $ctr\backslash E$ represents the continuations that X prefers. At any given state S' in $ctr\backslash E$, an edge coming out of the node labeled by S' indicates the next move for X in accordance with the strategy $ctr\backslash E$. If the edge is not under X's control, then the next move for X is idling, *i.e.*, waiting for others to move.

To define fairness and other properties, we are interested in strategies in which the coalition X drives the protocol to a state in which some property holds:

Definition 2. *If there is a strategy $ctr\backslash E$ from S for coalition X such that all leaf nodes of $ctr\backslash E$ are labeled by states S' that satisfy some property $\phi(S')$, then X* has a strategy from S to reach a state in which ϕ holds.

The definition for A-silent strategies is similar.

Since the players' objective in the game is to obtain each other's signatures, we are interested in the states where A possesses B's signature and the ones where B possesses A's signature. Formally, B *possesses* some term u in a reachable state S if u is derivable, using the rules in **Possess**, from the terms in B's internal role state predicate B_i in S and B's additional memory in S given to him by the threat model. Possession is always monotonic. The definition for A is symmetric, except that the threat model does not have to be considered.

Definition 3. *If there is a strategy for coalition X such that all leaf nodes in the strategy are labeled by states in which A possesses B's signature, then X* has a strategy from S to give A B's signature. *Moreover, if $X = \{A\}$, then A is said to have a* strategy to obtain B's signature.

3.2 Fairness, Optimism, Timeliness, and Advantage

We now use the notion of strategy to define what it means for a contract signing protocol to be fair, optimistic, and timely, and what it means for a participant to enjoy an advantage.

The definitions are quite subtle. For example, we need to draw the distinction between a *strategy* for achieving some outcome, and a *possibility* that the outcome will happen under the right circumstances. This requires introduction of a four-valued variable to characterize the degree of each player's control over the protocol game.

Fairness. Fairness is the basic symmetry property of an exchange protocol. There is a known impossibility result [14,24] demonstrating that no deterministic two-party protocol can be fair. Therefore, fairness requires introduction of at least one other party, *e.g.*, the trusted third party T. Our definition is equivalent to a common definition of fairness in terms of state reachability [18,11]. Intuitively, a protocol is fair for the honest signer A, if, whenever B has obtained A's signature, A has a strategy in coalition with T to obtain B's signature.

Definition 4. *A protocol is* fair *for honest A if, for each state S reachable by honest A such that B possesses A's signature in S, the coalition of A and T has an honest strategy from S to give A B's signature for all bounds on the number of moves that a dishonest B makes.*

Advantage. Intuitively, fairness says that either both players obtain what they want, or neither does. This is not always sufficient, however. A player's ability to decide unilaterally *whether* the transaction happens or not can be of great value in scenarios where resource commitment is important, such as online trading and auction bidding.

To characterize the degree to which each participant controls the outcome of the protocol in a given state, we now define a pair of values $rslv_A, rslv_B$ associated with each reachable state. We are interested in what a participant *may* do in the worse possible case. Therefore, despite our assumption that A is honest, we will consider A's dishonest moves when reasoning about A's ability to control the outcome.

Definition 5. *Define $rslv_A$ for any reachable state S as follows:*

$rslv_A(S) = 2$, *if A has a strategy to obtain B's signature for all bounds on the number of dishonest moves of B,*

$= 1$, *if $rslv_A(S) \neq 2$, but A has a B-silent strategy to reach state S' such that $rslv_A(S') = 2$,*

$= \frac{1}{2}$, *if $rslv_A(S) \neq \{1, 2\}$, but there is state S' reachable from S such that $rslv_A(S') = 2$, and no transition on the $S \to S'$ path is in $\mathbf{B} \cup \mathbf{B_{threat}}$,*

$= 0$, *otherwise.*

The strategies need not be honest. Definition of $rslv_B$ is symmetric.

Intuitively, $rslv_B(S) = 2$ if B can obtain A's signature no matter what A does, 1 if B can obtain A's signature provided A stops communicating and remains silent, $\frac{1}{2}$ if there is a possibility (but no strategy) for B to obtain A's signature when A is silent, and 0 means that B cannot obtain A's signature without A's involvement. The difference between 1 and $\frac{1}{2}$ is essential. For example, $rslv_B(S) = 1$ if B can obtain A's signature by sending a message to T as long as A is silent, while $rslv_B(S) = \frac{1}{2}$ if A is silent, but some previously sent message is already on the channel to T, and the outcome of the protocol depends on the race condition between this message and B's message.

Given an initial state S_0, we assume that $rslv_A(S_0) = rslv_B(S_0) = 0$. The signature exchange problem is not meaningful otherwise.

Definition 6. *B has an* abort strategy *in S if B has a strategy to reach a state S' such that $rslv_A(S') = 0$. B has a* resolve strategy *in S if B has a strategy to reach a state S'' such that $rslv_B(S'') = 2$. B has an* advantage *in S if B has both an abort strategy and a resolve strategy.*

If B has an advantage in S, then A does not have an advantage in S, and vice versa.

Optimism. Intuitively, a protocol is optimistic if it enables two honest parties to exchange signatures without involving the trusted third party, assuming they do not time out waiting for each other's messages. Such protocols potentially provide a practical means of fair exchange between mistrusting agents without relying on a third party in most instances.

We say that *A does not send a message to T* in the transition from S to S' if (i) the transition is an application of a rule in $\mathbf{A} \cup \mathbf{A_{threat}}$, and (ii) no fact created by the transition matches a term in the left hand side of a rule in $\mathbf{T}$.

Definition 7. *A fair protocol is* optimistic *for B if, assuming A is honest and B controls the timeouts of both A and B, B has an honest strategy at S_0 such that*
1) no messages are sent by any signer to T;
2) every leaf node is labeled by a state in which B possesses A's signature;
3) there is a trace from S_0 to a leaf node that involves only the transitions in $\mathbf{A} \cup \mathbf{B}$.

Any trace in this strategy is an optimistic trace. *Definition of* optimistic for A *is symmetric. A protocol is optimistic if it is optimistic for both signers.*

Our definition of optimism implies that the protocol specification does not permit honest participants to contact T nondeterministically, *i.e.*, every rule that results in a message sent to T is conditional on some timer timing out.

Timeliness. We now formalize the following intuition [3]: "one player cannot force the other to wait for any length of time — a fair and timely termination can always be forced by contacting the third party." Timeliness has been emphasized by the designers of fair exchange protocols, since it is essential for practical use. In any state of the protocol, each participant should be able to terminate the exchange unilaterally. If he has not been able to obtain the other's signature, he can always reach a terminal state where he can stop and be sure that the opponent will not be able to obtain his signature, either.

Definition 8. *A fair, optimistic protocol is* timely *for B if in every state on an optimistic trace B has an A-silent strategy to reach a state S' such that $rslv_A(S') = 0$ or $rslv_B(S') = 2$. A protocol is timely if it is timely for both signers.*

To illustrate the importance of timeliness, consider a protocol that is *not* timely, *e.g.*, Boyd-Foo protocol [8]. In this protocol, originator O releases some information that can be used by responder R to obtain O's signature from T at some later point. If R stops communicating, O is at his mercy. He may have to wait, possibly forever, before he learns whether the exchange has been successful.

For the rest of this paper, we assume that the protocol is fair, timely, and optimistic for both signers.

4 Impossibility of Balance in Optimistic Protocols

As explained in the introduction, optimistic contract signing protocols are only valuable insofar as they offer benefit to an optimistic participant. We say that the honest participant A is *optimistic* if, in any state where he is permitted by the protocol specification to contact trusted third party T, he waits for B's response before contacting T.

The propensity of the optimistic participant to wait for the opponent's response before contacting T can be exploited by the opponent. Recall that definition 7 implies that an honest participant only contacts T after some timer times out. We use this to model optimism of A by giving B the ability to schedule the timeout rules of A by an "out-of-band" signal. In any implementation of the protocol, B does not actually schedule A's timers. This is simply a technical device to restrict the set of execution traces under consideration to those that may occur when one of the participants is optimistic.

Definition 6 can thus be extended to cases where A is optimistic by permitting B's strategy to include control over timeouts of A and B. If B does not have a strategy for reaching a state where he has an advantage over an optimistic A, we say that the protocol is *balanced* for an optimistic A. As we will now show, balance cannot be achieved by any fair, timely, optimistic protocol.

The first observation underlying our proof is that, in the interleaving semantics of concurrency used by our model, the order of application of state transition rules that affect independent parts of the system can be commuted. The second observation is that the strategies available to the dishonest player are not negatively affected by messages sent to him by the honest player or by the honest player's timeouts because the dishonest player is free to ignore both.

We start with an auxiliary proposition, which follows directly from definition 5.

Proposition 1. *Let $S \to S'$ be a state transition* not *in $\mathbf{B} \cup \mathbf{B}_{\mathbf{threat}}$. If $rslv_B(S) = 2$, then $rslv_B(S') = 2$. If $rslv_A(S) = 0$, then $rslv_A(S') = 0$.*

Proposition 1 implies that if $rslv_A(S) = 0$ and $rslv_A(S') > 0$, then the $S \to S'$ transition must be in $\mathbf{B} \cup \mathbf{B}_{\mathbf{threat}}$. Similarly, if $rslv_B(S) = 0$ and $rslv_B(S') > 0$, then $S \to S'$ is in $\mathbf{A} \cup \mathbf{A}_{\mathbf{threat}}$. Intuitively, a player acquires some degree of control over the outcome of the protocol for the first time only because of the other player's move.

Just like we defined $ctr_{[A]}$ to be the tree obtained from ctr by removing all edges in $\mathbf{A} \cup \mathbf{A}_{\mathbf{threat}}$, we define $ctr_{[A+]}$ to be the tree obtained from ctr by removing all edges in $\mathbf{A} \cup \mathbf{A}_{\mathbf{threat}} \cup \mathbf{A}_{\mathbf{timeouts}}$. If E is a selection of edges in $ctr_{[A+]}$ under B's control, then $ctr_{[A+]} \backslash E$ is a strategy available to B if A remains silent *and* no timers time out. We will call such a strategy *weak A-silent strategy*.

Proposition 2. *Let $S \to S'$ be a state transition in $\mathbf{A}_{\mathbf{timeouts}}$. B has a weak A-silent abort [resolve] strategy at S' if and only if B has a weak A-silent abort [resolve] strategy at S.*

The proof of proposition 2 relies on the fact that the moves of B and T that constitute a weak A-silent strategy cannot depend on the state of A's timers.

Proposition 3. *B has an A-silent abort [resolve] strategy at S if and only if B has a weak A-silent abort [resolve] strategy at S.*

In the proof, we use proposition 2 to construct an A-silent strategy from a weak A-silent strategy by induction on the height of the continuation tree. Proposition 3 establishes that the strategies available to the dishonest player are not negatively affected by the honest player's timeouts. We now show that they are not affected by the honest player's messages to the dishonest player.

Lemma 1. *Let $S \to S'$ be a transition in $\mathbf{A} \cup \mathbf{A_{threat}}$. If B has an A-silent abort [resolve] strategy in S, and A does not send a message to T in the $S \to S'$ transition, then B has an A-silent abort [resolve] strategy in S'.*

Proof. The proof, illustrating our general proof techniques, is in appendix B.

We use lemma 1 to show that for each strategy conditional on A remaining silent, there is an equivalent strategy in which A is not silent, but B simply ignores A's messages. The strategy works as long as A does not try to talk to T.

Lemma 2. *If B has an A-silent abort [resolve] strategy at S, and A does not send any messages to T, then B has an abort [resolve] strategy.*

Proof. (Omitted for space reasons).

We now show that a strategy conditional on A not talking to T works against an optimistic A since he waits for B's messages instead of trying to contact T.

Lemma 3. *Let S be a state that does not contain $Z(k, timed\_out)$ for any timer predicate Z. If B has an A-silent abort [resolve] strategy in state S, then B has an abort [resolve] strategy against optimistic A in S.*

Proof. (Sketch) Definition 7 implies that an optimistic A contacts T only when some timer times out. B controls the timeouts of an optimistic A. Hence B can prevent A from sending any message to T. We then apply lemma 2.

Theorem 1 (Impossibility of Balance). *Let $\mathbf{P}$ be a fair, optimistic, timely protocol between signers A and B. If A is optimistic, then there is a non-initial state S^* such that B has an advantage against an optimistic A at S^*.*

Proof. (Sketch) By definition 7, there is an optimistic trace from the initial state S_0 which contains only the transitions in $\mathbf{A} \cup \mathbf{B}$ and leads to S' such that $rslv_B(S') = 2$. Consider the first transition $S \to S^*$ on this trace such that $rslv_B(S) = 0, rslv_B(S^*) > 0$. Proposition 1 implies that this must be a transition in $\mathbf{A} \cup \mathbf{A_{threat}}$. By definition 7, A does not send a message to T anywhere in the trace, including this transition.

By definition 8, B has an A-silent strategy to reach a state S'' such that $rslv_A(S'') = 0$ or $rslv_B(S'') = 2$. Since $rslv_B(S) = 0$, it must be the case that $rslv_A(S'') = 0$, *i.e.*, B has an A-silent abort strategy. By lemma 1, B has an A-silent abort strategy in S^*. Therefore, by lemma 3, B has an abort strategy against optimistic A in S^*.

By definition 7, B has a strategy at S_0 to obtain A's signature since B controls the timeouts of A and B. Because S^* is reached a part of this strategy (recall that the $S \to S^*$ transition is on an optimistic trace), B also has a strategy to obtain A's signature at S^*. Hence B has a resolve strategy against optimistic A in S^*. Since B has both abort and resolve strategies, B has an advantage against an optimistic A in S^*. □

We'd like to emphasize that the result of theorem 1 is *not* a trivial "second-mover" advantage. A and B are not protocol roles, but simply notation for the honest and dishonest participant, respectively. An optimistic participant is at a disadvantage *regardless* of the role he plays in the protocol. Even if he chooses the responder role, he will lose control over the outcome of the protocol at some point as long as he remains optimistic. For example, in Garay *et al.*'s abuse-free contract signing protocol [18], the originator enjoys the advantage over the responder, even though the responder is the first to receive information that potentially enables him to obtain the originator's signature.

5 Abuse-Free Protocols and Provable Advantage

Theorem 1 states that any fair, optimistic, timely protocol necessarily provides a dishonest participant with control over the outcome against an optimistic opponent. The problem may be alleviated by ensuring that no participant can *prove* to an outside party that he controls the outcome. Such protocols have been called *abuse-free* in the literature [18], and concrete protocols [3,18] have been constructed using zero-knowledge cryptographic techniques such as verifiable signature escrows and designated-verifier proofs. To formalize "ability to prove," we rely on a knowledge-theoretic framework borrowed from epistemic logic [19,16].

Reasoning about knowledge. Given a participant P and a reachable state S, let P's *view* of S be the submultiset of S containing all the facts corresponding to role states in the role theory of P, timers of P and messages on P's channels to other participants. Intuitively, this set represents all that P may observe in S. Given a trace tr from the initial state S_0 to S, construct a new labeled chain by relabeling the nodes by P's view of S. Relabel the edges not associated with P by ϵ, which indicates that somebody other than P may have moved. Since P cannot observe other players' moves, insert an ϵ between any two consecutive edges labeled by rules of P (duplicate the node connecting these edges) as well as at the start and end of the trace. If there are two or more consecutive ϵ edges, but P's view does not change when moving across one of them, then delete that edge. The resulting chain tr' is called P's *observation* of the protocol, $Obsv_P(S, tr)$. Intuitively, P's observation is just P's own history in the trace.

In the spirit of algorithmic knowledge [16,22], observations $Obsv_P(S, tr)$ and $Obsv_P(S^*, tr^*)$ are equivalent if they are computationally indistinguishable by P.

Definition 9. *Given a trace tr from S_0 ending in S, we say that P* knows *in (S, tr) that logical formula F is true if*
i) F is true in S, and
ii) for each trace tr^* from S_0 to S^* such that $Obsv_P(S^*, tr^*)$ is indistinguishable by P from $Obsv_P(S, tr)$, F is true in S^*.

Intuitively, P *knows* that F is true if F holds in all possible executions of the protocol consistent with P's observations.

Abuse-freeness. To reason about abuse-freeness, we augment the protocol with an outside party C and consider his knowledge at different stages of the protocol. C does not possess the signers' or the third party's private keys, and obtains all of his evidence about the protocol from one of the protocol participants, *e.g.*, B, who forwards arbitrary messages to C in an attempt to cause C to *know* that A is participating in the protocol.

Definition 10. *B has* provable advantage *against A in state S if*
i) B has an advantage over A at S, and
ii) B can provide information, derived from the protocol execution up to S, that causes C to know *that A is participating in the protocol.*
A protocol is abuse-free *for A if B has no provable advantage in any reachable state.*

Definition 10 is weaker than one might expect. If B enjoys an advantage at S, then in order for B to enjoy provable advantage, B merely has to prove A's participation in the protocol. B may succeed even if his protocol with A is already over. But since we are concerned with making the protocol as safe as possible for an optimistic A, the weaker definition is acceptable since it makes abuse-freeness (its negation) stronger. Combining theorem 1 and definition 10, we obtain

Corollary 1. *In any fair, optimistic, timely, abuse-free protocol between A and B, there is a trace tr from S_0 to state S such that*
i) B has an advantage over optimistic A at S,
ii) C does not know in (S, tr) that A is participating in the protocol, i.e., *there is another trace tr^* from S_0 to some S^* such that $Obsv_C(S^*, tr^*)$ is indistinguishable by C from $Obsv_C(S, tr)$, and A is not participating in tr^*.*

6 Related Work

Previous game-theoretic approaches to the study of fair exchange [11,20,21] focused on formalizing fairness for the strongest possible honest player without taking optimism into account. In [20], fairness is formalized as the existence of a defense strategy for the honest player, which is not sufficient if the honest player faces nondeterministic choices in the protocol, as is the case in the abuse-free protocol of Garay *et al.* [18]. Another game-theoretic model was developed in [9], but it focuses mainly on economic equilibria in fair exchange. Cryptographic proofs of correctness by protocol designers [2,3,18] focus on basic fairness and ignore the issues of optimism and fundamental asymmetry of communication between the signers and the trusted third party.

To the best of our knowledge, we are the first to apply an epistemic logic framework to formalize the "ability to prove" and thus abuse-freeness. In [27], belief logic SVO is used to reason about correctness of the non-repudiation protocol [26], but it is not clear how belief logics might apply to fairness and abuse-freeness. [21] models advantage, but not the concepts of proof and knowledge, which we believe provide a more compelling characterization of abuse-freeness.

7 Conclusions and Further Work

We have studied contract signing protocols in a game-theoretic model, giving precise, formal definitions of properties such as fairness and timeliness. We characterized optimism of honest protocol participants using a form of out-of-band signal that forces the optimistic player to wait for the opponent. While the out-of-band signal does not correspond to any realistic mechanism in distributed computation, it accurately reduces the set of protocol traces to those where the optimistic player waits for the opponent instead of contacting the trusted third party.

Our main result is that in any fair, optimistic, timely protocol, an optimistic player yields an advantage to his opponent. This means that the opponent has both a strategy to complete the signature exchange and a strategy to keep the players from obtaining each other's signatures. Since the protocol is fair, the outcome for both players is the same, but the player with an advantage can choose what this outcome is. This holds regardless of whether the optimistic player is the first or second mover.

Since advantage cannot be eliminated, the best a protocol can do to protect optimistic participants is prevent the opponent from proving to any outside party that he has reached a position of advantage. This property is known as abuse-freeness. We define abuse-freeness using the concept of algorithmic knowledge adapted from epistemic logic to formalize what it means to "prove" something to an outside observer.

One direction for further investigation involves the notion of trusted third party accountability. The relationship between our definitions and the cryptographic definitions of fairness [3] may also merit further study. Finally, we believe that our techniques will prove useful for investigating multi-party contract signing protocols.

Acknowledgments. We are particularly grateful to D. Malkhi for pointing out the vulnerability of optimistic players in fair exchange. We also thank I. Cervesato, S. Even, D. Gollmann, S. Kremer, J.F. Raskin, C. Meadows, and J. Millen for interesting and helpful discussions.

References

1. M. Abadi and A. Gordon. A calculus for cryptographic protocols: the spi-calculus. *Information and Computation*, 143:1–70, 1999.
2. N. Asokan, M. Schunter, and M. Waidner. Optimistic protocols for fair exchange. In *Proc. 4th ACM Conf. on Computer and Communications Security*, pages 7–17, 1997.
3. N. Asokan, V. Shoup, and M. Waidner. Optimistic fair exchange of digital signatures. *IEEE Journal on Selected Areas in Communications*, 18(4):593–610, 2000.
4. J. Banatre and D. Le Metayer. Computing by multiset transformation. *Communications of the ACM (CACM)*, 36(1):98–111, 1993.
5. M. Ben-Or, O. Goldreich, S. Micali, and R. L. Rivest. A fair protocol for signing contracts. *IEEE Transactions on Information Theory*, 36(1):40–46, 1990.
6. G. Berry and D. Boudol. The chemical abstract machine. *Theoretical Computer Science*, 96(1):217–248, 1992.
7. D. Boneh and M. Naor. Timed commitments and applications. In *Proc. CRYPTO 'OO*, pages 236–254, 2000.
8. C. Boyd and E. Foo. Off-line fair payment protocols using convertible signatures. In *Proc. ASIACRYPT '98*, pages 271–285, 1998.
9. L. Buttyán and J.-P. Hubaux. Toward a formal model of fair exchange — a game theoretic approach. Technical Report SSC/1999/39, Swiss Federal Institute of Technology (EPFL), Lausanne, Switzerland, December 1999.

10. I. Cervesato, N. Durgin, P. D. Lincoln, J. C. Mitchell, and A. Scedrov. A meta-notation for protocol analysis. In *Proc. 12th IEEE Computer Security Foundations Workshop*, pages 55–69, 1999.
11. R. Chadha, M. Kanovich, and A. Scedrov. Inductive methods and contract signing protocols. In *Proc. 8th ACM Conf. on Computer and Communications Security*, pages 176–185, 2001.
12. I. B. Damgård. Practical and provably secure release of a secret and exchange of signatures. *J. Cryptology*, 8(4):201–222, 1995.
13. D. Dolev and A. Yao. On the security of public-key protocols. In *Proc. 22nd Annual IEEE Symposium on Foundations of Computer Science*, pages 350–357, 1981.
14. S. Even and Y. Yacobi. Relations among public key signature schemes. Technical Report 175, Computer Science Dept. Technion, Israel, March 1980.
15. F.J. Thayer Fábrega, J. Herzog, and J. Guttman. Strand spaces: Why is a security protocol correct? In *Proc. IEEE Symposium on Security and Privacy*, pages 160–171, 1998.
16. R. Fagin, J. Halpern, Y. Moses, and M. Vardi. *Reasoning about Knowledge*. MIT Press, 1995.
17. M. Fischer, N. Lynch, and M. Patterson. Impossibility of distributed consensus with one faulty process. *JACM*, 32(2):374–382, 1985.
18. J. Garay, M. Jakobsson, and P. MacKenzie. Abuse-free optimistic contract signing. In *Proc. CRYPTO '99*, pages 449–466, 1999.
19. J. Hintikka. *Knowledge and Belief*. Cornell University Press, 1962.
20. S. Kremer and J.-F. Raskin. A game-based verification of non-repudiation and fair exchange protocols. In *Proc. CONCUR '01*, pages 551–565, 2001.
21. S. Kremer and J.-F. Raskin. Game analysis of abuse-free contract signing. In *Proc. 15th IEEE Computer Security Foundations Workshop*, pages 206–220, 2002.
22. R. Pucella and J. Halpern. Modeling adversaries in a logic for security protocol analysis. In *Formal Aspects of Security, 2002 (FASec '02)*.
23. O. Markowitch and S. Saeednia. Optimistic fair exchange with transparent signature recovery. In *Proc. 5th International Conf. on Financial Cryptography*, pages 339–350, 2001.
24. H. Pagnia and F. Gaertner. On the impossibility of fair exchange without a trusted third party. Technical Report TUD-BS-1999-02, Department of Computer Science, Darmstadt University of Technology, Germany, March 1999.
25. T.Y.C. Woo and S.S. Lam. A semantic model for authentication protocols. In *Proc. IEEE Symposium on Security and Privacy*, pages 178–194, 1993.
26. J. Zhou and D. Gollmann. A fair non-repudiation protocol. In *Proc. IEEE Symposium on Security and Privacy*, pages 55–61, 1996.
27. J. Zhou and D. Gollmann. Towards verification of non-repudiation protocols. In *Proc. International Refinement Workshop and Formal Methods Pacific*, pages 370–380, 1998.

A Role and Protocol Theories

We assume that the vocabulary contains the following basic sorts: PK (for public keys), M (for messages), C (for pre-agreed contract texts), PI (for protocol instances), and UI (for globally unique instance identifiers, since we assume that each protocol instance has such an identifier). We also assume a function $\langle \_, \_, \_, \_, \_ \rangle : PK \times PK \times PK \times C \times UI \to PI$, *i.e.*, a protocol instance is determined by the signers' public key, the key of the trusted third party, pre-agreed contract text, the and unique identifier. For example, $p = \langle k_o, k_r, k_t, m, n \rangle$ describes a protocol instance, identified as n, in which signers with public keys k_o and k_r exchange signatures on the pre-agreed text m with the help of the trusted third party whose key is k_t.

Definition 11. *Theory* **A** *is a* role theory for participant A with public key k_a, *where* k_a *is a constant of the sort* PK, *if it satisfies the following:*
i) A *includes a finite list of predicates* $A_0, \ldots, A_n$, *called* role state predicates, *and a finite list of timer predicates, called* timers *of A. The two lists are disjoint.*
ii) A_0 *is a binary predicate whose arguments are of the sort* PK *and* PI, *respectively. We call* A_0 *the* initial role state predicate.
iii) For each rule $l \to r$ *in* **A**,

1. *There is exactly one occurrence of a role state predicate in* l, *say* A_i, *and exactly one occurrence of a role state predicate in* r, *say* A_j. *Furthermore, it is the case that* $i < j$. *If* A_0 *occurs in* l, *then* $A_0(k_a, p) \in l$ *for some term* p *of the sort* PI.
2. *If* A_j *is a k-ary role state predicate occurring in* l, *and* A_j *is an* m*-ary role state predicate occurring in* r, *then* $m > k$. *Furthermore, if* $A_i(u_1, \ldots, u_k) \in l$ *and* $A_j(v_1, \ldots, v_m) \in l$, *then* u_q *and* v_q *are the same terms for all* $1 \leq q \leq k$.
3. *Let* $A_i(u_1, \ldots, u_m) \in l$, $A_j(v_1, \ldots, v_m) \in r$. *Let* $\mathcal{MSG}$ *be the set of terms* u *such that* $N(u)$ *or* $tc(k_1, k_2, u) \in l$ *for some TTPchannel predicate tc. For each* q, v_q *is derivable from* $u_1, \ldots, u_m$ *and* $\mathcal{MSG}$ *using the rules in* **Possess**.
4. *For each timer* Z *of* A,
   i) l *and* r *each contain at most one occurrence of* Z. *Occurrences are of the form* $Z(k_a, ts)$, *where* ts *is a constant of the sort* $timer\_state$. *If* Z *occurs in* r, *then it occurs in* l.
   ii) If $Z(k_a, unset) \in l$, *then either* $Z(k_a, unset) \in r$, *or* $Z(k_a, set) \in r$.
   iii) If $Z(k_a, set) \in l$, *then* $Z(k_a, set) \in r$.
   iv) If $Z(k_a, timed\_out) \in l$, *then* $Z(k_a, timed\_out) \in r$.
5. *If* $N(u) \in l$, *where* N *is a network predicate and* u *is term of the sort* M, *then* $N(u) \in r$. *If* $tc(k_1, k_2, u) \in l$, *where* tc *is a TTPchannel predicate, and terms* k_1, k_2, u *are of the sort* PK, PK, M, *respectively, then* $tc(k_1, k_2, u) \in r$.
6. *For any predicate* $\mathcal{P}$ *other than a role state, timer, network, or TTPchannel predicate, atomic formula* $\mathcal{P}(t_1, ..., t_n)$ *has the same occurrences in* l *as in* r.

Definition 12. *If* Z *is a timer of the participant with public key* k_a, *then* $Z(k_a, set) \to Z(k_a, timed\_out)$ *is the* timeout rule *of* Z.

Definition 13. *Theory* **P** *is a* protocol theory *for signers* O *and* R *and trusted third party* T *with public keys* k_o, k_r, k_t, *respectively, where* k_o, k_r, k_t *are constants of the sort* PK, *if* $\mathbf{P} = \mathbf{O} \uplus \mathbf{R} \uplus \mathbf{T_0} \uplus \mathbf{O_{timeouts}} \uplus \mathbf{R_{timeouts}} \uplus \mathbf{T_{timeouts}}$, *where*

1. $\mathbf{O}, \mathbf{R}, \mathbf{T_0}$ *are role theories for, respectively,* O, R, T *with public keys* k_o, k_r, k_t.
2. *At most one TTPchannel predicate, say* tc_o, *occurs in* **O**. *Each occurrence of* tc_o *is of the form* $tc_o(k_o, k_t, m)$, *where* m *is of the sort* M, *and* tc_o *may not occur in* **R**.
3. *At most one TTPchannel predicate, say* tc_r, *occurs in* **R**. *Each occurrence of* tc_r *is of the form* $tc_r(k_r, k_t, m)$, *where* m *is of the sort* M, *and* tc_r *may not occur in* **O**.
4. *If some TTPchannel predicate occurs in* $\mathbf{T_0}$, *then it also occurs in* **O** *or* **R**.
5. *The role state predicates and the timers of* O *(respectively,* R*) do not occur in* **R** *(respectively,* **O***) and* $\mathbf{T_0}$. *The role state predicates and the timers of* T *do not occur in* **O** *or* **R**.
6. $\mathbf{O_{timeouts}}$, $\mathbf{R_{timeouts}}$, *and* $\mathbf{T_{timeouts}}$ *are the sets of timeout rules of all timers of* O, R, *and* T, *respectively.*

B Proof of Lemma 1

Proof. We rely on the observation that state transition rules affecting independent parts of the system may be commuted. Intuitively, moves of B and T are independent of A's internal state. Therefore, as long as A does not send any messages to T, B may ignore any message sent to him by A and follow the same strategy in S' as in S. In light of proposition 3, all we need to show is that B has a weak A-silent abort [resolve] strategy at S' if B has a weak A-silent abort [resolve] strategy at S. We prove this by induction on the height of the continuation tree at S.
Base case: The height of the continuation tree at S is 0. The lemma is vacuously true.
Induction hypothesis: Suppose the lemma is true for all states S such that the height of the continuation tree at S is $\leq n$.
Induction step: Consider state S such that i) the height of the continuation tree at S is $n+1$, and ii) B has a weak A-silent abort [resolve] strategy at S.

Consider the continuation tree at S', and remove all edges that are in $\mathbf{A} \cup \mathbf{A_{threat}} \cup \mathbf{A_{timeouts}}$ along with their descendants. For each remaining edge e from S' to some state S'', let t be the state transition rule labeling e and consider the following cases:

Case 1: $t \in \mathbf{T}$. Since no message is sent to T in the $S \to S'$ transition, t can be applied at S as well, resulting in some state $\hat{S}$. Observe that:
i) the height of the continuation tree at $\hat{S}$ is $\leq n$;
ii) B has a weak A-silent strategy at $\hat{S}$;
iii) S'' can be obtained from $\hat{S}$ by the same transition that labels $S \to S'$: simply commute $S \to S'$ and $S' \to S''$ transitions.
By the induction hypothesis, B has a weak A-silent strategy at S''. Replace the continuation tree at S'' by this strategy.

Case 2: $t \in \mathbf{B} \cup \mathbf{B_{threat}}$. There are three possibilities:
2.1) t cannot be applied at S. Remove edge e along with its descendants.
2.2) t can be applied at S, but it is not a part of the A-silent strategy at S. Remove edge e along with its descendants.
2.3) t can be applied at S, and it is a part of the A-silent strategy at S. Then, as in Case 1, replace the continuation tree at S'' by this strategy.

Case 3: $t \in \mathbf{B_{timeouts}}$. If t is not a part of the A-silent strategy at S, remove edge e along with its descendants. If it is a part of the A-silent strategy, replace the continuation tree at S'' by this strategy.

By constructing the right continuation tree for any immediate descendant of S', we have constructed a weak A-silent strategy at S'. It remains to show that it is indeed an abort [resolve] strategy. There are two possibilities :

Case A: The height of the constructed strategy is 0. From the construction, it follows that the height of the weak A-silent abort [resolve] strategy at S is also 0. Therefore, $rslv_A(S) = 0$ [$rslv_B(S) = 2$]. By proposition 1, $rslv_A(S') = 0$ [$rslv_B(S') = 2$].

Case B: The height of the constructed strategy is > 0. By construction, all leaf nodes are labeled by states S^* such that $rslv_A(S^*) = 0$ [$rslv_B(S^*) = 2$].

We conclude that B has a weak A-silent abort [resolve] strategy at S', which completes the induction. □

Full Abstraction for HOPLA

Mikkel Nygaard[1] and Glynn Winskel[2]

[1] BRICS*, University of Aarhus
[2] Computer Laboratory, University of Cambridge

Abstract. A fully abstract denotational semantics for the higher-order process language HOPLA is presented. It characterises contextual and logical equivalence, the latter linking up with simulation. The semantics is a clean, domain-theoretic description of processes as downwards-closed sets of computation paths: the operations of HOPLA arise as syntactic encodings of canonical constructions on such sets; full abstraction is a direct consequence of expressiveness with respect to computation paths; and simple proofs of soundness and adequacy shows correspondence between the denotational and operational semantics.

1 Introduction

HOPLA (Higher-Order Process LAnguage [19]) is an expressive language for higher-order nondeterministic processes. It has a straightforward operational semantics supporting a standard bisimulation congruence, and can directly encode calculi like CCS, higher-order CCS and mobile ambients with public names. The language came out of work on a linear domain theory for concurrency, based on a categorical model of linear logic and associated comonads [4,18], the comonad used for HOPLA being an exponential ! of linear logic.

The denotational semantics given in [19] interpreted processes as presheaves. Here we consider a "path semantics" for HOPLA which allows us to characterise operationally the distinguishing power of the notion of computation path underlying the presheaf semantics (in contrast to the distinguishing power of the presheaf structure itself). Path semantics is similar to trace semantics [10] in that processes denote downwards-closed sets of computation paths and the corresponding notion of process equivalence, called *path equivalence*, is given by equality of such sets; computation paths, however, may have more structure than traditional traces. Indeed, we characterise contextual equivalence for HOPLA as path equivalence and show that this coincides with logical equivalence for a fragment of Hennessy-Milner logic which is characteristic for simulation equivalence in the case of image-finite processes [8].

To increase the expressiveness of HOPLA (for example, to include the type used in [24] for CCS with late value-passing), while still ensuring that every operation in the language has a canonical semantics, we decompose the "prefix-sum" type $\Sigma_{\alpha \in A} \alpha . \mathbb{P}_\alpha$ in [19] into a sum type $\Sigma_{\alpha \in A} \mathbb{P}_\alpha$ and an anonymous action

* Basic Research in Computer Science (`www.brics.dk`), funded by the Danish National Research Foundation.

R. Amadio, D. Lugiez (Eds.): CONCUR 2003, LNCS 2761, pp. 383–398, 2003.

prefix type $!\mathbb{P}$. The sum type, also a product, is associated with injection ("tagging") and projection term constructors, βt and $\pi_\beta t$ for $\beta \in A$. The prefix type is associated with constructions of prefixing $!t$ and prefix match $[u > !x \Rightarrow t]$, subsuming the original terms $\beta.t$ and $[u > \beta.x \Rightarrow t]$ using $\beta!t$ and $[\pi_\beta u > !x \Rightarrow t]$.

In Sect. 2 we present a domain theory of path sets, used in Sect. 3 to give a fully abstract denotational semantics to HOPLA. Section 4 presents the operational semantics of HOPLA, essentially that of [19], and relates the denotational and operational semantics with pleasingly simple proofs of soundness and adequacy. Section 5 concludes with a discussion of related and future work.

2 Domain Theory from Path Sets

In the path semantics, processes are intuitively represented as collections of their computation paths. Paths are elements of preorders $\mathbb{P}, \mathbb{Q}, \ldots$ called *path orders* which function as process types, each describing the set of possible paths for processes of that type together with their sub-path ordering. A process of type $\mathbb{P}$ is then represented as a downwards-closed subset $X \subseteq \mathbb{P}$, called a *path set.* Path sets $X \subseteq \mathbb{P}$ ordered by inclusion form the elements of the poset $\widehat{\mathbb{P}}$ which we'll think of as a domain of meanings of processes of type $\mathbb{P}$.

The poset $\widehat{\mathbb{P}}$ has many interesting properties. First of all, it is a complete lattice with joins given by union. In the sense of Hennessy and Plotkin [7], $\widehat{\mathbb{P}}$ is a "nondeterministic domain", with joins used to interpret nondeterministic sums of processes. Accordingly, given a family $(X_i)_{i \in I}$ of elements of $\widehat{\mathbb{P}}$, we sometimes write $\Sigma_{i \in I} X_i$ for their join. A typical finite join is written $X_1 + \cdots + X_k$ while the empty join is the empty path set, the inactive process, written $\varnothing$.

A second important property of $\widehat{\mathbb{P}}$ is that any $X \in \widehat{\mathbb{P}}$ is the join of certain "prime" elements below it; $\widehat{\mathbb{P}}$ is a *prime algebraic complete lattice* [16]. Primes are down-closures $y_{\mathbb{P}} p = \{p' : p' \leq_{\mathbb{P}} p\}$ of individual elements $p \in \mathbb{P}$, representing a process that may perform the computation path p. The map $y_{\mathbb{P}}$ reflects as well as preserves order, so that $p \leq_{\mathbb{P}} p'$ iff $y_{\mathbb{P}} p \subseteq y_{\mathbb{P}} p'$, and $y_{\mathbb{P}}$ thus "embeds" $\mathbb{P}$ in $\widehat{\mathbb{P}}$. We clearly have $y_{\mathbb{P}} p \subseteq X$ iff $p \in X$ and prime algebraicity of $\widehat{\mathbb{P}}$ amounts to saying that any $X \in \widehat{\mathbb{P}}$ is the union of its elements:

$$X = \bigcup_{p \in X} y_{\mathbb{P}} p \ . \qquad (1)$$

Finally, $\widehat{\mathbb{P}}$ is characterised abstractly as the *free join-completion* of $\mathbb{P}$, meaning (i) it is join-complete and (ii) given any join-complete poset C and a monotone map $f : \mathbb{P} \to C$, there is a unique join-preserving map $f^\dagger : \widehat{\mathbb{P}} \to C$ such that the diagram on the left below commutes.

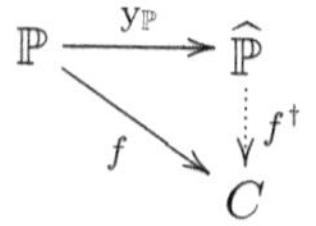

$$f^\dagger X = \bigcup_{p \in X} f p \ . \qquad (2)$$

We call $f^\dagger$ *the extension of* f *along* $y_{\mathbb{P}}$. Uniqueness of $f^\dagger$ follows from (1).

Notice that we may instantiate C to any poset of the form $\widehat{\mathbb{Q}}$, drawing our attention to join-preserving maps $\widehat{\mathbb{P}} \to \widehat{\mathbb{Q}}$. By the freeness property (2), join-preserving maps $\widehat{\mathbb{P}} \to \widehat{\mathbb{Q}}$ are in bijective correspondence with monotone maps $\mathbb{P} \to \widehat{\mathbb{Q}}$. Each element Y of $\widehat{\mathbb{Q}}$ can be represented using its "characteristic function", a monotone map $f_Y : \mathbb{Q}^{\mathrm{op}} \to \mathbf{2}$ from the opposite order to the simple poset $0 < 1$ such that $Y = \{q : f_Y q = 1\}$ and $\widehat{\mathbb{Q}} \cong [\mathbb{Q}^{\mathrm{op}}, \mathbf{2}]$. Uncurrying then yields the following chain:

$$[\mathbb{P}, \widehat{\mathbb{Q}}] \cong [\mathbb{P}, [\mathbb{Q}^{\mathrm{op}}, \mathbf{2}]] \cong [\mathbb{P} \times \mathbb{Q}^{\mathrm{op}}, \mathbf{2}] = [(\mathbb{P}^{\mathrm{op}} \times \mathbb{Q})^{\mathrm{op}}, \mathbf{2}] \cong \widehat{\mathbb{P}^{\mathrm{op}} \times \mathbb{Q}} \ . \tag{3}$$

So the order $\mathbb{P}^{\mathrm{op}} \times \mathbb{Q}$ provides a function space type. We'll now investigate what additional type structure is at hand.

2.1 Linear and Continuous Categories

Write **Lin** for the category with path orders $\mathbb{P}, \mathbb{Q}, \ldots$ as objects and join-preserving maps $\widehat{\mathbb{P}} \to \widehat{\mathbb{Q}}$ as arrows. It turns out **Lin** has enough structure to be understood as a categorical model of Girard's linear logic [5,22]. Accordingly, we'll call arrows of **Lin** *linear* maps.

Linear maps are represented by elements of $\widehat{\mathbb{P}^{\mathrm{op}} \times \mathbb{Q}}$ and so by downwards-closed subsets of the order $\mathbb{P}^{\mathrm{op}} \times \mathbb{Q}$. This relational presentation exposes an involution central in understanding **Lin** as a categorical model of classical linear logic. The involution of linear logic, yielding $\mathbb{P}^{\perp}$ on an object $\mathbb{P}$, is given by $\mathbb{P}^{\mathrm{op}}$; clearly, downwards-closed subsets of $\mathbb{P}^{\mathrm{op}} \times \mathbb{Q}$ correspond to downwards-closed subsets of $(\mathbb{Q}^{\mathrm{op}})^{\mathrm{op}} \times \mathbb{P}^{\mathrm{op}}$, showing how maps $\mathbb{P} \to \mathbb{Q}$ correspond to maps $\mathbb{Q}^{\perp} \to \mathbb{P}^{\perp}$ in **Lin**. The tensor product of $\mathbb{P}$ and $\mathbb{Q}$ is given by the product of preorders $\mathbb{P} \times \mathbb{Q}$; the singleton order $\mathbb{1}$ is a unit for tensor. Linear function space $\mathbb{P} \multimap \mathbb{Q}$ is then obtained as $\mathbb{P}^{\mathrm{op}} \times \mathbb{Q}$. Products $\mathbb{P} \,\&\, \mathbb{Q}$ are given by $\mathbb{P} + \mathbb{Q}$, the disjoint juxtaposition of preorders. An element of $\widehat{\mathbb{P} \,\&\, \mathbb{Q}}$ can be identified with a pair (X, Y) with $X \in \widehat{\mathbb{P}}$ and $Y \in \widehat{\mathbb{Q}}$, which provides the projections $\pi_1 : \mathbb{P} \,\&\, \mathbb{Q} \to \mathbb{P}$ and $\pi_2 : \mathbb{P} \,\&\, \mathbb{Q} \to \mathbb{Q}$ in **Lin**. More general, not just binary, products $\&_{i \in I} \mathbb{P}_i$ with projections π_j, for $j \in I$, are defined similarly. From the universal property of products, a collection of maps $f_i : \mathbb{P} \to \mathbb{P}_i$, for $i \in I$, can be tupled together to form a unique map $\langle f_i \rangle_{i \in I} : \mathbb{P} \to \&_{i \in I} \mathbb{P}_i$ with the property that $\pi_j \circ \langle f_i \rangle_{i \in I} = f_j$ for all $j \in I$. The empty product is given by the empty order $\mathbb{O}$ and, as the terminal object, is associated with unique maps $\varnothing_{\mathbb{P}} : \mathbb{P} \to \mathbb{O}$, constantly $\varnothing$, for any path order $\mathbb{P}$. All told, **Lin** is a $*$-autonomous category, so a symmetric monoidal closed category with a dualising object, and has finite products as required by Seely's definition of a model of linear logic [22].

In fact, **Lin** also has all coproducts, also given on objects $\mathbb{P}$ and $\mathbb{Q}$ by the juxtaposition $\mathbb{P} + \mathbb{Q}$ and so coinciding with products. Injection maps $in_1 : \mathbb{P} \to \mathbb{P} + \mathbb{Q}$ and $in_2 : \mathbb{Q} \to \mathbb{P} + \mathbb{Q}$ in **Lin** derive from the obvious injections into the disjoint sum of preorders. The empty coproduct is the empty order $\mathbb{O}$ which is then a zero object. This collapse of products and coproducts highlights that **Lin** has arbitrary *biproducts*. Via the isomorphism $\mathbf{Lin}(\mathbb{P}, \mathbb{Q}) \cong \widehat{\mathbb{P}^{\mathrm{op}} \times \mathbb{Q}}$, each homset of **Lin** can be seen as a commutative monoid with neutral element the always $\varnothing$ map, itself written $\varnothing : \mathbb{P} \to \mathbb{Q}$, and sum given by union, written $+$. Composition in **Lin** is bilinear in that, given $f, f' : \mathbb{P} \to \mathbb{Q}$ and $g, g' : \mathbb{Q} \to \mathbb{R}$, we have $(g + g') \circ (f + f') = g \circ f + g \circ f' + g' \circ f + g' \circ f'$. Further, given a family of objects $(\mathbb{P}_\alpha)_{\alpha \in A}$, we have for each $\beta \in A$ a diagram

$$\mathbb{P}_\beta \underset{in_\beta}{\overset{\pi_\beta}{\rightleftarrows}} \Sigma_{\alpha\in A}\mathbb{P}_\alpha \quad \text{such that} \quad \begin{aligned} \pi_\beta \circ in_\beta &= 1_{\mathbb{P}_\beta} \ , \\ \pi_\beta \circ in_\alpha &= \varnothing \text{ if } \alpha \neq \beta, \text{ and} \\ \Sigma_{\alpha \in A}(in_\alpha \circ \pi_\alpha) &= 1_{\Sigma_{\alpha\in A}\mathbb{P}_\alpha} \ . \end{aligned} \tag{4}$$

Processes of type $\Sigma_{\alpha\in A}\mathbb{P}_\alpha$ may intuitively perform computation paths in any of the component path orders $\mathbb{P}_\alpha$.

We see that **Lin** is rich in structure. But linear maps alone are too restrictive. Being join-preserving, they in particular preserve the empty join. So, unlike e.g. prefixing, linear maps always send the inactive process $\varnothing$ to itself. Looking for a broader notion of maps between nondeterministic domains we follow the discipline of linear logic and consider *non-linear* maps, i.e. maps whose domain is under an exponential, !. One choice of a suitable exponential for **Lin** is got by taking $!\mathbb{P}$ to be the preorder obtained as the free finite-join completion of $\mathbb{P}$. Concretely, $!\mathbb{P}$ can be defined to have finite subsets of $\mathbb{P}$ as elements with ordering given by $\preceq_\mathbb{P}$, defined for arbitrary subsets X, Y of $\mathbb{P}$ as follows:

$$X \preceq_\mathbb{P} Y \iff_{\text{def}} \forall p \in X. \exists q \in Y. p \leq_\mathbb{P} q \ . \tag{5}$$

When $!\mathbb{P}$ is quotiented by the equivalence induced by the preorder we obtain a poset which is the free finite-join completion of $\mathbb{P}$. By further using the obvious inclusion of this completion into $\widehat{\mathbb{P}}$, we get a map $i_\mathbb{P} : !\mathbb{P} \to \widehat{\mathbb{P}}$ sending a finite set $\{p_1, \ldots, p_n\}$ to the join $y_\mathbb{P} p_1 + \cdots + y_\mathbb{P} p_n$. Such finite sums of primes are the finite (isolated, compact) elements of $\widehat{\mathbb{P}}$. The map $i_\mathbb{P}$ assumes the role of $y_\mathbb{P}$ above. For any $X \in \widehat{\mathbb{P}}$ and $P \in !\mathbb{P}$, we have $i_\mathbb{P} P \subseteq X$ iff $P \preceq_\mathbb{P} X$, and X is the directed join of the finite elements below it:

$$X = \bigcup_{P \preceq_\mathbb{P} X} i_\mathbb{P} P \ . \tag{6}$$

Further, $\widehat{\mathbb{P}}$ is the *free directed-join completion* of $!\mathbb{P}$ (also known as the ideal completion of $!\mathbb{P}$). This means that given any monotone map $f : !\mathbb{P} \to C$ for some directed-join complete poset C, there is a unique directed-join preserving (i.e. Scott continuous) map $f^\ddagger : \widehat{\mathbb{P}} \to C$ such that the diagram below commutes.

$$\begin{array}{ccc} !\mathbb{P} & \xrightarrow{i_\mathbb{P}} & \widehat{\mathbb{P}} \\ & {\scriptstyle f}\searrow & \vdots {\scriptstyle f^\ddagger} \\ & & C \end{array} \qquad f^\ddagger X = \bigcup_{P \preceq_\mathbb{P} X} f P \ . \tag{7}$$

Uniqueness of $f^\ddagger$, called the *extension of f along $i_\mathbb{P}$*, follows from (6). As before, we can replace C by a nondeterministic domain $\widehat{\mathbb{Q}}$ and by the freeness properties (2) and (7), there is a bijective correspondence between linear maps $!\mathbb{P} \to \mathbb{Q}$ and continuous maps $\widehat{\mathbb{P}} \to \widehat{\mathbb{Q}}$.

We define the category **Cts** to have path orders $\mathbb{P}, \mathbb{Q}, \ldots$ as objects and continuous maps $\widehat{\mathbb{P}} \to \widehat{\mathbb{Q}}$ as arrows. These arrows allow more process operations, including prefixing, to be expressed. The structure of **Cts** is induced by that of **Lin** via an adjunction between the two categories.

2.2 An Adjunction

As linear maps are continuous, **Cts** has **Lin** as a sub-category, one which shares the same objects. We saw above that there is a bijection

$$\mathbf{Lin}(!\mathbb{P}, \mathbb{Q}) \cong \mathbf{Cts}(\mathbb{P}, \mathbb{Q}) \ . \tag{8}$$

This is in fact natural in $\mathbb{P}$ and $\mathbb{Q}$ so an adjunction with the inclusion $\mathbf{Lin} \hookrightarrow \mathbf{Cts}$ as right adjoint. Via (7) the map $\mathrm{y}_{!\mathbb{P}} : !\mathbb{P} \to \widehat{!\mathbb{P}}$ extends to a map $\eta_{\mathbb{P}} = \mathrm{y}_{!\mathbb{P}}^{\ddagger} : \mathbb{P} \to !\mathbb{P}$ in **Cts**. Conversely, $i_{\mathbb{P}} : !\mathbb{P} \to \widehat{\mathbb{P}}$ extends to a map $\varepsilon_{\mathbb{P}} = i_{\mathbb{P}}^{\dagger} : !\mathbb{P} \to \mathbb{P}$ in **Lin** using (2). These maps are the unit and counit, respectively, of the adjunction:

$$\eta_{\mathbb{P}} X = \textstyle\bigcup_{P \preceq_{\mathbb{P}} X} \mathrm{y}_{!\mathbb{P}} P \qquad \varepsilon_{\mathbb{P}} X = \textstyle\bigcup_{P \in X} i_{\mathbb{P}} P \tag{9}$$

The left adjoint is the functor $! : \mathbf{Cts} \to \mathbf{Lin}$ given on arrows $f : \mathbb{P} \to \mathbb{Q}$ by $(\eta_{\mathbb{Q}} \circ f \circ i_{\mathbb{P}})^{\dagger} : !\mathbb{P} \to !\mathbb{Q}$. The bijection (8) then maps $g : !\mathbb{P} \to \mathbb{Q}$ in **Lin** to $\bar{g} = g \circ \eta_{\mathbb{P}} : \mathbb{P} \to \mathbb{Q}$ in **Cts** while its inverse maps $f : \mathbb{P} \to \mathbb{Q}$ in **Cts** to $\bar{f} = \varepsilon_{\mathbb{Q}} \circ !f$ in **Lin**. We call $\bar{g}$ and $\bar{f}$ the *transpose* of g and f, respectively; of course, transposing twice yields back the original map. As **Lin** is a sub-category of **Cts**, the counit is also a map in **Cts**. We have $\varepsilon_{\mathbb{P}} \circ \eta_{\mathbb{P}} = 1_{\mathbb{P}}$ and $1_{!\mathbb{P}} \leq \eta_{\mathbb{P}} \circ \varepsilon_{\mathbb{P}}$ for all objects $\mathbb{P}$.

Right adjoints preserve products, and so **Cts** has products given as in **Lin**. Hence, **Cts** is a symmetric monoidal category like **Lin**, and in fact, our adjunction is symmetric monoidal. In detail, there are isomorphisms of path orders,

$$k : \mathbb{1} \cong !\mathbb{O} \quad \text{and} \quad m_{\mathbb{P},\mathbb{Q}} : !\mathbb{P} \times !\mathbb{Q} \cong !(\mathbb{P} \,\&\, \mathbb{Q}) \ , \tag{10}$$

with $m_{\mathbb{P},\mathbb{Q}}$ mapping a pair $(P, Q) \in !\mathbb{P} \times !\mathbb{Q}$ to the union $in_1 P \cup in_2 Q$; any element of $!(\mathbb{P} \,\&\, \mathbb{Q})$ can be written on this form. These isomorphisms induce isomorphisms with the same names in **Lin** with m natural. Moreover, k and m commute with the associativity, symmetry and unit maps of **Lin** and **Cts**, such as $s^{\mathbf{Lin}}_{\mathbb{P},\mathbb{Q}} : \mathbb{P} \times \mathbb{Q} \cong \mathbb{Q} \times \mathbb{P}$ and $r^{\mathbf{Cts}}_{\mathbb{Q}} : \mathbb{Q} \,\&\, \mathbb{O} \cong \mathbb{Q}$, making ! symmetric monoidal. It then follows [13] that the inclusion $\mathbf{Lin} \hookrightarrow \mathbf{Cts}$ is symmetric monoidal as well, and that the unit and counit are monoidal transformations. Thus, there are maps

$$l : \mathbb{O} \to \mathbb{1} \quad \text{and} \quad n_{\mathbb{P},\mathbb{Q}} : \mathbb{P} \,\&\, \mathbb{Q} \to \mathbb{P} \times \mathbb{Q} \tag{11}$$

in **Cts**, with n natural, corresponding to k and m above; l maps $\varnothing$ to $\{*\}$ while $n_{\mathbb{P},\mathbb{Q}}$ is the extension $h^{\ddagger}$ of the map $h(in_1 P \cup in_2 Q) = i_{\mathbb{P}} P \times i_{\mathbb{Q}} Q$. Also, the unit makes the diagrams below commute and the counit satisfies similar properties.

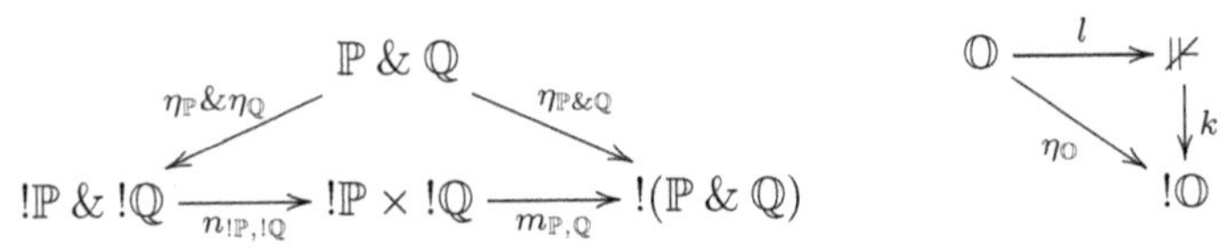

(12)

The diagram on the left can be written as $str_{\mathbb{P},\mathbb{Q}} \circ (1_{\mathbb{P}} \,\&\, \eta_{\mathbb{Q}}) = \eta_{\mathbb{P}\&\mathbb{Q}}$ where *str*, the *strength* of ! viewed as a monad on **Cts**, is the natural transformation

$$\mathbb{P} \,\&\, !\mathbb{Q} \xrightarrow{\eta_{\mathbb{P}}\&1_{!\mathbb{Q}}} !\mathbb{P} \,\&\, !\mathbb{Q} \xrightarrow{n_{!\mathbb{P},!\mathbb{Q}}} !\mathbb{P} \times !\mathbb{Q} \xrightarrow{m_{\mathbb{P},\mathbb{Q}}} !(\mathbb{P} \,\&\, \mathbb{Q}) \ . \tag{13}$$

Finally, recall that the category **Lin** is symmetric monoidal closed so that the functor $(\mathbb{Q} \multimap -)$ is right adjoint to $(- \times \mathbb{Q})$ for any object $\mathbb{Q}$. Together with the natural isomorphism m this provides a right adjoint $(\mathbb{Q} \to -)$, defined by $(!\mathbb{Q} \multimap -)$, to the functor $(- \,\&\, \mathbb{Q})$ in **Cts** via the chain

$$\begin{aligned}\mathbf{Cts}(\mathbb{P} \,\&\, \mathbb{Q}, \mathbb{R}) &\cong \mathbf{Lin}(!(\mathbb{P} \,\&\, \mathbb{Q}), \mathbb{R}) \cong \mathbf{Lin}(!\mathbb{P} \times !\mathbb{Q}, \mathbb{R}) \\ &\cong \mathbf{Lin}(!\mathbb{P}, !\mathbb{Q} \multimap \mathbb{R}) \cong \mathbf{Cts}(\mathbb{P}, !\mathbb{Q} \multimap \mathbb{R}) = \mathbf{Cts}(\mathbb{P}, \mathbb{Q} \to \mathbb{R})\end{aligned} \tag{14}$$

—natural in $\mathbb{P}$ and $\mathbb{R}$. This demonstrates that **Cts** is cartesian closed, as is well known. The adjunction between **Lin** and **Cts** now satisfies the conditions put forward by Benton for a categorical model of intuitionistic linear logic, strengthening those of Seely [1,22]; see also [13] for a recent survey of such models.

3 Denotational Semantics

HOPLA is directly suggested by the structure of **Cts**. The language is typed with types given by the grammar

$$\mathbb{T} ::= \mathbb{T}_1 \to \mathbb{T}_2 \mid \Sigma_{\alpha \in A} \mathbb{T}_\alpha \mid !\mathbb{T} \mid T \mid \mu_j \vec{T}.\vec{\mathbb{T}} \ . \tag{15}$$

The symbol T is drawn from a set of type variables used in defining recursive types; closed type expressions are interpreted as path orders. Using vector notation, $\mu_j \vec{T}.\vec{\mathbb{T}}$ abbreviates $\mu_j T_1, \ldots, T_k.(\mathbb{T}_1, \ldots, \mathbb{T}_k)$ and is interpreted as the j-component, for $1 \le j \le k$, of "the least" solution to the defining equations $T_1 = \mathbb{T}_1, \ldots, T_k = \mathbb{T}_k$, in which the expressions $\mathbb{T}_1, \ldots, \mathbb{T}_k$ may contain the T_j's. We shall write $\mu \vec{T}.\vec{\mathbb{T}}$ as an abbreviation for the k-tuple with j-component $\mu_j \vec{T}.\vec{\mathbb{T}}$, and confuse a closed expression for a path order with the path order itself. Simultaneous recursive equations for path orders can be solved using information systems [21,12]. Here, it will be convenient to give a concrete, inductive characterisation based on a language of *paths*:

$$p, q ::= P \mapsto q \mid \beta p \mid P \mid abs\, p \ . \tag{16}$$

Above, P ranges over finite sets of paths. We use $P \mapsto q$ as notation for pairs in the function space $(!\mathbb{P})^{\mathrm{op}} \times \mathbb{Q}$. The language is complemented by formation rules using judgements $p : \mathbb{P}$, meaning that p belongs to $\mathbb{P}$, displayed below on top of rules defining the ordering on $\mathbb{P}$ using judgements $p \le_{\mathbb{P}} p'$. Recall that $P \preceq_{\mathbb{P}} P'$ means $\forall p \in P. \exists p' \in P'.\ p \le_{\mathbb{P}} p'$.

$$\frac{P : !\mathbb{P} \quad q : \mathbb{Q}}{P \mapsto q : \mathbb{P} \to \mathbb{Q}} \qquad \frac{p : \mathbb{P}_\beta \quad \beta \in A}{\beta p : \Sigma_{\alpha \in A} \mathbb{P}_\alpha} \qquad \frac{p_1 : \mathbb{P} \cdots p_n : \mathbb{P}}{\{p_1, \ldots, p_n\} : !\mathbb{P}} \qquad \frac{p : \mathbb{T}_j[\mu \vec{T}.\vec{\mathbb{T}}/\vec{T}]}{abs\, p : \mu_j \vec{T}.\vec{\mathbb{T}}}$$

$$\frac{P' \preceq_{\mathbb{P}} P \quad \le_{\mathbb{Q}} q'}{P \mapsto q \le_{\mathbb{P} \to \mathbb{Q}} P' \mapsto q'} \qquad \frac{p \le_{\mathbb{P}_\beta} p'}{\beta p \le_{\Sigma_{\alpha \in A} \mathbb{P}_\alpha} \beta p'} \qquad \frac{P \preceq_{\mathbb{P}} P'}{P \le_{!\mathbb{P}} P'} \qquad \frac{p \le_{\mathbb{T}_j[\mu \vec{T}.\vec{\mathbb{T}}/\vec{T}]} p'}{abs\, p \le_{\mu_j \vec{T}.\vec{\mathbb{T}}} abs\, p'}$$

Using information systems as in [12] yields the same representation, except for the tagging with *abs* in recursive types, done to help in the proof of adequacy in Sect. 4.1. So rather than the straight equality between a recursive type and its unfolding which we are used to from [12], we get an isomorphism $abs : \mathbb{T}_j[\mu\vec{T}.\vec{\mathbb{T}}/\vec{T}] \cong \mu_j\vec{T}.\vec{\mathbb{T}}$ whose inverse we call *rep*.

As an example consider the type of CCS processes given in [19] as the path order $\mathbb{P}$ satisfying $\mathbb{P} = \Sigma_{\alpha\in A}!\mathbb{P}$ where A is a set of CCS actions. The elements of $\mathbb{P}$ then have the form $abs(\beta P)$ where $\beta \in A$ and P is a finite set of paths from $\mathbb{P}$. Intuitively, a CCS process can perform such a path if it can perform the action β and, following that, is able to perform each path in P.

The raw syntax of HOPLA terms is given by

$$t, u ::= x \,|\, rec\, x.t \,|\, \Sigma_{i\in I} t_i \,|\, \lambda x.t \,|\, t\ u \,|\, \beta t \,|\, \pi_\beta t \,|\, !t \,|\, [u > !x \Rightarrow t] \,|\, abs\, t \,|\, rep\, t\ . \qquad (17)$$

The variables x in the terms $rec\, x.t$, $\lambda x.t$, and $[u > !x \Rightarrow t]$ are binding occurrences with scope t. We shall take for granted an understanding of free and bound variables, and substitution on raw terms.

Let $\mathbb{P}_1, \ldots, \mathbb{P}_k, \mathbb{Q}$ be closed type expressions and assume that the variables $x_1, \ldots, x_k$ are distinct. A syntactic judgement $x_1 : \mathbb{P}_1, \ldots, x_k : \mathbb{P}_k \vdash t : \mathbb{Q}$ stands for a map $[\![x_1 : \mathbb{P}_1, \ldots, x_k : \mathbb{P}_k \vdash t : \mathbb{Q}]\!] : \mathbb{P}_1 \& \cdots \& \mathbb{P}_k \to \mathbb{Q}$ in **Cts**. We'll write Γ, or Λ, for an environment list $x_1 : \mathbb{P}_1, \ldots, x_k : \mathbb{P}_k$ and most often abbreviate the denotation to $\mathbb{P}_1 \& \cdots \& \mathbb{P}_k \xrightarrow{t} \mathbb{Q}$, or $\Gamma \xrightarrow{t} \mathbb{Q}$, or even $[\![t]\!]$, suppressing the typing information. When the environment list is empty, the corresponding product is the empty path order $\mathbb{O}$.

The term-formation rules are displayed below alongside their interpretations as constructors on maps of **Cts**, taking the maps denoted by the premises to that denoted by the conclusion (cf. [2]). We assume that the variables in any environment list which appears are distinct.

Structural rules. The rules handling environment lists are given as follows:

$$\frac{}{x : \mathbb{P} \vdash x : \mathbb{P}} \qquad \frac{}{\mathbb{P} \xrightarrow{1_\mathbb{P}} \mathbb{P}} \qquad (18)$$

$$\frac{\Gamma \vdash t : \mathbb{Q}}{\Gamma, x : \mathbb{P} \vdash t : \mathbb{Q}} \qquad \frac{\Gamma \xrightarrow{t} \mathbb{Q}}{\Gamma \& \mathbb{P} \xrightarrow{t \& \varnothing_\mathbb{P}} \mathbb{Q} \& \mathbb{O} \xrightarrow{r^{\mathbf{Cts}}_\mathbb{Q}} \mathbb{Q}} \qquad (19)$$

$$\frac{\Gamma, y : \mathbb{Q}, x : \mathbb{P}, \Lambda \vdash t : \mathbb{R}}{\Gamma, x : \mathbb{P}, y : \mathbb{Q}, \Lambda \vdash t : \mathbb{R}} \qquad \frac{\Gamma \& \mathbb{Q} \& \mathbb{P} \& \Lambda \xrightarrow{t} \mathbb{R}}{\Gamma \& \mathbb{P} \& \mathbb{Q} \& \Lambda \xrightarrow{t \circ (1_\Gamma \& s^{\mathbf{Cts}}_{\mathbb{P},\mathbb{Q}} \& 1_\Lambda)} \mathbb{R}} \qquad (20)$$

$$\frac{\Gamma, x : \mathbb{P}, y : \mathbb{P} \vdash t : \mathbb{Q}}{\Gamma, z : \mathbb{P} \vdash t[z/x, z/y] : \mathbb{Q}} \qquad \frac{\Gamma \& \mathbb{P} \& \mathbb{P} \xrightarrow{t} \mathbb{Q}}{\Gamma \& \mathbb{P} \xrightarrow{1_\Gamma \& \Delta_\mathbb{P}} \Gamma \& \mathbb{P} \& \mathbb{P} \xrightarrow{t} \mathbb{Q}} \qquad (21)$$

In the formation rule for contraction (21), the variable z must be fresh; the map $\Delta_\mathbb{P}$ is the usual diagonal, given as $\langle 1_\mathbb{P}, 1_\mathbb{P} \rangle$.

Recursive definition. Since each $\widehat{\mathbb{P}}$ is a complete lattice, it admits least fixed-points of continuous maps. If $f : \widehat{\mathbb{P}} \to \widehat{\mathbb{P}}$ is continuous, it has a least fixed-point, *fix* $f \in \widehat{\mathbb{P}}$ obtained as $\bigcup_{n\in\omega} f^n(\varnothing)$. Below, *fix* f is the fixpoint in $\mathbf{Cts}(\Gamma,\mathbb{P}) \cong \widehat{\Gamma \to \mathbb{P}}$ of the continuous operation f mapping $g : \Gamma \to \mathbb{P}$ in **Cts** to the composition $[\![t]\!] \circ (1_\Gamma \,\&\, g) \circ \Delta_\Gamma$.

$$\frac{\Gamma, x : \mathbb{P} \vdash t : \mathbb{P}}{\Gamma \vdash rec\, x.t : \mathbb{P}} \qquad \frac{\Gamma \,\&\, \mathbb{P} \xrightarrow{t} \mathbb{P}}{\Gamma \xrightarrow{fix\, f} \mathbb{P}} \tag{22}$$

Nondeterministic sum. Each path order $\mathbb{P}$ is associated with a join operation, $\Sigma : \&_{i\in I}\, \mathbb{P} \to \mathbb{P}$ in **Cts** taking a tuple $\langle t_i \rangle_{i\in I}$ to the join $\Sigma_{i\in I} t_i$ in $\widehat{\mathbb{P}}$. We'll write $\varnothing$ and $t_1 + \cdots + t_k$ for finite sums.

$$\frac{\Gamma \vdash t_j : \mathbb{P} \quad \text{all } j \in I}{\Gamma \vdash \Sigma_{i\in I} t_i : \mathbb{P}} \qquad \frac{\Gamma \xrightarrow{t_j} \mathbb{P} \quad \text{all } j \in I}{\Gamma \xrightarrow{\langle t_i \rangle_{i\in I}} \&_{i\in I}\, \mathbb{P} \xrightarrow{\Sigma} \mathbb{P}} \tag{23}$$

Function space. As noted at the end of Sect. 2.2, the category **Cts** is cartesian closed with function space $\mathbb{P} \to \mathbb{Q}$. Thus, there is a 1-1 correspondence *curry* from maps $\mathbb{P} \,\&\, \mathbb{Q} \to \mathbb{R}$ to maps $\mathbb{P} \to (\mathbb{Q} \to \mathbb{R})$ in **Cts**; its inverse is called *uncurry*. We obtain application, $app : (\mathbb{P} \to \mathbb{Q}) \,\&\, \mathbb{P} \to \mathbb{Q}$ as $uncurry(1_{\mathbb{P}\to\mathbb{Q}})$.

$$\frac{\Gamma, x : \mathbb{P} \vdash t : \mathbb{Q}}{\Gamma \vdash \lambda x.t : \mathbb{P} \to \mathbb{Q}} \qquad \frac{\Gamma \,\&\, \mathbb{P} \xrightarrow{t} \mathbb{Q}}{\Gamma \xrightarrow{curry\, t} \mathbb{P} \to \mathbb{Q}} \tag{24}$$

$$\frac{\Gamma \vdash t : \mathbb{P} \to \mathbb{Q} \quad \Lambda \vdash u : \mathbb{P}}{\Gamma, \Lambda \vdash t\, u : \mathbb{Q}} \qquad \frac{\Gamma \xrightarrow{t} \mathbb{P} \to \mathbb{Q} \quad \Lambda \xrightarrow{u} \mathbb{P}}{\Gamma \,\&\, \Lambda \xrightarrow{t\&u} (\mathbb{P} \to \mathbb{Q}) \,\&\, \mathbb{P} \xrightarrow{app} \mathbb{Q}} \tag{25}$$

Sum type. The category **Cts** does not have coproducts, but we can build a useful sum type out of the biproduct of **Lin**. The properties of (4) are obviously also satisfied in **Cts**, even though the construction is universal only in the subcategory of linear maps because composition is generally not bilinear in **Cts**. We'll write $\mathbb{O}$ and $\mathbb{P}_1 + \cdots + \mathbb{P}_k$ for the empty and finite sum types. The product $\mathbb{P}_1 \,\&\, \mathbb{P}_2$ of [19] with pairing (t, u) and projection terms *fst* t, *snd* t can be encoded, respectively, as the type $\mathbb{P}_1 + \mathbb{P}_2$, and the terms $1t + 2u$ and $\pi_1 t$, $\pi_2 t$.

$$\frac{\Gamma \vdash t : \mathbb{P}_\beta \quad \beta \in A}{\Gamma \vdash \beta t : \Sigma_{\alpha\in A}\mathbb{P}_\alpha} \qquad \frac{\Gamma \xrightarrow{t} \mathbb{P}_\beta \quad \beta \in A}{\Gamma \xrightarrow{t} \mathbb{P}_\beta \xrightarrow{in_\beta} \Sigma_{\alpha\in A}\mathbb{P}_\alpha} \tag{26}$$

$$\frac{\Gamma \vdash t : \Sigma_{\alpha\in A}\mathbb{P}_\alpha \quad \beta \in A}{\Gamma \vdash \pi_\beta t : \mathbb{P}_\beta} \qquad \frac{\Gamma \xrightarrow{t} \Sigma_{\alpha\in A}\mathbb{P}_\alpha \quad \beta \in A}{\Gamma \xrightarrow{t} \Sigma_{\alpha\in A}\mathbb{P}_\alpha \xrightarrow{\pi_\beta} \mathbb{P}_\beta} \tag{27}$$

Prefixing. The adjunction between **Lin** and **Cts** provides a type constructor, $!(-)$, for which the unit $\eta_\mathbb{P} : \mathbb{P} \to\, !\mathbb{P}$ and counit $\varepsilon_\mathbb{P} :\, !\mathbb{P} \to \mathbb{P}$ may interpret term constructors and deconstructors, respectively. The behaviour of $\eta_\mathbb{P}$ with respect to maps of **Cts** fits that of an anonymous prefix operation. We'll say that $\eta_\mathbb{P}$ maps u of type $\mathbb{P}$ to a "prefixed" process $!u$ of type $!\mathbb{P}$; intuitively, the process

$!u$ will be able to perform an action, which we call !, before continuing as u.

$$\frac{\Gamma \vdash u : \mathbb{P}}{\Gamma \vdash !u : !\mathbb{P}} \qquad \frac{\Gamma \xrightarrow{u} \mathbb{P}}{\Gamma \xrightarrow{u} \mathbb{P} \xrightarrow{\eta_{\mathbb{P}}} !\mathbb{P}} \tag{28}$$

By the universal property of $\eta_{\mathbb{P}}$, if t of type $\mathbb{Q}$ has a free variable of type $\mathbb{P}$, and so is interpreted as a map $t : \mathbb{P} \to \mathbb{Q}$ in **Cts**, then the transpose $\bar{t} = \varepsilon_{\mathbb{Q}} \circ !t$ is the unique map $!\mathbb{P} \to \mathbb{Q}$ in **Lin** such that $t = \bar{t} \circ \eta_{\mathbb{P}}$. With u of type $!\mathbb{P}$, we'll write $[u > !x \Rightarrow t]$ for $\bar{t}u$. Intuitively, this construction "tests" or matches u against the pattern $!x$ and passes the results of successful matches for x on to t. Indeed, first prefixing a term u of type $\mathbb{P}$ and then matching yields a successful match u for x as $\bar{t}(\eta_{\mathbb{P}} u) = tu$. By linearity of $\bar{t}$, the possibly multiple results of successful matches are nondeterministically summed together; the denotations of $[\Sigma_{i \in I} u_i > !x \Rightarrow t]$ and $\Sigma_{i \in I}[u_i > !x \Rightarrow t]$ are identical.

The above clearly generalises to the case where u is an open term, but if t has free variables other than x, we need to make use of the strength map (13):

$$\frac{\Gamma, x : \mathbb{P} \vdash t : \mathbb{Q} \quad \Lambda \vdash u : !\mathbb{P}}{\Gamma, \Lambda \vdash [u > !x \Rightarrow t] : \mathbb{Q}} \qquad \frac{\Gamma \,\&\, \mathbb{P} \xrightarrow{t} \mathbb{Q} \quad \Lambda \xrightarrow{u} !\mathbb{P}}{\Gamma \,\&\, \Lambda \xrightarrow{1_\Gamma \& u} \Gamma \,\&\, !\mathbb{P} \xrightarrow{str_{\Gamma,\mathbb{P}}} !(\Gamma \,\&\, \mathbb{P}) \xrightarrow{\bar{t}} \mathbb{Q}} \tag{29}$$

Recursive types. Folding and unfolding recursive types is accompanied by term constructors *abs* and *rep*:

$$\frac{\Gamma \vdash t : \mathbb{T}_j[\mu \vec{T}.\vec{\mathbb{T}}/\vec{T}]}{\Gamma \vdash abs\, t : \mu_j \vec{T}.\vec{\mathbb{T}}} \qquad \frac{\Gamma \xrightarrow{t} \mathbb{T}_j[\mu \vec{T}.\vec{\mathbb{T}}/\vec{T}]}{\Gamma \xrightarrow{t} \mathbb{T}_j[\mu \vec{T}.\vec{\mathbb{T}}/\vec{T}] \xrightarrow{abs} \mu_j \vec{T}.\vec{\mathbb{T}}} \tag{30}$$

$$\frac{\Gamma \vdash t : \mu_j \vec{T}.\vec{\mathbb{T}}}{\Gamma \vdash rep\, t : \mathbb{T}_j[\mu \vec{T}.\vec{\mathbb{T}}/\vec{T}]} \qquad \frac{\Gamma \xrightarrow{t} \mu_j \vec{T}.\vec{\mathbb{T}}}{\Gamma \xrightarrow{t} \mu_j \vec{T}.\vec{\mathbb{T}} \xrightarrow{rep} \mathbb{T}_j[\mu \vec{T}.\vec{\mathbb{T}}/\vec{T}]} \tag{31}$$

3.1 Useful Equivalences

We provide some technical results about the path semantics which are used in the proof of soundness, Proposition 10. Proofs can be found in [20].

Lemma 1 (Substitution). *Suppose $\Gamma, x : \mathbb{P} \vdash t : \mathbb{Q}$ and $\Lambda \vdash u : \mathbb{P}$ with Γ and Λ disjoint. Then $\Gamma, \Lambda \vdash t[u/x] : \mathbb{Q}$ with denotation given by the composition $[\![t]\!] \circ (1_\Gamma \,\&\, [\![u]\!])$.*

Corollary 2. *If $\Gamma, x : \mathbb{P} \vdash t : \mathbb{P}$, then $\Gamma \vdash t[rec\, x.t/x] : \mathbb{P}$ and $[\![rec\, x.t]\!] = [\![t[rec\, x.t/x]]\!]$ so recursion amounts to unfolding.*

Corollary 3. *Application amounts to substitution. In the situation of the substitution lemma, we have $[\![(\lambda x.t)\, u]\!] = [\![t[u/x]]\!]$.*

Proposition 4. *From the properties of the biproduct and by linearity of injections and projections, we get:*

$$\begin{aligned} [\![\pi_\beta(\beta t)]\!] &= [\![t]\!] \\ [\![\pi_\alpha(\beta t)]\!] &= \varnothing \qquad \text{if } \alpha \neq \beta \\ [\![\Sigma_{\alpha\in A}\alpha(\pi_\alpha(t))]\!] &= [\![t]\!] \end{aligned} \qquad \begin{aligned} [\![\beta(\Sigma_{i\in I}t_i)]\!] &= [\![\Sigma_{i\in I}(\beta t_i)]\!] \\ [\![\pi_\beta(\Sigma_{i\in I}t_i)]\!] &= [\![\Sigma_{i\in I}(\pi_\beta t_i)]\!] \end{aligned} \tag{32}$$

Proposition 5. *The prefix match satisfies the properties*

$$\begin{aligned} [\![[!u > !x \Rightarrow t]]\!] &= [\![t[u/x]]\!] \\ [\![[\Sigma_{i\in I}u_i > !x \Rightarrow t]]\!] &= [\![\Sigma_{i\in I}[u_i > !x \Rightarrow t]]\!] \end{aligned} \tag{33}$$

3.2 Full Abstraction

We define a *program* to be a closed term t of type $!\mathbb{O}$. A $(\Gamma,\mathbb{P})$*-program context* C is a term with holes into which a term t with $\Gamma \vdash t : \mathbb{P}$ may be put to form a program $\vdash C(t) : !\mathbb{O}$. The denotational semantics gives rise to a type-respecting contextual preorder [15]:

Definition 6. *Suppose $\Gamma \vdash t_1 : \mathbb{P}$ and $\Gamma \vdash t_2 : \mathbb{P}$. We say that t_1 and t_2 are related by* contextual preorder, *written $t_1 \precsim t_2$, iff for all $(\Gamma,\mathbb{P})$-program contexts C, we have $[\![C(t_1)]\!] \neq \varnothing \implies [\![C(t_2)]\!] \neq \varnothing$. If both $t_1 \precsim t_2$ and $t_2 \precsim t_1$, we say that t_1 and t_2 are* contextually equivalent.

Contextual equivalence coincides with path equivalence:

Theorem 7 (Full abstraction). *For any terms $\Gamma \vdash t_1 : \mathbb{P}$ and $\Gamma \vdash t_2 : \mathbb{P}$,*

$$[\![t_1]\!] \subseteq [\![t_2]\!] \iff t_1 \precsim t_2 \ . \tag{34}$$

Proof. Suppose $[\![t_1]\!] \subseteq [\![t_2]\!]$ and let C be a $(\Gamma,\mathbb{P})$-program context with $[\![C(t_1)]\!] \neq \varnothing$. As $[\![t_1]\!] \subseteq [\![t_2]\!]$ we have $[\![C(t_2)]\!] \neq \varnothing$ by monotonicity, and so $t_1 \precsim t_2$ as wanted.

Now suppose that $t_1 \precsim t_2$. With $p : \mathbb{P}$ we define closed terms t_p of type $\mathbb{P}$ and $(\mathbb{O},\mathbb{P})$-program contexts C_p that respectively "realise" and "consume" the path p, by induction on the structure of p. We'll also need realisers t'_P and consumers C'_P of finite sets of paths:

$$\begin{aligned} t_{P\mapsto q} &\equiv \lambda x.[C'_P(x) > !x' \Rightarrow t_q] \\ t_{\beta p} &\equiv \beta t_p \\ t_P &\equiv !t'_P \\ t_{abs\,p} &\equiv abs\,t_p \end{aligned} \qquad \begin{aligned} C_{P\mapsto q} &\equiv C_q(-\ t'_P) \\ C_{\beta p} &\equiv C_p(\pi_\beta -) \\ C_P &\equiv [- > !x \Rightarrow C'_P(x)] \\ C_{abs\,p} &\equiv C_p(rep\ -) \end{aligned} \tag{35}$$

$$\begin{aligned} t'_{\{p_1,\dots,p_n\}} &\equiv t_{p_1} + \cdots + t_{p_n} \\ C'_{\{p_1,\dots,p_n\}} &\equiv [C_{p_1} > !x \Rightarrow \cdots \Rightarrow [C_{p_n} > !x \Rightarrow !\varnothing]\cdots] \end{aligned}$$

Note that $t'_\varnothing \equiv \varnothing$ and $C'_\varnothing \equiv !\varnothing$. Although the syntax of t'_P and C'_P depends on a choice of permutation of the elements of P, the semantics obtained for different

$$\frac{\mathbb{P} : t[rec\,x.t/x] \xrightarrow{a} t'}{\mathbb{P} : rec\,x.t \xrightarrow{a} t'} \qquad \frac{\mathbb{P} : t_j \xrightarrow{a} t'}{\mathbb{P} : \Sigma_{i \in I} t_i \xrightarrow{a} t'}\, j \in I \qquad \frac{\mathbb{Q} : t[u/x] \xrightarrow{a} t'}{\mathbb{P} \to \mathbb{Q} : \lambda x.t \xrightarrow{u \mapsto a} t'} \qquad \frac{\mathbb{P} \to \mathbb{Q} : t \xrightarrow{u \mapsto a} t'}{\mathbb{Q} : t\,u \xrightarrow{a} t'}$$

$$\frac{\mathbb{P}_\beta : t \xrightarrow{a} t'}{\Sigma_{\alpha \in A}\mathbb{P}_\alpha : \beta t \xrightarrow{\beta a} t'} \qquad \frac{\Sigma_{\alpha \in A}\mathbb{P}_\alpha : t \xrightarrow{\beta a} t'}{\mathbb{P}_\beta : \pi_\beta t \xrightarrow{a} t'} \qquad \frac{}{!\mathbb{P} : !t \xrightarrow{!} t} \qquad \frac{!\mathbb{P} : u \xrightarrow{!} u' \quad \mathbb{Q} : t[u'/x] \xrightarrow{a} t'}{\mathbb{Q} : [u > !x \Rightarrow t] \xrightarrow{a} t'}$$

$$\frac{\mathbb{T}_j[\mu \vec{T}.\vec{\mathbb{T}}/\vec{T}] : t \xrightarrow{a} t'}{\mu_j \vec{T}.\vec{\mathbb{T}} : abs\,t \xrightarrow{abs\,a} t'} \qquad \frac{\mu_j \vec{T}.\vec{\mathbb{T}} : t \xrightarrow{abs\,a} t'}{\mathbb{T}_j[\mu \vec{T}.\vec{\mathbb{T}}/\vec{T}] : rep\,t \xrightarrow{a} t'}$$

Fig. 1. Operational rules

permutations is the same. Indeed, we have (z being a fresh variable):

$$\begin{array}{ll} [\![t_p]\!] = y_{\mathbb{P}}p & [\![\lambda z.C_p(z)]\!] = y_{\mathbb{P}\to!\mathbb{O}}(\{p\} \mapsto \varnothing) \\ [\![t'_P]\!] = i_{\mathbb{P}}P & [\![\lambda z.C'_P(z)]\!] = y_{\mathbb{P}\to!\mathbb{O}}(P \mapsto \varnothing) \end{array} \tag{36}$$

Suppose t_1 and t_2 are closed. Given any $p \in [\![t_1]\!]$ we have $[\![C_p(t_1)]\!] \neq \varnothing$ and so using $t_1 \precsim t_2$, we get $[\![C_p(t_2)]\!] \neq \varnothing$, so that $p \in [\![t_2]\!]$. It follows that $[\![t_1]\!] \subseteq [\![t_2]\!]$.

As for open terms, suppose $\Gamma \equiv x_1 : \mathbb{P}_1, \dots, x_k : \mathbb{P}_k$. Writing $\lambda\vec{x}.t_1$ for the closed term $\lambda x_1.\cdots\lambda x_k.t_1$ and likewise for t_2, we get

$$t_1 \precsim t_2 \implies \lambda\vec{x}.t_1 \precsim \lambda\vec{x}.t_2 \implies [\![\lambda\vec{x}.t_1]\!] \subseteq [\![\lambda\vec{x}.t_2]\!] \implies [\![t_1]\!] \subseteq [\![t_2]\!] \ . \tag{37}$$

The proof is complete. □

4 Operational Semantics

HOPLA can be given an operational semantics using *actions* defined by

$$a ::= u \mapsto a \mid \beta a \mid\, ! \mid abs\,a \ . \tag{38}$$

We assign types to actions a using a judgement of the form $\mathbb{P} : a : \mathbb{P}'$. Intuitively, performing the action a turns a process of type $\mathbb{P}$ into a process of type $\mathbb{P}'$.

$$\frac{\vdash u : \mathbb{P} \quad \mathbb{Q} : a : \mathbb{P}'}{\mathbb{P} \to \mathbb{Q} : u \mapsto a : \mathbb{P}'} \qquad \frac{\mathbb{P}_\beta : a : \mathbb{P}' \quad \beta \in A}{\Sigma_{\alpha \in A}\mathbb{P}_\alpha : \beta a : \mathbb{P}'} \qquad \frac{}{!\mathbb{P} : ! : \mathbb{P}} \qquad \frac{\mathbb{T}_j[\mu\vec{T}.\vec{\mathbb{T}}/\vec{T}] : a : \mathbb{P}'}{\mu_j\vec{T}.\vec{\mathbb{T}} : abs\,a : \mathbb{P}'} \tag{39}$$

Notice that in $\mathbb{P} : a : \mathbb{P}'$, the type $\mathbb{P}'$ is unique given $\mathbb{P}$ and a. The operational rules of Fig. 1 define a relation $\mathbb{P} : t \xrightarrow{a} t'$ where $\vdash t : \mathbb{P}$ and $\mathbb{P} : a : \mathbb{P}'$.[1] An easy rule induction shows

Proposition 8. *If* $\mathbb{P} : t \xrightarrow{a} t'$ *with* $\mathbb{P} : a : \mathbb{P}'$, *then* $\vdash t' : \mathbb{P}'$.

Accordingly, we'll write $\mathbb{P} : t \xrightarrow{a} t' : \mathbb{P}'$ when $\mathbb{P} : t \xrightarrow{a} t'$ and $\mathbb{P} : a : \mathbb{P}'$.

[1] The explicit types in the operational rules were missing in the rules given in [19]. They are needed to ensure that the types of t and a agree in transitions.

4.1 Soundness and Adequacy

For each action $\mathbb{P} : a : \mathbb{P}'$ we define a linear map $a^* : \mathbb{P} \to !\mathbb{P}'$ which intuitively maps a process t of type $\mathbb{P}$ to a representation of its possible successors after performing the action a. In order to distinguish between, say, the successor $\varnothing$ and no successors, a^* embeds into the type $!\mathbb{P}'$ rather than using $\mathbb{P}'$ itself. For instance, the successors after action ! of the processes $!\varnothing$ and $\varnothing$ are, respectively, $!^*[\![!\varnothing]\!] = 1_{!\mathbb{P}}(\eta_{\mathbb{P}}\varnothing) = \eta_{\mathbb{P}}\varnothing$ and $!^*[\![\varnothing]\!] = 1_{!\mathbb{P}}\varnothing = \varnothing$. It will be convenient to treat a^* as a syntactic operation and so we define a term a^*t such that $[\![a^*t]\!] = a^*[\![t]\!]$:

$$\begin{aligned}
(u \mapsto a)^* &= a^* \circ \mathit{app} \circ (- \,\&\, [\![u]\!]) & (u \mapsto a)^*t &\equiv a^*(t\ u)\\
(\beta a)^* &= a^* \circ \pi_\beta & (\beta a)^*t &\equiv a^*(\pi_\beta t)\\
!^* &= 1_{!\mathbb{P}} & !^*t &\equiv t\\
(\mathit{abs}\, a)^* &= a^* \circ \mathit{rep} & (\mathit{abs}\, a)^*t &\equiv a^*(\mathit{rep}\, t)
\end{aligned} \tag{40}$$

The role of a^* is to reduce the action a to a prefix action. Formally the reduction is captured by the lemma below, proved by structural induction on a:

Lemma 9. $\mathbb{P} : t \xrightarrow{a} t' : \mathbb{P}' \iff !\mathbb{P}' : a^*t \xrightarrow{!} t' : \mathbb{P}'$.

Note that the reduction is done uniformly at all types using deconstructor contexts: application, projection, and unfolding. This explains the somewhat mysterious function space actions $u \mapsto a$. A similar use of labels to carry context information appears e.g. in [6].

Soundness says that the operational notion of "successor" is included in the semantic notion. The proof is by rule induction on the transition rules, see [20].

Proposition 10 (Soundness). *If* $\mathbb{P} : t \xrightarrow{a} t' : \mathbb{P}'$, *then* $\eta_{\mathbb{P}'}[\![t']\!] \subseteq a^*[\![t]\!]$.

We obtain a corresponding adequacy result using logical relations $X \trianglelefteq_{\mathbb{P}} t$ between subsets $X \subseteq \mathbb{P}$ and closed terms of type $\mathbb{P}$. Intuitively, $X \trianglelefteq_{\mathbb{P}} t$ means that all paths in X can be "operationally realised" by t. Because of recursive types, these relations cannot be defined by structural induction on the type $\mathbb{P}$ and we therefore employ a trick essentially due to Martin-Löf (see [23], Ch. 13). We define auxiliary relations $p \in_{\mathbb{P}} t$ between paths $p : \mathbb{P}$ and closed terms t of type $\mathbb{P}$, by induction on the structure of p:

$$\begin{aligned}
X \trianglelefteq_{\mathbb{P}} t &\iff_{\mathrm{def}} \forall p \in X.\ p \in_{\mathbb{P}} t\\
P \mapsto q \in_{\mathbb{P}\to\mathbb{Q}} t &\iff_{\mathrm{def}} \forall u.\ (P \trianglelefteq_{\mathbb{P}} u \implies q \in_{\mathbb{Q}} t\ u)\\
\beta p \in_{\Sigma_{\alpha\in A}\mathbb{P}_\alpha} t &\iff_{\mathrm{def}} p \in_{\mathbb{P}_\beta} \pi_\beta t\\
P \in_{!\mathbb{P}} t &\iff_{\mathrm{def}} \exists t'.\ !\mathbb{P} : t \xrightarrow{!} t' : \mathbb{P} \text{ and } P \trianglelefteq_{\mathbb{P}} t'\\
\mathit{abs}\, p \in_{\mu_j \vec{T}.\vec{\mathbb{T}}} t &\iff_{\mathrm{def}} p \in_{\mathbb{T}_j[\mu\vec{T}.\vec{\mathbb{T}}/\vec{T}]} \mathit{rep}\, t
\end{aligned} \tag{41}$$

The main lemma below is proved by structural induction on terms, see [20].

Lemma 11. *Suppose* $\vdash t : \mathbb{P}$. *Then* $[\![t]\!] \trianglelefteq_{\mathbb{P}} t$.

Proposition 12 (Adequacy). *Suppose* $\vdash t : \mathbb{P}$ *and* $\mathbb{P} : a : \mathbb{P}'$. *Then*

$$a^* [\![t]\!] \neq \varnothing \iff \exists t'.\ \mathbb{P} : t \xrightarrow{a} t' : \mathbb{P}' \tag{42}$$

Proof. The "$\Leftarrow$" direction follows from soundness. Assume $a^*[\![t]\!] \neq \varnothing$. Then because $a^*[\![t]\!]$ is a downwards-closed subset of $!\mathbb{P}'$ which has least element $\varnothing$, we must have $\varnothing \in a^*[\![t]\!]$. Thus $\varnothing \in_{!\mathbb{P}'} a^*t$ by Lemma 11, which implies the existence of a term t' such that $!\mathbb{P}' : a^*t \xrightarrow{!} t' : \mathbb{P}'$. By Lemma 9 we have $\mathbb{P} : t \xrightarrow{a} t' : \mathbb{P}'$. □

4.2 Full Abstraction w.r.t. Operational Semantics

Adequacy allows an operational formulation of contextual equivalence. If t is a program, we write $t\xrightarrow{!}$ if there exists t' such that $!\mathbb{O} : t \xrightarrow{!} t' : \mathbb{O}$. We then have $t\xrightarrow{!}$ iff $[\![t]\!] \neq \varnothing$ by adequacy. Hence, two terms t_1 and t_2 with $\Gamma \vdash t_1 : \mathbb{P}$ and $\Gamma \vdash t_2 : \mathbb{P}$ are related by contextual preorder iff for all $(\Gamma, \mathbb{P})$-program contexts C, we have $C(t_1)\xrightarrow{!} \implies C(t_2)\xrightarrow{!}$.

Full abstraction is often formulated in terms of this operational preorder. With t_1 and t_2 as above, the inclusion $[\![t_1]\!] \subseteq [\![t_2]\!]$ holds iff for all $(\Gamma, \mathbb{P})$-program contexts C, we have the implication $C(t_1)\xrightarrow{!} \implies C(t_2)\xrightarrow{!}$.

4.3 Simulation

The path semantics does not capture enough of the branching behaviour of processes to characterise bisimilarity (for that, the presheaf semantics is needed, see [11,19]). As an example, the processes $!\varnothing + !!\varnothing$ and $!!\varnothing$ have the same denotation, but are clearly not bisimilar. However, using Hennessy-Milner logic we can link path equivalence to simulation. In detail, we consider the fragment of Hennessy-Milner logic given by possibility and finite conjunctions; it is characteristic for simulation equivalence in the case of image-finite processes [8]. With a ranging over actions, formulae are given by

$$\phi ::= \langle a \rangle \phi \mid \textstyle\bigwedge_{i \leq n} \phi_i \ . \tag{43}$$

The empty conjunction is written $\top$. We type formulae using judgements $\phi : \mathbb{P}$, the idea being that only processes of type $\mathbb{P}$ should be described by $\phi : \mathbb{P}$.

$$\frac{\mathbb{P} : a : \mathbb{P}' \quad \phi : \mathbb{P}'}{\langle a \rangle \phi : \mathbb{P}} \qquad \frac{\phi_i : \mathbb{P} \quad \text{all } i \leq n}{\bigwedge_{i \leq n} \phi_i : \mathbb{P}} \tag{44}$$

The notion of *satisfaction*, written $t \vDash \phi : \mathbb{P}$, is defined by

$$t \vDash \langle a \rangle \phi : \mathbb{P} \iff \exists t'.\ \mathbb{P} : t \xrightarrow{a} t' : \mathbb{P}' \text{ and } t' \vDash \phi : \mathbb{P}' \tag{45}$$

$$t \vDash \textstyle\bigwedge_{i \leq n} \phi_i : \mathbb{P} \iff t \vDash \phi_i : \mathbb{P} \text{ for each } i \leq n \ . \tag{46}$$

Note that $\top : \mathbb{P}$ and $t \vDash \top : \mathbb{P}$ for all $\vdash t : \mathbb{P}$.

Definition 13. *Closed terms t_1, t_2 of the same type $\mathbb{P}$ are related by* logical preorder, *written $t_1 \lesssim_{\mathrm{L}} t_2$, iff for all formulae $\phi : \mathbb{P}$ we have $t_1 \vDash \phi : \mathbb{P} \implies t_2 \vDash \phi : \mathbb{P}$. If both $t_1 \lesssim_{\mathrm{L}} t_2$ and $t_2 \lesssim_{\mathrm{L}} t_1$, we say that t_1 and t_2 are* logically equivalent.

To each formula $\phi : \mathbb{P}$ we can construct a $(\mathbb{O}, \mathbb{P})$-program context C_ϕ with the property that

$$C_\phi(t) \xrightarrow{!} \iff t \vDash \phi : \mathbb{P} \ . \tag{47}$$

Define

$$\begin{aligned} C_{\langle u \mapsto a \rangle \phi} &\equiv C_{\langle a \rangle \phi}(-\ u) \ , & C_{\langle ! \rangle \phi} &\equiv [- > !x \Rightarrow C_\phi(x)] \ , \\ C_{\langle \beta a \rangle \phi} &\equiv C_{\langle a \rangle \phi}(\pi_\beta -) \ , & C_{\langle abs\, a \rangle \phi} &\equiv C_{\langle a \rangle \phi}(rep\ -) \ , \\ C_{\bigwedge_{i \leq n} \phi_i} &\equiv [C_{\phi_1} > !x \Rightarrow \cdots \Rightarrow [C_{\phi_n} > !x \Rightarrow !\varnothing] \cdots] \ . \end{aligned} \tag{48}$$

Corollary 14. *For closed terms t_1 and t_2 of the same type,*

$$t_1 \lesssim t_2 \iff t_1 \lesssim_{\mathrm{L}} t_2 \ . \tag{49}$$

Proof. The direction "⇒" follows from (47) and the remarks of Sect. 4.2. As for the converse, we observe that the program contexts C_p of the full abstraction proof in Sect. 3.2 are all subsumed by the contexts above. Thus, if $t_1 \lesssim_{\mathrm{L}} t_2$, then $[\![t_1]\!] \subseteq [\![t_2]\!]$ and so $t_1 \lesssim t_2$ by full abstraction. □

5 Related and Future Work

Matthew Hennessy's fully abstract semantics for higher-order CCS [9] is a path semantics, and what we have presented here can be seen as a generalisation of his work via the translation of higher-order CCS into HOPLA, see [19].

The presheaf semantics originally given for HOPLA is a refined version of the path semantics. A path set $X \in \widehat{\mathbb{P}}$ can be seen to give a "yes/no answer" to the question of whether or not a path $p \in \mathbb{P}$ can be realised by the process (cf. the representation in Sect. 2 of path sets as monotone maps $\mathbb{P}^{\mathrm{op}} \to \mathbf{2}$). A presheaf over $\mathbb{P}$ is a functor $\mathbb{P}^{\mathrm{op}} \to \mathbf{Set}$ to the category of sets and functions, and gives instead a set of "realisers", saying *how* a path may be realised. This extra information can be used to obtain refined versions of the proofs of soundness and adequacy, giving hope of extending the full abstraction result to a characterisation of bisimilarity, possibly in terms of *open maps* [11].

Replacing the exponential ! by a "lifting" comonad yields a model **Aff** of affine linear logic and an affine version of HOPLA, again with a fully abstract path semantics [20]. The tensor operation of **Aff** can be understood as a simple parallel composition of event structures [17]. Thus, the affine language holds promise of extending our approach to "independence" models like Petri nets or event structures in which computation paths are partial orders of events. Work is in progress to provide an operational semantics for this language together with results similar to those obtained here.

Being a higher-order process language, HOPLA allows process passing and so can express certain forms of mobility, in particular that present in the ambient calculus with public names [3,19]. Another kind of mobility, mobility of communication links, arises from name-generation as in the π-calculus [14]. Inspired by HOPLA, Francesco Zappa Nardelli and GW have defined a higher-order process language with name-generation, allowing encodings of full ambient calculus and π-calculus. Bisimulation properties and semantic underpinnings are being developed [25].

References

1. P.N. Benton. A mixed linear and non-linear logic: proofs, terms and models (extended abstract). In *Proc. CSL'94*, LNCS 933.
2. T. Bräuner. *An Axiomatic Approach to Adequacy.* Ph.D. Dissertation, University of Aarhus, 1996. BRICS Dissertation Series DS-96-4.
3. L. Cardelli and A.D. Gordon. Anytime, anywhere: modal logics for mobile ambients. In *Proc. POPL'00.*
4. G.L. Cattani and G. Winskel. Profunctors, open maps and bisimulation. Manuscript, 2000.
5. J.-Y. Girard. Linear logic. *Theoretical Computer Science*, 50(1):1–102, 1987.
6. A.D. Gordon. Bisimilarity as a theory of functional programming. In *Proc. MFPS'95*, ENTCS 1.
7. M.C.B. Hennessy and G.D. Plotkin. Full abstraction for a simple parallel programming language. In *Proc. MFCS'79*, LNCS 74.
8. M. Hennessy and R. Milner. Algebraic laws for nondeterminism and concurrency. *Journal of the ACM*, 32(1):137–161, 1985.
9. M. Hennessy. A fully abstract denotational model for higher-order processes. *Information and Computation*, 112(1):55–95, 1994.
10. C.A.R. Hoare. *A Model for Communicating Sequential Processes.* Technical monograph, PRG-22, University of Oxford Computing Laboratory, 1981.
11. A. Joyal, M. Nielsen, and G. Winskel. Bisimulation from open maps. *Information and Computation*, 127:164–185, 1996.
12. K.G. Larsen and G. Winskel. Using information systems to solve recursive domain equations effectively. In *Proc. Semantics of Data Types, 1984*, LNCS 173.
13. P.-A. Melliès. Categorical models of linear logic revisited. Submitted to *Theoretical Computer Science*, 2002.
14. R. Milner, J. Parrow and D. Walker. A calculus of mobile processes, parts I and II. *Information and Computation*, 100(1):1–77, 1992.
15. J.H. Morris. *Lambda-Calculus Models of Programming Languages.* PhD thesis, MIT, 1968.
16. M. Nielsen, G. Plotkin and G. Winskel. Petri nets, event structures and domains, part I. *Theoretical Computer Science*, 13(1):85–108, 1981.
17. M. Nygaard. Towards an operational understanding of presheaf models. Progress report, University of Aarhus, 2001.
18. M. Nygaard and G. Winskel. Linearity in process languages. In *Proc. LICS'02.*
19. M. Nygaard and G. Winskel. HOPLA—a higher-order process language. In *Proc. CONCUR'02*, LNCS 2421.

20. M. Nygaard and G. Winskel. Domain theory for concurrency. Submitted to *Theoretical Computer Science*, 2003.
21. D.S. Scott. Domains for denotational semantics. In *Proc. ICALP'82*, LNCS 140.
22. R.A.G. Seely. Linear logic, $*$-autonomous categories and cofree coalgebras. In *Proc. Categories in Computer Science and Logic*, 1987.
23. G. Winskel. *The Formal Semantics of Programming Languages*. MIT Press, 1993.
24. G. Winskel. A presheaf semantics of value-passing processes. In *Proc. CONCUR'96*, LNCS 1119.
25. G. Winskel and F. Zappa Nardelli. Manuscript, 2003.

Modeling Consensus in a Process Calculus*

Uwe Nestmann[1], Rachele Fuzzati[1], and Massimo Merro[2]

[1] EPFL, Switzerland
[2] University of Verona, Italy

Abstract. We give a process calculus model that formalizes a well-known algorithm (introduced by Chandra and Toueg) solving consensus in the presence of a particular class of failure detectors ($\diamond\mathcal{S}$); we use our model to formally prove that the algorithm satisfies its specification.

1 Introduction and Summary

This paper serves the following purposes: (1) to report on the first *formal* proof known to us of a Consensus algorithm developed by Chandra and Toueg using a particular style of failure detectors [CT96]; (2) to demonstrate the feasibility of using process calculi to carry out solid proofs for such algorithms; (3) to report on an operational semantics model for failure detectors that is easier to use in proofs than the original one based on so-called failure patterns.

Distributed Consensus. In the field of Distributed Algorithms, a widely-used computation model is based on *asynchronous* communication between a fixed number n of connected processes, where *no timing assumptions* can be made. Moreover, processes are subject to *crash-failure*: once crashed, they do not recover. The Distributed Consensus problem is well-known in this field: initially, each process proposes some value; eventually, all processes who do not happen to crash shall agree on one of the proposed values. More precisely, Consensus is specified by the following three properties on possible runs of a system.

Termination: Every correct process (eventually) decides some value.
Validity: If a process decides v, then v was proposed by some process.
Agreement: No two correct processes decide differently.

Here, a process is called *correct* in a given run, if it does not crash *in this run.* An important impossibility result states that Consensus cannot be solved in the aforementioned computation model when even a single process may fail [FLP85]. Since this impossibility result, several refinements of the computation model have been developed to overcome it. One of them is the addition of *unreliable failure detectors* (FD), i.e., modules attached to each process that can be locally queried to find out whether another process is currently locally suspected to

* Supported by the Swiss National Science Foundation, grant No. 21-67715.02, the Hasler Foundation, grant No. DICS 1825, an EPFL start-up grant, and the EU FET-GC project PEPITO. The full version is available at: `http://lamp.epfl.ch/~uwe/`

R. Amadio, D. Lugiez (Eds.): CONCUR 2003, LNCS 2761, pp. 399–414, 2003.

have crashed [CT96,CHT96]. FDs are unreliable in that they may have wrong suspicions, they may disagree among themselves, and they may change their suspicions at any time. To become useful, the behavior of FDs is constrained by abstract reliability properties about (i) the guaranteed suspicion of crashed processes, and (ii) the guaranteed non-suspicion of correct processes. Obviously, due to the run-based definition of correctness of processes, also these constraints are expressed over runs. A number of different combinations of FD-constraints were proposed in [CT96], one pair of which is commonly referred to as $\diamond\mathcal{S}$:

Strong Completeness (SC): Eventually every process that crashes is permanently suspected by (the FD of) every correct process.

Eventual Weak Accuracy (EWA): There is a time after which some correct process is never suspected by (the FD of) any correct process.

Chandra and Toueg also provide an algorithm – using pseudo-code, without formal semantics – in the context of FDs satisfying the reliability constraints of $\diamond\mathcal{S}$. The algorithm solves Consensus under the condition that a majority $\lceil\frac{n+1}{2}\rceil$ of processes are correct. It proceeds in *rounds* and is based on the *rotating coordinator* paradigm: for each round number, a single process is predetermined to play a coordinator role, while all other processes in this round play the role of participants. Each of the n processes counts rounds locally and knows at any time, who is the coordinator of its current round. Note that, due to asynchrony, any such system may easily reach states, in which all processes are in different rounds. Each round proceeds in four phases, in which (1) each participant sends to the coordinator of its current round its current estimate of the consensus value stamped with the round number at which it adopted this estimate; (2) the coordinator waits for sufficiently many estimates to arrive, selects one of those with the highest stamp; this is the round proposal that is distributed to the participants; (3) each participant either waits for the coordinator's round-proposal or, if this is currently permitted by its local FD, suspects the coordinator – in both cases, participants then send (positive or negative) acknowledgments to the coordinator and proceed to the next round; (4) the coordinator waits for sufficiently many acknowledgments; if they are all positive it proceeds to the decision, otherwise it proceeds to the next round. "Deciding on a value" means to send the value to all processes using Reliable Broadcast (RB). The reception of an RB-message is called RB-*delivery*; processes may RB-deliver independent of their current round and phase. On RB-delivery, a process "officially" decides on the broadcast value. Note that also the broadcast-initiator must perform RB-delivery. Since RB satisfies a termination property, every non-crashed process will eventually receive the broadcast messages.

Intuitively the algorithm works because coordinators always wait for a majority of messages before they proceed (which is why, to ensure the computation is non-blocking, strictly less than the majority are allowed to crash). Once a majority of processes have positively acknowledged in the same round, the coordinator's proposal of that round is said to be *locked*: if ever "after" another majority positively acknowledges, it will be for the very same value, thus satisfying Agreement. If some coordinator manages to get these acknowledgments and survives until RB-delivery, the algorithm also satisfies Termination. The interest in having a FD with $\diamond\mathcal{S}$ is the guarantee (EWA) that eventually there will be a correct process that is never again suspected, thus will be positively acknowledged when playing the coordinator. $\diamond\mathcal{S}$ also gives the guarantee (SC) that such a process will indeed be able to reach a round in which it plays the coordinator role. More detailed proofs of termination, validity, and agreement, are given in natural language and found in [CT96]. We found them reasonable, but hard to follow and very hard to formally verify.

Our first main criticism is that the pseudo-code does not have a formal semantics. Thus, there is no well-defined way to generate system runs, which are the base of the FD and Consensus properties. To tackle this problem, many years of research on concurrency theory provide us with a variety of decent formalisms that only need to be extended to also model failures and their detection.

Our second main criticism is more subtle. Some proofs of properties over runs make heavy reference to the concept of rounds, e.g., using induction on round numbers, although the relation between runs and asynchronous rounds is never clarified. This is problematic! Typically, such an induction starts with the smallest round in which some property X holds, e.g., in which a majority has positively acknowledged. In a given run, to find this starting point one may take the initial state and search from there for the first state in which X holds for some round. However, this procedure is not correct. It may well be that at a *later state* of the run, X holds for a *smaller round*! Accordingly, when the induction proceeds to a higher round, it might go backwards in time along a system run. Therefore, the concept of time – and of iteration along a run – is not fully compatible with the concept of asynchronous rounds. The solution, rather implicit in [CT96], is to consider runs as a whole, ignoring *when* events happen, just noting *that* they happened. In other words, we should pick a sufficiently advanced state of a given run (for example the last one in a finite run), and then find an appropriately abstract way to *reason about its possible past*. Summing up, the proofs would profit much from a global view on system states and their past that provides us with precise information about what processes *have been* in which round in the past, and what they precisely did when they were there.

Our Approach. We provide a process calculus setting that faithfully captures the asynchronous process model. We equip this model with an operational control over crash-failure and FD properties (§2). However, instead of $\diamond\mathcal{S}$, for which the algorithm was designed, we use the following FD [CHT96]:

Eventual Perpetual Uniform Trust (Ω) There is a time after which all the (correct) processes always trust the same correct process.

The FDs Ω and $\diamond\mathcal{S}$ are equivalent in the sense that one can be used to implement the other, and vice versa. Although Ω was introduced only to simplify the minimality proofs of [CHT96], it turns out to be more natural to develop our operational model for it rather than for $\diamond\mathcal{S}$. (Briefly, instead of keeping track of loads of useless unreliable suspicion information, Ω only requires to model small amounts of reliable trust information.) We then model the Consensus algorithm as a term in this calculus (§3), allowing us in principle to analyze its properties over runs generated by its local-view formal operational semantics. However, we do not do this as one might expect by iteration along system runs, showing the preservation of invariants. Instead, in order to formally deal with the round abstraction, we develop a global-view matrix-like representation of reachable states that contains the complete history of message sent "up to now" (§4). Also for this abstraction, we provide a formal semantics, and we use it instead of the local-view semantics to prove the Consensus properties (§5). The key justification for this approach is a very tight formal operational correspondence proof between the local-view process semantics and the global-view matrix semantics. It exploits slightly non-standard process calculus technology (see the full paper).

Contributions. One novelty is the operational modeling of FD properties.

However, the essential novelty is the formal global-view matrix representation of the reachable states of a Consensus system that formally captures the round abstraction. It allowed us to bridge the gap between the local-view code and semantics describing the algorithm on the one hand, and the round-based reasoning that enables comprehensible structured proofs on the other hand.

Another contribution is that some proofs of [CT96], especially for Agreement, can now be considered as being formalized. Instead of trying to directly formalize Termination, we came up with different proof ideas for it (Theorem 2).

Conclusion. The matrix semantics provides us with a tractable way to perform a formal analysis of this past, according to when and which messages have been sent in the various earlier rounds.

We use process calculus and operational semantics to justify proofs via global views that are based on the abstraction of rounds. In fact, this round-based global view of a system acts as a vehicle for many proofs about distributed algorithms, while to our knowledge it has never been formally justified and thus remained rather vague. Thus, we expect that our contribution will not only be valuable for this particular verification exercise, but also generally improve the understanding of distributed algorithms in asynchronous systems.

Related Work. We are only aware of a formal model and verification of *Randomized* Consensus using probabilistic I/O-automata [PSL00].

Future Work. Apart from this application-oriented work, we have also modeled the other failure detectors of [CT96]. We are currently working on the formal comparison of our representation to theirs. This work is independent of the language used to describe the algorithms that make use of failure detectors.

It would also be interesting to study extensions of our operational semantics setting for failure detectors towards more dynamic mobile systems.

Acknowledgments. We thank Sergio Mena, André Schiper, Pawel Wojciechowski, and Rachid Guerraoui for discussions on the Consensus problem; James Leifer, Peter Sewell, Holger Hermanns for discussions on proof techniques; Daniel

Bünzli and Aoife Hegarty for improving the presentation; and the anonymous reviewers.

2 The Process Calculus Model

We use a simple *distributed asynchronous value-passing* process calculus; name-passing is not needed for static process groups. We use an extension with *named sites* inspired by Berger and Honda [BH00], but unlike them we do not have to model message loss. Our notion of sites also resembles the locations of Dπ [RH01] and the Nomadic pi calculus [WS00]. For convenience, we also employ site-local purely signaling synchronous actions. We do not need the usual restriction operator, because we are going to study only internal transitions.

$$
\begin{array}{rcl}
v & ::= & x \mid i \mid \mathsf{t} \mid \mathsf{f} \mid \mathsf{f}(\tilde{v}) \mid \ldots \\
M & ::= & \overline{a}\langle \tilde{v} \rangle \\
\alpha & ::= & a(\tilde{x}) \mid \tau \mid \mathrm{susp}_j \mid a \mid \overline{a} \\
G & ::= & G{+}G \mid \alpha.P \mid \mathbf{0} \\
P & ::= & P|P \mid \mathsf{Y}\langle \tilde{v} \rangle \mid M \mid G \mid \mathsf{if}\ v\ \mathsf{then}\ P\ \mathsf{else}\ P \\
N & ::= & N|N \mid i[P] \mid M
\end{array}
$$

where process constants Y are associated with defining equations $\mathsf{Y}(\tilde{x}) := P$, which also gives us recursion. $\mathbf{I} \subseteq \mathbf{V}$ is a set of site identifiers (metavariables i, j, k, n), for which we simply take a subset of the natural numbers $\mathbf{Nat}$ equipped with standard operations like equality and modulo. $\{\mathsf{t}, \mathsf{f}\} \subseteq \mathbf{V}$ is the set of boolean values. The set $\mathbf{V}$ of value expressions (metavariable v) contains various operations on sets and lists, like addition, extraction, arity, and comparison. We also use a function eval that performs the deterministic evaluation of value expressions. By abuse of notation, we use all value metavariables (and x) also as input variables. Names $\mathbf{N}$ (metavariable a) are different from values ($\mathbf{N} \cap \mathbf{V} = \emptyset$).

We use $\mathbb{G}$, $\mathbb{P}$, and $\mathbb{N}$, to refer to the sets of terms generated by the respective non-terminal symbols for guards G, local processes P, and networks N. Sites $i[P]$ are named and may be syntactically distributed over terms; sometimes, we refer to them as *processes*. The interpretation of all operators is standard [BH00]. For actions susp_j, see the explanation and formal semantics later on. We include both synchronous signals $(a, \overline{a})$ and asynchronous messages M with matching receivers; for simplicity, we do not introduce separate syntactic categories for respective channels. As usual, parallel composition is associative and commutative; with finite indexing sets I we use $\prod_{i \in I} P_i$ as abbreviation for the arbitrarily ordered and nested parallel composition of the P_i, and similar for $\prod_{i \in I} N_i$.

Structural Equivalence. The relation $\Lleftarrow\!\!\Rrightarrow$ is defined as the smallest equivalence relation generated by the laws of the commutative monoids $(\mathbb{G}, +, \mathbf{0})$, $(\mathbb{P}, |, \mathbf{0})$, and $(\mathbb{N}, |, \mathbf{0})$, the law $i[P_1] \,|\, i[P_2] \Lleftarrow\!\!\Rrightarrow i[P_1|P_2]$ that defines the scope of sites, the straightforward laws induced by evaluation of value expressions:

- $\mathsf{if}\ v\ \mathsf{then}\ P_1\ \mathsf{else}\ P_2 \Lleftarrow\!\!\Rrightarrow P_1$ if $\mathrm{eval}(v) = \mathsf{t}$,

Table 1. Network transitions

$$(\text{TAU})\quad i[\,\tau.P+G\,] \xrightarrow{\tau@i} i[\,P\,] \qquad (\text{SUSPECT?})\quad i[\,\text{susp}_j\,.P+G\,] \xrightarrow{\text{susp}_j@i} i[\,P\,]$$

$$(\text{COM})\quad i[\;\overline{a}.P_1+G_1 \mid a.P_2+G_2\;] \xrightarrow{\tau@i} i[\;P_1 \mid P_2\;]$$

$$(\text{SND})\quad i[\,M\,] \xrightarrow{\tau@i} M \qquad (\text{RCV})\quad \overline{a}\langle\tilde{v}\rangle \mid i[\,a(\tilde{x}).P+G\,] \xrightarrow{\tau@i} i[\,P\{^{\tilde{v}}/_{\tilde{x}}\}\,]$$

$$(\text{STR})\ \frac{N \Lleftarrow\!\!\!\Rrightarrow \hat{N} \qquad \hat{N} \xrightarrow{\mu@i} \hat{N}' \qquad \hat{N}' \Lleftarrow\!\!\!\Rrightarrow N'}{N \xrightarrow{\mu@i} N'} \qquad (\text{PAR})\ \frac{N_1 \xrightarrow{\mu@i} N_1'}{N_1|N_2 \xrightarrow{\mu@i} N_1'|N_2}$$

- if v then P_1 else $P_2 \Lleftarrow\!\!\!\Rrightarrow P_2$ if $\text{eval}(v) = \mathsf{f}$,
- $\overline{a}\langle\tilde{v}\rangle \Lleftarrow\!\!\!\Rrightarrow \overline{a}\langle\text{eval}(\tilde{v})\rangle$, $\mathsf{Y}\langle\tilde{v}\rangle \Lleftarrow\!\!\!\Rrightarrow \mathsf{Y}\langle\text{eval}(\tilde{v})\rangle$;
- $\mathsf{Y}\langle\tilde{v}\rangle \Lleftarrow\!\!\!\Rrightarrow P\{^{\tilde{v}}/_{\tilde{x}}\}$ if $\mathsf{Y}(\tilde{x}) := P$,

and that is preserved within non-prefix contexts. The inclusion of conditional resolution and recursion unfolding within structural equivalence is to allow us to have the transition relation defined below to deal exclusively with interactions. However, an unconstrained use of $\Lleftarrow\!\!\!\Rrightarrow$ quickly leads to problems when applying equivalence laws in an unintended direction. Thus, for proofs, we replace the relation $\Lleftarrow\!\!\!\Rrightarrow$ and the rule (STR) of Table 1 with a directed *(normalized)* version.

Network Transitions. Transitions on networks are generated by the laws in Table 1. Each transition $\mu@i$ is labeled by the action $\mu \in \{\tau, \text{susp}_j\}$ and the site identifier i indicating the site required for the action. The communication of asynchronous messages takes two steps: once they are sent, i.e., appear at top-level on a site, they need to leave the sender site (SND) into the buffering "ether"; once in the ether, they may be received by a process on the target site (RCV).

Without definition (see the full paper for details), let $\xrightarrow{\mu@i}_{\text{n}}$ denote the *normalized* transition relation that we get when using a directed structural relation; this relation is defined on the subset of *normalized* network terms $\mathbb{N}^{\text{n}}$.

Environment Transitions. By adding an environment component Γ to networks, we model both failures and their detection, as well as "trust" in the sense of Ω. Environments $\Gamma := (\mathsf{TI}, \mathsf{C})$ contain (i) information about sites $i \in \mathsf{TI} \subseteq \mathbf{I}$ that have become *trusted* forever and *immortal*, so they can no longer be suspected nor crash, and (ii) information about sites $i \in \mathsf{C} \subseteq \mathbf{I}$ that have already crashed.

Environments are updated according to the rules in Table 2. Rule (TRUST) models the instant at which (according to Ω) processes become trusted – in our model they also become immortal: they will be "correct" in every possible future. Rule (CRASH) keeps track of processes that crash and is subject to an upper bound: for instance, the Consensus algorithm of [CT96] is supposed

Table 2. Environment transitions

$$(\text{TRUST})\ \frac{i \notin \mathsf{TI} \cup \mathsf{C}}{(\mathsf{TI}, \mathsf{C}) \ \rightarrow \ (\mathsf{TI} \cup \{i\}, \mathsf{C})} \qquad (\text{CRASH})\ \frac{i \notin \mathsf{TI} \cup \mathsf{C} \qquad |\mathsf{C}| \leq \lfloor \frac{n-1}{2} \rfloor}{(\mathsf{TI}, \mathsf{C}) \ \rightarrow \ (\mathsf{TI}, \mathsf{C} \cup \{i\})}$$

Table 3. System transitions

$$(\text{DETECT})\ \frac{\Gamma \rightarrow \Gamma'}{\Gamma \vdash N \rightarrow \Gamma' \vdash N} \qquad (\text{ACT})\ \frac{i \notin \mathsf{C} \qquad N \xrightarrow{\tau @ i} N'}{(\mathsf{TI}, \mathsf{C}) \vdash N \rightarrow (\mathsf{TI}, \mathsf{C}) \vdash N'}$$

$$(\text{SUSPECT!})\ \frac{j \neq i \notin \mathsf{C} \qquad N \xrightarrow{\text{susp}_j @ i} N' \qquad j \notin \mathsf{TI}}{(\mathsf{TI}, \mathsf{C}) \vdash N \rightarrow (\mathsf{TI}, \mathsf{C}) \vdash N'}$$

to work correctly only under the constraint that at most $\lfloor \frac{n-1}{2} \rfloor$ processes may crash.

System Transitions. Configurations are pairs of the form $\Gamma \vdash N$. Their transitions come either from the environment Γ (DETECT), modeling the unconstrained occurrence of its transitions, or they come from the network N. In this case, the environment must explicitly permit the network actions. Rule (ACT) guarantees that only non-crashed sites may act. Rule (SUSPECT!) provides the model for suspicions: a site j may only be suspected by a process on another (different) non-crashed site i and – which is crucial – the *suspected site must not be trusted.* Note that suspicions in this model are "very unreliable" since every non-trusted site may be suspected from within any non-crashed site at any time.

Runs. FD properties are based on the notion of *run*. In our language, runs are complete (in)finite sequences of transitions (denoted by $\rightarrow^*$) starting in some initial configuration $(\emptyset, \emptyset) \vdash N$. According to [CT96], a process is called *correct* in a given run, if it does not crash in that run. There is a close relation between this notion and the environment information in states of system runs.

Lemma 1 (Correctness in System Runs).

1. *If R is the run $(\emptyset,\emptyset) \vdash N_0 \rightarrow^* (\mathsf{TI},\mathsf{C}) \vdash N \not\rightarrow$ then:*
   - *$i \in \mathsf{TI}$ iff i is correct in R; $i \in \mathsf{C}$ iff i is not correct in R.*
   - *$|\mathsf{TI}| \geq n - \lfloor \frac{n-1}{2} \rfloor$, $|\mathsf{C}| \leq \lfloor \frac{n-1}{2} \rfloor$, and $\mathsf{TI} \uplus \mathsf{C} = \{1..,n\}$.*
2. *If R is the run $(\emptyset,\emptyset) \vdash N_0 \rightarrow^* (\mathsf{TI},\mathsf{C}) \vdash N \rightarrow^{*/\omega}$ then:*
   - *If $i \in \mathsf{TI}$, then i is correct in R. If $i \in \mathsf{C}$, then i is not correct in R.*
   - *$|\mathsf{TI}| \geq n - \lfloor \frac{n-1}{2} \rfloor$, $|\mathsf{C}| \leq \lfloor \frac{n-1}{2} \rfloor$, and $\mathsf{TI} \uplus \mathsf{C} \subseteq \{1..,n\}$.*

Proof (Sketch). By the rules of Table 2 and rule (SUSPECT!) in Table 3. □

For finite runs, Lemma 1(1) states that in final states all decisions concerning "life" and "death" are taken. For intermediate states of infinite runs, Lemma 1(2) provides us with only partial but nevertheless reliable information.

Our operational representation of the FD Ω consists of two parts: (i) the above rule (SUSPECT!), and (ii) a condition on runs that at least one site must eventually become trusted and immortal (for the current run) such that it cannot be suspected afterwards and will turn out to be correct.

Definition 1 (Ω-Runs). *Let R be a run starting in $(\emptyset, \emptyset) \vdash N_0$.*
R is called Ω-run if $(\emptyset, \emptyset) \vdash N_0 \rightarrow^* (\mathsf{TI}, \mathsf{C}) \vdash N$ is a prefix of R with $\mathsf{TI} \neq \emptyset$.

The condition $\mathsf{TI} \neq \emptyset$ means that, for at least one transition in the run R, the rule (TRUST) must have been applied. In Ω-runs, it is sufficient to check a syntactic condition on states that guarantees the absence of subsequent unpermitted suspicions. In contrast, the original FD model requires to carefully check that after some hypothetical (not syntactically indicated) time all occurrences of suspicion steps do not address a particular process that happens to be correct in this run by analyzing every single step of the run. Thus, our operational FD model considerably simplifies the analysis of runs. The formal comparison of operational models and the original history-based models is ongoing work, in which we also address the remaining failure detector classes introduced in [CT96].

3 Solving Consensus with Ω-Detection

Table 4 shows the Consensus algorithm of [CT96] represented as the process calculus term $Consensus_{(v_1..,v_n)}$. When no confusion is possible, we may omit the initial values $(v_1.., v_n)$. We use the notation $\mathsf{Y}_i^{\tilde{v}}$ as an abbreviation for both $\mathsf{Y}_i(i, \tilde{v})$ and $\mathsf{Y}_i\langle i, \tilde{v}\rangle$, so the subscript is part of the constant while the superscripts represent formal/actual parameters. The subscript must, in fact, also be considered part of the parameters, because we will access it in the body, but since we never change this parameter, we omit it in the abbreviation.

Let n be the number of processes, and $\mathrm{crd}(r) := ((r-1) \bmod n)+1$ denote the coordinator of round r. $\mathsf{Y}_i^{r,v,s,L}$ represents participant i in round r with current estimate v dating back to round s, and a list L of messages previously received from other participants (see below). Y_i itself ranges over $\mathsf{P1}_i, \mathsf{P2}_i, \mathsf{P4}_i, \mathsf{R}_i, \mathsf{Z}_i$ for $i = \mathrm{crd}(r)$, and over $\mathsf{P1}_i, \mathsf{P3}_i, \mathsf{R}_i$ for $i \neq \mathrm{crd}(r)$. D_i is part of the RB-protocol: it is the component that "decides" and re-broadcasts on RB-delivery.

All protocol participants are interconnected; we use separate channel names $(c1_i, c2_i, c3_i)$ for the messages sent in the first three phases, and further channel names for broadcasting (b_i) and announcing decisions $(decide_i)$. For convenience, we use site-indexed channel names, but note that the indices i are only virtual: they are considered to be part of the indivisible channel name. In addition to these $5{*}n$ asynchronous channels, we use n synchronous channels $(undecided_i)$, also "indexed". We use the latter to conveniently avoid fairness conditions on runs concerning the reception of the otherwise asynchronous signals. We include

Table 4. Consensus

$$\mathit{Consensus}_{(v_1..,v_n)} \stackrel{\text{def}}{=} \prod_{i=1}^{n} i\Big[\, \mathsf{P1}_i^{1,v_i,0,\emptyset} \,|\, \mathsf{D}_i \,\Big]$$

$$\mathsf{P1}_i^{r,v,s,L} \stackrel{\text{def}}{=} \overline{c1_{\mathrm{crd}(r)}}\langle i, r, v, s\rangle \mid \mathsf{if}\ i{=}\mathrm{crd}(r)\ \mathsf{then}\ \mathsf{P2}_i^{r,v,s,L}\ \mathsf{else}\ \mathsf{P3}_i^{r,v,s,L}$$

$$\begin{aligned}\mathsf{P2}_i^{r,v,s,L} \stackrel{\text{def}}{=}\ & \mathsf{if}\ |L_1^r| < \lceil \tfrac{n+1}{2} \rceil \\ & \mathsf{then}\ c1_i\,(\tilde{x})\,.\,\mathsf{P2}_i^{r,v,s,(1,\tilde{x})::L} \\ & \mathsf{else}\ \tau.\Big(\prod_{i \neq k=1}^{n} \overline{c2_k}\langle k, r, \mathrm{best}(L_1^r)\rangle \mid \mathsf{P4}_i^{r,\mathrm{best}(L_1^r),r,L} \Big)\end{aligned}$$

$$\begin{aligned}\mathsf{P3}_i^{r,v,s,L} \stackrel{\text{def}}{=}\ & \mathsf{if}\ L_2^r = \emptyset \\ & \mathsf{then}\ \Big(c2_i\,(\tilde{x})\,.\,\mathsf{P3}_i^{r,v,s,(2,\tilde{x})::L} \ +\ \mathsf{susp}_{\mathrm{crd}(r)}\,.\,\big(\overline{c3_{\mathrm{crd}(r)}}\langle i, r, \mathsf{f}\rangle \mid \mathsf{R}_i^{r,v,s,L} \big)\Big) \\ & \mathsf{else}\ \tau.\big(\overline{c3_{\mathrm{crd}(r)}}\langle i, r, \mathsf{t}\rangle \mid \mathsf{R}_i^{r,\mathrm{val}(L_2^r),r,L} \big)\end{aligned}$$

$$\begin{aligned}\mathsf{P4}_i^{r,v,s,L} \stackrel{\text{def}}{=}\ & \mathsf{if}\ |L_3^r| < \lceil \tfrac{n+1}{2} \rceil - 1 \\ & \mathsf{then}\ c3_i\,(\tilde{x})\,.\,\mathsf{P4}_i^{r,v,s,(3,\tilde{x})::L} \\ & \mathsf{else\ if}\ \bigwedge_{l \in L_3^r} \mathrm{bool}(l)\ \mathsf{then}\ \tau.\Big(\prod_{k=1}^{n} \overline{b_k}\langle i, k, 1, r, v\rangle \mid \mathsf{Z}_i^{r,v,s,L} \Big)\ \mathsf{else}\ \mathsf{R}_i^{r,v,r,L}\end{aligned}$$

$$\mathsf{Z}_i^{r,v,s,L} \stackrel{\text{def}}{=} \mathbf{0}$$

$$\mathsf{R}_i^{r,v,s,L} \stackrel{\text{def}}{=} \mathit{undecided}_i\,.\,\mathsf{P1}_i^{r+1,v,s,L}$$

$$\mathsf{D}_i \stackrel{\text{def}}{=} \overline{\mathit{undecided}_i}\,.\,\mathsf{D}_i \ +\ b_i\,(j,\cdot,m,r,v)\,.\,\Big(\overline{\mathit{decide}_i}\langle j, i, m, r, v\rangle \mid \prod_{k=1}^{n} \overline{b_k}\langle i, k, 2, r, v\rangle \Big)$$

some redundant information (gray-shaded in Table 4) within messages – especially about the sender and receiver identification – such that we can easily and uniquely distinguish messages. We also add some τ-steps, which are only there to facilitate the presentation of some of the proofs.

Behaviors. In the 1st phase, we ($\mathsf{P1}_i^{r,v,s,L}$) send our current estimate and depending on whether we are coordinator of our round, we move to phase 2 or 3.

In the 2nd phase, we ($\mathsf{P2}_i^{r,v,s,L}$) wait for sufficiently many 3rd-phase estimate messages for our current round r. Once we have them, we determine the best one among them (see below), and we impose *its* value as the one to adopt in the round r by sending it to everybody else. (As a slight optimization of [CT96], we do not send the proposal to ourselves, and also we do not send an acknowledgment to ourselves, assuming that we agree with our own proposal.) Remembering the just proposed value, we then move to phase 4.

In the 3rd phase, we ($\mathsf{P3}_i^{r,v,s,L}$) are waiting for the proposal from the coordinator of our current round r. As soon as it arrives, we positively acknowledge it, and (try to) restart and move to the next round. As long as it has not yet arrived, we may also have the possibility to suspect the coordinator in order to move on; in this case, we continue with our old value and stamp.

In the 4th phase, we ($\mathsf{P4}_i^{r,v,s,L}$) wait for sufficiently many 3rd-phase acknowledgment messages for our current round r. Once we have them, we check whether they are all positive. If yes, then we launch reliable broadcast by sending our decision value v on all b_k; it becomes reliable only through the definition of D_k on the receiver side of the b_k. If no, then we simply try to restart.

If we ($\mathsf{R}_i^{r,v,s,L}$) want to restart, we must get the explicit permission from our broadcast controller process D_i along the local synchronous channel $undecided_i$. This permission will never again be given as soon as we (at site i) have "delivered", i.e., received the broadcast along b_i and subsequently have decided.

When halting a coordinator, we do not just let it become $\mathbf{0}$ or disappear, but use a specific constant Z_i to denote the final state. The reason is that we can keep accessibly within the term the final information of halted processes ($\mathsf{Z}_i^{r,v,s,L}$), which would otherwise disappear as well.

Data Structures. The parameter $L \in \mathbb{L}$ is a heterogeneous list of elements in $\mathbb{L}_1$ for 1st-phase messages, $\mathbb{L}_2$ for 2nd-phase messages, and $\mathbb{L}_3$ for 3rd-phase messages. By L_1, L_2, L_3, we denote the various homogeneous sublists of L for the corresponding phases. By $|L|$, we denote the length of a list L. By $l{::}L$, we denote the addition of element l to L. For each homogeneous type of sublist, we provide some more notation. For convenience, we allow ourselves to use component access via "logical" names rather than "physical" projections. For example, in all types, one component represents a round number. By $L^r := \{\, l \in L \mid \mathrm{round}(l) = r \,\}$, we extract all elements of list L that apparently belong to round r. Similarly, the function $\mathrm{val}(l)$ extracts the value field of list element l.

Elements of $\mathbb{L}_1$ ($\{1\} \times I \times \mathbb{N} \times \mathbf{V} \times \mathbb{N}$), like 1st-phase messages, consist of a site identifier ($\in I$), a round number ($\in \mathbb{N}$), an estimate value ($\in \mathbf{V}$), and a stamp ($\in \mathbb{N}$). Let $L \in \mathbb{L}_1^*$. By $\mathrm{max\_s}(L) := \max\{\, \mathrm{stamp}(l) \mid l \in L \,\}$ we extract the maximal stamp occurring in the elements of L. By $\mathrm{best}(L) := \mathrm{val}(\mathrm{min\_i}\{\, l \in L \mid \mathrm{stamp}(l) = \mathrm{max\_s}(L) \,\})$, we extract among all the elements of L that have the highest stamp the one element that is smallest with respect to the site identifier, and return the value of it.

Elements of $\mathbb{L}_2$ ($\{2\} \times I \times \mathbb{N} \times \mathbf{V}$), like 2nd-phase messages, consist of a site identifier ($\in I$), a round number ($\in \mathbb{N}$), and an estimated value ($\in \mathbf{V}$).

Elements of $\mathbb{L}_3$ ($\{3\} \times I \times \mathbb{N} \times \mathbb{B}$), like 3rd-phase messages, consist of a sender site identifier ($\in I$), a round number ($\in \mathbb{N}$), and a boolean value ($\in \mathbb{B}$). Let $l \in \mathbb{L}_3$. By $\mathrm{bool}(l)$, we extract the boolean component of list element l.

4 A Global Message-Oriented View: Matrices

By analysis of Chandra and Toueg's proofs of the Consensus properties [CT96], we observe that they become feasible *only* if we manage to argue formally and globally about the contributions of processes to individual rounds. To this aim, we design an alternative representation of the reachable state of *Consensus*: message *matrices* $\mathbb{M}$. In fact, matrices contains precisely the same information as

Table 5. From messages M to matrix entries $\mathbf{x}$ ... and back

$M := \mathcal{E}_{\mathrm{M}}^{-1}(\mathbf{x})$	$l := \mathcal{E}_{\mathrm{L}}^{-1}(\mathbf{x})$	snd	rcv	rnd	$\mathcal{E}_t(M) =: \mathbf{x} := \mathcal{E}_t(l)$	tag($\mathbf{x}$)
$\overline{c1_{\mathrm{crd}(r)}}\langle i, r, v, s\rangle$	$(1, i, r, v, s)$	i	crd(r)	r	$(i, r) \overset{1}{\mapsto} (v, s, t)$	t
$\overline{c2_i}\langle i, r, v\rangle$	$(2, i, r, v)$	crd(r)	i	r	$(i, r) \overset{2}{\mapsto} (v, t)$	t
$\overline{c3_{\mathrm{crd}(r)}}\langle i, r, z\rangle$	$(3, i, r, z)$	i	crd(r)	r	$(i, r) \overset{3}{\mapsto} (z, t)$	t
$\overline{b_i}\langle j, i, m, r, v\rangle$		j	i	r	$(i, j, m) \overset{\mathrm{b}}{\mapsto} (r, v, t)$	t
$\overline{decide_i}\langle j, i, m, r, v\rangle$		i	–	r	$(i) \overset{\mathrm{d}}{\mapsto} (j, m, r, v, t)$	t

terms: we can freely move between the two representations via formal mappings:

$$\mathbb{N} \underset{\mathcal{N}[\![\ ]\!] \text{ using } \mathcal{E}^{-1}()}{\overset{\mathcal{M}[\![\ ]\!] \text{ using } \mathcal{E}_t()}{\rightleftarrows}} \mathbb{M}$$

It is for this tight connection that we augmented the definition of *Consensus* in Table 4 with book-keeping data, never forgetting any message ever received.

$\mathcal{M}[\![\ ]\!]$*: From Networks to Matrices.* With any state reachable starting from *Consensus*, we associate a matrix structure containing all the asynchronous messages that have been sent "up to now", organized according to the round in which they were sent. For the 1st-, 2nd-, and 3rd-phase messages, the resulting structure is a specific kind of two-dimensional matrix (see column six of Table 5): one dimension for *process ids* (variable i ranging from 1 to n), one dimension for *round numbers* (variable r ranging unboundedly over natural numbers starting at 1). For broadcast- and decision-messages, which may only eventually occur for a single round per process, the format is slightly different.

For each message, we distinguish three *transmission states*:

- being *sent*, but not yet having left the sender site ($\surd$)
- being *in transit*, i.e., having left the sender site, but not yet arrived ($\surd\!\!\surd$)
- being *received*, i.e., appearing in the list L ($\surd\!\!\surd\!\!\surd$)

We usually let t range over the elements of the ordered set $\{\surd < \surd\!\!\surd < \surd\!\!\surd\!\!\surd\}$. For d-entries, aka: decision messages, there is no receiver and thus always $t \neq \surd\!\!\surd\!\!\surd$.

Networks can be mapped into matrices because our process representation memorizes the required information on past messages ($\surd\!\!\surd\!\!\surd$) in the state parameters L; messages that are sent and not yet received ($\surd, \surd\!\!\surd$) can be analyzed "directly" from the respective system state component. Table 5 lists the various entry types of matrices, and how they correspond to the formats found in networks, namely messages M and list entries $l \in L$. For better orientation, we include columns snd and rcv that indicate the respective sender and receiver.

We may view a matrix $\mathfrak{M}$ as the heterogeneous superposition of five homogeneous parts. Each part can be regarded either as a set of elements $\mathfrak{M}.\mathbf{x}$ as in column six of Table 5, or as a function according to the domain of x, ranging over $\{\mathfrak{M}.1_i^r, \mathfrak{M}.2_i^r, \mathfrak{M}.3_i^r, \mathfrak{M}.\mathrm{b}_{ij}^m, \mathfrak{M}.\mathrm{d}_i\}$; we use $\top$ and $\bot$ to denote defined and undefined images. Matrix update $\mathfrak{M}\{\mathrm{x} := \tilde{v}\}$ is overriding.

Table 6. Example matrix

Rounds Number	Phase	Processes 1	2	3	4	5
1	1	$(v_1,0,\surd)$	$(v_2,0,\surd\!\!\surd)$	$(v_3,0,\surd\!\!\surd\!\!\surd)$	$(v_4,0,\surd\!\!\surd\!\!\surd)$	$(v_5,0,\surd\!\!\surd\!\!\surd)$
	2	—	$(v_3,\surd)$	$(v_3,\surd)$	$(v_3,\surd\!\!\surd)$	$(v_3,\surd\!\!\surd\!\!\surd)$
	3	—		$(f,\surd\!\!\surd\!\!\surd)$	$(f,\surd\!\!\surd\!\!\surd)$	$(t,\surd\!\!\surd)$
2	1	$(v_3,1,\surd)$		$(v_3,0,\surd\!\!\surd)$	$(v_4,0,\surd)$	$(v_3,1,\surd\!\!\surd)$
	2		—			
	3		—	$(f,\surd\!\!\surd)$	$(f,\surd)$	$(f,\surd\!\!\surd)$
3	1			$(v_3,0,\surd\!\!\surd\!\!\surd)$	$(v_4,0,\surd\!\!\surd\!\!\surd)$	$(v_3,1,\surd\!\!\surd\!\!\surd)$
	2	$(v_3,\surd\!\!\surd)$	$(v_3,\surd)$	—	$(v_3,\surd\!\!\surd)$	$(v_3,\surd\!\!\surd\!\!\surd)$
	3			—	$(f,\surd\!\!\surd\!\!\surd)$	$(f,\surd\!\!\surd\!\!\surd)$
4	1			$(v_3,3,\surd)$	$(v_4,0,\surd\!\!\surd\!\!\surd)$	$(v_3,1,\surd\!\!\surd)$
	2				—	
	3			$(f,\surd)$	—	

Matrix Semantics. The initial matrix of *Consensus* is denoted by

$$\mathfrak{Consensus}_{(v_1..,v_n)} := \mathcal{M}[\![\, Consensus_{(v_1..,v_n)} \,]\!] = \emptyset\{\, \forall i : 1_i^1 := (v_i, 0, \surd) \,\}$$

In order to simulate the behavior of networks at the level of matrices, we propose an operational semantics that manipulates matrices precisely mimicking the behavior of their corresponding networks. As with networks, the rules in Tables 2 and 3, where networks and their transitions are replaced by matrices and their (equally labeled) transitions, allow us to completely separate the treatment of behavior in the context of crashes from the description of the behavior of messages in the matrix. The rules are given in the full paper. Here, we just look at an example of a matrix for $n = 5$ (Table 6) that is reachable by using the matrix semantics. For instance, to be a valid matrix, coordinators can only have proceeded to the next round if they received ($\surd\!\!\surd\!\!\surd$) a majority−1 of 3rd-phase messages; c.f. the coordinators of rounds 1 and 3. Also, participants proceed with the value of the previous round if they *nack* (f), or with the proposed value of the previous coordinator if they *ack* (t); c.f. process 5 in its rounds 2 and 4.

Some transitions that are enabled from within the example matrix are: messages with tag $\surd$ may be released to the network and get tag $\surd\!\!\surd$; process 4 may receive 1st-phase messages from process 5, from either round 2 or 4. Many other requirements like these are represented by the 12 rules of the matrix semantics.

$\mathcal{N}[\![\ ]\!]$*: From Matrices to Networks.* We only note here that the presence of all previously sent messages, distinguishing all their transmission states, allows us to uniquely reconstruct the term counterpart, i.e., for every site i we may uniquely determine its phase $\mathrm{phs}_i(\mathfrak{M}) {:=} \mathsf{Y}_i \in \{\, \mathsf{P1}_i, \mathsf{P2}_i, \mathsf{P3}_i, \mathsf{P4}_i, \mathsf{R}_i, \mathsf{Z}_i \,\}$ with accompanying parameters $r {:=} \mathrm{rnd}_i(\mathfrak{M}), v, s, L$ and its decision state $\mathrm{dec}_i(\mathfrak{M}) \in \{\, \mathsf{D}_i, \mathbf{0} \,\}$.

The matrix semantics mimics the network semantics very closely.

Proposition 1 (Operational Correspondence). *Let* $Consensus^{\mathrm{n}} \rightarrow_{\mathrm{n}}^{*} N$.

1. *If* $N \xrightarrow{\mu@i}_{\mathrm{n}} N'$, *then* $\mathcal{M}[\![\, N \,]\!] \xrightarrow{\mu@i} \mathcal{M}[\![\, N' \,]\!]$.

2. If $\mathcal{M}[\![N]\!] \xrightarrow{\mu@i} \mathfrak{M}$, *then* $N \xrightarrow{\mu@i}_{\rm n} \Lleftarrow\!\!\!\Rrightarrow \mathcal{N}[\![\mathfrak{M}]\!]$.

Normalized network runs can then straightforwardly be translated step-by-step into matrix runs using $\mathcal{M}[\![\ ]\!]$, and vice versa using $\mathcal{N}[\![\ ]\!]$. If a network run is infinite, then its corresponding matrix run is infinite as well. Or, conversely, if a corresponding matrix run is finite, then the original network run must have been finite as well. Furthermore, since we produce system runs – where the distributed algorithm is embedded into our failure-sensitive environments – with either networks or matrices, the correspondence carries over also to the system level. Therefore, we may use the matrix semantics instead of the original network semantics to reason about the Consensus algorithm and its properties.

5 Properties of the Algorithm: Consensus

In this section, we prove the three required Consensus properties – validity, agreement, and termination – using the matrix structures. As the graphical sketches in Table 7 show, we heavily exploit the fact that the matrix abstraction allows us to analyze message patterns that have been sent in the past. We do not need to know precisely in which order all the messages have been sent, but we do need to have some information about the order in which they *cannot* have been sent. Our formal matrix semantics provides us with precisely this kind of information.

We conclude this section by transferring the results back to networks.

Validity. From the definition, every decided value has initially been proposed.

Proposition 2 (Validity). *Let* $\mathfrak{Consensus}_{(v_1..,v_n)} \rightarrow^* \mathfrak{M}$.
If $\mathfrak{M}.\mathrm{d}_i = (j, m, r, v, t)$, *then there is* $k \in \{1..,n\}$ *with* $v = v_k$.

Agreement. We call $\mathrm{val}^r(\mathfrak{M})$ the value that the coordinator of round r in $\mathfrak{M}$ tried to impose in its second phase; it may be undefined. In the Introduction, we said that a value gets *locked* as soon as enough processes have, in the same round, positively acknowledged to the coordinator of this round. This condition translates into matrix terminology, as follows:

Definition 2. *A value v is called* locked *for round r in matrix $\mathfrak{M}$, written* $\mathfrak{M} \overset{r}{\mapsto} v$, *if* $\#\{\, j \mid \mathfrak{M}.3^r_j = (\mathrm{t}, \cdot) \,\} \geq \lceil \frac{n+1}{2} \rceil - 1$.

Note the convenience of the matrix abstraction to access the messages that were sent in the past, without having to look at the run leading to the current state. Now, if $\mathfrak{M} \overset{r}{\mapsto} v$ then $v = \mathrm{val}^r(\mathfrak{M})$. Also, broadcast is always for a locked value.

Lemma 2. *If* $\mathfrak{M}.\mathrm{b}^m_{ij} = (r, v, \cdot)$, *then* $\mathfrak{M} \overset{r}{\mapsto} v$.

Lemma 3. *If* $\mathfrak{M} \overset{r}{\mapsto} v_1$ *and* $\mathfrak{M} \overset{r}{\mapsto} v_2$, *then* $v_1 = v_2$.

The key idea is to compare lockings in two different rounds.

Table 7. Matrix proofs

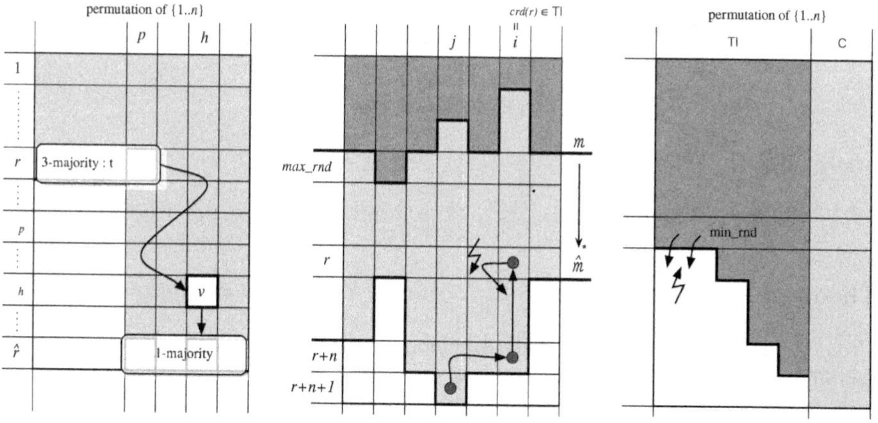

(I) Pre-Agreement (II) Ω-Finiteness (III) Termination

Proposition 3 (Pre-Agreement). *If* $\mathfrak{M} \overset{r_1}{\mapsto} v_1$ *and* $\mathfrak{M} \overset{r_2}{\mapsto} v_2$, *then* $v_1 = v_2$.

Note that both lockings have already happened in the past of $\mathfrak{M}$.

Proof (Sketch). Suppose that $\mathfrak{M} \overset{r}{\mapsto} v$, so $v = \mathrm{val}^r(\mathfrak{M})$. We prove by course-of-value induction that for all $\hat{r} > r$, if $\mathfrak{M} \overset{\hat{r}}{\mapsto} \hat{v}$, then $v = \hat{v}$.

First, in both rounds r and $\hat{r}$, a majority is responsible for the locking. In Table 7(I), we make explicit (by permutation) that there is a process p that belongs to both majorities. Then, let h be the process that won the first phase of round $\hat{r}$ in that $\mathrm{crd}(\hat{r})$ chose h's estimate as its round-proposal. Using the matrix semantics, we identify the rounds $\hat{s}_p$ and $\hat{s}_h$, in which p and h acknowledged the estimate that they still believe in at round $\hat{r}$. By a number of of auxiliary lemmas on matrices we conclude that $r \leq \hat{s}_p \leq \hat{s}_h < \hat{r}$.

Now, if $r = \hat{r}$, then trivially $\mathrm{val}^r(\mathfrak{M}) = \mathrm{val}^{\hat{r}}(\mathfrak{M})$ (Lemma 3).

If $\hat{r} > r$ then, by induction, we have $\mathrm{val}^r(\mathfrak{M}) = \mathrm{val}^{\hat{s}_h}(\mathfrak{M})$, and since h preserves the value it adopted in $\hat{s}_h$ until it reaches $\hat{r}$, where it "wins", also $\mathrm{val}^{\hat{s}_h}(\mathfrak{M}) = v = \mathrm{val}^{\hat{r}}(\mathfrak{M})$, we conclude $\mathrm{val}^r(\mathfrak{M}) = \mathrm{val}^{\hat{r}}(\mathfrak{M})$. □

Theorem 1 (Agreement). *If* $\mathfrak{M}.\mathrm{d}_i = (\cdot,\cdot,\cdot,v_i,\cdot)$ *and* $\mathfrak{M}.\mathrm{d}_j = (\cdot,\cdot,\cdot,v_j,\cdot)$, *then* $v_i = v_j$.

Proof (Sketch). If $\mathfrak{M}.\mathrm{d}_i = (k_i, m_i, r_i, v_i, \cdot)$, then by the only matrix rule to generate d_i-entries there must (have) be(en) r_i with $\mathfrak{M}.\mathrm{b}^{m_i}_{ik_i} = (r_i, v_i, \cdot)$. Analogously for j: if $\mathfrak{M}.\mathrm{d}_j = (k_j, m_j, r_j, v_j, \cdot)$, there must (have) be(en) r_j with $\mathfrak{M}.\mathrm{b}^{m_j}_{jk_j} = (r_j,_j\,,\cdot)$. By Lemma 2, both $\mathfrak{M} \overset{r_i}{\mapsto} v_i$ and $\mathfrak{M} \overset{r_j}{\mapsto} v_j$. By Proposition 3, we conclude $v_i = v_j$. □

Termination. In an infinite run, every round is reached.

Lemma 4 (Infinity). *Let R denote an infinite system run of $\mathfrak{Consensus}$. Then, for all $r > 0$, there is a prefix of R of the form*

$$(\emptyset, \emptyset) \vdash \mathfrak{Consensus} \rightarrow^* \Gamma \vdash \mathfrak{M}$$

where $\mathfrak{M}.1_i^r = \top$ for some i.

Proof (Sketch). By combinatorics on the number of steps per round. □

Theorem 2 (Ω-Finiteness). *All Ω-runs of $\mathfrak{Consensus}$ are finite.*

Proof (Sketch). Assume, by contradiction, to have an infinite Ω-run. The bold line $\mathfrak{M}$ in Table 7(II), marks the global state at instant t, when process i becomes $\in \mathsf{TI}$. Call max_rnd the greatest round at time t, and $r > max\_rnd$ the first round in which $i = \mathrm{crd}(r)$. Since the run is infinite, with Lemma 4 there is a time $\hat{t} > t$, where we reach state $\hat{\mathfrak{M}}$, where round $r{+}n{+}1$ is populated by some j. Since $i \in \mathsf{TI}$, j can reach $r{+}n{+}1$ only by positively acknowledging i in round $r{+}n$. So i was in $r{+}n$, therefore also in r. Since i was in r and has gone further, it has been suspected. But here we get a contradiction because in round r already $i \in \mathsf{TI}$ and no process was allowed to suspect it, while the matrix $\mathfrak{M}$ evolved into $\hat{\mathfrak{M}}$. So, no Ω-run can be infinite. □

Theorem 3 (Termination). *All Ω-runs of $\mathfrak{Consensus}$ are of the form*

$$(\emptyset, \emptyset) \vdash \mathfrak{Consensus} \rightarrow^* (\mathsf{TI}, \mathsf{C}) \vdash \mathfrak{M} \not\rightarrow$$

with $\mathsf{TI} \uplus \mathsf{C} = \{1..,n\}$ and $i \in \mathsf{TI} \neq \emptyset$ implies that $\mathfrak{M}.\mathrm{d}_i = \top$.

Proof (Sketch). We first show that if there was $i \in \mathsf{TI}$ with $\mathfrak{M}.\mathrm{d}_i = \bot$, then actually $\mathfrak{M}.\mathrm{d}_j = \bot$ for all $j \in \mathsf{TI}$. Since $\mathfrak{M} \not\rightarrow$, we may thus call all processes in $j \in \mathsf{TI}$ as being in deadlock. Then, we proceed by contradiction. We concentrate on the non-empty set $Min \subseteq \mathsf{TI}$ of processes in the currently minimal round. The contradiction arises, as in Table 7(III), by using the matrix semantics to show that Min must be empty, otherwise contradicting that $\mathfrak{M} \not\rightarrow$. □

Back to the Process Calculus. With Table 5, we observe that the definedness of an entry $\mathfrak{M}.\mathrm{d}_i = \top$ corresponds to a message $\overline{decide_i}\langle\cdots\rangle$ having been sent. Therefore, and with the operational correspondence (Proposition 1), which closely resembles strong bisimulation, all the previous results carry over to networks.

References

[BH00] M. Berger and K. Honda. The Two-Phase Commitment Protocol in an Extended pi-Calculus. In L. Aceto and B. Victor, eds, *Proceedings of EXPRESS '00*, volume 39.1 of *ENTCS*. Elsevier Science Publishers, 2000.

[CHT96] T.D. Chandra, V. Hadzilacos and S. Toueg. The Weakest Failure Detector for Solving Consensus. *Journal of the ACM*, 43(4):685–722, 1996.

[CT96] T.D. Chandra and S. Toueg. Unreliable Failure Detectors for Reliable Distributed Systems. *Journal of the ACM*, 43(2):225–267, 1996.

[FLP85] M.J. Fisher, N. Lynch and M. Patterson. Impossibility of Distributed Concensus with One Faulty Process. *Journal of the ACM*, 32(2):374–382, 1985.

[PSL00] A. Pogosyants, R. Segala and N. Lynch. Verification of the Randomized Consensus Algorithm of Aspnes and Herlihy: a Case Study. *Distributed Computing*, 13(3):155–186, 2000.

[RH01] J. Riely and M. Hennessy. Distributed Processes and Location Failures. *Theoretical Computer Science*, 226:693–735, 2001.

[WS00] P. Wojciechowski and P. Sewell. Nomadic Pict: Language and Infrastructure Design for Mobile Agents. *IEEE Concurrency*, 8(2):42–52, 2000.

Linear Forwarders

Philippa Gardner[1], Cosimo Laneve[2], and Lucian Wischik[2]

[1] Imperial College, London
pg@doc.ic.ac.uk
[2] University of Bologna, Italy
laneve@cs.unibo.it
lu@wischik.com

Abstract. A *linear forwarder* is a process which receives one message on a channel and sends it on a different channel. Such a process allows for a simple implementation of the asynchronous pi calculus, by means of a direct encoding of the pi calculus' *input capability* (that is, where a received name is used as the subject of subsequent input). This encoding is fully abstract with respect to barbed congruence.
Linear forwarders are actually the basic mechanism of an earlier implementation of the pi calculus called the *fusion machine*. We modify the fusion machine, replacing fusions by forwarders. The result is more robust in the presence of failures, and more fundamental.

1 Introduction

Distributed interaction has become a necessary part of modern programming languages. We regard the asynchronous pi calculus as a basis for such a language. In the pi calculus, a program (or process) has a collection of channels, and it runs through interaction over these channels. A possible distributed implementation is to let each channel belong to a single location. For instance, there is one location for the channels u, v, w and another for x, y, z, and the input resource $u(a).P$ goes in the first location. If an output $\overline{u}\,x$ should arise anywhere else in the system, it knows where it can find a matching input. This basic scheme is used in the join calculus [7], in the $\pi_{1\ell}$ calculus [3], and in the fusion machine [8]. (A different approach is taken in Dπ [2], in nomadic pict [19], and in the ambient calculus [6], where agent migration is used for remote interaction.)

We immediately face the problem of **input capability**, which is the ability in the pi calculus to receive a channel name and subsequently accept input on it. Consider the example $x(u).u(v).Q$. This program is located at (the location of) x, but upon reaction with $\overline{x}\,w$ it produces the continuation $w(v).Q\{^w/_u\}$ – and this continuation is still at x, whereas it should actually be at w. Solving the problem of input capability is the key challenge in distributing the pi calculus.

The point of this paper is to solve the problem of input capability with a language that is "just right" – it neither disallows more features than necessary (as does the join calculus), nor adds more implementation work than is necessary (as does the fusion machine). One measure of our solution is that we obtain full abstraction with the asynchronous pi calculus, up to weak barbed congruence.

R. Amadio, D. Lugiez (Eds.): CONCUR 2003, LNCS 2761, pp. 415–430, 2003.

First of all, let us consider in more detail the other solutions to input capability. The join calculus and localised pi calculus [13] simply disallow it: that is, in a term $x(u).P$, the P may not contain any inputs on channel u. The problem now is how to encode input capability into such a *localised* calculus. An encoding is possible, but awkward: when the term $x(u).u(v).Q \mid \overline{x}\,w$ is encoded and then performs the reaction, it does not perform the substitution $\{^w/_u\}$, but rather encodes this substitution as a persistent forwarder between w and u. Next, a firewall is needed to protect the protocol used by these forwarders. (The forwarder is called a "merged proxy pair" in the join calculus).

The fusion machine instead implements input capability through the runtime migration of code. In our example, $w(v).Q\{^w/_u\}$ would migrate from x over to w after the interaction. The migration is costly however when the continuation Q is large. In addition, code migration requires an elaborate infrastructure. To mitigate this, a large amount of the work on the fusion machine involved an encoding of arbitrary programs into *solos* programs (ones which have only simple continuations) without incurring a performance penalty. But the encoding used fusions, implemented through persistent trees of forwarders, which seem awkward and fragile in the presence of failures.

The solution presented in this paper is to disallow general input capability, and to introduce instead a limited form of input, the **linear forwarder**. A linear forwarder $x{\multimap}y$ is a process which allows just one x to be turned into a y. The essential point is that this limited form can be used to easily encode general input capability. For example, consider the pi calculus term $x(u).u(v).Q$. We will encode it as

$$x(u).(u')(u{\multimap}u' \mid u'(v).Q')$$

where the input $u(v)$ has been turned into a local input $u'(v)$ at the same location as x, and where the forwarder allows one output on u to interact with u' instead. The encoding has the property that if the forwarder $u{\multimap}u'$ exists, then there is guaranteed to be an available input on u'. We remark that linearity is crucial: if the forwarder persisted, then the guarantee would be broken; any further u turned into u' would become inert since there are no other inputs on u'.

One might think of a linear forwarder $x{\multimap}y$ as the pi calculus agent $x(u).\overline{y}\,u$ located at x. This agent would be suitable for a point-to-point network such as the Internet. But we have actually turned forwarders into first-class operators in order to abstract away from any particular implementation. This is because other kinds of networks benefit from different implementations of linear forwarders. In a broadcast network, $x{\multimap}y$ might be located at y; whenever it hears an offer of $\overline{x}\,u$ being broadcast, the machine at y can take up the offer. Another possibility is to use a shared tuple-space such as Linda [9], and ignore all linearity information. (The fusion machine also amounts to a shared state which ignores linearity).

In this paper we show how to encode the pi calculus into a linear forwarder calculus. Conversely, we also show how linear forwarders can be encoded into the pi calculus. We therefore obtain full abstraction with respect to barbed congruence.

We also describe a *linear forwarder machine*. It is a simplified form of our earlier fusion machine, and more robust with respect to failures. This machine gives an implementation of distributed rendezvous which can be performed locally. In this respect it is different from Facile [10], which assumes a three-party handshake. This handshake is a protocol for interaction, and so prevents full abstraction. We prove full abstraction between the machine and the linear forwarder calculus, with respect to barbed congruence.

Related Work. Forwarders have already been studied in detail by the pi community. Much work centres around the πI calculus [17] – a variant of the pi calculus in which only private names may be emitted, as in $(x)\overline{u}\,x$. Boreale uses forwarders to encode the emission of free names [4]: the reaction $u(a).Q \mid \overline{u}\,x$ does not perform the substitution $\{^x/a\}$, but instead encodes it as a persistent forwarder between a and x. The same technique is used by Merro and Sangiorgi [13] in proofs about the localised pi calculus; and both are inspired by Honda's *equators* [11], which are bidirectional forwarders. Something similar is also used by Abadi and Fournet [1]. When channels are used linearly, Kobayashi et al. [12] show that a linear forwarder can simulate a substitution.

We remark upon some differences. If substitutions are encoded as persistent forwarders, then the ongoing execution of a program will create steadily more forwarders. In contrast, we perform substitution directly, and in our setting the number of forwarders decreases with execution. More fundamentally, the πI calculus uses forwarders to effect the substitution of data, and they must be persistent (nonlinear) since the data might be used arbitrarily many times by contexts. We use forwarders to effect the input capability of code, and this is linear because a given piece of source code contains only finitely many input commands. Our proofs are similar in structure to those of Boreale, but are much simpler due to linearity.

Structure. The structure of this paper is as follows. Section 2 gives the linear forwarder calculus, and shows how to encode the pi calculus (with its input mobility) into this calculus. Section 3 gives bisimulations for the linear forwarder calculus, and Section 4 proves full abstraction of the pi calculus encoding. Section 5 describes a distributed abstract machine for implementing the linear forwarder calculus, and Section 6 proves full abstraction for this implementation. We outline future developments in Section 7.

2 The Linear Forwarder Calculus

We assume an infinite set $\mathcal{N}$ of *names* ranged over by $u, v, x, \ldots$. Names represent communication channels, which are also the values being transmitted in communications. We write $\widetilde{x}$ for a (possibly empty) finite sequence $x_1 \cdots x_n$ of names. *Name substitutions* $\{\widetilde{y}/\widetilde{x}\}$ are as usual.

Definition 1 (Linear forwarder calculus). *Terms are given by*

$$P \ ::= \ \mathbf{0} \ \mid \ \overline{x}\,\widetilde{y} \ \mid \ x(\widetilde{y}).P \ \mid \ (x)P \ \mid \ P|P \ \mid \ !P \ \mid \ x \multimap y$$

Structural congruence $\equiv$ *is the smallest equivalence relation satisfying the following and closed with respect to contexts and alpha-renaming:*

$$P|\mathbf{0} \equiv P \qquad P|Q \equiv Q|P \qquad P|(Q|R) \equiv (P|Q)|R \qquad !P \equiv P|!P$$
$$(x)(y)P \equiv (y)(x)P \qquad (x)(P|Q) \equiv P \mid (x)Q \quad \textit{if } x \notin \mathrm{fn}\, P$$

Reaction is the smallest equivalence satisfying the following and closed under $\equiv$, $(x)\_$ *and* $\_ \mid \_$*:*

$$u(\widetilde{x}).P \mid \overline{u}\,\widetilde{y} \;\rightarrow\; P\{\widetilde{y}/\widetilde{x}\} \qquad \overline{x}\,\widetilde{u} \mid x \multimap y \;\rightarrow\; \overline{y}\,\widetilde{u}$$

The operators in the syntax are all standard apart from the linear forwarder $x \multimap y$. This allows one output on x to be transformed into one on y, through the second reaction rule. In the output $\overline{x}\,\widetilde{y}$ and the input $x(\widetilde{y}).P$, the name x is called the *subject* and the names $\widetilde{y}$ are the *objects*. In the restriction $(x)P$, the name x is said to be *bound*. Similarly, in $x(\widetilde{y}).P$, the names $\widetilde{y}$ are bound in P. The *free names* in P, denoted $\mathrm{fn}(P)$, are the names in P with a non-bound occurrence. We write $(x_1 \cdots x_n)P$ for $(x_1)\cdots(x_n)P$.

Next we make a *localised* sub-calculus, by adding the *no-input-capability* constraint. It is standard from the π_L calculus [13] and the join calculus that such a constraint makes a calculus amenable to distributed implementation.

Definition 2 (Localised calculus). *The* localised *linear forwarder calculus, which we abbreviate* $L\ell$, *is the sub-calculus of the linear forwarder calculus which satisfies the* no-input-capability *constraint: in* $x(\widetilde{u}).P$, *the* P *has no free occurrence of* $\widetilde{u}$ *as the subject of an input.*

We remark that the no-input-capability constraint is preserved by structural congruence and by reaction.

The asynchronous pi calculus [5] is a sub-calculus of the linear forwarder calculus, obtained by dropping linear forwarders. We give an encoding of the asynchronous pi calculus into the *localised* linear forwarder calculus $L\ell$, showing that the input capability can be expressed using forwarders and local inputs. Henceforth, when we refer to the pi calculus, we mean the asynchronous pi calculus.

Definition 3 (Encoding pi). *The encoding* $[\![\cdot]\!]$ *maps terms in the pi calculus into terms in the* $L\ell$ *calculus as follows. (In the input and restriction cases, assume that the bound names do not clash with* $\widetilde{u}$*.) Define* $[\![P]\!] = [\![P]\!]_{\emptyset}$, *where*

$$\begin{aligned}
[\![x(\widetilde{y}).P]\!]_{\widetilde{u}} &= \begin{cases} x(\widetilde{y}).[\![P]\!]_{\widetilde{u}\widetilde{y}} & \textit{if } x \notin \widetilde{u} \\ (u_i')(u_i \multimap u_i' \mid u_i'(\widetilde{y}).[\![P]\!]_{\widetilde{u}\widetilde{y}}) & \textit{if } x = u_i, u_i \in \widetilde{u} \end{cases} \\
[\![(x)P]\!]_{\widetilde{u}} &= (x)([\![P]\!]_{\widetilde{u}}) \\
[\![P|Q]\!]_{\widetilde{u}} &= [\![P]\!]_{\widetilde{u}} \mid [\![Q]\!]_{\widetilde{u}} \\
[\![!P]\!]_{\widetilde{u}} &= ![\![P]\!]_{\widetilde{u}} \\
[\![\overline{x}\,\widetilde{y}]\!]_{\widetilde{u}} &= \overline{x}\,\widetilde{y} \\
[\![\mathbf{0}]\!]_{\widetilde{u}} &= \mathbf{0}
\end{aligned}$$

To understand the encoding, note that we use "primed" names to denote local copies of names. So the encoding of $x(u).u(y).P$ will use a new channel u' and a process $u'(y).P$, both at the same location as x. It will also create exactly one forwarder $u \multimap u'$, from the argument passed at runtime to u'. Meanwhile, any output use of u is left unchanged.

To illustrate the connection between the reactions of a pi term and of its translation, we consider the pi calculus reduction $\overline{u}\, y \mid u(x).P \;\rightarrow\; P\{^y\!/\!_x\}$. By translating we obtain:

$$\begin{aligned}
[\![\overline{u}\, y \mid u(x).P]\!]_u &= \overline{u}\, y \mid (u')(u \multimap u' \mid u'(x).[\![P]\!]_{xu}) \\
&\rightarrow (u')(\overline{u}'y \mid u'(x).[\![P]\!]_{xu}) \\
&\rightarrow (u')([\![P]\!]_{xu}\{^y\!/\!_x\}) \\
&\equiv [\![P]\!]_{xu}\{^y\!/\!_x\}
\end{aligned}$$

Note that the final state of the translated term is subscripted on x and u, not just on u. In effect, the translated term ends up with some garbage that was not present in the original. Because of this garbage, it is not in general true that $Q \rightarrow Q'$ implies $[\![Q]\!] \rightarrow^* [\![Q']\!]$; instead we must work up to some behavioural congruence. The following section deals with barbed congruence.

We remark that linearity is crucial in the translation. For instance, consider a non-linear translation where forwarders are replicated:

$$[\![u(x).P]\!]_u \;=\; (u')(!u \multimap u' \mid u'(y).P)$$

Then consider the example

$$\begin{aligned}
[\![u().P \mid u().Q \mid \overline{u} \mid \overline{u}]\!]_u &= (u')(!u \multimap u' \mid u'().P) \mid (u'')(!u \multimap u'' \mid u''().Q) \mid \overline{u} \mid \overline{u} \\
&\rightarrow (u')(P \mid !u \multimap u') \mid (u'')(!u \multimap u'' \mid u''().Q) \mid \overline{u} \\
&\rightarrow (u')(P \mid \overline{u}' \mid !u \multimap u') \mid (u'')(!u \multimap u'' \mid u''().Q)
\end{aligned}$$

Here, both outputs were forwarded to the local name u', even though the resource $u'().P$ had already been used up by the first one. This precludes the second one from reacting with Q – a reaction that would have been possible in the original pi calculus term. We need linearity to prevent the possibility of such dead ends.

3 Bisimulation and Congruence

We use barbed congruence [15] as our semantics for the $L\ell$ calculus.

Definition 4 (Barbed congruence). *The observation relation $P \downarrow u$ is the smallest relation generated by*

$$\begin{array}{llll}
\overline{u}\,\widetilde{x} \downarrow u & & P|Q \downarrow u & \textit{if } P \downarrow u \textit{ or } Q \downarrow u \\
(x)P \downarrow u & \textit{if } P \downarrow u \textit{ and } u \neq x & !P \downarrow u & \textit{if } P \downarrow u
\end{array}$$

We write $\Downarrow$ for $\rightarrow^*\downarrow$ and $\Rightarrow$ for $\rightarrow^*$. A symmetric relation $\mathcal{R}$ is a weak barbed bisimulation *if whenever $P \;\mathcal{R}\; Q$ then*

1. $P \Downarrow u$ implies $Q \Downarrow u$
2. $P \to P'$ implies $Q \Rightarrow Q'$ such that $P' \mathcal{R} Q'$

Let $\dot{\approx}$ be the largest weak barbed bisimulation. Two terms P and Q are weak barbed congruent *in the $L\ell$ calculus when, for every C, then $C[P] \dot{\approx} C[Q]$, where $C[P]$ and $C[Q]$ are assumed to be terms in the $L\ell$ calculus. Let $\approx$ be the least relation that relates all congruent terms.*

We remark that barbed bisimulation $\dot{\approx}$ is defined for the linear forwarder calculus. However, the weak barbed *congruence* $\approx$ is predicated upon the $L\ell$ subcalculus. Similar definitions may be given for the pi calculus, and, with abuse of notation, we keep $\dot{\approx}$ and $\approx$ denoting the corresponding semantic relations.

As an example of $\approx$ congruent terms in the $L\ell$ calculus, we remark that

$$u(x).P \quad \approx \quad u(x').(x)(!x{\multimap}x' \mid !x'{\multimap}x \mid P). \tag{1}$$

This is a straightforward variant of a standard result for equators [14], and we use it in Lemma 9.

Our overall goal is to prove that the encoding $[\![\cdot]\!]$ preserves the $\approx$ congruence. The issue, as described near the end of the previous section, is that an encoded term may leave behind garbage. To show that it is indeed garbage, we must prove that $[\![P]\!]_u$ and $[\![P]\!]_{ux}$ are congruent. But the barbed semantics offer too weak an induction hypothesis for this proof. A standard alternative technique (used for example by Boreale [4]) is to use barbed semantics as the primary definition, but then to use in the proofs a labelled transition semantics and its corresponding bisimulation – which is stronger than barbed congruence. The remainder of this section is devoted to the labelled semantics.

Definition 5 (Labelled semantics). *The labels, ranged over by μ, are the standard labels for interaction $\xrightarrow{\tau}$, input $\xrightarrow{u(\widetilde{x})}$ and possibly-bound output $\xrightarrow{(\widetilde{z})\overline{u}\,\widetilde{x}}$ where $\widetilde{z} \subseteq \widetilde{x}$. The bound names $\mathrm{bn}(\mu)$ of these input and output labels are $\widetilde{x}$ and $\widetilde{z}$ respectively.*

$$u(\widetilde{x}).P \xrightarrow{u(\widetilde{x})} P \qquad \overline{u}\,\widetilde{x} \xrightarrow{\overline{u}\,\widetilde{x}} \mathbf{0} \qquad u{\multimap}v \xrightarrow{u(\widetilde{x})} \overline{v}\,\widetilde{x}$$

$$\frac{P \xrightarrow{\mu} P' \quad y \notin \mu}{(y)P \xrightarrow{\mu} (y)P'} \qquad \frac{P \xrightarrow{(\widetilde{z})\overline{u}\,\widetilde{x}} P' \quad y \neq u,\ y \in \widetilde{x} \backslash \widetilde{z}}{(y)P \xrightarrow{(y\widetilde{z})\overline{u}\,\widetilde{x}} P'}$$

$$\frac{P|!P \xrightarrow{\mu} P'}{!P \xrightarrow{\mu} P'} \qquad \frac{P \xrightarrow{\mu} P' \quad \mathrm{bn}(\mu) \cap \mathrm{fn}(Q) = \emptyset}{P|Q \xrightarrow{\mu} P'|Q}$$

$$\frac{P \xrightarrow{(\widetilde{z})\overline{u}\,\widetilde{y}} P' \quad Q \xrightarrow{u(\widetilde{x})} Q' \quad \widetilde{z} \cup \mathrm{fn}(Q) = \emptyset}{P|Q \xrightarrow{\tau} (\widetilde{z})(P'|Q'\{\widetilde{y}/\widetilde{x}\})}$$

The transitions of $P|Q$ have mirror cases, which we have omitted. We implicitly identify terms up to alpha-renaming $\equiv_\alpha$: that is, if $P \equiv_\alpha \xrightarrow{\mu} P'$ then $P \xrightarrow{\mu} P'$. We write $\xRightarrow{\mu}$ for $\xrightarrow{\tau}{}^* \xrightarrow{\mu}$.

A symmetric relation $\mathcal{R}$ is a weak labelled bisimulation if whenever $P \mathcal{R} Q$ then $P \xrightarrow{\mu} P'$ implies $Q \stackrel{\mu}{\Longrightarrow} Q'$. Let $\approx_\ell$ be the largest labelled bisimulation.

This definition is given for terms in the full linear forwarder calculus. It is a standard result that $\approx_\ell$ is a congruence with respect to contexts in the full calculus, and hence also with respect to contexts in the $L\ell$ calculus and the pi calculus. The connection between labelled and barbed semantics is also standard:

Lemma 6. *In the $L\ell$ calculus,*

1. $P \to P'$ *iff* $P \xrightarrow{\tau}\equiv P'$.
2. $P \downarrow u$ *iff* $P \xrightarrow{(\widetilde{z})\overline{u}\,\widetilde{x}} P'$.
3. $\approx_\ell \subset \approx$.

The bisimulation $\approx_\ell$ allows for some congruence properties to be proved trivially: (the first will be used in Proposition 10)

$$\begin{aligned} u{\multimap}v \;\approx_\ell\; u(\widetilde{x}).\overline{v}\,\widetilde{x} \;&\approx_\ell\; (u')(u{\multimap}u' \mid u'(\widetilde{x}).\overline{v}\,\widetilde{x}). \\ u(x).P \;&\approx_\ell\; (u')(u{\multimap}u' \mid u'(x).P). \end{aligned} \tag{2}$$

4 Full Abstraction for the Pi Calculus Encoding

The $L\ell$ calculus is fully abstract with respect to the pi calculus encoding: $P \approx Q$ in pi if and only if $[\![P]\!] \approx [\![Q]\!]$ in $L\ell$. Informally, this is because the pi calculus input capability can be encoded with linear forwarders (as in Definition 3); and conversely a linear forwarder $x{\multimap}y$ can be encoded as the pi calculus term $x(\widetilde{u}).\overline{y}\,\widetilde{u}$. This section builds up to a formal proof of the result.

The structure of the proof follows that of Boreale ([4], Definition 2.5 to Proposition 3.6). However, the proofs are significantly easier in our setting. We begin with a basic lemma about the encoding $[\![\cdot]\!]$.

Lemma 7. *In the linear forwarder calculus,*

1. $[\![P]\!]_{\widetilde{z}} \approx_\ell [\![P]\!]_{x\widetilde{z}}$.
2. $[\![P]\!]_{\widetilde{z}}\{y/x\} \approx_\ell [\![P\{y/x\}]\!]_{\widetilde{z}}$.
3. $[\![P]\!]_{\widetilde{z}}\{\widetilde{y}/\widetilde{x}\} \approx_\ell (u')(\overline{u}'\widetilde{y} \mid u'(\widetilde{x}).[\![P]\!]_{\widetilde{z}\widetilde{x}})$.

Proof. The first two are trivial inductions on P. The last one follows directly. □

We draw attention to the first part of Lemma 7. This is an important simplifying tool. It means that, even though the encoding $[\![u(x).P]\!]_u = (u')(u'{\multimap}u \mid u'(x).[\![P]\!]_{ux})$ involves progressively more subscripts, they can be ignored up to behavioural equivalence. Thus, although a context C might receive names $\widetilde{x}$ in input, we can ignore this fact: $C\big[[\![P]\!]_{\widetilde{x}}\big] \approx_\ell C\big[[\![P]\!]\big]$ in the linear forwarder calculus. (Notice that, given a localised term $C\big[[\![P]\!]_{\widetilde{x}}\big]$, it is not necessary the case that $C\left[\llbracket P \rrbracket\right]$ is also localised; hence the result does not carry over to $\approx$, which is only defined for $L\ell$ contexts). Part 1 does not hold for Boreale, and so his equivalent of Part 3 uses a significantly longer (5-page) alternative proof. The essential difference is that Boreale produces forwarders upon reaction; we consume them.

Note that this section has even simpler proofs, using the property $P \approx_\ell \llbracket P \rrbracket$ which is deduced directly from Lemma 7 and the definition of $\llbracket\cdot\rrbracket$. However, this property relates terms from two different sub-calculi of the linear forward calculus, which some readers found inelegant – so we have avoided it.

The following proposition is equivalent to Boreale's Propositions 3.5–3.6:

Proposition 8. *For P, Q in the pi calculus, $P \dot\approx Q$ if and only if $\llbracket P \rrbracket \dot\approx \llbracket Q \rrbracket$.*

We will also need the following lemma. It generalises Lemma 7.1 to apply to barbed congruence rather than just labelled bisimulation. Effectively, it implies that a non-localised context can be transformed into a localised one.

Lemma 9. *For P, Q in the pi calculus, $\llbracket P \rrbracket \approx \llbracket Q \rrbracket$ implies $\llbracket P \rrbracket_{\widetilde{z}} \approx \llbracket Q \rrbracket_{\widetilde{z}}$.*

Proof. From Lemma 7.1 we get $\llbracket P \rrbracket_{\widetilde{z}} \approx_\ell \llbracket P \rrbracket \approx \llbracket Q \rrbracket \approx_\ell \llbracket Q \rrbracket_{\widetilde{z}}$. The result follows by Lemma 6.3 and the transitivity of $\approx$. (We thank an anonymous reviewer for this simpler proof.) □

We are now ready to establish full abstraction for the encoding of the pi calculus into the $L\ell$ calculus.

Theorem 10 (Full abstraction). *For P, Q in the pi calculus, $P \approx Q$ if and only if $\llbracket P \rrbracket \approx \llbracket Q \rrbracket$ in the $L\ell$ calculus.*

Proof. We show that (1) $P \not\approx Q$ implies $\llbracket P \rrbracket \not\approx \llbracket Q \rrbracket$ and (2) $\llbracket P \rrbracket \not\approx \llbracket Q \rrbracket$ implies $P \not\approx Q$. We write C_π and $C_{L\ell}$ to range over contexts such that $C_\pi[P], C_\pi[Q]$ are terms in the pi calculus, and $C_{L\ell}\left[\llbracket P \rrbracket\right], C_{L\ell}\left[\llbracket Q \rrbracket\right]$ are terms in the $L\ell$ calculus.

To establish (1), extend the translation $\llbracket\cdot\rrbracket$ to contexts in the obvious way. Since the translation $\llbracket\cdot\rrbracket$ is compositional, we get $\llbracket C[P] \rrbracket = \llbracket C \rrbracket\left[\llbracket P \rrbracket_{\widetilde{z}}\right]$ and $\llbracket C[Q] \rrbracket = \llbracket C \rrbracket\left[\llbracket Q \rrbracket_{\widetilde{z}}\right]$ for some $\widetilde{z}$ determined by C. Next, we reason by contradiction: we prove that $P \not\approx Q$ and $\llbracket P \rrbracket \approx \llbracket Q \rrbracket$ is false. Assuming $P \not\approx Q$, there exists a context $C_\pi[\,]$ such that $C_\pi[P] \not\dot\approx C_\pi[Q]$. By $\llbracket P \rrbracket \approx \llbracket Q \rrbracket$ and Lemma 9, we also have $\llbracket P \rrbracket_{\widetilde{z}} \approx \llbracket Q \rrbracket_{\widetilde{z}}$. Therefore, in particular $\llbracket C_\pi \rrbracket\left[\llbracket P \rrbracket_{\widetilde{z}}\right] \dot\approx \llbracket C_\pi \rrbracket\left[\llbracket Q \rrbracket_{\widetilde{z}}\right]$ and, by the above equalities, $\llbracket C_\pi[P] \rrbracket \dot\approx \llbracket C_\pi[Q] \rrbracket$. By Proposition 8, this latter bisimulation contradicts $C_\pi[P] \not\dot\approx C_\pi[Q]$.

To establish (2), we show that pi contexts are as expressive as linear forwarder contexts, by exhibiting a pi implementation of linear forwarders. To this end, we define $\widehat{\cdot}$, which translates $x \multimap y$ into $x(\widetilde{u}).\overline{y}\,\widetilde{u}$ and leaves all else unchanged. Similarly to (1), we prove that $\llbracket P \rrbracket \not\approx \llbracket Q \rrbracket$ and $P \approx Q$ are contradictory. We are given a context $C_{L\ell}[\,]$ such that $C_{L\ell}[\llbracket P \rrbracket] \not\dot\approx C_{L\ell}[\llbracket Q \rrbracket]$. Consider the agent $\llbracket \widehat{C_{L\ell}[P]} \rrbracket$, which by definition is equal to $\llbracket \widehat{C_{L\ell}} \rrbracket\left[\llbracket P \rrbracket_{\widetilde{z}}\right]$ for some $\widetilde{z}$. By Lemma 7 this is $\approx_\ell$-bisimilar to $[\![\widehat{C_{L\ell}}]\!]\big[[\![P]\!]\big]$. Now we consider the double translation $\widehat{[\![\cdot]\!]}$; it will convert each forwarder $u{-\!\circ}v$ into either $u(\widetilde{x}).\overline{v}\,\widetilde{x}$ or $(u')(u{-\!\circ}u' \mid u'(\widetilde{x}).\overline{v}\,\widetilde{x})$. Thanks to Equation 2, $[\![\widehat{C_{L\ell}}]\!]\big[[\![P]\!]\big] \approx_\ell C_{L\ell}\big[[\![P]\!]\big]$. And, with similar reasoning, the same holds for Q. The proof follows from these results. From the assumption that $P \approx Q$ we get $\widehat{C_{L\ell}[P]} \mathrel{\dot{\approx}} \widehat{C_{L\ell}[Q]}$. By Proposition 8, $[\![\widehat{C_{L\ell}[P]}]\!] \mathrel{\dot{\approx}} [\![\widehat{C_{L\ell}[Q]}]\!]$. Now we focus on P. $[\![\widehat{C_{L\ell}[P]}]\!] = [\![\widehat{C_{L\ell}}]\!]\big[[\![P]\!]_{\widetilde{z}}\big]$ (by definition of $[\![\cdot]\!]$); $\approx_\ell [\![\widehat{C_{L\ell}}]\!]\big[[\![P]\!]\big]$ (by Lemma 7.1); $\approx_\ell C_{L\ell}\big[[\![P]\!]\big]$ (by Equation 2). Doing the same to Q, we obtain $C_{L\ell}\big[[\![P]\!]\big] \mathrel{\dot{\approx}} C_{L\ell}\big[[\![Q]\!]\big]$, contradicting $[\![P]\!] \not\approx [\![Q]\!]$ and so proving the result. □

5 A Linear Forwarder Machine

In this section we develop a distributed machine for the $L\ell$ calculus, suitable for a point-to-point network such as the Internet. This machine is actually very similar to the fusion machine [8], but with linear forwarders instead of fusions (trees of persistent forwarders). We first give a diagrammatic overview of the machine. Then we provide a formal syntax, and prove full abstraction with respect to barbed congruence.

We assume a set of locations. Each channel belongs to a particular location. For instance, channels u, v, w might belong to ℓ_1 and x, y to ℓ_2. The structure of a channel name u might actually be the pair (IP:TCP), giving the IP number and port number of a channel-manager service on the Internet. Every input process is at the location of its subject channel. Output processes may be anywhere. For example,

$\ell_1{:}uvw$	$\ell_2{:}xy$
$u(x).(x')(x{-\!\circ}x' \mid x'(z).P)$	$\overline{u}\,y \mid \overline{y}\,w$

In a point-to-point network such as the Internet, the output message $\overline{u}\,y$ would be sent to u to react; in a broadcast network such as wireless or ethernet, the offer of output would be broadcast and then ℓ_1 would accept the offer. In both cases, the result is a reaction and a substitution $\{y/x\}$ as follows:

	$\ell_1{:}uvw\,x'$	$\ell_2{:}xy$
$\rightarrow$	$y{-\!\circ}x' \mid x'(z).P\{y/x\}$	$\overline{y}\,w$

The overall effect of the linear forwarder $y{-\!\circ}x'$ will be to turn the $\overline{y}\,w$ into $\overline{x}'w$. In a point-to-point network this can be implemented by migrating the $y{-\!\circ}x'$ to ℓ_2, there to *push* the $\overline{y}\,w$ to x', as shown below. (In the following diagrams, some steps are shown as heating transitions $\rightharpoonup$; these steps were abstracted away in the $L\ell$ calculus).

	$\ell_1{:}uvw\,x'$	$\ell_2{:}xy$
$\rightharpoonup$	$x'(z).P\{y/x\}$	$y{-\!\circ}x' \mid \overline{y}\,w$

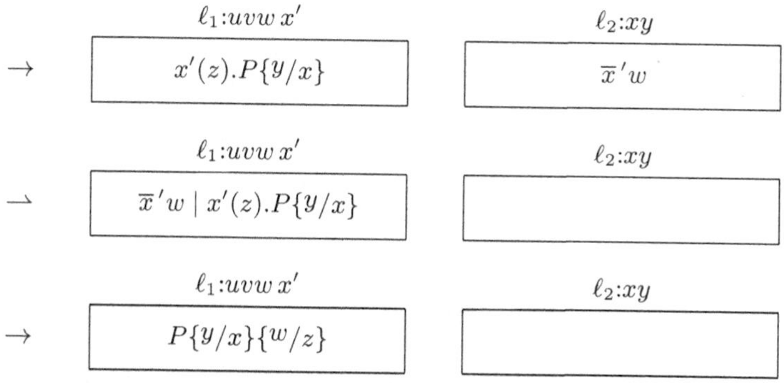

In a broadcast network, the linear forwarder $y{\multimap}x'$ would instead stay at ℓ_1; later, when the offer of $\overline{y}\,a$ is broadcast, the linear forwarder can *grab* the offer.

We remark that the above machine basically avoids code migration: after an input, the continuation remains in the same place (with the minor exception that forwarders $x{\multimap}x'$ and outputs $\overline{x}\,y$ may migrate, which is easy to implement). Because code does not migrate, there is no need for a run-anywhere infrastructure such as Java, and it is possible to compile into CPU-specific machine code.

Distributed Choice. A well-known operator in process calculi is the input-guarded choice $x(u).P + y(v).Q$. In the case where x and y are at separate locations, the choice is awkward to implement, but we can compile it into an (easily implementable) localised choice as follows:

$$\begin{aligned}\llbracket x(u).P + y(v).Q \rrbracket \;&=\; (x'y')(\; x{\multimap}x' \mid y{\multimap}y' \\ &\qquad \mid\; x'(u).(P \mid y'{\multimap}y) + y'(v).(Q \mid x'{\multimap}x)\;\;)\end{aligned}$$

To understand this encoding, note that the new names x' and y' will be created at the same location, and so the choice between x' and y' will a local one. Next, $\overline{x}\,v$ may be forwarded to x', or $\overline{y}\,v$ to y', or both. If the reaction with x' is taken, this yields $y'{\multimap}y$ to "undo" the effect of the forwarder that was not taken. (It undoes it up to weak barbed congruence, in the sense that $(y')(y{\multimap}y' \mid y'{\multimap}y) \approx \mathbf{0}$.) Note that even if the location of x should fail, then the y option remains open, and vice versa.

Failure. We briefly comment on the failure model for the linear forwarder machine. It is basically the same as failure in the join calculus: either a message can be lost, or an entire location can fail. If a linear forwarder $x{\multimap}y$ should fail, the effect is the same as if a single future message $\overline{x}\,\widetilde{u}$ should fail. A command '*iffail(u) then P*' might be added to determine if u's location is unresponsive. The current theory was largely prompted by criticisms of the fragility of our earlier fusion machine.

5.1 The Machine Calculus

Definition 11 (Linear forwarder machine). *Localised linear forwarder machines M are given by the following grammar, where P ranges over terms in the $L\ell$ calculus (Definition 2):*

$$M \quad ::= \quad \mathbf{0} \quad | \quad x[P] \quad | \quad (\!|x|\!)[P] \quad | \quad M, M$$

The presentation here is similar to that given for the fusion machine [8]. The *basic channel-manager* $x[P]$ denotes a channel-manager at channel x containing a *body* P. The *local channel-manager* $(\!|x|\!)[P]$ denotes a channel-manager where the name x is not visible outside the machine. We write $\mathsf{chan}\, M$ to denote the set of names of all channel-managers in the machine, and $\mathsf{Lchan}\, M$ for the names of only the local channel-managers.

We assume a *co-location* equivalence relation L on channels. We write $x@y$ to mean that $(x, y) \in L$, with the intended meaning that the two channels are at the same location. It is always possible to create a fresh channel at an existing location: therefore let each equivalence class in L be infinitely large. In the machine calculus, we generally assume L rather than writing it explicitly.

There are a number of well-formedness conditions on machines:

(1) *Localised.* All code is in the right place, and does not need to be moved at runtime. Formally, in every channel $u[P]$, every free input $v(\widetilde{x}).Q$ satisfies $u@v$. (Also, no received name is used as the subject of input; this already holds from the $L\ell$ calculus).

(2) *Singly-defined.* There is exactly one channel-manager per channel. Formally, a machine $x_1[B_1], \cdots, x_n[B_n]$ is singly-defined when $i \neq j$ implies $x_i \neq x_j$ (x_i or x_j may be local).

(3) *Complete.* It does not make sense to write a program that refers to channels which do not exist. We say that a machine is complete when it has no such references. Formally, the free names of a machine must be contained in $\mathsf{chan}\, M$.

A machine is *well-formed* when it is localised, singly-defined and complete. In the following, we consider only well-formed machines.

Definition 12 (Dynamics). *The* structural congruence *for well-formed machines $\equiv$ is the smallest equivalence relation satisfying the following laws:*

$$M, \mathbf{0} \equiv M \qquad M_1, M_2 \equiv M_2, M_1 \qquad M_1, (M_2, M_3) \equiv (M_1, M_2), M_3$$
$$P \equiv Q \;\; \textit{implies} \;\; u[P] \equiv u[Q] \;\textit{and}\; (\!|u|\!)[P] \equiv (\!|u|\!)[Q]$$

The reduction step $\rightarrow$ and the heating step $\rightharpoonup$ are the smallest relations satisfying the rules below, and closed with respect to structural congruence. Each rule addresses generically both free and local channel-managers.

$$u[\overline{u}\,\widetilde{y} \mid u(\widetilde{x}).P \mid R] \;\rightarrow\; u[P\{\widetilde{y}/\widetilde{x}\} \mid R] \qquad \text{(react)}$$
$$u[u \multimap v \mid \overline{u}\,\widetilde{x} \mid R] \;\rightarrow\; u[\overline{v}\,\widetilde{x} \mid R] \qquad \text{(fwd)}$$

$$u[(x)P \mid R] \;\rightharpoonup\; u[P\{x'/x\} \mid R],\ (\!|x'|\!)[\,],\quad x' \textit{ fresh},\ x'@u \qquad \text{(dep.new)}$$

$$u[x\multimap y \mid R_1],\ x[R_2] \ \rightharpoonup\ u[R_1],\ x[x\multimap y \mid R_2], \quad \text{if } u \neq x \qquad \text{(dep.fwd)}$$
$$u[\overline{x}\,\widetilde{y} \mid R_1],\ x[R_2] \ \rightharpoonup\ u[R_1],\ x[\overline{x}\,\widetilde{y} \mid R_2], \quad \text{if } u \neq x \qquad \text{(dep.out)}$$
$$u[x(\widetilde{y}).P \mid R_1],\ x[R_2] \ \rightharpoonup\ u[R_1],\ x[x(\widetilde{y}).P \mid R_2], \quad \text{if } u \neq x,\ u@x \qquad \text{(dep.in)}$$

For every transition rule above, we close it under contexts:

$$\frac{M \rightarrow M', \quad \mathsf{chan}\, M' \cap \mathsf{chan}\, N = \emptyset}{M, N \rightarrow M', N} \qquad \frac{M \rightharpoonup M', \quad \mathsf{chan}\, M' \cap \mathsf{chan}\, N = \emptyset}{M, N \rightharpoonup M', N}$$

We draw attention to two of the rules. The rule *(dep.new)* picks a fresh channel-name x' and this channel is deemed to be at the location where the command was executed. The rule *(dep.in)* will only move an input command from one channel u to another channel x, if the two channels are co-located; hence, there is no "real" movement. In the current presentation we have used arbitrary replication $!P$, but in a real machine we would instead use guarded replication [18], as is used in the fusion machine. All rules preserve well-formedness.

Definition 13 (Bisimulation). *The observation $M \downarrow u$ is the smallest relation satisfying $u[P] \downarrow u$ if $P \downarrow u$, and $M_1, M_2 \downarrow u$ if $M_1 \downarrow u$ or $M_2 \downarrow u$. Write $\Rightarrow$ for $(\rightarrow^* \rightharpoonup^*)^*$, and $M \Downarrow u$ for $M \Rightarrow\downarrow u$. A weak barbed bisimulation $\mathcal{R}$ between machines is a symmetric relation such that if $M\ \mathcal{S}\ N$ then*

1. *$M \Downarrow u$ implies $N \Downarrow u$*
2. *$M \Rightarrow M'$ implies $N \Rightarrow N'$ such that $M'\ \mathcal{R}\ N'$*

Let $\dot{\approx}$ be the largest barbed bisimulation. Two machines M_1 and M_2 are weak barbed equivalent, written $M_1 \simeq M_2$, when for every machine N, then $N, M_1 \dot{\approx} N, M_2$. (Note that N, M_1 and N, M_2 are assumed to be well-formed, so $\mathsf{chan}\, N \cap \mathsf{chan}\, M_1 = \mathsf{chan}\, N \cap \mathsf{chan}\, M_2 = \emptyset$.)

We will prove correctness using a translation $\mathsf{calc}\, M = (\mathsf{Lchan}\, M)\widehat{M}$ from machines to terms in the $L\ell$ calculus, where

$$\widehat{0} = \mathbf{0} \qquad \widehat{u[P]} = P \qquad \widehat{(u)[P]} = P \qquad \widehat{M_1, M_2} = \widehat{M_1} \mid \widehat{M_2}$$

One might prefer to prove correctness of a "compiling" translation, which takes a term and compiles it into a machine – rather than the reverse translation calc. However, different compilers are possible, differing in their policy for which location to upload code into. We note that all correct compilers are contained in the inverse of calc, so our results are more general.

The correctness of the forwarder machine relies on the following lemma

Lemma 14 (Correctness). *$M_1 \approx M_2$ if and only if $\mathsf{calc}\, M_1 \approx \mathsf{calc}\, M_2$*

Proof. It is clear that machine operations are reflected in the calculus: $M \equiv M'$ implies $\mathsf{calc}\, M \equiv \mathsf{calc}\, M'$, and $M \rightharpoonup M'$ implies $\mathsf{calc}\, M \equiv \mathsf{calc}\, M'$, and $M \rightarrow M'$ implies $\mathsf{calc}\, M \rightarrow \mathsf{calc}\, M'$, and $M \downarrow u$ implies $\mathsf{calc}\, M \downarrow u$.

The reverse direction is more difficult. We wish to establish that

1. $\mathsf{calc}\, M \downarrow u$ implies $M \rightharpoonup^* \downarrow u$, and
2. $\mathsf{calc}\, M \to P'$ implies $\exists M' : M \rightharpoonup^* \to M'$ and $P' \equiv \mathsf{calc}\, M'$.

Both parts share a similar proof; we focus on part 2. Given the machine M, there is also a *fully-deployed* machine M' such that $M \rightharpoonup^* M'$ and M' has no heating transitions: that is, all unguarded restrictions have been used to create fresh channels, and all outputs and forwarders are at the correct location. Therefore $\mathsf{calc}\, M \equiv \mathsf{calc}\, M' \to P'$. The structure of $\mathsf{calc}\, M'$ has the form $(\mathsf{Lchan}\, M')\widehat{M'}$. The reaction must have come from an output $\overline{u}\,\widetilde{y}$ and an input $u(\widetilde{x}).P$ in M' (or from an output and a forwarder). Because M' is fully-deployed, it must contain $u[\overline{u}\,\widetilde{y} \mid u(\widetilde{x}).P]$. Therefore it too must allow the corresponding reaction. The bisimulation result follows directly from the above. □

We now prove full abstraction: that two machines are barbed *equivalent* if and only if their corresponding $L\ell$ calculus terms are barbed *congruent*. There is some subtlety here. With the $L\ell$ calculus, $\approx$ is closed under restriction, input prefix and parallel contexts. With the abstract machine, $\simeq$ is only closed under the partial machine composition: contexts can add new channel managers to the machine, but not additional programs to existing ones. We defined machine equivalence around this partial closure, because this is the most natural equivalence in the machine setting. (It is not surprising that contexts gain no additional discriminating power through restriction and input-prefixing, since the same holds in the asynchronous pi calculus without matching [14]: $P \approx Q$ if and only if $R|P \dot{\approx} R|Q$ for every R.)

It has been suggested that a weaker simulation result would suffice. But we believe that full abstraction shows our machine to be a natural implementation of the pi calculus, in contrast to Facile and Join. Practically, full abstraction means that a program can be debugged purely at source-level rather than machine-level.

Theorem 15 (Full abstraction). $M_1 \simeq M_2$ *if and only if* $\mathsf{calc}\, M_1 \approx \mathsf{calc}\, M_2$.

Proof. The reverse direction is straightforward, because machine contexts are essentially parallel compositions. In the forwards direction, it suffices to prove $R|\,\mathsf{calc}\, M_1 \dot{\approx} R|\,\mathsf{calc}\, M_2$ for every R. By contradiction suppose the contrary: namely, there exists an R such that the two are not barbed bisimilar. Expanding the definition of calc we obtain that $R|(\mathrm{lchan}\, M_1)\widehat{M_1} \not\dot{\approx} R|(\mathrm{lchan}\, M_2)\widehat{M_2}$.

We now show how to construct a machine context M_R such that $M_R, M_1 \not\dot{\approx} M_R, M_2$, thus demonstrating a contradiction. Without loss of generality, suppose that R does not clash with the local names $\mathrm{lchan}\, M_1$ or $\mathrm{lchan}\, M_2$. This gives $(\mathrm{lchan}\, M_1)(R|\widehat{M_1}) \not\dot{\approx} (\mathrm{lchan}\, M_2)(R|\widehat{M_2})$. In order to ensure well-formedness of M_R, M_1 and M_R, M_2, let $\widetilde{z} = \mathrm{chan}\, M_1 \cup \mathrm{chan}\, M_2$. By Lemma 7 we get $[\![R]\!]_{\widetilde{z}} \approx_\ell R$, and by definition $[\![R]\!]_{\widetilde{z}}$ contains no inputs on $\widetilde{z}$, so satisfying the localised property. Now assume without loss of generality that R contains no top-level restrictions. Let $M_R = u[R]$ for a fresh name u such that, for every free input $u_i(\widetilde{x}).R'$ in R, then $u_i@u$. Hence $\mathrm{lchan}\, M_R = \emptyset$ and $\widehat{M_R} = [\![R]\!]_{\widetilde{z}}$. This yields calc $M_R, M_1 = (\text{lchan}\, M_1)([\![R]\!]_{\tilde{z}}|\widehat{M_1}) \approx_\ell (\text{lchan}\, M_1)(R|\widehat{M_1})$, and similarly for M_2. And finally, by construction, both M_R, M_1 and M_R, M_2 are singly-defined and complete. □

6 Further Issues

The point of this paper is to provide a distributed implementation of the input capability of the pi calculus. We have shown that a limited form of input capability (linear forwarders) is enough to easily express the full input capability. We have expressed this formally through a calculus with linear forwarders, and a proof of its full abstraction with respect to the pi calculus encoding.

The calculus in this paper abstracts away from certain details of implementation (such as the choice between a point-to-point or broadcast network). Nevertheless, thanks to its *localisation* property, it remains easy to implement.

Coupled Bisimulation. There is an interesting subtlety in the encoding of input capability. Our first attempt at an encoding gave, for example,

$$[\![u(x).(x().P \mid x().Q)]\!] \quad = \quad u(x).(x')(x{\multimap}x' \mid x'().P \mid x{\multimap}x' \mid x'().Q)$$

That is, we tried to reuse the same local name x' for all bound inputs. But then, a subsequent reaction $x{\multimap}x' \mid \overline{x}\,z \rightarrow \overline{x}\,'z$ would be a commitment to react with one of $x'().P$ or $x'().Q$, while ruling out any other possibilities. This *partial* commitment does not occur in the original pi calculus expression, and so the encoding does not even preserve behaviour. An equivalent counterexample in the pi calculus is that $\tau.P|\tau.Q|\tau.R$ and $\tau.P|\tau.(\tau.Q|\tau.R)$ are not weak bisimilar. We instead used an encoding which has a fresh channel for each bound input:

$$[\![u(x).(x().P \mid x().Q)]\!] \quad = \quad u(x).\big(\,(x')(x{\multimap}x' \mid x'().P) \mid (x'')(x{\multimap}x'' \mid x''().Q)\,\big)$$

Now, any reaction with a forwarder is a *complete* rather than a partial commitment. In fact, both encodings are valid. The original encoding, although not a bisimulation, is still a *coupled bisimulation* [16]. (Coupled bisimulation is a less-strict form of bisimulation that is more appropriate for an implementation, introduced for the same reasons as given here.) In this paper we chose the normal bisimulation and the repaired encoding, because they are simpler.

The Join Calculus and Forwarders. We end with some notes on the difference between the join calculus [7] and the $L\ell$ calculus. The core join calculus is

$$P \;::=\; \mathbf{0} \quad\big|\quad \overline{x}\,\widetilde{u} \quad\big|\quad P|P \quad\big|\quad \text{def}\; x(\widetilde{u})|y(\widetilde{v}) \rhd P \;\text{in}\; Q$$

The behaviour of the *def* resource is, when two outputs $\overline{x}\,\widetilde{u}'$ and $\overline{y}\,\widetilde{v}'$ are available, then it consumes them to yield a copy $P\{\widetilde{u}'\widetilde{v}'/\widetilde{u}\widetilde{v}\}$ of P. Note that x and y are bound by *def*, and so input capability is disallowed by syntax. The core join calculus can be translated into the pi calculus (and hence $L\ell$) as follows [7]:

$$[\![\mathbf{0}]\!] = \mathbf{0} \qquad [\![\overline{x}\,\widetilde{u}]\!] = \overline{x}\,\widetilde{u} \qquad [\![P|Q]\!] = [\![P]\!] \mid [\![Q]\!]$$
$$[\![\text{def } x(\widetilde{u})|y(\widetilde{v}) \triangleright P \text{ in } Q]\!] = (xy)(\,[\![Q]\!] \mid !x(\widetilde{u}).y(\widetilde{v}).[\![P]\!]\,)$$

If a join program is translated into the linear forwarder machine and then executed, then the result has exactly the same runtime behaviour (i.e. same number of messages) as the original join program. Additionally, we can provide the same distribution of channels through the co-location operator discussed above.

A reverse translation is more difficult, because of linear forwarders. One might try to translate $x{\multimap}y \mid R$ into def $x(\widetilde{u}){\triangleright}\overline{y}\,\widetilde{u}$ in $[\![R]\!]$, analogous with the translation of a forwarder into the pi calculus that was used in Proposition 10. But the $L\ell$ calculus allows a received name to be used as the source of a forwarder, as in $u(x).(x{\multimap}y \mid P)$, and the same is not possible in the join calculus. Therefore contexts in the $L\ell$ calculus are strictly more discriminating than contexts in the join calculus. (As an example, def $x(u) \triangleright \overline{y}\,u$ in $\overline{z}\,x$ is equivalent to $\overline{z}\,y$ in the join calculus, but the context $z(a).(a{\multimap}b \mid \overline{a} \mid \_)$ can distinguish them in the $L\ell$ calculus.)

References

1. M. Abadi and C. Fournet. Mobile values, new names, and secure communication. In *POPL 2001*, pages 104–115. ACM Press.
2. R. Amadio, G. Boudol, and C. Lhoussaine. The receptive distributed pi-calculus (extended abstract). In *FSTTCS 1999*, LNCS 1738:304–315.
3. R. Amadio. An asynchronous model of locality, failure, and process mobility. In *COORDINATION 1997*, LNCS 1282:374–391.
4. M. Boreale. On the expressiveness of internal mobility in name-passing calculi. *Theoretical Computer Science*, 195(2):205–226, 1998.
5. G. Boudol. Asynchrony and the π-calculus (note). Rapport de Recherche 1702, INRIA Sophia-Antipolis, 1992.
6. L. Cardelli and A. D. Gordon. Mobile ambients. *Theoretical Computer Science*, 240(1):177–213, 2000.
7. C. Fournet and G. Gonthier. The reflexive chemical abstract machine and the join-calculus. In *POPL 1996*, pages 372–385. ACM Press.
8. P. Gardner, C. Laneve, and L. Wischik. The fusion machine. In *CONCUR 2002*, LNCS 2421:418–433.
9. D. Gelernter, N. Carriero, S. Chandran, and S. Chang. Parallel programming in Linda. In *ICPP 1985*, pages 255–263. IEEE.
10. A. Giacalone, P. Mishra, and S. Prasad. Facile: A symmetric integration of concurrent and functional programming. *International Journal of Parallel Programming*, 18(2):121–160, 1989.
11. K. Honda and N. Yoshida. On reduction-based process semantics. *Theoretical Computer Science*, 152(2):437–486, 1995.
12. N. Kobayashi, B. C. Pierce, and D. N. Turner:. Linearity and the pi-calculus. *ACM Transactions on Programming Languages and Systems*, 21(5):914–947, 1999.
13. M. Merro and D. Sangiorgi. On asynchrony in name-passing calculi. In *ICALP 1998*, LNCS 1443:856–867.

14. M. Merro. On equators in asynchronous name-passing calculi without matching. In *EXPRESS 1999*, volume 27 of *Electronic Notes in Theoretical Computer Science*. Elsevier Science Publishers.
15. R. Milner and D. Sangiorgi. Barbed bisimulation. In *ICALP 1992*, LNCS 623:685–695.
16. J. Parrow and P. Sjödin. Multiway synchronization verified with coupled simulation. In *CONCUR 1992*, LNCS 630:518–533.
17. D. Sangiorgi. Pi-calculus, internal mobility and agent-passing calculi. *Theoretical Computer Science*, 167(1–2):235–275, 1996.
18. D. Sangiorgi. On the bisimulation proof method. *Mathematical Structures in Computer Science*, 8(5):447–479, 1998.
19. P. Sewell, P. Wojciechowski, and B. Pierce. Location independence for mobile agents. In *ICCL 1998*, LNCS 1686:1–31.

Abstract Patterns of Compositional Reasoning

Nina Amla[1], E. Allen Emerson[2], Kedar Namjoshi[3], and Richard Trefler[4]

[1] Cadence Design Systems
[2] Univ. of Texas at Austin*
[3] Bell Labs, Lucent Technologies
[4] Univ. of Waterloo**

Abstract. Compositional Reasoning – reducing reasoning about a concurrent system to reasoning about its individual components – is an essential tool for managing proof complexity and state explosion in model checking. Typically, such reasoning is carried out in an *assume-guarantee* manner: each component guarantees its behavior based on assumptions about the behavior of other components. Restrictions imposed on such methods to avoid unsoundness usually also result in incompleteness – i.e., one is unable to prove certain properties. In this paper, we construct an abstract framework for reasoning about process composition, formulate an assume-guarantee method, and show that it is sound and semantically complete. We then show how to instantiate the framework for several common notions of process behavior and composition. For these notions, the instantiations result in the first methods known to be complete for mutually inductive, assume-guarantee reasoning.

1 Introduction

A large system is typically structured as a composition of several smaller components that interact with one another. An essential tool for the formal analysis of such systems is a *compositional reasoning* method – one that reduces reasoning about the entire system to reasoning about its individual components. This is particularly important when applying model checking [10,25] to a concurrent composition of interacting, non-deterministic processes, where the full transition system can have size exponential in the number of components. This *state explosion* problem is one of the main obstacles to the application of model checking. Compositional reasoning techniques (see e.g., [12,11]) are particularly useful for ameliorating state explosion, since they systematically decompose the model checking task into smaller, more tractable sub-tasks. A typical *assume-guarantee* style of reasoning (cf. [9,17,2,6,21,23]), establishes that the composition of processes P_1 and P_2 refines the composition of Q_1 and Q_2 if P_1 composed with Q_2 refines Q_1, and Q_1 composed with P_2 refines Q_2. Here, Q_1 and Q_2 act as mutually inductive hypotheses.

* This author's research is supported in part by NSF grants CCR-009-8141 & ITR-CCR-020-5483, and SRC Contract No. 2002-TJ-1026.
** Sponsored in part by an Individual Discovery Grant from NSERC of Canada.

R. Amadio, D. Lugiez (Eds.): CONCUR 2003, LNCS 2761, pp. 431–445, 2003.

However, existing methods for compositional reasoning can be hard to apply, for a number of reasons. Firstly, they are often bound to a particular syntax for describing processes, and particular notions of behavior and composition. Thus, it is not clear if a reasoning pattern devised for one such choice also applies to another. Another key factor is that several methods are known to be incomplete [23]. The completeness failure can usually be traced back to restrictions placed to avoid unsound, semantically circular reasoning with assumptions and guarantees. As argued in [11], completeness is an important property for a proof method. An incomplete method can be a serious impediment in practice, since it can make it impossible to prove that a correct program is correct. Moreover, the completeness failures demonstrated in [23] all occur for simple and common programming patterns. Lastly, safety and liveness properties are handled differently by most methods. For instance, the above method is *not* sound if both Q_1 and Q_2 include liveness or fairness constraints. It appears that there is a delicate balance between adding enough restrictions to avoid unsound circular reasoning, while yet allowing enough generality to ensure that the method is complete for both safety and liveness properties.

This paper addresses these problems in the following way. First, we construct an abstract, algebraic framework to reason about processes and composition. We formulate a mutually inductive, assume-guarantee method that applies to both safety and liveness properties, and show that it is sound and complete, all within the abstract setting. The framework makes explicit all assumptions needed for these proofs, uses as few assumptions as possible, and clarifies the key ideas used to show soundness and completeness. Our proof method extends the one given above with a soundness check for liveness properties. We show that a simple extension of the proof method in [6], which replaces Q_1 in the second hypothesis with its safety closure, is also complete. The two methods are closely related, but we show that ours is more widely applicable.

We then show how the abstract framework can be concretized in several different ways, obtaining a sound and complete method for each instantiation. In this paper, we discuss interleaving and fully synchronous composition, and notions of process behavior that include liveness, fairness, branching and closure under stuttering. The resulting instantiations are the first mutually inductive, assume-guarantee methods known to be semantically complete for general properties. That such diverse notions of composition and behavior can be handled in a common framework may seem surprising. To a large extent, this is due to the key property that, in each case, composition is represented as a conjunction of languages (cf. [4,2]). The abstract framework thus provides a clean separation between the general axioms needed for soundness and completeness, and the assumptions needed for their validity in specific contexts. It simplifies and unifies a large body of work, and allows one to easily experiment with – and prove correct – different patterns of compositional reasoning.

Related Work: Methods for compositional reasoning about concurrent processes have been extensively studied for nearly three decades. Assume-guarantee reasoning was introduced by Chandy and Misra [9] and Jones [16] for analyzing safety properties. These methods were extended to some progress properties in the following decade (e.g., [24]; the book [11] has a comprehensive historical survey). More recently, Abadi and Lamport [2] and McMillan [21] extended the methods to temporal liveness properties. However, as shown by Namjoshi and Trefler [23], these extensions are not complete, usually for liveness properties of simple programs. Building on McMillan's formulation, they present a complete method for model checking of linear time temporal logic properties.

The methods presented in this paper apply to a process refinement methodology. In this setting, the method of [2] for asynchronous composition is incomplete [23]. Our methods are complete for asynchronous composition, both with and without closure under stuttering. Alur and Henzinger propose a method in [6] for the Reactive Modules language. We show that our new formulation, and a slight extension of their method are complete for this setting. Henzinger et. al. [15] showed how the same pattern of reasoning applies also to simulation-based refinement of Moore machines. Our proof method is different, and applies somewhat more generally (e.g., to Mealy machines). A major contribution of this paper, we believe, is the demonstration that all of these instantiations can be obtained from a single abstract pattern of reasoning. There is work by Abadi and Plotkin [4], Abadi and Merz [3], Viswanathan and Viswanathan [26], and Maier [18] on similar abstract formulations, but none of these result in complete methods. For a (non-standard) notion of completeness, Maier [19] shows that sound, circular, assume-guarantee rules cannot be complete, and that complete rules (in the standard sense) must use auxiliary assertions.

2 Abstract Compositional Reasoning

Notation. We use a notation popularized by Dijkstra and Scholten [13]. In the term $(\mathcal{Q}x : r(x) : p(x))$, $\mathcal{Q}$ is a quantifier, $r(x)$ is the range for variable x, and $p(x)$ is the term being operated on. The operator $[\phi]$ (read as "box") universally quantifies over the free variables of ϕ. Proof steps are linked by a transitive connective such as $\equiv$ or $\Rightarrow$, with an associated hint. For convenience, we move freely between set-based and predicate-based notations. For instance, $a \in S$ may be written as the predicate $S(a)$, and $[A \Rightarrow B]$ represents $A \subseteq B$.

2.1 Processes, Closure, and Composition

The abstract space of *processes* is denoted by $\mathcal{P}$. The set of abstract process *behaviors*, $\mathcal{B}$, is assumed to be equipped with a partial order $\preceq$ (read as "prefix"), and partitioned into non-empty subsets of finite behaviors, $\mathcal{B}_*$, and infinite behaviors, $\mathcal{B}_\infty$. We make the following assumptions about the behavior space.

WF $\mathcal{B}_*$ is downward closed under $\preceq$, and $\prec$ is well-founded on $\mathcal{B}_*$.

By downward closure, we mean that any prefix of a behavior in $\mathcal{B}_*$ is also in $\mathcal{B}_*$. An *initial* behavior is a finite behavior with no strict prefix. The set of initial elements, which is non-empty by the well-foundedness assumption, is denoted by $\mathcal{B}_0$. In our concretizations, behaviors are either computations or computation trees, under the standard prefix ordering, so that the **WF** assumption is satisfied. Initial behaviors then correspond to initial states of a process. The semantics of an abstract process P is a subset of $\mathcal{B}$. We call this the *language* of P and denote it by $\mathcal{L}(P)$. The finite behaviors in the language are denoted by $\mathcal{L}_*(P)$ and the infinite behaviors by $\mathcal{L}_\infty(P)$. These subsets must satisfy the following condition.

L1 Every finite prefix of a behavior in $\mathcal{L}(P)$ is a finite behavior of P.

This condition can be expressed succinctly by the notion of the limit of a set of behaviors. For a set S, $lim\,(S) = \{x \mid (\forall y : y \preceq x \wedge y \in \mathcal{B}_* : y \in S)\}$. The condition **L1** is $[\mathcal{L}(P) \;\Rightarrow\; lim\,\mathcal{L}_*(P)]$. The *closure* of a subset S of behaviors, denoted by $cl(S)$, is the set $\{x \mid (\forall y : y \in \mathcal{B}_* \wedge y \preceq x : (\exists z : y \preceq z : z \in S))\}$. I.e., an element x is in $cl(S)$ iff every finite prefix y of x has an "extension" z that is in S. It is not hard to show that cl is monotonic, idempotent, and weakening. We call a set S where $[cl(S) \;\equiv\; S]$ a *safety* property, and a set S where $[cl(S) \;\equiv\; true]$ a *liveness* property in analogy with the definitions of temporal safety and liveness in [5].

Lemma 0. *(cf. [5,20]) Any set of behaviors, S, can be expressed as the intersection of the safety property $cl(S)$ and the liveness property $(\neg cl(S) \vee S)$.* □

The main process *composition* operator is denoted by //, and maps a finite subset of $\mathcal{P}$ to $\mathcal{P}$. The *process closure* operator, CL, has signature $CL : \mathcal{P} \to \mathcal{P}$. The process *choice* operator, $+$, also maps a finite subset of $\mathcal{P}$ to $\mathcal{P}$. We say that process P *refines* process Q, written as $P \models Q$, provided that $[\mathcal{L}(P) \;\Rightarrow\; \mathcal{L}(Q)]$. We assume that these operators enjoy the following properties.

P1 *Composition is conjunction of languages*: $[\mathcal{L}(/\!/\, i : P_i) \;\equiv\; (\wedge\, i : \mathcal{L}(P_i))]$. This implies the corresponding assertions for $\mathcal{L}_*$ and $\mathcal{L}_\infty$.
P2 *Choice is disjunction of languages*: $[\mathcal{L}(+i : P_i) \;\equiv\; (\vee\, i : \mathcal{L}(P_i))]$.
P3 *Closure represents language closure*: $[\mathcal{L}(CL(P)) \;\equiv\; cl(\mathcal{L}(P))]$.

Thus, // and + are associative and commutative. To state the circular reasoning method, we also need the concepts of behavior equivalence and non-blocking.

Behavior Equivalence: For each process P, we assume the existence of an equivalence relation $\sim_P$ on $\mathcal{B}_*$. This relation is used to state when two behaviors are to be considered equivalent relative to P – for example, in a concrete setting, two computations that agree on values of the external variables of P would be equivalent. We define the closure function, $\langle \sim_P \rangle$, induced by this relation as $\langle \sim_P \rangle(S) = \{x \mid (\exists y : y \in S \wedge x \sim_P y)\}$, for any subset S of behaviors. This must have the property below.

BEQ For any process P, $\mathcal{L}(P)$ is closed under $\sim_P$: $[\langle \sim_P \rangle(\mathcal{L}(P)) \;\Rightarrow\; \mathcal{L}(P)]$.

Non-blocking: Process Q *does not block* a process P iff

(a) for every initial behavior of P there is a matching initial behavior of $P /\!/ Q$. Formally, $[\mathcal{B}_0 \wedge \mathcal{L}_*(P) \;\Rightarrow\; \langle \sim_P \rangle(\mathcal{B}_0 \wedge \mathcal{L}_*(P /\!/ Q))]$ and

```
P1::   var x: (0,1);  trans. x' = y
P2::   var y: (0,1);  trans. y' = x
T::    var x,y:(0,1);  fair inf. often (x=1) and inf. often (y=1)
Q1::   var x:(0,1);  fair inf. often (x=1)
Q2::   var y:(0,1);  fair inf. often (y=1)
```

Fig. 1. Unsoundness for liveness

(b) every extension by P of a finite behavior of $P \,/\!\!/\, Q$ has a matching extension by $P \,/\!\!/\, Q$. Formally, for any x, y, if $x \in \mathcal{L}_*(P)$ and $y \prec x$ and $y \in \mathcal{L}_*(P \,/\!\!/\, Q)$, then $x \in \langle \sim_P \rangle(\mathcal{L}_*(P \,/\!\!/\, Q))$.

Note: This completes the list of postulates. It may also be interesting to know what we do *not* postulate: we do not assume that process languages exhibit either *machine-closure* or *receptivity*, and we do not require the behavior equivalence relations to be a congruence relative to $\preceq$. It is the lack of the latter restriction that lets us concretize the framework to stuttering closed languages.

2.2 Compositional Reasoning

The aim of a compositional reasoning method is to provide a systematic way of breaking down a proof that a composition $(/\!\!/\, i : P_i)$ refines a target process T into proof steps that reason individually about each P_i. This is done by abstracting the other processes into an environment for P_i. To uniformly handle such environments, we generalize the original problem into showing that $(/\!\!/\, i : P_i)$ refines T when constrained by an environment process, E. The choice composition operator can be handled quite simply: $E \,/\!\!/ (+i : P_i) \models T$ if, and only if, (from **P2**), for every i, $E \,/\!\!/\, P_i \models T$. We assume that, in the typical case, the composition $(/\!\!/\, i : P_i)$ is too large to be handled directly, by model checking methods, but that T and E are small. Consider the two process case: $E \,/\!\!/\, P_1 \,/\!\!/\, P_2 \models T$.

A non-circular reasoning method shows that $E \,/\!\!/\, P_1 \,/\!\!/\, P_2 \models T$ by finding Q such that (a) $E \,/\!\!/\, P_1 \models Q$, and (b) $Q \,/\!\!/\, P_2 \models T$. Soundness follows immediately from **P1** and the transitivity of $\Rightarrow$. This method is most useful when the interaction between P_1 and P_2 is in one direction only. For bi-directional interaction, it is often convenient to reason in a mutually inductive manner, as in the following (syntactically) circular reasoning method, where Q_1 and Q_2 supply the mutually inductive hypotheses. For $i \in \{1, 2\}$, we write $\hat{\imath}$ for the other index.

Circular Reasoning I: To show that $E \,/\!\!/\, P_1 \,/\!\!/\, P_2 \models T$, find Q_1, Q_2 such that (a) for some i, Q_i does not block $P_{\hat{\imath}}$, and $[\sim_{P_i} \Rightarrow \sim_{Q_i}]$, (b) $P_1 \,/\!\!/\, Q_2 \models Q_1$, (c) $Q_1 \,/\!\!/\, P_2 \models Q_2$, and (d) $E \,/\!\!/\, Q_1 \,/\!\!/\, Q_2 \models T$.

This method is easily shown to be complete for processes P_1 and P_2 such that one does not block the other: given that $E \,/\!\!/\, P_1 \,/\!\!/\, P_2 \models T$, choose P_1 for Q_1 and P_2 for Q_2. It is sound for finite behaviors – the non-blocking hypothesis (a) enables a mutual induction on finite behaviors using (b) and (c).

However, this method is *not sound in general* – it is unsound when both Q_1 and Q_2 define liveness properties. Since it is difficult to exhibit the unsoundness in the abstract, consider the concrete setting of Fig. 1, where $/\!/$ is interpreted as synchronous composition, and the behavior of a process is the set of its computations. Each process is specified by its initial condition, a transition relation (a primed variable refers to its next value), and a fairness constraint on infinite computations (unspecified components are *true*, i.e., unconstrained).

The composition $P_1 /\!/ P_2$ *does not* refine T, since it has a computation where x and y are both always 0. However, consider the processes Q_1 and Q_2 in Fig. 1. As their initial and transition conditions are unconstrained, they are non-blocking, satisfying hyp. (a). The fairness condition of Q_2 requires that $y = 1$ is true infinitely often in any computation of $P_1 /\!/ Q_2$, so the update $x' = y$ in P_1 forces $x = 1$ to be true infinitely often, ensuring that hyp. (b) holds. Hyp. (c) holds for a similar reason. Hyp. (d) holds since Q_1 and Q_2 together satisfy T. Thus, the method leads to the *unsound* conclusion that $P_1 /\!/ P_2$ does refine T.

2.3 A Sound and Complete Method

The previous example shows that a method based on mutual induction is not always sound for liveness properties; however, it is hard to ascribe a reason for the failure. One possible explanation is that the problem arises from the lack of an inductive structure: in the example, both Q_1 and Q_2 restrict only the "future" of a computation. This observation has led to several methods (e.g., [2, 21,23]) where a temporal next-time operator supplies the necessary well-founded structure. Here, we adopt a different strategy, and augment the previous method with an additional check to ensure that the reasoning is sound.

Circular Reasoning II: To show that $E /\!/ P_1 /\!/ P_2 \models T$, find Q_1, Q_2 such that the conditions (a)-(d) from **Circular Reasoning I** hold, and additionally, (e) for some i, $E /\!/ P_i /\!/ CL(T) \models (T + Q_1 + Q_2)$.

To gain some intuition for the new hypotheses (e), consider the earlier soundness failure. The reasoning fails because the error computation π, where x and y are both always 0, is *ignored* by hypotheses (b) and (c), since neither Q_1 nor Q_2 contain π. The "missing" computations are those, such as π, that belong to $\mathcal{L}(P_1 /\!/ P_2)$, but not to either of $\mathcal{L}(Q_1)$ or $\mathcal{L}(Q_2)$. We want these computations to also be part of $\mathcal{L}(T)$. As is shown by the soundness proof below, hypotheses (a)-(d) ensure that all computations of $P_1 /\!/ P_2$ belong to the safety part of the language of T. Thus, the missing computations must belong to the liveness part of T. A direct statement of this condition is: $[\mathcal{L}(E) \wedge \mathcal{L}(P_1 /\!/ P_2) \wedge \neg(\mathcal{L}(Q_1) \vee \mathcal{L}(Q_2)) \Rightarrow (\neg cl(\mathcal{L}(T)) \vee \mathcal{L}(T))]$. However, this includes $P_1 /\!/ P_2$, which is just what we are trying to avoid reasoning about directly.

We show in the soundness proof that one can replace $P_1 /\!/ P_2$ with a *single* process, P_i, resulting in a stronger condition, but *without sacrificing completeness.* Rearranging the new condition, we get that, for some i, $[\mathcal{L}(E) \wedge \mathcal{L}(P_i) \wedge cl(\mathcal{L}(T)) \Rightarrow (\mathcal{L}(Q_1) \vee \mathcal{L}(Q_2) \vee \mathcal{L}(T))]$. Put in terms of the process operators using **P1-P3**, this is just condition (e). For our example, both $\mathcal{L}(E)$ and $cl(\mathcal{L}(T))$ are *true* (i.e., the set of all computations), and condition (e) does not hold – as expected – for either P_1 or P_2, because π belongs to both processes, but not to either of T, Q_1, or Q_2. We show completeness first, which is the simpler half of the proof.

Theorem 0. *(**Completeness**) For processes P_1, P_2 such that one of the processes does not block the other, and for any process T, if $E \,/\!\!/\, P_1 \,/\!\!/\, P_2 \models T$, then this can be shown using the **Circular Reasoning II** method.*

Proof. Choose $Q_1 = P_1$ and $Q_2 = P_2$. Condition (a) holds by the non-blocking assumption. Condition (b) becomes $P_1 \,/\!\!/\, P_2 \models P_1$ which, by the definition of $\models$ and assumption **P1**, is equivalent to $[\mathcal{L}(P_1) \wedge \mathcal{L}(P_2) \;\Rightarrow\; \mathcal{L}(P_1)]$, which is trivially true. Conditions (c) and (d) can be dealt with in a similar manner. Condition (e), for $i = 1$, simplifies, using **P1-P3**, to $[\mathcal{L}(E) \wedge \mathcal{L}(P_1) \wedge cl(\mathcal{L}(T)) \;\Rightarrow\; \mathcal{L}(T) \vee \mathcal{L}(P_1) \vee \mathcal{L}(P_2)]$, which is true trivially. □

Note: The completeness proof, as may be expected, considers the worst case, where $Q_1 = P_1$ and $Q_2 = P_2$. This is unavoidable in general, because it is a manifestation of the state explosion problem. However, the language of a process is usually given by its external input-output behavior, which can be much less complex than its full behavior. Thus, Q_i, which represents external behavior, can be smaller than P_i, and yet be an adequate replacement for it in the method.

Soundness requires a more complex proof. First, we show, using well-founded induction on $\prec$, and hypotheses (a)-(d), that the finite language of $P_1 \,/\!\!/\, P_2$ is contained in the finite language of $Q_1 \,/\!\!/\, Q_2$. This is used to prove that any behavior of $E \,/\!\!/\, P_1 \,/\!\!/\, P_2$ is a safe behavior of T. We then utilize hypothesis (e) to conclude that all behaviors of $E \,/\!\!/\, P_1 \,/\!\!/\, P_2$ are behaviors of T.

Lemma 1. *For processes P_1, P_2, Q_1, Q_2, and T satisfying the conditions of the **Circular Reasoning II** method, $[\mathcal{L}_*(P_1 \,/\!\!/\, P_2) \;\Rightarrow\; \mathcal{L}_*(Q_1 \,/\!\!/\, Q_2)]$.*

Proof. This proof is based on well-founded induction on the set of finite behaviors. Both the base and inductive cases make use of the non-blocking hypothesis (a), and the mutual induction supplied by hypotheses (b) and (c) on the processes Q_1 and Q_2. Without loss of generality, we assume that hyp. (a) holds for the pair (P_1, Q_2); the proof is symmetric in the other case.

(Base case: initial elements) For any initial behavior x, x is in $\mathcal{L}_*(P_1 \,/\!\!/\, P_2)$ iff (by **P1**), it belongs to both $\mathcal{L}_*(P_1)$ and $\mathcal{L}_*(P_2)$. As Q_2 does not block P_1, x is in $\langle \sim_{P_1} \rangle(\mathcal{B}_0 \wedge \mathcal{L}_*(P_1 \,/\!\!/\, Q_2))$. From Hyp. (b), x belongs to $\langle \sim_{P_1} \rangle(\mathcal{B}_0 \wedge \mathcal{L}_*(Q_1))$. As $[\sim_{P_1} \Rightarrow \sim_{Q_1}]$ from Hyp. (a), using **BEQ** and the monotonicity of $\langle \sim_P \rangle$, x belongs to $\mathcal{L}_*(Q_1)$. Since x belongs to both $\mathcal{L}_*(P_2)$ and $\mathcal{L}_*(Q_1)$, using **P1** and Hyp. (c), it belongs to $\mathcal{L}_*(Q_2)$, and thus, (by **P1**) to $\mathcal{L}_*(Q_1 \,/\!\!/\, Q_2)$.

(Inductive case) Consider any non-initial behavior x in $\mathcal{L}_*(P_1 \,/\!\!/\, P_2)$. Let y be a strict prefix of x. By **WF**, y is a finite behavior, so, by **L1**, it belongs to $\mathcal{L}_*(P_1 \,/\!\!/\, P_2)$. From the inductive hypothesis, y belongs to $\mathcal{L}_*(Q_1 \,/\!\!/\, Q_2)$. It follows from **P1**, that y belongs to $\mathcal{L}_*(P_1 \,/\!\!/\, Q_2)$. As Q_2 does not block P_1 (Hyp. (a)), and x belongs to $\mathcal{L}_*(P_1)$, it follows that x belongs to $\langle \sim_{P_1} \rangle(\mathcal{L}_*(P_1 \,/\!\!/\, Q_2))$. Reasoning now as in the earlier case, it follows that x belongs to $\mathcal{L}_*(Q_1 \,/\!\!/\, Q_2)$. □

Lemma 2. *For any process P, $[cl(\mathcal{L}(P)) \;\equiv\; lim\, \mathcal{L}_*(P)]$.*

Proof. For any x, $x \in cl(\mathcal{L}(P))$ iff (by definition of closure) $(\forall y : y \in \mathcal{B}_* \wedge y \preceq x : (\exists z : y \preceq z \wedge z \in \mathcal{L}(P)))$. By condition **L1**, as $[\mathcal{L}_*(P) \Rightarrow \mathcal{L}(P)]$, this is equivalent to $(\forall y : y \in \mathcal{B}_* \wedge y \preceq x : y \in \mathcal{L}_*(P))$ – i.e., x is in $lim\, \mathcal{L}_*(P)$. □

Theorem 1. *(**Soundness**) For processes E, P_1, P_2, Q_1, Q_2, and T satisfying the conditions of **Circular Reasoning II**, $[\mathcal{L}(E /\!/ P_1 /\!/ P_2) \Rightarrow \mathcal{L}(T)]$.*

Proof. The decomposition of $\mathcal{L}(T)$ into safety and liveness components gives us a natural decomposition of this proof into safety and liveness proofs. The safety proof shows that $[\mathcal{L}(E /\!/ P_1 /\!/ P_2) \Rightarrow cl(\mathcal{L}(T))]$, while the liveness proof shows that $[\mathcal{L}(E /\!/ P_1 /\!/ P_2) \Rightarrow (\neg cl(\mathcal{L}(T)) \vee \mathcal{L}(T))]$.

(Safety)

$\mathcal{L}(E /\!/ P_1 /\!/ P_2)$

$\Rightarrow cl(\mathcal{L}(E /\!/ P_1 /\!/ P_2))$ (cl is weakening)

$\equiv lim\, \mathcal{L}_*(E /\!/ P_1 /\!/ P_2)$ (Lemma 2)

$\Rightarrow lim\, \mathcal{L}_*(E /\!/ Q_1 /\!/ Q_2)$ (Lemma 1; monotonicity of lim)

$\equiv cl(\mathcal{L}(E /\!/ Q_1 /\!/ Q_2))$ (Lemma 2)

$\Rightarrow cl(\mathcal{L}(T))$ (by hyp. (d); monotonicity of cl)

(Liveness)

$\mathcal{L}(E /\!/ P_1 /\!/ P_2) \wedge cl(\mathcal{L}(T))$

$\equiv \mathcal{L}(E /\!/ P_1 /\!/ P_2) \wedge \mathcal{L}(CL(T))$ (by **P3**)

$\equiv \mathcal{L}(E) \wedge \mathcal{L}(P_1) \wedge \mathcal{L}(P_2) \wedge \mathcal{L}(CL(T))$ (by **P1**)

$\Rightarrow \mathcal{L}(E) \wedge \mathcal{L}(P_1) \wedge \mathcal{L}(P_2) \wedge \mathcal{L}(T + Q_1 + Q_2)$
(by hyp. (e) (pick i))

$\equiv \mathcal{L}(E) \wedge \mathcal{L}(P_1) \wedge \mathcal{L}(P_2) \wedge (\mathcal{L}(T) \vee \mathcal{L}(Q_1) \vee \mathcal{L}(Q_2))$
(by **P2**)

$\Rightarrow \mathcal{L}(T) \vee (\mathcal{L}(E) \wedge \mathcal{L}(P_2) \wedge \mathcal{L}(Q_1)) \vee (\mathcal{L}(E) \wedge \mathcal{L}(P_1) \wedge \mathcal{L}(Q_2))$
($\wedge$ over $\vee$; dropping conjuncts)

$\Rightarrow \mathcal{L}(T) \vee (\mathcal{L}(E) \wedge \mathcal{L}(Q_1) \wedge \mathcal{L}(Q_2))$ (by **P1**; hyp. (b) and (c))

$\Rightarrow \mathcal{L}(T)$ (by hyp. (d))

□

Note: One can replace hyp. (e) with (e'): for some i, $E /\!/ P_i /\!/ CL(Q_1) /\!/ CL(Q_2) \models (T + Q_1 + Q_2)$, without losing either soundness or completeness. Hyp. (e') is weaker (by hyp. (d)) than (e), so it is more likely to hold. However, it might also be more difficult to check in practice as $CL(Q_1) /\!/ CL(Q_2)$ could have size larger than $CL(T)$. Hyp. (e') holds if either $\mathcal{L}(Q_1)$ or $\mathcal{L}(Q_2)$ are safety languages, showing that the first circular reasoning rule is sound in this case.

The new hypothesis also provides a direct link to the reasoning method proposed by Alur and Henzinger in [6]. In our notation, it reads as follows: to show that $P_1 /\!/ P_2 \models Q_1 /\!/ Q_2$, prove that (i) $P_1 /\!/ Q_2 \models Q_1$, and (ii) $P_2 /\!/ CL(Q_1) \models Q_2$ (non-blocking is assured by the language definition). This is incomplete as stated, but it can be generalized to showing that $P_1 /\!/ P_2 \models T$ by making a *choice* of Q_1, Q_2 such that (i) and (ii) hold, as well as (iii) $Q_1 /\!/ Q_2 \models T$. This is sound, by arguments in [6], as well as complete (choose P_i for Q_i).

Both methods are complete, but we may go further and compare their solution sets for $\{Q_i\}$ (i.e., those that satisfy the hypotheses), for fixed $\{P_i\}$ and T. In one direction, if (i)-(iii) hold for $\{Q_i\}$, so do the hypotheses (a)-(e') of **Circular Reasoning II**, for the same choice. Hyp. (a) is the non-blocking property. Hyp. (b),(d) are identical to (i),(iii), respectively. Hyp. (c) holds using (ii) and the weakening property of *cl*. Finally, Hyp. (e') holds for P_2 using (ii). The converse "almost" holds (assuming non-blocking): from (e) (for P_2) and (c), one can derive that $P_2 \mathbin{/\!/} CL(Q_1) \models Q_2 + T$, which is weaker than hyp. (ii). It turns out, in fact, that there are specific $\{P_i\}$, $\{Q_i\}$, and T for which the hypotheses of our method hold, but those of the generalized [6] method do not. This implies that our method is more widely applicable. Indeed, we had noticed in earlier work [7, 8] that it was not possible to use the generalized Alur-Henzinger rule for Q_i with liveness constraints that were *automatically* generated from property T. In [7], we showed that conditions (a), (d), and (e) *always* hold for the generated Q_i's, thus leaving one to (safely) apply the mutually inductive checks (b) and (c).

3 Concrete Compositional Reasoning

In this section, we show how the abstract framework can be instantiated for a variety of composition operators and language definitions in shared-variable concurrency. For each choice, we show that assumptions **WF**, **L1**, **P1-P3**, and **BEQ** are met. For lack of space, detailed proofs are left to the full version.

We assume available a set of variable names, with associated domains of values. For a variable w, the primed variable w' is used to denote the value of w in the next state. A *process*, P, is given by a tuple (z, ι, τ, Φ), where: z is a finite set of *variables*, partitioned into *local* variables (x), and *interface* variables (y), $\iota(z)$ is an *initial* condition, $\tau(z, z')$ is a *transition* relation, relating current and next values of z, and $\Phi(z)$ is a temporal fairness formula defining a *fairness* condition on the states. A *state* of P is a function giving values to its variables. We write states as pairs (a, b), where a is the value of x and b the value of y. A *path*, π, is a finite or infinite sequence of states where for each i such that $0 < i < |\pi|$, $\tau(\pi_{i-1}, \pi_i)$ holds; $|\pi|$ is the length of π (the number of states on it). A state t is *reachable* from state s by a path π of length n iff $\pi_0 = s$ and $\pi_{n-1} = t$.

A process exhibits *finite local branching* (cf. [1]) if (i) for each b, there are finitely many a such that (a, b) is an initial state, and (ii) for any state (a, b) and any b', there are finitely many a' such that (a', b') is reachable from (a, b) by a path where the value of y is b for all non-final states. We assume that all processes satisfy this restriction. We use temporal logic to define the languages of processes. We only need the operators G (always) and $\overset{\infty}{\mathsf{G}}$ (from some point on). Formally, for a sequence π, and a predicate q on states (transitions), $\mathsf{G}(q)$ holds of π iff q holds for every state (transition) on π. For an infinite sequence π, and a predicate q, $\overset{\infty}{\mathsf{G}}(q)$ holds of π if, from some $k \in \mathbf{N}$, the suffix of π starting at point k satisfies $\mathsf{G}(q)$. The path operator A quantifies over all paths from a state.

3.1 Linear-Time Interleaving Semantics

In the interleaving semantics, the *language* of a process, P_i, is defined relative to an *external context* for P_i. The context is represented by a set of variables, w_i, disjoint from z_i (we write z_i instead of z_{P_i} for clarity). The language is a subset of the set of sequences, both finite and infinite, of valuations to $z_i \cup w_i$. We denote the language by the expression $\mathcal{L}^{[]}(P_i)(z_i; w_i)$, where ";" is used to separate the process' variables from its context. The set of finite computations ($\mathcal{L}^{[]}_*(P_i)(z_i; w_i)$) is defined by the temporal logic formula below, where, for a set X of variables, $unch(X) \equiv (\forall x : x \in X : x' = x)$.

$$\iota_i(z_i) \wedge \mathsf{G}((\tau_i(z_i, z_i') \wedge unch(w_i)) \vee unch(x_i))$$

This formula ensures that the first state is an initial one, every transition is either that of P_i and leaves the context w_i unchanged, or is an "environment" transition that leaves the local variables of P_i unchanged. The set of infinite sequences in the language, denoted by $\mathcal{L}^{[]}_\infty(P_i)(z_i; w_i)$, is defined by the same formula (interpreted over infinite sequences), together with the constraint $\Phi_i(z_i)$ which ensures that the fairness condition of P_i holds. The full language, $\mathcal{L}^{[]}(P_i)(z_i; w_i)$, is given by $\mathcal{L}^{[]}_*(P_i)(z_i; w_i) \cup \mathcal{L}^{[]}_\infty(P_i)(z_i; w_i)$. The *external language* of process P_i for context w_i is denoted by $\mathcal{L}^{[]}_{ext}(P_i)(y_i; w_i)$. It is defined as the projection of the language of P_i on its interface and context variables: i.e., $\mathcal{L}^{[]}_{ext}(P_i)(y_i; w_i) \equiv (\exists x_i : \mathcal{L}^{[]}(P_i)(z_i; w_i))$. Here, existential quantification refers to the choice of a *sequence* of values for x_i. Two computations are *behavior equivalent* relative to P_i iff their projections on $y_i \cup w_i$ are identical. The ordering $\preceq$ is defined as the prefix ordering on finite sequences. This choice satisfies **WF**. From the language definitions, every finite prefix of a sequence in $\mathcal{L}^{[]}_\infty(P)$ satisfies the initial and transition conditions, and is thus in $\mathcal{L}^{[]}_*(P)$, so that **L1** holds for $\mathcal{L}^{[]}_{ext}(P)$. As $\mathcal{L}^{[]}_{ext}(P)$ is defined over the non-local variables of P, it satisfies **BEQ**.

Interleaving Composition. For a set of processes $\{P_i\}$, their *interleaving composition* Q is denoted as $([]i : P_i)$. It is defined provided that the local variables of each process are disjoint from the variables of other processes; i.e., $x_i \cap z_j = \emptyset$ for $i \neq j$. The process Q has local variables $(\cup i : x_i)$, interface variables $(\cup i : y_i)$, initial condition $(\wedge i : \iota_i(z_i))$, and fairness condition $(\wedge i : \Phi_i(z_i))$.

For any i, let X_i be the set of local variables of the other processes; formally, $X_i = x_Q \setminus x_i$. Let Y_i be the set of interface variables of other processes that are *not shared* with the interface variables of process P_i; formally, $Y_i = y_Q \setminus y_i$. Let $Z_i = X_i \cup Y_i$. The transition relation of Q is defined to be $(\vee i : \tau_i(z_i, z_i') \wedge unch(Z_i))$. This definition implies that a transition of process P_i, leaves unchanged the values of all variables that are not shared with P_i.

It is important to note that we do not distinguish between the usage of shared variables, such as read-only or write-only variables. All shared variables can be both read and written to by the sharing processes. Since this is the most general case, our results apply to more specific situations as well.

Non-blocking: The condition $[\sim_{P_i} \Rightarrow \sim_{Q_i}]$ in hyp. (a) of the **Circular Reasoning II** method is equivalent to saying that y_{Q_i} is a subset of y_{P_i}. Furthermore, only the first part of the non-blocking definition (for initial states) is needed, because, for any x, y: if x is a finite computation in $\mathcal{L}^{\Box}_{ext}(P)$, $y \prec x$ and $y \in \mathcal{L}^{\Box}_{ext}(P \Box Q)$ then y can be extended to a computation z in $\mathcal{L}^{\Box}_{ext}(P \Box Q)$ such that z and x agree on the external variables of P by following the transitions in x while keeping the local variables of Q unchanged.

Theorem 2. *(**P1** for $\mathcal{L}^{\Box}$) Let $\{P_i\}$ be a set of processes such that $Q = (\Box i : P_i)$ is defined, and let w be a context for Q. Then, $[\mathcal{L}^{\Box}(Q)(z_Q; w) \equiv (\wedge i : \mathcal{L}^{\Box}(P_i)(z_i; w \cup Z_i))]$.* $\Box$

The previous theorem shows that asynchronous composition corresponds to conjunction of languages. We are usually interested only in the externally visible behavior of a process. Abadi and Lamport showed in [2] that a similar theorem applies to the external languages, under restrictions which ensure that the external language is closed under stuttering (i.e., finite repetition of states). However, there are applications such as the analysis of timing diagrams [8], where one wants to count the number of events in an asynchronous system, and this is not a stuttering-closed language. We therefore show the analogous theorem under a different non-interference assumption on composition.

The *mutex* Assumption: A set of processes $\{P_i\}$ satisfies this assumption if in any jointly enabled transition, at most one process changes its local state. This may be realized in practice by using a turn-based interleaving scheduler, possibly implemented by the processes themselves, and guarding each transition with a check that it is the process' turn.

Theorem 3. *(**P1** for $\mathcal{L}^{\Box}_{ext}$) Let $\{P_i\}$ be a set of processes such that $Q = (\Box i : P_i)$ is defined, and let w be a context for Q. Under the mutex assumption for $\{P_i\}$, $[\mathcal{L}^{\Box}_{ext}(Q)(y_Q; w) \equiv (\wedge i : \mathcal{L}^{\Box}_{ext}(P_i)(y_i; w \cup Y_i))]$.*$\Box$

Stuttering Equivalence: Stuttering refers to finite repetition of identical values. Two sequences are stuttering equivalent if they are identical up to such finite repetition. The language of a process is closed relative to stuttering; however, its external language is not, as existential quantification does not preserve closure in general. The stuttering closed external language of P_i for context w is denoted by $\tilde{\mathcal{L}}^{\Box}_{ext}(P_i)(y_i; w)$, and is defined as $(\tilde{\exists} x_i : \mathcal{L}^{\Box}(P_i)(z_i; w))$. Here, $\tilde{\exists}$ is a stuttering closed version of $\exists$, defined by: for a sequence π over a set of variables W, $\pi \in (\tilde{\exists} X : S(X, W))$ iff there exists a sequence ξ defined over $X \cup W$ such that π and ξ are stuttering equivalent on X, and $\xi \in S$.

The *Sequencing* Assumption: This is a semantic formulation of the syntactic constraints considered by Abadi and Lamport in [2]. It holds if, for every jointly enabled transition t of $Q = (\Box i : P_i)$, there is a finite sequence σ that satisfies the transition condition of Q on all transitions, and σ and t are stuttering equivalent relative to the external variables $y_Q \cup w$.

Theorem 4. *(**P1** for $\tilde{\mathcal{L}}^{\Box}_{ext}$) (cf. [2]) Let $\{P_i\}$ be a set of processes such that $Q = (\Box i : P_i)$ is defined, and let w be a context for Q. Under the sequencing* *assumption, and if Φ_Q is stuttering-closed relative to $y_Q \cup w$,* $[\tilde{\mathcal{L}}^{\square}_{ext}(Q)(y_Q; w) \equiv (\wedge i : \tilde{\mathcal{L}}^{\square}_{ext}(P_i)(y_i; w \cup Y_i))]$. $\square$

Process Choice. For a set of processes $\{P_i\}$, the process $C = (+i : P_i)$ is defined whenever $(\square i : P_i)$ is defined, with local variables $(\cup i : x_i) \cup \{c\}$, where c is a fresh variable not in $(\cup i : x_i)$, interface variables $(\cup i : y_i)$, initial condition $(\vee i : c = i \wedge \iota_i(z_i))$ transition relation $(c' = c) \wedge (\wedge i : c = i \Rightarrow \tau_i(z_i, z_i'))$, and fairness condition $(\vee i : \overset{\infty}{\mathsf{G}}(c = i) \wedge \Phi_i(z_i))$. The variable c records the initial (fixed) choice of $\{P_i\}$. The following theorem is an easy consequence.

Theorem 5. *(**P2** for $\mathcal{L}^{\square}_{ext}$ and $\tilde{\mathcal{L}}^{\square}_{ext}$) For a set of processes $\{P_i\}$ such that $C = (+i : P_i)$ is defined, and a context w for C, (i) $[\mathcal{L}^{\square}_{ext}(C)(y_C; w) \equiv (\vee i : \mathcal{L}^{\square}_{ext}(P_i)(y_i; w))]$, and (ii) $[\tilde{\mathcal{L}}^{\square}_{ext}(C)(y_C; w) \equiv (\vee i : \tilde{\mathcal{L}}^{\square}_{ext}(P_i)(y_i; w))]$.* $\square$

Process Closure. For a process Q, let $CL(Q)$ be the process $(z_Q, \iota_Q, \tau_Q, true)$. I.e., $CL(Q)$ has the same transition structure as Q, but has a trivial fairness constraint. The proof of the following theorem relies on the finite local branching requirement and König's Lemma.

Theorem 6. *(**P3** for $\mathcal{L}^{\square}_{ext}$, and for $\tilde{\mathcal{L}}^{\square}_{ext}$ (cf. [1])) For any process Q and context w for Q, $[\mathcal{L}^{\square}_{ext}(CL(Q))(y_Q; w) \equiv cl(\mathcal{L}^{\square}_{ext}(Q)(y_Q; w))]$, and $[\tilde{\mathcal{L}}^{\square}_{ext}(CL(Q))(y_Q; w) \equiv cl(\tilde{\mathcal{L}}^{\square}_{ext}(Q)(y_Q; w))]$.* $\square$

3.2 Synchronous Semantics

In the synchronous semantics, all processes in a composition make a transition simultaneously. Synchronous semantics are appropriate for modeling hardware systems, where several components are controlled by a single clock. The *language* of a process, P_i, is a subset of the set of sequences, both finite and infinite, of valuations to z_i. The finite part of the language, denoted by $\mathcal{L}^{||}_{*}(P_i)(z_i)$, is given by the temporal formula $\iota_i(z_i) \wedge \mathsf{G}(\tau_i(z_i, z_i'))$. The infinite part, denoted by $\mathcal{L}^{||}_{\infty}(P_i)(z_i)$ is given by sequences satisfying the same formula, but with the additional fairness constraint, $\Phi_i(z_i)$. The full language, denoted by $\mathcal{L}^{||}(P_i)(z_i)$ is $\mathcal{L}^{||}_{*}(P_i)(z_i) \cup \mathcal{L}^{||}_{\infty}(P_i)(z_i)$. The external language of the process, denoted by $\mathcal{L}^{||}_{ext}(P_i)(y_i)$, is defined, as before, by $(\exists x_i : \mathcal{L}^{||}(P_i)(z_i))$.

Synchronous Composition. For a set of processes $\{P_i\}$, their *synchronous composition* Q is denoted as $(|| \; i : P_i)$. The components of Q are defined as for interleaving composition, except the transition relation, which is defined as $(\wedge i : \tau_i(z_i, z_i'))$, since transitions are simultaneous. Process choice and closure are defined exactly as in the asynchronous case. Similarly, the behavior equivalence relationship from Hyp. (a) of the method is given by $y_{Q_i} \subseteq y_{P_i}$. Completely specified Moore machines are non-blocking by definition. Mealy machines can be made non-blocking by syntactically preventing combinational cycles between the input-output variables of the composed processes (e.g., as in Reactive Modules [6]). In [7], it is shown that these definitions enjoy the properties **P1-P3** for $\mathcal{L}^{||}_{ext}$; thus, the **Circular Reasoning II** method is applicable.

3.3 Refinement through Fair Simulation

In this part of the paper, we interpret the relation $\models$ as refinement through a simulation relation that includes the effect of fairness constraints cf. [15]. Composition is synchronous, with the operators defined as in the previous section. The difference is that the language of a process is now a set of trees instead of sequences. We assume that each process has the finite branching property.

Tree Languages: A *tree* is a prefix-closed subset of $\mathbf{N}^*$. We consider only finite branching trees. Finite behaviors are finite depth trees. The $\preceq$ ordering on trees is subtree inclusion. The finite branching condition ensures that this ordering is well-founded on finite behaviors, since each finite-depth tree has finitely many nodes. A *labeled tree*, u, is a triple (U, f, Σ), where U is a tree and $f : U \rightarrow \Sigma$.

A *computation tree* of a process P is obtained, informally, by unrolling its transition relation starting from an initial state. A *computation tree fragment* (CTF, for short) of process P is obtained from a computation tree of P by either dropping or duplicating some subtrees of the computation tree, while satisfying its fairness constraint along all infinite branches. Formally, a CTF is a labeled tree $u = (U, f, \Sigma)$ where: Σ is the state space of P, $f(\lambda)$ is an initial state of P, $(f(x), f(x.i))$ is a transition of P for each $x \in \mathbf{N}^*$ and $i \in \mathbf{N}$ such that $x, x.i \in U$, and the fairness condition of P holds along all infinite branches of U. Tree fragments represent the result of composing P with arbitrary, nondeterministic environments. The set of CTF's of P forms its (tree) *language*, denoted by $\mathcal{L}^T(P)$. The finite language of P, $\mathcal{L}^T_*(P)$, is therefore, the set of finite-depth trees satisfying the CTL-like formula $\iota_P(z_P) \wedge \mathsf{AG}(\tau_P(z_P, z'_P))$. The infinite language of P, $\mathcal{L}^T_\infty(P)$, is the set of infinite trees satisfying, additionally, $\mathsf{A}(\Phi_P(z_P))$. Let $\mathcal{L}^T(P) = \mathcal{L}^T_*(P) \cup \mathcal{L}^T_\infty(P)$. The external language of M, denoted by $\mathcal{L}^T_{ext}(P)$, is obtained, as before, by projecting out the local variable component of each node label: formally, this is represented as $(\exists x_P : \mathcal{L}^T(P))$.

In [14], it is shown (with different notation) that for non-fair processes P and Q, Q simulates P (in the standard sense [22]) iff $[\mathcal{L}^T_{ext}(P) \Rightarrow \mathcal{L}^T_{ext}(Q)]$. Based on this correspondence, they propose using language inclusion as the definition of simulation under fairness. We show that this choice satisfies conditions **P1-P3**. Conditions **P1** and **P2** follow quite directly from the CTL formulation of process language. From the prefix ordering, a tree is in the closure of a set of trees S iff every finite depth subtree can be extended to a tree in S. This is called *finite closure* in [20], where it is used to define "universally safe" branching properties. The proof of **P3** relies on the finite-branching property, and König's lemma.

Theorem 7. *(**P1,P2,P3** for $\mathcal{L}^T_{ext}$) If $Q = (\| i : P_i)$ is defined, then (i) $[\mathcal{L}^T_{ext}(Q) \equiv (\wedge i : \mathcal{L}^T_{ext}(P_i))]$, (ii) $[\mathcal{L}^T_{ext}((+i : P_i)) \equiv (\vee i : \mathcal{L}^T_{ext}(P_i))]$, and (iii) for any P, $[cl(\mathcal{L}^T_{ext}(P)) \equiv \mathcal{L}^T_{ext}(CL(P))]$.*

4 Conclusions

This paper develops a general framework for designing sound and complete methods for mutually inductive, assume-guarantee reasoning. The key challenge is in balancing, on one hand, the restrictions needed to avoid unsound circular reasoning with liveness properties and, on the other, the generality necessary to ensure completeness. Furthermore, we show how to instantiate this framework for linear-time interleaving composition, and extend it to stuttering closed external languages. We then outline the instantiation for synchronous composition in both linear and branching time semantics. We believe that the resulting rules can be adapted without much difficulty to specific modeling languages.

References

1. M. Abadi and L. Lamport. The existence of refinement mappings. In *LICS*, 1988.
2. M. Abadi and L. Lamport. Conjoining specifications. *ACM Trans. on Programming Languages and Systems (TOPLAS)*, May 1995.
3. M. Abadi and S. Merz. An abstract account of composition. In *MFCS*, volume 969 of *LNCS*. Springer Verlag, 1995.
4. M. Abadi and G. Plotkin. A logical view of composition and refinement. In *POPL*, 1991.
5. B. Alpern and F. Schneider. Defining liveness. *Information Processing Letters*, 21(4), 1985.
6. R. Alur and T. Henzinger. Reactive modules. In *LICS*, 1996.
7. N. Amla, E.A. Emerson, K.S. Namjoshi, and R.J. Trefler. Assume-guarantee based compositional reasoning for synchronous timing diagrams. In *TACAS*, volume 2031 of *LNCS*, 2001.
8. N. Amla, E.A. Emerson, K.S. Namjoshi, and R.J. Trefler. Visual specifications for modular reasoning about asynchronous systems. In *FORTE*, volume 2529 of *LNCS*, 2002.
9. K.M. Chandy and J. Misra. Proofs of networks of processes. *IEEE Transactions on Software Engineering*, 7(4), 1981.
10. E.M. Clarke and E. A. Emerson. Design and synthesis of synchronization skeletons using branching time temporal logic. In *Workshop on Logics of Programs*, volume 131 of *LNCS*, 1981.
11. W-P. de Roever, F. de Boer, U. Hannemann, J. Hooman, Y. Lakhnech, M. Poel, and J. Zwiers. *Concurrency Verification: Introduction to Compositional and Non-compositional Proof Methods.* Cambridge University Press, 2001.
12. W-P. de Roever, H. Langmaack, and A. Pnueli, editors. *Compositionality: The Significant Difference*, volume 1536 of *LNCS*. Springer-Verlag, 1997.
13. E.W. Dijkstra and C.S. Scholten. *Predicate Calculus and Program Semantics.* Springer Verlag, 1990.
14. T.A. Henzinger, O. Kupferman, and S. Rajamani. Fair simulation. In *CONCUR*, volume 1243 of *LNCS*, 1997.
15. T.A. Henzinger, S. Qadeer, S.K. Rajamani, and S. Tasiran. An assume-guarantee rule for checking simulation. *ACM Trans. on Programming Languages and Systems (TOPLAS)*, January 2002.
16. C.B. Jones. *Development methods for computer programs including a notion of interference.* PhD thesis, Oxford University, 1981.
17. R.P. Kurshan. *Computer-Aided Verification of Coordinating Processes: The Automata-Theoretic Approach.* Princeton University Press, 1994.
18. P. Maier. A set-theoretic framework for assume-guarantee reasoning. In *ICALP*, volume 2076 of *LNCS*, 2001.

19. P. Maier. Compositional circular assume-guarantee rules cannot be sound and complete. In *FoSSaCS*, volume 2620 of *LNCS*, 2003.
20. P. Manolios and R. J. Trefler. Safety and liveness in branching time. In *LICS*, 2001.
21. K.L. McMillan. Circular compositional reasoning about liveness. In *CHARME*, volume 1703 of *LNCS*, 1999.
22. R. Milner. An algebraic definition of simulation between programs. In *2nd IJCAI*, 1971.
23. K.S. Namjoshi and R.J. Trefler. On the completeness of compositional reasoning. In *CAV*, volume 1855 of *LNCS*, 2000.
24. P. Pandya and M. Joseph. P-A logic - a compositional proof system for distributed programs. *Distributed Computing*, 1991.
25. J.P. Queille and J. Sifakis. Specification and verification of concurrent systems in CESAR. In *Proc. of the 5th Intl. Symp. on Programming*, volume 137 of *LNCS*, 1982.
26. M. Viswanathan and R. Viswanathan. Foundations for circular compositional reasoning. In *ICALP*, volume 2076 of *LNCS*. Springer Verlag, 2001.

Relating Fairness and Timing in Process Algebras*

F. Corradini[1], M.R. Di Berardini[1], and W. Vogler[2]

[1] Dipartimento di Informatica
Universitá di L'Aquila
{flavio,mdiberar}@di.univaq.it
[2] Institut für Informatik
Universität Augsburg
vogler@informatik.uni-augsburg.de

Abstract. This paper contrasts two important features of parallel system computations: *fairness* and *timing*. The study is carried out at specification system level by resorting to a well-known process description language. The language is extended with labels which allow to filter out those process executions that are not (weakly) fair (as in [5,6]), and with upper time bounds for the process activities (as in [2]).
We show that fairness and timing are closely related. Two main results are stated. First, we show that each everlasting (or non-Zeno) timed process execution is fair. Second, we provide a characterization for fair executions of untimed processes in terms of timed process executions. This results in a finite representation of fair executions using regular expressions.

1 Introduction

In the theory and practice of parallel systems, fairness and timing play an important role when describing the system dynamics. Fairness requires that a system activity which is continuously enabled along a computation will eventually proceed; this is usually a necessary requirement for proving liveness properties of the system. Timing gives information on when actions are performed and can serve as a basis for considering efficiency.

We will show that fairness and timing are somehow related - although they are used in different contexts. Our comparison is conducted at system specification level by resorting to a standard (CCS-like) process description language. We consider two extensions of this basic language. The first extension permits to isolate the fair system executions and follows the approach of Costa and Stirling [5,6]. The second one adds upper time bounds for the execution time of system activities and follows the approach taken in [2].

* This work was supported by MURST project 'Sahara: Software Architectures for Heterogeneous Access Networks infrastructures' and by the Center of Eccellence for Research 'DEWS: Architectures and Design Methodologies for Embedded Controllers, Wireless Interconnect and System-on-chip'.

R. Amadio, D. Lugiez (Eds.): CONCUR 2003, LNCS 2761, pp. 446–460, 2003.

Costa and Stirling distinguish between fairness of actions (also called events), which they study in [5] for a CCS-like language without restriction, and fairness of components [6]. In both cases they distinguish between weak and strong fairness. Weak fairness requires that if an action (a component, resp.) can *almost always* proceed then it must eventually do so, and in fact it must proceed infinitely often, while strong fairness requires that if an action (a component) can proceed *infinitely often* then it must proceed infinitely often. Differences between fairness of actions and fairness of components and between weak and strong fairness are detailed in [6]; for the purpose of this paper, we are interested in weak fairness of actions. A useful result stated in [5,6] characterizes fair computations as the concatenation of certain finite sequences, called LP-steps in [6].

Regarding timing, we follow the approach taken in the timed process algebra PAFAS (Process Algebra for Faster Asynchronous Systems). Based on ideas first studied for Petri nets e.g. in [10], this new process description language has been proposed as a useful tool for comparing the worst-case efficiency of asynchronous systems (see [9,2] for the general theory and [3] for an application). PAFAS is a CCS-like process description language [12] where basic actions are atomic and instantaneous but have an associated time bound (which is 1 or 0 for simplicity) as a maximal time delay for their execution.[1] When, for an action with time bound 1, this idle-time of 1 has elapsed, the action becomes *urgent* (i.e. its time bound becomes 0) and it must be performed (or be deactivated) before time may pass further – unless it has to wait for a synchronization, with another component, which either does not offer synchronization on this action at all or at least can still delay the synchronization.

We prove two main results relating timed computations of PAFAS processes and weak fairness of actions. First, we prove that all everlasting (or non-Zeno) computations [2] are fair. This result shows that timing with upper time bounds imposes fairness among the different system activities. Intuitively, when one time unit passes, the active actions become urgent and must be performed (or be deactivated) before time may pass further; this clearly ensures that an activated action does not wait forever in a computation with infinitely many time steps.

As a second main result we show that LP-steps – defined for untimed processes – coincide in the timed setting with sequences of basic actions between two consecutive time steps. As a consequence of this lemma we have that non-Zeno process computations fully characterize fair computations.

Besides providing a formal comparison between fairness and timing, our timed characterization of fair executions results in a representation with technical advantages compared to the approach of [5,6]. In order to keep track of the different instances of system activities along a system execution, Costa and Stirling associate labels to actions, and the labels are essential in the definition of fair

---

[1] As discussed in [2], due to these upper time bounds time can be used to evaluate efficiency, but it does not influence functionality (which actions are performed); so compared to CCS, also PAFAS treats the full functionality of asynchronous systems.

[2] A process computation is a Zeno computation when infinitely many actions happen in finite time.

computations. New labels are created dynamically during the system evolution with the immediate effect of changing the syntax of process terms; thus, cycles in the transition system of a process are impossible and even finite-state processes (according to the ordinary operational semantics) usually become infinite-state. From the maximal runs of such a transition system, Costa and Stirling filter out the unfair computations by a criterion that considers the processes and their labels on a maximal run. Our timed semantics also provides such a two-level description: we also change the syntax of processes – in our case by adding timing information –, but this is much simpler than the labels of [5,6], and it leaves finite-state processes finite-state. Then we apply a simpler filter, which does not consider the processes: we simply require that infinitely many time steps occur in a run. As a small price, we have to project away these time steps in the end.

As mentioned above, Costa and Stirling give a one-level characterization of fair computations with an SOS-semantics defining so-called LP-steps; these are (finite, though usually unbounded) sequences of actions leading from ordinary processes to ordinary processes, with the effect that even finite-state transition systems for LP-steps usually have infinitely many transitions – although they are at least finite-state. In contrast, our time-based operational semantics defines steps with single actions (or unit time steps), and consequently a finite-state transition system is really finite. Finally, using standard automata-theoretic techniques, we can get rid of the time steps in such a finite-state transition system by constructing another finite-state transition system with regular expressions as arc labels; maximal runs in this transition system are exactly the fair runs. This way we also arrive at a one-level description, and ours is truly finite. See [4] for full details.

2 A Process Algebra for Faster Asynchronous Systems

In this section we give a brief description of PAFAS, a process algebra introduced in [2] to consider the functional behaviour and the temporal efficiency of asynchronous systems. The PAFAS transitional semantics is given by two sets of SOS-rules. One describes the functional behaviour and is very similar to the SOS-rules for standard CCS [12]. The other describes the temporal behaviour and is based on a notion of refusal sets.

2.1 PAFAS Processes

We use the following notation: $\mathbb{A}$ is an infinite set of basic actions. An additional action τ is used to represent internal activity, which is unobservable for other components. We define $\mathbb{A}_\tau = \mathbb{A} \cup \{\tau\}$. Elements of $\mathbb{A}$ are denoted by $a, b, c, \ldots$ and those of $\mathbb{A}_\tau$ are denoted by $\alpha, \beta, \ldots$. Actions in $\mathbb{A}_\tau$ can let time 1 pass before their execution, i.e. 1 is their maximal delay. After that time, they become *urgent* actions written $\underline{a}$ or $\underline{\tau}$; these have maximal delay 0. The set of urgent actions is denoted by $\underline{\mathbb{A}}_\tau = \{\underline{a} \mid a \in \mathbb{A}\} \cup \{\underline{\tau}\}$ and is ranged over by $\underline{\alpha}, \underline{\beta}, \ldots$. Elements of $\mathbb{A}_\tau \cup \underline{\mathbb{A}}_\tau$ are ranged over by μ.

$\mathcal{X}$ is the set of process variables, used for recursive definitions. Elements of $\mathcal{X}$ are denoted by $x, y, z, \ldots$. $\Phi : \mathbb{A}_\tau \to \mathbb{A}_\tau$ is a *general relabelling function* if the set $\{\alpha \in \mathbb{A}_\tau \mid \emptyset \neq \Phi^{-1}(\alpha) \neq \{\alpha\}\}$ is finite and $\Phi(\tau) = \tau$. Such a function can also be used to define *hiding*: P/A, where the actions in A are made internal, is the same as $P[\Phi_A]$, where the relabelling function Φ_A is defined by $\Phi_A(\alpha) = \tau$ if $\alpha \in A$ and $\Phi_A(\alpha) = \alpha$ if $\alpha \notin A$.

We assume that time elapses in a discrete way.[3] Thus, an action prefixed process $a.P$ can either do action a and become process P (as usual in CCS) or can let one time step pass and become $\underline{a}.P$; $\underline{a}$ is called *urgent* a, and $\underline{a}.P$ as a stand-alone process cannot let time pass, but can only do a to become P.

Definition 1. (*timed process terms*)
The set $\tilde{\mathbb{P}}$ of the timed process terms is generated by the following grammar:

$$P ::= \mathsf{nil} \mid x \mid \alpha.P \mid \underline{\alpha}.P \mid P + P \mid P \|_A P \mid P[\Phi] \mid \mathsf{rec}\, x.P$$

where $x \in \mathcal{X}$, $\alpha \in \mathbb{A}_\tau$, Φ is a general relabelling function and $A \subseteq \mathbb{A}$ possibly infinite. We assume that the recursion is *(time-)guarded*, i.e. for $\mathsf{rec}\, x.P$ variable x only appears in P within the scope of a prefix $\alpha.()$ with $\alpha \in \mathbb{A}_\tau$. A term P is *guarded* if each occurrence of a variable is guarded in this sense. A significant subset of $\tilde{\mathbb{P}}$ is $\tilde{\mathbb{P}}_1$, the set of *initial* timed process terms. These are $\tilde{\mathbb{P}}$-terms where every action is in $\mathbb{A}_\tau$; they correspond to ordinary CCS-like processes.

The set of closed (i.e., every variable x in a timed process term P is bound by the corresponding $\mathsf{rec}\, x$-operator) timed process terms in $\tilde{\mathbb{P}}$ and $\tilde{\mathbb{P}}_1$, simply called *processes* and *initial processes* resp., is denoted by $\mathbb{P}$ and $\mathbb{P}_1$ resp.[4]

A brief description of the (PAFAS) operators now follows. nil is the Nil-process; it cannot perform any action, but may let time pass without limit. A trailing nil will often be omitted, so e.g. $a.b + c$ abbreviates $a.b.\mathsf{nil} + c.\mathsf{nil}$. $\mu.P$ is action prefixing, known from CCS, with the timed behaviour as explained above. $P_1 + P_2$ models the choice between two conflicting processes P_1 and P_2. $P_1 \|_A P_2$ is the parallel composition of two processes P_1 and P_2 that run in parallel and have to synchronize on all actions from A; this synchronization discipline is inspired from TCSP. $P[\Phi]$ behaves as P but with the actions changed according to Φ. $\mathsf{rec}\, x.P$ models a recursive definition.

2.2 The Functional Behaviour of PAFAS Processes

The transitional semantics describing the functional behaviour of PAFAS processes indicates which basic actions they can perform; timing information can be disregarded, since we only have *upper* time bounds; see the two PREF-rules below.

[3] PAFAS is not time domain dependent, meaning that the choice of discrete or continuous time makes no difference for the testing-based semantics of asynchronous systems, see [2] for more details.

[4] As shown in [3], $\mathbb{P}_1$ processes do not have time-stops; i.e. every finite process run can be extended such that time grows unboundedly.

Definition 2. (*Functional operational semantics*) The following SOS-rules define the transition relations $\xrightarrow{\alpha} \subseteq (\tilde{\mathbb{P}} \times \tilde{\mathbb{P}})$ for $\alpha \in \mathbb{A}_\tau$, the *action transitions*.

As usual, we write $P \xrightarrow{\alpha} P'$ if $(P, P') \in \xrightarrow{\alpha}$ and $P \xrightarrow{\alpha}$ if there exists a $P' \in \tilde{\mathbb{P}}$ such that $(P, P') \in \xrightarrow{\alpha}$, and similar conventions will apply later on.

$$\text{PREF}_{a1}\frac{}{\alpha.P \xrightarrow{\alpha} P} \qquad \text{PREF}_{a2}\frac{}{\underline{\alpha}.P \xrightarrow{\alpha} P}$$

$$\text{SUM}_a\frac{P_1 \xrightarrow{\alpha} P_1'}{P_1 + P_2 \xrightarrow{\alpha} P_1'}$$

$$\text{PAR}_{a1}\frac{\alpha \notin A,\ P_1 \xrightarrow{\alpha} P_1'}{P_1\|_A P_2 \xrightarrow{\alpha} P_1'\|_A P_2} \qquad \text{PAR}_{a2}\frac{\alpha \in A,\ P_1 \xrightarrow{\alpha} P_1',\ P_2 \xrightarrow{\alpha} P_2'}{P_1\|_A P_2 \xrightarrow{\alpha} P_1'\|_A P_2'}$$

$$\text{REL}_a\frac{P \xrightarrow{\alpha} P'}{P[\Phi] \xrightarrow{\Phi(\alpha)} P'[\Phi]} \qquad \text{REC}_a\frac{P\{\mathsf{rec}\,x.P/x\} \xrightarrow{\alpha} P'}{\mathsf{rec}\,x.P \xrightarrow{\alpha} P'}$$

Additionally, there are symmetric rules for Par_{a1} and Sum_a for actions of P_2.

2.3 The Temporal Behaviour of PAFAS Processes

We are now ready to define the refusal traces of a term $P \in \tilde{\mathbb{P}}$. Intuitively a refusal trace records, along a computation, which actions process P can perform ($P \xrightarrow{\alpha} P'$, $\alpha \in \mathbb{A}_\tau$) and which actions P can refuse to perform when time elapses ($P \xrightarrow{X}_r P'$, $X \subseteq \mathbb{A}$).

Definition 3. (*Refusal transitional semantics*)

The following inference rules define $\xrightarrow{X}_r \subseteq (\tilde{\mathbb{P}} \times \tilde{\mathbb{P}})$, where $X \subseteq \mathbb{A}$.

$$\text{NIL}_r\frac{}{\mathsf{nil} \xrightarrow{X}_r \mathsf{nil}}$$

$$\text{PREF}_{r1}\frac{}{\alpha.P \xrightarrow{X}_r \underline{\alpha}.P} \qquad \text{PREF}_{r2}\frac{\alpha \notin X \cup \{\tau\}}{\underline{\alpha}.P \xrightarrow{X}_r \underline{\alpha}.P}$$

$$\text{PAR}_r\frac{P_i \xrightarrow{X_i}_r P_i' \text{ for } i = 1,2,\ X \subseteq (A \cap (X_1 \cup X_2)) \cup (X_1 \cap X_2) \backslash A}{P_1\|_A P_2 \xrightarrow{X}_r P_1'\|_A P_2'}$$

$$\text{SUM}_r\frac{\forall\, i = 1,2\ P_i \xrightarrow{X}_r P_i'}{P_1 + P_2 \xrightarrow{X}_r P_1' + P_2'}$$

$$\text{REL}_r\frac{P \xrightarrow{\Phi^{-1}(X \cup \{\tau\}) \backslash \{\tau\}}_r P'}{P[\Phi] \xrightarrow{X}_r P'[\Phi]} \qquad \text{REC}_r\frac{P\{\mathsf{rec}\,x.P/x\} \xrightarrow{X}_r P'}{\mathsf{rec}\,x.P \xrightarrow{X}_r P'}$$

Rule PREF_{r1} says that a process $\alpha.P$ can let time pass and refuse to perform any action while rule PREF_{r2} says that a process P prefixed by an urgent action $\underline{\alpha}$, can let time pass but action α cannot be refused. Process $\underline{\tau}.P$ cannot let time pass and cannot refuse any action; also in any context, $\underline{\tau}.P$ has to perform τ as explained by Rule PREF_{a2} in Definition 2 before time can pass further. Another rule worth noting is PAR_r which defines which actions a parallel composition can refuse during a time-step. The intuition is that $P_1\|_A P_2$ can refuse an action α if either $\alpha \notin A$ (P_1 and P_2 can perform α independently) and both P_1 and P_2 can refuse α, or $\alpha \in A$ (P_1 and P_2 are forced to synchronize on α) and at least one of P_1 and P_2 can refuse α, i.e. can delay it. Thus, an action in a parallel composition is urgent (cannot be further delayed) only when all synchronizing 'local' actions are urgent. The other rules are as expected.

A transition like $P \xrightarrow{X}_r P'$ is called a (partial) *time-step*. The actions listed in X are not urgent; hence P is justified in not performing them, but performing a time step instead. This time step is partial because it can occur only in contexts that can refuse the actions not in X. If $X = \mathbb{A}$ then P is fully justified in performing this time-step; i.e., P can perform it independently of the environment. If $P \xrightarrow{\mathbb{A}}_r P'$ we write $P \xrightarrow{1} P'$ and say that P performs a *full-time* step.

In [2], it is shown that refusal traces characterize an efficiency preorder, which is intuitively justified by a testing scenario. In the present paper, we need partial time steps only to set up the following SOS-semantics; our real interest is in runs where all time steps are full. We let λ range over $\mathbb{A}_\tau \cup \{1\}$.

3 Fairness and PAFAS

In this section we briefly describe our theory of fairness. It closely follows Costa and Stirling's theory of (weak) fairness. The main ingredients of the theory are:

- *A Labelling for Process Terms.* This allows to detect during a transition which action is actually performed; e.g., for process $P = \mathsf{rec}\, x.\alpha.x$, we need additional information to detect whether the left hand side instance of action α or the right hand one is performed in the transition $P\|_\emptyset P \xrightarrow{\alpha} P\|_\emptyset P$. When an action is performed, we speak of an *event*, which corresponds to a label – or actually a tuple of labels as we will see.
- *Live Events.* An action of a process term is live if it can currently be performed. In $a.b.\mathsf{nil}\|_{\{b\}}\, b.\mathsf{nil}$ only a can be performed while b cannot, momentarily. Such a live action corresponds to a possible event, i.e. to a label.
- *Fair Sequences.* A maximal sequence is fair when no event in a process term becomes live and then remains live throughout.

These items sketch the general methodology used by Costa and Stirling to define and isolate fair computations in [5,6]. It has to be noted, however, that in [6] Costa and Stirling concentrate on fairness of process components; i.e., along a fair computation, there cannot exist any subprocess that could always contribute some action but never does so. In contrast, we will require fairness for actions. In the setting of [5], i.e. with CCS-composition but without restriction, these two views coincide.

To demonstrate the difference, consider $a\|_{\{a\}}\mathsf{rec}\, x.(a.x + b.x)$ and a run consisting of infinitely many b's. This run is not fair to the component a, since this component is enabled at every stage, but never performs its a. In our view, this run is fair for the synchronized *action* a, since the second component offers always a fresh a for synchronization. (Another intuitive explanation is that action a is not possible *while* b is performed.) Correspondingly, the label for such a synchronization (called an event label) is a pair of labels, each stemming from one of the components; such a pair is a live event, and it changes with each transition. It is not clear how our approach could deal with fairness for components.

We now describe the three items in more detail. Most of the definitions in the rest of this section are taken from [6] with the obvious slight variations due to the different language we are using. We also take from [6] those results that are language independent.

3.1 A Labelling for Process Terms

Costa and Stirling associate labels with basic actions and operators inside a process. Along a computation, labels are unique and, once a label disappears, it will not reappear in the process anymore.

The set of *labels* is $\mathsf{LAB} = \{1,2\}^*$ with ε as the empty label and $u, v, w, \ldots$ as typical elements; $\leq$ is the *prefix preorder* on LAB. We have that $u \leq v$ if there is $u' \in \mathsf{LAB}$ such that $v = uu'$ (and $u < v$ if $u' \in \{1,2\}^+$). We also use the following notation:

- (Set of Tuples) $\mathcal{N} = \{\langle v_1, \ldots, v_n\rangle \mid n \geq 1,\ v_1, \ldots, v_n \in \mathsf{LAB}\}$;
- (Composition of Tuples) $s_1 \times s_2 = \langle v_1, \ldots, v_n, w_1, \ldots, w_m\rangle$, where $s_1, s_2 \in \mathcal{N}$ and $s_1 = \langle v_1, \ldots, v_n\rangle$, $s_2 = \langle w_1, \ldots, w_m\rangle$;
- (Composition of Sets of Tuples) $N \times M = \{s_1 \times s_2 \mid s_1 \in N \text{ and } s_2 \in M\}$, where $N, M \subseteq \mathcal{N}$. Note that $N = \emptyset$ or $M = \emptyset$ implies $N \times M = \emptyset$.

All PAFAS operators and variables will now be labelled in such a way that no label occurs more than once in an expression. As indicated above, an action being performed might correspond to a pair or more generally to a tuple of labels, cf. the definition of live events below (6); therefore, we call tuples of labels *event labels.*

Labels (i.e. elements of LAB) are assigned systematically following the structure of PAFAS terms usually as indexes and in case of parallel composition as upper indexes. Due to recursion the labelling is dynamic: the rule for rec generates new labels.

Definition 4. (*labelled process algebra*)

The labelled process algebra $\mathsf{L}(\tilde{\mathbb{P}})$ (and similarly $\mathsf{L}(\tilde{\mathbb{P}}_1)$ etc.) is defined as $\bigcup_{u\in\mathsf{LAB}} \mathsf{L}_u(\tilde{\mathbb{P}})$, where $\mathsf{L}_u(\tilde{\mathbb{P}}) = \bigcup_{P\in\tilde{\mathbb{P}}} \mathsf{L}_u(P)$ and $\mathsf{L}_u(P)$ is defined inductively as follows:

Nil, Var : $\mathsf{L}_u(\mathsf{nil}) = \{\mathsf{nil}_u\}$, $\mathsf{L}_u(x) = \{x_u\}$,
In examples, we will often write nil for nil_u, if the label u is not relevant.

Pref: $\mathsf{L}_u(\mu.P) = \{\mu_u.P' \mid P' \in \mathsf{L}_{u1}(P)\}$

Sum: $\mathsf{L}_u(P + Q) = \{P' +_u Q' \mid P' \in \mathsf{L}_{u1}(P),\ Q' \in \mathsf{L}_{u2}(Q)\}$

Par: $\mathsf{L}_u(P\|_A Q) = \{P' \|_A^u Q' \mid P' \in \mathsf{L}_{u1v}(P),\ Q' \in \mathsf{L}_{u2v'}(Q)$ where $v, v' \in \mathsf{LAB}\}$

Rel: $\mathsf{L}_u(P[\Phi]) = \{P'[\Phi_u] \mid\ P' \in \mathsf{L}_{u1v}(P)$ where $v \in \mathsf{LAB}\}$

Rec: $\mathsf{L}_u(\mathsf{rec}\, x.P) = \{\mathsf{rec}\, x_u.P' \mid P' \in \mathsf{L}_{u1}(P)\}$

We assume that, in $\mathsf{rec}\, x_u.P$, $\mathsf{rec}\, x_u$ binds all free occurrences of a labelled x. We let $\mathsf{L}(P) = \bigcup_{u \in \mathsf{LAB}} \mathsf{L}_u(P)$ and $\mathsf{LAB}(P)$ is the set of labels occurring in P.

The unicity of labels must be preserved under derivation. For this reason in the rec rule the standard substitution must be replaced by a substitution operation which also changes the labels of the substituted expression. The new substitution operation, denoted by $\{\!| \_ |\!\}$, is defined on $\mathsf{L}(\tilde{\mathbb{P}})$ using the following operators:

$()^{+v}$: If $P \in \mathsf{L}_u(\tilde{\mathbb{P}})$, then $(P)^{+v}$ is the term in $\mathsf{L}_{vu}(\tilde{\mathbb{P}})$ obtained by prefixing v to all labels in P.
$()_\varepsilon$: If $P \in \mathsf{L}_u(\tilde{\mathbb{P}})$, then $(P)_\varepsilon$ is the term in $\mathsf{L}_\varepsilon(\tilde{\mathbb{P}})$ obtained by removing the prefix u from all labels in P. (Note that u is the unique prefix-minimal label in P.)

Suppose $P, Q \in \mathsf{L}(\tilde{\mathbb{P}})$ and $x_u, \dots, x_v$ are all free occurrences of a labelled x in P then $P\{\!| Q/x |\!\} = P\{((Q)_\varepsilon)^{+u}/x_u, \dots, ((Q)_\varepsilon)^{+v}/x_v\}$. In such a way in $P\{\!|Q/x|\!\}$ each substituted Q inherits the label of the x it replaces.

The behavioural and temporal operational semantics of the labelled PAFAS are obtained by replacing the rules Rec_a and Rec_r in Definition 2 and 3 with:

$$\mathrm{Rec}_a\frac{P\{\!| \mathsf{rec}\, x_u.P/x |\!\} \xrightarrow{\alpha} P'}{\mathsf{rec}\, x_u.P \xrightarrow{\alpha} P'} \qquad \mathrm{Rec}_r\frac{P\{\!| \mathsf{rec}\, x_u.P/x |\!\} \xrightarrow{X}_r P'}{\mathsf{rec}\, x_u.P \xrightarrow{X}_r P'}$$

and the rules Pref_{a1} and Pref_{a2} in Definition 2 with the rules:

$$\mathrm{Pref}_{a1}\frac{}{\alpha_u.P \xrightarrow{\alpha} P} \qquad \mathrm{Pref}_{a2}\frac{}{\underline{\alpha}_u.P \xrightarrow{\alpha} P}$$

because we assume that labels are not observable when actions are performed. The other rules are unchanged.

The following proposition shows that labels are just annotations that do not interfere with transitions of PAFAS and labelled PAFAS processes. Let R be the operation of removing labels from a labelled PAFAS term.

Proposition 5. Let $P \in \mathsf{L}_u(\tilde{\mathbb{P}})$ and $A \subseteq \mathbb{A}$. Then:

i. $P \xrightarrow{\alpha} Q$ ($P \xrightarrow{X}_r Q$) implies $\mathsf{R}(P) \xrightarrow{\alpha} \mathsf{R}(Q)$ ($\mathsf{R}(P) \xrightarrow{X}_r \mathsf{R}(Q)$) in unlabelled PAFAS;
ii. if $P' \xrightarrow{\alpha} Q'$ ($P' \xrightarrow{X}_r Q'$) in unlabelled PAFAS and $P' = \mathsf{R}(P)$, then for some Q with $Q' = \mathsf{R}(Q)$, we have $P \xrightarrow{\alpha} Q$ ($P \xrightarrow{X}_r Q$).

An immediate consequence of the labelling are the following facts that have been proven in [6]: No label occurs more than once in a given process $P \in \mathsf{L}_u(\tilde{\mathbb{P}})$ and $w \in \mathsf{LAB}(P)$ implies $u \leq w$. Moreover, central to labelling is the persistence and disappearance of labels under derivation. In particular, once a label disappears it can never reappear. It is these features which allow us to recognize when a component contributes to the performance of an action.

Throughout the rest of this section we assume the labelled calculus.

3.2 Live Events

To capture the fairness constraint for execution sequences, we need to define the *live events*. For $\alpha_u.\mathsf{nil} \|_{\{\alpha\}} \alpha_v.\mathsf{nil}$ (with labels u and v), there is only one live action. This is action α; i.e. there is only one α-event, which we will identify with the tuple $\langle u, v\rangle$, i.e. with the tuple of labels of 'local' α's that synchronize when the process performs α; recall that we call such tuples *event labels*.[5] In a similar way, there is only one live action in $\alpha_u.\beta_v.\mathsf{nil} \|_{\{\beta\}} \beta_y.\mathsf{nil}$ (action α corresponding to tuple $\langle u\rangle$) because the parallel composition prevents the instance of β labelled by $\langle y\rangle$ from contributing an action. However, note that $\langle v, y\rangle$ becomes live, once action α is performed. We now define $\mathsf{LE}(P, A)$ as the set of live events of P (when the execution of actions in A are prevented by the environment).

Definition 6. (*live events*)
Let $P \in \mathsf{L}(\tilde{\mathbb{P}})$ and $A \subseteq \mathbb{A}$. The set $\mathsf{LE}(P, A)$ is defined by induction on P.

Var, Nil: $\mathsf{LE}(x_u, A) = \mathsf{LE}(\mathsf{nil}_u, A) = \emptyset$

Pref: $\mathsf{LE}(\mu_u.P, A) = \begin{cases} \{\langle u\rangle\} & \text{if } \mu = \alpha \text{ or } \mu = \underline{\alpha} \text{ and } \alpha \notin A \\ \emptyset & \text{otherwise} \end{cases}$

Sum: $\mathsf{LE}(P +_u Q, A) = \mathsf{LE}(P, A) \cup \mathsf{LE}(Q, A)$

Par: $\mathsf{LE}(P \|_B^u Q, A) = \mathsf{LE}(P, A \cup B) \cup \mathsf{LE}(Q, A \cup B) \cup$
$\bigcup_{\alpha \in B \setminus A} (\mathsf{LE}(P, \mathbb{A} \setminus \{\alpha\}) \times \mathsf{LE}(Q, \mathbb{A} \setminus \{\alpha\}))$

Rel: $\mathsf{LE}(P[\Phi_u], A) = \mathsf{LE}(P, \Phi^{-1}(A))$
Rec: $\mathsf{LE}(\mathsf{rec}\, x_u.P, A) = \mathsf{LE}(P, A)$

The *set of live events* in P is defined as $\mathsf{LE}(P, \emptyset)$ which we abbreviate to $\mathsf{LE}(P)$.

[5] Since Costa and Stirling deal with fairness of components, they have no need for tuples.

The set A represents the restricted actions. Then, $\mathsf{LE}(a_u.P,\{a\})$ must be empty because the action a is prevented. Note that, in the Par-case, $\mathsf{LE}(P, A \cup B) \cup \mathsf{LE}(Q, A \cup B)$ is the set of the labels of the live actions of P and Q, when the environment prevents actions from A and from the synchronization set B – corresponding to those actions that P and Q can perform independently. To properly deal with synchronization, for all $\alpha \in B \backslash A$ we combine each live event of P corresponding to α with each live event of Q corresponding to α, getting tuples of labels. The other rules are as expected.

3.3 Fair Execution Sequences

We can now define the (weak) fairness constraint. The following definitions and results are essentially borrowed from [6], and just adapted to our notions of fairness and labelling. First of all, for a process P_0, we say that a sequence of transitions $\gamma = P_0 \xrightarrow{\lambda_0} P_1 \xrightarrow{\lambda_1} \ldots$ with $\lambda_i \in \mathbb{A}_\tau \cup \{1\}$ is a *timed execution sequence* if it is an infinite sequence of action transitions and full time steps[6].

A timed execution sequence is *everlasting* in the sense of having infinitely many time steps if and only if it is *non-Zeno*; a Zeno run would have infinitely many actions in a finite amount of time, which in a setting with discrete time means that it ends with infinitely many action transitions without a time step.

For an *initial* process P_0, we say that a sequence of transitions $\gamma = P_0 \xrightarrow{\alpha_0} P_1 \xrightarrow{\alpha_1} \ldots$ with $\alpha_i \in \mathbb{A}_\tau$ is an *execution sequence* if it is a maximal sequence of action transitions; i.e. it is infinite or ends with a process P_n such that $P_n \not\xrightarrow{\alpha}$ for any action α. Now we formalize fairness by calling a (timed) execution sequence *fair*, if no event becomes live and then remains live throughout.

Definition 7. (*fair execution sequences*)

Let $\gamma = P_0 \xrightarrow{\lambda_0} P_1 \xrightarrow{\lambda_1} \ldots$ be an execution sequence or a timed execution sequence; we will write '(timed) execution sequence' for such a sequence. We say that γ is *fair* if

$$\neg(\exists\, s\, \exists\, i\, .\, \forall\, k \geq i\, :\, s \in \mathsf{LE}(P_k))$$

Following [6], we now present an alternative, more local, definition of fair computations which will be useful to prove our main statements.

Definition 8. (*B-step*)

We say that $P_0 \xrightarrow{\lambda_0} P_1 \xrightarrow{\lambda_1} \ldots \xrightarrow{\lambda_{n-1}} P_n$ with $n > 0$ is a *timed B-step* when: (i) B is a finite set of event labels, and (ii) $B \cap \mathsf{LE}(P_0) \cap \ldots \cap \mathsf{LE}(P_n) = \emptyset$. If $\lambda_i \in \mathbb{A}_\tau$, $i = 0, \ldots, n-1$, then the sequence is a *B-step*. If $P_0 \xrightarrow{\lambda_0} P_1 \xrightarrow{\lambda_1} \ldots \xrightarrow{\lambda_{n-1}} P_n$ is a (timed) B-step and $v = \lambda_0 \ldots \lambda_{n-1}$ we write $P_0 \xrightarrow{v}_B P_{n+1}$.

In particular, a (timed) $\mathsf{LE}(P)$-step from P is "locally" fair: all live events of P lose their liveness at some point in the step.

[6] Note that a maximal sequence of such transitions/steps is never finite, since for $\gamma = P_0 \xrightarrow{\lambda_0} P_1 \xrightarrow{\lambda_1} \ldots \xrightarrow{\lambda_{n-1}} P_n$, we have $P_n \xrightarrow{\alpha}$ or $P_n \xrightarrow{1}$ (this is because processes are guarded, see [4]).

Definition 9. (*fair-step sequences*)
A *(timed) fair-step sequence* from P_0 is any maximal sequence of (timed) steps of the form $P_0 \xrightarrow{v_0}_{\mathsf{LE}(P_0)} P_1 \xrightarrow{v_1}_{\mathsf{LE}(P_1)} \cdots$

A fair-step sequence is simply a concatenation of locally fair steps. If δ is a (timed) fair-step sequence, then its *associated* (timed) execution sequence is the sequence which drops all references to the sets $\mathsf{LE}(P_i)$.

Now we have the expected result stating that fair execution sequences and fair-step sequences are essentially the same.

Theorem 10. A (timed) execution sequence is fair if and only if it is the sequence associated with a (timed) fair-step sequence.

4 Fairness and Timing

This section is the core of the paper. It relates fairness and timing in a process algebraic setting.

4.1 Fairness of Everlasting Sequences

The following proposition is a key statement for proving our main results. It states that each sequence of functional transitions between two full-time steps is actually an $\mathsf{LE}(P)$-step.

Proposition 11. Let $P \in \mathsf{L}(\tilde{\mathbb{P}})$ and $v, w \in (\mathbb{A}_\tau)^*$.

1. If $P \xrightarrow{1} P_1 \xrightarrow{v} P_2 \xrightarrow{1}$ then $P \xrightarrow{1v}_{\mathsf{LE}(P)} P_2$;
2. If $P \xrightarrow{v} Q \xrightarrow{1} Q_1 \xrightarrow{w} Q_2 \xrightarrow{1}$ then $P \xrightarrow{v1w}_{\mathsf{LE}(P)} Q_2$.

Then, we prove that each everlasting timed execution sequences of PAFAS processes is fair.

Theorem 12. Each everlasting timed execution sequence, i.e. each timed execution sequence of the form

$$\gamma = P_0 \xrightarrow{v_0} P_1 \xrightarrow{1} P_2 \xrightarrow{v_1} P_3 \xrightarrow{1} P_4 \xrightarrow{v_2} P_5 \xrightarrow{1} \ldots$$

with $v_0, v_1, v_2 \ldots \in (\mathbb{A}_\tau)^*$ is fair.

Observe that an everlasting timed execution sequence, by its definition, does not depend on the labelling, i.e. it is a notion of the *unlabelled* PAFAS calculus.

4.2 Relating Timed Executions and Fair Executions

While in the previous section we have shown that every everlasting timed execution is fair, we show in this section that everlasting timed execution sequences of initial PAFAS processes in fact *characterize* the fair untimed executions of these processes. Observe that the latter is a notion of an ordinary *labelled untimed* process algebra (like CCS or TCSP), while the former is a notion of our *unlabelled timed* process algebra.

A key statement for proving this shows that whenever an initial process P can perform a sequence v of basic actions and this execution turns out to be an $\mathsf{LE}(P)$-step, then P can alternatively let time pass (perform a 1-time step) and then perform the sequence of basic actions v, and vice versa.

To relate processes in fair-step sequences and execution sequences, we define the set of processes $\mathcal{UU}(P)$ obtained by *unfolding* recursive terms in a given process P (that are not action guarded) and then by making some of the (initial) actions be *urgent*. In the following definition, if I and J are set of processes and op is a PAFAS binary operator then I op J is defined as $\{P \text{ op } Q | P \in I \text{ and } Q \in J\}$; analogously, we deal with the unary operators $[\Phi]$.

Definition 13. (*urgency and unfolding*)
Let $P \in \mathsf{L}(\tilde{\mathbb{P}}_1)$ be a labelled term. Define $\mathcal{UU}(P)$ by induction on P as follows:

Var, Stop, Nil:	$\mathcal{UU}(x_u) = \{x_u\},\ \mathcal{UU}(\mathsf{nil}_u) = \{\mathsf{nil}_u\}$
Pref:	$\mathcal{UU}(\alpha_u.P, A) = \{\alpha_u.P\} \cup \{\underline{\alpha}_u.P\}$
Sum:	$\mathcal{UU}(P +_u Q) = \mathcal{UU}(P) +_u \mathcal{UU}(Q)$
Par:	$\mathcal{UU}(P \parallel_B^u Q) = \mathcal{UU}(P) \parallel_B^u \mathcal{UU}(Q)$
Rel:	$\mathcal{UU}(P[\Phi_u]) = (\mathcal{UU}(P))[\Phi_u]$
Rec:	$\mathcal{UU}(\mathsf{rec}\, x_u.P) = \{\mathsf{rec}\, x_u.P\} \cup \mathcal{UU}(P)\{\!\vert \mathsf{rec}\, x_u.P/x \vert\!\}$

In [4] we present a key proposition relating LE-steps and temporal transitions. More in detail, we relate initial processes to processes that can perform a full time step; an initial process Q_0 can perform an $\mathsf{LE}(Q_0)$-step v to Q_n if and only if any process $R \in \mathcal{UU}(Q_0)$ that can perform a full time step can perform the action sequence v afterwards to another process $R_n \in \mathcal{UU}(Q_n)$ that can again perform a full time step. Note that this statement would not hold if we would define (as in [6]), $\mathsf{LE}(P +_u Q, A) = \{u\}$ if $\mathsf{LE}(P, A) \cup \mathsf{LE}(Q, A) \neq \emptyset$ and $\mathsf{LE}(P +_u Q, A) = \emptyset$ otherwise. Indeed, for the initial process $P = (a_{111}.\mathsf{nil} \parallel_\emptyset^{11} b_{112}.\mathsf{nil}) +_1 c_{12}.\mathsf{nil}$ we would have $P \xrightarrow{a}_{\mathsf{LE}(P)} \mathsf{nil} \parallel_\emptyset^{11} b_{112}.\mathsf{nil}$, but $P \xrightarrow{1} (\underline{a}_{111}.\mathsf{nil} \parallel_\emptyset^{11} \underline{b}_{112}.\mathsf{nil}) +_1 \underline{c}_{12}.\mathsf{nil} \xrightarrow{a} \mathsf{nil} \parallel_\emptyset^{11} \underline{b}_{112}.\mathsf{nil}$, where the latter process cannot let one time unit pass.

We now provide a characterization result for fair execution sequences in terms of timed execution sequences.

Theorem 14. (*Characterization of fair timed execution sequences*)
Let $P \in \mathsf{L}(\mathbb{P}_1)$ and $\alpha_0, \alpha_1, \alpha_2 \ldots \in \mathbb{A}_\tau$. Then:

1. For any fair execution sequence from P

$$P = P_0 \xrightarrow{\alpha_0} P_1 \xrightarrow{\alpha_1} P_2 \ldots P_i \xrightarrow{\alpha_i} P_{i+1} \cdots$$

there exists a timed execution sequence in unlabelled PAFAS

$$\mathsf{R}(P) = S_{i_0} \xrightarrow{1} S'_{i_0} \xrightarrow{v_{i_0}} S_{i_1} \xrightarrow{1} S'_{i_1} \xrightarrow{v_{i_1}} S_{i_2} \ldots S_{i_j} \xrightarrow{1} S'_{i_j} \xrightarrow{v_{i_j}} S'_{i_{j+1}} \cdots$$

where $i_0 = 0$, $v_{i_j} = \alpha_{i_j}\alpha_{i_j+1}\ldots\alpha_{i_{j+1}-1}$, $S_{i_j} \in \mathcal{UU}(\mathsf{R}(P_{i_j}))$ and $j \geq 0$. A similar result holds if $P_{i+1} \not\xrightarrow{\alpha}$, for any $\alpha \in \mathbb{A}_\tau$; then $S'_{i_j} \xrightarrow{1} S''_{i_j} \xrightarrow{1} S''_{i_j} \xrightarrow{1}$.

2. For any timed execution sequence from $\mathsf{R}(P)$ in unlabelled PAFAS

$$\mathsf{R}(P) = S_{i_0} \xrightarrow{1} S'_{i_0} \xrightarrow{v_{i_0}} S_{i_1} \xrightarrow{1} S'_{i_1} \xrightarrow{v_{i_1}} S_{i_2} \ldots S_{i_j} \xrightarrow{1} S'_{i_j} \xrightarrow{v_{i_j}} S'_{i_{j+1}} \cdots$$

where $i_0 = 0$, $v_{i_j} = \alpha_{i_j}\alpha_{i_j+1}\ldots\alpha_{i_{j+1}-1}$, for every $j \geq 0$, there exists a fair execution sequence

$$P = P_0 \xrightarrow{\alpha_0} P_1 \xrightarrow{\alpha_1} P_2 \ldots P_i \xrightarrow{\alpha_i} P_{i+1} \ldots$$

where $S_{i_j} \in \mathcal{UU}(\mathsf{R}(P_{i_j}))$, for every $j \geq 0$. Again, there is a variation where $S'_{i_j} \xrightarrow{1} S''_{i_j} \xrightarrow{1} S''_{i_j} \xrightarrow{1}$ implies $P_{i+1} \not\xrightarrow{\alpha}$, for any $\alpha \in \mathbb{A}_\tau$.

4.3 Fair Execution Sequences and Finite State Processes

We call an initial process $P \in \mathsf{L}(\mathbb{P}_1)$ (i.e. a standard untimed process) *finite state*, if only finitely many processes are *action-reachable*, i.e. can be reached according to the functional operational semantics, i.e. with transitions $\xrightarrow{\alpha}$.

For the definition of fair executions, we followed Costa and Stirling and introduced two semantic levels: one level (the positive) prescribes the finite and infinite execution sequences of labelled processes disregarding their fairness, while the other (the negative) filters out the unfair ones. The labels are notationally heavy, and keeping track of them is pretty involved. Since the labels evolve dynamically along computations, the transition system defined for the first level is in general infinite state even for finite state processes (as long as they have at least one infinite computation). Also the filtering mechanism is rather involved, since we have to check repeatedly what happens to live events along the computation, and for this we have to consider the processes passed in the computation.

With the characterization results of the previous subsection, we have not only shown a conceptional relationship between timing and fairness. We have also given a much lighter description of the fair execution sequences of a process $P \in \mathsf{L}(\mathbb{P}_1)$ via the transition system of processes *timed-reachable* (i.e. with transitions $\xrightarrow{\alpha}$ and $\xrightarrow{1}$) from P, which we will denote by $\mathcal{TT}(P)$: the marking of some actions with underlines is easier than the labelling mechanism, and the filtering simply requires infinitely many time steps, i.e. non-Zeno behaviour; hence, for filtering one does not have to consider the processes passed. Furthermore, the transition system $\mathcal{TT}(P)$ is finite for finite state processes.

Theorem 15. If $P \in \mathsf{L}(\mathbb{P}_1)$ is finite state, then $\mathcal{TT}(P)$ is finite.

The main result in [5,6] is a characterization of fair execution sequences with only one (positive) level: SOS-rules are given that describe all transitions $P \xrightarrow{v} Q$

with $v \in (\mathbb{A}_\tau)^*$ such that $P \overset{v}{\rightarrow}_{\mathsf{LE}(P)} Q$. This is conceptionally very simple, since there is only one level and there is no labelling or marking of processes: the corresponding transition system for P only contains processes reachable from P. In particular, the transition system is finite-state if P is finite-state. The drawback is that, in general, P has infinitely many $\mathsf{LE}(P)$-steps (namely, if it has an infinite computation), and therefore the transition system is infinitely branching and has infinitely many arcs. (Observe that this drawback is not shared by our transition system of timed-reachable processes.)

As a second main result, we will now derive from $\mathcal{TT}(P)$ for a finite-state process P a finite transition system with finitely many arcs that describes the fair execution sequences in one level: the essential idea is that the arcs are inscribed with regular expressions (and not just with sequences as in [5,6]). The states of this transition system are the states Q of $\mathcal{TT}(P)$ such that $Q \overset{1}{\rightarrow} Q'$; if R is another such state, we have an arc from Q to R labelled with a regular expression e. This expression is obtained by taking $\mathcal{TT}(P)$ with Q' as initial state and R as the only final state, deleting all transitions $\overset{1}{\rightarrow}$ and applying the well-known Kleene construction to get an (equivalent) regular expression from a nondeterministic automaton. (The arc can be omitted, if e describes the empty set.) Clearly, such an arc corresponds to a set of B-steps which are also present in the one-level characterization of Costa and Stirling, but there is one exception: if $Q' \overset{1}{\rightarrow}$, then Q and Q' cannot perform any action; hence, there will only be an ε-labelled arc from Q' to itself and, if $Q \neq Q'$, from Q to Q'.

Thus, we can obtain exactly the action sequences performed in fair execution sequences of P by taking the infinite paths from P in the new transition system and replacing each regular expression e by a sequence in the language of e.

Observe that P is a state of the new transition system, but not all states are initial processes. It would be nice to have only arcs $P \overset{e}{\rightarrow} Q$ such that P and Q are initial processes and for each v belonging to e one has $P \overset{v}{\rightarrow} Q$; to get this, we still have a technical problem corresponding to the fact that some recursions may be unfolded in a process $R \in \mathcal{UU}(Q)$ compared to Q.

Another interesting line of research regards the notion of fairness for components and the possibility of characterizing this different fairness notion in our setting. As already discussed in Sect. 3, the current timed operational semantics of initial terms is not appropriate for this purpose. It would be nice to find a suitable operational semantics and to check if it corresponds to some notion of time.

A very recent paper shows some similarities with our work. In [1], Stephen Brookes gives a denotational trace semantics for CSP processes to describe a weak notion of fairness close to ours. In his setting, processes explicitly declare to the external environment the actions that are waiting for a synchronization on the current state. Thus, besides input actions $a?$ and output actions $a!$ (Brookes uses a CCS-like synchronization mechanism), processes can perform transitions like $P \overset{\delta_X}{\longrightarrow} P$, where X is a set of actions which – in our terms – are live. The achievement is a notion of traces to describe fair behavior which gives a compositional semantics and which can be defined denotationally and operationally, such that the same notion of trace can be used both for synchronous and asynchronous communicating processes. The resulting fairness notion is different from ours in that Brookes only cares about fairness of internal actions. We also have different issues. We start with a notion of timed traces that has a meaning of its own, and show how these traces are related to and can be used for easier descriptions of Costa and Stirling's fair traces.

Acknowledgments. We would like to thank Colin Stirling for useful discussions during a preliminary stage of this work.

References

1. S. Brookes. Traces, Pomsets, Fairness and Full Abstarctions for Communicating Processes. In *Concur'02*, LNCS 2421, pp. 466–482, Springer, 2002.
2. F. Corradini, W. Vogler and L. Jenner. Comparing the Worst-Case Efficiency of Asynchronous Systems with PAFAS. *Acta Informatica* **38**, pp. 735–792, 2002.
3. F. Corradini, M.R. Di Berardini and W. Vogler. PAFAS at work: Comparing the Worst-Case Efficiency of Three Buffer Implementations. In Proc. of 2nd Asia-Pacific Conference on Quality Software, APAQS 2001, pp. 231–240, IEEE, 2001.
4. F. Corradini, M.R. Di Berardini and W. Vogler. Relating Fairness and Timing in Process Algebra. RR 03-2003, Univ. of L'Aquila, 2003. Available from: http://www.di.univaq.it/flavio.
5. G. Costa, C. Stirling. A Fair Calculus of Communicating Systems. *Acta Informatica* **21**, pp. 417–441, 1984.
6. G. Costa, C. Stirling. Weak and Strong Fairness in CCS. *Information and Computation* **73**, pp. 207–244, 1987.
7. R. De Nicola and M.C.B. Hennessy. Testing equivalence for processes. *Theoret. Comput. Sci.* **34**, pp. 83–133, 1984.
8. C.A.R. Hoare. *Communicating Sequential Processes.* Prentice Hall, 1985.
9. L. Jenner, W. Vogler. Comparing the Efficiency of Asynchronous Systems. In Proc. of AMAST Workshop on Real-Time and Probabilistic Systems, LNCS 1601, pp. 172–191, 1999. Modified full version as [2].
10. L. Jenner and W. Vogler. Fast asynchronous systems in dense time. *Theoret. Comput. Sci.*, 254:379–422, 2001. Extended abstract in Proc. ICALP 96, LNCS 1099.
11. N. Lynch. *Distributed Algorithms.* Morgan Kaufmann Publishers, 1996.
12. R. Milner. *Communication and Concurrency.* Prentice Hall, 1989.
13. W. Vogler. Efficiency of asynchronous systems, read arcs, and the MUTEX-problem. *Theoret. Comput. Sci.* 275, pp. 589–631, 2002.
14. W. Vogler. Modular Construction and Partial Order Semantics of Petri Nets. LNCS 625, Springer, 1992.

A Compositional Semantic Theory for Synchronous Component-Based Design*

Barry Norton[1], Gerald Lüttgen[2], and Michael Mendler[3]

[1] Department of Computer Science, University of Sheffield, UK
b.norton@dcs.shef.ac.uk
[2] Department of Computer Science, University of York, UK
gerald.luettgen@cs.york.ac.uk
[3] Informatics Theory Group, University of Bamberg, Germany
michael.mendler@wiai.uni-bamberg.de

Abstract. Digital signal processing and control (DSPC) tools allow application developers to assemble systems by connecting predefined components in signal-flow graphs and by hierarchically building new components via encapsulating sub-graphs. Run-time environments then dynamically schedule components for execution on some embedded processor, typically in a synchronous cycle-based fashion, and check whether one component jams another by producing outputs faster than can be consumed. This paper develops a process-algebraic model of coordination for synchronous component-based design, which directly lends itself to compositionally formalising the monolithic semantics of DSPC tools. By uniformly combining the well-known concepts of abstract clocks, maximal progress and clock-hiding, it is shown how the DSPC principles of dynamic synchronous scheduling, isochrony and encapsulation may be captured faithfully and compositionally in process algebra, and how observation equivalence may facilitate jam checks at compile-time.

1 Introduction

One important domain for embedded-systems designers are *digital signal processing and control* (DSPC) applications. These involve dedicated software for control and monitoring problems in industrial production plants, or software embedded in engineering products. The underlying programming style within this domain relies on *component-based design*, based on the rich repositories of pre-compiled and well-tested software components (PID-controllers, FIR-filters, FFT-transforms, etc.) built by engineers over many years. Applications are simply programmed by interconnecting components, which frees engineers from most of the error-prone low-level programming tasks. Design efficiency is further aided by the fact that DSPC programming tools, including LabView [9], iConnect [15] and Ptolemy [10], provide a graphical user interface that supports hierarchical extensions of signal-flow graphs. These permit the *encapsulation* of sub-systems into single components, thus enabling the reuse of system designs.

* Research supported by EPSRC grant GR/M99637.

R. Amadio, D. Lugiez (Eds.): CONCUR 2003, LNCS 2761, pp. 461–476, 2003.

While the visual signal-flow formalism facilitates the structural design of DSPC applications, the behaviour of a component-based system manifests itself only once its components are scheduled on an embedded processor. This *scheduling* is often handled dynamically by run-time environments, as is the case in LabView and iConnect, in order to achieve more efficient and adaptive real-time behaviour. The scheduling typically follows a cycle-based execution model with the phases *collect input* (I), *compute reaction* (R) and *deliver output* (O). At the top level, the scheduler continuously iterates between executing the *source components* that produce new inputs, e.g., by reading sensor values, and executing *computation components* that transform input values into output values, which are then delivered to the system environment, e.g., via actuators. Each phase obeys the *synchrony principle*, i.e., in (I) all source components are given a chance to collect input from the environment before any computation component is executed, in (R) every computation component whose inputs are available will be scheduled for execution, and in (O) all generated outputs will be delivered before the current cycle ends. The constraint in phase (O), which is known as *isochrony* [6], implies that each output signal will be 'simultaneously' and 'instantaneously' received at each connected input. This synchronous scheme can be applied in a hierarchical fashion, abstracting a sequence of RO-steps produced by a sub-system into a single RO-step (cf. Sect. 2).

Like in synchronous programming, the implicit synchrony hypothesis of IRO scheduling assumes that the reaction of a (sub-)system is always faster than its environment issues execution requests. If a component cannot consume its input signals at the pace at which they arrive, a *jam* occurs [15], indicating a serious real-time problem (cf. Sect. 2). Unfortunately, in existing tools, there are no compile-time checks for detecting jams, thereby forcing engineers to rely on extensive simulations for validating their applications before delivery. Moreover, there is no formal model of IRO scheduling for DSPC programming systems that can be used for the static analysis of jams, and the question of how to distribute the monolithic IRO scheduler into a uniform model of coordination has not been addressed in the literature either.

The objective of this paper is to show that a relatively small number of standard concepts studied in concurrency theory provides the key to *compositionally* formalising the semantics of component-based DSPC designs, and to enabling static jam checks. The most important concepts from the process-algebra toolbox are *handshake* synchronisation from CCS [12] and *abstract clocks* in combination with *maximal progress* as investigated in temporal process algebras, specifically TPL [7], PMC [1] and CSA [3]. We use handshake synchronisation for achieving serialisation and maximal-progress clocks for enforcing synchrony. Finally, given maximal progress, synchronous encapsulation may be captured naturally in terms of *clock-hiding*, similar to hiding in CSP [8]. We will uniformly integrate all three concepts into a single process language (cf. Sect. 3), to which we refer as *Calculus for Synchrony and Encapsulation* (CaSE) and which conservatively extends CCS in being equipped with a behavioural theory based on *observation equivalence* [12].

As our main contribution we will formally establish that CaSE is expressive enough for faithfully modelling the principles of IRO scheduling and for capturing jams (cf. Sect. 4). First, using a single clock and maximal progress we will show how one may derive a decentralised description of the synchronous scheduler. Second, we prove that isochrony across connections can be modelled via multiple clocks and maximal progress. Third, the subsystems-as-components principle is captured by the clock-hiding operator. Moreover, we will argue that observation equivalence lends itself for statically detecting jams by reducing jam checking to timelock checking. In this way, our modelling in CaSE yields a *model of coordination* for synchronous component-based design, whose virtue is its compositional style for specifying and reasoning about DSPC systems and its support for the static capture of semantic properties of DSPC programs. Thus, CaSE provides a foundation for developing new-generation DSPC tools that offer the compositional, static analysis techniques desired by engineers.

2 An Example of DSPC Design

Our motivating example is a *digital spectrum analyser* whose hierarchical signal-flow graph is sketched in Fig. 1. The task is to analyse an audio signal and continually show an array of bar-graphs representing the intensity of the signal in disjoint sections of the frequency range. Our spectrum analyser is designed with the help of components Soundcard, Const, Element and BarGraph. Each instance c1, c2, ... of Element, written as ck:Element or simply ck, for $k = 1, 2, \ldots$, is responsible for assessing the intensity of one frequency range, which is then displayed by component instance dk:BarGraph. The first input port ei_{k1} of ck:Element is connected to the output port $\overline{\mathrm{so}}$ of the single instance s0:Soundcard, which generates the audio signal and provides exactly one audio value each time it is scheduled. As can be seen by the wire stretching from output port $\overline{\mathrm{co}}_k$ to input port ei_{k2}, ck:Element is also connected to instance sk:Const of component Const, which initialises ck:Element by providing filter parameters when it is first scheduled. In contrast to components Soundcard and Const, Element is not a basic but a hierarchical component. Indeed, every ck encapsulates one instance of Filter, ck1:Filter, and one of Quantise, ck2:Quantise, as shown in Fig. 1 on the right-hand side.

Scheduling. According to IRO scheduling, our example application will be serialised as follows within each IRO-cycle. First, each source component instance gets the chance to execute. In the first cycle, this will be s0:Soundcard and all sk:Const, which will be interleaved in some arbitrary order. In all subsequent cycles, only s0:Soundcard will request to be scheduled, since sk:Const can only produce a value once. Each produced sound value will be instantaneously propagated from output port $\overline{\mathrm{so}}$ of s0 to the input port ei_{k1} of each ck:Element, for all $k \geq 1$, according to the principle of isochronic broadcast discussed below. The scheduler then switches to scheduling computation components. Since all necessary inputs of each ck are available in each IRO-cycle, every ck will request to be scheduled. The scheduler will serialise these requests, each ck will

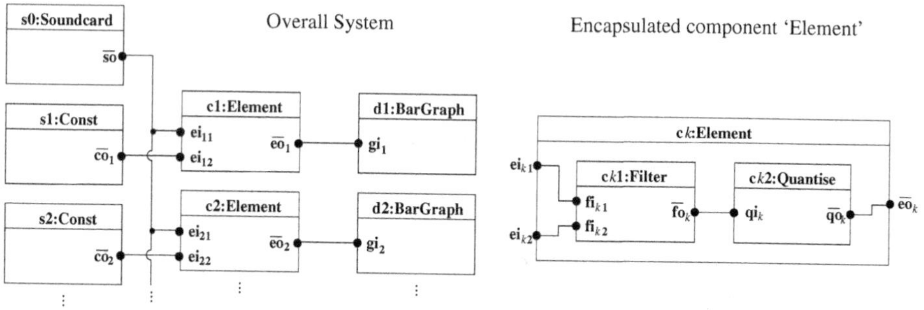

Fig. 1. Example: Digital spectrum analyser

execute accordingly, and the synthesised frequency-strength signal will be emitted by component $ck2$:Quantise via port $\overline{\text{qo}}_k$ and propagated by ck through port $\overline{\text{eo}}_k$. Upon reception of this signal by dk:BarGraph at port gi_k, this computation component instance will also request to be scheduled and, according to the synchrony hypothesis, granted execution within the same IRO-cycle. When all components dk have executed, the current IRO-cycle ends since these do not generate outputs that need to be propagated to the system environment.

It is important to note that, since each ck encapsulates further computation component instances, its execution is non-trivial and involves a sub-scheduler that will schedule $ck1$:Filter and $ck2$:Quantise in such a way that an RO-cycle of these instances will appear atomic outside of ck. This ensures that the scheduling of the inner $ck1$ and $ck2$ will not be interleaved with the execution of any sibling instance cl of ck, for $l \neq k$, or any component instance dk.

Isochronic Output. Whenever s0:Soundcard is scheduled in our example system, it generates an audio signal whose value is propagated via a wire from port $\overline{\text{so}}$, which forks to port ei$k$1 of each instance ck:Element, for $k \geq 1$. In order for the array of bar-graphs to display a consistent state synchronous with the environment, all ck must have received the new value from s0:Soundcard before any cl:Element may be scheduled. Thus, s0:Soundcard and all ck:Element, for $k \geq 1$, must synchronise to transmit sound values instantaneously. This form of synchronisation is called *isochrony* [6] in hardware, where it is the weakest known synchronisation principle from which non-trivial sequential behaviour can be implemented safely, without internal real-time glitches.

Jams. Let us now consider what happens if instances s0:Soundcard and s1:Const are accidently connected the wrong way around, i.e., output port $\overline{\text{so}}$ is connected to input port ei_{12}, and output port $\overline{\text{co}}_1$ of s1:Const to input port ei_{11} of c1:Element. Recall that c11:Filter within c1:Element will only read a value, an initialisation value, from port ei_{12} in the first IRO-cycle and never again afterwards. Thus, when the value of s0:Soundcard produced in the second cycle is propagated to port ei_{12} and further to fi_{12}, the system *jams*. This is because the value that has been produced in the second IRO-cycle and stored at this latter port, has not yet been read by c11:Filter. Observe that a jam is different from a deadlock; indeed, our example system does not deadlock since all instances of Element other than c1:Element continue to operate properly.

3 CaSE: Calculus for Synchrony and Encapsulation

This section presents our process calculus CaSE, which serves as a framework for deriving our formal model of coordination for DSPC design in Sect. 4. The purpose here is not to develop yet another process algebra, but to tailor several well-studied semantic concepts for addressing a specific application domain. CaSE is inspired by Hennessy and Regan's TPL [7], which is an extension of Milner's CCS [12] with regard to syntax and operational semantics. In addition to CCS, TPL includes (i) a *single abstract clock* σ that is interpreted not quantitatively as some number, but qualitatively as a recurrent global synchronisation event; (ii) a *timeout operator* $\lfloor P \rfloor \sigma(Q)$, where the occurrence of σ deactivates process P and activates Q; (iii) the concept of *maximal progress* that implements the synchrony hypothesis by demanding that a clock can only tick within a process, if the process cannot engage in any internal activity τ.

CaSE further extends TPL by (i) allowing for *multiple clocks* $\sigma, \rho, \dots$ as in PMC [1] and CSA [3], while, in contrast to PMC and CSA, maintaining the global interpretation of maximal progress; (ii) explicit *timelock* operators Δ and Δ_σ that prohibit the ticking of all clocks and of clock σ, respectively; (iii) *clock-hiding* operators P/σ that internalise all clock ticks σ of process P. Clock hiding is basically hiding as in CSP [8], i.e., hidden actions are made non-observable. In combination with maximal progress, this has the important effect that all inner clock ticks become included within the synchronous cycle of an outer clock. This is the essence of synchronous encapsulation, as is required for modelling isochronous broadcast and the subsystems-as-components principle. Finally, in contrast to TPL and similar to CCS and CSA, we will equip CaSE with a bisimulation-based semantic theory [12].

Syntax and Operational Semantics. We let $\Lambda = \{a, b, \dots\}$ be a countable set of *input actions* and $\overline{\Lambda} = \{\overline{a}, \overline{b}, \dots\}$ be the set of complementing *output actions*. As in CCS [12], an action a communicates with its complement $\overline{a}$ to produce the *internal action* τ. The symbol $\mathcal{A}$ denotes the set of all actions $\Lambda \cup \overline{\Lambda} \cup \{\tau\}$. Moreover, CaSE is parameterised in a set $\mathcal{T} = \{\sigma, \rho, \dots\}$ of *abstract clocks*, or clocks for brief. The syntax of CaSE is defined by the following BNF:

$$P \ ::= \ \mathbf{0} \mid \Delta \mid \Delta_\sigma \mid x \mid \alpha.P \mid P{+}P \mid P|P \mid P\backslash L \mid P/\sigma \mid \lfloor P \rfloor \sigma(P) \mid \mu x.P \ ,$$

where x is a *variable* taken from some countably infinite set, and $L \subseteq \mathcal{A} \setminus \{\tau\}$ is a *restriction set*. Further, we use the standard definitions for *static* and *dynamic* operators, *free* and *bound* variables, *open* and *closed* terms, and *guarded* terms. We refer to closed and guarded terms as *processes*, collected in the set $\mathcal{P}$. For convenience, we write $\overline{L}$ for the set $\{\overline{a} \mid a \in L\}$, where $\overline{\overline{a}} =_{\mathrm{df}} a$, and $x \stackrel{\mathrm{def}}{=} P$ for the process $\mu x.P$.

The *operational semantics* of a CaSE process P is given by a labelled transition system $\langle \mathcal{P}, \mathcal{A} \cup \mathcal{T}, \longrightarrow, P \rangle$, where $\mathcal{P}$ is the set of states, $\mathcal{A} \cup \mathcal{T}$ the alphabet,

Table 1. Operational semantics of CaSE

Act $$\frac{-}{\alpha.P \xrightarrow{\alpha} P}$$

tAct $$\frac{-}{\alpha.P \xrightarrow{\sigma} \alpha.P}\ \alpha \neq \tau$$

Sum1 $$\frac{P \xrightarrow{\alpha} P'}{P+Q \xrightarrow{\alpha} P'}$$

tNil $$\frac{-}{\mathbf{0} \xrightarrow{\sigma} \mathbf{0}}$$

tStall $$\frac{-}{\Delta_\sigma \xrightarrow{\rho} \Delta_\sigma}\ \sigma \neq \rho$$

Sum2 $$\frac{Q \xrightarrow{\alpha} Q'}{P+Q \xrightarrow{\alpha} Q'}$$

tSum $$\frac{P \xrightarrow{\sigma} P' \quad Q \xrightarrow{\sigma} Q'}{P+Q \xrightarrow{\sigma} P'+Q'}$$

Res $$\frac{P \xrightarrow{\alpha} P'}{P \backslash L \xrightarrow{\alpha} P' \backslash L}\ \alpha \notin L \cup \overline{L}$$

tRes $$\frac{P \xrightarrow{\sigma} P'}{P \backslash L \xrightarrow{\sigma} P' \backslash L}$$

Par1 $$\frac{P \xrightarrow{\alpha} P'}{P|Q \xrightarrow{\alpha} P'|Q}$$

tPar $$\frac{P \xrightarrow{\sigma} P' \quad Q \xrightarrow{\sigma} Q'}{P|Q \xrightarrow{\sigma} P'|Q'}\ P|Q \not\xrightarrow{\tau}$$

Par2 $$\frac{Q \xrightarrow{\alpha} Q'}{P|Q \xrightarrow{\alpha} P|Q'}$$

tHid1 $$\frac{P \xrightarrow{\sigma} P'}{P/\sigma \xrightarrow{\tau} P/\sigma}$$

Par3 $$\frac{P \xrightarrow{a} P' \quad Q \xrightarrow{\overline{a}} Q'}{P|Q \xrightarrow{\tau} P'|Q'}$$

tHid2 $$\frac{P \xrightarrow{\rho} P'}{P/\sigma \xrightarrow{\rho} P'/\sigma}\ \sigma \neq \rho,\ P \not\xrightarrow{\sigma}$$

Hid $$\frac{P \xrightarrow{\alpha} P'}{P/\sigma \xrightarrow{\alpha} P'/\sigma}$$

tTO1 $$\frac{-}{\lfloor P \rfloor \sigma(Q) \xrightarrow{\sigma} Q}\ P \not\xrightarrow{\tau}$$

TO $$\frac{P \xrightarrow{\alpha} P'}{\lfloor P \rfloor \sigma(Q) \xrightarrow{\alpha} P'}$$

tTO2 $$\frac{P \xrightarrow{\rho} P'}{\lfloor P \rfloor \sigma(Q) \xrightarrow{\rho} \lfloor P' \rfloor \sigma(Q)}\ \sigma \neq \rho$$

Rec $$\frac{P[\mu x.P/x] \xrightarrow{\alpha} P'}{\mu x.P \xrightarrow{\alpha} P'}$$

tRec $$\frac{P[\mu x.P/x] \xrightarrow{\sigma} P'}{\mu x.P \xrightarrow{\sigma} P'}$$

$\longrightarrow$ the transition relation and P the start state. We refer to transitions with labels in $\mathcal{A}$ as *action transitions* and to those with labels in $\mathcal{T}$ as *clock transitions*. The transition relation $\longrightarrow \subseteq \mathcal{P} \times (\mathcal{A} \cup \mathcal{T}) \times \mathcal{P}$ is defined in Table 1 using operational rules. We write γ for a representative of $\mathcal{A} \cup \mathcal{T}$, as well as $P \xrightarrow{\gamma} P'$ for $\langle P, \gamma, P' \rangle \in \longrightarrow$ and $P \xrightarrow{\gamma}$ for $\exists P' \in \mathcal{P}.\, P \xrightarrow{\gamma} P'$. Note that, despite the negative side conditions of some rules, the transition relation is well-defined for guarded processes. Our semantics obeys the following properties, for all clocks $\sigma \in \mathcal{T}$: (i) *maximal progress*, i.e., $P \xrightarrow{\sigma}$ implies $P \not\xrightarrow{\tau}$; (ii) *time determinacy*, i.e., $P \xrightarrow{\sigma} P'$ and $P \xrightarrow{\sigma} P''$ implies $P' = P''$. It is time determinacy that distinguishes clock ticks from CSP broadcasting [8].

Intuitively, the *nil* process $\mathbf{0}$ permits all clocks to tick, while the *timelock* processes Δ and Δ_σ prohibit the ticking of any clock and of clock σ, respectively. Process $\alpha.P$ may engage in action α and then behave like P. If $\alpha \neq \tau$, it may also idle for each clock σ; otherwise, all clocks are stopped, thus respecting maximal progress. The *summation operator* $+$ denotes nondeterministic choice, i.e., process $P + Q$ may behave like P or Q. Because of time determinacy, time has to proceed equally on both sides of summation. Process $P|Q$ stands for the *parallel composition* of P and Q according to an interleaving semantics with synchronised communication on complementary actions resulting in the internal action τ. Again, time has to proceed equally on both sides of the operator, and the side condition of Rule (tPar) ensures maximal progress. The *restriction operator* $\backslash L$ prohibits the execution of actions in $L \cup \overline{L}$ and thus permits the scoping of actions. The *clock-hiding operator* $/\sigma$ within a process P/σ turns every tick of clock σ in P into the internal action τ. This not only hides clock σ

but also pre-empts all other clocks ticking in P at the same states as σ, by Rule (tHid2). Process $\lfloor P \rfloor \sigma(Q)$ behaves as process P, and it can perform a σ-transition to Q, provided P cannot engage in an internal action as is reflected in the side condition of Rule (tTO1). The timeout operator disappears as soon as P engages in an action transition, but persists along clock transitions. Finally, $\mu x. P$ denotes *recursion* and behaves as a distinguished solution of the equation $x = P$.

Our interpretation of prefixes $\alpha.P$ adopted above, for $\alpha \neq \tau$, is *relaxed* [7], i.e., we allow this process to idle on clock ticks. In the remainder, *insistent prefixes* $\underline{\alpha}.P$ [1], which do not allow clocks to tick, will prove convenient as well. These can be expressed in CaSE by $\underline{\alpha}.P =_{\text{df}} \alpha.P + \Delta$. Similarly, one may define a prefix that only lets clocks not in T tick, for $T \subseteq \mathcal{T}$, by $\underline{\alpha}_T.P =_{\text{df}} \alpha.P + \Delta_T$, where $\Delta_T =_{\text{df}} \sum_{\sigma \in T} \Delta_\sigma$. As usual, $\sum$ denotes the indexed version of operator $+$, with the empty summation understood to be process $\mathbf{0}$. For convenience, we abbreviate $\lfloor \mathbf{0} \rfloor \sigma(P)$ by $\sigma.P$, and $\lfloor \Delta \rfloor \sigma(P)$ by $\underline{\sigma}.P$. We also write $P/\{\sigma_1, \sigma_2, \ldots, \sigma_k\}$ for $P/\sigma_1/\sigma_2 \cdots /\sigma_k$, if the order in which clocks are hidden is inessential. Moreover, for finite $A \subseteq \mathcal{A} \setminus \{\tau\}$ and process P, we let $A.P$ stand for the recursively defined process $\sum_{a \in A} a.(A \setminus \{a\}).P$, if $A \neq \emptyset$ and P, otherwise. Finally, instead of relabelling as in CCS [12] we use syntactic substitution, e.g., $P[a'/a, b'/b]$ relabels all occurrences of actions a, $\overline{a}$, b, $\overline{b}$ in P by a', $\overline{a}'$, b', $\overline{b}'$, respectively.

Temporal Observation Equivalence and Congruence. This section equips CaSE with a bisimulation-based semantics [12]. For the purposes of this paper we will concentrate on *observation equivalence* and *congruence*. The straightforward adaptation of strong bisimulation to our calculus immediately leads to a behavioural congruence, as can easily be verified by inspecting the format of our operational rules and by applying well-known results for structured operational semantics [16]. Observation equivalence is a notion of bisimulation in which any sequence of τ's may be skipped. For $\gamma \in \mathcal{A} \cup \mathcal{T}$ we define $\hat{\gamma} =_{\text{df}} \epsilon$ if $\gamma = \tau$ and $\hat{\gamma} =_{\text{df}} \gamma$, otherwise. Further, let $\stackrel{\epsilon}{\Rightarrow} =_{\text{df}} \stackrel{\tau}{\rightarrow}^*$ and $P \stackrel{\gamma}{\Rightarrow} P'$ if there exist processes P'' and P''' such that $P \stackrel{\epsilon}{\Rightarrow} P'' \stackrel{\gamma}{\rightarrow} P''' \stackrel{\epsilon}{\Rightarrow} P'$.

Definition 1. *A symmetric relation* $\mathcal{R} \subseteq \mathcal{P} \times \mathcal{P}$ *is a* temporal weak bisimulation *if* $P \stackrel{\gamma}{\rightarrow} P'$ *implies* $\exists Q'. Q \stackrel{\hat{\gamma}}{\Rightarrow} Q'$ *and* $\langle P', Q' \rangle \in \mathcal{R}$, *for every* $\langle P, Q \rangle \in \mathcal{R}$ *and for* $\gamma \in \mathcal{A} \cup \mathcal{T}$. *We write* $P \approx Q$ *if* $\langle P, Q \rangle \in \mathcal{R}$ *for some temporal weak bisimulation* $\mathcal{R}$.

Temporal observation equivalence $\approx$ is compositional for all operators except summation and timeout. However, for proving compositionality regarding parallel composition and hiding, the following proposition is central.

Proposition 1. *If* $P \approx Q$ *and* $P \stackrel{\sigma}{\rightarrow} P'$, *then* $\exists Q', Q'', Q'''. Q \stackrel{\epsilon}{\Rightarrow} Q'' \stackrel{\sigma}{\rightarrow} Q''' \stackrel{\epsilon}{\Rightarrow} Q'$, $P \approx Q''$, $P' \approx Q'$ *and* $\{\gamma \in \mathcal{A} \cup \mathcal{T} \mid P \stackrel{\gamma}{\rightarrow}\} = \{\gamma \in \mathcal{A} \cup \mathcal{T} \mid Q'' \stackrel{\gamma}{\rightarrow}\}$.

The validity of this proposition is due to the maximal-progress property in CaSE. To identify the largest equivalence contained in $\approx$, the summation fix of CCS is not sufficient. As in other work in temporal process algebras [3], the deterministic nature of clocks implies the following definition.

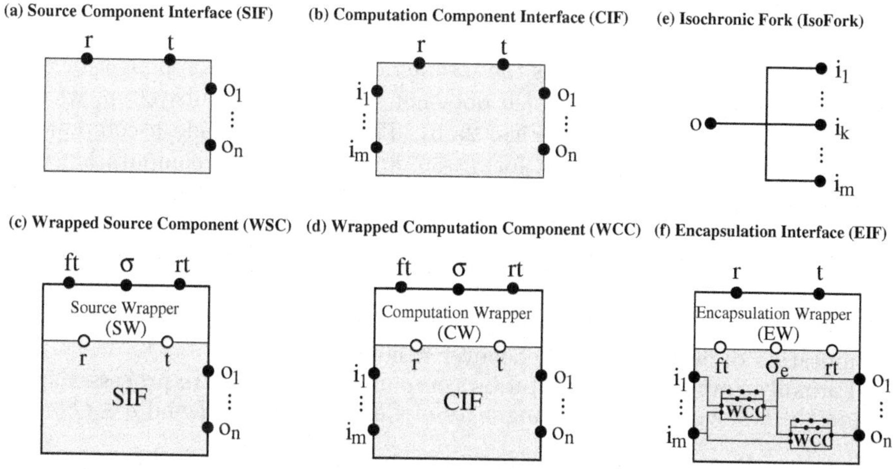

Fig. 2. Illustration of our modelling toolbox

Definition 2. *A symmetric relation* $\mathcal{R} \subseteq \mathcal{P} \times \mathcal{P}$ *is a* temporal observation congruence *if for every* $\langle P, Q \rangle \in \mathcal{R}$, $\alpha \in \mathcal{A}$ *and* $\sigma \in \mathcal{T}$*:*

1. $P \stackrel{\alpha}{\rightarrow} P'$ *implies* $\exists Q'. Q \stackrel{\alpha}{\Rightarrow} Q'$ *and* $P' \approx Q'$.
2. $P \stackrel{\sigma}{\rightarrow} P'$ *implies* $\exists Q'. Q \stackrel{\sigma}{\rightarrow} Q'$ *and* $\langle P', Q' \rangle \in \mathcal{R}$.

We write $P \cong Q$ *if* $\langle P, Q \rangle \in \mathcal{R}$ *for some temporal observation congruence* $\mathcal{R}$.

Theorem 1. *The equivalence* $\cong$ *is the largest congruence contained in* $\approx$.

CCS [12] can be identified in terms of syntax, operational semantics and bisimulation semantics as the sub-calculus of CaSE that is obtained by setting $\mathcal{T} = \emptyset$.

4 A Synchronous Coordination Model with Encapsulation

This section presents our model of coordination for DSPC applications on the basis of our process calculus CaSE. As illustrated in Fig. 2 we will successively model the key ingredients of a DSPC application: the behaviour of its source and computation components towards its environment (Figs. 2a,b), a compositional version of the centralised scheduler which is distributed to 'wrap' each component instance (Figs. 2c,d), the application's isochronous forks connecting output and input ports (Fig. 2e), and the facility to encapsulate several computation components (Fig. 2f). Having these ingredients at hand, a CaSE model of a DSPC application can then be built systematically along the structure of hierarchical signal-flow graphs, which we will illustrate by way of the digital-spectrum-analyser example introduced in Sect. 2. A particular emphasis will be given on showing how our modelling may facilitate static jam analysis.

4.1 Component Interfaces

A *component interface* describes the interaction of a source component or a basic computation component, which does not encapsulate a subsystem, with its environment via its ports (cf. Figs. 2a,b). These ports include a component's *output* ports, $O = \{o_1, \dots, o_n\}$, for $n \geq 0$, and, in case of a computation component, its *input* ports, $I = \{i_1, \dots, i_m\}$, for $m \geq 1$. Note that we abstract from values carried by signals through ports. In addition, each component interfaces to the system scheduler via port r, over which a component sends *request-to-be-scheduled* messages, and port t via which a *token* is passed between the scheduler and the component, with the intention that a component can go ahead with its computation of output signals whenever it holds the token.

Formally, source and computation component interfaces are processes specified in the following CCS sub-languages of CaSE, where $i \in I$ and $o \in O$:

Source Component Interface

$$\begin{aligned} \mathrm{SIF} &::= \mathbf{0} \mid x \mid \mathrm{SR} \mid \mu x.\,\mathrm{SR} \\ \mathrm{SR} &::= \overline{r}.t.\tau.\mathrm{SO} \\ \mathrm{SO} &::= \overline{o}.\mathrm{SO} \mid \mathrm{SO} + \mathrm{SO} \mid \overline{t}.\mathrm{SIF} \end{aligned}$$

Computation Component Interface

$$\begin{aligned} \mathrm{CIF} &::= \mathbf{0} \mid x \mid \mathrm{CI} \mid \mu x.\,\mathrm{CI} \\ \mathrm{CI} &::= i.\mathrm{CI} \mid \mathrm{CI} + \mathrm{CI} \mid i.\mathrm{CR} \\ \mathrm{CR} &::= \overline{r}.t.\tau.\mathrm{CO} \\ \mathrm{CO} &::= \overline{o}.\mathrm{CO} \mid \mathrm{CO} + \mathrm{CO} \mid \overline{t}.\mathrm{CIF} \end{aligned}$$

Intuitively, after reading its inputs in case of a computation component, a component instance (i) requests to be scheduled (action $\overline{r}$), (ii) waits for receiving the scheduler's token (action t), which indicates that the request has been granted and ensures serialisation on the underlying single processor, (iii) computes the output signal values (internal action τ), (iv) outputs these signal values over the corresponding output ports, and (v) returns the token to the scheduler (action $\overline{t}$). The interfaces of the source and basic computation component instances of our example system can then be specified as follows:

$$\begin{aligned} \mathrm{SIF}_{\mathrm{so}} &\stackrel{\mathrm{def}}{=} \overline{r}.t.\tau.\overline{so}.\overline{t}.\mathrm{SIF}_{\mathrm{so}} & \mathrm{SIF}_{\mathrm{sk}} &\stackrel{\mathrm{def}}{=} \overline{r}.t.\tau.\overline{co}_k.\overline{t}.\mathbf{0} \\ \mathrm{CIF}_{ck1} &\stackrel{\mathrm{def}}{=} fi_{k1}.fi_{k2}.\overline{r}.t.\tau.\overline{fo}_k.\overline{t}.\mathrm{CIF}'_{ck1} & \mathrm{CIF}'_{ck1} &\stackrel{\mathrm{def}}{=} fi_{k1}.\overline{r}.t.\tau.\overline{fo}_k.\overline{t}.\mathrm{CIF}'_{ck1} \\ \mathrm{CIF}_{ck2} &\stackrel{\mathrm{def}}{=} gi_k.\overline{r}.t.\tau.\overline{qo}_k.\overline{t}.\mathrm{CIF}_{ck2} & \mathrm{CIF}_{\mathrm{d}k} &\stackrel{\mathrm{def}}{=} gi_k.\overline{r}.t.\tau.\overline{t}.\mathrm{CIF}_{\mathrm{d}k} \end{aligned}$$

Note that sk:Const produces an output $\overline{\mathrm{so}}$ during the first cycle only, while ck:Element reads an input from port fi_{k1} during the first cycle only, as desired.

4.2 Component Instances and Scheduling

As seen above, a component uses its ports r and t to negotiate its execution with a scheduler. From the point of view of the component, it does not matter whether it communicates with a centralised or a distributed scheduler. In this section we develop a concept of *wrappers* for harnessing component instances with enough local control so they participate coherently in a global IRO-scheduling scheme, without the presence of a global scheduler (cf. Figs. 2c,d). Indeed all wrappers added together will represent a distributed version of an imagined central IRO scheduler.

Before introducing our distributed scheduler we present, for reference, an abstract model of the global centralised scheduler, as employed in the DSPC tool iConnect [15]. It uses an abstract clock σ that reflects the phase clock inherent in IRO scheduling. This clock organises the strict alternation between source and computation phases and, by way of maximal progress, implements run-to-completion within each phase. The global scheduler is defined via the following two sets of process equations, namely CSC that models the computation phase and CSS that models the source phase. They are stated relative to the sets $\mathcal{S}$ of source component instances and $\mathcal{C}$ of computation component instances within the signal-flow graph under consideration.

$$\begin{aligned} \mathrm{CSC}(W,\sigma) &\stackrel{\mathrm{def}}{=} \lfloor C(W,\sigma) \rfloor \sigma (\mathrm{CSS}(\emptyset,\emptyset,\sigma)) \\ C(W,\sigma) &\stackrel{\mathrm{def}}{=} (\sum_{c \in \mathcal{C} \setminus W} r_c.\mathrm{CSC}(W \cup \{c\},\sigma)) + (\sum_{c \in W} \bar{t}_c.\underline{t_c}_\sigma.\mathrm{CSC}(W \setminus \{c\},\sigma)) \\ \mathrm{CSS}(W,D,\sigma) &\stackrel{\mathrm{def}}{=} \lfloor S(W,D,\sigma) \rfloor \sigma (\mathrm{CSC}(\emptyset,\sigma)) \\ S(W,D,\sigma) &\stackrel{\mathrm{def}}{=} (\sum_{s \in \mathcal{S} \setminus (W \cup D)} r_s.\mathrm{CSS}(W \cup \{s\},D,\sigma)) + (\sum_{s \in W} \bar{t}_s.\underline{t_s}_\sigma.\mathrm{CSS}(W \setminus \{s\},D \cup \{s\},\sigma)) \end{aligned}$$

The process equations are parameterised in the phase clock σ, as well as the set W of component instances that are waiting for their scheduling request to be granted and the set D of source component instances that have already executed during the current source phase. Recall that each source component instance can execute at most once during each source phase, while each computation component instance may execute several times during a computation phase. While there are component instances that request to be scheduled or wait for being scheduled, the scheduler remains in the current phase, as is enforced by maximal progress. Otherwise, the phase clock may tick and switch phases.

To distribute this centralised scheduler over each component instance, all we assume is that the single embedded processor, on which the DSPC application is scheduled, provides some facility to ensure mutual exclusion. This may be modelled via a single token that the processor passes on to the component instance that may execute next: $\mathrm{CPUtoken} \stackrel{\mathrm{def}}{=} \overline{ft}.rt.\mathrm{CPUtoken}$, where ft stands for *fetch token* and rt for *release token*. Now, we may define the wrapping of computation and source component instances via meta-processes WCC and WSC, respectively. They are parameterised in the computation (source) component interface CIF_c (SIF_s) of a given computation (source) component instance c (s), as well as in the phase clock σ.

$$\begin{aligned} &\mathrm{WCC}(\mathrm{CIF}_c,\sigma) \stackrel{\mathrm{def}}{=} (\mathrm{CIF}_c \,|\, \mathrm{CW}(\sigma)) \setminus \{r,t\} \quad \mathrm{CW}(\sigma) \stackrel{\mathrm{def}}{=} \lfloor r.ft.\bar{t}.\underline{t}_\sigma.\overline{rt}.\mathrm{CW}(\sigma) \rfloor \sigma(\sigma.\mathrm{CW}(\sigma)) \\ &\mathrm{WSC}(\mathrm{SIF}_s,\sigma) \stackrel{\mathrm{def}}{=} (\mathrm{SIF}_s \,|\, \sigma.\mathrm{SW}(\sigma)) \setminus \{r,t\} \; \mathrm{SW}(\sigma) \stackrel{\mathrm{def}}{=} \lfloor r.ft.\bar{t}.\underline{t}_\sigma.\overline{rt}.\sigma.\sigma.\mathrm{SW}(\sigma) \rfloor \sigma(\sigma.\mathrm{SW}(\sigma)) \end{aligned}$$

Consider process $\mathrm{WCC}(\mathrm{CIF}_c,\sigma)$, which runs the wrapping process $\mathrm{CW}(\sigma)$ alongside the computation component interface CIF_c. Both synchronise via the now internalised channels r and t. If the component instance c signals its desire to be scheduled via a communication on channel r, the wrapping process $\mathrm{CW}(\sigma)$ waits until it may fetch the CPU token (action ft), passes on the token via the internal channel t, waits until the token has been passed back via the same channel, i.e., until the execution of c is complete, and then surrenders the token to the CPU (action $\overline{rt}$). If no computation component instance wishes to be scheduled, process $\mathrm{CW}(\sigma)$ may time out, thus allowing the overall system to switch to the source phase. In this state, component instance c must wait until the clock ticks again, i.e., until the scheduling has returned to the computation phase. The behaviour of $\text{WSC}(\text{SIF}_s, \sigma)$ wrapping source component instances is similar, except that those may only be scheduled once during each source phase. Thus, the source wrapper process $\text{SW}(\sigma)$ makes sure that two clock ticks have to pass before a request of the wrapped source component instance is considered again. Moreover, note that the initial σ-prefix in front of the wrapping process $\text{SW}(\sigma)$ ensures that the first source phase begins with the first ticking of σ. The following theorem shows that our compositional approach to scheduling coincides with the centralised one, where $\Pi_{k \in K} P_k$ stands for the parallel composition of processes P_k, for a finite index set K.

Theorem 2. *Let $\mathcal{S}$ ($\mathcal{C}$) be a finite set of source (computation) components with interfaces SIF_s (CIF_c), for $s \in \mathcal{S}$ ($c \in \mathcal{C}$), let σ be the phase clock, and let $R =_{df} \{r_s, t_s \mid s \in \mathcal{S}\} \cup \{r_c, t_c \mid c \in \mathcal{C}\}$. Then*

$$(\Pi_{s \in \mathcal{S}} WSC(SIF_s, \sigma) \mid \Pi_{c \in \mathcal{C}} WCC(CIF_c, \sigma) \mid CPUtoken) \setminus \{ft, rt\} \approxeq$$
$$(\Pi_{s \in \mathcal{S}} SIF_s[r_s/r, t_s/t] \mid \Pi_{c \in \mathcal{C}} CIF_c[r_c/r, t_c/t] \mid CSC(\emptyset, \sigma))) \setminus R\,.$$

4.3 Isochronic Forks

Before encoding isochronous forks in CaSE we present their naive modelling in CCS. To do so, we introduce a new output prefix $\overline{o}; P$ and assume that output port $\overline{o}$ shall be connected to input ports $I = \{i_1, i_2, \dots, i_m\}$ via an isochronic fork, as sketched in Fig. 2e. We need to ensure that the signal transmitted via $\overline{o}$ reaches all i_k, for $1 \leq i \leq m$, before process P executes. To model this, we define $\overline{o}; P =_{\text{df}} \overline{o}.f_o.P$ and $\text{ForkWire}(\overline{o}, I) =_{\text{df}} o.\overline{I}.\overline{f}_o.\text{ForkWire}(\overline{o}, I)$. Here, $\text{ForkWire}(\overline{o}, I)$ models the forking wire between port $\overline{o}$ and the ports in I. This wire distributes messages from the output port to all input ports and, once finished, signals this via the distinguished action $\overline{f}_o$. The sending process $\overline{o}; P$ has to wait for synchronisation on f_o before it can proceed with P, whence ensuring isochrony. While this solution is feasible, it requires that the number of intended recipients of a broadcasted signal is fixed up front and cannot grow as components are added to a signal-flow graph.

To overcome this problem we employ isochronic wires that connect the output port with exactly one input port, and use a fresh clock σ_o under maximal progress for synchronisation between sender and receivers of a broadcast signal. In analogy to the above we define the new isochronic output prefix $\overline{o}{:}P =_{\text{df}} C_{\overline{o},P}$ with $C_{\overline{o},P} \stackrel{\text{def}}{=} \lfloor \underline{\overline{o}}.C_{\overline{o},P} \rfloor \sigma_o(P)$ and an isochronic wire connecting $\overline{o}$ to input port i by $\text{IsoWire}(\overline{o}, i) =_{\text{df}} \underline{o}_{\sigma_o}.\underline{\overline{i}}_{\sigma_o}.\underline{\sigma}_o.\text{IsoWire}(\overline{o}, i)$. Thus, for a broadcast request $\overline{o}$, an arbitrary number of copies of the signal will be communicated on $\overline{o}$ until clock σ_o, which defines the isochronous instant in which the communication occurs, ticks and ends that instant. Because of maximal progress, σ_o can only tick when there are no further receivers listening on o. In this way signal $\overline{o}$ obtains maximal distribution, and one can add further receiving ports j later on by simply including a new isochronic wire from $\overline{o}$ to j without having to change the existing model. The following theorem shows that our compositional approach to isochronic broadcast faithfully models isochronous forks.

Theorem 3. *Let $\overline{o} \in \overline{\Lambda}$, $I \subseteq_{fin} \Lambda$ and $P \in \mathcal{P}$. Then*

$$(\overline{o}{:}P \mid \Pi_{i \in I} \mathit{IsoWire}(\overline{o}, i)) \backslash \{o\} / \sigma_o \;\approxeq\; (\overline{o};P \mid \mathit{ForkWire}(\overline{o}, I)) \backslash \{o, f_o\} \mid \Delta_{\sigma_o} .$$

The parallel component Δ_{σ_o} caters for the fact that the clock hiding operator $/\sigma_o$ eliminates clock σ_o. From now on we assume that all action prefixes $\overline{o}.P$ referring to the output ports of our component interfaces are replaced by isochronic ones $\overline{o}{:}P$, e.g., SIF_{s0} becomes $\overline{r}.t.\tau.\overline{so}{:}\overline{t}.\mathrm{SIF}_{s0}$.

Note that isochronous wiring cannot be modelled faithfully and compositionally in Hoare's CSP [8] or Prasad's CBS [13]. While the broadcasting primitive in CSP ignores the direction in which information is propagated, the one in CBS does not force receivers to synchronise with the sender.

4.4 Encapsulation

Hierarchical signal-flow graphs allow system designers to encapsulate several interconnected computation components, i.e., a subsystem, into a single computation component. As depicted in Fig. 2f, a subsystem is a tuple $\langle \mathcal{C}_e, W_e, I, O, W_I, W_O \rangle$ that consists of (i) a finite set $\mathcal{C}_e \subseteq \mathcal{C}$ of computation components, with disjoint sets of input ports I_e and sets of output ports O_e, (ii) a set of internal isochronic wires connecting output ports in O_e with input ports in I_e, (iii) a set of input ports $I = \{i_1, \dots, i_m\}$, (iv) a set of output ports $O = \{o_1, \dots, o_n\}$, (v) a set $W_I \subseteq I \times I_e$ of isochronic wires connecting the input ports of the subsystem with the input ports of the encapsulated components, and (vi) a set $W_O \subseteq O_e \times O$ of isochronic wires connecting the output ports of the encapsulated components with the output ports of the subsystem. In the example of Fig. 1 we have ck:Element $=$ $\langle \{ck1{:}\mathrm{Filter}, ck2{:}\mathrm{Quantise}\}, \{\langle \overline{\mathrm{fo}}_k, \mathrm{qi}_k \rangle\}, \{\mathrm{ei}_{k1}, \mathrm{ei}_{k2}\}, \{\overline{\mathrm{eo}}_k\}, \{\langle \mathrm{ei}_{k1}, \mathrm{fi}_{k1} \rangle, \langle \mathrm{ei}_{k2}, \mathrm{fi}_{k2} \rangle\}, \{\langle \overline{\mathrm{qo}}_k, \overline{\mathrm{eo}}_k \rangle\} \rangle$. The CaSE model of this subsystem is given by

$$\begin{aligned}\mathrm{Element}_k(\sigma_e) \stackrel{\mathrm{def}}{=} (&\Pi_{c \in \mathcal{C}_e} \mathrm{WCC}'(\mathrm{CIF}_c, \sigma_e) \mid \Pi_{\langle \overline{o}_e, i_e \rangle \in W_e} \mathrm{IsoWire}(\overline{o}_e, i_e) \mid \\ &\Pi_{\langle i, i_e \rangle \in W_I} \mathrm{IsoWire}(\overline{i}, i_e) \mid \Pi_{\langle \overline{o}_e, \overline{o} \rangle \in W_O} \mathrm{IsoWire}(\overline{o}_e, o)) \backslash I_e \backslash O_e / \sigma_{O_e} ,\end{aligned}$$

where $\sigma_{O_e} =_{\mathrm{df}} \{\sigma_{o_e} \mid o_e \in O_e\}$ contain the clocks governing the encapsulated isochronic wires. Also, $\mathrm{WCC}'(\mathrm{CIF}_c, \sigma_e) \stackrel{\mathrm{def}}{=} (\mathrm{CIF}_c \mid \mathrm{CW}'(\sigma_e)) \backslash \{r, t\}$ is an updated version of our instantiation wrapper given in Sect. 4.2, with $\mathrm{CW}'(\sigma_e) \stackrel{\mathrm{def}}{=} \lfloor r.(ft.\overline{t}.\underline{t}_{\sigma_e}.\overline{rt}.\mathrm{CW}'(\sigma_e) + \overline{r}_e.ft.\overline{t}.\underline{t}_{\sigma_e}.\overline{rt}.\mathrm{CW}'(\sigma_e)) \rfloor \sigma_e(\sigma_e.\mathrm{CW}'(\sigma_e))$. As subsystems must be executed atomically, the first encapsulated computation component that is ready to execute needs to request the mutual-exclusion token from its environment (action $\overline{r}_e$), i.e., from the subsystem at the next higher hierarchy level. Our modelling of encapsulation must then ensure that the token is only passed up to the environment once all computation components within the subsystem, which are able to execute, have actually executed. This is achieved via an *encapsulation wrapper* $\mathrm{EW}(\mathrm{SS}, I, O, \sigma_e)$ that is parameterised in the CaSE model SS of the subsystem under consideration, with input ports I, output ports O and subsystem clock σ_e. The encapsulation wrapper essentially translates back the scheduling interface $\{ft, rt, \sigma_e\}$ into $\{r, t\}$, which is the scheduling interface of a basic component.

$$\begin{aligned}
\mathrm{EW}(\mathrm{SS}, I, O, \sigma_e) &\stackrel{\mathrm{def}}{=} (\mathrm{SS}[i'_1/i_1, \ldots, i'_m/i_m, o'_1/o_1, \ldots, o'_n/o_n] \,|\, \mathrm{EI}(I, \sigma_e) \,|\, \mathrm{EO}(O, \sigma_e)) \\
&\qquad \setminus\{i'_1, \ldots, i'_m, o'_1, \ldots, o'_n, r_e\} / \sigma_I / \sigma_e \\
\mathrm{EI}(I, \sigma_e) &\stackrel{\mathrm{def}}{=} \sum_{i \in I} i.\overline{i}' {:} \mathrm{EI}(I, \sigma_e) \;+\; \underline{r_e}_{\sigma_e}.\underline{\overline{r}}_{\sigma_e}.\underline{t}_{\sigma_e}.\mathrm{EI}'(I, \sigma_e) \\
\mathrm{EI}'(I, \sigma_e) &\stackrel{\mathrm{def}}{=} \lfloor \overline{ft}.rt.\mathrm{EI}'(I, \sigma_e) \rfloor \sigma_e (\underline{\overline{t}}_{\sigma_e}.\mathrm{EI}(I, \sigma_e)) \\
\mathrm{EO}(O, \sigma_e) &\stackrel{\mathrm{def}}{=} \sum_{\overline{o} \in O} o'.\overline{o} {:} \mathrm{EO}(O, \sigma_e) \,,
\end{aligned}$$

where all i', for $i \in I$, and $\overline{o}'$, for $\overline{o} \in O$, are fresh port names not used in SS, and where $\sigma_I =_{\mathrm{df}} \{\sigma_i \,|\, i \in I\}$. The wrapper process $\mathrm{EI}(I, \sigma_e)$ propagates all input signals entering the subsystem to the desired receiving components, within the same cycle of the subsystem clock σ_e. Once an inner component requests to be scheduled (action r_e), the wrapper process forwards this request via port $\overline{r}$ to the next upper hierarchy level and waits for the token, indicating granted access to the embedded processor, to be passed down via port t. In this state, the encapsulation wrapper essentially behaves as process CPUtoken has done before, i.e., engaging in a communication cycle between ports $\overline{ft}$ and rt, until no further encapsulated component wishes to execute, i.e., until clock σ_e triggers a timeout and the token is passed back up (action $\overline{t}$). The outputs produced by components within the subsystem are instantaneously propagated to the subsystem's environment via the parallel process $\mathrm{EO}(O, \sigma_e)$, which is part of the encapsulation wrapper. Note that our encapsulation wrapper hides the inner clock σ_e, whose ticking thus appears like an internal, unobservable computation, from the point of view of components outside the subsystem under consideration. The following theorem puts the subsystems-as-components principle on a formal footing.

Theorem 4. *Let SS be the CaSE model of a subsystem* $\langle \mathcal{C}_e, W_e, I, O, W_I, W_O \rangle$ *using* σ_e *as subsystem clock. Then, there exists a computation component c with input ports I, output ports O and component interface* CIF_c *such that* $EW(SS, I, O, \sigma_e) \approx EW(WCC(CIF_c, \sigma_e), I, O, \sigma_e) \,|\, \Delta_{\sigma_I \cup \sigma_{O_e}}$.

We now have all tools of our DSPC modelling toolbox to complete the overall CaSE model $\mathrm{DSA}(\rho)$ of the digital spectrum analyser of Fig. 1, under phase clock ρ and given the component interfaces provided in Sect. 4.1: $\mathrm{DSA}(\rho) \stackrel{\mathrm{def}}{=}$

$$\begin{aligned}
&(\mathrm{WSC}(\mathrm{SIF}_{\mathrm{s}0}, \rho) \,| \\
&\;\Pi_{k \geq 1}(\mathrm{WSC}(\mathrm{SIF}_{\mathrm{s}k}, \rho) \,|\, \mathrm{WCC}(\mathrm{EW}(\mathrm{Element}_k(\sigma_k), \{\mathrm{ei}_{k1}, \mathrm{ei}_{k2}\}, \{\overline{\mathrm{eo}}_k\}, \sigma_k), \rho) \,| \\
&\qquad\quad \mathrm{WCC}(\mathrm{CIF}_{\mathrm{d}k}, \rho) \,|\, \mathrm{IsoWire}(\overline{\mathrm{so}}, \mathrm{ei}_{k1}) \,|\, \mathrm{IsoWire}(\overline{\mathrm{co}}_k, \mathrm{ei}_{k2}) \,|\, \mathrm{IsoWire}(\overline{\mathrm{eo}}_k, \mathrm{gi}_k)) \\
&)\setminus\{\mathrm{co}_k, \mathrm{ei}_{k1}, \mathrm{ei}_{k2}, \mathrm{eo}_k, \mathrm{gi}_k \,|\, k \geq 1\} \setminus \{\mathrm{so}\} / \{\sigma_{\mathrm{co}_k}, \sigma_{\mathrm{eo}_k} \,|\, k \geq 1\} / \sigma_{\mathrm{so}}
\end{aligned}$$

Observe that our modelling proceeds along the structure of the hierarchical signal-flow graph of Fig. 1.

4.5 Jam Analysis. A jam occurs when an output signal value produced by one component cannot be consumed by an intended receiving component within the same IRO-cycle. In current DSPC tools, jams are detected by the run-time system; upon detection of a jam, a DSPC application is simply terminated.

In our model of coordination we will encode jams in such a way that a jam manifests itself as a timelock regarding the overall system clock ρ. Such a timelock will occur when an isochronic wire is unable to pass on the value it holds. This can be achieved by modifying processes $\mathrm{IsoWire}(\overline{o}, i)$ throughout, such that clock ρ is stopped when the wire already stores a signal value but has not yet been able to pass it on to port i; formally, $\mathrm{IsoWire}(\overline{o}, i) =_{\mathrm{df}} \underline{o}_{\sigma_o}.\underline{\overline{i}}_{\{\rho,\sigma_o\}}.\underline{\sigma}_o.\mathrm{IsoWire}(\overline{o}, i)$. Consequently, the local 'jam' condition is turned into a timing flaw, which is a global condition that stops the complete system, as desired. The next theorem makes this mathematically precise; note that our model of coordination of a DSPC system does not possess any infinite τ-computations, as long as the system does not contain some computation components that are wired-up in feedback loops in which these components continuously trigger themselves.

Theorem 5. *Let P be a process that possesses only τ- and ρ-transitions and no infinite τ-computations, and let Check$=_{df} \mu x.\lfloor \Delta \rfloor \rho(x)$. Then $P \approx$ Check if and only if $\nexists P' \,\nexists s \in \{\tau, \rho\}^*.\ P \stackrel{s}{\rightarrow} P' \stackrel{\rho}{\not\Rightarrow}$.*

Hence, when considering that our model of coordination for an arbitrary hierarchical signal-flow graph can be automatically constructed from the flow graph's given component interfaces, one may statically check for jams by employing well-known algorithms for computing temporal observation equivalence [4].

5 Related Work

To the best of our knowledge, our process-algebraic model of coordination is the first formal model of the synchronous and hierarchical scheduling discipline behind DSPC tools. It complements existing work in *distributed object-oriented systems* and in *architectural description languages*. There, the focus is on distributed software rather than on embedded centralised systems, and consequently on asynchronous rather than on synchronous component behaviour.

In object-oriented systems, process-algebraic frameworks have been studied, where processes model the life-cycle of objects [14]. Within these frameworks, one may reason at compile-time whether each invocation of an object's method at run-time is permissible. This semantic analysis is different from jam analysis in DSPC applications, but similar to the *compatibility analysis* of interface automata [5], which we will discuss below. In architectural description languages, the formalism of process algebra has been studied by Bernardo et al. [2]. Their approach rests on the use of CSP-style broadcast communication together with asynchronous parallel composition. Like in our application domain of DSPC design, the intention is to identify communication problems, but these are diagnosed in terms of deadlock behaviour. As illustrated earlier, deadlock is a more specific property than the jam property investigated by us: a jam in one component jams the whole system, but a deadlock in one component does not necessarily result in a system deadlock.

From a practical point of view we envision our model of coordination based on the process calculus CaSE to play the role of a *reactive-types* language. This would enable designers to specify the intended interactions between a given component and its environment as a type, and permit tool implementations to reduce type checking to temporal observation-equivalence checking. This idea is somewhat similar to the one of *behavioural types* in the Ptolemy community [11]. Behavioural types are based on the formalism of *interface automata* [5] and employed for checking the *compatibility* property between components. However, interface automata are not expressive enough to reason about jams, which Ptolemy handles by linear-algebra techniques for the restricted class of synchronous data-flow (SDF) models. In contrast, CaSE's semantic theory is more general than SDF and lends itself to checking jams at compile-time.

6 Conclusions and Future Work

This paper presented a novel compositional model of coordination for the synchronous component-based design of and reasoning about DSPC applications. We demonstrated that the semantic concepts underlying the IRO principle of DSPC tools, namely dynamic synchronous scheduling, isochrony and encapsulation, can be captured by uniformly combining the process-algebraic concepts of abstract clocks, maximal progress and clock hiding, which have been studied in the concurrency-theory community. The standard notion of temporal observation equivalence then facilitates the desired static reasoning about jams in DSPC applications. Future work should integrate our work in DSPC tools in the form of a reactive-types system. A prototype written in Haskell is currently being implemented in Sheffield.

Acknowledgements. We thank the anonymous referees, as well as Rance Cleaveland and Matt Fairtlough for their valuable comments and suggestions.

References

1. H.R. Andersen and M. Mendler. An asynchronous process algebra with multiple clocks. In *ESOP '94*, volume 788 of *LNCS*, pages 58–73, 1994.
2. M. Bernardo, P. Ciancarini, and L. Donatiello. Detecting architectural mismatches in process algebraic descriptions of software systems. In *WICSA 2001*, pages 77–86. IEEE Comp. Soc. Press, 2001.
3. R. Cleaveland, G. Lüttgen, and M. Mendler. An algebraic theory of multiple clocks. In *CONCUR '97*, volume 1243 of *LNCS*, pages 166–180, 1997.
4. R. Cleaveland and S. Sims. The NCSU Concurrency Workbench. In *CAV '96*, volume 1102 of *LNCS*, pages 394–397, 1996.
5. L. de Alfaro and T.A. Henzinger. Interface automata. In *ESEC/FSE 2001*, volume 26, 5 of *Softw. Eng. Notes*, pages 109–120. ACM Press, 2001.
6. S. Hauck. Asynchronous design methodologies: An overview. *Proc. of the IEEE*, 83(1):69–93, 1995.

7. M. Hennessy and T. Regan. A process algebra for timed systems. *Inform. and Comp.*, 117:221–239, 1995.
8. C.A.R. Hoare. *Communicating Sequential Processes.* Prentice Hall, 1985.
9. G.W. Johnson and R. Jennings. *LabView Graphical Programming.* McGraw, 2001.
10. E.A. Lee. Overview of the Ptolemy project. Technical Report UCB/ERL M01/11, Univ. of California at Berkeley, 2001.
11. E.A. Lee and Y. Xiong. Behavioral types for component-based design. Technical Report UCB/ERL M02/29, Univ. of California at Berkeley, 2002.
12. R. Milner. *Communication and Concurrency.* Prentice Hall, 1989.
13. K.V.S. Prasad. Programming with broadcasts. In *CONCUR '93*, volume 715 of *LNCS*, pages 173–187, 1993.
14. F. Puntigam. Type specifications with processes. In *FORTE '95*, volume 43 of *IFIP Conf. Proc.* Chapman & Hall, 1995.
15. A. Sicheneder et al. Tool-supported software design and program execution for signal processing applications using modular software components. In *STTT '98*, BRICS Notes Series NS-98-4, pages 61–70, 1998.
16. C. Verhoef. A congruence theorem for structured operational semantics with predicates and negative premises. *Nordic J. of Computing*, 2(2):274–302, 1995.

Conditional Expectation and the Approximation of Labelled Markov Processes

Vincent Danos[1,*], Josée Desharnais[2], and Prakash Panangaden[3]

[1] Université Paris 7 & CNRS
[2] Université Laval, Québec
[3] McGill University, Montréal

Abstract. We develop a new notion of approximation of labelled Markov processes based on the use of conditional expectations. The key idea is to approximate a system by a coarse-graining of the state space and using averages of the transition probabilities. This is unlike any of the previous notions where the approximants are simulated by the process that they approximate. The approximations of the present paper are customizable, more accurate and stay within the world of LMPs. The use of averages and expectations may well also make the approximations more robust. We introduce a novel condition – called "granularity" – which leads to unique conditional expectations and which turns out to be a key concept despite its simplicity.

1 Introduction

Labelled Markov Processes (LMPs) are probabilistic transition systems where the state space might be any general measurable space, in particular this includes situations where the state space may be continuous. They are essentially traditional discrete-time Markov processes enriched with the process-algebra based notion of interaction by synchronization on labels. These have been studied intensively in the last few years ([6,7,8,15]). This is because they embody simple probabilistic interactive behaviours, and yet are rich enough to encompass many examples and to suggest interesting mathematics.

The initial motivation was the inclusion of continuous state spaces with a view towards eventual applications involving stochastic hybrid systems. An unexpected benefit of this additional generality has been the discovery that a simple temporal probabilistic logic, $\mathcal{L}_0$, captures a natural notion of equivalence between such processes, namely strong bisimulation. Remarkably this logic needs neither infinite conjunction, even though the systems may have even uncountable branching, nor negation nor any kind of negative construct (like the "must" modality). With this logical view, it became natural to think of the interplay between discrete structures (the logic) and the continuous mathematics of LMPs

* *Corresponding author:* Équipe PPS, Université Paris 7 Denis Diderot, Case 7014, 2 Place Jussieu 75251 Paris Cedex 05, Vincent.Danos@pps.jussieu.fr

R. Amadio, D. Lugiez (Eds.): CONCUR 2003, LNCS 2761, pp. 477–491, 2003.

(measure and probability theory). This led to the important question of understanding what it means to be an approximation of a given LMP and especially of a "finite" approximant.

The approximation theory has developed along two lines. Desharnais et. al. [7] have developed a metric between LMPs which can be viewed as a "relaxation" of the notion of strong bisimulation. This metric can be used to say that one LMP "comes close to" behaving like another. The other direction was to develop a notion of "finite" approximant [8,9] and cast this in a domain theoretic setting. The papers just cited established that even a system with an uncountable state space could be approximated by a family of finite state processes. The family of approximants converge to the system being approximated in both metric and domain-theoretic senses. The approximations interact smoothly with the logic in the following sense. Any formulas of $\mathcal{L}_0$ that are satisfied by any approximant of P are satisfied by the process P itself and any formula satisfied by P is satisfied by some approximant.

In a recent paper Danos and Desharnais [5] have developed a variant of the approximation that has two important advantages. First, the approximations can be "guided" by a family of formulas of interest. In other words, if there is a set of formulas of particular interest one can construct a specific finite approximant geared towards these formulas. One can then be sure that the process in question satisfied a formula of interest if and only if the approximant did. Second, a much more compact representation was used so that loops were not unwound and convergence was attained more rapidly. A disadvantage was that the approximations obtained were not LMPs because the transition "probabilities" are not measures. Instead they were capacities [2]. Capacities are not additive but they have instead a continuity property and are sub (or super) additive. The variants of LMPs obtained by using capacities instead of measures are called pre-LMPs.

In the present paper we show that we can have the best of both worlds in the sense that we can have the flexibility of a customizable approach to approximation and stay within the realm of LMPs. The approach is based on a radical departure from the ideas of the previous approaches [5,9]. In these approaches one always approximated a system by ensuring that the transition probabilities in the approximant were below the corresponding transition in the full system. Here we approximate a system by taking a coarse-grained discretization (pixellization) of the state space and then using *average* values. This new notion is not based on the natural simulation ordering between LMPs as were the previous approaches.

Instead we use *conditional expectation.* This is a traditional construction in probability theory which given a probability triple (S, Σ, p) (sample space), a Σ-measurable random variable X (observation) and a sub-σ algebra Λ (pixellization of the sample space), returns the conditional expectation of X with respect to p and Λ, written $\mathbb{E}_p(X|\Lambda)$, which in some suitable sense is the 'best' possible Λ-measurable approximation of X. The best will prove to be enough in our case, in that conditional expectations will construct for us low-resolution averages of any given LMP. Furthermore, an LMP will be known completely, up to bisimilarity, from its finite-resolution (meaning finite state) averages.

Moreover the new construction gives closer approximants in a sense that we will have to make precise later. They are also likely to be more robust to numerical variations in the system that one wants to approximate, since they are based on averages. Of course this is a speculative remark and needs to be thrashed out in subsequent work. To summarize, the new approximants are customizable, probabilistic and more accurate and possibly more robust as well.

Beyond the construction given here, we would like to convey the idea that probability theory and its toolkit – especially the uses of averages and expectation values – are remarkably well adapted to a computationally-minded approach to probabilistic processes. It has a way of meshing finite and continuous notions of computations which is not unlike domain-theory. We expect far more interaction in the future between these theories than what is reported here. Work on probabilistic powerdomains [12] and integration on domains [10,11] provides a beginning. Curiously enough the bulk of work in probabilistic process algebra rarely ever mentions averages or expectation values. We hope that the present paper stimulates the use of these methods by others.

Outline. First we recall the definitions of our two basic objects of concern, LMPs and conditional expectations. Then we identify circumstances in which the conditional expectation is actually defined pointwise and not only "almost everywhere". We construct an adaptation of Lebesgue measure on any given LMP that will serve as the ambient probability which we need to drive the construction home. With all this in place we may turn to the definition of approximants. We show they are correct both by a direct argument and by showing the precise relation in which they stand with the order-theoretic approximants given in [5].

2 Preliminaries

2.1 Measurable Spaces and Probabilities

A *measurable space* is a pair (S, Σ) where S is a set and $\Sigma \subset 2^S$ is a σ*-algebra* over S, that is, a set of subsets of S, containing S and closed under countable intersection and complement. Well-known examples are $[0,1]$ and $\mathbb{R}$ equipped with their respective *Borel* σ-algebras generated by the intervals which we will both denote by $\mathcal{B}$.

A map f between two measurable spaces (S, Σ) and (S', Σ') is said to be *measurable* if for all $Q' \in \Sigma'$, $f^{-1}(Q') \in \Sigma$. Writing $\sigma(f)$ for the σ-algebra generated by f, namely the set of sets of the form $f^{-1}(Q')$ with $Q' \in \Sigma'$, one can rephrase this by saying $\sigma(f) \subseteq \Sigma$. The set of measurable maps from (S, Σ) to $(\mathbb{R}, \mathcal{B})$ will be denoted $m\Sigma$.

A *subprobability* on (S, Σ) is a map $p : \Sigma \to [0,1]$, such that for any countable collection (Q_n) of pairwise disjoint sets in Σ, $p(\bigcup_n Q_n) = \sum_n p(Q_n)$. An actual probability is when in addition $p(S) = 1$. The condition on p is called σ*-additivity* and can be conveniently broken in two parts:

— *additivity*: $p(Q \cup Q') = p(Q) + p(Q')$, for Q, Q' disjoint,
— *continuity*: $\forall \uparrow Q_n \in \Sigma : p(\cup Q_n) = \sup_n p(Q_n)$.[1]

Let (S, Σ, p) be a probability triple, that is to say a measurable space (S, Σ) together with a probability p. A subset $N \subset S$ is said to be *negligible* if there exists a $Q \in \Sigma$ such that $N \subseteq Q$ and $p(Q) = 0$.

We write $\mathcal{N}_p$ for p-negligible subsets. Two functions X, Y on (S, Σ, p) are said to be *almost surely equal*, written $X = Y$ a.s., if $\{s \in S \mid X(s) \neq Y(s)\} \in \mathcal{N}_p$. Sometimes we say p-a.s. equal if we wish to emphasize which measure we are talking about.

The subset of $m\Sigma$ consisting of the functions that are integrable with respect to p will be denoted by $\mathcal{L}^1(S, \Sigma, p)$. A last piece of notation that we will use is to write $X_n \uparrow X$ when X_ns and X are in $m\Sigma$, meaning that $X_n \leq X_{n+1}$ with respect to the pointwise ordering and X_n converges pointwise to X.

2.2 Labelled Markov Processes

We need to define the objects of study:

Definition 1 (LMP). $\mathcal{S} = (S, \Sigma, h : L \times S \times \Sigma \to [0, 1])$ *is a Labelled Markov Process (LMP) if* (S, Σ) *is a measurable space, and:*
— *for all* $a \in L$, $Q \in \Sigma$, $h(a, s, Q)$ *is* Σ*-measurable as a function of* s*;*
— *for all* $s \in S$, $h(a, s, Q)$ *is a subprobability as a function of* Q.

Some particular cases: 1) when S is finite and $\Sigma = 2^S$ we have the familiar probabilistic transition system, 2) when $h(a, s, Q)$ does not depend on s or on a we have the familiar (sub)probability triple. An example of the latter situation is $([0, 1], \mathcal{B}, h)$ with $h(a, s, B) = \lambda(B)$ with λ the Lebesgue measure on the collection $\mathcal{B}$ of Borel sets.

Second we see that equivalently LMPs can be defined as follows:

Definition 2 (LMP2). *A Labelled Markov Process consists of a measurable space* (S, Σ) *and a family of* Σ*-measurable functions* $(h(a, Q))_{a \in L, Q \in \Sigma}$ *with values in* $[0, 1]$*, such that:*
— *additivity: for all disjoint* Q, Q': $h(a, Q \cup Q') = h(a, Q) + h(a, Q')$;
— *continuity: for all increasing sequence* $\uparrow Q_n$: $h(a, \bigcup Q_n) = \sup h(a, Q_n)$.

From the definition follows that for all a, s one has $h(a, S)(s) \leq 1$.

In this second definition we see an LMP as a Σ-indexed family of Σ-measurable functions, namely the random variables "probability of jumping to Q in one step labelled with a", instead of an S-indexed family of probabilities on Σ. Both definitions are related by $h'(a, s, Q) = h(a, Q)(s)$. The functions h, h' are commonly referred to as transition probability functions or Markov kernels (or stochastic kernels).

In previous treatments [6] LMPs were required to have an analytic state space. This was needed for the proof of the logical characterization of bisimulation. We will not mention this again in the present paper since we will not need the analytic structure. In fact it is hard to give examples of spaces that are not analytic, let alone one that might be useful in an example.

[1] Where $\uparrow Q_n$ denotes an increasing sequence of sets Q_n, *i.e.*, for all n, $Q_n \subseteq Q_{n+1}$.

2.3 Conditional Expectation

The expectation $\mathbb{E}_p(X)$ of a random variable X is the average computed by $\int X dp$ and therefore it is just a number. The *conditional* expectation is not a mere number but a random variable. It is meant to measure the expected value in the presence of additional information. The conditional expectation is typically thought of in the form: "if I know in advance that the outcome is in the set Q then my revised estimate of the expectation is $\mathbb{E}_p(X)$ is $\mathbb{E}_p(X|Q)$." However additional information may take a more subtle form than merely stating that the result is in or not in a set.

The additional information takes the form of a sub-σ algebra, say Λ, of Σ. In what way does this represent "additional information"? The idea is that an experimenter is trying to compute probabilities of various outcomes of a random process. The process is described by (S, Σ, p). However she may have partial information in advance by knowing that the outcome is in a measurable set Q. Now she may try to recompute her expectation values based on this information. To know that the outcome is in Q also means that it is *not* in Q^c. Note that $\{\varnothing, Q, Q^c, S\}$ is in fact a (tiny) sub-σ-algebra of Σ. Thus one can generalize this idea and say that for some given sub-σ-algebra Λ of Σ she knows for every $Q \in \Lambda$ whether the outcome is in Q or not. Now she can recompute the expectation values given this information.

How can she actually express this revised expectation when the σ-algebra Λ is large. It is presented as a density function so that for every Λ-measurable set B one can compute the conditional expectation by integration over B. Thus instead of a number we get a Λ-measurable function called the *conditional expectation given* Λ and is written $\mathbb{E}_p(\_|\Lambda)$.[2]

It is not at all obvious that such a function should exist and is indeed a fundamental result of Kolmogorov (see for instance [16], p.84).

Theorem 1 (Kolmogorov). *Let (S, Σ, p) be a probability triple, X be in $\mathcal{L}^1(S, \Sigma, p)$ and Λ be a sub-σ-algebra of Σ, then there exists a $Y \in \mathcal{L}^1(S, \Lambda, p)$ such that*

$$\forall B \in \Lambda. \int_B X dp = \int_B Y dp. \tag{1}$$

Not only does the conditional expectation exist, but it has a lot of properties. As a functional of type:

$$\mathbb{E}_p(\_|\Lambda) : \mathcal{L}^1(S, \Sigma, p) \rightarrow \mathcal{L}^1(S, \Lambda, p)$$

it is *linear*, *increasing* with respect to the pointwise ordering and *continuous* in the sense that for any sequence (X_n) with $0 \leq X_n \uparrow X$ and $X_n, X \in \mathcal{L}^1(S, \Sigma, p)$, then $\mathbb{E}_p(X_n|\Lambda) \uparrow \mathbb{E}_p(X|\Lambda) \ldots$ but it is *not* uniquely defined !

[2] Take note that, in the same way as $\mathbb{E}_p(X)$ is constant on S, the conditional expectation will be constant on every "pixel" or smallest observable set in Λ. In the above "tiny" sub-σ-algebra, this means constant on both Q and Q^c. This will turn out to be exactly what we need later when pixels are defined by sets of formulas.

All candidate conditional expectations are called *versions* of the conditional expectation. It is easy to prove that any two Λ-measurable functions satisfying the characteristic property (1) given above may differ only on a set of p-probability zero.

2.4 The Finite Case

As we have said before, the basic intuition of $\mathbb{E}_p(X|\Lambda)$ is that it averages out all variations in X that are below the resolution of Λ, *i.e.* which do not depend on Λ. In particular, if X is independent of Λ, then $\mathbb{E}_p(X|\Lambda) = \mathbb{E}_p(X)$,[3] and X is completely averaged out. On the other hand, if X is fully dependent on Λ, in other words if X is Λ-measurable, then $\mathbb{E}_p(X|\Lambda) = X$.[4]

Actually this intuition is exact in the case that the sample space S is *finite*. We may suppose then that $\Sigma = 2^S$, and Λ will be generated by a set of equivalence classes. But then $Y = \mathbb{E}_p(X|\Lambda)$ has to be constant on equivalence classes (else it is not Λ-measurable) and by the characteristic property, with B an equivalence class $[s]$, we get:

$$Y(s).p([s]) = \int_{[s]} Y dp = \int_{[s]} X dp = \sum_{t \in [s]} X(t)p(\{t\})) = \mathbb{E}(1_{[s]}X),$$

where $1_{[s]}$ is the characteristic function of the measurable set $[s]$.

When $p([s]) > 0$ we see that Y is exactly the *p-average* of X over equivalence classes associated to Λ:

$$Y(s) = \frac{1}{p([s])} \cdot \mathbb{E}(1_{[s]}X)$$

2.5 The Example That Says It All

Now that it is understood that in the finite state-space case conditional expectations are averages over equivalence classes, we can consider a revealing example. Put $S = \{x, y, 0, 1\}$, $\Sigma = 2^S$, $L = \{a\}$ (there is only one label, so we will not even bother to write a in the kernels); $h(\{0\})(x) = h(\{1\})(y) = 1$ and every other state-to-state transition is of probability zero. Suppose Λ identifies x and y, and call the resulting class z.

One can conceive of three ways to define a kernel k on the quotient space $\{z, 0, 1\}$. One can define the kernel as the *infimum* over $\{x, y\}$ or dually one can

[3] Recall that in this equation the left-hand side is a function while the right-hand side is a number; we mean to say that the function on the left is a constant function whose value is given by the right-hand side.

[4] Given a probability triple (S, Σ, p), a random variable $X \in m\Sigma$ is said to be independent of a sub-σ-algebra Λ if for any event $A \in \sigma(X)$ and $B \in \Lambda$, $p(A \cap B) = p(A)p(B)$. In particular, as one can easily verify, X is always independent of the trivial σ-algebra $\Lambda_0 = \{\varnothing, S\}$ and by the remark above, $\mathbb{E}_p(X|\Lambda_0) = \mathbb{E}_p(X)$ the ordinary unconditional expectation of X.

take it to be the *supremum*:

$$k_i(\{0\})(z) = 0,\ k_i(\{1\})(z) = 0,\ k_i(\{0,1\})(z) = 1,$$
$$k_s(\{0\})(z) = 1,\ k_s(\{1\})(z) = 1,\ k_s(\{0,1\})(z) = 1,$$

or else one can *average* (using here the uniform probability):

$$k_a(\{0\})(z) = 1/2,\ k_a(\{1\})(z) = 1/2,\ k_a(\{0,1\})(z) = 1.$$

As we said in the introduction, the use of the infimum results in super-additive kernels while the use of a supremum results in sub-additive kernels:

$$k_i(\{0,1\})(z) = 1 > k_i(\{0\})(z) + k_i(\{1\})(z) = 0$$
$$k_s(\{0,1\})(z) = 1 < k_s(\{0\})(z) + k_s(\{1\})(z) = 2$$

Of the three options, only using averages preserve additivity:

$$k_a(\{0,1\})(z) = 1 = k_a(\{0\})(z) + k_a(\{1\})(z).$$

Besides we observe that, perhaps not surprisingly, in all cases the kernel obtained by using averages is sandwiched between the others, *e.g.* :

$$0 = k_i(\{0\})(z) \leq k_a(\{0\})(z) = 1/2 \leq k_s(\{0\})(z) = 1.$$

The rest of the paper is essentially about structuring this nice concrete notion of approximant by averages as a general construction and explaining in what sense these approximants are actually approximating what they are supposed to be approximants of.

2.6 Logic and Metric

The other goal of having approximants that are customizable with respect to formulas of interest will be achieved by using the notion of expectation above with Λ a σ-algebra generated by a set of formulas of a suitable logic. We will prove that the approximant satisfies exactly the same formulas of the given set as does the process being approximated.

The following logic $\mathcal{L}_0$ is a central tool for asserting properties of LMPs, since it characterizes strong bisimulation between them [6].

$$\theta := \top,\ \theta \wedge \theta,\ \langle a \rangle_r \theta.$$

The parameter r above can be any rational in $[0, 1]$.

Definition 3. *Given an LMP $\mathcal{S}$, one defines inductively the map $[\![.]\!]_\mathcal{S} : \mathcal{L}_0 \to \Sigma$ as:*

— $[\![\top]\!]_\mathcal{S} = S$,
— $[\![\theta_0 \wedge \theta_1]\!]_\mathcal{S} = [\![\theta_0]\!]_\mathcal{S} \cap [\![\theta_1]\!]_\mathcal{S}$,
— $[\![\langle a \rangle_r \theta]\!]_\mathcal{S} = \{s \in S \mid h(a, [\![\theta]\!]_\mathcal{S})(s) \geq r\}$.

Let $\mathcal{S}$ be an LMP, one says $s \in S$ satisfies θ, written $s \models \theta$, if $s \in [\![\theta]\!]_{\mathcal{S}}$; one says $\mathcal{S}$ satisfies θ, still written $\mathcal{S} \models \theta$, if there exists an $s \in S$ such that $s \models \theta$. Finally given another LMP $\mathcal{S}'$ and $\mathcal{F}$ a subset of formulas of $\mathcal{L}_0$, we write $\mathcal{S} \approx_{\mathcal{F}} \mathcal{S}'$ if $\mathcal{S}$ and $\mathcal{S}'$ satisfy the same formulas of $\mathcal{F}$.

As we have already said, we will take the simplifying stance that two LMPs are bisimilar iff they satisfy the same formulas. This was proven in the case of analytic state spaces and that is a general enough class to encompass any conceivable physical example.

In [7] a family of metrics, d^c for $c \in (0,1)$, has been introduced that is closely related to this logic. Indeed one can think of the metric as measuring the complexity of the distinguishing formula between two states if any.

We do not need to give the precise definition of these metrics here, but we do want to use it to show convergence of approximants. This will be done using the following result which is a direct consequence of results relating the logic and the metric that can be found in [7].

Proposition 4. *Let $(\mathcal{F}_i)_{i\in\mathbf{N}}$ be an increasing sequence of sets of formulas converging to the set of all formulas of $\mathcal{L}_0$. Let $\mathcal{S}$ be an LMP and $(\mathcal{S}_i)_{i\in\mathbf{N}}$ a sequence of LMPs. Then if $\mathcal{S}_i \approx_{\mathcal{F}_i} \mathcal{S}$ for every set $\mathcal{F}_i$ of formulas of $\mathcal{L}_0$, then for all $c \in (0,1)$:*

$$d^c(\mathcal{S}_i, \mathcal{S}) \longrightarrow_{i\to\infty} 0.$$

3 When $\mathbb{E}_p(\_|\Lambda)$ Is Unique

There is one thing we have to confront. As we noted before, conditional expectations are unique only "almost surely." Now we want to use them to average our family of $h(a,Q)$ and, from the definition of an LMP, we need these averages to be defined *pointwise*, not only up to p. Yet, in the case of finite systems, one option is to choose for p the uniform probability on S, in which case "almost surely" actually means "surely," since only the empty set is in $\mathcal{N}_p$. This, intuitively, is because points are big enough chunks to be seen by the probability distribution. This leads to the following two definitions.

Definition 5 (pixels). *Let (S,Σ) be a measurable space, one says s and $t \in S$ are Σ-indistinguishable if $\forall Q \in \Sigma$, $s \in Q \leftrightarrow t \in Q$.*

This is an equivalence on S and we write $[s]_\Sigma$ or sometimes simply $[s]$ to denote the equivalence class of s. One has $[s]_\Sigma = \cap\{Q \mid s \in Q \in \Sigma\}$ so equivalence classes might not be measurable themselves unless Σ is countably generated, which is the case we are interested in.

Definition 6 (granularity). *Let (S,Σ,p) be a probability triple and $\Lambda \subseteq \Sigma$ be a sub-σ-algebra of Σ; p is said to be* **granular** *over Λ if for all $s \in S$, $[s]_\Lambda \notin \mathcal{N}_p$.*

In other words, p is granular over Λ if no Λ equivalence class is negligible. What this means intuitively is that the "pixellization" of Λ is always seen by p. It may be instructive to point out that there are at most countably many equivalence classes in this case.

As an example, we can take the probability triple $([0,1]^2, \mathcal{B}_2, \lambda_2)$, where λ_2 is the Lebesgue measure on the square, and $\Lambda = \mathcal{B} \times [0,1]$. Then $[s]_\Lambda = \{s\} \times [0,1] \in \Lambda$ and $\lambda_2([s]) = 0$ so our p is not granular over this Λ. The measurable sets of Λ are very thin strips. They are too fine to be granular. But if we take a cruder Λ, namely that containing the squares $[k/n, k+1/n] \times [h/n, h+1/n]$ for k, $h < n$ (with n fixed), then $[s]_\Lambda$ is such a square of λ_2-measure $1/n^2$, so here p is granular.

The big payoff of granularity is the following

Lemma 7 (Uniqueness lemma). *Let (S, Σ, p) be a probability triple, $\Lambda \subseteq \Sigma$, p granular over Λ, X and Y both Λ-measurable, then:*

$$X = Y \ a.s. \Rightarrow X = Y.$$

So in this case "almost surely" does mean "surely !"

Proof. Set $Q := \{s \in S \mid X(s) = \alpha \wedge Y(s) = \beta\}$ and $t \in Q$. One has $Q \in \Lambda$, by Λ-measurability of X and Y, but then $[t]_\Lambda \subseteq Q$ (otherwise Q splits $[t]_\Lambda$). So by granularity $p(Q) > 0$ (else $[t]_\Lambda$ is negligible), and therefore $\alpha = \beta$ or else X and Y differ on a non negligible set Q. □

So in this favourable circumstances we can do away with versions. If $X \in \mathcal{L}^1(S, \Sigma, p)$, and p is granular over Λ:

$$\mathbb{E}_p(X|\Lambda) : \mathcal{L}^1(S, \Sigma, p) \to \mathcal{L}^1(S, \Lambda, p)$$

is uniquely defined and we can proceed to the main definition.

4 Projecting LMPs

Definition 8 (projection of an LMP). *Given (S, Σ) a measurable space, Λ a sub-σ-algebra of Σ, p a probability on (S, Σ) granular over Λ, and $\mathcal{S} = (h(a, Q))_{a \in L, Q \in \Sigma}$ an LMP on (S, Σ), one defines the* p-projection *of $\mathcal{S}$ on Λ, written $(\mathcal{S}|\Lambda)_p$ as:*

$$h'(a, Q) = \mathbb{E}_p(h(a, Q)|\Lambda), \ \textit{for}\ a \in L, \ Q \in \Lambda.$$

Take note that this is *the* version of the conditional expectation. Existence follows from the fact that the $h(a, Q)$ evidently are integrable with respect to p (they are measurable, positive and bounded by 1), in other words they are in $\mathcal{L}^1(S, \Sigma, p)$.

Proposition 9 (Staying within LMPs). *$(\mathcal{S}|\Lambda)_p$ is an LMP.*

Proof. All maps $h'(a, Q)$ are Λ-measurable by definition of the conditional expectatioin; additivity is because $\mathbb{E}_p(\_|\Lambda)$ is linear; continuity follows because $\mathbb{E}_p(\_|\Lambda)$ is continuous as can be seen by using the conditional form of the monotone convergence theorem. □

We may now round off the construction by changing the state space.

Let us write $[\text{-}]_\Lambda : S \to [S]_\Lambda$ for the canonical surjection to the set of equivalence classes and denote accordingly the quotient σ-algebra by $[\Lambda]_\Lambda$. Then one can define the *quotient* LMP $([S]_\Lambda, [\Lambda]_\Lambda, k)$ with:

$$k(a, B)([s]_\Lambda) := h'(a, \cup B)(t) := \mathbb{E}_p(h(a, \cup B)|\Lambda)(t),$$

with $t \in [s]$. Take note that the right hand side is independent of the choice of $t \in [s]_\Lambda$ since $h'(a, Q)$ is Λ-measurable, and therefore $h'(a, Q)$ has to be constant on $[s]_\Lambda$ (else the equivalence is split by an event in Λ). Moreover, $[\text{-}]_\Lambda$ is a bisimulation morphism (which was formerly called a "zig-zag" [6]) from $(\mathcal{S}|\Lambda)_p$ to $([S]_\Lambda, [\Lambda]_\Lambda, k)$ and as such it preserves all $\mathcal{L}_0$ properties.

So far we have a quotient theory for LMPs when pixels are big enough, but everything hinges on the choice of an ambient p. This is the second problem we have to deal with.

5 A "Uniform" Probability on $(S, \sigma(\mathcal{L}_0))$

The key is to construct an appropriate measure, and we will use $\mathcal{L}_0$ to do this. So, given an LMP $\mathcal{S} = (S, \Sigma, h)$, and a *fixed* enumeration (θ_n) of $\mathcal{L}_0$, we first define a sequence (S, Λ_n) of measurable spaces:[5]

$$\Lambda_0 := \{\varnothing, S\},\ \Lambda_n := \sigma([\![\theta_i]\!]_{\mathcal{S}}; i < n).$$

Then for each n, we set $\tau_n := 1_{[\![\theta_n]\!]_{\mathcal{S}}}$ and define $\alpha_n : \{0,1\}^n \to \Lambda_n$ as:

$$\alpha_n(\boldsymbol{x}) = \cap_{i<n}\{s \mid \tau_i(s) = \boldsymbol{x}_i\},$$

with the convention that $\{0,1\}^0 = \{*\}$ and $\alpha_0(*) = S$.

Each Λ_n is a finite boolean algebra and so has atoms (non empty sets in Λ_n with no proper subsets); each atom of Λ_n is the image by α_n of a unique sequence $\boldsymbol{x} \in \{0,1\}^n$, but not all sequences are mapped to atoms, some are mapped to the empty set.

Now the idea is to construct p stagewise and at each stage to divide evenly the mass of an atom $\alpha_n(\boldsymbol{x}) \in \Lambda_n$ between its proper subsets in Λ_{n+1} if there are some. Specifically, we define inductively p_n on Λ_n-atoms as:

$$\begin{array}{l} p_0(\varnothing) = 0,\ p_0(S) = 1 \\ \alpha_{n+1}(\boldsymbol{x}0) \neq \varnothing,\ \alpha_{n+1}(\boldsymbol{x}1) \neq \varnothing \Rightarrow p_{n+1}(\alpha_{n+1}(\boldsymbol{x}0)) = p_{n+1}(\alpha_{n+1}(\boldsymbol{x}1)) = \frac{1}{2} \cdot p_n(\alpha_n(\boldsymbol{x})) \\ \alpha_{n+1}(\boldsymbol{x}0) = \varnothing,\ \alpha_{n+1}(\boldsymbol{x}1) \neq \varnothing \Rightarrow p_{n+1}(\alpha_{n+1}(\boldsymbol{x}0)) = 0,\ p_{n+1}(\alpha_{n+1}(\boldsymbol{x}1)) = p_n(\alpha_n(\boldsymbol{x})) \\ \alpha_{n+1}(\boldsymbol{x}0) \neq \varnothing,\ \alpha_{n+1}(\boldsymbol{x}1) = \varnothing \Rightarrow p_{n+1}(\alpha_{n+1}(\boldsymbol{x}0)) = p_n(\alpha_n(\boldsymbol{x})),\ p_{n+1}(\alpha_{n+1}(\boldsymbol{x}1)) = 0 \end{array}$$

Clearly each p_n extends to a unique probability on (S, Λ_n) since it is defined on Λ_n-atoms and the p_n are compatible in the sense that $p_{n+1} \upharpoonright \Lambda_n = p_n$; the sequence p_n converges to some "skewed" Lebesgue measure p on $\sigma(\mathcal{L}_0)$, the σ-algebra generated by our temporal formulas.[6]

[5] For each n, $\Lambda_n \subseteq \Lambda_{n+1}$, this is usually called a filtration.

[6] To be exact, by $\sigma(\mathcal{L}_0)$ we mean $\sigma([\![\theta]\!]_{\mathcal{S}}; \theta \in \mathcal{L}_0)$.

First, we have to remind for future use that for any finite set of formulas $\mathcal{F} \subset \mathcal{L}_0$ and $\Lambda_{\mathcal{F}}$ the associated σ-algebra:

$$p([s]_{\Lambda_{\mathcal{F}}}) \geq 2^{-N} \tag{2}$$

where $N = \max\{i \mid \theta_i \in \mathcal{F}\}$.

Second, we observe that the p obtained here will depend on the original enumeration, and we leave for future investigation the question of whether there is a principled way of choosing p. In our case, all choices will work equally well.

As an example we can consider the transition sytem with only state s, only one letter a and $h(a, \{s\})(s) = 1/2$. Then $s \models \theta$ iff all coefficients used in θ are below $1/2$. In this case, and as with all one-state systems, at any stage there will be at most one atom namely $\{s\}$ and therefore $p(\{s\}) = 1$.

5.1 Compressing Σ

But the reader might protest that to apply the projection, one needs a probability on an arbitrary Σ not just on $\sigma(\mathcal{L}_0)$. Well, in fact, it is enough to consider the latter case because:

Proposition 10. *$\sigma(\mathcal{L}_0)$ is the smallest σ-algebra which is closed under the shifts:*

$$\langle a \rangle_r(Q) = \{s \mid h(a, s)(Q) \geq r\}$$

That it is the smallest is obvious, but that it is closed is not [3].

Therefore, $\sigma(\mathcal{L}_0)$ is always included in Σ, since Σ has to be stable by shifts (this is equivalent to asking that $h(a, Q)$ are all Σ-measurable) and one can always 'compress' an LMP to $\sigma(\mathcal{L}_0)$. The obtained LMP is obviously bisimilar to the first since by construction states are the same and their temporal properties remain the same as well. Without loss of generality, we may and will suppose thereafter that $\Sigma = \sigma(\mathcal{L}_0)$.

6 Approximations

Now we can complete the approximation construction.

6.1 Finite-State Approximants

Let $\mathcal{S}$ be a compressed LMP $\mathcal{S} = (S, \Sigma, h)$ with $\Sigma = \sigma(\mathcal{L}_0)$, and $\mathcal{F} \subseteq \mathcal{L}_0$ be a *finite* set of formulas, set Λ to be the σ-algebra, $\sigma(\mathcal{F})$, generated by $\mathcal{F}$ on S.

We observe that by inequation (2), p is granular over Λ, so the machinery gets us a *finite-state* LMP approximant:

$$\mathcal{S} = (S, \Sigma, h) \xrightarrow{[\cdot]_\Lambda} \mathcal{S}_{\mathcal{F}} = ([S]_\Lambda, [\Lambda]_\Lambda, k)$$

which is the quotient constructed above after the appropriate projection.

There are at most $2^{|\mathcal{F}|}$ states in $\mathcal{S}_{\mathcal{F}}$, in particular it is a finite-state probabilistic transition sytem.

6.2 Convergence

We need to say how the obtained $\mathcal{S}_\mathcal{F}$ approximates $\mathcal{S}$. In the previous approach [8], approximants were always below the approximated process and hence simulated by it. It was shown that they converge in the domain of all LMPs. It is not the case here since approximants are neither above nor below $\mathcal{S}$. However, $\mathcal{S}_\mathcal{F}$ does converge to $\mathcal{S}$.

Proposition 11. *For every finite subformula-closed set of formulas $\mathcal{F} \subset \mathcal{L}_0$: $\mathcal{S}_\mathcal{F} \approx_\mathcal{F} \mathcal{S}$.*

Proof. We prove something stronger, namely that if $\theta \in \mathcal{F}$, then $\cup[\![\theta]\!]_{\mathcal{S}_\mathcal{F}} = [\![\theta]\!]_\mathcal{S}$ or equivalently that $[\![\theta]\!]_{\mathcal{S}_\mathcal{F}} = [[\![\theta]\!]_\mathcal{S}]_\Lambda$ (recall that $\Lambda = \sigma(\mathcal{F})$). This is done by induction on the structure of formulas in $\mathcal{F}$, which is why we ask $\mathcal{F}$ to be closed by subformulas.
The only interesting case is when $\theta = \langle a \rangle_r \phi$. If all states in class $[s]$ satisfy θ (equivalently if one state in $[s]$ satisfies θ), that is to say if $h(a, [\![\phi]\!]_\mathcal{S})(t) \geq r$ for all $t \in [s]$, then obviously the conditional expectation is also above r and hence $[s]$ satisfies θ since:

$$k(a, [\![\phi]\!]_{\mathcal{S}_\mathcal{F}})[s] := \mathbb{E}_p(h(a, \cup[\![\phi]\!]_{\mathcal{S}_\mathcal{F}})|\Lambda)(t) = \mathbb{E}_p(h(a, [\![\phi]\!]_\mathcal{S})|\Lambda)(t) \geq r,$$

where the first equation is by definition of k, and the second equation is by induction hypothesis. Conversely, if the conditional expectation on $[s]$ (recall that it is constant on this set) is $\geq r$, then at least one $t \in [s]$ must satisfy $h(a, [\![\phi]\!]_\mathcal{S})(t) \geq r$. Since all states in $[s]$ satisfy the same formulas of $\mathcal{F}$, then they all satisfy formula θ, as required. □

Notice that this proposition is also true for a logic extended with a greatest fixpoint operator [4,5].[7]

From Proposition 4, it follows now easily that:

Theorem 2. *If $(\mathcal{F}_i)$ is an increasing sequence of subformula-closed sets of formulas converging to the set of all formulas $\mathcal{L}_0$, then for all $c \in (0, 1)$:*

$$d^c(\mathcal{S}_{\mathcal{F}_i}, \mathcal{S}) \longrightarrow_{i\to\infty} 0.$$

We could have taken another route to prove Proposition 11. As the example 2.5 suggested, quotients constructed with conditional expections do lie between the inf- and the sup- approximants [5]:

$$\begin{aligned}
k(a, [Q])([s]_\Lambda) &:= h'(a, Q)(s) \\
&= \tfrac{1}{p([s]_\Lambda)} \textstyle\int_{[s]_\Lambda} h'(a, Q)\, dp && h'(a, Q) \text{ constant on } [s]_\Lambda \\
&= \tfrac{1}{p([s]_\Lambda)} \textstyle\int_{[s]_\Lambda} h(a, Q)\, dp && [s]_\Lambda \in \Lambda \\
&\geq \inf_{t \in [s]_\Lambda} h(a, Q)
\end{aligned}$$

[7] The proof can be found in a survey of LMP approximation (to appear in ENTCS).

The second equation holds both because $h'(a, Q)$ is constant on equivalence classes *and* because p is granular and therefore $p([s]_\Lambda) > 0$. The third equation is the characteristic property of conditional expectations. A similar type of argument allows one to reason analogously for the supremum case.

Thus another, indirect, way to prove the previous proposition, is to use this sandwiching effect and the fact that the infimum and supremum were proven to give approximations in the same sense as proposition 11 [5]. This also makes clear that the average-based approximants are better than the order-theoretic ones.

7 Conclusion

We have given an approximation technique for LMPs that has a number of good properties. It can be customized in the sense that if one is interested in a special set of formulas one can arrange the approximation so that one obtains a finite system (assuming that one had finitely many formulas) with the property that the formulas of interest are satisfied by the original system if and only if they are satisfied by the finite approximant. This brings the work much closer to the goal of using automated verification tools on continuous state space systems. This property is shared by the infima technique [5] however, unlike that result, we can also stay within the framework of traditional LMPs and avoid having to work with capacities.

The results of this paper give yet another approximation construction and one may well wonder if this is just one more in a tedious sequence of constructions that are of interest only to a small group of researchers. In fact, we feel that there are some new directions in this work whose significance extends beyond the properties of the construction. First, the idea of granularity is, we feel, significant. One of the big obstacles to the applicability of modern probability theory on general spaces to the computational setting has been the curse of non uniqueness embodied in the phrases "almost everywhere" and "almost surely" seen almost everywhere in probability theory. One can even argue that the bulk of the computer science community has worked with discrete systems to try and avoid this non uniqueness. Our use of granularity shows a new sense in which the discrete can be used to dispel the non uniqueness that arises in measure theory.

The second important direction that we feel should be emphasized is the use of averages rather than infima. This should lead to better numerical properties. More striking than that however is the fact that the simulation order is not respected by the approximants. Perhaps it suggests that some sort of non monotone approximation occurs. Similar phenomena have been observed by Martin [13] – which was the first departure from Scott's ideas of monotonicity as being one of the key requirements of computability – and also in the context of non determinate dataflow [14].

One might ask why we do not mention any analytic space property contrarily to what is done in previous papers on LMPs. In fact, analyticity is needed if one wants to use the fact that the relational definition of bisimulation is characterized by the logic. If one is happy with only the logic or the metric in order to compare or work with LMPs, there is no need for analyticity of the state space in the definition. However, if one indeed needs the analytic property of processes, the results of the present paper carry through since the quotient of an analytic space under countably many conditions is analytic, as reported in [9]. This follows essentially from well known facts about analytic spaces, see for example chapter 3 of "Invitation to C^*-algebras" by Arveson [1].

Acknowledgements. Josée Desharnais would like to acknowledge the support of NSERC and FCAR. Prakash Panangaden would like to thank NSERC for support.

References

1. W. Arveson. *An Invitation to C^*-Algebra.* Springer-Verlag, 1976.
2. G. Choquet. Theory of capacities. *Ann. Inst. Fourier (Grenoble)*, 5:131–295, 1953.
3. Vincent Danos and Josée Desharnais. Note sur les chaînes de Markov étiquetées. Unpublished (in french), 2002.
4. Vincent Danos and Josée Desharnais. A fixpoint logic for labeled Markov Processes. In Zoltan Esik and Igor Walukiewicz, editors, *Proceedings of an international Workshop FICS'03 (Fixed Points in Computer Science)*, Warsaw, 2003.
5. Vincent Danos and Josée Desharnais. Labeled Markov Processes: Stronger and faster approximations. In *Proceedings of the 18th Symposium on Logic in Computer Science*, Ottawa, 2003. IEEE.
6. J. Desharnais, A. Edalat, and P. Panangaden. Bisimulation for labelled Markov processes. *Information and Computation*, 179(2):163–193, Dec 2002. Available from `http://www.ift.ulaval.ca/~jodesharnais`.
7. J. Desharnais, V. Gupta, R. Jagadeesan, and P. Panangaden. Metrics for labeled Markov processes. In *Proceedings of CONCUR99*, Lecture Notes in Computer Science. Springer-Verlag, 1999.
8. J. Desharnais, V. Gupta, R. Jagadeesan, and P. Panangaden. Approximating continuous Markov processes. In *Proceedings of the 15th Annual IEEE Symposium On Logic In Computer Science, Santa Barbara, Californie, USA*, 2000. pp. 95-106.
9. J. Desharnais, V. Gupta, R. Jagadeesan, and P. Panangaden. Approximating labeled Markov processes. *Information and Computation*, 2003. To appear. Available from `http://www.ift.ulaval.ca/~jodesharnais`.
10. Abbas Edalat. Domain of computation of a random field in statistical physics. In C. Hankin, I. Mackie, and R. Nagarajan, editors, *Theory and Formal Methods 1994: Proceedings of the second Imperial College Department of Computing Workshop on Theory and Formal Methods*, pages 11–14. IC Press, 1994.
11. Abbas Edalat. Domain theory and integration. *Theoretical Computer Science*, 151:163–193, 1995.
12. C. Jones and G. D. Plotkin. A probabilistic powerdomain of evaluations. In *Proceedings of the Fourth Annual IEEE Symposium On Logic In Computer Science*, pages 186–195, 1989.
13. Keye Martin. The measurement process in domain theory. In *International Colloquium on Automata, Languages and Programming*, pages 116–126, 2000.

14. P. Panangaden and V. Shanbhogue. The expressive power of indeterminate dataflow primitives. *Information and Computation*, 98(1):99–131, 1992.
15. Franck van Breugel, Michael Mislove, Joël Ouaknine, and James Worrell. Probabilistic simulation, domain theory and labelled Markov processes. To appear in the Proceedings of FOSSACS'03, 2003.
16. David Williams. *Probability with Martingales*. CUP, Cambridge, 1991.

Comparative Branching-Time Semantics for Markov Chains
(Extended Abstract)

Christel Baier[1], Holger Hermanns[2,3], Joost-Pieter Katoen[2], and Verena Wolf[1]

[1] Institut für Informatik I, University of Bonn
Römerstraße 164, D-53117 Bonn, Germany
[2] Department of Computer Science, University of Twente
P.O. Box 217, 7500 AE Enschede, The Netherlands
[3] Department of Computer Science
Saarland University, D-66123 Saarbrücken, Germany

Abstract. This paper presents various semantics in the branching-time spectrum of discrete-time and continuous-time Markov chains (DTMCs and CTMCs). Strong and weak bisimulation equivalence and simulation pre-orders are covered and are logically characterised in terms of the temporal logics PCTL and CSL. Apart from presenting various existing branching-time relations in a uniform manner, our contributions are: (i) weak simulation for DTMCs is defined, (ii) weak bisimulation equivalence is shown to coincide with weak simulation equivalence, (iii) logical characterisation of weak (bi)simulations are provided, and (iv) a classification of branching-time relations is presented, elucidating the semantics of DTMCs, CTMCs and their interrelation.

1 Introduction

Equivalences and pre-orders are important means to compare the behaviour of transition systems. Prominent branching-time relations are bisimulation and simulation. Bisimulations [36] are equivalences requiring related states to exhibit identical stepwise behaviour. Simulations [30] are preorders requiring state s' to mimic s in a stepwise manner, but not necessarily the reverse, i.e., s' may perform steps that cannot be matched by s. Typically, strong and weak relations are distinguished. Whereas in *strong* (bi)simulations, each individual step needs to be mimicked, in *weak* (bi)simulations this is only required for observable steps but not for internal computations. Weak relations thus allow for stuttering.

A plethora of strong and weak (bi)simulations for labelled transition systems has been defined in the literature, and their relationship has been studied by process algebraists, most notably by van Glabbeek [22,23]. These "comparative" semantics have been extended with logical characterisations. Strong bisimulation, for instance, coincides with CTL-equivalence [13], whereas strong simulation agrees with a "preorder" on the universal (or existential) fragment of CTL [15]. Similar results hold for weak (bi)simulation where typically the next operator is omitted, which is not compatible with stuttering.

For probabilistic systems, a similar situation exists. Based on the seminal works of [31, 35], notions of (bi)simulation (see, e.g., [2,7,8,11,12,24,27,28,32,38,40,41]) for models with and without nondeterminism have been defined during the last decade, and various

R. Amadio, D. Lugiez (Eds.): CONCUR 2003, LNCS 2761, pp. 492–507, 2003.

logics to reason about such systems have been proposed (see e.g., [1,4,10,26]). This holds for both discrete probabilistic systems and variants thereof, as well as systems that describe continuous-time stochastic phenomena. In particular, in the discrete setting several slight variants of (bi)simulations have been defined, and their logical characterisations studied, e.g., [3,17,21,19,40]. Although the relationship between (bi)simulations is fragmentarily known, a clear, concise classification is – in our opinion – lacking. Moreover, continuous-time and discrete-time semantics have largely been developed in isolation, and their connection has received scant attention, if at all.

This paper attempts to study the comparative semantics of branching-time relations for probabilistic systems that do not exhibit any nondeterminism. In particular, time-abstract (or discrete-time) fully probabilistic systems (FPS) and continuous-time Markov chains (CTMCs) are considered. Strong and weak (bi)simulation relations are covered together with their characterisation in terms of the temporal logics PCTL [26] and CSL [4,10] for the discrete and continuous setting, respectively. Apart from presenting various existing branching-time relations and their connection in a uniform manner, several new results are provided. For FPSs, weak bisimulation [7] is shown to coincide with $\text{PCTL}_{\setminus X}$-equivalence, weak simulation is introduced whose kernel agrees with weak bisimulation, and the preorder weakly preserves a safe (live) fragment of $\text{PCTL}_{\setminus X}$. In the continuous-time setting, strong simulation is defined and is shown to coincide with a preorder on CSL. These results are pieced together with various results known from the literature, forming a uniform characterisation of the semantic spectrum of FPSs, CTMCs and of their interrelation.

Organisation of the Paper. Section 2 provides the necessary background. Section 3 defines strong and weak (bi)simulations. Section 4 introduces PCTL and CSL and presents the logical characterisations. Section 5 presents the branching-time spectrum. Section 6 concludes the paper. Some proofs are included in this paper; for remaining proofs, see [9].

2 Preliminaries

This section introduces the basic concepts of the Markov models considered within this paper; for a more elaborate treatment see e.g., [25,33,34]. Let *AP* be a fixed, finite set of atomic propositions.

Definition 1. *A* fully probabilistic system *(FPS) is a tuple* $\mathcal{D} = (S, \mathbf{P}, L)$ *where:*

- *S is a countable set of states*
- *$\mathbf{P} : S \times S \to [0,1]$ is a probability matrix satisfying $\sum_{s' \in S} \mathbf{P}(s,s') \in [0,1]$ for all $s \in S$*
- *$L : S \to 2^{AP}$ is a labelling function which assigns to each state $s \in S$ the set $L(s)$ of atomic propositions that are valid in s.* ■

If $\sum_{s' \in S} \mathbf{P}(s,s') = 1$, state s is called stochastic, if this sum equals zero, state s is called absorbing; otherwise, s is called sub-stochastic.

Definition 2. *A (labelled)* DTMC *is an FPS where any state is either stochastic or absorbing, i.e., $\sum_{s' \in S} \mathbf{P}(s,s') \in \{0,1\}$ for all $s \in S$.* ■

For $C \subseteq S$, $\mathbf{P}(s,C) = \sum_{s' \in C} \mathbf{P}(s,s')$ denotes the probability for s to move to a C-state. For technical reasons, $\mathbf{P}(s,\perp) = 1 - \mathbf{P}(s,S)$. Intuitively, $\mathbf{P}(s,\perp)$ denotes the probability to stay forever in s without performing any transition; although $\perp$ is not a "real" state (i.e., $\perp \notin S$), it may be regarded as a deadlock. In the context of simulation relations later on, $\perp$ is treated as an auxiliary state that is simulated by any other state. Let $S_\perp = S \cup \{\perp\}$. $\mathsf{Post}(s) = \{\, s' \mid \mathbf{P}(s,s') > 0 \,\}$ denotes the set of direct successor states of s, and $\mathsf{Post}_\perp(s) = \{\, s' \in S_\perp \mid \mathbf{P}(s,s') > 0 \,\}$, i.e., $\mathsf{Post}(s) \cup \{\, \perp \mid \mathbf{P}(s,\perp) > 0 \,\}$.

We consider FPSs and therefore also DTMCs as *time-abstract* models. The name DTMC has historical reasons. A (discrete-)timed interpretation is appropriate in settings where all state changes occur at equidistant time points. For weak relations the time-abstract view will be decisive. In contrast, CTMCs are considered as *time-aware*, as they have an explicit reference to (real-)time, in the form of transition rates which determine the stochastic evolution of the system in time.

Definition 3. *A (labelled) CTMC is a tuple $\mathcal{C} = (S, \mathbf{R}, L)$ with S and L as before, and* rate matrix *$\mathbf{R} : S \times S \to I\!R_{\geqslant 0}$ such that the exit rate $E(s) = \sum_{s' \in S} \mathbf{R}(s,s')$ is finite.* ■

As in the discrete case, $\mathsf{Post}(s) = \{\, s' \mid \mathbf{R}(s,s') > 0 \,\}$ denotes the set of direct successor states of s, and for $C \subseteq S$, $\mathbf{R}(s,C) = \sum_{s' \in C} \mathbf{R}(s,s')$ denotes the rate of moving from state s to C via a single transition.

The meaning of $\mathbf{R}(s,s') = \lambda > 0$ is that with probability $1 - e^{-\lambda \cdot t}$ the transition $s \to s'$ is enabled within the next t time units (provided that the current state is s). If $\mathbf{R}(s,s') > 0$ for more than one state s', a *race* between the outgoing transitions from s exists. The probability of s' winning this race before time t is $\frac{\mathbf{R}(s,s')}{E(s)} \cdot (1 - e^{-E(s) \cdot t})$. With $t \to \infty$ we get the time-abstract behaviour by the so-called embedded DTMC:

Definition 4. *The* embedded *DTMC of CTMC $\mathcal{C} = (S, \mathbf{R}, L)$ is given by $emb(\mathcal{C}) = (S, \mathbf{P}, L)$, where $\mathbf{P}(s,s') = \mathbf{R}(s,s')/E(s)$ if $E(s) > 0$ and $\mathbf{P}(s,s') = 0$ otherwise.* ■

A CTMC is called *uniformised* if all states in $\mathcal{C}$ have the same exit rate. Each CTMC can be transformed into a uniformised CTMCs by adding self-loops [39]:

Definition 5. *Let $\mathcal{C} = (S, \mathbf{R}, L)$ be a CTMC and let (uniformisation rate) E be a real such that $E \geqslant \max_{s \in S} E(s)$. Then, $unif(\mathcal{C}) = (S, \overline{\mathbf{R}}, L)$ is a uniformised CTMC with $\overline{\mathbf{R}}(s,s') = \mathbf{R}(s,s')$ for $s \neq s'$, and $\overline{\mathbf{R}}(s,s) = \mathbf{R}(s,s) + E - E(s)$.* ■

In *unif*$(\mathcal{C})$ all rates of self-loops are "normalised" with respect to E, such that state transitions occur with an average "pace" of E, uniform for all states of the chain. We will later see that $\mathcal{C}$ and *unif*$(\mathcal{C})$ are related by weak bisimulation.

Paths and the probability measures on paths in FPSs and CTMCs are defined by a standard construction, e.g., [25,33,34], and are omitted here.

3 Bisimulation and Simulation

We will use the subscript "d" to identify relations defined in the discrete setting (FPSs or DTMCs), and "c" for the continuous setting (CTMCs).

Definition 6. [33,35,32,24] *Let $\mathcal{D} = (S, \mathbf{P}, L)$ be a FPS and R an equivalence relation on S. R is a* strong bisimulation *on $\mathcal{D}$ iff for $s_1 \, R \, s_2$: $L(s_1) = L(s_2)$ and $\mathbf{P}(s_1, C) = \mathbf{P}(s_2, C)$ for all C in S/R. s_1 and s_2 in $\mathcal{D}$ are strongly bisimilar, denoted $s_1 \sim_d s_2$, if there exists a strong bisimulation R on $\mathcal{D}$ with $s_1 \, R \, s_2$.* ■

Definition 7. [14,28] *Let $\mathcal{C} = (S, \mathbf{R}, L)$ be a CTMC and R an equivalence relation on S. R is a* strong bisimulation *on $\mathcal{C}$ if for $s_1 \, R \, s_2$: $L(s_1) = L(s_2)$ and $\mathbf{R}(s_1, C) = \mathbf{R}(s_2, C)$ for all C in S/R. s_1 and s_2 in $\mathcal{C}$ are strongly bisimilar, denoted $s_1 \sim_c s_2$, if there exists a strong bisimulation R on $\mathcal{C}$ with $s_1 \, R \, s_2$.* ■

As $\mathbf{R}(s, C) = \mathbf{P}(s, C) \cdot E(s)$, the condition on the cumulative rates can be reformulated as (i) $\mathbf{P}(s_1, C) = \mathbf{P}(s_2, C)$ for all $C \in S/R$ and (ii) $E(s_1) = E(s_2)$. Hence, $\sim_c$ agrees with $\sim_d$ in the embedded DTMC provided that exit rates are treated as additional atomic propositions. By the standard construction, it can be shown that $\sim_d$ and $\sim_c$ are the coarsest strong bisimulations.

Proposition 1. *For CTMC $\mathcal{C} = (S, \mathbf{R}, L)$:*

1. *$s_1 \sim_c s_2$ implies $s_1 \sim_d s_2$ in emb($\mathcal{C}$), for any state $s_1, s_2 \in S$.*
2. *if $\mathcal{C}$ is uniformised then $\sim_c$ coincides with $\sim_d$ in emb($\mathcal{C}$).*

Definition 8. *A* distribution *on set S is a function $\mu : S \to [0, 1]$ with $\sum_{s \in S} \mu(s) \leqslant 1$.* ■

We put $\mu(\perp) = 1 - \sum_{s \in S} \mu(s)$. *Distr*$(S)$ denotes the set of all distributions on S. Distribution μ on S is called stochastic if $\mu(\perp) = 0$. For simulation relations, the concept of weight functions is important.

Definition 9. [29,31] *Let S be a set, $R \subseteq S \times S$, and $\mu, \mu' \in Distr(S)$. A* weight function *for μ and μ' with respect to R is a function $\Delta : S_\perp \times S_\perp \to [0, 1]$ such that:*

1. *$\Delta(s, s') > 0$ implies $s \, R \, s'$ or $s = \perp$*
2. *$\mu(s) = \sum_{s' \in S_\perp} \Delta(s, s')$ for any $s \in S_\perp$*
3. *$\mu'(s') = \sum_{s \in S_\perp} \Delta(s, s')$ for any $s' \in S_\perp$*

We write $\mu \sqsubseteq_R \mu'$ (or simply $\sqsubseteq$, if R is clear from the context) iff there exists a weight function for μ and μ' with respect to R. $\sqsubseteq_R$ is the lift of R to distributions. ■

Definition 10. [31] *Let $\mathcal{D} = (S, \mathbf{P}, L)$ be a FPS and $R \subseteq S \times S$. R is a* strong simulation *on $\mathcal{D}$ if for all $s_1 \, R \, s_2$: $L(s_1) = L(s_2)$ and $\mathbf{P}(s_1, \cdot) \sqsubseteq_R \mathbf{P}(s_2, \cdot)$. s_2 strongly simulates s_1 in $\mathcal{D}$, denoted $s_1 \precsim_d s_2$, iff there exists a strong simulation R on $\mathcal{D}$ such that $s_1 \, R \, s_2$.* ■

It is not difficult to see that $s_1 \sim_d s_2$ implies $s_1 \precsim_d s_2$. For a DTMC without absorbing states, $\precsim_d$ is symmetric and coincides with $\sim_d$, see [31].

Proposition 2. [5,16] *For any FPS, $\precsim_d \cap \precsim_d^{-1}$ coincides with $\sim_d$.*

Definition 11. *Let $\mathcal{C} = (S, \mathbf{R}, L)$ be a CTMC and $R \subseteq S \times S$. R is a* strong simulation *on $\mathcal{C}$ if for all $s_1 \, R \, s_2$: $L(s_1) = L(s_2)$, $\mathbf{P}(s_1, \cdot) \sqsubseteq_R \mathbf{P}(s_2, \cdot)$ and $E(s_1) \leqslant E(s_2)$. s_2 strongly simulates s_1 in $\mathcal{C}$, denoted $s_1 \precsim_c s_2$, iff there exists a strong simulation R on $\mathcal{C}$ such that $s_1 \, R \, s_2$.* ■

Proposition 3. *For any CTMC $\mathcal{C}$:*

1. *$s_1 \sim_c s_2$ implies $s_1 \precsim_c s_2$, for any state $s_1, s_2 \in S$.*
2. *$s_1 \precsim_c s_2$ implies $s_1 \precsim_d s_2$ in emb($\mathcal{C}$), for any state $s_1, s_2 \in S$.*
3. *$\precsim_c \cap \precsim_c^{-1}$ coincides with $\sim_c$.*
4. *if $\mathcal{C}$ is uniformised then $\precsim_c$ is symmetric and coincides with $\sim_c$.*

Weak Bisimulation. In this paper, we only consider weak bisimulation which relies on branching bisimulation in the style of van Glabbeek and Weijland and only abstracts from stutter-steps inside the equivalence classes. While for ordinary transition systems branching bisimulation is strictly finer than Milner's observational equivalence, they agree for FPSs [7], and thus for CTMCs.

Let $\mathcal{D} = (S, \mathbf{P}, L)$ be a DTMC and $R \subseteq S \times S$ an equivalence relation. Any transition $s \to s'$ where s and s' are R-equivalent is an R-silent move. Let Silent_R denote the set of states $s \in S$ for which $\mathbf{P}(s, [s]_R) = 1$, i.e., all stochastic states that do not have a successor state outside their R-equivalence class. For any state $s \notin \mathsf{Silent}_R$, $s' \in S$ with $s' \notin [s]_R$:

$$\mathbf{P}(s, s' \mid \text{no } R\text{-silent move}) = \frac{\mathbf{P}(s, s')}{1 - \mathbf{P}(s, [s]_R)}$$

denotes the conditional probability to move from s to s' via a single transition under the condition that from s *no* transition inside $[s]_R$ is taken. Thus, either a transition is taken to another equivalence class under R or, for sub-stochastic states, the system deadlocks. For $C \subseteq S$ with $C \cap [s]_R = \varnothing$ let $\mathbf{P}(s, C \mid \text{no } R\text{-silent move}) = \sum_{s' \in C} \mathbf{P}(s, s' \mid \text{no } R\text{-silent move})$.

Definition 12. [7] *Let $\mathcal{D} = (S, \mathbf{P}, L)$ be a FPS and R an equivalence relation on S. R is a* weak bisimulation *on $\mathcal{D}$ if for all $s_1 \, R \, s_2$:*

1. $L(s_1) = L(s_2)$
2. *If $s_1, s_2 \notin \mathsf{Silent}_R$ then:* $\mathbf{P}(s_1, C \mid \text{no } R\text{-silent move}) = \mathbf{P}(s_2, C \mid \text{no } R\text{-silent move})$ *for all $C \in S/R$, $C \neq [s_1]_R$.*
3. *If $s_1 \in \mathsf{Silent}_R$ and $s_2 \notin \mathsf{Silent}_R$ then s_1 can reach a state $s' \in [s_1]_R \setminus \mathsf{Silent}_R$ with positive probability.*

s_1 and s_2 in $\mathcal{D}$ are weakly bisimilar, denoted $s_1 \approx_d s_2$, iff there exists a weak bisimulation R on $\mathcal{D}$ such that $s_1 \, R \, s_2$. ■

By the third condition, for any R-equivalence class C, either all states in C are R-silent (i.e., $\mathbf{P}(s, C) = 1$ for $s \in C$) or for $s \in C$ there is a path fragment that ends in an equivalence class that differs from C.

Example 1. For the following DTMC (where equally shaded states are equally labeled) the reachability condition is needed to establish a weak bisimulation for states s_1 and s_2:

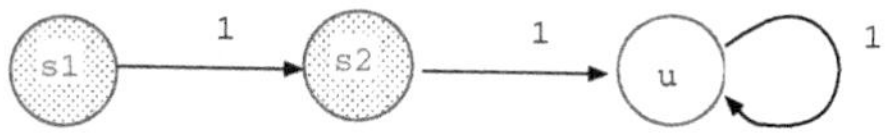

We have $s_1 \approx_d s_2$, and s_1 is $\approx_d$-silent while s_2 is not. Here, the reachability condition is obviously fulfilled. This condition can, however, not be dropped: otherwise s_1 and s_2 would be weakly bisimilar to an absorbing state with the same labeling. ■

Definition 13. [12] *Let $\mathcal{C} = (S, \mathbf{R}, L)$ be a CTMC and R an equivalence relation on S. R is a* weak bisimulation *on $\mathcal{C}$ if for all $s_1 \, R \, s_2$: $L(s_1) = L(s_2)$ and $\mathbf{R}(s_1, C) = \mathbf{R}(s_2, C)$ for all $C \in S/R$ with $C \neq [s_1]_R$. s_1 and s_2 in $\mathcal{C}$ are weakly bisimilar, denoted $s_1 \approx_c s_2$, iff there exists a weak bisimulation R on $\mathcal{C}$ such that $s_1 \, R \, s_2$.* ■

Proposition 4. *For any CTMC $\mathcal{C}$:*

1. *$\sim_c$ is strictly finer than $\approx_c$.*
2. *if $\mathcal{C}$ is uniformised then $\approx_c$ coincides with $\sim_c$.*
3. *$\approx_c$ coincides with $\approx_c$ in unif$(\mathcal{C})$.*

The last result can be strengthened as follows. Any state s in $\mathcal{C}$ is weakly bisimilar to s considered as a state in *unif*$(\mathcal{C})$. (For this, consider the disjoint union of $\mathcal{C}$ and *unif*$(\mathcal{C})$ as a single CTMC.)

Proposition 5. *For CTMC $\mathcal{C}$ with $s_1, s_2 \in S$: $s_1 \approx_c s_2$ implies $s_1 \approx_d s_2$ in emb$(\mathcal{C})$.*

Proof. Let R be a weak bisimulation on $\mathcal{C}$. We show that R is a weak bisimulation on *emb*$(\mathcal{C})$ as follows. First, observe that all R-equivalent states have the same labelling. Assume $s_1 \, R \, s_2$ and $B = [s_1]_R = [s_2]_R$. Distinguish two cases. (i) s_1 is R-silent, i.e., $\mathbf{P}(s_1, B) = 1$. Hence, $\mathbf{R}(s_1, B) = E(s_1)$ and therefore $0 = \mathbf{R}(s_1, C) = \mathbf{R}(s_2, C)$ for all $C \in S/R$ with $C \neq B$. So, $\mathbf{P}(s_2, B) = 1$. (ii) Neither s_1 nor s_2 is R-silent, i.e., $\mathbf{P}(s_i, B) < 1$, for i=1, 2. Note that:

$$E(s_i) = \sum_{\substack{C \in S/R \\ C \neq B}} \mathbf{R}(s_i, C) \; + \; \mathbf{R}(s_i, B)$$

As $s_1 \approx_c s_2$, $\mathbf{R}(s_1, C) = \mathbf{R}(s_2, C)$ for all $C \in S/R$ with $C \neq B$. Hence, $\sum_{C \in S/R, C \neq B} \mathbf{R}(s_1, C) = \sum_{C \in S/R, C \neq B} \mathbf{R}(s_2, C)$ and therefore $E(s_1) - \mathbf{R}(s_1, B) = E(s_2) - \mathbf{R}(s_2, B)$ (*). For any $C \in S/R$ with $C \neq B$ we derive:

$$\mathbf{P}(s_1, C \mid \text{no } R\text{-silent move}) \overset{def}{=} \frac{\mathbf{P}(s_1, C)}{1 - \mathbf{P}(s_1, B)} = \frac{E(s_1) \cdot \mathbf{P}(s_1, C)}{E(s_1) - E(s_1) \cdot \mathbf{P}(s_1, B)}$$

$$\overset{def. \mathbf{R}}{=} \frac{\mathbf{R}(s_1, C)}{E(s_1) - \mathbf{R}(s_1, B)} \overset{(*), s_1 \approx_c s_2}{=} \frac{\mathbf{R}(s_2, C)}{E(s_2) - \mathbf{R}(s_2, B)} = \frac{\mathbf{P}(s_2, C)}{1 - \mathbf{P}(s_2, B)}$$

which, by definition, equals $\mathbf{P}(s_2, C \mid \text{no } R\text{-silent move})$. So, $s_1 \approx_d s_2$. ■

Remark 1. Prop. 1.2 states that for a uniformised CTMC, $\sim_c$ coincides with $\sim_d$ on the embedded DTMC. The analogue for $\approx_c$ does not hold, as, e.g., in the uniformised CTMC of Example 1 we have $s_1 \approx_d s_2$ but $s_1 \not\approx_c s_2$ as $\mathbf{R}(s_1, [u]) \neq \mathbf{R}(s_2, [u])$. Intuitively, although s_1 and s_2 have the same time-abstract behaviour (up to stuttering) they have distinct timing behaviour. s_1 is "slower than" s_2 as it has to perform a stutter step prior to an observable step (from s_2 to u) while s_2 can immediately perform the latter step. Note that by Prop 4.2 and Prop. 1.2, $\approx_c$ coincides with $\sim_d$ for uniformised CTMCs. In fact, Prop. 5 can be strengthened in the following way: $\approx_c$ is the coarsest equivalence finer than $\approx_d$ such that $s_1 \approx_c s_2$ implies $\mathbf{R}(s_1, S \setminus [s_1]) = \mathbf{R}(s_2, S \setminus [s_2])$. ■

Weak Simulation. Weak simulation on FPSs is inspired by our work on CTMCs [8]. Roughly speaking, $s_1 \precsim s_2$ if the successor states of s_1 and s_2 can be grouped into subsets U_i and V_i (assume, for simplicity, $U_i \cap V_i = \varnothing$). All transitions from s_i to V_i are viewed as stutter-steps, i.e., internal transitions that do not change the labelling and respect $\precsim$. To that end, any state in V_1 is required to be simulated by s_2 and, symmetrically, any state in V_2 simulates s_1. Transitions from s_i to U_i are regarded as visible steps. Accordingly, we require that the distributions for the conditional probabilities $u_1 \mapsto \mathbf{P}(s_1, u_1)/K_1$ and $u_2 \mapsto \mathbf{P}(s_2, u_2)/K_2$ to move from s_i to U_i are related via a weight function (as for $\precsim_d$). K_i denotes the total probability to move from s_i to a state in U_i in a single step. For technical reasons, we allow $\perp \in U_i$ and $\perp \in V_i$.

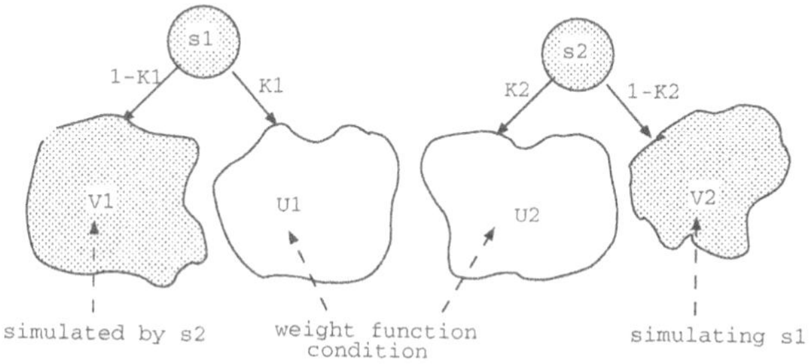

Definition 14. *Let $\mathcal{D} = (S, \mathbf{P}, L)$ be a FPS and $R \subseteq S \times S$. R is a* weak simulation *on $\mathcal{D}$ iff for $s_1 \, R \, s_2$: $L(s_1) = L(s_2)$ and there exist functions $\delta_i : S_\perp \to [0,1]$ and sets $U_i, V_i \subseteq S_\perp$ (i=1, 2) with*

$$U_i = \{ u_i \in \mathsf{Post}_\perp(s_i) \mid \delta_i(u_i) > 0 \} \text{ and } V_i = \{ v_i \in \mathsf{Post}_\perp(s_i) \mid \delta_i(v_i) < 1 \}$$

such that:

1. *(a) $v_1 \, R \, s_2$ for all $v_1 \in V_1$, $v_1 \neq \perp$, and (b) $s_1 \, R \, v_2$ for all $v_2 \in V_2$, $v_2 \neq \perp$*
2. *there exists a function $\Delta : S_\perp \times S_\perp \to [0,1]$ such that:*
   a) $\Delta(u_1, u_2) > 0$ implies $u_1 \in U_1$, $u_2 \in U_2$ and either $u_1 \, R \, u_2$ or $u_1 = \perp$,
   b) if $K_1 > 0$ and $K_2 > 0$ then for all states $w \in S$:

$$K_1 \cdot \sum_{u_2 \in U_2} \Delta(w, u_2) = \delta_1(w) \cdot \mathbf{P}(s_1, w), \quad K_2 \cdot \sum_{u_1 \in U_1} \Delta(u_1, w) = \delta_2(w) \cdot \mathbf{P}(s_2, w)$$

   where $K_i = \sum_{u_i \in U_i} \delta_i(u_i) \cdot \mathbf{P}(s_i, u_i)$ for i=1, 2
3. *for $u_1 \in U_1$, $u_1 \neq \perp$ there exists a path fragment $s_2, w_1, \ldots, w_n, u_2$ such that $n \geqslant 0$, $s_1 \, R \, w_j$, $0 < j \leqslant n$, and $u_1 \, R \, u_2$.*

s_2 weakly simulates s_1 in $\mathcal{D}$, denoted $s_1 \precsim_d s_2$, iff there exists a weak simulation R on $\mathcal{D}$ such that $s_1 \, R \, s_2$. ■

Note the correspondence to $\approx_d$ (cf. Def. 12), where $[s_1]_R$ plays the role of V_1, while the successors outside $[s_1]_R$ play the role of U_1, and the same for s_2, V_2 and U_2.

Example 2. In the following FPS we have $s_1 \precsim_d s_2$:

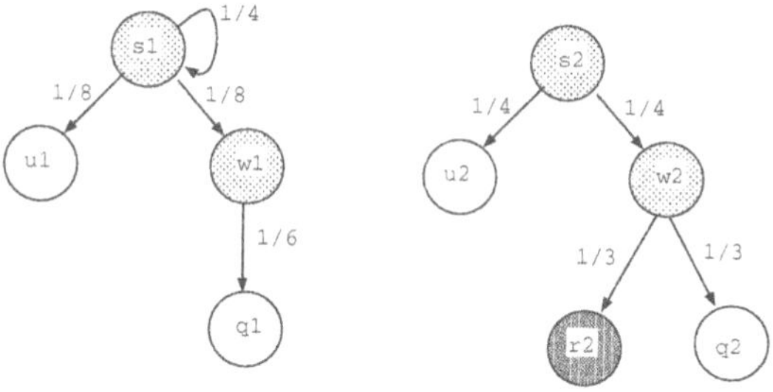

First, observe that $w_1 \precsim_d w_2$ since $R = \{ (q_1, q_2), (w_1, w_2) \}$ is a weak simulation, as we may deal with

- δ_1 the characteristic function of $U_1 = \{ q_1, \bot \}$ (and, thus, $V_1 = \varnothing$ and $K_1 = 1$)
- δ_2 the characteristic function of $U_2 = \{ r_2, q_2, \bot \}$ (and $V_2 = \varnothing$ and $K_2 = 1$)

and the weight function $\Delta(q_1, q_2) = \Delta(\bot, q_2) = \frac{1}{6}, \Delta(\bot, r_2) = \Delta(\bot, \bot) = \frac{1}{3}$. To establish a weak simulation for (s_1, s_2) consider the relation:

$$R = \{ (s_1, s_2), \ (u_1, u_2), \ (w_1, w_2), \ (q_1, q_2) \}$$

and put $V_1 = \{ \bot, s_1 \}$ and $V_2 = \varnothing$ while $U_i = \{ u_i, w_i, \bot \}$ where $\delta_1(\bot) = 1/2$, $\delta_i(u_i) = \delta_i(w_i) = \delta_2(\bot) = 1$. Then, $K_1 = \frac{1}{8} + \frac{1}{8} + \frac{1}{2} \cdot \frac{1}{2} = \frac{1}{2}$, $K_2 = \frac{1}{4} + \frac{1}{4} + \frac{1}{2} = 1$. This yields the following distribution for the U-successors of s_1 and s_2: $u_1 : \frac{1}{4}$, $w_1 : \frac{1}{4}$, $\bot : \frac{1}{2}$, $u_2 : \frac{1}{4}$, $w_2 : \frac{1}{4}$, and $\bot : \frac{1}{2}$. Note that, e.g., $\frac{\delta_1(u_1) \cdot \mathbf{P}(s_1, u_1)}{K_1} = \frac{1}{4}$ and $\frac{\delta_1(\bot) \cdot \mathbf{P}(s_1, \bot)}{K_1} = \frac{1}{2}$. Hence, an appropriate weight function is: $\Delta(u_1, u_2) = \Delta(w_1, w_2) = \frac{1}{4}$, $\Delta(\bot, \bot) = \frac{1}{2}$, and $\Delta(\cdot) = 0$ for the remaining cases. Thus, according to Def. 14, R is a weak simulation. ■

Proposition 6. *For any FPS $\mathcal{D}$: $s_1 \approx_d s_2$ implies $s_1 \precsim\!\!\approx_d s_2$, and $s_1 \precsim_d s_2$ implies $s_1 \precsim\!\!\approx_d s_2$.*

Definition 15. [8] *Let $\mathcal{C} = (S, \mathbf{R}, L)$ be a CTMC and $R \subseteq S \times S$. R is a* weak simulation *on $\mathcal{C}$ iff for $s_1 \, R \, s_2$: $L(s_1) = L(s_2)$ and there exist $\delta_i : S \to [0, 1]$ and $U_i, V_i \subseteq S$ (i=1, 2) satisfying conditions 1. and 2. of Def. 14 (ignoring $\bot$) and the rate condition:*

$$\sum_{u_1 \in U_1} \delta_1(u_1) \cdot \mathbf{R}(s_1, u_1) \leqslant \sum_{u_2 \in U_2} \delta_2(u_2) \cdot \mathbf{R}(s_2, u_2)$$

s_2 weakly simulates s_1 in $\mathcal{C}$, denoted $s_1 \precsim\!\!\approx_c s_2$, iff there exists a weak simulation R on $\mathcal{C}$ such that $s_1 \, R \, s_2$. ■

The condition on the rates which replaces the reachability condition in FPSs states that s_2 is "faster than" s_1 in the sense that the total rate to move from s_2 to (the δ_2-part of) the U_2-states is at least the total rate to move from s_1 to (the δ_1-part of) the U_1-states. Note that $K_i \cdot E(s_i) = \sum_{u_i \in U_i} \delta_i(u_i) \cdot \mathbf{R}(s_i, u_i)$. Hence, the condition in Def. 15 can be rewritten as $K_1 \cdot E(s_1) \leqslant K_2 \cdot E(s_2)$. In particular, $K_2 = 0$ implies $K_1 = 0$. Therefore, a reachability condition as for weak simulation on FPSs is not needed here.

Proposition 7. *For CTMC $\mathcal{C}$ and states $s_1, s_2 \in S$:*

1. *$s_1 \precsim\!\!\approx_c s_2$ implies $s_1 \precsim\!\!\approx_d s_2$ in emb($\mathcal{C}$).*
2. *$s_1 \approx_c s_2$ implies $s_1 \precsim\!\!\approx_c s_2$.*
3. *$\precsim\!\!\approx_c$ coincides with $\precsim_c$ in unif($\mathcal{C}$).*

A few remarks are in order. Although $\precsim_c$ and $\precsim_d$ coincide for uniformised CTMCs (as $\precsim_c$ agrees with $\sim_c$, $\sim_c$ agrees with $\sim_d$, and $\sim_d$ agrees with $\precsim_d$), this does not hold for $\precsim\!\!\approx_d$ and $\precsim\!\!\approx_c$. For example, in:

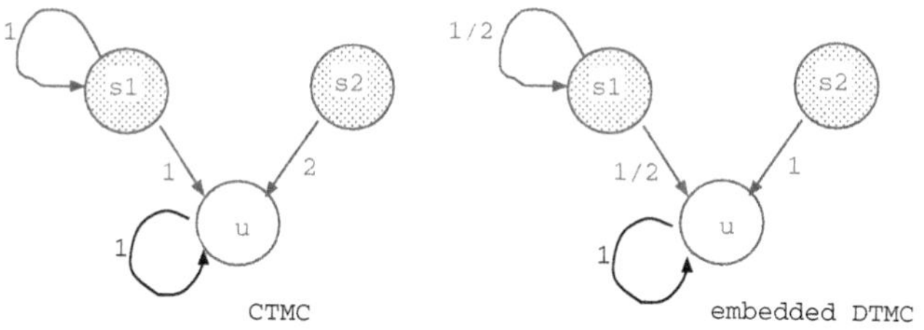

$s_2 \precapprox_d s_1$ in the embedded DTMC (on the right), but $s_2 \not\precapprox_c s_1$ in the CTMC (on the left), as the rate condition in Def. 15 is violated. Secondly, note that the analogue of Prop. 7.3 for $\precapprox_c$ does not hold. This can be seen by considering the above embedded DTMC (on the right) as a uniformised CTMC. Finally, we note that although for uniformised CTMCs, $\sim_c$ and $\approx_c$ agree, a similar result for the simulation preorders does not hold. An example CTMC for which $s_1 \precapprox_c s_2$ but $s_1 \not\precsim_c s_2$ is depicted on the left. The fact that $s_1 \not\precsim_c s_2$ follows from the weight function condition in Def. 11. To see that $s_1 \precapprox_c s_2$, consider the reflexive closure R of $\{(s_1, s_2)\}$ and the partitioning $V_1 = \{s_2\}$, $V_2 = \{s_1\}$ and $U_1 = U_2 = \{u\}$ for which the conditions of a weak simulation are fulfilled.

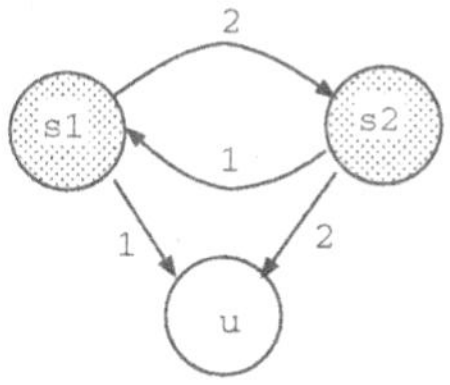

Theorem 1.

1. *For any FPS, weak simulation equivalence $\precapprox_d \cap \precapprox_d^{-1}$ coincides with $\approx_d$.*
2. *For any CTMC, weak simulation equivalence $\precapprox_c \cap \precapprox_c^{-1}$ coincides with $\approx_c$.*

4 Logical Characterisations

PCTL. In Probabilistic CTL (PCTL) [26], state-formulas are interpreted over states of a FPS and path-formulas are interpreted over paths (i.e., sequences of states) in a FPS. The syntax of PCTL is as follows[1], where $\trianglelefteq \in \{\leqslant, \geqslant\}$:

$$\Phi ::= \mathrm{tt} \;\Big|\; a \;\Big|\; \Phi \wedge \Phi \;\Big|\; \neg\Phi \;\Big|\; \mathcal{P}_{\trianglelefteq p}(X\Phi) \;\Big|\; \mathcal{P}_{\trianglelefteq p}(\Phi\,\mathcal{U}\,\Phi) \;\Big|\; \mathcal{P}_{\trianglelefteq p}(\Phi\,\mathcal{W}\,\Phi)$$

where $p \in [0,1]$ and $a \in AP$. The satisfaction relation $\models$ is similar to CTL, where $s \models \mathcal{P}_{\trianglelefteq p}(\varphi)$ iff $\Pr(s,\varphi) \trianglelefteq p$. Here, $\Pr(s,\varphi)$ denotes the probability measure of the set of paths starting in state s fulfilling path-formula φ. As in CTL, X is the next-step operator, and the path-formula $\Phi\,\mathcal{U}\,\Psi$ asserts that Ψ will eventually be satisfied and that at all preceding states Φ holds (strong until). $\mathcal{W}$ is its weak counterpart, and does not require Ψ to eventually become true. The until-operator and the weak until-operator are closely related. For any PCTL-formula Φ and Ψ the following two formulae are equivalent:

$$\mathcal{P}_{\geqslant p}(\Phi\,\mathcal{W}\,\Psi) \;\equiv\; \mathcal{P}_{\leqslant 1-p}((\neg\Psi)\,\mathcal{U}\,\neg(\Phi \vee \Psi)).$$

A similar equivalence holds when the weak until- and the until-operator are swapped.

[1] The bounded until-operator [26] is omitted here as for weak relations, FPSs are viewed as being time-abstract. For the strong relations on FPSs, this operator could, however, be considered without any problem.

CSL. Continuous Stochastic Logic (CSL) [10] is a variant of the (identically named) logic by Aziz *et al.* [4] and extends PCTL by operators that reflect the real-time nature of CTMCs: a time-bounded until-operator and a steady-state operator. We focus here on a fragment of CSL where the time bounds of (weak) until are of the form "$\leqslant t$"; other time bounds can be handled by mappings on this case, cf. [6]. The syntax of CSL is, for real t, or $t = \infty$:

$$\Phi ::= \mathrm{tt} \;\Big|\; a \;\Big|\; \Phi \wedge \Phi \;\Big|\; \neg\Phi \;\Big|\; \mathcal{P}_{\trianglelefteq p}(X^{\leqslant t}\Phi) \;\Big|\; \mathcal{P}_{\trianglelefteq p}(\Phi\,\mathcal{U}^{\leqslant t}\,\Phi) \;\Big|\; \mathcal{P}_{\trianglelefteq p}(\Phi\,\mathcal{W}^{\leqslant t}\,\Phi) \;\Big|\; \mathcal{S}_{\trianglelefteq p}(\Phi)$$

To have a well-defined steady-state operator it is assumed that the steady-state probabilities in the CTMC do exist for any starting state. Intuitively, $\mathcal{S}_{\trianglelefteq p}(\Phi)$ asserts that on the long run, the probability for a Φ-state meets the bound $\trianglelefteq p$. The path-formula $\Phi\,\mathcal{U}^{\leqslant t}\,\Psi$ asserts that Ψ is satisfied at some time instant before t and that at all preceding time instants Φ holds (strong until). The connection between the until-operator and the weak until-operator is as in PCTL.

Logical Characterisation of Bisimulation. In both the discrete and the continuous setting, strong bisimulation ($\sim_d$ and $\sim_c$) coincides with logical equivalence (in PCTL and CSL, respectively) [3,6,19]. For weak bisimulation, the next-step operator is ignored, as it is not invariant with respect to stuttering. Let PCTL$_{\setminus X}$ denote the fragment of PCTL without the next-step operator; similarly, CSL$_{\setminus X}$ is defined. PCTL$_{\setminus X}$-equivalence (denoted $\equiv_{\text{PCTL}_{\setminus X}}$) and CSL$_{\setminus X}$-equivalence ($\equiv_{\text{CSL}_{\setminus X}}$) are defined in the obvious way.

Theorem 2. *For any FPS: $\approx_d$ coincides with PCTL$_{\setminus X}$-equivalence.*

Proof. By structural induction on the syntax of PCTL$_{\setminus X}$-formulae. We only consider the until operator. Let $\varphi = \Phi_1\,\mathcal{U}\,\Phi_2$. By the induction hypothesis we may assume that $Sat(\Phi_i)$ for $i{=}1,2$ is a disjoint union of equivalence classes under $\approx_d$. Let $B = [s]_{\approx_d}$. Then, $B \cap Sat(\Phi_i) = \varnothing$ or $B \subseteq Sat(\Phi_i)$. Only the cases $B \subseteq Sat(\Phi_1)$ and $B \cap Sat(\Phi_2) = \varnothing$ are of interest; otherwise, $\Pr(s_1, \varphi) = \Pr(s_2, \varphi) \in \{0, 1\}$ for all $s_1, s_2 \in B$. Let S' be the set of states that reach a Φ_2-state via a (non-empty) Φ_1-path, i.e., $S' = \{\, s \in Sat(\Phi_1) \setminus Sat(\Phi_2) \mid \Pr(s, \varphi) > 0 \,\}$. It follows that S' is the disjoint union of equivalence classes under $\approx_d$.

We first observe the following. For $s \notin S'$, $\Pr(s, \varphi) \in \{0, 1\}$. For $s \in S'$, the vector $\big(\Pr(s, \varphi)\big)_{s \in S'}$ is the *unique* solution of the equation system:

$$x_s = \mathbf{P}(s, Sat(\Phi_2)) + \sum_{s' \in Sat(\Phi_1) \setminus Sat(\Phi_2)} \mathbf{P}(s, s') \cdot x_{s'} \tag{1}$$

For any $\approx_d$-equivalence class $B \subseteq S'$, select $s_B \in B$ such that $\mathbf{P}(s_B, B) < 1$. Such state is guaranteed to exist, since if $\mathbf{P}(s, B)$ would equal 1 for any $s \in B$ then none of the B-states can reach a Φ_2-state, contradicting being in S'. Now consider the unique solution $(x_B)_{B \in S/\approx_d, B \subseteq S'}$ of the equation system:

$$x_B = \mathbf{P}(s_B, Sat(\Phi_2)) + \sum_{\substack{C \in S/\approx_d \\ C \subseteq S'}} \mathbf{P}(s_B, C) \cdot x_C.$$

A calculation shows that the vector $(x_s)_{s \in S'}$ where $x_s = x_B$ if $s \in B$ is a solution to (1). Hence, $x_B = \Pr(s, \varphi)$ for all states $s \in B$.

The fact that PCTL$_{\setminus X}$-equivalence implies $\approx_d$ is proven as follows. W.l.o.g. we assume S to be finite and that any equivalence class C under $\equiv_{\text{PCTL}_{\setminus X}}$ is represented by a PCTL$_{\setminus X}$-formula Φ_C. (for infinite-state CTMCs approximations of master-formulae can be used). For $\text{PCTL}_{\setminus X}$ equivalence classes B and C with $B \neq C$, consider the path formulae $\varphi = \Phi_B \,\mathcal{U}\, \Phi_C$ and $\psi = \diamond\neg\Phi_B$. Then, $\Pr(s_1, \varphi) = \Pr(s_2, \varphi)$ and $\Pr(s_1, \psi) = \Pr(s_2, \psi)$ for any $s_1, s_2 \in B$. In particular, if $\mathbf{P}(s, B) < 1$ for some $s \in B$ then $\Pr(s, \psi) > 0$. Hence, for any $s' \in B$ there exists a path leading from s' to a state not in B. Assume that $s_1, s_2 \in B$ and that $\mathbf{P}(s_i, B) < 1$ for i=1, 2. Then:

$$\Pr(s_i, \varphi) \;=\; \frac{\mathbf{P}(s_i, C)}{1 - \mathbf{P}(s_i, B)}.$$

This is justified as follows. If $\Pr(s_i, \varphi) = 0$, then obviously $\mathbf{P}(s_i, C) = 0$. Otherwise, by instantiating the equation system in (1) with $S' = B$, $\Phi_2 = \Phi_C$, $\Phi_1 = \Phi_B$ it can easily be verified that the vector with the values $x_s = \frac{\mathbf{P}(s,C)}{1-\mathbf{P}(s,B)}$ (for $s \in B$) is a solution. ∎

Proposition 8. *For CTMC $\mathcal{C}$, s in $\mathcal{C}$, and $\text{CSL}_{\setminus X}$-formula Φ: $s \models \Phi$ iff $s \models \Phi$ in unif$(\mathcal{C})$.*

Proof. By induction on the syntax of Φ. For the propositional fragment the result is obvious. For the $\mathcal{S}$- and $\mathcal{P}$-operator, we exploit the fact that steady-state and transient distributions in $\mathcal{C}$ and *unif*$(\mathcal{C})$ are identical, and that the semantics of $\mathcal{U}^{\leqslant t}$ and $\mathcal{W}^{\leqslant t}$ agrees with transient distributions [6]. ∎

Proposition 9. *For any uniformised CTMC: $\equiv_{\text{CSL}}$ coincides with $\equiv_{\text{CSL}_{\setminus X}}$.*

Proof. The direction "⇒" is obvious. We prove the other direction. Assume CTMC $\mathcal{C}$ is uniformised and s_1, s_2 be states in $\mathcal{C}$. From Prop. 1.1 and the logical characterisations of $\sim_c$ and $\sim_d$ it follows:

$$s_1 \equiv_{\text{CSL}} s_2 \;\text{ iff }\; s_1 \sim_c s_2 \;\text{ iff }\; s_1 \sim_d s_2 \;\text{ iff }\; s_1 \equiv_{\text{PCTL}} s_2.$$

Hence, it suffices to show that $\equiv_{\text{CSL}_{\setminus X}}$ implies $\equiv_{\text{PCTL}_{\setminus X}}$ (for uniformised CTMC). This is done by structural induction on the syntax of PCTL-formulae. Clearly, only the next step operator is of interest. Consider PCTL-path formula $\varphi = X\Phi$. By induction hypothesis $Sat(\Phi)$ is a (countable) union of equivalence classes of $\equiv_{\text{CSL}_{\setminus X}}$. In the following, we establish for $s_1 \equiv_{\text{CSL}_{\setminus X}} s_2$:

$$\mathbf{P}(s_1, Sat(\Phi)) = \mathbf{P}(s_2, Sat(\Phi)) \text{ that is } \Pr(s_1, X\Phi) = \Pr(s_2, X\Phi).$$

Let $B = [s_1]_{\equiv_{\text{CSL}_{\setminus X}}} = [s_2]_{\equiv_{\text{CSL}_{\setminus X}}}$. First observe that $\mathbf{P}(s_1, B) = \mathbf{P}(s_2, B)$; otherwise, if, e.g., $\mathbf{P}(s_1, B) < \mathbf{P}(s_2, B)$ one would have $\Pr(s_1, \diamond^{\leqslant t}\neg\Phi_B) < \Pr(s_2, \diamond^{\leqslant t}\neg\Phi_B)$ for some sufficiently small t, contradicting $s_1 \equiv_{\text{CSL}_{\setminus X}} s_2$. As in the proof of Theorem 2 we assume a finite state space and that any $\equiv_{\text{CSL}_{\setminus X}}$-equivalence class C can be characterised by $\text{CSL}_{\setminus X}$ formula Φ_C. Distinguish:

- $\mathbf{P}(s_1, B) = \mathbf{P}(s_2, B) < 1$. Using the same arguments as in the proof of Theorem 2 we obtain:

$$\Pr(s_i, \Phi_B \,\mathcal{U}\, \Phi) \;=\; \frac{\mathbf{P}(s_i, Sat(\Phi))}{1 - \mathbf{P}(s_1, B)}, \quad i = 1, 2.$$

  As $s_1 \equiv_{\text{CSL}_{\setminus X}} s_2$ and $\Phi_B \,\mathcal{U}\, \Phi$ is a $\text{CSL}_{\setminus X}$-path formula we get: $\Pr(s_1, \Phi_B \,\mathcal{U}\, \Phi) = \Pr(s_2, \Phi_B \,\mathcal{U}\, \Phi)$. Since $\mathbf{P}(s_1, B) = \mathbf{P}(s_2, B)$, it follows $\mathbf{P}(s_1, Sat(\Phi)) = \mathbf{P}(s_2, Sat(\Phi))$.
- $\mathbf{P}(s_1, B) = \mathbf{P}(s_2, B) = 1$. As $Sat(\Phi)$ is the union of equivalence classes under $\equiv_{\text{CSL}_{\setminus X}}$, the intersection with B is either empty or equals B. For $i = 1, 2$: $\mathbf{P}(s_i, Sat(\Phi)) = 1$ if $B \subseteq Sat(\Phi)$ and 0 if $B \cap Sat(\Phi) = \varnothing$. Hence, $\mathbf{P}(s_1, Sat(\Phi) = \mathbf{P}(s_2, Sat(\Phi))$.

Thus, $s_1 \equiv_{\text{PCTL}} s_2$. ∎

Theorem 3. *For any CTMC: $\approx_c$ coincides with $CSL_{\setminus X}$-equivalence.*

Proof.

$$
\begin{array}{lll}
& s_1 \approx_c^{\mathcal{C}} s_2 & \\
\text{iff} & s_1 \approx_c^{unif(\mathcal{C})} s_2 & \text{(by Prop. 4.3)} \\
\text{iff} & s_1 \sim_c^{unif(\mathcal{C})} s_2 & \text{(by Prop. 4.2)} \\
\text{iff} & s_1 \equiv_{\mathrm{CSL}}^{unif(\mathcal{C})} s_2 & \text{(since } \sim_c \text{ and CSL-equivalence coincide)} \\
\text{iff} & s_1 \equiv_{\mathrm{CSL}_{\setminus X}}^{unif(\mathcal{C})} s_2 & \text{(by Prop. 9)} \\
\text{iff} & s_1 \approx_{\mathrm{CSL}_{\setminus X}}^{\mathcal{C}} s_2 & \text{(by Prop. 8)}
\end{array}
$$
■

Logical Characterisation of Simulation. $\precsim_d$ for DTMCs without absorbing states equals $\sim_d$ [31], and hence, equals $\equiv_{\mathrm{PCTL}}$. For FPS where $\precsim_d$ is non-symmetric and strictly coarser than $\sim_d$, a logical characterisation is obtained by considering a fragment of PCTL in the sense that $s_1 \precsim_d s_2$ iff all PCTL-safety properties that hold for s_2 also hold for s_1. A similar result can be established for $\precsim_c$ and a safe fragment of CSL.

Safe and Live Fragments of PCTL and CSL. In analogy to the universal and existential fragments of CTL, safe and live fragments of PCTL and CSL are defined as follows. We consider formulae in positive normal form, i.e., negations may only be attached to atomic propositions. In addition, only a restriced class of probability bounds is allowed in the probabilistic operator. The syntax of PCTL-safety formulae (denoted by Φ_S) is as follows:

$$\mathrm{tt} \;\Big|\; \mathrm{ff} \;\Big|\; a \;\Big|\; \neg a \;\Big|\; \Phi_S \wedge \Phi_S \;\Big|\; \Phi_S \vee \Phi_S \;\Big|\; \mathcal{P}_{\leqslant p}(X\Phi_L) \;\Big|\; \mathcal{P}_{\geqslant p}(\Phi_S \,\mathcal{W}\, \Phi_S) \;\Big|\; \mathcal{P}_{\leqslant p}(\Phi_L \,\mathcal{U}\, \Phi_L)$$

PCTL-liveness formulae (denoted by Φ_L) are defined as follows:

$$\mathrm{tt} \;\Big|\; \mathrm{ff} \;\Big|\; a \;\Big|\; \neg a \;\Big|\; \Phi_L \wedge \Phi_L \;\Big|\; \Phi_L \vee \Phi_L \;\Big|\; \mathcal{P}_{\leqslant p}(X\Phi_L) \;\Big|\; \mathcal{P}_{\geqslant p}(\Phi_L \,\mathcal{W}\, \Phi_L) \;\Big|\; \mathcal{P}_{\leqslant p}(\Phi_S \,\mathcal{U}\, \Phi_S)$$

As a result of the aforementioned relationship between $\mathcal{U}$ and $\mathcal{W}$, there is a duality between safety and liveness properties for PCTL, i.e., for any formula Φ_S there is a liveness property equivalent to $\neg\Phi_S$, and the same applies to liveness property Φ_L. Safe and live fragments of CSL are defined in an analogous way, where the steady-state operator is not considered, see [8].

Logical Characterisation of Simulation. Let $s_1 \precsim_{\mathrm{PCTL}}^{safe} s_2$ iff for all PCTL-safety formulae Φ_S: $s_2 \models \Phi_S$ implies $s_1 \models \Phi_S$. Likewise, $s_1 \precapprox_{\mathrm{PCTL}_{\setminus X}}^{safe} s_2$ iff this implication holds for all $\mathrm{PCTL}_{\setminus X}$-safety formulae. The preorders $\precsim_{\mathrm{PCTL}}^{live}$ and $\precapprox_{\mathrm{PCTL}_{\setminus X}}^{live}$ are defined similarly, and the same applies for the preorders corresponding to the safe and live fragments of CSL and $\mathrm{CSL}_{\setminus X}$. The first of the following results follows from a result by [17] for a variant of Hennessy-Milner logic. The fourth result has been reported in [8]. The same proof strategy can be used to prove the second and third result [9]. We conjecture that the converse of the third and fourth result also holds.

Theorem 4.

1. *For any FPS:* $\precsim_d$ *coincides with* $\precsim^{safe}_{PCTL}$ *and with* $\precsim^{live}_{PCTL}$.
2. *For any CTMC:* $\precsim_c$ *coincides with* $\precsim^{safe}_{CSL}$ *and with* $\precsim^{live}_{CSL}$.
3. *For any FPS:* $\precapprox_d \subseteq \precapprox^{safe}_{PCTL\setminus X}$ *and* $\precapprox_d \subseteq \precapprox^{live}_{PCTL\setminus X}$.
4. *For any CTMC:* $\precapprox_c \subseteq \precapprox^{safe}_{CSL\setminus X}$ *and* $\precapprox_c \subseteq \precapprox^{live}_{CSL\setminus X}$.

5 The Branching-Time Spectrum

Summarising the results obtained in the literature together with our results in this paper yields the 3-dimensional spectrum of branching-time relations depicted in Fig. 1. All strong bisimulation relations are clearly contained within their weak variants, i.e., $\sim_d \subseteq \approx_d$ and $\sim_c \subseteq \approx_c$. The plane in the "front" (black arrows) represents the continuous-time setting, whereas the plane in the "back" (light blue or gray arrows) represents the discrete-time setting. Arrows connecting the two planes (red or dark gray) relate CTMCs and their embedded DTMCs. $R \longrightarrow R'$ means that R is finer than R', while $R \not\longrightarrow R'$ means that R is not finer than R'. The dashed arrows in the continuous setting refer to uniformised CTMCs, i.e., if there is a dashed arrow from R to R', R is finer than R' for uniformised CTMCs. In the discrete-time setting the dashed arrows refer to DTMCs without absorbing states. Note that these models are obtained as embeddings of uniformised CTMCs (except for the pathological CTMC where all exit rates are 0, in which case all relations in the picture

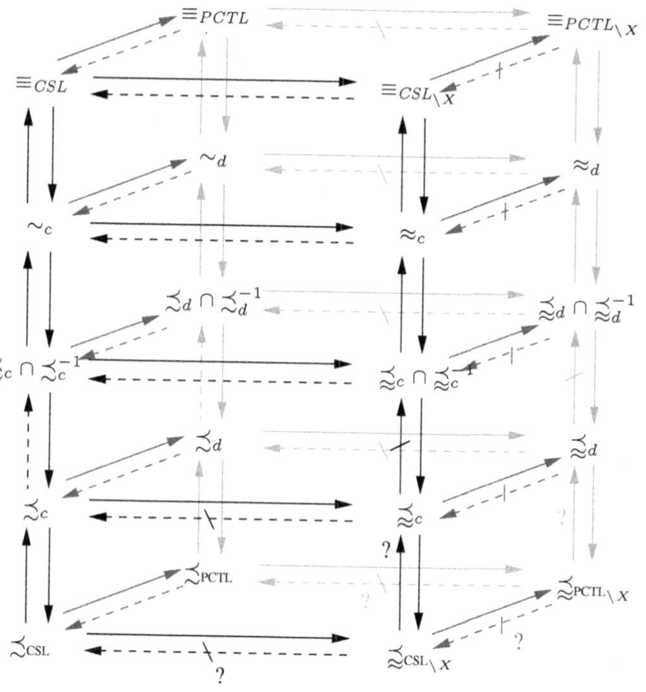

Fig. 1. Spectrum of branching-time relations for CTMCs and DTMCs

agree). If a solid arrow is labeled with a question mark, we claim the result, but have no proof (yet). For negated dashed arrows with a question mark, we claim that the implication does not hold even for uniformised CTMCs (DTMCs without absorbing states). The only difference between the discrete and continuous setting is that weak and strong bisimulation equivalence agree for uniformised CTMCs, but not for DTMCs without absorbing states.

The weak bisimulation proposed in [2] is strictly coarser than $\approx_d$, and thus does not preserve $\equiv_{\mathrm{PCTL}_{\setminus X}}$. The ordinary, non-probabilistic branching-time spectrum is more diverse, because there are many different weak bisimulation-style equivalences [23]. In the setting considered here, the spectrum spanned by Milner-style observational equivalence and branching bisimulation equivalence collapses to a single "weak bisimulation equivalence" [7]. Another difference is that for ordinary transition systems, simulation equivalence is strictly coarser than bisimulation equivalence. Further, in this non-probabilistic setting weak relations have to be augmented with aspects of divergence to obtain a logical characterisation by $\mathrm{CTL}_{\setminus X}$ [37]. In the probabilistic setting, divergence occurs with probability 0 or 1, and does not need any distinguished treatment.

6 Concluding Remarks

This paper has explored the spectrum of strong and weak (bi)simulation relations for countable fully probabilistic systems as well as continuous-time Markov chains. Based on a cascade of definitions in a uniform style, we have studied strong and weak (bi)simulations, and have provided logical characterisations in terms of fragments of PCTL and CSL. The definitions have three ingredients: (1) a condition on the labelling of states with atomic propositions, (2) a time-abstract condition on the probabilistic behaviour, and (3) a model-dependent condition: a rate condition for CTMCs (on the exit rates in the strong case, and on the total rates of "visible" moves in the weak case), and a reachability condition on the "visible" moves in the weak FPS case. The strong FPS case does not require a third condition.

As the rate conditions imply the corresponding reachability condition, the "continuous" relations are finer than their "discrete" counterparts, and the continuous-time setting excludes the possibility to abstract from stuttering occurring with probability 1.[2] While weak bisimulation in CTMCs (and FPSs) is a rather fine notion, it is the best abstraction preserving all properties that can be specified in CSL (PCTL) without next-step.

References

1. L. de Alfaro. Temporal logics for the specification of performance and reliability. *Symp. on Th. Aspects of Comp. Sc.*, LNCS 1200, pp. 165–176, 1997.
2. S. Andova and J. Baeten. Abstracion in probabilistic process algebra. *Tools and Algorithms for the Construction and Analysis of Systems*, LNCS 2031, pp. 204–219, 2001.
3. A. Aziz, V. Singhal, F. Balarin, R. Brayton and A. Sangiovanni-Vincentelli. It usually works: the temporal logic of stochastic systems. *CAV*, LNCS 939, pp. 155–165, 1995.
4. A. Aziz, K. Sanwal, V. Singhal and R. Brayton. Verifying continuous time Markov chains. *Computer-Aided Verification*, LNCS 1102, pp. 269–276, 1996.

[2] In process-algebraic terminology, the reachability condition guarantees the law $\tau.P = P$ for FPS. This law cannot hold for CTMCs due to the advance of time while stuttering (performing τ).

5. C. Baier. On algorithmic verification methods for probabilistic systems. Habilitation thesis, University of Mannheim, 1998.
6. C. Baier, B.R. Haverkort, H. Hermanns and J.-P. Katoen. Model checking continuous-time Markov chains by transient analysis. *CAV*, LNCS 1855, pp. 358–372, 2000.
7. C. Baier, H. Hermanns. Weak bisimulation for fully probabilistic processes. *Computer-Aided Verification*, LNCS 1254, pp. 119–130. 1997.
8. C. Baier, J.-P. Katoen, H. Hermanns and B. Haverkort. Simulation for continuous-time Markov chains. *Concurrency Theory*, LNCS 2421, pp. 338–354, 2002.
9. C. Baier, H. Hermanns, J.-P. Katoen and V. Wolf. Comparative branching-time semantics for Markov chains. Tech. Rep., Univ. of Twente, 2003.
10. C. Baier, J.-P. Katoen and H. Hermanns. Approximate symbolic model checking of continuous-time Markov chains. *Concurrency Theory*, LNCS 1664, pp. 146–162, 1999.
11. M. Bernardo and R. Gorrieri. Extended Markovian process algebra. *Concurrency Theory*, LNCS 1119, pp. 315–330, 1996.
12. M. Bravetti. Revisiting interactive Markov chains. *3rd Workshop on Models for Time-Critical Systems*, BRICS Notes NP-02-3, pp. 68–88, 2002.
13. M. Brown, E. Clarke, O. Grumberg. Characterizing finite Kripke structures in propositional temporal logic. *Th. Comp. Sc.*, **59**: 115–131, 1988.
14. P. Buchholz. Exact and ordinary lumpability in finite Markov chains. *J. of Appl. Prob.*, **31**: 59–75, 1994.
15. E. Clarke, O. Grumberg and D.E. Long. Model checking and abstraction. *ACM Tr. on Progr. Lang. and Sys.*, **16**(5): 1512–1542, 1994.
16. J. Desharnais. *Labelled Markov Processes.* PhD Thesis, McGill University, 1999.
17. J. Desharnais. Logical characterisation of simulation for Markov chains. *Workshop on Probabilistic Methods in Verification*, Tech. Rep. CSR-99-8, Univ. of Birmingham, pp. 33–48, 1999.
18. J. Desharnais, A. Edalat and P. Panangaden. A logical characterisation of bisimulation for labelled Markov processes. In *IEEE Symp. on Logic in Comp. Sc.*, pp. 478–487, 1998.
19. J. Desharnais and P. Panangaden. Continuous stochastic logic characterizes bisimulation of continuous-time Markov processes. *J. of Logic and Alg. Progr.*, **56**: 99–115, 2003.
20. J. Desharnais, V. Gupta, R. Jagadeesan and P. Panangaden. Approximating labelled Markov processes. In *IEEE Symp. on Logic in Comp. Sc.*, pp. 95–106, 2000.
21. J. Desharnais, V. Gupta, R. Jagadeesan and P. Panangaden. Weak bisimulation is sound and complete for PCTL*. *Concurrency Theory*, LNCS 2421, pp. 355–370, 2002.
22. R.J. van Glabbeek. The linear time – branching time spectrum I. The semantics of concrete, sequential processes. Ch. 1 in *Handbook of Process Algebra*, pp. 3–100, 2001.
23. R.J. van Glabbeek. The linear time – branching time spectrum II. The semantics of sequential processes with silent moves. *Concurrency Theory*, LNCS 715, pp. 66–81, 1993.
24. R.J. van Glabbeek, S.A. Smolka and B. Steffen. Reactive, generative, and stratified models of probabilistic processes. *Inf. & Comp.*, **121**: 59–80, 1995.
25. W. Feller. *An Introduction to Probability Theory and its Applications.* John Wiley, 1968.
26. H. Hansson and B. Jonsson. A logic for reasoning about time and reliability. *Form. Asp. of Comp.* **6**: 512–535, 1994.
27. H. Hermanns. *Interactive Markov Chains.* LNCS 2428, 2002.
28. J. Hillston. *A Compositional Approach to Performance Modelling.* Cambr. Univ. Press, 1996.
29. C. Jones and G. Plotkin. A probabilistic powerdomain of evaluations. In *IEEE Symp. on Logic in Computer Science*, pp. 186–195, 1989.
30. B. Jonsson. Simulations between specifications of distributed systems. *Concurrency Theory*, LNCS 527, pp. 346–360, 1991.
31. B. Jonsson and K.G. Larsen. Specification and refinement of probabilistic processes. In *IEEE Symp. on Logic in Comp. Sc.*, pp. 266–277, 1991.
32. C.-C. Jou and S.A. Smolka. Equivalences, congruences, and complete axiomatizations for probabilistic processes. *Concurrency Theory*, LNCS 458, pp. 367–383, 1990.

33. J.G. Kemeny and J.L. Snell. *Finite Markov Chains*. Van Nostrand, 1960.
34. V.G. Kulkarni. *Modeling and Analysis of Stochastic Systems*. Chapman & Hall, 1995.
35. K.G. Larsen and A. Skou. Bisimulation through probabilistic testing. *Inf. and Comp.*, **94**(1): 1–28, 1991.
36. R. Milner. *Communication and Concurrency*. Prentice-Hall, 1989.
37. R. De Nicola and F. Vaandrager. Three logics for branching bisimulation (extended abstract). In *IEEE Symp. on Logic in Comp. Sc.*, pp. 118–129, 1992.
38. A. Philippou, I. Lee, and O. Sokolsky. Weak bisimulation for probabilistic systems. *Concurrency Theory*, LNCS 1877, pp. 334–349, 2000.
39. M.L. Puterman. *Markov Decision Processes: Discrete Stochastic Dynamic Programming*. John Wiley & Sons, 1994.
40. R. Segala and N.A. Lynch. Probabilistic simulations for probabilistic processes. *Nordic J. of Computing*, **2**(2): 250–273, 1995.
41. M.I.A. Stoelinga. *Verification of Probabilistic, Real-Time and Parametric Systems*. PhD Thesis, University of Nijmegen, 2002.

Quantitative Relations and Approximate Process Equivalences

Alessandra Di Pierro[1,⋆], Chris Hankin[2,⋆⋆], and Herbert Wiklicky[2]

[1] Dipartimento di Informatica, Universitá di Pisa, Italy
[2] Department of Computing, Imperial College London, UK

Abstract. We introduce a characterisation of probabilistic transition systems (PTS) in terms of linear operators on some suitably defined vector space representing the set of states. Various notions of process equivalences can then be re-formulated as abstract linear operators related to the concrete PTS semantics via a probabilistic abstract interpretation. These process equivalences can be turned into corresponding approximate notions by identifying processes whose abstract operators "differ" by a given quantity, which can be calculated as the norm of the difference operator. We argue that this number can be given a statistical interpretation in terms of the tests needed to distinguish two behaviours.

1 Introduction

We study the notion of *relation* on a set X in terms of linear operators on a space representing the elements in X. In this setting classical relations corresponds to 0/1 matrices. By considering matrices with generic (numerical) entries, we generalise the classical notion by introducing *quantitative relations*. We will concentrate on a special type of quantitative relations, namely probabilistic transition relations. These represent a central notion in probabilistic process algebra [19], where process semantics and thus the various process equivalences are defined in terms of probabilistic transition systems (PTS).

We introduce a technique for defining approximated versions of various process equivalences, which exploits the operator algebraic view of quantitative relations. The fact that these quantities correspond in a PTS to probabilities allows for a statistical interpretation of the approximation according to the "button-pushing experiments" view of process semantics [22,31].

The technique is based on the definition of a PTS as a continuous linear operator on a Hilbert space built out of the states and actions. Process equivalences are special linear operators which correspond to some probabilistic abstractions of the PTS semantics. By using some appropriate operator norm we are then able to quantify equivalences, and use the resulting measure ε to define corresponding notions of approximate equivalences. These ε-relations are no longer equivalence relations but instead they approximate equivalence relations.

⋆ Supported by Progetto MEFISTO (Metodi Formali per la Sicurezza e il Tempo).
⋆⋆ Partly funded by the EU FET open project SecSafe.

R. Amadio, D. Lugiez (Eds.): CONCUR 2003, LNCS 2761, pp. 508–522, 2003.

We will illustrate our approach on two process semantics, namely graph isomorphism and a generic notion of probabilistic bisimulation, which we will characterise by using the Probabilistic Abstract Interpretation framework introduced in [13,14]. The possibility of reasoning in terms of a non-exact semantics is important for program analysis, where it is often more realistic to consider a margin of tolerance in the identification of two processes. For example, in the area of security, approximate versions of process equivalences can be used to define security properties which reflect more closely the various security problems which occur in practice. For the approximate version of bisimulation, which we call ε-bisimulation, we will mention possible applications in this area. This approach has been adopted in [12,11], where an approximate notion of observational equivalence is considered to address the problem of confidentiality.

2 Quantitative Relations

Standard models in semantics are usually based on a *qualitative* concept of a relation $R \subseteq X \times X$, which states whether two elements are related or not. We are concerned here with *quantitative* (more precisely probabilistic) relations. Such relations not only specify which elements in X are related, but also how "strong" this relation is. As an example, probabilistic transition relations are quantitative relations which specify how likely it is that one state is reachable from another. We begin with an investigation of the general notion of quantitative relation, which we characterise as a linear operator; we then apply these general results to the special case of probabilistic transition relations, which are at the base of the process equivalences we will study in the following.

Definition 1. **(i)** *A* quantitative *or* weighted relation *R over a space X with weights in $\mathbb{W}$ is a subset $R \subseteq X \times \mathbb{W} \times X$.*
(ii) *A* labelled quantitative relation *L is a subset $L \subseteq X \times A \times \mathbb{W} \times X$, where A is a set of labels.*
(iii) *A* probabilistic relation *P is a quantitative relation with $\mathbb{W} = [0,1]$, i.e. $P \subseteq X \times [0,1] \times X$, where for each $x \in X$ the function $\mu_x : X \mapsto [0,1]$ defined by $\mu_x(y) = p$ for $(x,p,y) \in P$ is a distribution, i.e. for a fixed $x \in X$: $\sum_{y \in X} \mu_x(y) = \sum_{(x,p,y)\in P} p = 1$.*

We will consider here only quantitative relations over *countable* sets X and *finite* sets of labels A. Furthermore we will assume complex weights, i.e. $\mathbb{W} = \mathbb{C}$, as we can embed the other common weight sets, e.g. $\mathbb{Z}, \ldots, \mathbb{R}$, easily in $\mathbb{C}$.

Note that for numerical weights – i.e. for $\mathbb{W}$ a ring, field, etc. – we can interpret $R \subseteq X \times \mathbb{W} \times X$ as a function $R : X \times X \to \mathbb{W}$ by adding all the weights associated to the same pair $(x,y) \in X \times X$, i.e. $R(x,y) = \sum_{(x,w,y)\in R} w$.

2.1 Linear Representations

Qualitative as well as quantitative relations have a simple representation as linear operators. In order to define the matrix associated to a relation on a set X, we first have to lift X to a vector space.

Definition 2. *The vector space $\mathcal{V}(X)$ over a set X is the space of formal linear combinations of elements in X with coefficients in some field $\mathbb{W}$ (e.g. $\mathbb{W} = \mathbb{C}$) which are represented by sequences of elements in $\mathbb{W}$ indexed by elements in X:*

$$\mathcal{V}(X) = \{(c_x)_{x \in X} \mid c_x \in \mathbb{W}\}.$$

We associate to each relation $R \subseteq X \times X$ a 0/1-matrix, i.e. a linear operator $\mathbf{M}(R)$ on $\mathcal{V}(X)$ defined by:

$$(\mathbf{M}(R))_{xy} = \begin{cases} 1 \text{ iff } (x, y) \in R \\ 0 \text{ otherwise} \end{cases}$$

where $x, y \in X$, and $(\mathbf{M}(R))_{xy}$ denotes the entry in column x and row y in the matrix representing $\mathbf{M}(R)$. Analogously, the matrix representing a quantitative relation $R \subseteq X \times \mathbb{W} \times X$ is defined by:

$$(\mathbf{M}(R))_{xy} = \begin{cases} w \text{ iff } (x, w, y) \in R \\ 0 \text{ otherwise} \end{cases}$$

Note that these definitions rely on the interpretation of (numerical) quantitative relations as functions mentioned above. For probabilistic relations, where $\mathbb{W} = [0, 1]$, we obtain a *stochastic matrix*, that is a positive matrix where the entries in each row sum up to one.

For finite sets X the representation of (quantitative) relations as linear operators on $\mathcal{V}(X) \simeq \mathbb{C}^n$ is rather straightforward: since all finite dimensional vector spaces are isomorphic to the n-dimensional complex vector space $\mathbb{C}^n$ for some $n < \infty$, their topological structure is unique [18, 1.22] and every linear operator is automatically continuous. For infinite (countable) sets, however, the algebra of infinite matrices which we obtain this way is topologically "unstable". The algebra of infinite matrices has no universal topological structure and the notions of linearity and continuity do not coincide. It is therefore difficult, for example, to define the limit of a sequence of infinite matrices in a general way. In [15] Di Pierro and Wiklicky address this problem by concentrating on relations which can be represented as elements of a C*-algebra, or concretely as elements in $\mathcal{B}(\ell^2)$, i.e. the algebra of bounded, and therefore continuous linear operators on the standard Hilbert space $\ell^2(X) \subseteq \mathcal{V}(X)$. This is the space of infinite vectors:

$$\ell^2 = \ell^2(X) = \{(x_i)_{i \in X} \mid x_i \in \mathbb{C} : \sum_{i \in X} |x_i|^2 < \infty\}.$$

The algebraic structure of a C*-algebra allows for exactly one norm topology and thus offers the same advantages as the linear algebra of finite dimensional matrices. A formal justification for this framework is given in [15]. We just mention here that the representation of (probabilistic transition) relations as operators on $\ell^2(X)$ – and not for example on $\ell^1(S)$ (which a priori might seem to be a more appropriate structure, e.g. [20]) – allows us to treat "computational states" and "observables" as elements of the same space (as Hilbert spaces are *self-dual*). Furthermore, this approach is consistent with the well established study of *(countable) infinite graphs* via their adjacency operator as an element in $\mathcal{B}(\ell^2)$, e.g. [23].

2.2 Probabilistic Transition Relations

A labelled transition system specifies a class of sequential processes $\mathcal{P}$ on which binary predicates $\xrightarrow{a}$ are defined for each action a a process is capable to perform. Probabilistic Transition Systems (PTS) are labelled transition systems with a probabilistic branching: a process p can be in a relation $\xrightarrow{a}$ with any p' in a set S of possible successors with a given probability $\mu(p')$ such that μ forms a distribution over the set S [19,21].

Given a countable set S, we call a function $\pi : S \mapsto [0,1]$ a *distribution* on S iff $\sum_{s\in S} \pi(s) = 1$. We denote by $Dist(S)$ the set of all distributions on S. Every distribution corresponds to a vector in the vector space $\mathcal{V}(S)$. Furthermore as $x^2 \leq x$ for $x \in [0,1]$ we have $\sum_{s\in S} \pi(s)^2 \leq \sum_{s\in S} \pi(s) = 1$, i.e. every distribution corresponds to a vector in $\ell^2(S) \subseteq \mathcal{V}(S)$.

Given an equivalence relation $\sim$ on S and a distribution π on S, the *lifting* of π to the set of equivalence classes of $\sim$ in S, $S/_\sim$, is defined for each equivalence class $[s] \in S/_\sim$ by $\pi([s]) = \sum_{s'\in[s]} \pi(s')$. It is straightforward to show that this is indeed a distribution on $S/_\sim$ (e.g. [19, Def 1 & Thm 1]). We write $\pi \sim \varrho$ if the lifting of π and ϱ coincide.

Definition 3. *A* probabilistic transition system *is a tuple* $(S, A, \longrightarrow, \pi_0)$*, where:*

- S *is a non-empty, countable set of* states,
- A *is a non-empty, finite set of* actions,
- $\longrightarrow \subseteq S \times A \times Dist(S)$ *is a* transition relation, *and*
- $\pi_0 \in Dist(S)$ *is an* initial distribution *on* S.

For $s \in S$, $\alpha \in A$ and $\pi \in Dist(S)$ we write $s \xrightarrow{\alpha} \pi$ for $(s, \alpha, \pi) \in \longrightarrow$. By $s \xrightarrow{\alpha}_{\pi(t)} t$ we denote the transition to individual states t with probability $\pi(t)$.

The above definition of a PTS allows for fully probabilistic as well as non-deterministic transitions as there might be more than one distribution associated to a state s and an action α. In this paper we will concentrate on fully probabilistic models where a non-deterministic choice never occurs.

Definition 4. *Given a probabilistic transition system* $X = (S, A, \longrightarrow, \pi_0)$*, we define its* matrix *or* operator representation $\mathbf{X} = (\mathbf{M}(X), \mathbf{M}(\pi_0))$ *as the direct sum of the operator representations of the transition relations for each* $\alpha \in A$*:*

$$\mathbf{M}(X) = \bigoplus_{\alpha\in A} \mathbf{M}(\xrightarrow{\alpha}),$$

and $|A|$ *copies of the vector representing* π_0*:* $\mathbf{M}(\pi_0) = \bigoplus_{\alpha\in A} \boldsymbol{\pi}_0$.

In the following we will denote $\mathbf{M}(\xrightarrow{\alpha})$ by $\mathbf{M}_\alpha$.

Given a set $\{\mathbf{M}_i\}_{i=1}^k$ of $n_i \times m_i$ matrices, then the direct sum of these matrices is given by the $(\sum_{i=1}^k n_i) \times (\sum_{i=1}^k m_i)$ matrix:

$$\mathbf{M} = \bigoplus_i \mathbf{M}_i = \begin{pmatrix} \mathbf{M}_1 & 0 & 0 \dots & 0 \\ 0 & \mathbf{M}_2 & 0 \dots & 0 \\ \vdots & \vdots & \vdots \ddots & \vdots \\ 0 & 0 & 0 \dots & \mathbf{M}_k \end{pmatrix}$$

Distributions are represented by vectors in the vector space $\ell^2(S)\oplus\ldots\oplus\ell^2(S) = (\ell^2)^{|A|} \subseteq \mathcal{V}(S)^{|A|}$. The matrix $\mathbf{M}(X)$ represents a linear operator on this space.

It is easy to see that starting with $\mathbf{M}(\pi_0)$ and applying $\mathbf{M}(X)$ repeatedly for n steps we get the distributions corresponding to the n-step closure of $\longrightarrow$ (by summing up the factors in the direct sum). More precisely:

- Take an initial $\pi_0 \in Dist(S)$ and represent it as a vector $\mathbf{M}(\pi_0) \in \mathcal{V}(S)$
- Combine $|A|$ copies of $\mathbf{M}(\pi_0)$ to obtain $\mathbf{M}(\pi_0)^{|\mathcal{A}|} = \bigoplus_{\alpha\in\mathcal{A}} \mathbf{M}(\pi_0)$.
- Apply $\mathbf{M}(X) = \bigoplus_{\alpha\in\mathcal{A}} \mathbf{M}_\alpha$ to this vector.
- Obtain $(\bigoplus_{\alpha\in\mathcal{A}} \mathbf{M}_\alpha)(\mathbf{M}(\pi_0)^{|\mathcal{A}|}) = \bigoplus_{\alpha\in\mathcal{A}} \mathbf{M}_\alpha(\mathbf{M}(\pi_0))$.
- Denote the factors by $\mathbf{M}(\pi'_\alpha) = \mathbf{M}_\alpha(\mathbf{M}(\pi_0))$.
- Construct the compactification $\mathbf{M}(\pi_1) = \sum_{\alpha\in\mathcal{A}} \mathbf{M}(\pi'_\alpha)$.
- Restart the iteration process with π_1.

For the sake of simplicity we will denote by $\mathbf{PX}$ the multiplication of a direct sum $\bigoplus_\alpha \mathbf{P}$ of the same matrix $\mathbf{P}$ with the matrix $\mathbf{X} = \bigoplus_\alpha \mathbf{X}_\alpha$. By the properties of the direct sum this is the same as $\bigoplus_\alpha(\mathbf{PX}_\alpha)$.

Given a PTS $X = (S, A, \longrightarrow, \pi_0)$ and a state $p \in S$, we denote by $S_p \subseteq S$ the set of all states reachable from p, by $T(p)$ the transition system induced on the restricted state space S_p, and by $\mathbf{M}(p)$ the matrix representation of $T(p)$.

3 Probabilistic Abstract Interpretation

Probabilistic Abstract Interpretation was introduced in [13,14] as a probabilistic version of the classical *abstract interpretation* framework by Cousot & Cousot [5,6]. This framework provides general techniques for the analysis of programs which are based on the construction of *safe* approximations of concrete semantics of programs via the notion of *Galois connection* [7,25]. Probabilistic abstract interpretation re-casts these techniques in a probabilistic setting, where linear spaces replace the classical order-theoretic based domains, and the notion of *Moore-Penrose pseudo-inverse* of a linear operator replaces the classical notion of a Galois connections. It is thus essentially different from approaches applying classical abstract interpretation to probabilistic domains [24].

By a *probabilistic domain* we mean a space which represents the distributions $Dist(S)$ on the state space S of a PTS, i.e. in our setting the Hilbert space $\ell^2(S)$. For finite state spaces we can identify $\mathcal{V}(S) \simeq \ell^2(S)$.

Definition 5. *Let $\mathcal{C}$ and $\mathcal{D}$ be two probabilistic domains. A* probabilistic abstract interpretation *is a pair of bounded linear operators $\mathbf{A} : \mathcal{C} \to \mathcal{D}$ and $\mathbf{G} : \mathcal{D} \to \mathcal{C}$, between (the concrete domain) $\mathcal{C}$ and (the abstract domain) $\mathcal{D}$, such that $\mathbf{G}$ is the Moore-Penrose pseudo-inverse of $\mathbf{A}$, and vice versa.*

The Moore-Penrose pseudo-inverse is usually considered in the context of so-called *least-square approximations* as it allows the definition of an optimal generalised solution of linear equations. The Moore-Penrose pseudo-inverse of a linear map between two Hilbert spaces is defined as follows (for further details see e.g. [4], or [3]):

Definition 6. *Let $\mathcal{C}$ and $\mathcal{D}$ be two Hilbert spaces and $\mathbf{A} : \mathcal{C} \mapsto \mathcal{D}$ a linear map between them. A linear map $\mathbf{A}^\dagger = \mathbf{G} : \mathcal{D} \mapsto \mathcal{C}$ is the* Moore-Penrose pseudo-inverse *of $\mathbf{A}$ iff*

$$\mathbf{A} \circ \mathbf{G} = \mathbf{P}_A \quad \textit{and} \quad \mathbf{G} \circ \mathbf{A} = \mathbf{P}_G$$

where $\mathbf{P}_A$ and $\mathbf{P}_G$ denote orthogonal projections onto the ranges of $\mathbf{A}$ and $\mathbf{G}$.

A simple method for constructing a probabilistic abstract interpretation which we will use in this paper is as follows: given a linear operator Φ on some Hilbert space $\mathcal{V}$ expressing the probabilistic semantics of a concrete system, and a linear abstraction function $\mathbf{A} : \mathcal{V} \mapsto \mathcal{W}$ from the concrete domain into an abstract domain $\mathcal{W}$, we compute the Moore-Penrose pseudo-inverse $\mathbf{G} = \mathbf{A}^\dagger$ of $\mathbf{A}$. The abstract semantics can then be defined as the linear operator on the abstract domain $\mathcal{W}$:

$$\Psi = \mathbf{A} \circ \Phi \circ \mathbf{G}.$$

Moore-Penrose inverses always exist for operators on finite dimensional vector spaces [3]. For operator algebras, i.e. operators over infinite dimensional Hilbert spaces, the following theorem provides conditions under which the existence of Moore-Penrose inverses is guaranteed [3, Thm 4.24]:

Theorem 1. *An operator $\mathbf{A} : \mathcal{C} \to \mathcal{D}$ between two Hilbert spaces is Moore-Penrose invertible if and only if it is* normally solvable, *i.e. if its range $\{\mathbf{A}x \mid x \in \mathcal{C}\}$ is closed.*

For the special case of operators $\mathbf{A}$ which are defined via an approximating sequence $(\mathbf{A}_n)_n$ of finite-dimensional operators, we are not only guaranteed that the Moore-Penrose pseudo-inverse exists, but we can also construct it via an approximation sequence provided that the sequence $(\mathbf{A}_n)_n$ and the sequence $(\mathbf{A}_n^*)_n$ of their adjoints converges *strongly* to $\mathbf{A}$ and $\mathbf{A}^*$ [3, Cor 4.34]. In the strong operator topology a sequence of operators $(\mathbf{A}_n)_n$ converges *strongly* if there exists an $\mathbf{A} \in \mathcal{B}(\ell^2)$ such that for all $\boldsymbol{x} \in \ell^2$: $\lim_{n\to\infty} \|\mathbf{A}_n\boldsymbol{x} - \mathbf{A}\boldsymbol{x}\| = 0$.

Proposition 1. *Let $\mathbf{A} : \mathcal{C} \to \mathcal{D}$ be an operator between two separable Hilbert spaces. If there is a sequence $\mathbf{A}_n$ of finite dimensional operators with $\sup_n \|\mathbf{A}_n\| < \infty$ and such that $\mathbf{A}_n \to \mathbf{A}$ and $\mathbf{A}_n^* \to \mathbf{A}^*$ strongly, then $\mathbf{A}$ is normally solvable and $\mathbf{A}_n^\dagger \to \mathbf{A}^\dagger$ strongly.*

This construction is sufficient for most cases as it can be shown that the operational or collecting semantics of finitely branching processes can always be approximated in this way [15].

4 Approximate Process Equivalences

In the classical approaches process equivalences are qualitative relations. Alternatively, process equivalences can be seen as a kind of quantitative relations, namely probabilistic relations. One advantage of having a quantity (the probability) attached to a relation is that we can calculate the behavioural difference between two processes and use the resulting quantity to define *approximate* notions of equivalences. The latter weaken strict equivalences by identifying processes whose behaviour is "the same up to ε", ε being the approximation error.

The ε versions of process equivalences are closely related to approaches which aim to distinguish probabilistic processes by statistical testing. A general setting for a statistical interpretation of ε is provided by the concept of *hypothesis testing*, see e.g. [28]. The problem can be stated as follows: given two processes A and B let us assume that one of these is executed as a black-box process X, i.e. we know that either $X = A$ or $X = B$. The idea is to formulate two (exclusive) hypotheses $H_0 : X$ is A and $H_1 : X$ is B. The aim is to determine the probability that either H_0 or H_1 holds based on a number of statistical tests performed on X. The number ε gives us a direct measure for how many tests we have to perform in order to accept H_0 or H_1 with a certain confidence. In essence: the smaller the ε, the more tests we have to perform in order to obtain the same level of confidence.

The details of the exact relation between the number of required tests n to distinguish H_0 and H_1 with a certain confidence α are not easy to be worked out in general, but can in principle be achieved using methods from mathematical statistics. More details for a concrete case – applied to the problem of probabilistic confinement related to the simple notion of process equivalence based on input/output observables – can be found in [12,11].

Approximate equivalences turn out to be very useful in program analysis where they can be used to define approximate and yet more realistic analyses of programs properties, such as confinement [12,11,10], which are directly defined in terms of some process equivalences.

In order to define approximate process equivalences we first look at relations as linear operators; then using an appropriate operator norm we measure the "distance" between relations. In this way we are able to define a relation which is ε-close to the strict (original) equivalence. For the characterisation of equivalence relations as linear operators we use the framework of probabilistic abstract interpretation. In particular, we will show that each equivalence on a given system corresponds to a pair of Moore-Penrose pseudo-inverses which define a probabilistic abstract interpretation of the system.

4.1 Graph Equivalence

To illustrate our basic strategy for approximating process equivalences let us first look at the strongest – in some sense too strong [31, Fig 1] – notion of process equivalence, that is tree equivalence. Following [31, Def 1.3] the graph associated to a process p of a labelled transition system with actions A is a directed graph rooted in p whose edges are labelled by elements in A. Two processes are *tree equivalent* if their associated graphs are isomorphic. Graph isomorphism is defined as follows ([31, Def 1.3,Def 1.4], [17, p2]):

Definition 7. *An* isomorphism *between directed graphs* (V_1, E_1) *and* (V_2, E_2) *is a bijection* $\varphi : V_1 \mapsto V_2$ *such that* $\langle v, w\rangle \in E_1 \Leftrightarrow \langle \varphi(v), \varphi(w)\rangle \in E_2$.

In the usual way, we define the *adjacency operator* $\mathbf{A}(X)$ of a directed graph $X = (V, E)$ as an operator on $\ell^2(V)$ representing the edge-relation E [23]. Then the notion of isomorphism between (finite) graphs can be re-stated in terms of permutation matrices.

An $n \times n$-matrix $\mathbf{P}$ is called a *permutation matrix* if there exists a permutation $\pi : \{1, \dots, n\} \to \{1, \dots, n\}$ such that $\mathbf{P}_{ij} = 1$ iff $j = \pi(i)$ and otherwise $\mathbf{P}_{ij} = 0$. This notion can easily be extended to *permutation operators* for infinite structures.

We denote by $\mathcal{P}(n)$ the set of all $n \times n$ permutation matrices and by $\mathcal{P}(\mathcal{H})$ the set of permutation operators on $\mathcal{H}$; obviously we have $\mathcal{P}(n) = \mathcal{P}(\mathbb{C}^n)$.

Proposition 2. *For any permutation operator $\mathbf{P} \in \mathcal{P}(\mathcal{H})$ the following holds: $\mathbf{P}^{-1} = \mathbf{P}^* = \mathbf{P}^T = \mathbf{P}^\dagger$, i.e.* inverse, adjoint, transpose, *and* pseudo-inverse *of permutation operators coincide.*

We then have the following result [17, Lemma 8.8.1]:

Proposition 3. *Let $X = (V, E_1)$ and $Y = (V, E_2)$ be two directed graphs on the same set of nodes V. Then X and Y are isomorphic if and only if there is a permutation operator $\mathbf{P}$ such that the following holds: $\mathbf{P}^T\mathbf{A}(X)\mathbf{P} = \mathbf{A}(Y)$.*

By using these notions and the operator representation of (probabilistic) transition systems (cf. Definition 4) we can reformulate tree-equivalence of processes as follows.

Proposition 4. *Given the operator representations $\mathbf{X}$ and $\mathbf{Y}$ of two probabilistic transition systems $X = (S, A, \longrightarrow, s_0)$ and $Y = (S', A, \longrightarrow', s'_0)$ with $|S| = |S'|$, then X and Y are tree-equivalent iff there exists $\mathbf{P} \in \mathcal{P}(\ell^2(S)) = \mathcal{P}(\ell^2(S'))$, such that:*

$$\mathbf{P}^T\mathbf{X}\mathbf{P} = \mathbf{Y},$$

i.e. for all $\alpha \in A$ we have $\mathbf{P}^T\mathbf{M}(\xrightarrow{\alpha})\mathbf{P} = \mathbf{M}(\xrightarrow{\alpha}{}')$ and $\mathbf{P}^T\pi_0\mathbf{P} = \pi'_0$.

Therefore, tree equivalence of two systems X and Y corresponds to the existence of an abstraction operator (the operator $\mathbf{P}$) which induces a probabilistic abstract interpretation Y of X.

Approximate Graph Equivalence. In the case where there is no $\mathbf{P}$ which satisfies the property in Proposition 4, i.e. X and Y are definitely not isomorphic, we could still ask how close X and Y are to being isomorphic. The most direct way to define a kind of "isomorphism defect" would be to look at the difference $\mathbf{X} - \mathbf{Y}$ between the operators representing X and Y and then measure in some way, e.g. using a norm, this difference.

Obviously, this is not the idea we are looking for: it is easy to see that the same graph – after enumerating its vertices in a different ways – has different adjacency operators; it would thus have a non-zero "isomorphism defect" with itself. To remedy this we have to allow first for a reordering of vertices before we measure the difference between the operators representing two probabilistic transition systems. This is the underlying idea behind the following definition.

Definition 8. *Let $X = (S, A, \longrightarrow, \pi_0)$ and $Y = (S', A, \longrightarrow', \pi'_0)$ be probabilistic transition systems over the same set of actions A, and let $\mathbf{X}$ and $\mathbf{Y}$ be their operator representations. We say that X and Y are ε-*graph equivalent*, denoted by $X \sim_i^\varepsilon Y$, iff*

$$\inf_{\mathbf{P} \in \mathcal{P}} \|\mathbf{P}^T \mathbf{X} \mathbf{P} - \mathbf{Y}\| = \varepsilon$$

where $\|.\|$ denotes an appropriate norm.

Note that, in the case of finite probabilistic transition systems, for $\varepsilon = 0$ we recover the original notion of (strict) graph equivalence, i.e. $\sim_i = \sim_i^0$.

Proposition 5. *An ε-isomorphism for $\varepsilon = 0$, i.e. $\sim_i^0$, of finite transition systems is an isomorphism.*

We believe that a similar proposition can be stated for infinite PTS's too. However, this would require the development of a more elaborate operator algebraic framework for modelling PTS's than the one presented in this paper, and we refer to [15] for a more detailed treatment of this case.

4.2 Probabilistic Bisimulation Equivalence

The finest process equivalence is bisimulation equivalence [31]. Bisimulation is a relation on processes, i.e. states of a labelled transition system. Alternatively, it can be seen as a relation between the *transition graphs* associated to the processes. The classical notion of bisimulation equivalence for labelled transition systems is as follows, e.g. [31, Def 12]:

Definition 9. *A* bisimulation *is a binary relation $\sim_b$ on states of a labelled transition system satisfying for all $\alpha \in A$:*

$$p \sim_b q \text{ and } p \xrightarrow{\alpha} p' \Rightarrow \exists q' : q \xrightarrow{\alpha} q' \text{ and } p' \sim_b q',$$
$$p \sim_b q \text{ and } q \xrightarrow{\alpha} q' \Rightarrow \exists p' : p \xrightarrow{\alpha} p' \text{ and } q' \sim_b p'.$$

Given two processes p and q, we say that they are *bisimilar* if there exists a bisimulation relation $\sim_b$ such that $p \sim_b q$. Bisimulations are equivalence relations [31, Prop 8.1].

The standard generalisation of this notion to probabilistic transition systems, i.e. *probabilistic bisimulation*, is due to [21]. We will concentrate here on fully probabilistic systems or reactive systems in the terminology of [19]. In this model all states $s \in S$ are deterministic in the sense that for each action $\alpha \in A$, there is only one distribution π such that $s \xrightarrow{\alpha} \pi$.

Definition 10. *[19, Def 4][9, Def 3.2] A* probabilistic bisimulation *is an equivalence relation $\sim_b$ on states of a probabilistic transition system satisfying for all $\alpha \in A$:*

$$p \sim_b q \text{ and } p \xrightarrow{\alpha} \pi \Rightarrow q \xrightarrow{\alpha} \varrho \text{ and } \pi \sim_b \varrho.$$

We now introduce the notion of a classification operator which we will use to define a probabilistic bisimulation equivalence via a probabilistic abstract interpretation. A *classification matrix* or *classification operator* is given by an (infinite) matrix containing only a single non zero entry equal to one in each row, and no column with only zero entries. Classification operators simply represent a classical relation, i.e. is a $0/1$ matrix, which happens to be a (surjective) function from one state space into another.

Classification matrices and operators are thus particular kinds of stochastic matrices and operators. We denote by $\mathcal{C}(n,m)$ the set of all $n \times m$-classification matrices, and by $\mathcal{C}(\mathcal{H}_1,\mathcal{H}_2)$ the set of classification operators; again we have $\mathcal{C}(n,m) = \mathcal{C}(\mathbb{C}^n,\mathbb{C}^m)$.

Obviously, every permutation matrix is also a classification matrix: $\mathcal{P}(n) \subseteq \mathcal{C}(n,n)$, and similarly $\mathcal{P}(\mathcal{H}) \subseteq \mathcal{C}(\mathcal{H},\mathcal{H})$. Furthermore, the multiplication of two classification operators gives again a classification operator. These properties follow easily from the following correspondence between classification operators and equivalence relations:

Proposition 6. *Let X be a countable set. Then for each equivalence relation $\approx$ on X there exists a classification operator $\mathbf{K} \in \mathcal{C}(\ell^2(X),\ell^2(X/_\approx))$ and vice versa.*

For finite sets with $|X| = n$ and $|X/_\approx| = m$ we get a classification matrix in $\mathcal{C}(n,m)$.

Proposition 7. *The pseudo-inverse of a classification operator $\mathbf{K}$ corresponds to its normalised transpose or adjoint (these operations coincide for real $\mathbf{K}$).*

The *normalisation* operation $\mathcal{N}$ is defined for a matrix $\mathbf{A}$ by $\mathcal{N}(\mathbf{A})_{ij} = \frac{\mathbf{A}_{ij}}{a_j}$ if $a_j = \sum_i \mathbf{A}_{ij} \neq 0$ and $\mathcal{N}(\mathbf{A})_{ij} = 0$ otherwise. Although the classification operator $\mathbf{K}$ represents a classical function, i.e. corresponds to an (infinite) $0/1$-matrix, the pseudo-inverse will in general not be an (infinite) $0/1$-matrix.

It is easy to see that a probabilistic bisimulation equivalence $\sim$ on a PTS $T = (S, A, \rightarrow, \pi_0)$ defines a probabilistic abstract interpretation of T. In fact, by Proposition 6, there is a classification matrix $\mathbf{K} \in \mathcal{C}(\ell^2(S),\ell^2(S')$, for some S' which represents $\sim$. If $\mathbf{M}(T)$ is the operator representation of T then $\mathbf{K}^\dagger\mathbf{M}(T)\mathbf{K}$ is the abstract operator induced by $\mathbf{K}$. Intuitively, this is an operator which abstracts the original system T by encoding only the transitions between equivalence classes instead of the ones between single states.

Consider now two processes $p, q \in S$ and their operator representations $\mathbf{M}(p)$ and $\mathbf{M}(q)$. The restrictions of $\mathbf{K}$ to these two sets of nodes, which we call $\mathbf{K}_p$ and $\mathbf{K}_q$, are the abstraction operators for the two processes p and q and allow us to express exactly the condition for the probabilistic bisimilarity of p and q:

Proposition 8. *Given the operator representation $\mathbf{M}(p)$ and $\mathbf{M}(q)$ of two probabilistic processes p and q, then p and q are bisimilar iff there exists a $\mathbf{K}_p \in \mathcal{C}(\ell^2(S_p),\ell^2(S))$ and $\mathbf{K}_q \in \mathcal{C}(\ell^2(S_q),\ell^2(S))$ for some set S such that*

$$\mathbf{K}_p^\dagger\mathbf{M}(p)\mathbf{K}_p = \mathbf{K}_q^\dagger\mathbf{M}(q)\mathbf{K}_q.$$

Corollary 1. *Given the matrix representation* $\mathbf{M}(p)$ *and* $\mathbf{M}(q)$ *of two processes* p *and* q. *Then* p *and* q *are bisimilar, i.e.* $p \sim_b q$, *iff there exists a PTS* x *which is the probabilistic abstract interpretation of both* p *and* q.

Example 1. Consider the following two processes A and B from [21, Fig.4]:

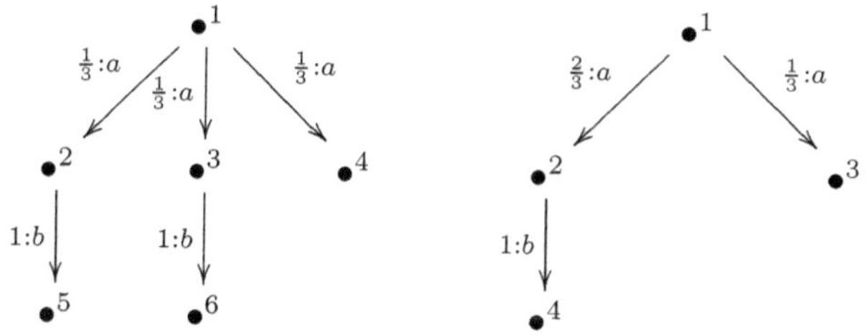

The corresponding matrices are:

$$\mathbf{M}(A) = \mathbf{M}_a(A) \oplus \mathbf{M}_b(A) = \begin{pmatrix} 0 & \frac{1}{3} & \frac{1}{3} & \frac{1}{3} & 0 & 0 \\ 0 & 0 & 0 & 0 & 0 & 0 \\ 0 & 0 & 0 & 0 & 0 & 0 \\ 0 & 0 & 0 & 0 & 0 & 0 \\ 0 & 0 & 0 & 0 & 0 & 0 \\ 0 & 0 & 0 & 0 & 0 & 0 \end{pmatrix} \oplus \begin{pmatrix} 0 & 0 & 0 & 0 & 0 & 0 \\ 0 & 0 & 0 & 0 & 1 & 0 \\ 0 & 0 & 0 & 0 & 0 & 1 \\ 0 & 0 & 0 & 0 & 0 & 0 \\ 0 & 0 & 0 & 0 & 0 & 0 \\ 0 & 0 & 0 & 0 & 0 & 0 \end{pmatrix}$$

and

$$\mathbf{M}(B) = \mathbf{M}_a(B) \oplus \mathbf{M}_b(B) = \begin{pmatrix} 0 & \frac{2}{3} & \frac{1}{3} & 0 \\ 0 & 0 & 0 & 0 \\ 0 & 0 & 0 & 0 \\ 0 & 0 & 0 & 0 \end{pmatrix} \oplus \begin{pmatrix} 0 & 0 & 0 & 0 \\ 0 & 0 & 0 & 1 \\ 0 & 0 & 0 & 0 \\ 0 & 0 & 0 & 0 \end{pmatrix}$$

The classification operators and their pseudo-inverses are given by:

$$\mathbf{K}_A = \begin{pmatrix} 1 & 0 & 0 & 0 \\ 0 & 1 & 0 & 0 \\ 0 & 1 & 0 & 0 \\ 0 & 0 & 1 & 0 \\ 0 & 0 & 0 & 1 \\ 0 & 0 & 0 & 1 \end{pmatrix} \qquad \mathbf{K}_A^\dagger = \begin{pmatrix} 1 & 0 & 0 & 0 & 0 & 0 \\ 0 & \frac{1}{2} & \frac{1}{2} & 0 & 0 & 0 \\ 0 & 0 & 0 & 1 & 0 & 0 \\ 0 & 0 & 0 & 0 & \frac{1}{2} & \frac{1}{2} \end{pmatrix}$$

and $\mathbf{K}_B$ and $\mathbf{K}_B^\dagger$ are simply 4×4 identity matrices. We then get:

$$\mathbf{K}_A^\dagger \cdot \mathbf{M}_a(A) \cdot \mathbf{K}_A = \mathbf{M}_a(B)$$
$$\mathbf{K}_A^\dagger \cdot \mathbf{M}_b(A) \cdot \mathbf{K}_A = \mathbf{M}_b(B)$$

which shows that A and B are probabilistically bisimilar.

The matrix formulation of (probabilistic) bisimulation makes it also easy to see how graph and bisimulation equivalence are related, as $\mathcal{P}(n) \subset \mathcal{C}(n,n)$ we have:

Proposition 9. *If $p \sim_i q$ then $p \sim_b q$.*

Note that probabilistic bisimulation is only related to a particular kind of probabilistic abstract interpretation: we consider only abstractions which are induced by classification matrices and not by more general ones. The relation between abstract interpretation and (bi)simulation has been recognised before in the classical Galois connection based setting ([8], [27]), but this appears to be the first investigation of such a relation in a probabilistic setting.

Approximate Bisimulation Equivalences. When it is not possible to find a bisimulation equivalence for two processes p and q of a PTS T, we can still identify them although only approximately. In order to do so, we introduce an ε-version of probabilistic bisimilarity. The intuitive idea is to find a classification operator $\mathbf{K}$ which is the closest one to a bisimulation relation in which p and q are equivalent. The difference between the abstract operators induced by $\mathbf{K}$ for the two processes will give us an estimate of the non-bisimilarity degree of p and q.

Definition 11. *Let $T = (S, A, \longrightarrow, \pi_0)$ be a probabilistic transition systems and let p and q be two states in S with operator representations $\mathbf{X}$ and $\mathbf{Y}$. We say that p and q are* ε-bisimilar, *denoted by $p \sim_b^\varepsilon q$, iff*

$$\min_{\mathbf{K}_p, \mathbf{K}_q \in C} \|\mathbf{K}_p^\dagger \mathbf{X} \mathbf{K}_p - \mathbf{K}_q^\dagger \mathbf{Y} \mathbf{K}_q\| = \varepsilon$$

where $\|.\|$ denotes an appropriate norm.

In determining the "degree of similarity" ε of two processes $\mathbf{X}$ and $\mathbf{Y}$ our aim is to identify two "abstract processes" $\mathbf{K}_p^\dagger \mathbf{X} \mathbf{K}_p$ and $\mathbf{K}_q^\dagger \mathbf{X} \mathbf{K}_q$ such that their behaviour is most similar. The concrete numerical value of ε depends on the norm we choose and the type of classification operators we consider. In particular, we can strengthen the above definition by restricting the number of "abstract states", i.e. the dimension of $\mathbf{K}_p$ and $\mathbf{K}_q$, in order to obtain an estimation ε relative to only those equivalences with a fixed number of classes.

Note that it is possible to use this definition also to introduce an approximate version of the classical notion of bisimulation. Furthermore, for $\varepsilon = 0$ we recover partially the original notion of strict (probabilistic)bisimulation:

Proposition 10. *An ε-bisimulation for $\varepsilon = 0$, i.e. $\sim_b^0$, is a (probabilistic) bisimulation for finite transition systems.*

For infinite PTS, the same remarks as for Proposition 5 apply.

Example 2. In this example we will use a more "probabilistic" form of PTS which are called *generative* in [26]. In this model the probability distribution on the branching takes into account the internal decision of the process to react to a given action. Thus the transition relation is a subset of $S \times Dist(A \times S)$.

Let us compare the following, obviously somehow "similar", processes:

$$A \equiv \mathit{fix}\ A.b : A +_{\frac{1}{2}} a : \mathbf{0}$$

$$B \equiv a : \mathbf{0} +_{\frac{3}{4}} (\mathit{fix}\ X.b : X +_{\frac{1}{2}} a : \mathbf{0})$$

$$C \equiv a : \mathbf{0} +_{\frac{1}{2}} (\mathit{fix}\ X.b : X +_{\frac{51}{100}} a : \mathbf{0})$$

Their transition graphs are given by:

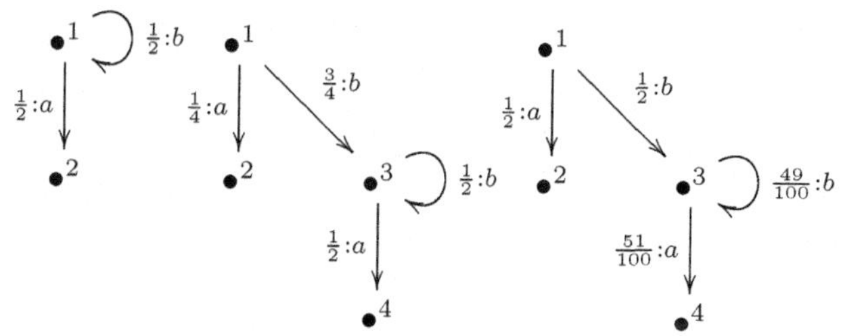

These processes are not probabilistically bisimilar. However one can try to determine how similar they are. The matrix representations are as follows:

$$\mathbf{A} = \mathbf{M}(A) = \mathbf{M}_a(A) \oplus \mathbf{M}_b(A) = \begin{pmatrix} 0 & \frac{1}{2} \\ 0 & 0 \end{pmatrix} \oplus \begin{pmatrix} \frac{1}{2} & 0 \\ 0 & 0 \end{pmatrix}$$

$$\mathbf{B} = \mathbf{M}(B) = \mathbf{M}_a(B) \oplus \mathbf{M}_b(B) = \begin{pmatrix} 0 & \frac{1}{4} & 0 & 0 \\ 0 & 0 & 0 & 0 \\ 0 & 0 & 0 & \frac{1}{2} \\ 0 & 0 & 0 & 0 \end{pmatrix} \oplus \begin{pmatrix} 0 & 0 & \frac{3}{4} & 0 \\ 0 & 0 & 0 & 0 \\ 0 & 0 & \frac{1}{2} & 0 \\ 0 & 0 & 0 & 0 \end{pmatrix}$$

$$\mathbf{C} = \mathbf{M}(C) = \mathbf{M}_a(C) \oplus \mathbf{M}_b(C) = \begin{pmatrix} 0 & \frac{1}{2} & 0 & 0 \\ 0 & 0 & 0 & 0 \\ 0 & 0 & 0 & \frac{51}{100} \\ 0 & 0 & 0 & 0 \end{pmatrix} \oplus \begin{pmatrix} 0 & 0 & \frac{1}{2} & 0 \\ 0 & 0 & 0 & 0 \\ 0 & 0 & \frac{49}{100} & 0 \\ 0 & 0 & 0 & 0 \end{pmatrix}$$

The problem is to find a $\mathbf{K}_A, \mathbf{K}_B$, and $\mathbf{K}_C \in \mathcal{C}$ such that the norm of the difference between $\mathbf{K}_A^\dagger \mathbf{A} \mathbf{K}_A$ and $\mathbf{K}_B^\dagger \mathbf{B} \mathbf{K}_B$ or $\mathbf{K}_C^\dagger \mathbf{C} \mathbf{K}_C$ is minimal. There is only a finite (though exponentially growing) number of possible classification operators $\mathbf{K} \in \mathcal{C}$. A brute force approach looking at all possible $\mathbf{K}$ allows us to determine the ε-bisimilarity of A and B, and of A and C. Interestingly the optimal $\mathbf{K} = \mathbf{K}_B = \mathbf{K}_C$ is coincidentally the same for both B and C:

$$\mathbf{K} = \begin{pmatrix} 1 & 0 \\ 0 & 1 \\ 1 & 0 \\ 0 & 1 \end{pmatrix} \quad \mathbf{K}^\dagger = \begin{pmatrix} \frac{1}{2} & 0 & \frac{1}{2} & 0 \\ 0 & \frac{1}{2} & 0 & \frac{1}{2} \end{pmatrix},$$

while for $\mathbf{K}_A$ we can take the identity.

Measuring the difference based on the operator norm leads to the following:

$$\inf_{\mathbf{K} \in \mathcal{C}} \|\mathbf{A} - \mathbf{K}^\dagger \mathbf{B} \mathbf{K}\| = \frac{1}{8}, \qquad \inf_{\mathbf{K} \in \mathcal{C}} \|\mathbf{A} - \mathbf{K}^\dagger \mathbf{C} \mathbf{K}\| = \frac{1}{200}.$$

5 Conclusions

In this paper we have investigated *quantitative relations*, in particular probabilistic transition relations. We were able to extend the classical framework of Abstract Interpretation to a quantitative domain by taking the Moore-Penrose pseudo-inverse as an appropriate replacement for the order-theoretic concept of Galois connections. Based on this methodology of *Probabilistic Abstract Interpretation*, previously introduced only in a finite dimensional setting [13], we recast (probabilistic) process equivalences in terms of linear operators. This formulation has a very strong resemblance to notions of similarity in mathematical control theory, e.g. [29, Def 4.1.1]. Finally we were able to weaken strict process equivalences to approximate ones. This provides a novel approach towards the notion of approximative or ε-bisimilarity and adds new aspects to existing approaches, like those based on metrics [16] or pseudo-metrics [9,30]. In particular, our approach allows for a statistical interpretation of the approximation ε which relates this quantity to the number of tests we need to perform in order to accept a given hypothesis with a certain confidence in a "hypothesis testing" approach to statistical testing. This is particularly important in a security context; we are confident that these notions of approximate similarity can be fruitfully employed in security related applications, such as *approximate confinement*, which provided the original motivation for this work [12]. Aldini et al. adopted a similar approach to study probabilistic non-interference in a CSP-like language [1].

References

1. A. Aldini, M. Bravetti, and R. Gorrieri. A process algebraic approach for the analysis of probabilistic non-interference. *Journal of Computer Security*, 2003. To appear.
2. J.A. Bergstra, A. Ponse, and S.A. Smolka, editors. *Handbook of Process Algebra*. Elsevier Science, Amsterdam, 2001.
3. A. Böttcher and B. Silbermann. *Introduction to Large Truncated Toeplitz Matrices*. Springer Verlag, New York, 1999.
4. S.L. Campbell and D. Meyer. *Generalized Inverse of Linear Transformations*. Constable and Company, London, 1979.
5. P. Cousot and R. Cousot. Abstract Interpretation: A Unified Lattice Model for Static Analysis of Programs by Construction or Approximation of Fixpoints. In *Proceedings of POPL'77*, pages 238–252, Los Angeles, 1977.
6. P. Cousot and R. Cousot. Systematic Design of Program Analysis Frameworks. In *Proceedings of POPL'79*, pages 269–282, San Antonio, Texas, 1979.
7. P. Cousot and R. Cousot. Abstract Interpretation and Applications to Logic Programs. *Journal of Logic Programming*, 13(2–3):103–180, July 1992.
8. D. Dams, R. Gerth, and O. Grumberg. Abstract interpretation of reactive systems. *ACM Transactions on Programming Languages and Systems*, 19(2):253–291, 1997.
9. J. Desharnais, R. Jagadeesan, V. Gupta, and P.Panangaden. The metric analogue of weak bisimulation for probabilistic processes. In *Proceedings of LICS'02*, pages 413–422, Copenhagen, Denmark, 22–25 July 2002. IEEE.

10. A. Di Pierro, C. Hankin, and H. Wiklicky. Approximate confinement under uniform attacks. In *Proceedings of SAS'02*, volume 2477 of *Lecture Notes in Computer Science*. Springer Verlag, 2002.
11. A. Di Pierro, C. Hankin, and H. Wiklicky. Approximate non-interference. In *Proceedings of CSFW'02*, pages 3–17, Cape Breton, 24–26 June 2002. IEEE.
12. A. Di Pierro, C. Hankin, and H. Wiklicky. Approximate non-interference. *Journal of Computer Security (WITS '02 Issue)*, 2003. To appear.
13. A. Di Pierro and H. Wiklicky. Concurrent Constraint Programming: Towards Probabilistic Abstract Interpretation. In *Proceedings of PPDP'00*, pages 127–138, Montréal, Canada, 2000. ACM.
14. A. Di Pierro and H. Wiklicky. Measuring the precision of abstract interpretations. In *Proceedings of LOPSTR'00*, volume 2042 of *Lecture Notes in Computer Science*, pages 147–164. Springer Verlag, 2001.
15. A. Di Pierro and H. Wiklicky. A C*-algebraic approach to the operational semantics of programming languages. In preparation, 2003.
16. A. Giacalone, C.-C. Jou, and S.A. Smolka. Algebraic reasoning for probabilistic concurrent systems. In *Proceedings of the IFIP WG 2.2/2.3 Working Conference on Programming Concepts and Methods*, pages 443–458. North-Holland, 1990.
17. C. Godsil and G. Royle. *Algebraic Graph Theory*, volume 207 of *Graduate Texts in Mathematics*. Springer Verlag, New York – Heidelberg – Berlin, 2001.
18. W.H. Greub. *Linear Algebra*, volume 97 of *Grundlehren der mathematischen Wissenschaften*. Springer Verlag, New York, third edition, 1967.
19. B. Jonsson, W. Yi, and K.G. Larsen. *Probabilistic Extensions of Process Algebras*, chapter 11, pages 685–710. Elsevier Science, Amsterdam, 2001. see [2].
20. D. Kozen. Semantics for probabilistic programs. *Journal of Computer and System Sciences*, 22:328–350, 1981.
21. K.G. Larsen and A. Skou. Bisimulation through probabilistic testing. *Information and Computation*, 94:1–28, 1991.
22. R. Milner. *A Calculus of Communicating Systems*, volume 92 of *Lecture Notes in Computer Science*. Springer-Verlag, Berlin – New York, 1980.
23. B. Mohar and W. Woess. A survey on spectra of infinite graphs. *Bulletin of the London Mathematical Society*, 21:209–234, 1988.
24. D. Monniaux. Abstract interpretation of probabilistic semantics. In *Proceedings of SAS'00*, volume 1824 of *Lecture Notes in Computer Science*. Springer Verlag, 2000.
25. F. Nielson, H. Riis Nielson, and C. Hankin. *Principles of Program Analysis*. Springer Verlag, Berlin – Heidelberg, 1999.
26. S.A. Smolka R.J. van Glabbeek and B. Steffen. Reactive, Generative and Stratified Models of Probabilistic Processes. *Information and Computation*, 121:59–80, 1995.
27. D.A. Schmidt. Binary relations for abstraction and refinement. In *Workshop on Refinement and Abstraction*, Amagasaki, Japan, November 1999.
28. J. Shao. *Mathematical Statistics*. Springer Texts in Statistics. Springer Verlag, New York – Berlin – Heidelberg, 1999.
29. E.D. Sontag. *Mathematical Control Theory: Deterministic Finite Dimensional Systems*, volume 6 of *Texts in Applied Mathematics*. Springer Verlag, 1990.
30. F. van Breugel and J. Worrell. Towards quantitative verification of probabilistic transition systems. In *Proceedings of ICALP'01*, volume 2076 of *Lecture Notes in Computer Science*, pages 421–432. Springer Verlag, 2001.
31. R.J. van Glabbeek. *The Linear Time – Branching Time Spectrum I. The Semantics of Concrete, Sequential Processes*, chapter 1, pages 3–99. Elsevier Science, Amsterdam, 2001. see [2].

Author Index